15th Latin American Workshop on Plasma Physics (LAWPP 2014) and 21st IAEA TM on Research Using Small Fusion Devices (RUSFD)

Journal of Physics: Conference Series
Volume 591

San Jose, California, USA
27 - 31 January 2014

ISBN: 978-1-5108-0213-1
ISSN: 1742-6588

Printed from e-media with permission by:

Curran Associates, Inc.
57 Morehouse Lane
Red Hook, NY 12571

CURRAN ASSOCIATES INC. proceedings .com

Some format issues inherent in the e-media version may also appear in this print version.

Printed by Curran Associates, Inc. (2015)

Institute of Physics
Dirac House, Temple Back
Bristol BS1 6BE UK

Phone: 44 1 17 929 7481
Fax: 44 1 17 920 0979

techtracking@iop.org

Additional copies of this publication are available from:

Curran Associates, Inc.
57 Morehouse Lane
Red Hook, NY 12571 USA
Phone: 845-758-0400
Fax: 845-758-2634
Email: curran@proceedings.com
Web: www.proceedings.com

Table of contents

Volume 591

15th Latin American Workshop on Plasma Physics (LAWPP 2014) and 21st IAEA TM on Research Using Small Fusion Devices (RUSFD)
27–31 January 2014, San José, Costa Rica

Accepted papers received: 9 February 2015
Published online: 24 March 2015

Papers

Tokamaks

Stellarators

Alternative Confinement Concepts

Dense Magnetised Plasmas

Plasma-wall interactions

General Theory and Fundamental Processes

Laser Plasma Interactions

Numerical simulations

Plasmas Processing and Technological Applications

Journal of Physics: Conference Series **591** (2015) 011001 doi:10.1088/1742-6596/591/1/011001

Preface

Written contributions from participants of the Joint 15th Latin American Workshop on Plasma Physics (LAWPP 2014)-21st IAEA Technical Meeting on Research Using Small Fusion Devices (21st IAEA TM RUSFD).

The International Advisory Committees of the 15th Latin American Workshop on Plasma Physics (LAWPP 2014) and the 21st IAEA TM on Research Using Small Fusion Devices (RUSFD), agreed to carry out together this Joint LAWPP 2014 -21st RUSFD in San José, Costa Rica, on 27–31 January 2014. The Joint LAWPP 2014 -21st RUSFD meeting, organized by the Instituto Tecnológico de Costa Rica, Universidad Nacional de Costa Rica, and Ad Astra Rocket Company in collaboration with the International Atomic Energy Agency (IAEA).

The Latin American Workshop on Plasma Physics (LAWPP) is a series of events which has been held periodically since 1982, with the purpose of providing a forum in which the research of the Latin American plasma physics community can be displayed, as well as fostering collaborations among plasma scientists within the region and with researchers from the rest of the world. Recognized plasma scientists from developed countries are specially invited to the meeting to present the state of the art on several "hot" topics related to plasma physics. It is an open meeting, with an International Advisory Committee, in which the working language is English. It was firstly held in 1982 in Cambuquira, Brazil, followed by workshops in Medellín, Colombia (1985), Santiago de Chile, Chile (1988), Buenos Aires, Argentina (1990), Mexico City, Mexico (1992), Foz do Iguaçu, Brazil (1994, combined with the International Congress on Plasma Physics (ICPP)), Caracas, Venezuela (1997), Tandil, Argentina (1998), La Serena, Chile (2000), Sao Pedro, Brazil (2003), Mexico City, Mexico (2005), Caracas, Venezuela (2007), Santiago de Chile, Chile (2010, combined with the ICPP) and Mar de Plata, Argentina (2011).

The 21st IAEA TM on Research Using Small Fusion Devices is an ideal forum for small laboratory size fusion experiments, as compared to those of the larger laboratories, to report about their latest achievements working with medium size and small scale tokamaks, stellarators, compact tori, dense plasma focus, reversed field pinches, helical devices, linear machines, and other small plasma devices. The Technical Meeting aims at stimulating new synergies which can contribute to better streamline the research outputs to the mainstream fusion research. Previous meetings in the series were held in Budapest, Hungary (1985), Nagoya, Japan (1986), Nice, France (1988), Washington DC, USA (1990), Hefei, China (1991), Wuerzburg, Germany (1992), Campinas, Brazil (1993), Madrid, Spain (1994), Ahmedabad, India (1995), Prague, Czech Republic (1996), Cairo, Egypt (1997), Tokyo, Japan (1998) in Chengdu, China (1999), São Paulo, Brazil (2002), Vienna, Austria (2003) in Mexico City, Mexico (2005), Lisbon, Portugal (2007), in Alushta, Ukraine (2008), Kurchatov, Kazakhstan (2009) and Vienna, Austria (2011).

The 1st Costa Rican Summer School on Plasma Physics was held a week before the Joint LAWPP 2014 -21st IAEA TM RUSFD, and the 2nd Latin American Workshop on Industrial Applications of Plasma Technology (AITP) was organized in parallel with the it. The objective of the AITP Workshop is to enhance the regional academic and industrial cooperation in the field of plasma assisted surface technology.

The Joint LAWPP 2014 -21st IAEA TM RUSFD was held at the Crowne Plaza Corobici Hotel in San Jose from 27 to 31 January 2014. The LAWPP scientific programme, which was spread along the whole week, had 15 invited speakers, 126 participants from 20 countries around the world. It included 7 plenary talks, 8 invited talks and 12 oral contributed papers were chosen out of 92 submissions. 82 contributions in 25 topics were presented in poster sessions on Monday 27, Tuesday 28 and Thursday 30 January 2014. The 21st IAEA TM RUSFD was held along the LAWPP 2014 from 27 to 29 January 2014; was attended by 37 participants formally registered with the IAEA, who joined the LAWPP 2014 participants. Its separate scientific programme had two plenary talks, 12 oral presentations and 14 papers presented in poster sessions on Monday 27 and Tuesday 28 January 2014. The 2nd Workshop on Industrial Applications of Plasma Technology (2nd AITP) was held on 30 and 31 January 2014, had six invited speakers, which included 2 plenary talks, 4 invited talks, 11 oral presentations and 31 contributions in a single poster session on Thursday 30 January, 2014. Its proceedings have been merged with those of the joint meeting. Finally the 1st Costa Rican Summer School on Plasma Physics, held in Santa Clara, San Carlos on 20-24 January 2014, in the week previous to the meetings, had 80 participants, 40 international conferences on different plasma physics topics, and 12 professors.

The topics included in the programme of the Joint LAWPP 2014 -21st IAEA TM RUSFD were: space plasmas, dusty plasmas, nuclear fusion, nonthermal plasmas, plasma space propulsion, basic plasma processes, plasma simulation, and industrial plasma applications among others.

We are very grateful to the sponsors of the meetings:, the Instituto Tecnológico de Costa Rica, the International Atomic Energy Agency (IAEA), the Universidad Nacional de Costa Rica, and Ad Astra Rocket Company. We also want to thank our exhibitors and contributors: INTERCOVAMEX, Nuclear & Plasma Sciences Society, and the IEEE Costa Rica Chapter. The publication of the proceedings was fully supported by the International Atomic Energy Agency (IAEA).

The support of the International Advisory and the Local Organizing Committees, is also acknowledged in a heartfelt way. Finally, the Editors of this special issue are grateful to Jose Asenjo for his excellent work and cooperation for the preparation of the proceedings.

Iván Vargas-Blanco and J. Julio E. Herrera-Velázquez
Editors of the proceedings

Journal of Physics: Conference Series **591** (2015) 011001 doi:10.1088/1742-6596/591/1/011001

Participants of LAWPP 2014 -21st RUSFD

Journal of Physics: Conference Series **591** (2015) 012002 doi:10.1088/1742-6596/591/1/012002

International Programme Advisory Committee of the 21st IAEA Technical Meeting on Research using Small Fusion Devices (RUSFD)

Guido Van Oost, Belgium
Edson del Bosco, Brazil
Leopoldo Soto, Chile
Jan Stöckel, Czech Republic
A. Sagara, Japan
J. Julio E. Herrera, Mexico
Vladimir Gribkov, Russia
Ryzard Miklaszewski, Poland
Igor Garkusha, Ukraine
Mikhail Gryaznevich, United Kingdom
Paul Bellan, U.S.A.

International Advisory Committee of the 15th Latin American Workshop on Plasma Physics (LAWPP)

Luis Bilbao, Argentina
María Virginia Alves, Brazil
Ibere Caldas, Brazil
Leopoldo Soto, Chile
Henry Riascos, Colombia
Iván Vargas-Blanco, Costa Rica (Chair)
Ramón Rodríguez, Cuba
Kohnosuke Sato, Japan
Mayo Villagrán, Mexico
Luis Felipe Delgado-Aparicio, Peru
Carlos Varandas, Portugal
Edberto Leal-Quirós, Puerto Rico
Carlos Alejaldre, Spain
Michael Tendler, Sweden
Malcolm G. Haines, UK – (Deceased)
Anthony Peratt, USA
George Morales, USA
Julio Puerta, Venezuela

Journal of Physics: Conference Series **591** (2015) 012002 doi:10.1088/1742-6596/591/1/012002

Local Organizing Committee of the Joint RUSFD, LAWPP Meeting

Iván Vargas Blanco (Chairman: LAWPP 2014), ITCR
Saúl Guadamuz (Chairman: 2nd AITP), ITCR
Celso Ribeiro (Chairman: 21st RUSFD), ITCR
Giovanni Sáenz-Arce, UNA
José Javier Saavedra-Arias, UNA
Juan Ignacio del Valle, Ad Astra Rocket Company
Jorge Oguilve, Ad Astra Rocket Company
José Casto Nieto, Ad Astra Rocket Company
Julio Herrera-Velázquez, Scientific Program Committee coordinator, UNAM

Peer review statement

All papers published in this volume of *Journal of Physics: Conference Series* have been peer reviewed through processes administered by the proceedings Editors. Reviews were conducted by expert referees to the professional and scientific standards expected of a proceedings journal published by IOP Publishing.

Report on recent results obtained in TCABR

RMO Galvão[1], CHS Amador[2],WAH Baquero[1], F Borges[3], IL Caldas[1], NAM Cuevas[1], VN Duarte[1], AG Elfimov[1], JI Elizondo[1], AMM Fonseca[1], TM Germano[1], GG Grenfell[1] , ZO Guimarães-Filho[1], JL Jeronimo[1], YuK Kuznetsov[1], MAM Manrique[1], IC Nascimento[1], CJA Pires[1], PGP Puglia[1], AP Reis[1], G Ronchi[1], LF Ruchko[1], WP de Sá[1], RJF Sgalla[1], EK Sanada[1], JHF Severo[1], VC Theodoro[1] and D.L. Toufen[4]

[1] Instituto de Física da Universidade de São Paulo, São Paulo, Brasil.
[2] Universidade Tecnológica Federal do Paraná, Cornélio Procópio, Brasil.
[3] Instituto de Física, Universidade Federal Fluminense, Brasil.
[4] Instituto Federal de Educação, Ciência e Tecnologia de São Paulo, Guarulhos, Brasil.

E-mail: rgalvao@if.usp.br

Abstract. Recent results of experimental work and theoretical modeling carried out in the TCABR tokamak are reported on characterization of MHD instabilities, improved diagnostics of rotation of the plasma column, excitation of Alvén global modes, identification of GAMs, and the effect of rotation on their behavior. Detailed measurements of edge electrostatic perturbations and of magnetic island evolution and rotation indicate that the edge turbulence is substantially affected by the islands growth, leading to a strong modulation of the edge particle losses at the same frequency of the MHD activity. Measurements with spatial resolution also show that the growth of the MHD activity is due to nonlinear coupling of magnetic islands with different poloidal mode numbers, which increases the impurity influx. A new system of data acquisition and processing of the TCABR plasma rotation diagnostic was implemented. The system is based upon a single monochromator coupled with six photomultipliers tubes and allows one toroidal and two poloidal simultaneous rotation measurements. The excitation of Global Alfvén Waves - GAW has been investigated, using a new type of radio frequency amplifier. The GAW resonances are searched either by a pre-programmed density variation, at fixed generator frequency, or through three RF frequency sweeps from 2 to 4.5 MHz, at stationary density. GAW resonances have been found and their somewhat new characteristics are presented. The investigation of the effect of poloidal and toroidal rotation on the characteristics of the geodesic acoustic mode has been investigated, both theoretically and experimentally. It is found that the assumption of isothermal flux surfaces gives rise to a third branch of this mode. Detailed predictions coupled with experimental measurements are currently being carried out to investigate this question.

1. Introduction

The TCABR tokamak was designed and built at the *Centre de Recherches en Physique des Plasmas - CRPP*, École Politechnique Fédérale de Lausane, with the main objective of investigating plasma heating by Alfvén waves [1]. It was operated in CRPP from 1980 to 1992, approximately. A couple of years after being de-commissioned, its main components were transferred to the Plasma Physics Laboratory of the University of São Paulo, where the tokamak was rebuilt, with new systems of discharge control and of Alfvén wave excitation, and renamed TCABR, for Tokamak Chauffage Alfvén Bresilien. It started operating in São Paulo in 1999 and produced the first new results around a year later [2,3].

The main parameters of TCABR are major radius R = 0.61m, minor radius a = 0.18m, toroidal magnetic field B = 1.1T, plasma current $I_p \leq$ 100kA, plasma density $n \leq 3\times10^{19}\text{m}^{-3}$, and electron temperature $T_e \leq$ 650 eV. The discharge lasts approximately 100ms, with flat-top plasma current around 40 to 60ms. In this paper we discuss recent results obtained in TCABR on the characterization of MHD activities, measurement of the plasma column rotation, relation between long-range correlations and Geodesic Acoustic Modes - GAMs, and excitation of Global Alfvén Waves – GAWs.

2. Characterization of MHD Instabilities

One of the main lines of research in TCABR, as in many other 'small tokamaks', is the investigation of improved confinement regimes triggered by electrostatic polarization of the plasma edge by external electrodes [4]. Along this line, one of the relevant results obtained in TCABR was on the effect of MHD activity on the triggering and maintenance of the improved confinement regime [5,6].

Although strong MHD activity can entirely suppress improved confinement [6], some conditions have been found in which this regime can survive to somewhat milder activity. An example can be seen in figure 1. In discharge 24126, the polarization pulse has been applied after the occurrence of strong MHD activity and a density rise combined with a little decrease of H-alpha emission is observed, as usual in the TCABR improved confinement regime [5]. The discharge programming was then slightly modified to delay the occurrence of the MHD activity, in order to make it coincide with the application of the polarization pulse, discharge 24128. In this case the MHD activity is not as strong as in cases previously reported [6]. Although only a short decrease in the H-alpha is observed and disappears when the MHD activity sets in, the density remains high during the polarization pulse.

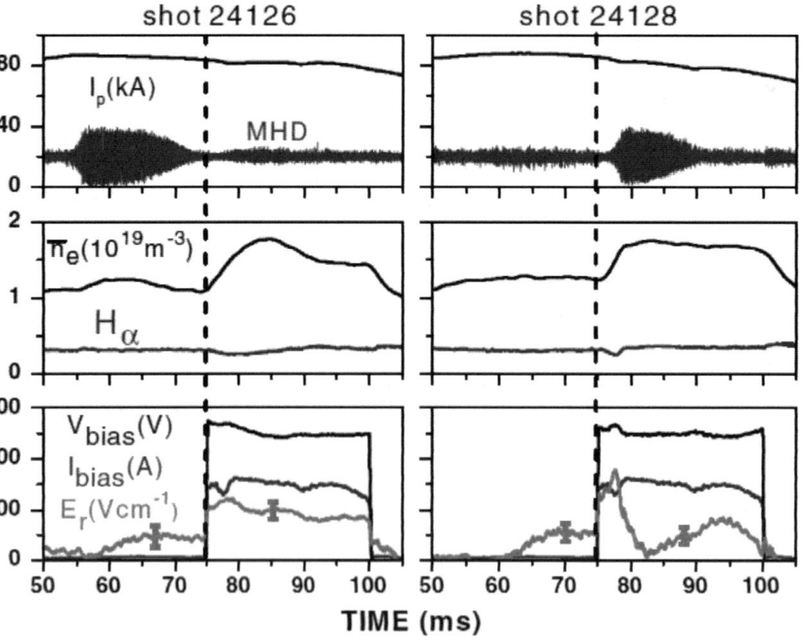

Figure 1 Time traces of plasma current (I_p), Mirnov coil signal (MHD), line average plasma density at the center of the plasma column ($\bar{n}_e$), H-alpha emission, electrode polarization voltage (V_{bias}) and current (I_{bias}), and edge radial electric field (E_r) measured with a radial array of Langmuir probes for discharges 24126 and 24128 of TCABR.

The MHD activity may growth in TCABR up to a saturated stage with large magnetic islands even in the Ohmic phase, as in the case of discharge 24126. For another TCABR discharge with high MHD activity, figure 2 shows the quite similar spectral behavior of the magnetic oscillations and turbulent fluctuations, which indicates that the edge turbulence is modulated by the magnetic islands during the high MHD phase of the discharge. Moreover, figure 2(c) shows that the particle losses are also synchronized with the MHD signals.

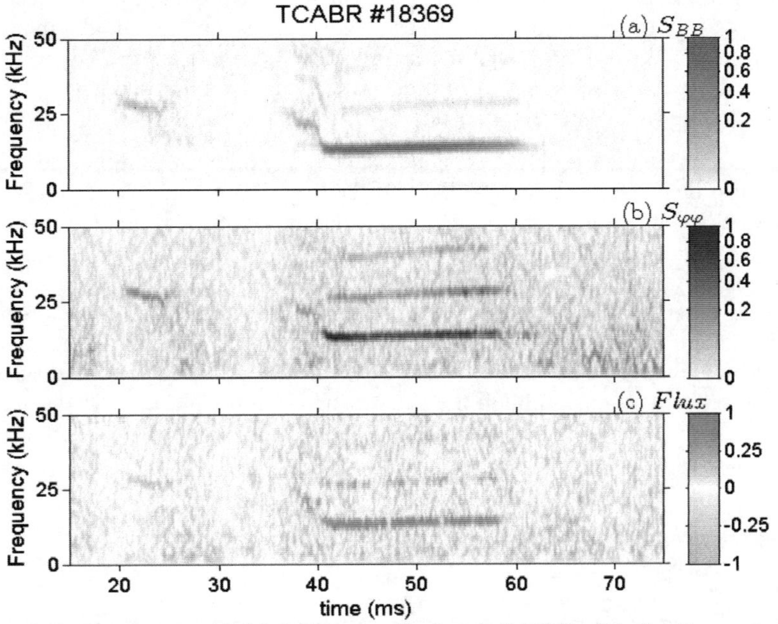

Figure 2 Spectrograms of (a) Mirnov coil signal (MHD), (b) floating potential fluctuations measured with Langmuir probes, and (c) the spectral contribution for the turbulence-driven radial particle flux for discharge 18369 of TCABR.

To better investigate the dynamics of the growth of the MHD activity and their effect on the confinement, we have combined the external magnetic measurements (twenty-four Mirnov coils in a poloidal cross-section closely surrounding the circular plasma column) with soft X-rays and total radiation measurements. The soft X-ray array views the plasma column laterally, from the low-field side, along twenty chords, while the bolometer array views the plasma column vertically, from top to bottom, along twenty-four chords, as shown in figures 3(a) and 3(b), respectively.

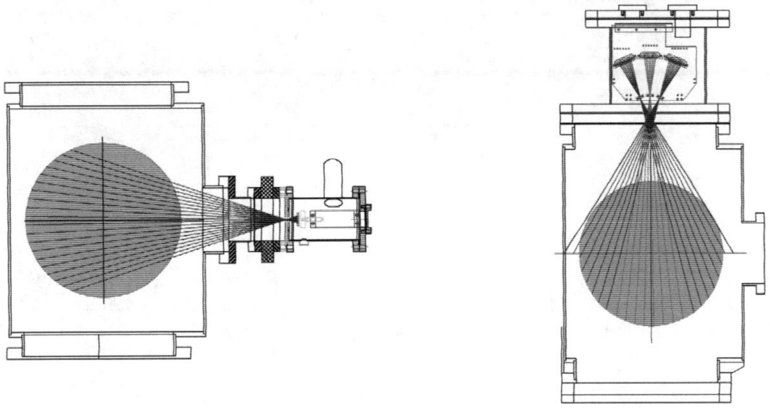

(a) (b)

Figure 3 Layout of the soft X-ray (a) and bolometer (b) arrays in TCABR

We have found out that during the growth of the MHD activity a large impurity influx associated with coupling of islands of different mode numbers is often observed. This is illustrated for discharge 26880, shown in figures 4 and 5. In this case, two MHD modes are observed in the spectrogram in the time interval $54 \leq t \leq 69$ ms, approximately. These modes correspond to poloidal mode numbers $m = 2$ ($f \approx 17$ kHz) and $m = 3$ ($f \approx 20$ kHz). At $t \approx 68$ ms, there is a programmed slight increase in the plasma current and the two modes merge, becoming a dominantly $m = 2$ mode with substantially larger intensity and a somewhat smaller frequency, $f \approx 12$ kHz, than before merging.

As the value of the cylindrical safety factor, indicated by the pink trace in the spectrograms for the mode number, is just below four, it is expected that the $m = 3$ is located rather close to the plasma boundary. Therefore, when it couples with the $m = 2$, a larger saturated island is formed [7], bringing in part of the edge plasma and, possibly, impurities. This effect is indeed corroborated by the radial profiles of soft X-ray and bolometer emissions and by a drop in the central electron temperature, as shown in figure 4. When the two modes merge, at $t \approx 68$ ms, there is a large broadening of the emission observed in the central chords of both the X-ray and radiation diagnostics, concurrent with the drop in the central electron temperature measured with ECE. Although in this discharge the mode coupling was triggered on purpose, through the slight increase in the plasma current, it seems that the same effect occurred in the high MHD discharge 18369 shown in figure 2 and in the improved confinement examples of discharges 24126 and 24128, figure 1.

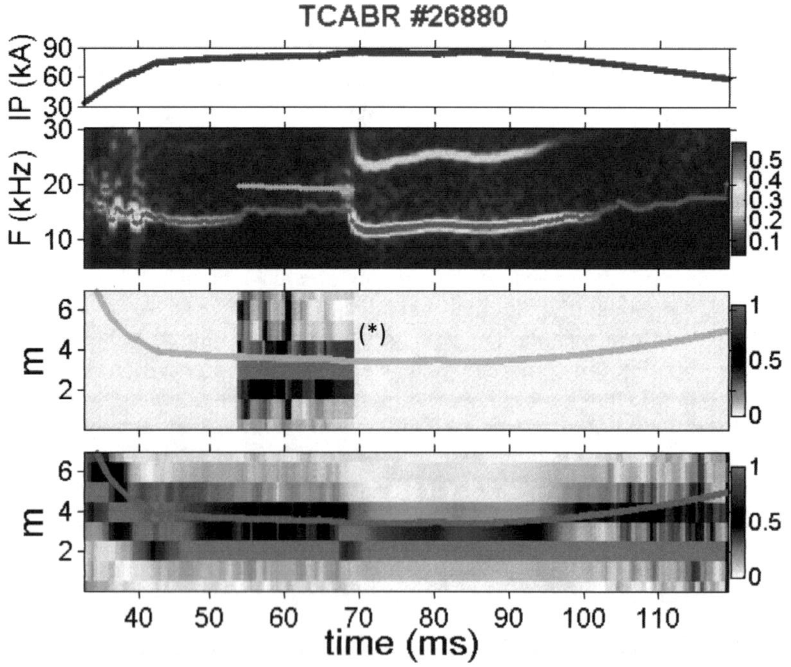

Figure 4 Plasma current, I_p, spectrogram of MHD activity measured with Mirnov coils, and two spectrograms of the poloidal mode number for discharge 26880 in TCABR. Spectogram (*) corresponds to the mode with higher frequency observed in the interval The pink trace in the last two figures corresponds to the value of the edge safety factor.

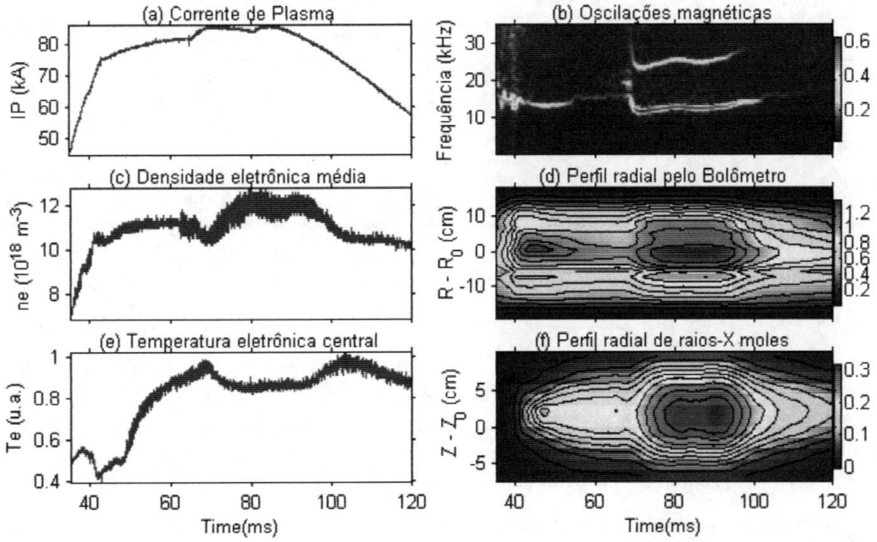

Figure 5 Plasma current, I_p, plasma density, n_e, central electron temperature, T_e, spectrogram of MHD activity measured with Mirnov coils, and spectrogram of the emission measured by the bolometer and X-ray diagnostics for discharge 26880 in TCABR.

3. Measurement of plasma rotation

A somewhat original spectroscopic method to measure the rotation of the plasma column in TCABR was developed sometime time ago and it has been under permanent improvement ever since [8-11]. The method is based upon measuring simultaneously two distinct bands of the profile of an appropriate impurity emission line to estimate the Doppler shift due to rotation. The two main results obtained so far with this diagnostic were to show that the value of the poloidal plasma rotation agrees with the neoclassical predictions in the collisional regime [8,9] and that the toroidal rotation follows the same scaling verified in large tokamaks [12]. According to this scaling, the product of the plasma current by the maximum variation of the radial profile of the toroidal velocity is proportional to the same variation of the ion temperature profile.

Recently the system has been substantially improved to allow simultaneous measurements of the toroidal and poloidal components of the rotation velocity at one and two points, respectively. Details of the new system were presented in another paper in another paper in this conference [12]. With this system it was possible to confirm that direction of the toroidal component is mostly opposite to the plasma current inside the plasma, reverting direction close to the plasma boundary, whereas the poloidal component is in the electron diamagnetic drift velocity, as indicate in figure 6. The maximum value of the toroidal and poloidal components can reach 15 km/s and 6 km/s, respectively; actual experimental values are given in [12].

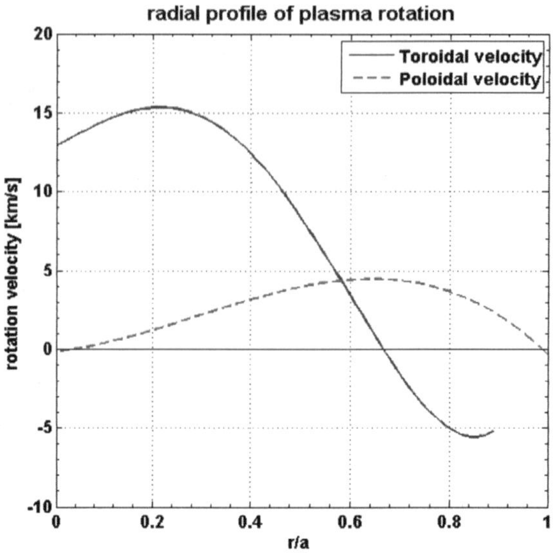

Figure 6 Sketch of the standard radial profiles of the toroidal and poloidal components of the plasma rotation velocity in TCABR discharges. For the toroidal component, positive means in the direction of the toroidal field (opposite to the plasma current) and for the poloidal component positive means in the electron diamagnetic drift velocity.

4. Long-range correlations and GAMs

It is currently accepted that zonal flows- ZFs and geodesical acoustic modes – GAMs play a quite relevant role on reducing the associate anomalous transport caused by turbulence [13,14]. For this reason there has been intensive theoretical and experimental investigation on the characteristics of these modes under different conditions and on their effects [15,16]. In standard fluid theory, the zonal flow and the geodesic acoustic mode in fact correspond to the zero and non-zero frequencies of the same dispersion relation, respectively. The frequency of the GAM is proportional to the ion sound speed and, for standard tokamak equilibria, it represents a continuum of singular stable perturbations on the magnetic surfaces.

Since both the ZFs and the GAMs have constant phases in the toroidal direction, one of the standard techniques to identify them experimentally, close to the plasma boundary, is to search for electrostatic perturbations with long-range correlations in the toroidal direction [17]. This method is being intensively used in TCABR using different types of electrostatic probes, located in different positions, as described Kuznetsov et al [18]. In this work it was clearly shown that electrostatic potential fluctuations with long distance correlations are substantially amplified during the transition to the improved confinement regime, triggered by electrostatic biasing of the plasma edge. Furthermore, for long toroidal distances between probes, the cross-spectrum is concentrated at low frequencies $f < 60$ kHz with three peaks observed at $f < 5$ kHz, $f = 13–15$ kHz and $f \approx 40$ kHz and low wave numbers with a maximum at $k = 0$. Careful analysis of data has shown that these peaks are indeed associated with electrostatic fluctuations, and not the result of MHD activity.

This result is somewhat surprising because one should expect only two peaks in the correlation spectra, one at rather low frequency, associated with ZFs, and another in the range of frequencies of the second peak, associated with GAMs. Although the precision in the determination of the lowest frequency peak is not good enough, due to the limitation on the maximum number of data points, it is

Journal of Physics: Conference Series **591** (2015) 012001 doi:10.1088/1742-6596/591/1/012001

definitely above 1 kHz. Nevertheless, a reasonable qualitative explanation for the occurrence of the three peaks can be put forward by taking into account recent theoretical developments on the effect of plasma rotation on the dynamics of GAMs.

Indeed, Elfimov and collaborators have shown that, for axisymmetric magnetic confinement configurations with isothermal flux surfaces and toroidal and poloidal rotation, there appears three solutions of the dispersion relation for GAMs [19]. The highest branch ω_1 corresponds to the standard GAM with a frequency up-shift,

$$\omega_1^2 = \frac{c_s^2}{R_0^2}\left[2 + \frac{1}{q^2} + 4M_t^2 - 4M_pM_t\left(1 - \frac{1}{q^2}\right) + M_p^2\left(2 - \frac{1}{q^2} + \frac{2}{q^4}\right)\right], \tag{1}$$

where c_s is the ion sound speed, R_0 is the major radius of the plasma column, q is the inverse rotational transform, and M_t and M_p are the toroidal and poloidal Mach numbers, respectively. The second branch,

$$\omega_2^2 = \frac{c_s^2}{q^2R_0^2}\left[1 - 4M_pM_t + M_p^2\left(3 - \frac{2}{q^2}\right)\right], \tag{2}$$

is driven by toroidal rotation, so that it does not exist in the limit $M_t \to 0$, due to a degeneracy in the dispersion relation.

These modes had already been obtained by other authors assuming different models for the flux surfaces [20]. However, a third low frequency mode is found for the isothermal flux surfaces, with frequency given by

$$\omega_3^2 = \frac{c_s^2}{q^2R_0^2}\left[M_p^2(1 - M_d)^2 + \frac{q^2(\gamma - 1)}{2(1 + 2q^2)}\left(M_t^2 - 2M_tM_p + 2M_p^2\right)M_t^2\right], \tag{3}$$

where M_d is a quantity that depends on the heat flux at the magnetic surface to warrant constant temperature [19].

The radial profile of these frequencies have been calculated from the relevant quantities measured for TCABR, in particular the rotation velocities, and the results are shown in figure 7. We see that, before the application of the biasing pulse on the external electrode, that is, in the ohmic phase of the discharge, only the MHD oscillations at f $\approx$ 16 kHz and its harmonics are observed. When biasing is applied, there appears three distinct peaks concentrated around the toroidal wave number k $\approx$ 0, i.e., quite distinct from the MHD modes have non zero k. The frequencies of these peaks agree well with the predictions from the theory presented in [19], as schematically indicated by the arrows.

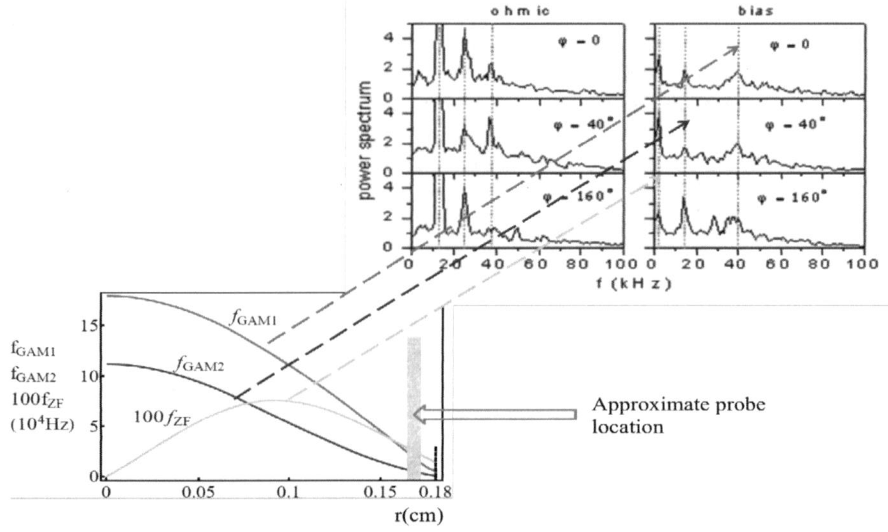

Figure 7 The three roots of the dispersion relation for GAMs, equations (1), (2), and (3), for actual parameters of the TCABR discharge, in an experiment with edge electrostatic biasing. The insert on the top is taken from figure 8 of [18]. The angle φ gives the toroidal location of the probes for which the cross-correlations are calculated.

5. Excitation of Global Alfvén Waves

The Global Alfvén Waves – GAWs are discrete modes that can be excited in axisymmetric magnetic confinement configurations below the Alfvén continuum [21,22]. They have been experimentally observed in PRETEXT [23] and TCA [24] and lately proposed to be used as a diagnostic tool to determine the inverse rotational transform q and the effective mass number in tokamaks [25]. However they are somewhat difficult to be excited in large tokamaks, due to effects of the Hall parameter (ω/ω_{ci}), where ω_{ci} is the cyclotron frequency, plasma pressure, and shear at the minimum value of q [26]. On the other, the so-called Toroidal Alfvén Eigenmodes – TAEs, which are also discrete modes but excited in the gaps of the continuum introduced by toroidal mode coupling, can be excited and have a deleterious effect on the confinement of alpha particles in large tokamaks [27].

The excitation of GAWs in carefully investigated in TCABR with two main purposes: its application as a diagnostic tool and to acquire know-how on the excitation and detection of TAEs in large tokamaks. The modes are excited using the same antennae used for Alfvén wave heating, but at low power ($\leq$ 1 kW) and are detected by two magnetic probes. The experimental setup is described in detail in another paper presented at this meeting [28]. The main results are shown in figure 8, which is the same of that work. The eigenmodes appear in four distinct bands, corresponding to different combinations of the poloidal n and toroidal m modes numbers, which are being carefully identified. The frequency in each band decreases inversely proportional to the square root of the density, as predicted by theory.

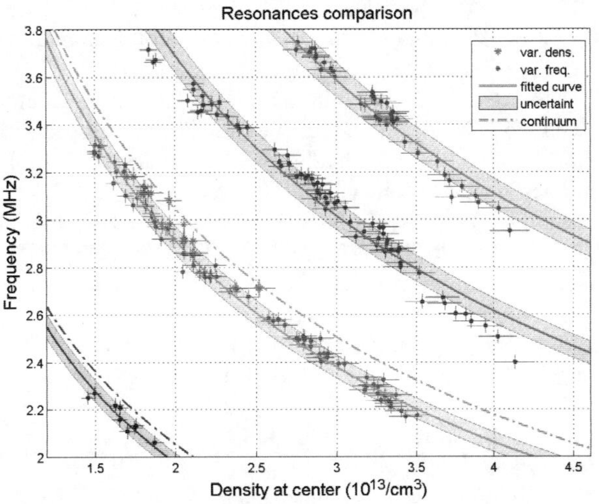

Figure 8. Summary of identified resonant frequencies over the estimated density at the center of the plasma. The dot-dash lines represent the Alfvén continuum minimum for effective mass number A=1.36 and density 1.15 times higher as the value shown on axis n/m = -1/-1 (blue) and -2/-1 (green).

6. Conclusions

The results of four lines of experimental work being currently carried out in TCABR have been summarized in this work. The effect of coupling between MHD modes on the influx of impurity, which affects the performance of improved confinement regimes triggered by edged electrostatic biasing, has been clarified. The radial profiles of the poloidal and toroidal rotation of the plasma column have been determined and it has been shown that, in the presence of rotation, the three branches of the GAMs that were theoretically predicted agree well with experimental measurements of long distance electrostatic potential fluctuations at the edge of the plasma column. Finally, a detailed investigation of the excitation of GAWs has been carried out and the results are in good agreement with theoretical predictions.

Acknowledgements

This work was supported by FAPESP (Research Foundation of the State of São Paulo), grant 2011/50773-0, and by CNPq (National Council for Scientific and Technological Development), grants 307511/2011-1 and 480733/2013-9.

References

[1] TCA Team 1985 *Nuclear Fusion* **25** 1041

[2] Campos D O, Elizondo, J I, Fagundes A N, Ferreira A A, Galvão R M O, Kuznetsov Yu K, Nascimento I C, Ruchko L F, Severo J H F and Vannucci A 2000 *Proceedings of the 2000 International Congress on Plasma Physics - Contributed Papers. Quebec*

[3] Galvão R M O, Bellintani V, Bengtson R D, Elfimov A G, Elizondo J I, Fagundes A N, A Ferreira A A, Fonseca A M M, Kuznetsov Y K, Lerche E A, Nascimento I C, Ruchko L F, de Sa W P, Saettone E A S, Sanada E, Severo J H F, da Silva R P, Tsypin V S, Usuriaga O C and Vannucci A 2001 *Plasma Phys. Control. Fusion* **43** A299

[4] Weynants R R, Jachmichz S and G Van Oost 1998 *Plasma Phys. Control. Fusion* **40** 635

[5] Nascimento I C, Kuznetsov Yu K, Severo J H F, Fonseca A M M, Elfimov A G, Bellintani V, Machida M, Heller M VA P, Galvão R M O, Sanada E and Elizondo, J I 2005 *Nuclear Fusion* **45** 796

[6] Nascimento I C, Kuznetsov Yu K, Guimarães-Filho Z O, El Chamma-Neto, Usuriaga O, Fonseca A M M, Galvão R M O, Caldas I L, Severo J H F, Semenov I B, Ribeiro C, Heller M VA P, Bellintani V, Elizondo, J I and Sanada E 2007 *Nuclear Fusion* **47** 1570

[8] Severo J H F, Nascimento I C, Tsypin V S and Galvão R M O 2003 *Nuclear Fusion* **43** 1047

[9] Severo J H F, Nascimento I C, Tsypin V S, Kuznetsov Yu K, Saettone E A S, Vannucci A, Galvão R M O, Tendler M and Mikhailovskii A B 2004 *Physics of Plasmas* **11** 846

[10] Severo J H F, Nascimento I C, Kuznetsov Yu K, Tsypin V S, Galvão R M O and Tendler M 2007 *Review of Scientific Instruments* **78** 043509

[11] Severo J H F, Nascimento I C, Kuznetsov Yu K, Galvão R M O, Guimarães-Filho Z O, Borges F O, Usuriaga O C, Elizondo, J I, de Sa W P, Sanada E K, and Tendler M 2009 *Nuclear Fusion* **49** 115026

[12] Ronchi G, Severo J H F, Sá W P and Galvão R M O 2014 *Data Acquisition and Automation for Plasma Rotation Diagnostic in the TCABR Tokamak*, submited to J. Phys.: Conf. Series.

[13] Lin Z, Hahm T S, Lee W W, Tang W M and White R B 1998 *Science* **281** 1835

[14] Diamond P H, Itoh S-I, Itoh K and Hahm T S 2005 *Plasma Phys. Control. Fusion* **47** R35

[15] Burrell K.H. 1997 *Phys. Plasmas* **4** 1499

[16] Fujisawa A 2009 *Nucl. Fusion* **49** 013001

[17] Xu Y, Jachmich S, Weynants R. R, Van Schoor M, Vergote M, Kraemer-Flecken A, Schmitz O, Unterberg B and Hidalgo C 2009 *Phys. Plasmas* **16** 110704

[18] Kuznetsov Yu K, Nascimento I C, Silva C, Figueiredo H, Guimarães-Filho Z O, Caldas I L, Galvão R M O, Severo J H F, Toufen D L, Ruchko L F, Elfimov A G, Elizondo, J I, de Sa W P, Usuriaga O C, *et al* 2012 *Nucl. Fusion* **52** 063004

[19] Elfimov A G, Galvão R M O and Sgalla R J F 2011 *Plasma Phys. Control. Fusion* **53** 105003

[20] Ilgisonis V I, Lakin V P, Smolyakov A I and Sorokina E A 2011 *Plasma Phys. Control. Fusion* **53** 065008

[21] Appert K, Balet B, Vaclavik J 1982 *Physics Letters A* 233

[22] Ross D W, Chen G L and Mahajan S M 1982 *Phys. Fluids* **25**, 652

[23] Evans T E, Valanj P M, Benesch J F, Bengston R D, Li Y M, Mahajan S M, Oakes M E, Ross, D W, Wang X Z and Watkins J G 1984 *Phys. Rev. Lett.* **53** 1743

[24] Collins G A, Hofmann F, Joye B, Keller R, Lietti A, Lister J B and Pochelon A 1986 *Physics of Fluids* **29**, 2260

[25] Holties H A, Fasoli A, Goedbloed J P, Huysmans G T A and Kerner W 1997 *Phys. Plasmas* **4**, 709

[26] Elfimov A G 2008 *Phys. Plasmas* **15**, 074503

[27] Fasoli A, Borba D, Gormezano C, Heeter R, Jaun A, Jacquinot J, Kerner W, King Q, Lister J B, Sharapov S, Start D and Villard L 1997 *Plasma Phys. Control. Fusion* **39**, B287

[28] Puglia P G P, Elfimov A G, Ruchko L F, , Galvão R M O, Guimarães-Filho Z O, Ronchi G, Fonseca A M M, , Kuznetsov Yu K, Nascimento I C, Reis A P, de Sá W P, Sanada E K, Severo J H F, Theodoro V C and Elizondo J I O 2014 *Excitation of Global Alfvén Waves by Low RF Power on TCABR*, submited to J. Phys.: Conf. Series.

Journal of Physics: Conference Series **591** (2015) 012002 doi:10.1088/1742-6596/591/1/012002

Excitation of Global Alfvén Waves by Low RF Power on TCABR

P G P Puglia, A G Elfimov, L Ruchko, R M O Galvão, Z O Guimarães-Filho, G Ronchi, A M M Fonseca, Yu K Kuznetsov, I C Nascimento, A P Reis, W P de Sá, E K Sanada, J H F Severo, V C Theodoro and J I Elizondo

Plasma Physics Laboratory, Institute of Physics at University of São Paulo, São Paulo, Brazil

E-mail: ppuglia@if.usp.br

Abstract. Recent results of Global Alfvén wave (GAW) excited by external antenna and fed by low radio-frequency (RF) power ($\leq$1kW) in tokamak TCABR are presented. The goal of this work is to develop a diagnostic tool based on the excitation of GAW using low power RF generators when the waves can be excited without perturbing the basic plasma parameters in TCABR. This method named MHD diagnostics has already been developed on other tokamaks for toroidicity induced Alfvén eigenmodes as well for GAW. Two magnetic probes are used for measurements of the magnetic field perturbations in two regimes of excitation. In the first one, the fixed RF frequency is combined with gas-puffing induced density rise and relaxation (1-2)$\times 10^{13}$ /cm^3 in order to meet the GAW resonance, while in the second excitation method RF frequency sweeps 2-4 MHz are applied to the relatively stable plasma density. The plasma discharges are accompanied by saw-tooth (ST) oscillations registered by Soft-X Ray diagnostics that helps in an identification of the GAW resonances due to ST density oscillations. It is found that the fixed frequency GAW resonance appears during density rise, as well during slow density reducing after stop of gas puffing. The effective mass number A_{eff}=1.36 is found.

1. Introduction

Global Alfvén waves (GAW) discovered theoretically [1, 2] and observed experimentally in tokamaks [3, 4] were later proposed to use for diagnostic purpose. This method named as "MHD diagnostics" has already been proposed in [5, 6, 7, 8, 9] to identify the mass number and security factor profile and was tested in other tokamaks for toroidicity induced Alfvén eigenmodes as well for GAW. The basic Alfvén wave (AW) continuum equation is presented as follows:

$$\omega_A = k_{\|}V_A = \frac{B}{\sqrt{\mu_0 n m_H A_{eff}}}\frac{1}{R}\left(N + \frac{M}{q}\right) \quad (1)$$

Here, ω_A is the Alfvén wave frequency, $k_{\|}$ is the wave vector parallel to the magnetic field, V_A is the Alfvén speed, B is the magnetic field strength, n is the electron density, m_H is the hydrogen nuclei mass, A_{eff} is the effective ion mass, R is the major tokamak radius, (N,M) the toroidal and poloidal wave numbers and q the safety factor. Typically, GAW eigenmodes may appear at the extreme points of the AW continuum to avoid strong continuum dissipation. As we can see in equation (1), the A_{eff} is inversely proportional to the Alfvén frequency in square giving us the possibility to try to estimate that parameter based on the identified GAW frequency, density profile, mode numbers and consideration of q profile.

In this work, we are going to use the excitation of Global Alfvén waves as a diagnostic tool to identify mass number in TCABR. The machine has the following main parameters: major radius $R_0 =$ 0.615m, minor radius a = 0.18m, toroidal magnetic field $B0$ = 1.07T, plasma current $IP \approx$80- 85kA, line averaged electron density $\bar{n}$= 1.2– 2.4×10^{19}m^{-3}, and central electron temperature $T_{e0}\approx$430-480eV, in the discussed ohm discharges. We find that the typical plasma, which is similar to the old TCA [10], stays in a mixed state of diffusion, a plateau diffusion for the hydrogen ions and electrons and Pfirsch-Schlüter regime for the impurities, that may cause an accumulation of impurities on the plasma core [11] changing the electronic density profile. If we assume a parabolic profile for the impurity free plasma, we end up with a density profile in the form [11]

$$n_e = n_{0H}(1 - r^2/a^2)^{\alpha} + n_{0Z}e^{-\frac{r^2}{(0.4a)^2}} \quad (2)$$

Where n_{0H} is the hydrogen density at the plasma center, a is the plasma minor radius, r the radial position, α a profile parameter and n_{0Z} the impurity density at the plasma center. The impurity accumulation creates a higher density on the plasma core, so we can expect a higher value for the effective ion mass on the center of the plasma because of the impurity accumulation.

Contrary to previous high RF power experiments [12], it is proposed to apply the low power (≤1kW) of RF generators to excite waves without disturbing the basic plasma parameters. Here, we are more concerned about the identification of the Alfvén waves, to perform an estimate of the plasma ion effective mass, and the ions relative concentration. The results presented in this paper are only a rough estimate and will be more explored in future work. The GAW excitation is proposed in the TCABR conditions with sawtooth oscillation when q-value at center is near to one. For monotonic q-profile typical for TCABR, the GAW eigenmodes may be exited at the plasma core for the toroidal modes N≥2. By identifying the frequency of excited waves and their mode numbers, therefore, it is possible to obtain information about the effective mass of ions.

2. Experimental Setup

To excite GAW in TCABR, we use the four strap antenna developed for the AW heating of the plasma [13]. Lower power is applied and the waves are detected with a pair of magnetic probes separated by the distance 25cm in toroidal direction. The scheme of our experimental setup is shown in figure 1, where we may observe two antenna modules composed of four straps each on opposite sides of the tokamak. In the presented experiments, we only use one strap of one antenna module to excite the GAW.

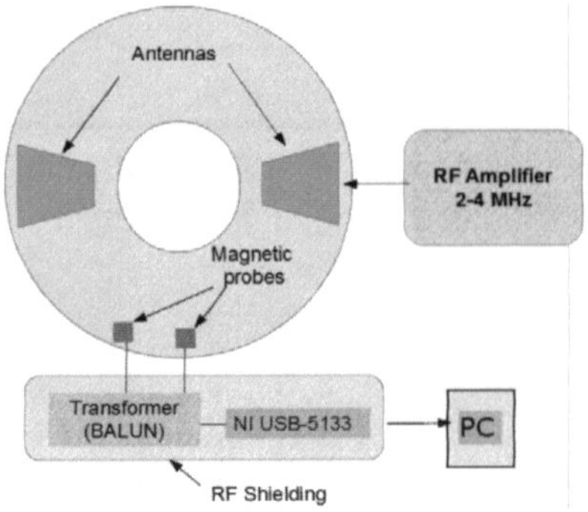

Figure 1. Schematic of the setup used for the experiments. Highlighting the principal equipment used for GAW excitation and detection.

Beyond the antenna, we have two magnetic probes, each probe may be used to measure the poloidal or toroidal magnetic field perturbations. Only the poloidal field is considered in the presented work. The RF antenna current is measured with a Rogowski coil. To collect the data we use a digitizer from National Instruments (NI) that transfers the collected raw data to the PC via USB connection. The amplifier used to feed the antenna is a class-D RF amplifier with MOSFET that operates in the frequency range 2-4 MHz.

2.1. Magnetic probes and BALUN

To measure the magnetic fields excited by the antenna in the plasma we use two poloidal magnetic probes [13] separated by a toroidal angle of approximately ~ 20° and on the same poloidal angle. The magnetic probe signal first goes through a BALUN (BALanced to UNbalanced transformer) [14] so we can ground one of the sides of the signal for measurement, the BALUN also works like a choke, eliminating common modes. The setup presented in figure 2 is also known as a choke BALUN. To digitize the signals we use two NI USB-5133 modules connected to a PC, this equipment is from National Instruments. We set the sampling frequency at 50MS/s, as we are interested in frequencies up to 4 MHz only, this gives sufficient number of points in a period for analyzes.

The PC, BALUN and NI USB-5133 are placed inside a shielded cabinet, so we have some kind of additional insulation from RF interference. The cables used to send the signals from the magnetic probes to the BALUN are BNO (twin-axial cabling).

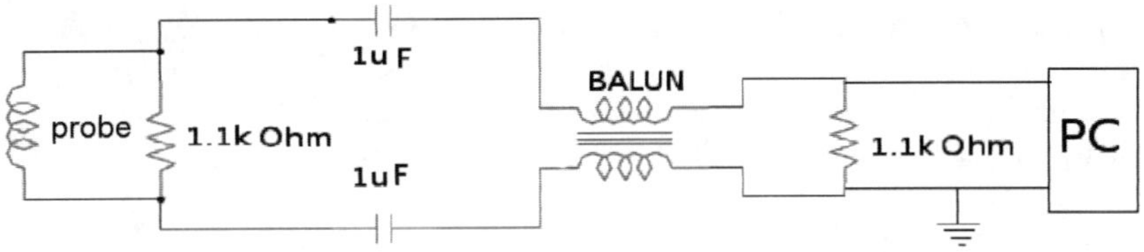

Figure 2. Schematic of the equipment use for measurement of magnetic probe signals.

The system has four input channels of 8 bits. Besides the probe signals, we also measured the antenna RF current using a Rogowski coil. There is care not to do ground loops as we use different systems, the use of the BALUN, as it is shown on the scheme figure 2, helps prevent it.

2.2. RF Amplifier and antennas

To induce the waves in the plasma we use a class-D RF amplifier with MOSFET, which operates in the frequency range 2-4 MHz it allows frequency scanning or fixed frequency currents. The amplitude of the RF current stays stable enough during ~20ms in the antenna with a Faraday screen. Each antenna set consists of four straps, two on top of the tokamak and two on the bottom, but only one strap was used to excite the waves in these experiments. The amplifier is a switching amplifier, and it needs an external function generator to switch the MOSFETs at the desired frequency to excite the GAW.

In figure 3, we present the time evolution of the RF current in the two different detection modes (fixed and swept frequencies). In the first case, gas injection is used to stimulate a ramp up of the plasma density, due to that the resonant condition may be met. In the second case, the density stays almost constant when the series of frequency sweeps of the RF current is induced.

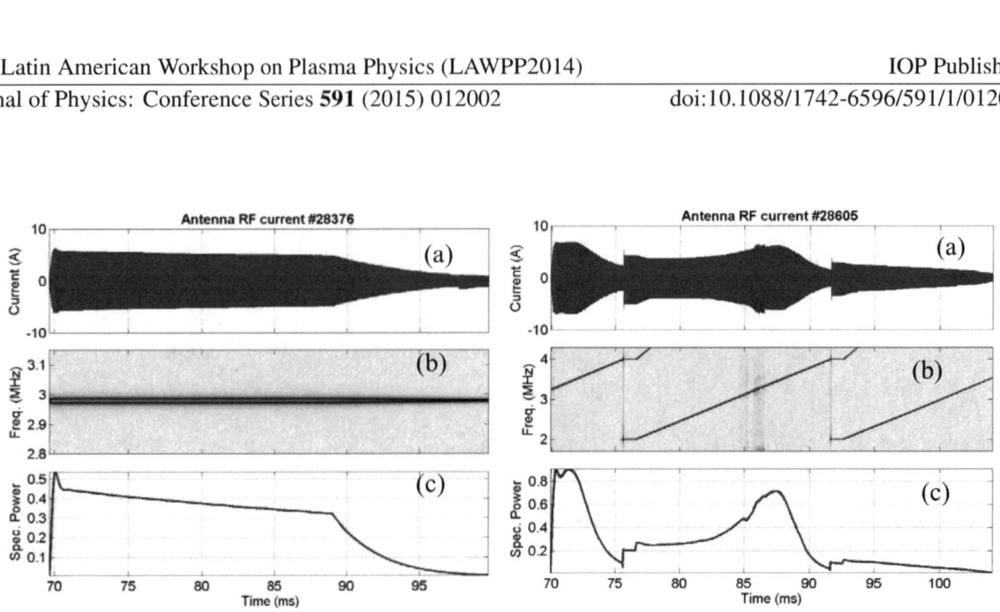

Figure 3. Typical characteristic of the RF current excited by the amplifier at the antenna. With fixed frequency (left) and frequency scanning (right). There is the raw data (a), the spectrogram (b) and spectral power on the RF frequency (c).

3. Results and Discussion

As it was mentioned the previous section, we used two methods to identify the resonances. To illustrate these methods, we selected two different shots that are typical for each case. The plasma parameters for each type of shots are shown in figure 4. In the shot #28376, a RF current of fixed frequency with a ramp up in the plasma density is used, while a sweep of the RF power frequency with a fixed plasma density is used in the shot #28605.

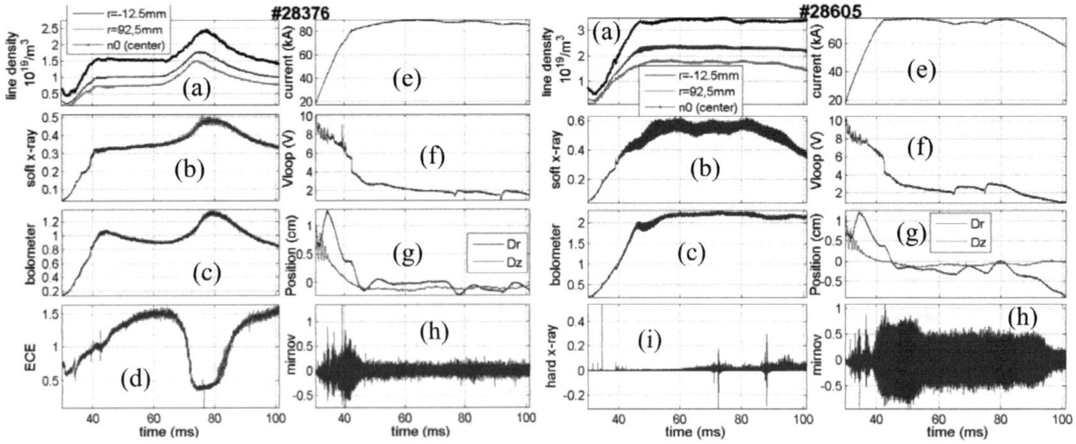

Figure 4. Basic plasma parameters for TCABR shots #28376 (left) and #28605 (right). The signals presented are the plasma electronic density (a), soft X-ray emission from core (b), bolometer signal from core, ECE signal (d), plasma current (e), loop voltage (f), plasma position (g), magnetic oscillations (h) and hard X-ray emission (i).

3.1. Shot #28376

We begin the discussion of RF experiments related to the shot #28376 shown in figure 4 (left) and the regime of the RF current to excite the GAW is presented in figure 3. In this case, gas injection is performed in the middle of the discharge and RF power was applied during the formed density ramp up. Assuming a parabolic density profile $\propto n_0(1 - r^2/a^2)^\alpha$, we calculate the density n_0 at the plasma center and the α-power using two line integrated density data from our interferometer. Those have the chord positions r=-1.25 and +9.25 cm with respect to the chamber center. It should be noted that the position of the plasma column was stable during the density ramp up and the estimated Shafranov shift is ~ 0.8 cm [15]. Due to that, we may adjust the central density and the α-parameter, those are necessary for A_{eff} in equation (1) because the GAW depends strongly on the central plasma density.

Using the respective probe measurements, in figure 5 we present the phase difference between the signals of the probes and the spectral power respective to the same frequency of the RF current. Two GAW resonances, one for the rising density and the other for the falling density, are identified by the abrupt changes in the phase and spectral power signals. The estimated density at the center of the plasma is the same for both resonances, $1.97 \times 10^{13}/cm^3$ and $1.93 \times 10^{13}/cm^3$ respectively, indicating the same kind of excited mode.

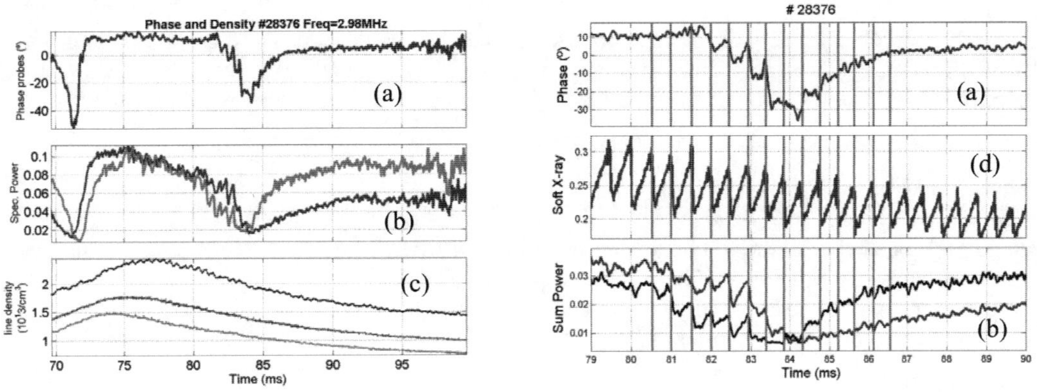

Figure 5. Probe signal for shot #28376 compared with the plasma electronic density (left) and with saw-tooth oscillations measured by soft X-ray (right). We have the phase between the probes (a) and their spectral power at the RF frequency (b), the electronic density measured by interferometer and estimation of central density (c) and saw-tooth oscillations measured by soft X-ray (d).

Beating of the probe signals by the strong saw-tooth (ST) oscillations measured with the soft x-ray diagnostic observed during the slow falling down density phase is shown in figure 5. Due to the slower variation, there are many saw-tooth oscillations that occur during the necessary time to pass the GAW resonant condition that reveals some specific details of the GAW resonances. We propose that this strong beating is related to the accumulation of impurities on the plasma core during the saw-tooth formation and its release after the saw-tooth crash [16]. The impurities accumulated on the plasma center changes the electron density and effective ion mass that consequently changes the GAW resonant frequency. This accumulation explains the beating, as well as the phase inversion of the saw-tooth oscillations at the minimum phase difference between the probes in figure 5. The phenomenon of the GAW frequency modulation had already been demonstrated in other experiments [7], but the phase inversion effect was not observed.

The ST phase inversion can be considered as the exact moment of the GAW resonant condition. Before that moment, the ST density rise gets closer to the GAW resonance, then, it abruptly moves away from the GAW resonant condition in each saw-tooth crash. After passing of the GAW resonance, situation is changed to inverse. Due to the ST effect and the localization of the q=1 position at $r \approx 4$ cm, we can expect that the GAW stays at the plasma core. Finally, the GAW mode numbers may be identified as *N/M*=-2/-1 because this mode has the q-minimum position at the magnetic axes for the standard discharges in TCABR for monotonic q-profile in accordance with equation (1).

As it was earlier mentioned, the Pfirsch-Schlüter regime may causes an accumulation of impurities on the plasma core [9] changing the density profile equation (2). This accumulation creates a higher density at the plasma core, so a higher value for the effective ion mass on the center of the plasma should be expected. To confirm that, we compare the cut off density of the ECE signal that is at $n_{cut}=2.22\text{x}10^{13}\text{cm}^{-3}$ with the one estimated from the interferometer data for the already mentioned parabolic profile $n_0 \approx 2.\text{x}10^{13}\text{cm}^{-3}$. That indicates that the electronic density may have a "Mexican hat" profile [16,17] due to impurity accumulation. Using the density value $n_0=2.22\text{x}10^{13}\text{cm}^{-3}$, we estimate $A_{eff} \approx 1.36$ with 10% relative error. To complete this analysis for the impurity composition (carbon + iron), we use scaling law of the old TCA [10].

$$Z_{eff}(0) = 1.84 \times 10^{-4}\left(T_{e0}^{1.5} V_{Loop}/B_T\right) - 0.755 \tag{3}$$

Where $T_{e0} \approx 480\text{eV}$ is the plasma temperature at the center, $V_{loop} \approx 1.8\text{V}$ the loop voltage and $B_T=1.07\text{T}$ the toroidal magnetic field. Using equation (3) we calculate $Z_{eff}=2.5$ for the data of discharge #28376. Finally, that gives us $A_{eff}=1.34$ considering 3% of light ion impurities (totally striped carbon), and 0.4% of heavy ion impurities, that is, iron (Fe^{+15}) in our case.

3.2. Shot#28605

Another method to excite the GAW modes is shown on figure 6. In this case, we try to maintain a flat plasma density together with the frequency sweeps from 2-4 MHZ lasting 15 ms. Opposed to the other shot, in this one we try to maintain the plasma density as flat as possible. The respective RF current used to excite the modes is shown in the right panel of figure 3 and the basic plasma parameters from this shot can be seen on the right panel of figure 4.

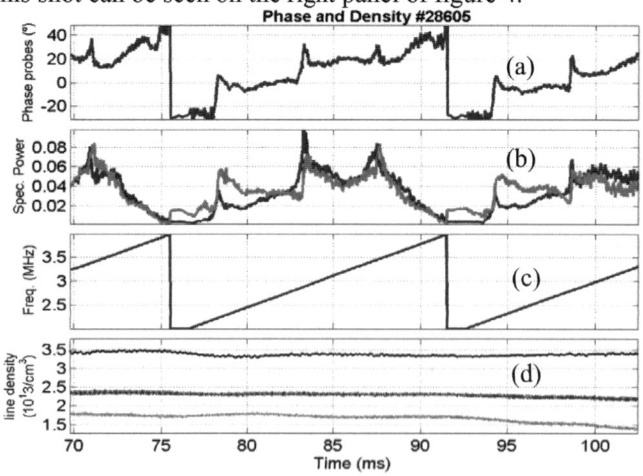

Figure 6. Time traces of the phase difference (a), spectral power (b) of the probe signals in comparison with the frequency sweeps (c) of the RF antenna current and the plasma electronic density (d) in the shot #28605.

The result of the signal treatment similar to the previous case is presented in figure 6. We can observe three distinct resonant spikes in the phase difference and spectral power of the probe signals during one sweep (75.4-91.5 ms). Each identified resonance corresponds to a different GAW mode number. Different from shot #28376, we cannot observe the beating with saw-tooth oscillations in figure 6 because the sweeping time for the frequency range is too fast to resolve the ST period.

3.3. Data obtained from series of shots

In figure 7, we present a summary of the GAW frequency dependence from central electronic density, based on data collected from about 50 shots with detected GAW resonances. The GAW modes form four distinct groups, due to different combination of the toroidal and poloidal mode numbers (*N/M*), and in each group the frequencies are inversely proportional to the square root of the density, as it comes from equation (1). It is difficult to get the spectrum of the higher frequencies, as we have large number of *M/N* combinations to satisfy the obtained data.

The presented data in figure 7 are fed with error bars from the measurements, we estimated an error of about 4% for the central density and 30 kHz for the frequency value. Regarding the mode numbers of the data presented, the lower blue mode should be the mode *N/M*=-1/-1, as this is the lowest possible mode number [1, 2]. The mode marked by the red dots has *N/M*=-2/-1 as mode numbers, the identification is due to the ST localization at the plasma center and its beating with the magnetic probe data, as explained previously. Identification of other modes is doubtful, because there are more than one combination of *N/M* that set the experimental data, and what we can say is only the most probable *N+M* combination for each specific band. This way we have *N+M*=-2 (blue), *N+M*=-3 (green), *N+M*=-4 (wine) and *N+M*=-5 (pink).

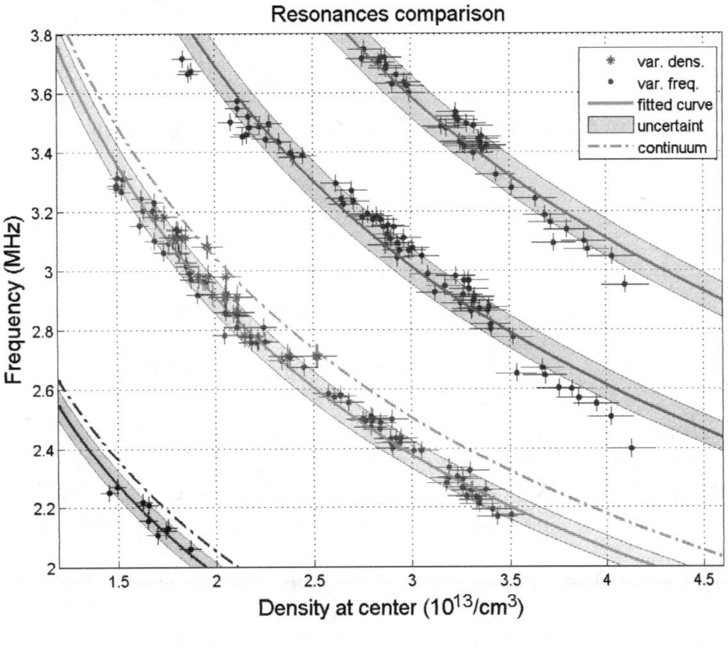

Figure7. Summary of identified resonant frequencies over the estimated density at the center of the plasma for resonances with N+M equals to -2 (blue), -3 (green), -4 (red) and -5 (pink). The dot-dash lines represent the Alfvén continuum minimum for effective mass number A=1.36 and density 1.15 times higher as the value shown on axis *N/M*=-1/-1 (blue) and -2/-1 (green).

The theoretical dependence of the frequency from density from equation (1) is also plotted (dash-dot lines) for the *N/M*=-1/-1 (blue) and -2/-1 (green) modes. The effective ion mass used was of

A_{eff}=1.36, taking into account 13% correction of the central density value, as we have found from ECE cut off for the shot #28376.

Furthermore, we note that the GAW frequency relation with the density is the same for the density ramp or frequency scan methods. It can be seen for the N+M=3 modes in figure 7 that the respective data points marked by the red stars and green dots statistically coincide. As well, we note that there is the distance $\Delta\omega/\omega_A \approx$2-5% between the continuum and main frequency for the *N/M*=-2/-1 eigenmode [1, 2, 18], this distance depends on Hall effect and magnetic shear.

4. Conclusion

The experiments with the low RF power excitation of the GAW and their detection by the magnetic probes in TCABR confirm that it may be a very effective method for diagnostics for the ion effective mass. Using the magnetic probes, we can identify the GAW resonances in TCABR and their behavior is found to be predicted by the Alfvén continuum equation (1). Also, it is demonstrated that there is no difference in results in identification of the GAW resonances excited by the fixed frequency during the density ramp up or RF current frequency sweep methods. The typical mass number is estimated to be at A_{eff}=1.36 and Z_{eff}=2.5, this corresponds to a mixture of 3% of the full stripped carbon impurities and 0.3% of iron Fe^{+15} impurities. For better results, it is necessary additional information of the central density detected by other diagnostics, and it should be noted that the presented method for A_{eff} identification is very sensitive to the considered value of the plasma density.

Acknowledgements

This work was supported by FAPESP (Research Foundation of the State of São Paulo), grant 2011/50773-0.

References

[1] Appert K, Gruber R, Troyuon F and Vaclavik J 1982 Excitation of global eigenmodes of the Alfven wave in Tokamaks *Plasma Phys.* **24** 1147

[2] Ross D W, Chen G L and Mahajan S M 1982 Kinetic description of Alfvén wave heating *Phys. Fluids* **25**, 652

[3] Evans T E et al 1984 Direct Observation of the Structure of Global Alfvén Eigenmodes in a Tokamak Plasma *Phys. Rev. Lett.* **53** 1743- pretex

[4] Collins G A, Hofmann F, Joye B, Keller R, Lietti A, Lister J B and Pochelon A 1986 The Alfvén wave spectrum as measured on a tokamak *Physics of Fluids* **29**, 2260-2272

[5] Holties H A, Fasoli A, Goedbloed J P, Huysmans G T A and Kerner W 1997 Determination of local tokamak parameters by magnetohydrodynamic spectroscopy, *Phys. Plasmas* **4**, 709-719

[6] Fasoli A, Dobbing A, Gormezano C, Jacquinot J, Lister J B, Sharapov S E and Sibley A 1996 Alfven eigenmode excitation by ICRH beat waves, *Nucl. Fusion*, **36**, 258

[7] Collins G A, Howling A A, Lister J B and Marmillod P 1987 Central mass and current density measurements in Tokamaks using the discrete Alfven wave spectrum *Plasma Phys. Control. Fusion* **29** 323

[8] Descamps P, Wassenhovel G, Koch R, Messiaen A M and Vandenplas P E 1990 Determination of central q and effective mass on textor based on discrete Alfvén wave (DAW) spectrum measurements *Phys. Lett.* A, **143**, 311

[9] Elfimov A G, Ruchko L F, Galvão R M O,Elizondo J I, Sanada E, Kuznetsov Y, Fagundes A N, de Sá W P, Varandas C A F, Manso M E C, Varela P, Silva A, Ivanov A A 2006 Identification of local Alfvén wave resonances with reflectometry as a diagnostic tool in tokamaks *Nucl. Fusion* **46** S722

[10] De Chambrier A, Collins G A, Hollenstein C, Joye B, Lister J B, Moret, J-M, Nowak S,

Pochelon A and Simm W 1984 *Target plasma conditions in TCA* Center for Research In Plasma Physics, École polytechnique fédérale de Lausanne

[11] Fussmann G, Field A R, Kallenbach A, Krieger K, Steuer K -H and The ASDEX Team 1991 *Plasma Phys. Control. Fusion* **33** 1677

[12] Ruchko L F, Ozono E, Galvão R M O, Nascimento I C, Degasperi F T and Lerche E 1998 Advanced antenna system for Alfvén wave plasma heating and current drive in TCABR tokamak. *Fusion Engineering and Design* **43** 15

[13] Lerche E 2003 *Aquecimento do plasma por ondas de Alfvén no tokamak TCABR*, . Universidade de São Paulo: Instituto de Física. online: http://www.teses.usp.br/teses/disponiveis/43/43134/tde-17072012-141903/.

[14] Sevick J 2006 *Transmission Line Transformers* (Raleigh: Scitech Publishing)

[15] Pereverzev G, Yushmanov P N 2002 Astra Automated System for Transport Analysis in a Tokamak *IPP-Report* 5/98 Max-Plank-Institute für Plasmaphysik

[16] Nicolas T, Sabot R, Garbet X, Lütjens H, Luciani J -F, Guimaraes-Filho Z, Decker J and Merle A 2012 *Phys. Plasmas* **19** 112305

[17] Nicolas T, Sabot R, Garbet X, Lütjens H, Luciani J -F, Sirinelli A, Decker J, Merle A and JET-EFDA Contributors 2013 Particle Flow during Sawtooth Reconnection: Numerical Simulations of Experimental Observations *Plasma and Fusion Research* **8** 2402131

[18] Amarante-Segundo G, Elfimov A G, Galvão, R M O, Ross D W and Nascimento I C 2001 Calculations of Alfvén wave driving forces, plasma flow, and current drive in the Tokamak Chauffage Alfvén wave experiment in Brazil (TCABR) *Phys. Plasmas* **8** 210

Correlation properties of Geodesic Acoustic Modes in the T-10 tokamak

A V Melnikov[1, 2 *], L G Eliseev[1], S E Lysenko[1], S V Perfilov[1], D A Shelukhin[1], V A Vershkov[1], V N Zenin[1], L I Krupnik[3], N K Kharchev[4] and HIBP team[3]

[1]National Research Centre "Kurchatov Institute", 123182, Moscow, Russia
[2] National Research Nuclear University MEPhI, 115409, Moscow, Russia
[3]Institute of Plasma Physics, NSC "Kharkov Institute of Physics and Technology", Kharkov, Ukraine
[4]Institute of General Physics, Russian Academy of Science, Moscow, Russia

E-mail: melnikov_07@yahoo.com

Abstract. Geodesic acoustic mode (GAM) of electrostatic potential and density fluctuations are simultaneously measured by Heavy Ion Beam Probe (HIBP) and Correlation Reflectometry (CR). The regimes with Ohmic and electron cyclotron resonance heating (ECRH) were studied (B = 2.2 T, I_p = 210 – 240 kA, n_e = (2.3 – 3.7) × 10^{19} m^{-3}, P_{EC} < 0.6 MW). GAMs are more pronounced during ECRH, when the typical frequencies were seen in the narrow band from 22 to 27 kHz for the main peak and 25-30 kHz for the higher frequency satellite peak. The local values of electric potential and density fluctuations show the significant coherency and constant phase shift at GAM frequency range. The existence of the long-distance (one quarter of the torus) correlations of core electric potential and density for GAM implying that GAM is a global mode, was shown for the first time in tokamaks.

1. Introduction

It is believed that the transport processes in the toroidal plasmas are dominated by the turbulence. The studies of the turbulent transport processes and the phenomena of the turbulence self-regulations like Zonal Flows (ZF) and Geodesic Acoustic Modes (GAMs) are of crucial importance to understand the physical picture of anomalous transport. It has been recently observed experimentally the correlation between GAM and high frequency turbulence, unless the level of the modulation was not found to be high [1-3]. The correlation properties of GAMs are important issues in the transport and turbulence studies. The most direct and powerful tool to study GAM in the core plasmas is HIBP [4], which is routinely used in T-10 for both mean potential profiles and oscillations [5]. It has been recently shown by HIBP that GAM poloidal mode number m = 0 [6]. The first experimental characterization of the correlation properties of ZF in stellarator plasma was done with dual HIBP in CHS [7].

The paper is dedicated to the first result of the core correlation measurements of the GAM oscillations made by different diagnostics in a tokamak. It reports the characterization of the GAMs on T-10, the correlation between the local values of potential and density, and the long-distance potential/density correlation.

* To whom any correspondence should be addressed.

2. Experimental set-up

Geodesic acoustic modes (GAM) were investigated in the T-10 tokamak using Heavy Ion Beam Probe (HIBP) [8, 9], Multipin Langmuir Probe (MLP) and Correlation Reflectometry (CR) [10] diagnostics. Regimes with Ohmic heating and with on- and off-axis ECR heating were studied ($B = 2.2$ T, $I_p = 210$ - 240 kA, $n_e = 2.3 - 3.7 \times 10^{19}$ m^{-3}). HIBP observed upper quadrant of the plasma column cross-section at the Low Field Side of the torus with radial position of Sample Volumes r_{SV}, varying as wide as $16<r_{SV}<30$ cm. One CR antenna was located at the High Field Side of the same cross section as HIBP, another one has a toroidal shift of one quarter of the torus at the Low Field Side. This layout was focused to the future study of the spatial mode structure of GAM.

3. Experimental results

3.1. GAM observation

HIBP is a powerful diagnostics to study GAMs in various toroidal devices [11]. It is able to get simultaneously the oscillatory components of two parameters: plasma electric potential ϕ by the energy of secondary ions and electron density n_e by total secondary beam current, I_t, the latter works, if the beam attenuation does not affect the signal (path integral effect) [12, 13]. This is the case of low density, which was studied here. It was shown that the GAMs are more pronounced in the plasma potential rather than in density, as presented in figure 1. Figure shows the potential and density power spectra, obtained by HIBP at the same time. It is clearly seen that GAM peak dominates the potential spectra, while MHD m=2 mode peak dominates the density spectra. Note that the GAM peak has dual structure, the lower frequency main peak and the higher frequency satellite, which is very typical for T-10 [2, 10].

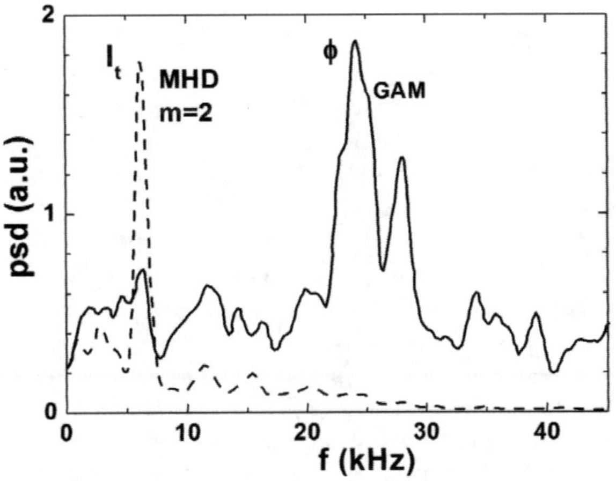

Figure 1. Power density spectra measured by HIBP for potential ϕ and total current I_t (or n_e) in discharge with ECRH. GAMs are more pronounced on the potential than on the density. In contrast, the amplitude of 7 kHz MHD m=2 mode oscillations is larger on the density and much smaller on the potential.

3.2. GAM intermittent character in OH and ECRH plasmas

Earlier observation had shown that the GAM frequency increases with T_e after the switching on the ECRH according to the GAM frequency scaling [14]. The recent experiments in OH regimes with auxiliary ECR heating shows that GAM might have a complex structure, not similar to harmonic oscillations with a single frequency. The wavelet spectrogram, presented in figure 2 (a) shows that GAM has an intermittent character presenting the stochastic sequence of the wave packages. Note that this feature holds for both the main GAM peak and the higher frequency satellite.

Journal of Physics: Conference Series **591** (2015) 012003 doi:10.1088/1742-6596/591/1/012003

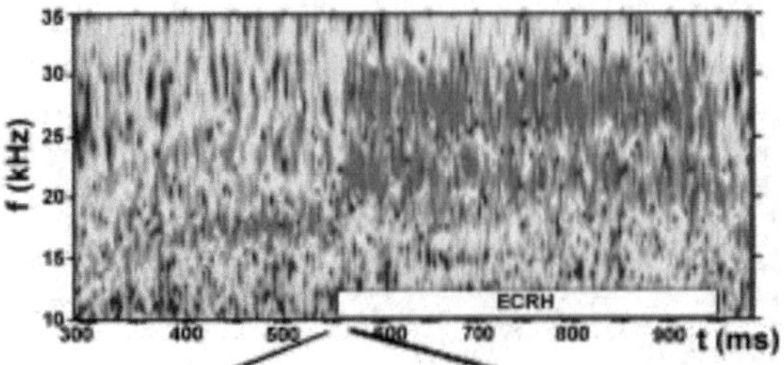

Figure 2 (a) Wavelet spectrogram for HIBP potential φ.

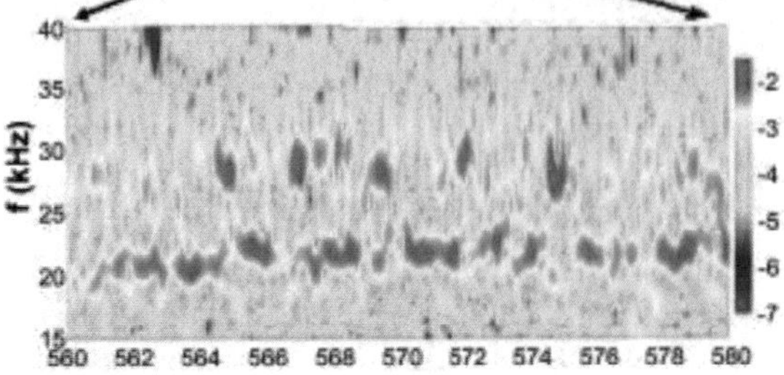

Figure 2 (b) Example of the GAM intermittent behaviour for ECR heated plasma

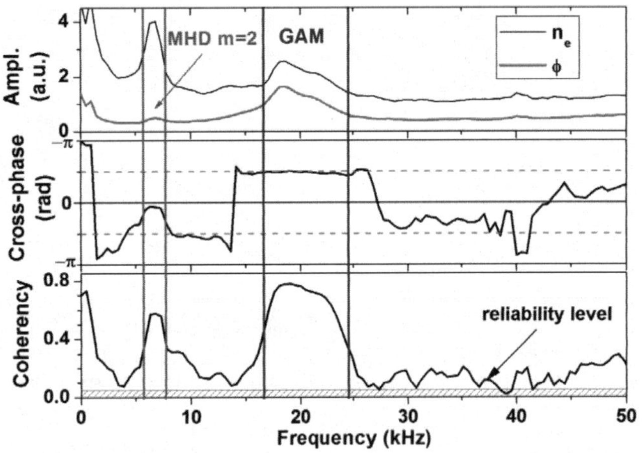

Figure 3. Local potential-density correlation. HIBP shows the high correlation at the GAM frequency range and the permanent cross-phase. For 7 kHz MHD mode m=2, the cross-phase differs from that of GAM. #44166, B=2.2 T, E_{beam} = 210 keV, r_{SV} = 22.1 cm.

For the considered T-10 experimental conditions the "lifetime" of the package lies in a range of 0.5–2 ms. This data shows that, the wavelet analysis used to be the most direct data processing tool to study the GAM properties. Figure 2 shows that in comparison with OH plasma, GAMs are more pronounced in ECRH plasmas, where the typical frequencies of the wave packages are observed in a narrow interval from 20-30 kHz at the outer one third of the plasma minor radius, as shown in figure 2 (a, b).

3.3. GAM potential-density cross-phase. Local HIBP correlations

Fourier coherence analysis with long time series (> 200 ms) shows clear coherence between local values of potential and density simultaneously measured by HIBP at the sample volume. The phase shift is found to be $\pi/2$ for GAMs in the presented example. In contrast, figure 3 shows that for MHD m=2 peak the phase shift is zero. Fourier coherence analysis with short time series (10 ms) shows the constant frequency and a bursty character of the coherence between HIBP potential and density at the GAM frequency, figure 4.

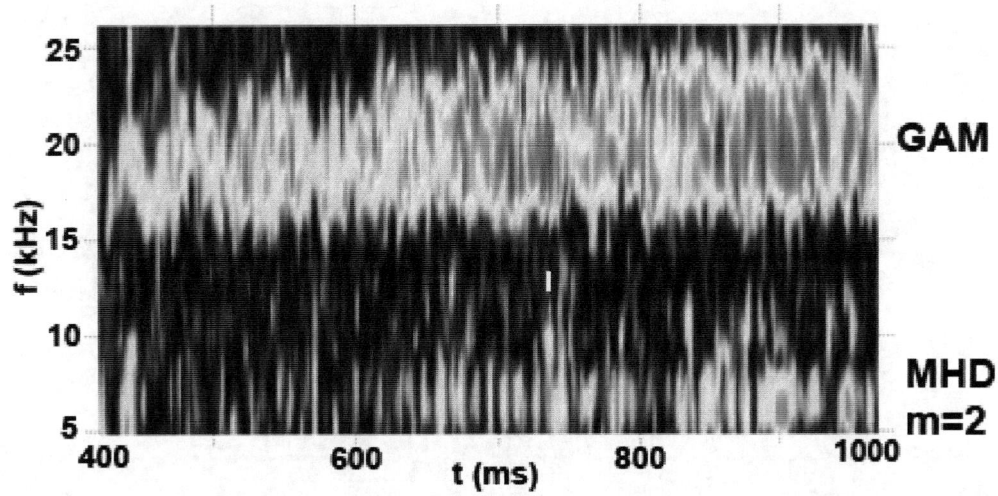

Figure 4. Fourier spectrogram for local HIBP potential-density coherency. Example of the GAM intermittent behaviour for Ohmic heated plasma. GAM frequency is around 20 kHz. MHD m=2 mode frequency is 7 kHz. #44160, B=2.2 T, $E_{\rm beam}$ = 250 keV, $r_{\rm SV}$ = 16.1 cm.

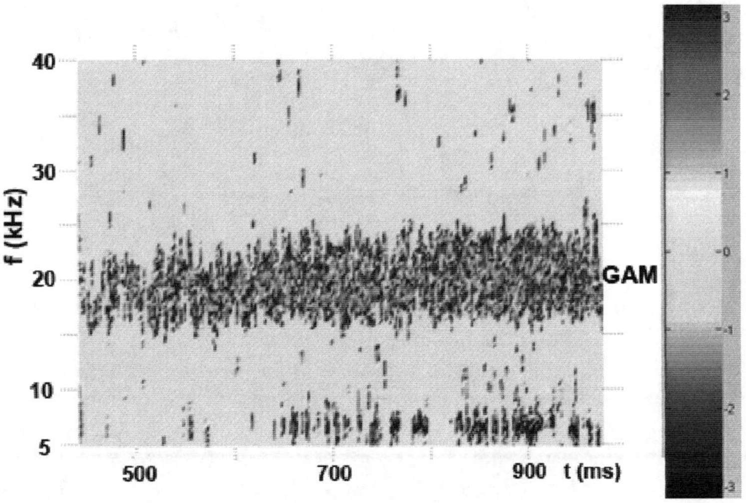

Figure 5. Time evolution of the cross-phase for local HIBP potential-density correlation, shown in figure 4. The threshold coherency $coh_t = 0.3$. If coherency > 0.3, the cross-phase is marked in colors, if coherency <0.3, the cross-phase is marked in green (zero).

The cross-phase between potential and density presents generally stochastic behavior over the observed frequency range. However, for quasicoherent modes, like MHD m=2 mode and GAMs the cross-phase is more systematic, see figure 5. To make the figure more clear and free from stochastic component, the only phase, which corresponds to the coherency exceeding some limit is presented with its color. For the lower coherency the cross-phase was marked as zero (green color in figure 5). To analyze the cross-phase for GAM, the histogram was made for the values, exceeding the threshold coherency coh. The result is shown in figure 6. The histogram method gives the same value as the long time series, as shown in figures 3 and 6.

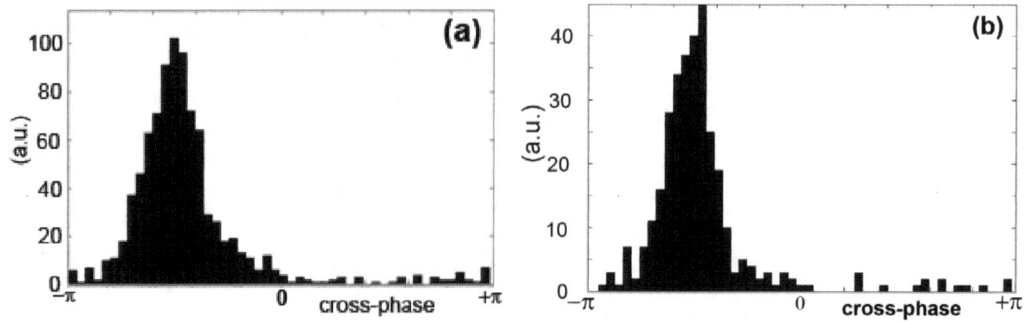

Figure 6. The examples of the cross-phase histogram for data taken from figures 4 and 5. The most frequent value for GAMs is $\pi/2$. (a) t = 760-780, (b) t = 920-940 ms.

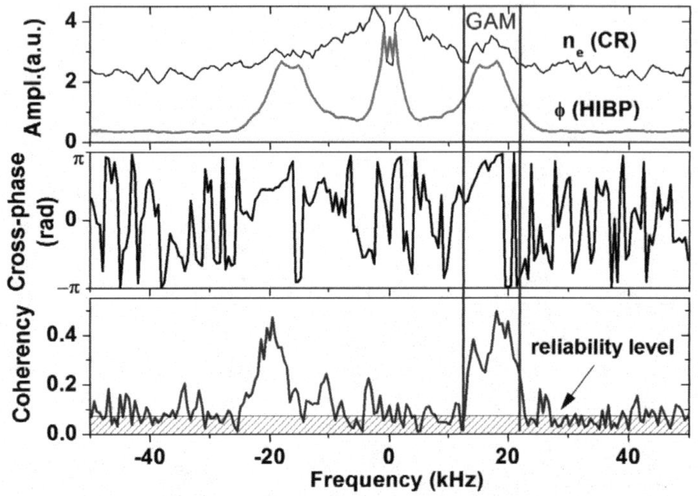

Figure 7. Long-distance potential-density coherency between HIBP and CR. #43285. B=2.2 T, E_{beam} = 250 keV, r_{SV} = 25±0.5 cm.

3.4. GAM potential-density cross-phase. Global HIBP-CR correlations

To study the long-distance correlations the following experiment was performed: both HIBP and CR were located at the same fixed radial position. Fourier correlation analysis with long time series (>200 ms) reliably shows a clear correlation between HIBP potential and CR density at the GAM frequency, see figure 7. This observation suggests a global character of the GAMs.

Fourier correlation analysis with short time series (4 ms) shows the constant frequency and a bursty character of the correlation between HIBP potential and CR density at the GAM frequency, figure 8.To study the radial range of the long-distance correlations the following experiment was performed: HIBP was located at the fixed position, while CR observation are (reflection layer) varied during a shot with the decay of the local density, presented in figure 9 (a) at 500 ms < t < 1000 ms. Figure 9 (b) shows the reliable existence of the correlation between the HIBP potential and CR density at the GAM frequency during all the shot. The coherency remains almost unchanged around 1 cm radial variation of CR, while HIBP position was 1 cm shifted outwards. This observation shows that the radial correlation length for GAM is higher than 2 cm. This agrees with our earlier CR estimation of k_r = 3-5 cm [2] and more recent observations of the large radial extend of the GAM [15].

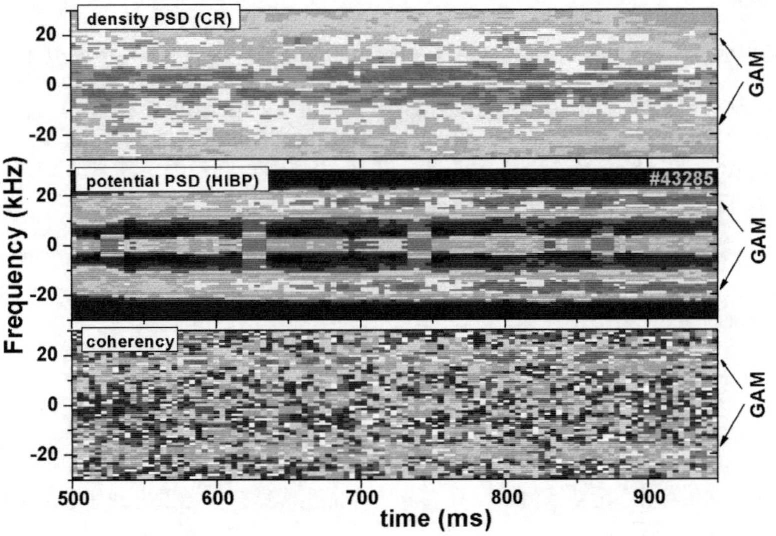

Figure 8. Time evolution of long-distance potential-density correlations. HIBP versus CR.

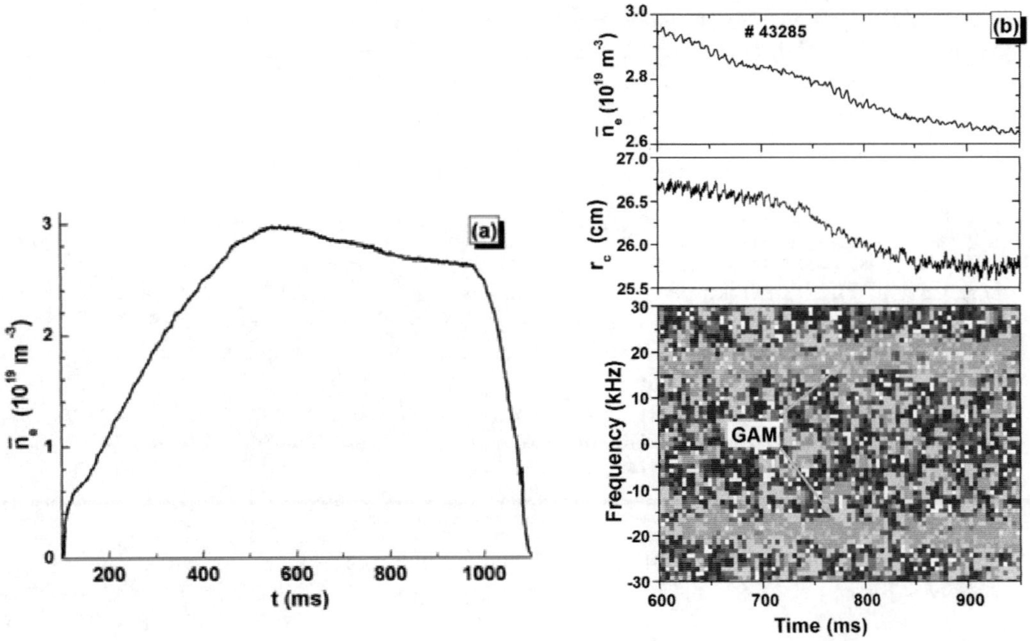

Figure 9. Radial evolution of the long-distance potential-density correlations. (a) Time trace of the central chord line-averaged density evolution in the OH discharge. (b) Potential-density correlation for different radii for one shot, HIBP versus CR. Top box: evolution of line-averaged density. Middle box: corresponding evolution of the CR observation radius. Bottom box: potential-density coherency. Density in this figure is normalized to the radius of vacuum chamber. Actual density should be enhanced by factor 4/3. HIBP radial position is $r_{SV} = 25 \pm 0.5$ cm.

4. Summary

GAM correlation study was performed by HIBP and CR. It shows that GAM is mainly manifested in the plasma potential and much less pronounced on the plasma density fluctuations. GAM has an intermittent character in amplitude and frequency, potential and density presents high correlations with a constant phase shift for GAMs. The existence of the long-distance correlations for GAM was shown for the first time in the core tokamak plasmas. This suggests that GAM is a global mode. Radial correlation length for GAM has a range of a few cm at least.

Acknowledgments

Authors acknowledge the support of IAEA RC No:18033/R0. The work was funded by Russian Science Foundation project 14-22-00193.

References

[1] Vershkov V A, Shelukhin D A, Soldatov S V, Urazbaev A O, Grashin S A, Eliseev L G, Melnikov A V and the T-10 team, 2005 *Nucl. Fusion* **45** S203-26

[2] Melnikov A V Vershkov V A, Eliseev L G, Grashin S A, Gudozhnik A V, Krupnik L I, Lysenko S E, Mavrin V A, Perfilov S V, Shelukhin D A et al., 2006 *Plasma Phys. Control. Fusion* **48** S87-S110

[3] Ido T, Miura Y, Hoshino K, Kamiya K, Hamada Y, Nishizawa A, Kawasumi Y, Ogawa H, Nagashima Y, Shinohara K et al., 2006 *Nucl. Fusion* **46** 512-20

[4] Diamond P H, Itoh S-I, Itoh K and Hahm T S 2005 *Plasma Phys. Control. Fusion* **47** R35

[5] Melnikov A V, Hidalgo C, Eliseev L G, Ascasibar E, Chmyga A A, Dyabilin K S, Krasilnikov I A, Krupin V A, Krupnik L I, Khrebtov S M, Komarov A D, Kozachek A et al., 2011 *Nucl. Fusion* **51** 083043

[6] Zenin V N, Eliseev L G, Kozachek A S, Krupnik L I, Lysenko S E, Melnikov AV and Perfilov S V *Problems of Atomic Science and Technology* 2014, №6 (94) *Series: Plasma Physics* (20) p. 269-271

[7] Fujisawa A, Itoh K, Iguchi H, Matsuoka K, Okamura S, Shimizu A, Minami T, Yoshimura Y, Nagaoka K, Takahashi C et al., 2004 *Phys. Rev. Lett.* **93** 165002

[8] Dnestrovskij Yu N, Melnikov A V, Krupnik L I and Nedzelskij I S, 1994 *IEEE Trans. Plasma Science*, **22**, 310

[9] Melnikov A V, Bondarenko I S, Efremov S L, Kharchev N K, Khrebtov S M, Krupnik L I, Nedzelskij I S, Zimeleva L G and Trofimenko Yu V, 1995 *Rev. Sci. Instrum.*, **66**, 317

[10] Melnikov A V Eliseev L G, Perfilov S V, Lysenko S E, Mavrin V A, Shurygin R V, Shelukhin D A, Vershkov V A, Tilinin G N, Grashin S A et al., 2008 *Proc. 22nd IAEA Fusion Energy Conf. (Geneva, 13-18 October 2008*), EX/P5-36

[11] Fujisawa A, Ido T, Shimizu A, Okamura S, Matsuoka K, Iguchi H, Hamada Y, Nakano H, Ohshima S, Itoh K et al., 2006 *Nucl. Fusion* **45** S718

[12] Melnikov A V, Eliseev L G, Grashin S A, Gudozhnik A V, Lysenko S E, Mavrin V A, Perfilov S V, Soldatov S V, Shelukhin D A, Vershkov V A et al., 2003 *Proc. 30th EPS Conf. on Controlled Fusion and Plasma Physics* (*St. Petersburg, 7-11 July 2003*) ECA, Vol **27A**, P3-114

[13] Melnikov A V, Eliseev L G, Perfilov S V, Lysenko S E, Mavrin V A, Shurygin R V, Shelukhin D A, Vershkov V A, Tilinin G N, Grashin S A et al., 2007 *Proc. 34th EPS Conf. on Plasma Physics (Warsaw, 2-6 July 2007)* ECA, vol **31F** P-1.096

[14] Melnikov A V, Eliseev L G, Gudozhnik A V, Lysenko S E, Mavrin V A, Perfilov S V, Zimeleva L G, Ufimtsev M V, Krupnik L I and Schoch P M, 2005 *Czech. J. Phys.* **55** 349-60

[15] Melnikov A V, Eliseev L G, Lysenko S E, Perfilov S V, Shurygin R V, Krupnik L I, Kozachek A S and Smolyakov A I, 2014 JETP Letters **100**, vol 9, 633 – 638 (in Russian).

Heat flux effects on the dispersion relation for geodesic modes in rotating plasmas

V. N. Duarte, R. J. F. Sgalla, A. G. Elfimov and R. M. O. Galvão
Institute of Physics, University of São Paulo, São Paulo, SP, Brazil.
E-mail: vnduarte@if.usp.br

Abstract. The MHD theory of the effect of toroidal and poloidal rotations on the dynamics of Zonal Flows – ZFs and Geodesic Acoustic Modes – GAMs in axisymmetric magnetic confinement configurations is revisited. The MHD model has an arbitrariness regarding the energy conservation equation and previous works on the effect of rotation on ZFs and GAMs adopted an adiabatic law, or other simplifying assumptions, to treat this problem. However, in fusion grade plasmas, the heat transport along the magnetic field lines is rather fast and, therefore, a somewhat more appropriate model is to assume isothermal flux surfaces. This implies to take into account the heat transport equation in the model and, in the presence of rotation, this leads to an increase in the degree of the dispersion relation for these modes, giving rise to a low-frequency third branch of these modes. This has been previously obtained by Elfimov, Galvão and Sgalla [1] employing a model of circular flux surfaces from the outset. In this paper, the theoretical development is generalized by using flux coordinates, following the method of Ilgisonis *et al* [2]. This allows a better assessment of the applicability of the results and to investigate the relevance of the low frequency mode in non-circular tokamaks. Specific results for the TCABR tokamak are presented.

1. Introduction

Geodesic acoustic modes (GAMs) are plasma eigenmodes characterized by a perturbed electric field constrained to oscillate only in the radial direction (i.e., $m = n = 0$, where m and n are the poloidal and toroidal mode numbers, respectively) and by a perturbed density that varies with the first order poloidal harmonic, $m = \pm 1$. GAMs, whose existence was predicted in the 1960s [1], are stable oscillations over magnetic surfaces. Their generation mechanism is based on the effect of the (radial) perturbed electric field, constant and perperdicular to the magnetic surfaces, which induces a poloidal flow over the surface due to the drift velocity, $\mathbf{E} \times \mathbf{B}$. In a toroidal magnetic surface, the geodesic curvature of the field lines gives rise to a restoring force against that flux, thus inducing an oscillatory motion. GAMs are the branch of nonzero frequency of the same dispersion relation associated with zonal flows (ZFs), which are zero-frequency modes.

The nonlinear process behind the existence of the GAMs could be described as follows. Density and temperature gradients trigger drift micro-instabilities, which give rise to a turbulent spectrum of fluctuations with large convective cells that cause the anomalous transport. After certain thresholds, the correlation between the same fluctuations triggers GAMs and ZFs by means of the Reynolds stress tensor and other parametric mechanisms. Sheared flows, associated with these modes, tend to tear the convective cells, thus reducing the anomalous transport.

So, GAMs and ZFs are a fundamental mechanism of anomalous transport self-control in high-temperature magnetized plasmas.

Although very important, the plasma rotation effects on GAMs ans ZFs have been long neglected in theoretical approaches. Recently, it has been recognized that plasma rotation can play an important role on the frequency of these modes, not only modifying the existing dispersion relation but also introducing new branches of frequencies. Even in works which attempt to consider the effects of rotation, it is usual to neglect the mass flow in the poloidal direction. It turns out that the poloidal flux seem to play a major role in L/H (tokamak confinement mode) transition. Internal transport barriers (ITB), related to the very existence of the H mode, are characterized by discontinuities in density and pressure. Curiously, in JET discharges, the poloidal velocity can reach values as high as $75km/s$ inside the ITB and virtually zero immediately outside. Moreover, poloidal flow can subtantially modify the equilibrium and can be a cause of the elliptic/hyperbolic regime transition of the Grad-Shafranov equation [2].

The set of MHD equations has an arbitrariness with respect to the equation of state that provides closure. In most of the situations, the energy balance equation is simplified to the adiabatic condition, i.e., the divergence of the heat flux vector is set equal to zero. It can be used to describe collisional phenomena which occur in a time scale much faster than the characteristic time of heat transport. Fusion grade plasmas, however, have a very low collision rate and are embedded in strong magnetic fields, so that the particles motion is practically parallel to the field. This generates a large phase-mixing which rapidly thermalizes the magnetic surfaces and thus isothermal magnetic surfaces, considered to be the most appropriate to model fusion plasmas, will be used as a hypothesis in the present work.

2. MHD equilibrium equations

The aim of this part is to derive in detail the equations that completely describe the equilibrium in terms of the stream functions used to express the magnetic field and the fluid velocity. For that purpose, we will follow the approach of Ilgisonis *et al* [3]. The complete set of unperturbed one-fluid ideal MHD equations, using the assumption of isothermal magnetic surfaces and a Gaussian system of units, reads

Momentum conservation:

$$\rho_0(\mathbf{v_0}\cdot\nabla)\mathbf{v_0} = -\nabla p_0 + \frac{1}{c}\mathbf{j_0}\times\mathbf{B_0} \tag{1}$$

with Ampère's law: $\mathbf{j_0} = \frac{c}{4\pi}\nabla\times\mathbf{B_0}$

Ideal Ohm's law:

$$\nabla\times(\mathbf{v_0}\times\mathbf{B_0}) = 0 \tag{2}$$

Magnetic field with no sources or sinks:

$$\nabla\cdot\mathbf{B_0} = 0 \tag{3}$$

Continuity equation:

$$\nabla\cdot(\rho_0\mathbf{v_0}) = 0 \tag{4}$$

Isothermal magnetic surfaces condition ($T_0 = T_0(\psi)$):

$$(\mathbf{B_0}\cdot\nabla)\frac{p_0}{\rho_0} = 0 \tag{5}$$

During the calculations, we will use two coordinate systems. The first is the usual cylindrical one (R, φ, z) where R is the distance from the axis passing through the center of the torus, ϕ is the toroidal angle and z is the distance from the center, along the axis of symmetry. The second is the straight field line system (ψ, θ, φ), where ψ is the flux over a magnetic surface and θ is the poloidal angle. Owing to the symmetry, the equilibrium quantities do not depend on the toroidal angle, φ. It will be assumed that the magnetic surfaces are nested surfaces $\psi(R, z) = const$ around the magnetic axis, i.e., $\mathbf{B_0} \cdot \nabla\psi = 0$. From the divergence-free requirement, we can express the magnetic field as

$$\mathbf{B_0} = F(R, z)\nabla\varphi + \nabla\psi \times \nabla\varphi \tag{6}$$

which implies

$$\mathbf{j_0} = \frac{c}{4\pi}\left(\nabla F \times \nabla\varphi - \Delta^*\psi\nabla\varphi\right) \tag{7}$$

where $\Delta^*\psi \equiv R^2\nabla \cdot \left(\frac{\nabla\psi}{R^2}\right)$. Note that [1]

$$B_0^2 = \frac{F^2 + |\nabla\psi|^2}{R^2} \tag{8}$$

It is possible to express the fluid velocity in terms of stream functions if we integrate equation (2), i.e. $\mathbf{v_0} \times \mathbf{B_0} = c\nabla\phi$, where ϕ is the electrostatic potential. Hence, $\mathbf{B_0} \cdot \nabla\phi = 0$ and from equation (8), it follows that

$$\mathbf{v_0} = \lambda(R, z)\mathbf{B_0} + R^2\Omega(\psi)\nabla\varphi \tag{9}$$

where λ is an arbitrary scalar function and the definition $\Omega(\psi) \equiv c\frac{\partial\phi}{\partial\psi}$ has been used. In the forthcoming calculations, we will consider that F, λ, $R^2\Omega$ and ρ_0 do not depend on the toroidal angle, φ, i.e. $\nabla F \cdot \nabla\varphi = \nabla\lambda \cdot \nabla\varphi = \nabla(R^2\Omega) \cdot \nabla\varphi = \nabla\rho_0 \cdot \nabla\varphi = 0$. Substituting equation (9) in equation (4) and noting that $\nabla^2\varphi = 0$, we get

$$\nabla \cdot (\rho_0\mathbf{v_0}) = \nabla(\lambda\rho_0) \cdot \mathbf{B_0} = 0 \Longrightarrow \lambda = \frac{\kappa(\psi)}{\rho_0}$$

where κ is a function depending only on ψ.

2.1. Bernoulli-like equation

To obtain a Bernoulli-like equation, we project the force balance equation (1) onto the direction of $\mathbf{B}_0$:

$$\rho_0\mathbf{B}_0 \cdot \nabla\left[\frac{\kappa^2B_0^2}{2\rho_0^2} - \frac{R^2\Omega^2}{2}\right] + \mathbf{B}_0 \cdot \nabla p_0 = 0 \tag{10}$$

Assuming quasi-neutrality, the main contribution to the mass density is due to the ions. So, from equation (5) and the ideal gases law, it follows that

$$\frac{p_0}{\rho_0}(\psi) \simeq \frac{T_0(\psi)}{m_i} \tag{11}$$

It should be noted that McClements and Hole [2] use $\frac{p_0}{\rho_0}(\psi) \simeq \frac{T_0(\psi)}{2m_i}$ since they consider T_0 to be the sum of ionic and electronic temperatures. Here we consider T to be the average temperature. Plugging (11) into (10), we get

$$\frac{\kappa^2B_0^2}{2\rho_0^2} - \frac{R^2\Omega^2}{2} + \frac{T_0}{m_i}ln\frac{\rho_0}{\rho_{0\psi}} = H(\psi) \tag{12}$$

[1] In cylindrical coordinates, $\nabla\varphi = \left(\frac{\partial}{\partial R}, \frac{1}{R}\frac{\partial}{\partial\varphi}, \frac{\partial}{\partial z}\right)\varphi = \frac{1}{R}\hat{\varphi} \Longrightarrow |\nabla\varphi|^2 = \frac{1}{R^2}$.

where $\rho_{0\psi}$ is an arbitrary quantity that may depend on ψ and the enthalpy $H(\psi)$ is a function of ψ only. Hence,

$$\frac{\rho_0(\psi,\theta)}{\rho_{0\psi}(\psi)} = \frac{p_0(\psi,\theta)}{p_{0\psi}(\psi)} = exp\left[\left(R^2\Omega^2 - \frac{\kappa^2 B_0^2}{\rho_0^2}\right)\frac{m_i}{2T_0}\right] \tag{13}$$

A factor $-R_0^2\Omega^2\frac{m_i}{2T_0}$ is often introduced in the argument of the exponential, without loss of generality. $\rho_{0\psi}$ and $p_{0\psi}$ are arbitrary quantities that depend only on ψ or are constant.

3. Perturbed equilibrium

Let us allow the following perturbations with respect to the equilibrium (of the form $f' = f'(\psi,\theta)exp(-i\omega t)$)

$$\rho = \rho_0 + \rho', \ \mathbf{v} = \mathbf{v}_0 + \mathbf{v}', \ p = p_0 + p', \ \mathbf{j} = \mathbf{j}_0 + \mathbf{j}'$$

where each of the quantities with subscript 0 depend on both ψ and θ. The linearized perturbed MHD equations are

$$\rho_0\left[-i\omega\mathbf{v}' + (\mathbf{v}_0\cdot\nabla)\mathbf{v}' + (\mathbf{v}'\cdot\nabla)\mathbf{v}_0\right] + \rho'(\mathbf{v}_0\cdot\nabla)\mathbf{v}_0 = -\nabla p' + \mathbf{j}'\times\mathbf{B}_0 \tag{14}$$

$$\mathbf{v}'\times\mathbf{B}_0 = \nabla\phi' \tag{15}$$

$$-i\omega\rho' + \nabla\cdot\left(\rho_0\mathbf{v}' + \rho'\mathbf{v}_0\right) = 0 \tag{16}$$

$$\nabla\cdot\mathbf{j}' = 0 \tag{17}$$

From Ohm's law, equation (15), the perturbed velocity can be written

$$\mathbf{v}' = \frac{1}{B_0^2}\left(V' - F\frac{d\phi'}{d\psi}\right)\mathbf{B}_0 + R^2\frac{d\phi'}{d\psi}\nabla\varphi \tag{18}$$

where $V' = \mathbf{v}'\cdot\mathbf{B}_0$. In order to provide closure to the system of perturbed equations, an energy equation is needed. To obtain the perturbed equation of state, we shall use equation (6.36) of [4] in a simplified form, in which the ion thermal flux vector is regarded to be more important than dissipative effects,

$$\frac{3}{2}p\frac{d}{dt}ln\left(\frac{p}{\rho^\gamma}\right) = -\nabla\cdot\mathbf{q} \tag{19}$$

with

$$\mathbf{q} \simeq \frac{5}{2}p\frac{\hat{b}\times\nabla T}{eB} \tag{20}$$

where $\hat{b} = \mathbf{B}/B$. One should note that the situation of pure toroidal flow is a degenerated case since it is not necessary to take into account the heat flow vector. This can be seen by writing the total time derivative of equation (19) in an explicit form:

$$\frac{3}{2}p\left[\left(\frac{\partial}{\partial t} + v_\varphi\frac{\partial}{\partial\varphi}\right)ln\left(\frac{p}{\rho^\gamma}\right)\right]$$

The above expression is always identically zero for stationary ($\frac{\partial(...)}{\partial t} = 0$) and axisymmetric ($\frac{\partial(...)}{\partial\varphi} = 0$) equilibrium. This particular situation was considered by Lahkin *et al* [5] for the case of purely toroidal rotation and isochoric magnetic surfaces. In their work the perturbed energy equation does not contain heat flux vector terms either, even though this is not an adiabatic regime. It is worth mentioning that, from the continuity equation, the validity of the hypothesis $\rho_0 = \rho_0(\psi)$ is only assured for zero poloidal flow. Back to the general situation, with both toroidal and poloidal flows, if we perturb equations (19) and (20), the following first-order, linearized energy equation is obtained:

$$\frac{3}{2}\rho_0^\gamma \left\{ \mathbf{v}' \cdot \nabla \left(p_0 \rho_0^{-\gamma} \right) + \gamma \frac{\rho'}{\rho_0} \mathbf{v}_0 \cdot \nabla \left(p_0 \rho_0^{-\gamma} \right) + \left(\frac{\partial}{\partial t} + \mathbf{v}_0 \cdot \nabla \right) \left(\frac{p' - c_s^2 \rho'}{\rho_0^\gamma} \right) \right\} = -\nabla \cdot \mathbf{q}_1 \quad (21)$$

Since $T' = \left(p' - \frac{\rho' p_0}{\rho_0} \right) \frac{m_i}{\rho_0}$, it follows that

$$\mathbf{q}_1 = \frac{5m_i}{2eB} \left\{ p_0 \hat{b} \times \nabla \left[\left(p' - \frac{\rho' p_0}{\rho_0} \right) \frac{1}{\rho_0} \right] + p' \hat{b} \times \nabla \left(\frac{p_0}{\rho_0} \right) \right\} \quad (22)$$

Substituting the perturbed velocity (18) into the perturbed continuity equation (16) leads to

$$-i\omega\rho' + \mathbf{B}_0 \cdot \nabla \left\{ \frac{\rho_0}{B_0^2} \left(V' - cF \frac{d\phi'}{d\psi} \right) + \kappa \frac{\rho'}{\rho_0} \right\} = 0 \quad (23)$$

The projection of (14) onto $\mathbf{B}_0$ is

$$-i\omega\rho_0 V' + \rho_0 \mathbf{B}_0 \cdot \left[(\mathbf{v}_0 \cdot \nabla)\mathbf{v}' + (\mathbf{v}' \cdot \nabla)\mathbf{v}_0 \right] - \frac{\rho'}{\rho_0} \mathbf{B}_0 \cdot \nabla p_0 = -\mathbf{B}_0 \cdot \nabla p' \quad (24)$$

where the equilibrium equation $\rho_0 \mathbf{B}_0 \cdot (\mathbf{v}_0 \cdot \nabla)\mathbf{v}_0 = -\mathbf{B}_0 \cdot \nabla p_0$ has been used. The term in brackets can be transformed the same way as in Appendix A of the paper by Ilgisonis et al [3], resulting $\mathbf{B}_0 \cdot [(\mathbf{v}_0 \cdot \nabla)\mathbf{v}' + (\mathbf{v}' \cdot \nabla)\mathbf{v}_0] = \mathbf{B}_0 \cdot \nabla \left\{ \frac{\kappa}{\rho_0} \left(V' - cF \frac{d\phi'}{d\psi} \right) - c\Omega R^2 \frac{d\phi'}{d\psi} \right\}$. Using the ideal gases law $p_0 \simeq \frac{\rho_0 T_0(\psi)}{m_i}$ and eq. (13), the term involving the equilibrium pressure can be written as

$$\begin{aligned} \frac{\rho'}{\rho_0} \mathbf{B}_0 \cdot \nabla p_0 = \frac{T_0 \rho'}{m_i \rho_0} \mathbf{B}_0 \cdot \nabla \rho_0 = \frac{T_0 \rho'}{m_i \rho_0} \mathbf{B}_0 \cdot \nabla \left\{ \rho_{0\psi}(\psi) exp \left[\left(R^2 \Omega^2 - \frac{\kappa^2 B_0^2}{\rho_0^2} \right) \frac{m_i}{2T_0} \right] \right\} = \\ = \rho_{0\psi} exp \left[\left(R^2 \Omega^2 - \frac{\kappa^2 B_0^2}{\rho_0^2} \right) \frac{m_i}{2T_0} \right] \frac{T_0 \rho'}{m_i \rho_0} \frac{m_i}{2T_0} \left\{ \Omega^2 \mathbf{B}_0 \cdot \nabla R^2 - \frac{\kappa^2}{\rho_0^2} \mathbf{B}_0 \cdot \nabla B_0^2 + \frac{2\kappa^2 B_0^2}{\rho_0^3} \mathbf{B}_0 \cdot \nabla \rho_0 \right\} = \\ = \frac{\rho'}{2} \left\{ \Omega^2 \mathbf{B}_0 \cdot \nabla R^2 - \frac{\kappa^2}{\rho_0^2} \mathbf{B}_0 \cdot \nabla B_0^2 + \frac{2\kappa^2 B_0^2}{\rho_0^3} \mathbf{B}_0 \cdot \nabla \rho_0 \right\} \end{aligned}$$

So, the perturbed momentum equation (14) along $\mathbf{B}_0$ becomes

$$\begin{aligned} -i\omega\rho_0 V' + \rho_0 \mathbf{B}_0 \cdot \nabla \left\{ \frac{\kappa}{\rho_0} \left(V' - cF \frac{d\phi'}{d\psi} \right) - c\Omega R^2 \frac{d\phi'}{d\psi} \right\} = \\ \frac{\rho'}{2} \left\{ \Omega^2 \mathbf{B}_0 \cdot \nabla R^2 - \frac{\kappa^2}{\rho_0^2} \mathbf{B}_0 \cdot \nabla B_0^2 + \frac{2\kappa^2 B_0^2}{\rho_0^3} \mathbf{B}_0 \cdot \nabla \rho_0 \right\} - \mathbf{B}_0 \cdot \nabla p' \end{aligned} \quad (25)$$

which is clearly different from equation (53) of reference [3], in which the simple algebraic equation $p' = c_s^2 \rho'$ (adiabatic case) was used to write the RHS simply as $\mathbf{B}_0 \cdot \nabla \left(c_s^2 \rho' / \rho_0 \right)$.

3.1. The average over a magnetic surface
Let us introduce the average over a magnetic surface as

$$\langle ... \rangle = \frac{\int_0^{2\pi} ... d\theta / J}{\int_0^{2\pi} d\theta / J}$$

At this point, it is important to differentiate between a periodic quantity and a quantity with zero average. An important result is that if f is a periodic function of θ, then $\langle \mathbf{B}_0 \cdot \nabla f \rangle = 0$:

$$\langle \mathbf{B}_0 \cdot \nabla f \rangle = \frac{\int_0^{2\pi} J \frac{\partial f}{\partial \theta} d\theta / J}{\int_0^{2\pi} d\theta / J} = \frac{\int_0^{2\pi} df}{\int_0^{2\pi} d\theta / J} = \frac{f(2\pi) - f(0)}{\int_0^{2\pi} d\theta / J} = 0$$

The next step is to obtain and expression for $\mathbf{j}'$ using the momentum conservation equation (14) and substitute it in the perturbed quasineutrality condition, eq. (17). The procedure is exactly the same of Ilgisonis *et al*, [3]. We have performed all the calculations but here we will just quote the final result without the whole reproduction (that can be found in Appendix B of [3]).

$$\begin{aligned} &\frac{d}{d\psi}\left\{\int_0^{2\pi} \frac{d\theta}{J}\left(-i\omega \frac{c\rho_0 |\nabla\psi|^2}{B_0^2}\frac{d\phi'}{d\psi} + p'\mathbf{B}_0 \cdot \nabla\left(\frac{F}{B_0^2}\right) + \right.\right. \\ &+\frac{\rho_0}{B_0^2} v'_{\parallel} B_0 \left[\left(\Omega + \frac{\kappa B_0^2}{\rho_0 F}\right)\mathbf{B}_0 \cdot \nabla R^2 - \frac{\kappa B_0^2}{\rho_0 F}\mathbf{B}_0 \cdot \nabla\left(\frac{|\nabla\psi|^2}{B_0^2}\right)\right] + \\ &\left.\left.+\frac{\rho' F}{2B_0^2}\left[\left(\Omega + \frac{\kappa B_0^2}{\rho_0 F}\right)^2 \mathbf{B}_0 \cdot \nabla R^2 - \left(\frac{\kappa B_0^2}{\rho_0 F}\right)^2 \mathbf{B}_0 \cdot \nabla\left(\frac{|\nabla\psi|^2}{B_0^2}\right)\right]\right)\right\} = 0 \end{aligned} \tag{26}$$

Equations (21) (with $\mathbf{q}_1$ given by (22)), (23), (25) and (26) describe exactly the perturbed rotating equilibrium. In order to become tractable, they must be simplified, as shown in the following.

3.2. The large-aspect-ratio tokamak ordering

In order to simplify the equations, let us consider the approximations $\epsilon \ll 1$; $B_0^2 = \frac{F^2}{R^2}\left(1 + \mathcal{O}(\epsilon^2)\right)$; $f = \bar{f}(\psi) + \tilde{f}(\psi, \theta)$, where ϵ corresponds to the reciprocal of the tokamak aspect ratio and f stands for the equilibrium quantities p_0, ρ_0, F, J, Ω_P and Ω_T. Before proceeding with the calculations, we should first note that[2]

$$J = (\nabla\psi \times \nabla\varphi) \cdot \nabla\theta = \mathbf{B}_0 \cdot \nabla\theta$$

and, since $q(\psi) \equiv \frac{\mathbf{B}_0 \cdot \nabla\varphi}{\mathbf{B}_0 \cdot \nabla\theta} = \frac{F}{JR^2}$,

$$\Omega_P \equiv \mathbf{v}_0 \cdot \nabla\theta = \frac{\kappa(\psi) J}{\rho_0}$$

$$\Omega_T \equiv \mathbf{v}_0 \cdot \nabla\varphi = \Omega(\psi) + \frac{\kappa(\psi) F}{\rho_0 R^2} \Rightarrow \Omega_T = \Omega(\psi) + q\Omega_P$$

where equation (9) was used. Let us also define the dimensionless coefficients λ_ρ and λ_F as

$$\frac{1}{\bar{\rho}_0}\mathbf{B}_0 \cdot \nabla\tilde{\rho}_0 \equiv \frac{\lambda_\rho}{R_0^2}\mathbf{B}_0 \cdot \nabla R^2 \tag{27}$$

$$\frac{1}{\bar{F}}\mathbf{B}_0 \cdot \nabla\tilde{F} \equiv \frac{\lambda_F}{R_0^2}\mathbf{B}_0 \cdot \nabla R^2 \tag{28}$$

Equation (23) becomes

$$\left(-i\omega + \bar{\Omega}_P \frac{\partial}{\partial\theta}\right)\rho' + \frac{\bar{\rho}_0 \bar{J}}{\bar{B}_0^2}\frac{\partial V'}{\partial\theta} = \frac{c\bar{\rho}_0}{\bar{F}}\frac{d\phi'}{d\psi}\left(1 + \lambda_\rho\right)\mathbf{B}_0 \cdot \nabla R^2$$

[2] Note that when working with the system (r, θ, φ) instead of (ψ, θ, φ) one could write the Jacobian explicitly as $J = \left[r\left(R_0 + r cos\theta\right)\right]^{-1}$.

which is the same as equation (63) of [3]. Now let us try to simplify equation (25). Our calculations will differ form the ones of [3] because we do not have a simple expression of p' in terms of ρ'. The LHS of equation (25) can be transformed using

$$\mathbf{B}_0\cdot\nabla\left\{\frac{\kappa F}{\rho_0}+\Omega R^2\right\}=\kappa\bar{F}\mathbf{B}_0\cdot\nabla\left(\frac{1}{\rho_0}\right)+\Omega\mathbf{B}_0\cdot\nabla R^2=\left(-\frac{\kappa\bar{F}}{\bar{\rho}_0}\frac{\lambda_\rho}{R_0^2}+\Omega\right)\mathbf{B}_0\cdot\nabla R^2$$

Since $\frac{\kappa\bar{F}}{\rho_0 R_0^2}=q\bar{\Omega}_P$ and $\bar{\Omega}=\bar{\Omega}_T-q\bar{\Omega}_P$, it follows that

$$\mathbf{B}_0\cdot\nabla\left\{\frac{\kappa F}{\rho_0}+\Omega R^2\right\}=\left(\bar{\Omega}_T-q\bar{\Omega}_P-q\bar{\Omega}_P\lambda_\rho\right)\mathbf{B}_0\cdot\nabla R^2$$

$$\begin{aligned}\bar{\rho}_0\left(-i\omega+\bar{\Omega}_P\tfrac{\partial}{\partial\theta}\right)V'+-\bar{\rho}_0 c\tfrac{d\phi'}{d\psi}\left(\bar{\Omega}_T-q\bar{\Omega}_P-q\bar{\Omega}_P\lambda_\rho\right)\mathbf{B}_0\cdot\nabla R^2=\\=\tfrac{\rho'}{2}\left\{\Omega^2+\tfrac{\kappa^2\bar{F}^2}{R_0^4\bar{\rho}_0^2}+\tfrac{2\kappa^2\bar{F}^2}{\bar{\rho}_0^2}\tfrac{\lambda_\rho}{R_0^4}\right\}\mathbf{B}_0\cdot\nabla R^2-\mathbf{B}_0\cdot\nabla p'\end{aligned}$$

where $\mathbf{B}_0\cdot\nabla B_0^2\simeq-\frac{\bar{F}^2}{R_0^4}\mathbf{B}_0\cdot\nabla R^2$ and equation (27) were used. Let us now analyze the tokamak ordering in the energy equation, equation (22). The expression for the heat flux vector divergence (equation (21)) can be simplified using that

$$\nabla\cdot\left[\frac{5m_i}{2eB}\left\{p_0\hat{b}\times\nabla\left[\left(p'-\frac{\rho' p_0}{\rho_0}\right)\frac{1}{\rho_0}\right]\right\}\right]=\left\{\hat{b}\times\nabla\left[\left(p'-\frac{\rho' p_0}{\rho_0}\right)\frac{1}{\rho_0}\right]\right\}\cdot\nabla\left(\frac{5m_i}{2eB}p_0\right)$$

Similarly, the divergence of the second term of the RHS of equation (22) can be written as

$$\nabla\cdot\left[\frac{5m_i}{2eB}p'\hat{b}\times\nabla\left(\frac{p_0}{\rho_0}\right)\right]=\left\{\hat{b}\times\nabla\left[\frac{T_0}{m_i}\right]\right\}\cdot\nabla\left(\frac{5m_i}{2eB}p_0\right)$$

Noticing that $\hat{b}$ is mostly oriented in the toroidal direction, it follows that

$$\nabla\cdot\mathbf{q}_1\simeq\frac{5m_iR_0^2\bar{J}}{2e\bar{F}}\left[\frac{dT_0}{d\psi}\frac{\partial p'}{\partial\theta}\frac{1}{m_i}-\frac{1}{\bar{\rho}_0}\frac{\partial\bar{p}_0}{\partial\psi}\left(\frac{\partial p'}{\partial\theta}-\frac{\bar{p}_0}{\bar{\rho}_0}\frac{\partial\rho'}{\partial\theta}\right)\right]$$

Writing the velocities explicitly, equation (21) can be put in the form

$$\begin{aligned}\tfrac{3}{2}\bar{\rho}_0^{\gamma}\left\{\tfrac{1}{B_0^2}\left(V'-F\tfrac{d\phi'}{d\psi}\right)\mathbf{B}_0\cdot\nabla\left(p_0\rho_0^{-\gamma}\right)+\gamma\tfrac{\rho'}{\bar{\rho}_0}\tfrac{\kappa}{\bar{\rho}_0}\mathbf{B}_0\cdot\nabla\left(p_0\rho_0^{-\gamma}\right)+\right.\\\left.+\left(-i\omega+\tfrac{\kappa}{\bar{\rho}_0}\mathbf{B}_0\cdot\nabla\right)\left(\tfrac{p'-c_s^2\rho'}{\bar{\rho}_0^{\gamma}}\right)\right\}=-\nabla\cdot\mathbf{q}_1\end{aligned}\tag{29}$$

Using the tokamak ordering, we note that $\frac{|\nabla\psi|^2}{B_0^2}\simeq\frac{B_{pol}R^2}{B_{tor}}=\frac{r^2}{q^2}$ and equation (26) can be simplified to

$$\frac{d}{d\psi}\left[-2\pi i\omega\frac{c\bar{\rho}_0}{\bar{J}}\frac{r^2}{q^2}\frac{d\phi'}{d\psi}+\frac{1}{\bar{J}\bar{F}}\int_0^{2\pi}d\theta\left(p'+\bar{\rho}_0R_0^2V'\frac{\bar{\Omega}_T}{\bar{F}}+\rho' R_0^2\frac{\bar{\Omega}_T^2}{2}\right)\mathbf{B}_0\cdot\nabla R^2\right]=0\tag{30}$$

Equation (29) can be further simplified if we work the term $\mathbf{B}_0\cdot\nabla\left(p_0\rho_0^{-\gamma}\right)$ out, using equation (13) (with $\rho_{0r}(r)$ arbitrary),

$$\rho_0(r,\theta)=\rho_{0r}(r)exp\left[\left(R^2\Omega^2-\frac{\kappa^2B_0^2}{\rho_0^2}\right)\frac{m_i}{2T_0}\right]$$

$$\mathbf{B}_0\cdot\nabla\left(p_0\rho_0^{-\gamma}\right)=\frac{T_0}{m_i}\mathbf{B}_0\cdot\nabla\rho_0^{-\gamma+1}=(1-\gamma)\rho_0^{-\gamma}\frac{T_0}{m_i}\mathbf{B}_0\cdot\nabla\rho_0$$

Noting that $\mathbf{B}_0\cdot\nabla\rho_0=\rho_{0r}(r)\frac{m_i}{2T_0}\left[\Omega^2\mathbf{B}_0\cdot\nabla R^2-\kappa^2\mathbf{B}_0\cdot\nabla\left(\frac{B_0^2}{\rho_0^2}\right)\right]exp\left[\left(R^2\Omega^2-\frac{\kappa^2B_0^2}{\rho_0^2}\right)\frac{m_i}{2T_0}\right]$. Since $\mathbf{B}_0\cdot\nabla B_0^2\simeq-\frac{\bar{F}^2}{R_0^4}\mathbf{B}_0\cdot\nabla R^2$, it follows that

$$\mathbf{B}_0\cdot\nabla\left(p_0\rho_0^{-\gamma}\right)=\frac{(1-\gamma)}{2}\rho_0^{-\gamma-1}\left[\Omega^2+\frac{\kappa^2\bar{F}^2}{R_0^4\bar{\rho}_0^2}\left(1+2\lambda_\rho\right)\right]\mathbf{B}_0\cdot\nabla R^2$$

$\mathbf{B}_0\cdot\nabla R^2$ is present in all four equation which are left to be solved. It can be simplified using

$$\mathbf{B}_0\cdot\nabla R^2=\mathbf{B}_0\cdot\nabla\left(R_0+rcos\theta\right)^2\simeq-2rR_0\bar{J}sin\theta$$

So, we can write equation (29) in the form

$$-\frac{3}{2}\frac{(1-\gamma)}{\bar{\rho}_0}rR_0\bar{J}sin\theta\left(\frac{R_0^2}{\bar{F}^2}\left(V'-c\bar{F}\frac{d\phi'}{d\psi}\right)+\gamma\frac{\rho'}{\bar{\rho}_0}\frac{\kappa}{\bar{\rho}_0}\right)\left[\Omega^2+\frac{\kappa^2\bar{F}^2}{R_0^4}\left(1+2\lambda_\rho\right)\right]+\frac{3}{2}\left(-i\omega+\frac{\kappa\bar{J}}{\bar{\rho}_0}\frac{\partial}{\partial\theta}\right)\left(p'-c_s^2\rho'\right)=$$
$$=-\frac{5m_iR_0^2\bar{J}}{2e\bar{F}}\left[\frac{dT_0}{d\psi}\frac{\partial p'}{\partial\theta}\frac{1}{m_i}-\frac{1}{\bar{\rho}_0}\frac{\partial\bar{p}_0}{\partial\psi}\left(\frac{\partial p'}{\partial\theta}-\frac{\bar{p}_0}{\bar{\rho}_0}\frac{\partial\rho'}{\partial\theta}\right)\right]$$

We are left with the following four equations (they correspond to equations (16)-(19) of [3], with the difference that in this work we consider isothermal surfaces instead of isentropic ones):

$$\left(-i\omega+\bar{\Omega}_P\frac{\partial}{\partial\theta}\right)\rho'+\frac{\bar{\rho}_0\bar{J}}{\bar{B}_0^2}\frac{\partial V'}{\partial\theta}=-2rR_0\bar{J}sin\theta\frac{c\bar{\rho}_0}{\bar{F}}\frac{d\phi'}{d\psi}\left(1+\lambda_\rho\right)\tag{31}$$

$$\bar{\rho}_0\left(-i\omega+\bar{\Omega}_P\tfrac{\partial}{\partial\theta}\right)V'+\bar{\rho}_0c\tfrac{d\phi'}{d\psi}\left(\bar{\Omega}_T-q\bar{\Omega}_P-q\bar{\Omega}_P\lambda_\rho\right)2rR_0\bar{J}sin\theta=$$
$$=-\frac{\rho'}{2}\left\{\Omega^2+\frac{\kappa^2\bar{F}^2}{R_0^4\bar{\rho}_0^2}\left(1+2\lambda_\rho\right)\right\}2rR_0\bar{J}sin\theta-\bar{J}\frac{\partial p'}{\partial\theta}\tag{32}$$

$$-\frac{3}{2}\frac{(1-\gamma)}{\bar{\rho}_0}rR_0\bar{J}sin\theta\left(\frac{R_0^2}{\bar{F}^2}\left(V'-\bar{F}\frac{d\phi'}{d\psi}\right)+\gamma\frac{\rho'}{\bar{\rho}_0}\frac{\kappa}{\bar{\rho}_0}\right)\left[\Omega^2+\frac{\kappa^2\bar{F}^2}{R_0^4\bar{\rho}_0^2}\left(1+2\lambda_\rho\right)\right]+\frac{3}{2}\left(-i\omega+\frac{\kappa\bar{J}}{\bar{\rho}_0}\frac{\partial}{\partial\theta}\right)\left(p'-c_s^2\rho'\right)=$$
$$=-\frac{5m_iR_0^2\bar{J}}{2e\bar{F}}\left[\frac{dT_0}{d\psi}\frac{\partial p'}{\partial\theta}\frac{1}{m_i}-\frac{1}{\bar{\rho}_0}\frac{\partial\bar{p}_0}{\partial\psi}\left(\frac{\partial p'}{\partial\theta}-\frac{\bar{p}_0}{\bar{\rho}_0}\frac{\partial\rho'}{\partial\theta}\right)\right]\tag{33}$$

$$\frac{d}{d\psi}\left[-2\pi i\omega\frac{c\bar{\rho}_0}{\bar{J}}\frac{r^2}{q^2}\frac{d\phi'}{d\psi}-\frac{1}{\bar{F}}\int_0^{2\pi}d\theta\left(p'+\bar{\rho}_0R_0^2V'\frac{\bar{\Omega}_T}{\bar{F}}+\rho'R_0^2\frac{\bar{\Omega}_T^2}{2}\right)2rR_0sin\theta\right]=0\tag{34}$$

Equations (31), (32) and (33) will be used to obtain expressions for V', ρ' and p' and the last one will give the dispersion relation. In theoretical treatments with the intention of obtaining dispersion relations for GAMs, it is usual to consider only first-order effects since higher-order harmonics are responsible only for higher order (in ϵ) contributions. So, we may assume

$$\rho'=\rho_s sin\theta+\rho_c cos\theta$$
$$p'=p_s sin\theta+p_c cos\theta\tag{35}$$
$$V'=V_s sin\theta+V_c cos\theta$$

4. Linear system resolution

Plugging (35) into equations (31), (32) and (33) implies the handling of considerably extensive expressions so that the aid of a symbolic software is imperative. To this end we have employed *Wolfram Mathematica* 8.0. The results of the 6×6 system (for ρ_s, ρ_c, V_s, V_c, p_s and p_c) are too lengthy to be shown here. The next step is to use the aforementioned expressions in (34) in order to proceed with the obtaintion of a dispersion relation. [3] A command is used in order to bring together the whole expression in order to analyze the global numerator and denominator. Equation (34) will give the desired relation; however one must be careful if the denominator could possibly cancel some factor of the numerator, thus reducing the number of roots of ω^2. To simplify the analysis, we use the following definitions:

- The acoustic speed:

$$c_s^2 = \gamma \frac{p_0}{\rho_0} = R_0^2 \omega_s^2$$

- The poloidal and toroidal Mach numbers:

$$M_{P,T} = \frac{\omega_s}{\Omega_{P,T}}$$

- A parameter to characterize $\mathbf{B_0} \cdot \nabla \rho_0$ against $\mathbf{B_0} \cdot \nabla R^2$: $\lambda\rho = \frac{\mathbf{B_0} \cdot \nabla \rho_0}{\mathbf{B_0} \cdot \nabla R^2} \frac{R_0^2}{\rho_0}$. After substitution in the equilibrium MHD equations, one obtains

$$\lambda\rho = \frac{\gamma (M_P^2 - M_P M_T + M_T^2/2)}{1 - \gamma M_P^2}$$

- A parameter to characterize $\mathbf{B_0} \cdot \nabla F$ against $\mathbf{B_0} \cdot \nabla R^2$: $\lambda_F = \frac{\mathbf{B_0} \cdot \nabla F}{\mathbf{B_0} \cdot \nabla R^2} \frac{R_0^2}{F}$. After substitution in the equilibrium MHD equations, one obtains $\lambda_F \sim \lambda_\rho \epsilon^2$, thus being negligible.
- A parameter to quantify the contribution of the heat flow vector: $\tau = \frac{R^2 \omega_s \rho_0}{e p_0} \frac{dT_0}{d\psi}$. The equivalence between this parameter and M_d introduced in Ref. [6] is $M_d = \frac{5}{3\gamma^2} \frac{\tau}{M_P}$.

The numerator of the dispersion relation (with terms up to order $\lambda\rho^2$) is

$$\begin{gathered} 8M_P^3 M_T q^2 \left(3 - (3+2\gamma)\Omega^2 + \gamma\Omega^4\right) - 2M_P^4 \left(-2 - 3\Omega^2 + 2q^2 \left(4 - (2+3\gamma)\Omega^2 + \gamma\Omega^4\right)\right) - \\ \left(-1+\Omega^2\right) \left(8M_T^2 q^2 \Omega^2 - 2\Omega^2 \left(-1 - 2q^2 + \Omega^2\right) + M_T^4 q^2 \left(1 + \gamma\left(-1+\Omega^2\right)\right)\right) - 2M_P M_T q^2 \times \\ \left(4\Omega^2 \left(3 + M_d - \Omega^2\right) + M_T^2 \left(1 + 3\Omega^2 + \gamma\left(-1 + 3\Omega^2 - 2\Omega^4\right)\right)\right) - 2M_P^2 \times \\ 1 + M_d^2 \left(1 + 2q^2\right) + 3\Omega^4 + 2M_d \left(-1 + 3\Omega^2 + q^2\left(-2+\Omega^2\right)\right) + q^2 \times \\ \left(\left(2\left(1 - 5\Omega^2 + \Omega^4\right) + M_T^2 \left(3 + \gamma - 13\Omega^2 - 7\gamma\Omega^2 + 4\gamma\Omega^4\right)\right)\right) \end{gathered} \tag{36}$$

while the denominator is

$$\begin{gathered} \left[M_P^3 + \Omega - 3M_P^2\Omega - \Omega^3 + M_P\left(-1 + M_d + 3\Omega^2\right)\right] \times \\ \times \left[M_P^3 + 3M_P^2\Omega + \Omega\left(-1+\Omega^2\right) + M_P\left(-1 + M_d + 3\Omega^2\right)\right] \end{gathered} \tag{37}$$

We have verified that, as reported in Ref. [6], if $M_P, M_T \neq 0$ no cancellation between the numerator and the denominator takes place and the problem is analitically intractable. An insight shown in the same reference is that if $M_T \to 0$, a factor $\Omega^2 - M_P^2(1 - M_d)^2$ is common to both parts. That limit is considered in order to simplify the equation and to obtain the other two frequencies and then the next order correction is calculated for the first root. Performing

[3] Curiously, only sine-terms will give nonzero contributions (in the $m = \pm 1$ approach).

the polynomial division of (36) (still with $M_T \neq 0$) by $\Omega^2 - M_P^2(1-M_d)^2$, using the subroutines *PolynomialRemainder* and *PolynomialQuotient*, a fourth order (in Ω) equation is obtained:

$$\begin{gathered}\Omega^4 - 2\Omega^2[1 + M_P^2 + q^2(1 + 2M_T^2 + M_P^2 - 2M_PM_T)] + \\ +2q^2(1 + 2M_T^2 + 4M_P^2 - 6M_PM_T) + 1 - 2M_P^2 + S_4(\Omega) = R_4/\left(\Omega^2 - M_P^2(1-M_d)^2\right)\end{gathered} \tag{38}$$

where the quotient $S_4(\Omega)$ and the remainder R_4 are terms of order $\lambda\rho^2$. Up to first order in $\lambda\rho$, we are left with a factorized equation of the form

$$\left(\Omega^2 - \Omega_1^2\right)\left(\Omega^2 - \Omega_2^2\right) \simeq 0$$

which has the solutions

$$\omega_{GAM1}^2 = \left[2 + \frac{1}{q^2} + \left(2 - \frac{1}{q^2} + \frac{2}{q^4}\right)M_P^2 + 4M_T^2 - 4\left(1 - \frac{1}{q^2}\right)M_PM_T\right]\frac{c_s^2}{R_0^2} \tag{39}$$

$$\omega_{GAM2}^2 = \left[1 + \left(3 - \frac{2}{q^2}\right)M_P^2 - 4M_PM_T\right]\frac{c_s^2}{q^2R_0^2} \tag{40}$$

Finally we are left with the task of calculating the frequency increment of the first branch (located near $\Omega_3^2 \simeq M_P^2(1-M_d)^2$). It is necessary to work up to order $\lambda\rho^2$ in eq. (38), i.e., keeping all terms. Let us consider a frequency of the form

$$\Omega_3 = \Omega_{d3} + \Delta_3;\ \Delta_3 \ll \Omega_{d3}$$

Substituting in eq. (38), we get

$$\left(\Omega_3^2 - \Omega_1^2\right)\left(\Omega_3^2 - \Omega_2^2\right) + S_4(\Omega_3) = \frac{R_4}{\left[(\Omega_{d3} + \Delta_3)^2 - \Omega_{d3}^2\right]}$$

So, approximately (for $\Omega_1, \Omega_2 \gg \Omega_3$), we have

$$\Omega_1^2\Omega_2^2 \simeq \frac{R_4}{2\Delta_3\Omega_{d3}} \Rightarrow \Delta_3 \simeq \frac{R_4}{2\Omega_1^2\Omega_2^2\Omega_{d3}}$$

and

$$\Omega_3^2 \simeq \Omega_{d3}^2 + 2\Delta_3\Omega_{d3} \simeq \Omega_{d3}^2 + \frac{R_4}{\Omega_1^2\Omega_2^2}$$

which allows us to express the lowest frequency mode as

$$\omega_{ZF}^2 = \left[\left(1 - M_d^2\right)M_P^2 + \frac{q^2(\gamma - 1)}{1 + 2q^2}(M_P^2 - M_PM_T + M_T^2/2)M_T^2\right]\frac{c_s^2}{q^2R_0^2} \tag{41}$$

This mode is identified as a zonal flow since it has zero frequency in the case of zero rotation. Equations (39), (40) and (41) are exactly the same ones obtained in [6].

5. Analysis and discussions

It is quite instructive to calculate the frequencies obtained in eq. (39) for realistic parameters of a real device. This was performed for typical discharges of TCABR. For the sake of simplicity, we used $\gamma = 5/3$ and $Z_{eff} = 1$. The radial profiles for the temperatures were assumed to be $T_e(r) = 600eV(1 - (r/0.18)^2)^2$, $T_i(r) = 300eV(1 - (r/0.18)^2)^2$, where r is measured in meters ($a = 0.18m$ being the minor radius), and the density profile was regarded as $n_{i,e}(r) = 2.2\ 10^{19}m^{-3}(1 - (r/0.18)^2) + 10^{14}m^{-3}$ and the toroidal current density profile as

$\propto (1-(r/0.18)^2)^{2.3}$. The poloidal field is then calculated under the constraint that the current density integrated over the whole plasma cross section must be equal to $100kA$. These profiles, combined with a nearly constant $B_{tor} \simeq 1.1T$, are capable of reproducing a quite resonable safety factor q, which is close to 1 at the center and grows monotonically to a value of approximately 3 at the edge. The radial poloidal and toroidal rotation profiles were obtained from the results of [7], which were then interpolated. The resulting profiles for each of the three continua are shown in figure 1. Most interesting is the measured power spectrum of acoustic modes measured in TCABR (figure 9 of [8]). It is surprising that there is strong evidence of the existence of three modes (most clearly seen in the bias regime - in Ohmic discharges the effect of higher harmonics seems to mask the lowest-order phenomena). They can be identified by characteristic frequencies of order $1kHz$, $15kHz$ and $30kHz$, which could correspond to the frequencies of figure 1 close to the edge (typical locations of geodesic modes). Unfortunately, the lack of window resolution in the measurements prevents us to identify the frequency of the modes more accurately.

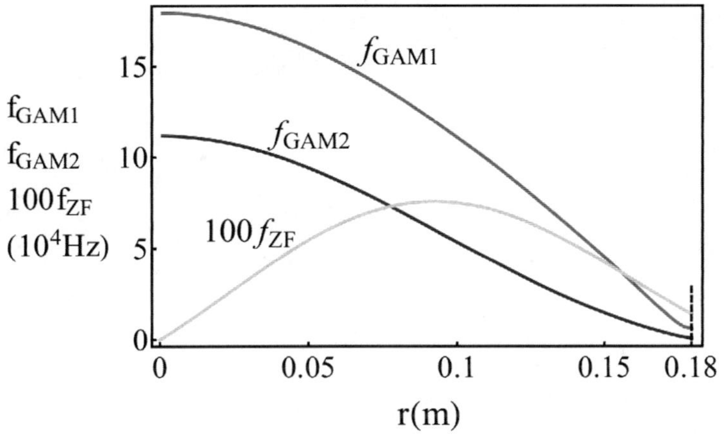

Figure 1. Estimation of the radial behavior of GAMs and ZF branches for typical TCABR tokamak parameters.

We note that the frequency of both branches of GAM modes were obtained in virtue of a factor cancellation in the numerator, which occurs in the formal limit $M_T \to 0$. We have analyzed in detail how fair is this approximation for TCABR tokamak rotating plasma parameters. To this purpose, the numerator (36) was solved numerically using the previously calculated safety factor $q(r)$. Then, equation (36) was solved numerically for a number of radial positions. It turned out that the analytical expressions for GAM frequencies are rather robust, with the analytical and numerical curves virtually overlaid, with a relative difference of less than $10^{-7}\%$. However, with increasing M_T, the error becomes more substantial. For example, if we double the toroidal velocity, the zonal flow branch obtained using the two methods present a 1% difference with respect to each other. It is interesting to note that, given the weak dependence of the GAM modes, equations (39) and (40), on the toroidal Mach number, they do not undergo severe changes when M_T is increased. However, the lowest-frequency mode, equation (41), has a forth-power dependence on M_T. For instance, if TCABR toroidal mass flow is multiplied by a factor of 5 the resulting error would be of order 10%. This kind of correction should be important for the case of an additional torque source, such as a neutral beam injection.

Acknowledgments
This work was supported by FAPESP (São Paulo Research Foundation) under Grant Nos. 2011/50773-0 and 2012/22830-2 and by CNPq (National Council for Scientific and Technological Development), under Grant Nos. 307511/2011-1 and 480733/2013-9.

References
[1] N. Winsor, J. L. Johnson, and J. M. Dawson; *Phys. Fluids* **11**, 2448 (1968).
[2] K. G. McClements and M. J. Hole, *Phys. Plasmas* **17**, 082509 (2010).
[3] V. I. Ilgisonis et al, *Plasma Phys. Controlled Fusion* **53**, 065008 (2011).
[4] S. L. Braginskii, in *Reviews of Plasma Physics*, vol. 1, p. 205 (Consultants Bureau, New York, 1965), ed. M. A. Leontovich.
[5] V. P. Lakhin, V. I. Ilgisonis and A. I. Smolyakov, *Phys. Lett. A* **374**, 4872 (2010).
[6] A. G. Elfimov, R. M. O. Galvão and R. J. F. Sgalla, *Plasma Phys. Controlled Fusion* **53**, 105003 (2011).
[7] J. H. F. Severo *et al*, *Rev. Sci. Instr.* **78**, 043509 (2007).
[8] Yu. K. Kuznetsov *et al*, *Nucl. Fusion* **52**, 063004 (2012).

Advancing Fusion by Innovations: Smaller, Quicker, Cheaper

M.P. Gryaznevich[1,2,3], V A Chuyanov[1], D Kingham[1], A Sykes[1] and Tokamak Energy Ltd

[1]Tokamak Energy Ltd, Culham Science Centre, Abingdon, OXON, OX143DB UK

[2]Imperial College, London, SW7 2AZ, UK

[3]Technical University of Denmark, Fysikvej DK-2800 Kgs. Lyngby, Denmark

mikhail.gryaznevich@tokamakenergy.co.uk

Abstract. On the path to Fusion power, the construction of ITER is on-going, however there is not much progress in performance improvements of tokamaks in the last 15 years, Fig.1. One possible reason for this stagnation is the lack of innovations in physics and technology that could be implemented with this approach in which progress is expected mainly from the increase in the size of a Fusion device. Such innovations could be easier to test and use in much smaller (and so cheaper and quicker to build) compact Fusion devices. In this paper we propose a new path to Fusion energy based on a compact high field Spherical Tokamak approach.

1. Introduction

We present a new concept of the exploration of Fusion power based on a compact high field Spherical Tokamak (ST). This concept differs from the mainstream path to the Fusion Power via big size devices ITER and DEMO. It allows significant reduction in costs and timescales and allows use of the most advanced technologies and innovations while still fully utilizing all developed Fusion technologies based on 50 years of the development of the Tokamak concept. Such an approach offers the potential to advance the fusion programme and provide a unique integrated system that will help to establish the expertise and techniques for designing and building future fusion power plants, as well as helping to prepare industry for the challenges of building them.

Construction of ITER is progressing, but does not go as fast and is more costly than expected. Certainly there are some subjective reasons for it, but there are also very objective reasons. High mechanical loads, tight tolerances and sizes at the limits of existing technologies have narrowed the industrial basis of ITER manufacturing. Only a small number of the most advanced industries of the world are able to participate in tendering. There is practically no competition and, as a result, very high monopoly prices. The participating companies have no incentive to develop new industrial capabilities and new technologies, because ITER is a single customer. Similar products will be not needed for anybody for next 10-20 years. Even in areas where the government investments have led to development of new products - $NbSn_3$ superconductivity, powerful short wave gyrotrons, MeV neutron beams – absence of customers in immediate future will lead to stagnation. In 20 years' time when the next fusion device will demand similar technologies the people who developed them will be retired and the know-

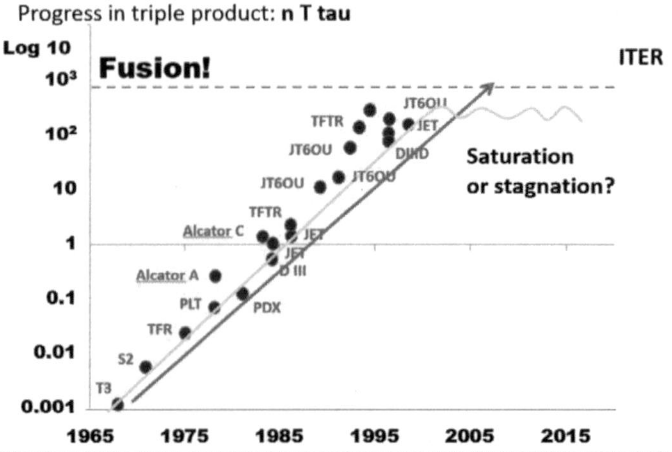

Figure 1. Progress in the triple product (n T tau) in magnetic fusion devices.

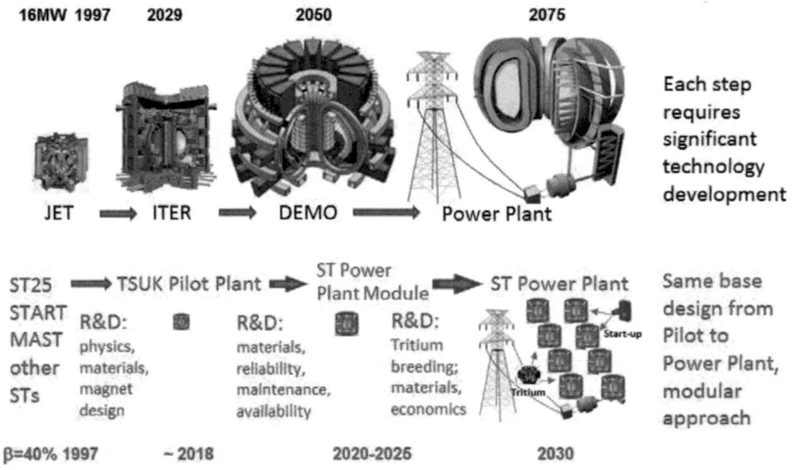

Figure 2. Comparison of roadmaps to Fusion Power: top/bottom – conventional/compact (modular) ST approach.

how will be forgotten. The development of a new technology with very big and expensive prototypes is inevitably slow and not efficient. The new proposed path, based on the compact high field ST approach, has several key features of an executable commercialization strategy [4]: a low-cost pilot plant that can attract commercial cost sharing with minimal financial risk and lead to power plants that are still small on an absolute scale. So many such devices can be constructed opening new sustainable market for the Fusion industry.

There is no question that the final fusion plant will be reasonably big, like all other power plants to day, at the level of 1000 MW(e). If it will be << 500 MW(e), the cost of its conventional part will become prohibitive. We propose another way. One energy conversion plant can serve many small fusion cores combined in one big fusion plant, so we need to build and test only one fusion core. If for example we can build a fusion core with 50 MW(th) fusion power, the total power station will consist of ~50 fusion cores and finally each core must cost < 65M Euros. Of course, the prototype may be several times more expensive and still fully prove solidness and competitiveness of the design. During the construction, the economy of scale will be substituted by economy of mass production. Building many fusion cores even for one station will justify development of new technologies for mass production. Typical learning curve during building of one station will lead to decrease of the mean cost of the first 40 cores below 50% of the first one. Low cost, short time of production and easy change of each core will lead to fast development of the technology and of its industrial base, Fig.2.

The recent advances in the development of high temperature superconductors (HTS) [1], and encouraging recent results on a strong favorable dependence of electron transport on higher toroidal field (TF) in Spherical Tokamaks (ST) [2], open new prospects for a high field ST as a compact fusion reactor. The combination of the high β (ratio of the plasma pressure to magnetic pressure), which has been achieved in STs, and the high TF that can be produced by HTS TF magnets opens a path to lower-volume fusion devices, in accordance with the fusion power scaling proportional to $\beta^2 Bt^4 V$. The proposed ST path to Fusion is described in Section 2. A compact ST can be also considered as an intense and efficient neutron source [3]. Some physics and engineering challenges of such devices will be discussed in Section 3. These include application of HTS in ST magnets, fast particle physics, effect of increase in B_t on micro-stability and plasma confinement, etc. Discussions on advantages of a modular approach are given in Section 4. The demonstration of reliable steady state operations in a compact ST even at the level of a few tens of MW Fusion output as a first step will significantly advance not only the mainstream Fusion for Energy research, but also the commercial exploitation of Fusion Power.

2. ST path to Fusion

We envisage three steps in the development of the ST path to Fusion: the first prototype, the ST Pilot plant and the Power plant module.

The goal of the first prototype is to demonstrate the possibility to produce burning plasma conditions, i.e. main plasma parameters to provide self-heating of the plasma that will be produced by alpha particles born in the DT reaction. As discussed in detail in [1], the proposed way to produce the necessary high toroidal field in a ST can be based on application of high temperature superconductors (HTS) in tokamak magnets. The recent progress in the development of HTS is remarkable and the first application of HTS in tokamak magnets [1] has already confirmed the feasibility of this approach. However the supply chain for production of high quality HTS in the needed quantities is very limited and although significant progress is expected, the timescale of our development path suggests an alternative or combined approach on the first stage of the prototype. This can be achieved in a compact high field ST with non-superconducting toroidal field magnet with HTS used for poloidal field coils only. We will also relax at this stage, requirements on the pulse duration, steady-state operations, full protection of the device from the neutron fluency and on protection of the device first wall and magnets from neutron damage. Short operations with tritium or with very small amounts of it are considered. This prototype will also be used as a research platform for conceptual studies, both in physics and technologies, and can be used as a demonstration of an intense Fusion Neutron Source [3].

Upgrades with high temperature superconducting magnets at the next stage will allow steady-state operations, and improved shielding will protect magnets from neutron damage and heating of magnets in this compact high-field ST Pilot Plant which will demonstrate the possibility of energy production in DT operations. The final stage, probably with an increased size, will provide full demonstration of a commercial Power Plant module.

Commercial opportunities of Fusion today may, or may not be, directly connected with the energy production and indeed the non-power producing applications are more feasible. There are three main opportunities for such commercial applications:

- R&D needed for the Fusion for Energy Research (ITER, DEMO, and support of the mainstream Fusion Research in areas that are not fully explored in presently operating devices);

- Potential application in the Nuclear Industry either as a Fusion core for subcritical systems (hybrid approach) or for waste management
- Commercial and research applications of Fusion neutrons at moderate level, i.e. using Fusion device as an intense Neutron Source.

The 2nd application is the most commercially attractive, but requires substantial development. A compact first step device will not only be a prototype of larger devices for the nuclear application, but readily serve to the 1st and 3rd applications. With the increased size that will allow more shielding and increase in the value of the magnetic field, a compact high field ST will allow demonstration of energy production. The main necessary engineering development for the use of Fusion both in the Nuclear Industry and for the energy production will be in the blanket design, which requires small scale prototypes for development and tests before been commercialized.

This approach is supported by the results of the previous extended studies of the ST path [4, 5] proposing to progress from a small prototype to high-power commercial devices without significant changes in physics and technology.

3. Some physics and engineering challenges of the ST path to Fusion

This new approach is stimulated by successful development of the Spherical Tokamak concept [6] in the last two decades and their obvious ability to produce high Fusion power in moderate size devices. 14 MeV fusion neutrons are produced when a deuterium (D) or deuterium-tritium (D-T) plasma becomes very hot so that the nuclei fuse together, releasing the fast neutrons. It is generally considered that the plasma needs to have high confinement time, high temperature, and high density to optimise this process. As mentioned above, the Fusion power in a reactor can be expressed as

$$P_{fus} \sim \beta_t^2 B_t^4 V \quad (1)$$

where $\beta_t = \langle p \rangle / B_t^2$, $\langle p \rangle$ is average plasmas pressure, B_t is toroidal field and V is the plasma volume.

A tokamak features a combination of strong toroidal field B_t (usually several Tesla) and high toroidal plasma current I_p (several mega-amps) and usually large plasma volume V and significant auxiliary heating, to provide a hot stable plasma so that fusion can occur. The auxiliary heating (usually via tens of megawatts of neutral beam injection of very high energy neutral H or D or T) is necessary to raise the temperature to sufficiently high values before transition to self-heated burning plasma regime occurs. The Lawson criterion requires the "triple product" $n\,T\,\tau_E \sim 3\times10^{21}$ kV sec/m^3 for such a regime at optimal conditions for T ~ 14 keV which requires considerably high energy confinement.

	B_t	I_p	n_e^{19}	k	tot
MAST	0.5	0.7	4	2.2	
Compact ST reactor	5	4	50	3	
τ_E gain	1.4	5.1	2.8	1.27	**25**

Table 1. Improvement in confinement in high-field ST following ITER IPB(y,2) scaling.

Increase in toroidal field, plasma current and plasma density will result in significant improvement in the plasma confinement. Table 1 shows this for a device similar to the MAST in size, but with increased parameters. Here ITER IPB(y,2) scaling was used. Assuming typical confinement in MAST of ~0.05 s [2], the predicted confinement will be 1.25 s. If it will be possible to achieve in this device temperatures above 10 keV, the Lawson criterion will be satisfied.

Although the ITER IPB(y,2) confinement scaling has been confirmed on spherical tokamaks in regimes acceptable for the data to be included in the ITER confinement database, results of recent more detailed studies from MAST and NSTX [2, 7] suggest different scaling which is much more favourable for a compact reactor dependence then with the ITER scaling :

$$\tau_E \sim B_T\text{^}0.9\text{-}1.4 \quad \text{vs} \quad \tau_{E98y,2} \sim B_T\text{^}0.15;$$
$$\tau_E \sim I_p\text{^}0.4 \quad \text{vs} \quad \tau_{E98y,2} \sim Ip\text{^}\ 0.93$$

If this scaling will be applicable for burning ST plasma conditions, parameters of a compact ST to achieve ignition cay be relaxed.

It is well known that stabilization of turbulence by sheared ExB flow is the most possible cause of transport barriers observed in tokamaks; and it has been pointed out that this effect would be very strong in the high shear of the ST leading to improved confinement. More recently, gyrokinetic studies by the NSTX team [8] suggest that the anomalous electron diffusivity decreases more rapidly for hotter, collissionless ST plasmas than for ITER-like plasmas. Gyrokinetic simulations are presently being undertaken [9] on a range of high-field ST (HFST) devices to investigate the possibility that the combination of high field and low-aspect-ratio could stabilize some forms of micro-instability and hence provide the improvement in confinement. They indicate that at low magnetic field the mixing length diffusivity is dominated by electromagnetic tearing modes; these are stabilised at higher B_t, diffusivity then being dominated by electrostatic twisting modes so significant improvement in confinement is expected.

The increase in toroidal field in ST is very favourable for achievement of burning phase. The required increase in the toroidal field can be achieved in STs, where the space for the central column is very limited, even with Cu magnets, however high power dissipation in the Cu magnet drives significant increase in the output Fusion power to achieve the positive economics balance. Fig.3 shows parameters of different superconductors that can replace Cu in the magnet to remove this issue. The left Figure shows dependence of the maximum possible irreversible field at different temperatures for different types of superconductors. The green circle represents the desired regime of operations for a compact high-field ST reactor. This regime can be easily achieved if the 2nd generation of HTS, YBCO tapes, is used. The right Figure shows low degradation of the highest achievable current density in superconductors at different magnetic field. The best low temperature superconductor, Nb_3Sn, shows significant degradation at fields above 25 T, while HTS does not show such degradation. Moreover, if the magnet is designed to position the YBCO tape parallel to the main field, its performance can be significantly enhanced.

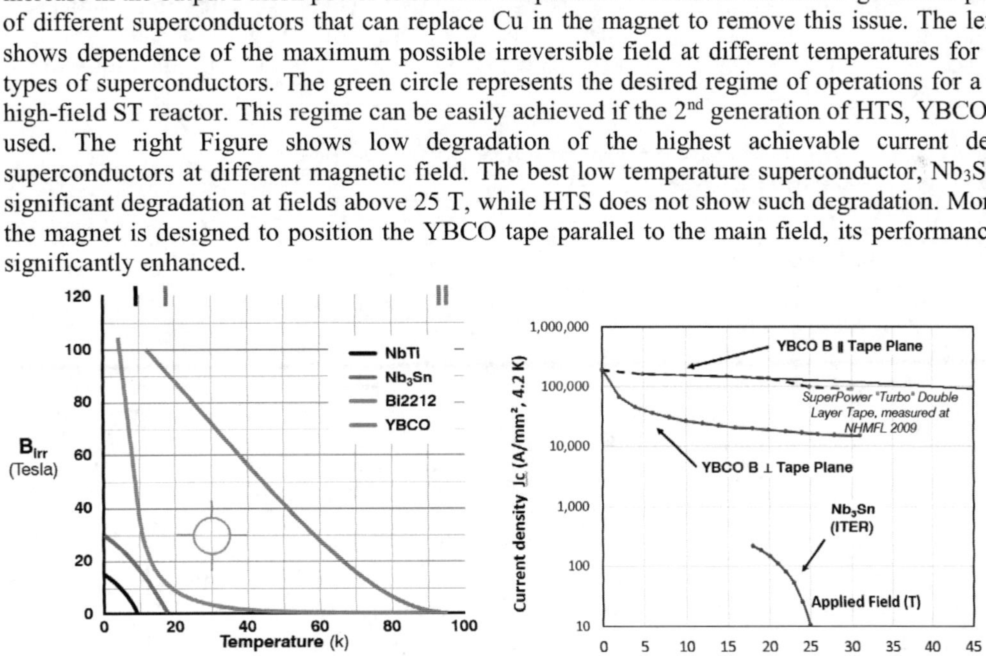

Figure 3. Comparison of different types of superconductors. Left: critical temperature at different magnetic field. The green circle indicates the area of interest for compact high-field ST. Right: critical current dependence on the magnetic field.

So, application of HTS in the ST magnet can allow achievement of high toroidal field needed for ignition, and also compactness of the HTS will provide more space for the necessary neutron shielding in the reactor to reduce requirements for cooling and protect the magnet from neutron damage. However, to achieve high Fusion power, alpha particles providing plasma self-heating must be confined.

In a conventional tokamak fusion reactor the value of the plasma current is determined by two requirements – necessary energy confinement to achieve ignition and necessary alpha particle confinement to heat plasma by fusion reactions. As for STs the energy confinement is a weak function of plasma current and the requirement of alpha-particle confinement may become the decisive one in the determination of the minimum plasma current. The simple estimates based on guiding centre approximation [10] show that to confine 3.7 MeV alpha particles, a plasma current of around 5-6 MA is needed and show a sharp reduction of alpha containment is observed when the plasma current is reduced below this level.

However, recent studies are more encouraging. Taking into account that the Larmor radius of alpha particle ($r \sim 0.16/B(T)$) is of the same order as the plasma minor radius in compact STs of consideration ($a \sim 0.3$-0.5 m) and the poloidal field is of the same order as the toroidal field, the most appropriate method to study alpha-particles confinement is a Monte-Carlo code with direct integration of gyro-orbital motion. Such calculations [11] show that in fact rather small plasma currents are sufficient for alpha confinement. Fig.4 presents results of Monte-Carlo modelling of alpha particles in a compact high-field ST. Alpha particle containment fraction and peak wall load dependences on the plasma current are shown at the left plot. The full gyro-orbit model is compared with guiding centre approximation. In the full-orbit case, the reduction of the containment and increase in the wall load are much less steep, because at large currents (even 8 MA) the gyrating particle - in contrast to the guiding centre particle - occupies a cylinder-like volume. This volume, characterized by the Larmor radius, touches the boundary at a distance of a Larmor radius already, thus producing more losses than the guiding centre particles. Thus more guiding centre particles than gyro - particles are lost. Results of full gyro-orbit simulation of alpha power deposition in the compact ST reactor show that the difference in the deposition for 6 MA and 4 MA cases is not significant.

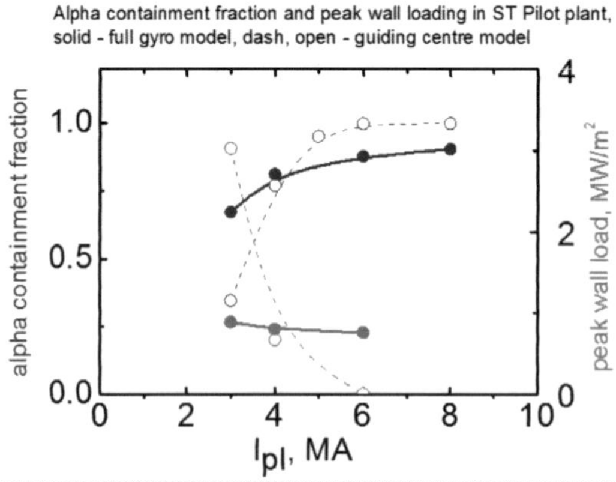

Figure 4. Results of Monte-Carlo modelling of alpha particles in a compact high-field ST. Alpha particle containment fraction and peak wall load dependences on the plasma current. Full gyro-orbit model is compared with guiding centre approximation.

The reduction in the plasma current may result not only in the reduction of the fusion power, but also in a significant increase in the wall load due to lost alpha particles. We have calculated this wall load (e.g. the additional wall load due to lost alphas) for different conditions and found that it is not negligible, but not too dangerous. For a device with the plasma current 4 MA, R/a = 0.8/0.5 m, B_t=5 T, k=3, the total wall load is 2.62 MW and the peak load is 0.8 MW/m^2. The conclusion of these studies is that to calculate losses and deposition profiles in a compact ST, the full-orbit model should be used, which shows less steep dependence on the plasma current. This is very promising, as any reduction in the plasma current will improve the economics of a compact ST reactor.

4. Discussions.

There is no question that the final fusion plant will be reasonably big, like all other power plants today, of a GWe level. However, modular approach does not require multi-GW reactors to be built to test their economics. It is sufficient to build and test only one fusion core module and a compact high field ST reactor is a good candidate for this. Such approach has several obvious advantages:

a. Reliability and availability of the power plant will be improved. If number of fusion cores N >>1, continuity of power can be achieved with a small additional capital cost ~1/N. The power of a modular power plant can be easily adjusted in accordance with grid requirements that change every hour. Such a flexible power station can have higher cost of electricity and still be acceptable by the grid.

b. The engineering costs (>20% for a complex project) will be shared between modules, so will be reduced ~1/N.

c. Internal parts of a fusion reactor become strongly activated during the first hours of operation. The change of the first wall and other repair works must be done by a remote handling equipment which in a single core power plant is idle during operation. Change of the first wall may be done in the most efficient way of changing a total fusion core with a smaller (~ 1/N) increase of capital cost in comparison with a single core power plant and without any loss of availability.

d. Manufacturing of many similar cores will justify development and preparation of special equipment for efficient production, which is usually not justified for production of a single unit (i.e. ITER). Smaller scale of multiple cores will broaden manufacturing basis for their production. Small companies will be able to participate in the tendering process broadening the competition.

For a new technology, big size of components represents not an advantage ("economy of scale"), but a disadvantage. Industries are not ready to prepare new big facilities for manufacturing a single new and risky device or system. As a result number of companies having proper equipment and able to participate in tendering is very small. It leads to a monopoly and high prices.

e. "Learning" of production can be realized during construction of the first module.

With a typical 15% industrial learning curve [12] (it is smaller than usually assumed in current fusion reactor studies when for a reactor N=10 the unit cost is 50% of the reactor N=1) one can expect that the mean cost per core for a reactor with N=40 cores will be below 50% of the cost of the first core. (See Fig 5). It is assumed, that the cost of the second core is equal 80% of the cost of the first on (a typical fission industry figure) and after that the learning rate is 15%, again the most probable industrial number. The prototype core may be at least 2 times more expensive that is needed for the competitiveness and still be acceptable.

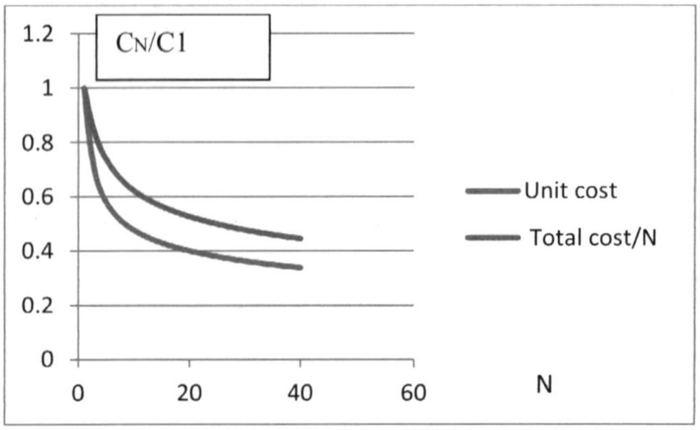

Fig 5. Normalized cost of a core and the mean cost of N cores as function of N.

f. "Economy of scale" in application to fusion devices may be reasonably expected when the "productivity" is proportional to the plasma volume and the "cost" is proportional to surface of the blanket surrounding the plasma to utilise the heat from neutrons for the energy production. Usually, all system codes calculate the cost of a component as the cost of materials and the cost of labour proportional to the cost of materials. The thickness of the fusion blanket and shield is practically constant and so the cost is proportional to the surface of the blanket. So it is beneficial to increase the wall loading up to a reasonable maximum, i.e. to reduce the surface-to-volume ratio, which is a great advantage of the ST geometry.

The cost of the magnet system in a tokamak is clearly depends on β and increase in β decrease the magnet capital cost, but for a fixed machine and a fixed toroidal field higher β means higher wall loading, which beyond some limit may be not desirable. To decrease the cost of the magnet one must use lower magnetic field. The value of magnetic field may be fixed by other considerations, like availability of RF generators for a given frequency, etc. The use of multiple cores permits to decouple β, magnetic field and wall loading by proper selection of the size of a core.

g. The auxiliary heating system of a fusion reactor - a very expensive part of the plant, must heat plasma to ignition point and also support BV ramp-up and increase in β to provide the increase in the bootstrap current fraction. With the modular design, the same system may be a common one for all (or some fraction) of the cores working consequently for each core during the start-up phase, or sharing power between different cores depending on the stage of the pulse.

h. In the modular approach, fusion cores may be pulsed. There is no necessity for steady-state current drive, which significantly increases power recirculation and capital cost of a fusion tokamak reactor. Constant output power may be supported by several pulsed cores with a small additional capital cost (~ 1/N or even << 1/N depending on ratio of pulse duration to dwell duration of a single core). This feature permits optimization of core operations and, for example, operations at high plasma densities for which the existing methods of current drive are very inefficient. For example, an operation regime can be suggested, when the plasma current is almost constant, but the fusion power is modulated by periodical changes of density. The current drive is efficiently performed at periods of low density and low fusion power. Loss of fusion power is compensated by other cores working in this moment of time at the highest plasma density and fusion power.

k. Fuelling/tritium systems for all cores may be combined to use the economy of scale. This common system may be tested separately from the fusion prototype. For the prototype a simplified fuelling/tritium system may be used. Blanket may be represented by just one module. Isotope separation system may be substituted by an external delivery. It will decrease the cost of the prototype even more.

5. Summary

The economy of scale concept can and must be substituted by economy of mass production and optimization, and any innovations could be easier to test and use in much smaller (and so cheaper and quicker to build) compact devices. The main problem of the fusion development is a high risk and a high cost of a prototype (DEMO) needed to justify construction of the power plant. The way to solve the problem is a combination of a big conventional energy conversion plant with multiple fusion cores – relatively small power fusion devices. Only they must be tested by construction of a prototype, significantly reducing the size and the cost of a DEMO. In this paper we propose a new path to the Fusion energy based on a compact high field Spherical Tokamak modular approach, which combines innovations in the Fusion science (ST concept), the Fusion technology (use of HTS magnets) and the Fusion economics (modular approach). We expect that application of these innovations will significantly advance the progress towards the Fusion energy.

References

[1] M Gryaznevich et al., "Progress in applications of High Temperature Superconductor in Tokamak Magnets". Fusion Eng. and Design. **88** (2013) 1593–1596
[2] M Valovic et al., Nuclear Fusion. **49** (2009) 075016
[3] M Gryaznevich et al., "Options for a Steady-State Compact Fusion Neutron Source". Fusion Science and Technology. **61** 1T (2012) 89
[4] R Stambaugh et al, "The Spherical Tokamak Path to Fusion Power", Fusion Technology. **33** (1998) 1
[5] M Gryaznevich et al. "The spherical tokamak path to fusion power, revisited". Proc. 40th EPS Conference, Helsinki, 1 - 5 July 2013, Europhysics Conferences Abstracts, Vol. 37F (2013) P2.111
[6] Y-K M Peng and D J Strickler, Nucl. Fusion. **26** (1986) 769
[7] S Kaye et al., Phys. Rev. Lett. **46** (2006) 848
[8] W Guttenfelder et al., "Simulation of microtearing turbulence in national spherical torus experiment" Physics of Plasmas. **19** (2012) 056119
[9] A Sykes, M Gryaznevich et al., "The Spherical Tokamak Path to Fusion Power – Revisited". Proceedings of 25th Symposium on Fusion Engineering (SOFE), 2013 IEEE , 10-14 June 2013, San Francisco, CA, p. 1 - 6
[10] G O Ludwig, M C R Andrade, M Gryaznevich, T N Todd. "Physics performance analysis of low-power tokamak reactors". Nuclear Fusion. **49** (2009) 085026
[11] M Gryaznevich, A Nicolai, P Buxton et al., "Fast particles in a steady-state compact FNS and compact ST reactor". Nuclear Fusion. **54** (2014), accepted for publication.
[12] A McDonald, L Schrattenholzer. "Learning rates for energy technologies". Energy Policy. **29** (2001) 255-261

Electron density profile reconstruction on the TCABR sweeping reflectometer

A.M.M. Fonseca[1], S. Hacquin[2], R.M.O. Galvão[1], J.I. Elizondo[1], P.G. Puglia[1], L.F. Ruchko[1], C. Amador[3], J.C. Raffaelli[4], J.H.F. Severo[1] and I.C. Nascimento[1]
[1]Instituto de Física da Universidade de São Paulo, Brazil
[2]Commissariat à l'Énergie Atomique - CEA , Cadarache, France
[3]Instituto Tecnológico Federal do Paraná, Brazil
[4]Instituto de Astronomia e Geofísica da Universidade de São Paulo, Brazil

E-mail: antonio.marques.fonseca@gmail.com

Abstract. A Frequency Sweeping Reflectometry System is under re-installation in the TCABR tokamak to provide the electron density profiles in this machine. This diagnostic operates in a frequency range between 18 GHz and 40 GHz, in Ordinary (O-mode) or in Extraordinary (X-mode) polarization modes. This diagnostic can cover electron densities between $0.4\times10^{19}m^{-3}$ and $2.0\times10^{19}m^{-3}$, when configured in O-mode. Electron density profiles can be inferred from the phase dependence of the probing frequency, $d\Phi(f)/df$, measured at the output of the reflectometer system. To evaluate the measuring capabilities and the requirements for using a frequency sweeping reflectometer, a numerical method to reconstruct the density profiles in the TCABR tokamak was implemented. The robustness of inversion method was tested through the following steps: i) Determination of $d\Phi(f)/df$ from a known density profile; ii) Reconstruction of the electron density profile by the method that will be presented here; iii) Comparison between the original and reconstructed electron density profiles to test the validity of the method. The method and analysis are applied for the X and O polarization modes. These computational analyses can be useful for the interpretation of reflectometer diagnostic data and the experimental data can be used to check the validity of the numerical model.

I – Introduction

Frequency sweeping reflectometry is a diagnostic technique frequently used to measure the density profiles in tokamak machines, successfully implemented for the O-mode [1] (when the electric field of the probing wave is parallel to the magnetic field in the plasma) and for the X-mode [2] (when the electric field of the probing wave is perpendicular to the magnetic field in the plasma). The density profiles can be inferred from the phase dependence of the probing frequency $d\Phi(f)/df$, measured at the output of the reflectometry system. In order to reduce the effect of turbulence in the density profile reconstruction, a fast sweeping local oscillator is used. In the TCABR tokamak, the sweeping reflectometer uses Hyperabrupt Varactor-Tuned (HTO) oscillator to provide density profiles in 8μs, faster than the typical period of the plasma fluctuations. During each shot, more than eleven thousand of sweeps are made and the output signals acquired in a dedicated Data

Acquisition System (DAS based in ATCA - *Advanced Telecom Computing Architecture*) with a sampling frequency up to 200 MHz.

To evaluate the measuring capabilities and the requirements for using a sweeping reflectometer, a numerical method to reconstruct the density profiles in the TCABR tokamak was implemented. The method and analysis are applied for the X and O modes of wave propagation in fusion plasmas. In this paper, a brief review of wave propagation conditions in cold plasma approximation and accessibility curves for the TCABR tokamak are presented in Section 2. The numerical method implemented for the reconstruction of density profiles is described in Section 3. The numerical result and comparison between the original and reconstructed electron density profile, to test the validity of the method, are also presented. In another example, the typical behaviors of phase and time of flight in the presence of a magnetic island are presented in Section 4. Finally, the main conclusions and future developments are discussed in Section 5.

II – Basic Principles

The phase delay of the reflected signal, measured at output of the reflectometer, is given by $\Phi_m = \Phi_r + \Phi_p$, where Φ_r is the reference phase (contribution from the waveguides and air/vacuum path lengths, which can be determined from calibration procedures or calculated numerically) and Φ_p is the phase contribution from plasma, given by:

$$\Phi_p(f) = 4\frac{\pi f}{c}\int_{x_0}^{x_c} N(f,x)dx\,, \tag{1}$$

where $N(f,x)$ is the wave refractive index and x_0 and x_c are the coordinates of the vacuum-plasma boundary (physical limiter in our case) and the cutoff layer, respectively. The plasma phase is obtained experimentally from the subtraction of the reference phase, Φ_r, of the measured phase of the reflected signal, Φ_m. The refractive indexes for the Ordinary and Extraordinary modes of propagation, in a cold plasma approximation and perpendicularly to the magnetic field [3], are given respectively by:

$$N_O = \left[1 - \left(\frac{f_{pe}}{f}\right)^2\right]^{1/2} \tag{2}$$

$$N_X = \left[1 - \left(\frac{f_{pe}}{f}\right)^2 \frac{f^2 - f_{pe}^2}{f^2 - f_{ce}^2 - f_{pe}^2}\right]^{1/2} \tag{3}$$

where $f_{pe} = 8.98(n_e(x))^{1/2}$ is the plasma frequency, $f_{ce} = 27.99 \times B$ is the electron cyclotron frequency (both in SI units), and B, in last expression, is the total magnetic field.

The total wave reflection inside the plasma column occurs when $N_{O,X} = 0$; for the O-mode the cutoff is given by $f = f_{pe}$ and, for the X-mode,

$$f_{R,L} = \frac{\pm f_{ce} + \left(f_{ce}^2 + 4f_{pe}^2\right)^{1/2}}{2} \tag{4}$$

where f_R is the Right-Hand cutoff (plus sign in the last expression) and f_L is the Left-Hand cutoff (minus sign in the last expression). The accessibility conditions on the equatorial plane of the TCABR tokamak (main parameters [4]: Major radius R_0 = 0.615m, Minor radius a = 0.18 m, Toroidal magnetic field B_0 = 1.0T – 1.1 T, Plasma current $I_p \le$ 120 kA, Plasma current duration $t_p \le$ 150 ms, Peak electron temperature $T_{e0} \le$ 600 eV, Peak electron density $n_{e0} \le 6.0\mathrm{x}10^{19}\ \mathrm{m}^{-3}$) , as function of major radius R, are presented in figure 1. The density [5,6] and temperature [7] profiles considered in the simulations are: $n_e(R) = n_{e0}\left[1 - (\frac{R-R_0}{a})^2\right]^{\beta}$ and $T_e(R) = T_{e0}\left[1 - (\frac{R-R_0}{a})^2\right]^{\beta Te}$, respectively. The total magnetic strength is given by $B = \sqrt{B_T^2 + B_p^2}$, where B_p is the poloidal magnetic field and $B_T = \frac{B_0 R_0}{r}$ is the toroidal magnetic field. The poloidal magnetic field is small, $^{B_0}/_{B_T} < 4\%$, and so the change in the orientation of the magnetic field of $\theta < 2.2^0$ is considered negligible.

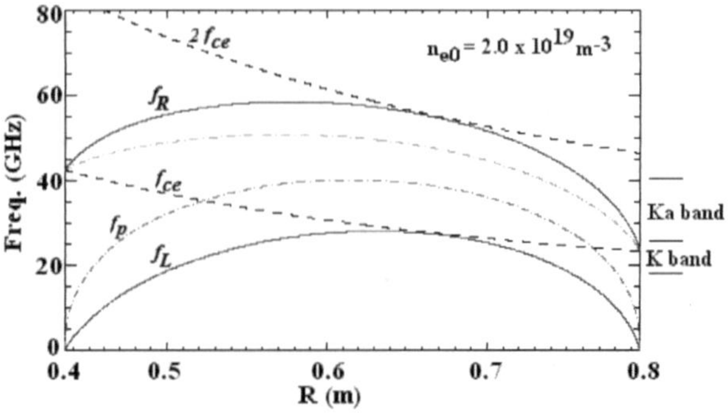

Figure 1: Accessibility conditions for typical density profiles in the TCABR tokamak: f_{ce} and 2. f_{ce} are the first and second Electron Cyclotron Emission, respectively, f_R is the Right-Hand cutoff, f_L is the Left-Hand cutoff and f_{pe} is the plasma frequency.

The phase shift of the wave, for the rays travelling inside the plasma column along the horizontal axis, can be obtained integrating numerically equation (1). The total phase, Φ_m, as function of the probing frequency, for two different central densities and for the O and X polarizations modes, are presented in figure 2(a) and figure 2(b), respectively. In this example, the reference phase, Φ_r (the distance between the antennas and the limiter is 18.5 cm), is represented by the blue color in figure 2.

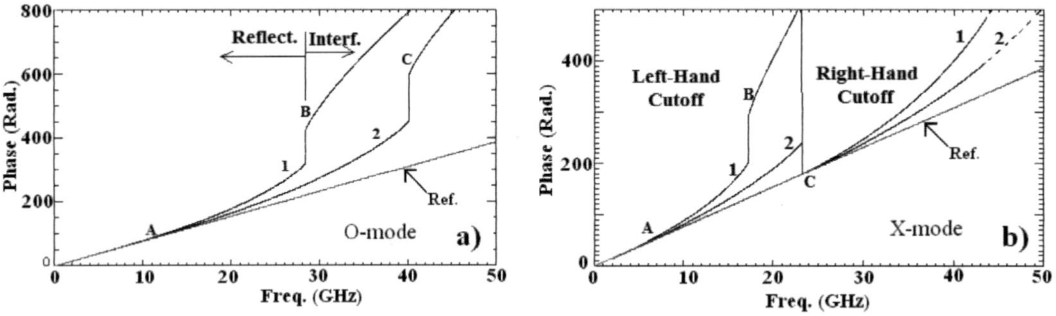

Figure 2: Total phase of the wave versus probing frequency, for peak densities $n_{e0} = 1.0 \times 10^{19}\ m^{-3}$ (curve 1) and $n_{e0} = 2.0 \times 10^{19}\ m^{-3}$ (curve 2), and for the O-mode (part a) and X-mode polarization (part b). The reference phase, ϕ_r, is shown by the blue color straight line.

The total phase diverges from the reference phase at $f > 11$ GHz for the O-mode, and $f > 6$ GHz for the X-mode, indicated by A in the figures 2.a) and 2.b). The points B and C in figure 2.a) (O-mode polarization) represent the phase jumps when the frequency of the probing wave reaches the maximum value of plasma frequency, i.e. at values of 28.4 GHz (peak density of $n_{e0} = 1.0 \times 10^{19}\ m^{-3}$) and 40.2 GHz (peak density of $n_{e0} = 2.0 \times 10^{19}\ m^{-3}$), and corresponds to the transition between Reflectometry to Interferometry (or Refractometry) operation mode. In X-mode polarization, the Reflectometry system can operate in Interferometry mode between the maximum of the Left-Hand cutoff (point B in figure 2.b)) and the minimum of the Right-Hand cutoff (point C in figure 2.b)). In the same figure, the point C corresponds also to the first fringe ($f \sim 23.2$ GHz) and this is very useful for the initialization of the density profile using the X-mode polarization. For frequencies higher than the second ECE harmonic at plasma edge, $f > 46.3$ GHz, the wave launched towards the plasma column in the X-mode polarization will be absorbed.

III – Profile inversion numerical method

To evaluate the measuring capabilities and the requirements for using a sweeping reflectometer, a numerical method to reconstruct the density profiles in the TCABR tokamak was implemented and based on the method described elsewhere [8]. The cutoff positions are determined interactively, using the following equation

$$X(i) = \frac{[A(0)\times X(0)+(i-1)\times mean[X(1:i-1)\times(A(1:i-1)-A(0:i-2))]-\Phi_p(i)]}{A(i-1)}, \quad (5)$$

where $A(i) = \frac{\pi\times f(i)\times[N(i)+N(i-1)]}{c}$ and N is the refractive index defined by the equations (2) and (3). The term $\Phi_p(i)$ is the contribution of the phase delay due to the plasma path. Once the cutoff position is determined, the electron density is easily obtained from well known relations (plasma frequency for the O-mode and equation (4) for the X-mode). To ensure numerical stability in the determination of the cutoff position, $X(i)$ was replaced by

$$X_i = \frac{X_i(i)+X_i(i-1)}{2}.$$

To test the method, in the next step we will consider a typical density profile in the TCABR tokamak. In figure 3.a) is represented the density profile reconstructed from the phase of the plasma, shown in figure 2.a). In the same simulation, the cutoff position as a function of the probing frequency is depicted in figure 3.b). In these simulations, the range of frequency probing considered was from 0.1 GHz up to 50 GHz.

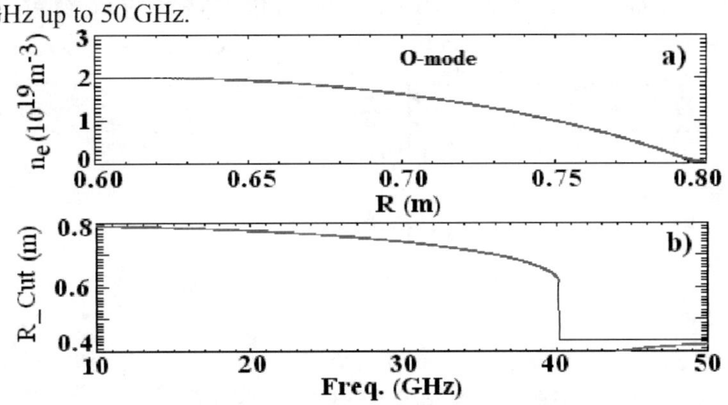

Figure 3: Expected (black curve) and reconstructed (red curve) obtained for the O-mode reconstruction method: a) Density profile; b) Cutoff position as function of probing frequency.

The density profile reconstructed from the plasma phase shift in X-mode (represented in figure 2.b)), is shown in figure 4.a). In this polarization mode, the density profile can be reconstructed from the edge of the plasma column (the first fringe occurs at $f = 23.2$ GHz) up to value for the second ECE frequency at edge. I n the X-mode polarization, the wave launched towards the plasma column will be absorbed at $f > 46.4$ GHz, that corresponds to the radial position $R < 73.4$ cm. The cutoff position as function of the probing frequency is shown in figure 4.b).

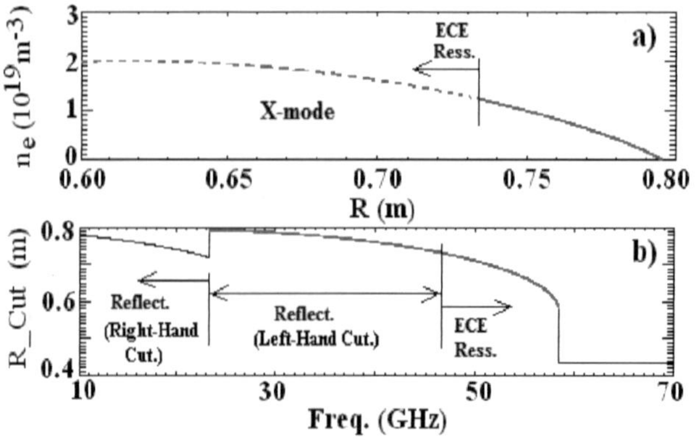

Figure 4: Reconstruction curves in X-mode: a) Density profile: Original (black curve) and reconstructed (red curve); b) Cutoff position as function of probing frequency: Expected (black curve) and reconstructed (red curve).

In the next figure are shown the errors in density and cutoff position for the O-mode (graphics at left) and X-mode (graphics at right).

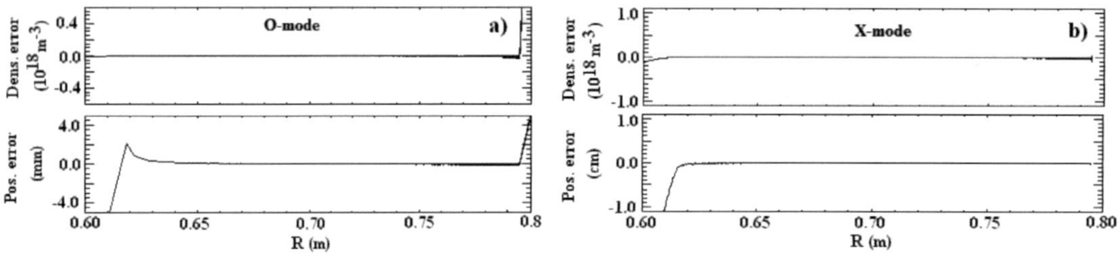

Figure 5: Error due to the reconstruction method for density profile and cutoff position, in the O-mode (part a) and X-mode (part b).

A very good agreement between the original and the reconstructed profiles, in both wave polarization modes, and from the plasma edge and close to the centre of plasma column, was found as seen in figure 5. The uncertainty in the cutoff positions rises quite fast, near the centre of the plasma column, due to flatness of the plasma frequency profile (for the O-mode) and in the Right-Hand cutoff for the X-mode.

The O-mode polarization has easier accessibility to plasma phase measurements than using the X-mode. The main disadvantage of using the O-mode polarization is the difficulty to estimate correctly the edge contribution required to initialize the density profiles [9].

The effect of the presence of a magnetic island in the density profile reconstruction will be presented in the next section.

IV – Effect of magnetic island in the reconstruction of the density profile

The perturbation in the density profile due to the presence of a magnetic island (red curve in the figure 6.) is given by the equation $\tilde{n}_{isl} = \tilde{n}_{e0}.\exp\left[\frac{|R-\tilde{x}|}{\tilde{W}/2}\right]^2$, where $\tilde{n}_{e0}$ is the maximum amplitude of perturbation, $\tilde{x}$ is the location of the rational surface and $\frac{\tilde{W}}{2}$ is the magnetic island half width. The values used in the density profile perturbation, presented in figure 6, are: $\tilde{n}_{e0} = 2 \times 10^{18} m^{-3}$, $\tilde{x} = 0.76\ m$ and $\widetilde{W} = 4.2\ cm$.

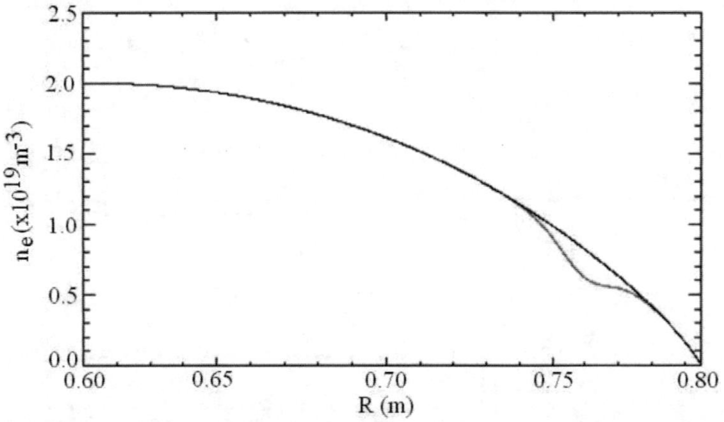

Figure 6: Non-perturbed (black curve) and perturbed density profiles by the presence of magnetic island (red curve).

The phase shift of the wave inside the plasma column as function of the probing frequency, in the presence of a magnetic island (blue curve) and non-perturbed profiles (red curve), for the X-mode polarization, is represented in right scale of figure 7. The cutoff position as function of the probing frequency, for a non-perturbed density profile (red curve) and in the presence of the magnetic island (blue curve), determined from the X-mode wave polarization reconstruction method, is presented in the left scale of figure 7. From this figure, we observe that the cutoff position recovers completely outside the magnetic island region but the same conclusion is not valid for the phase.

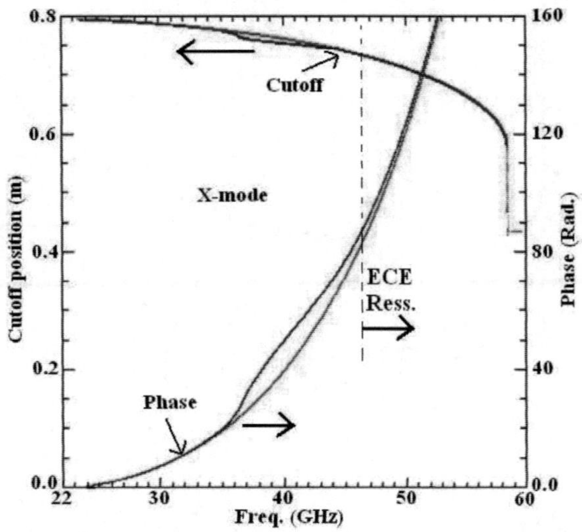

Figure 7: The cutoff position (left side in graphic) and phase delay inside the plasma column (right side in graphic), in X-mode polarization mode, for a non-perturbed (red curves) and perturbed density profiles by the presence of a magnetic island. The waves are absorbed by the plasma column for $f > 46.3$ GHz.

The density profile is perfectly reconstructed y numerical simulation, as shown in the figure 8.a). Also, the errors in the density and cutoff positions are negligible (see figures 8.b) and 8.c)).

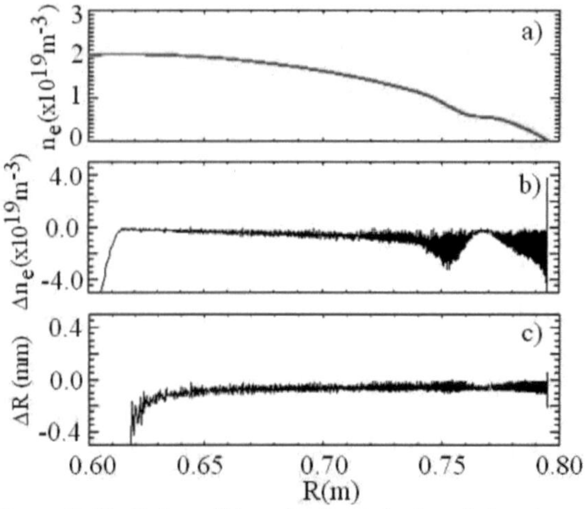

Figure 8: Radial profiles of: a) original and density profiles reconstructed; b) error in density and c) error in the cutoff position.

The time of flight, determined by $\tau_g = \frac{1}{2\pi}\frac{\delta\varphi}{\delta f}$ and, for the X-mode, is presented in the next figure.

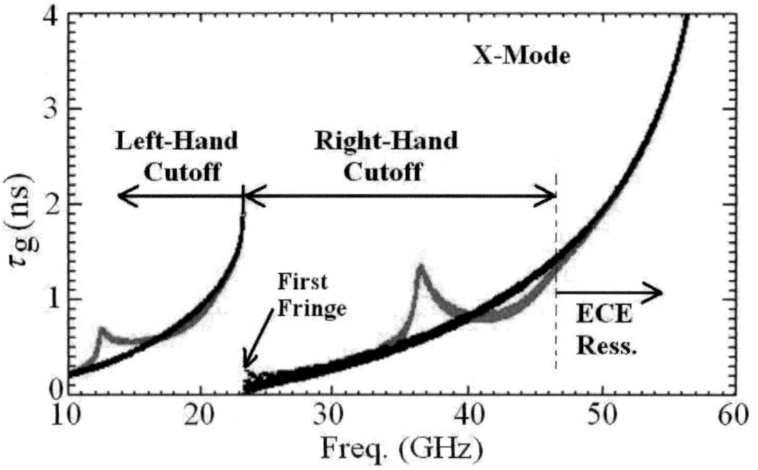

Figure 9: Time of flight, for the X-mode polarization wave, as function of probing frequency.

The effect of the magnetic island starts to appear at $f \sim 10$ GHz (in the Left-Hand cutoff) and at $f \sim 34$ GHz (in the Right-Hand cutoff), for the X-mode polarization, as shown in figure 9. The magnetic island effect in the time of flight reaches a maximum at $f \sim 12$ GHz and $f \sim 36.5$ GHz, for the Left-Hand cutoff and Right-Hand cutoff, respectively.

The time of flight as function of the cutoff position for the Left-Hand cutoff (curve A, occurs from the edge of the plasma column up to $R = 0.72$m) and for the Right-Hand cutoff (curve B, from the plasma edge up to R ~ 0.73m), is shown in figure 10. The red curves represent the effect of magnetic island in the time of flight. The X-mode polarized wave will be absorbed by the second ECE harmonic, for plasma positions with R ≤ 0.73m.

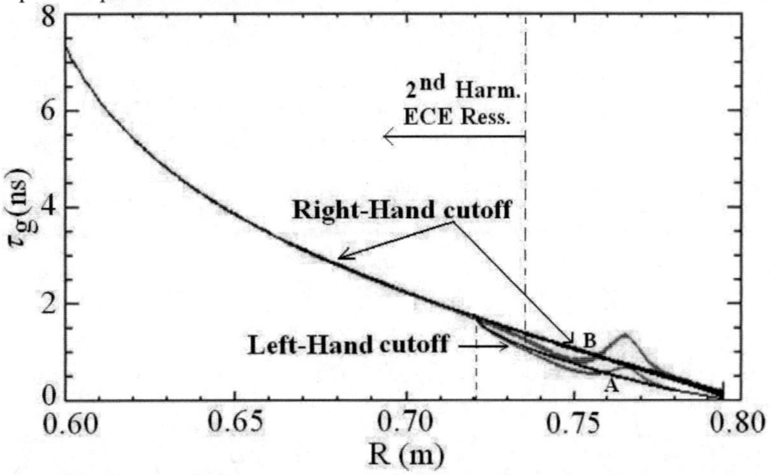

Figure 10: Time of flight as function of the cutoff position, in presence of a magnetic island (red curves) and in a non-perturbative density profile (black curve), obtained from the Left-Hand cutoff (curve A) and Right-Hand cutoff (curve B).

From figure10 on can observed that the effect of magnetic island in the time of flight appears between R ~ 72.0 cm and R ~78.0 cm, but in fact, the region with the presence of the magnetic island is between R ~ 74.0 cm and R ~ 78.0 cm (see figure 6). This happens because the wave is influenced by the presence of magnetic island before the wave be reflected in the cutoff layer.

The main conclusions and the next perspectives are presented in the next Section.

V – Conclusions and next steps

In this paper was presented a numerical method used for the reconstruction of density profiles from some predetermined density profiles. The robustness of the inversion method was tested to validate the numerical method and these analyses were applied for the X and O modes of wave propagation in fusion plasmas. The phase dependence of the probing frequency, $d\Phi(f)/df$, measured at the output of the reflectometer system, will be tested in the numerical method in order to obtain the reconstruction of density profiles in the TCABR tokamak.

Acknowledgments
This work was supported by CNPq "*Conselho Nacional de Desenvolvimento Científico e Tecnológico*" and FAPESP "*Fundação de Amparo à Pesquisa do Estado de São Paulo*" (Project 11/50773-0).

References

[1] Simonet F 1985 *Rev. Sci. Instrum.* **56** 664.
[2] Bottollier-Curtet H and Ichtchenko G 1987 *Rev. Sci. Instrum.* **58** 539
[3] Hutchinson I H 1987 *Principles of Plasma Diagnostics* (Cambridge University Press Cambridge).
[4] Nascimento I C, Galvão R M O, Kusnetsov Y, Ruchko L F, Elfimov A G, Sanada E, Cuevas N A M, Degasperi F T, Fagundes A N, Elizondo J I, da Silva R P, Vuolo J H,

Vannucci A, Fonseca A M M, Heller M V A P, dos Reis A P, de Sá W P, Campos D O, Chamaa I El, Ozono E, Amarante-Segundo G, Lerche E and Walder V S 1998 *Research Using Small Fusion Devices, VII Latin American Workshop on Plasma Phys.* Tandyl Argentina 15

[5] Fonseca A M M, da Silva R P, Galvão R M O, Kuznetzov Yu K, Elizondo J I, Calrderon E R, Ruchko L F and Vuolo J H 2004 *Braz. Journal of Physics*, **34** (4B) 1780

[6] Elizondo J I, Korneev D, Nascimento I C and de Sá W P 2002 *Braz. Journal of Physics* **32** 123

[7] da Silva R P, Fonseca A M M, Vuolo J H, Calderon E R, Galvão R M O, Kuznetzov Yu K and Raffaelli J C 2004 *Braz. Journal of Physics* **34** (4B) 1771

[8] Mazzucato 1998 E Rev. Sci. Instrum. **69** 2201

[9] Varela P, Manso M E, Silva A, Fernandes J and Silva F 1995 *Rev. Sci. Instrum.* **66** 4937

Data Acquisition and Automation for Plasma Rotation Diagnostic in the TCABR Tokamak

G Ronchi, J H F Severo, W P de Sá and R M O Galvão
Institute of Physics, University of São Paulo, São Paulo, SP, Brazil

E-mail: gronchi@if.usp.br

Abstract. In this work we describe the implementation of a full modular system of data acquisition and processing for the plasma rotation diagnostic in the TCABR tokamak. The experimental setup uses a single monochromator and six photomultipliers (PMT), in which pair of PMTs measures the light at slightly different wavelengths. Thus, it can measure the time evolution of the Doppler shift of the impurities emission lines coming from three spatial positions (one for toroidal rotation and two for poloidal rotation). The data acquisition and pre-analysis program were written with LabVIEW software and is capable of controlling the spectrometer wavelength, PMTs power supplies, data acquisition, and storage. All data are recorded in MDSplus trees that easily allow data visualization and post-processing analysis (both locally and remotely) via MATLAB, Python, Java and others programming languages. This system can run independently from other diagnostics and machine systems and can be integrated with the main tokamak control system by means of TCP/IP messages.

1. Introduction

It is well know that the plasma rotation in magnetic confinements experiments allow to stabilize large-scale MHD instabilities and to improve confinement reducing turbulent transport of particles and energy [1, 2, 3]. Due to its large size and to high density, effective momentum deposition by external neutral particles injection will be difficult to achieve in ITER. Therefore, if the spontaneous (intrinsic) rotation could be made large enough, it may play an important role in plasma stabilization. Furthermore, the role of plasma rotation in disruptions or in an H-mode transition, for instance, is not completely understood.

Thus we build a non-invasive diagnostic for measuring the time evolution of the intrinsic rotation in the TCABR tokamak (R = 0.615 m, a = 0.18 m, $I_p \leq 90$ kA, B_t = 1.07 T, $n_0 \leq 4 \times 10^{19}$ m^{-3}). The system employs a single monochromator with a beam splitter and two sets of PMTs for each radial position to record the light signal at slightly different wavelengths in order to estimate the Doppler shift due to the plasma velocity.

2. Plasma rotation diagnostic setup

The former experimental setup [2] could resolve temporarily the plasma rotation in just one optical port (toroidal or poloidal), the data acquisition was made using a Tektronix oscilloscope, and all high voltage power supplies for the photomultipliers (PMT) were manually adjusted, taking a lot of effort to proper store and record all relevant data for each tokamak discharge.

The upgraded system shown in the figures 1 and 2 is able to measure time-resolved and simultaneously two poloidal chords and one toroidal chord using the same technique. Currently the data acquisition is controlled by a LabVIEW program, and the data acquired are stored in MDSplus trees, as detailed in next sections.

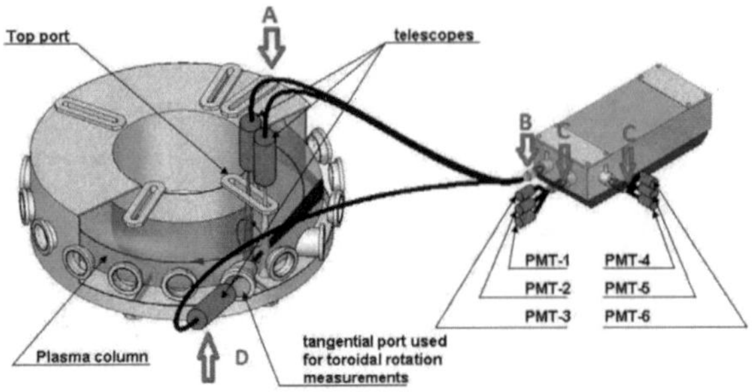

Figure 1. Experimental setup: the light from poloidal chords (A) and toroidal chord (D) arrive in the entrance slit (B) of the monochromator, and each signal is measured in the PMT at the axial and lateral output slit (C).

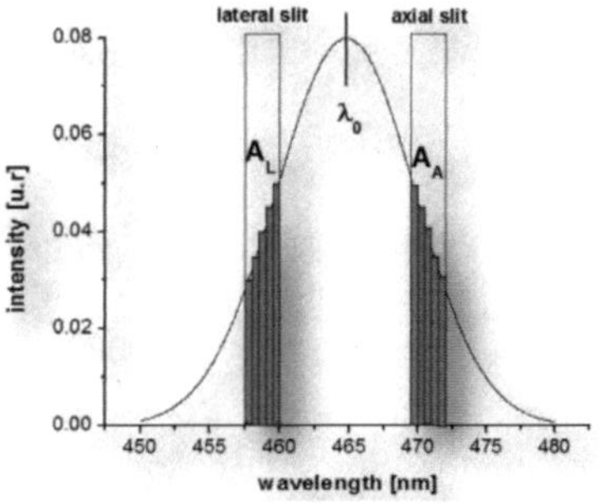

Figure 2. The light at the lateral and axial slits of the monochromator have slightly different wavelengths. So, from the ratio of the PMTs signals is possible to estimate the Doppler shift.

2.1. Doppler shift Calibration

The calibration is made with the optical ports looking at the central chord of the plasma poloidal cross section, where one expects to have no rotation along this line of sight (so, no effective Doppler shift). Then, the ratio of the signals of axial and lateral output slits are recorded from different wavelengths near the impurity emission, using tokamak discharges. The calibration is made separately for the Carbon III (464.7 nm) and Carbon VI (529.05 nm) lines because the Doppler broadening affects significantly the ratio of the outputs signals, as show in the figures 3 and 4. The optical chords are labelled as "central", "poloidal" and "toroidal" since the first one usually is keep looking the central chord of the plasma.

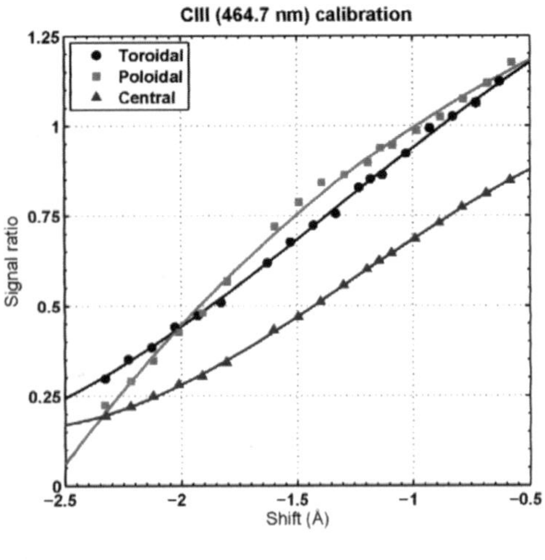

Figure 3. Calibration curve for C III.

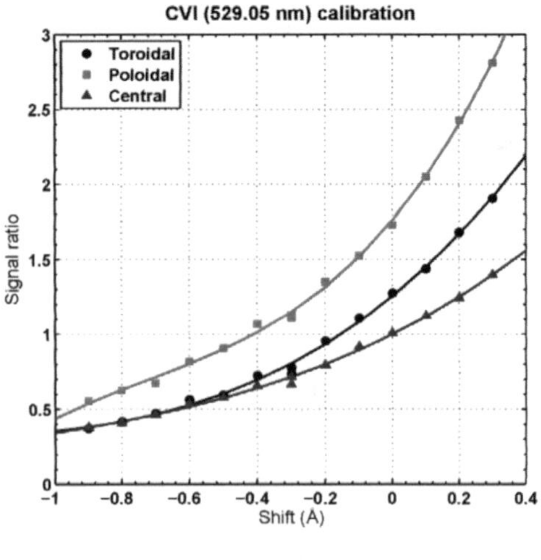

Figure 4. Calibration curve for C VI.

For the sake of simplicity we define zero "shift" in the calibration curves where the central chord ratio is 1. So, the velocity of rotation of the plasma can be estimated, taking as reference the ratio where there is no Doppler shift:

$$v = c\frac{\Delta\lambda_{\text{Doppler}}}{\lambda_0} \tag{1}$$

$$\Delta\lambda_{\text{Doppler}} = \text{shift}_i(R_i) - \text{shift}_i(R_{i\ \text{reference}}) \tag{2}$$

where v is the plasma velocity measured in the i-th chord, c is the speed of light, $\Delta\lambda_{\text{Doppler}}$ is the Doppler shift of the impurity emission line of wavelength λ_0, which is the difference of wavelength associated with the signal ratio of axial and lateral PMTs ($\text{shift}_i(R_i)$) and wavelength associated signal ratio with no Doppler shift (shift_i ($R_{i\ \text{reference}}$)) for that optical chord.

3. Experimental automation

During the experiments, it is necessary to control, monitor, and record the parameters of several instruments, like spectrometer position, PMT voltage, ADC gain, etc. The use of text-oriented programming language makes it hard the maintenance, and modifications would demand long time, specially if it uses any graphical user interface. Therefore we chose the LabVIEW graphical programming which has been used for wide applications in a multitude of instruments field like optics [4] and plasma physics [5] as well. The advantage of the LabVIEW platform over textual programming remains in its fast programming, readable source code, and instrumentation oriented interface due to its focus on instrument control.

The LabVIEW program controls the spectrometer wavelength by serial communication (RS-232), and the power supply for the PMTs by USBs ports. The data acquisition is made by NI PXI 5105 digitizers from National instrument, up to 60 MSamples/s and 12 bits resolution. The digitizers is connected to a PXI bus to the main controller, a NI PXIe-8108 with a Dual-Core processor running Windows 7.

The program running in the main controller takes also as input the radial position of the optical ports. The user interface of the program, for local setting, is shown in the figure 5 and 6.

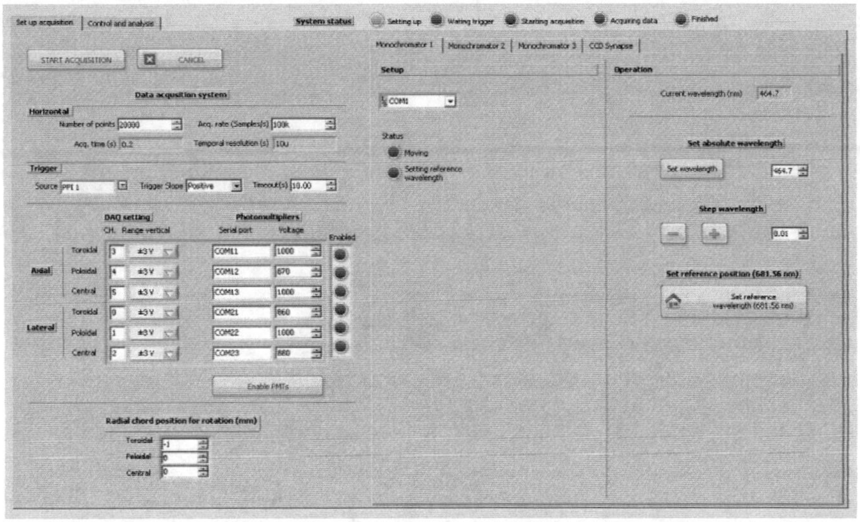

Figure 5. LabVIEW interface for instrumentation setup for local control.

Journal of Physics: Conference Series **591** (2015) 012007 doi:10.1088/1742-6596/591/1/012007

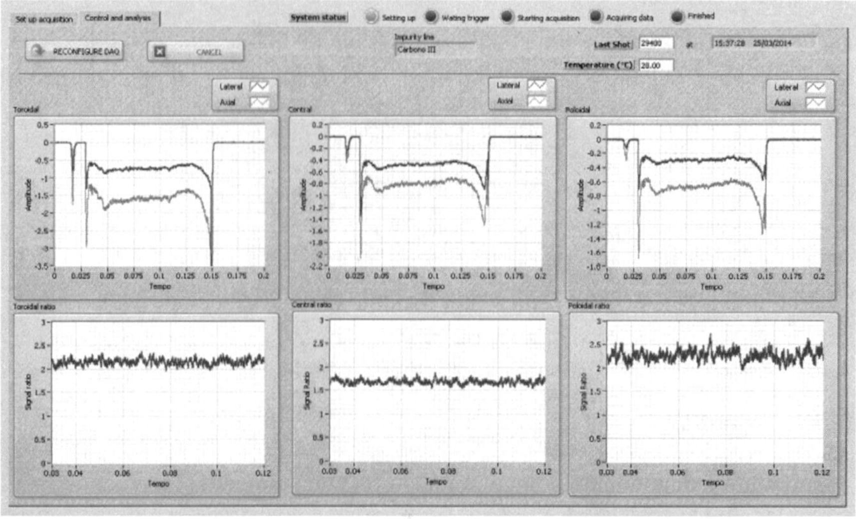

Figure 6. LabVIEW interface for post-analysis.

3.1. MDSplus

The MDSplus [6] is a set of programs and tools for experimental data acquisition, storage and control developed for managing the complicated environment of research as in plasma physics and nuclear fusion. Its focus in pulsed mode experiments (or intermittent events) makes it propitious for our experimental conditions. Besides, it is widely used in nuclear fusion experiments around the world [7]. Another important point is that the MDSplus provides application programming interfaces among different types of language (JAVA, Fortran, C, C++, Python) and programs such as MATLAB, IDL, OCTAVE [8].

Although MDSplus can also control the experimental setup and acquisition; in our case it acts just as a centralized data server in which the stored data is arranged hierarchically. After each plasma discharge, the LabVIEW Object Oriented programming interface (VLOOP) creates a new pulse file (an MDSplus Tree) and populates its nodes with all signals and parameters of the experiment. After this all data is already available for remote access. The full flowchart of the signals is shown in the figure 7.

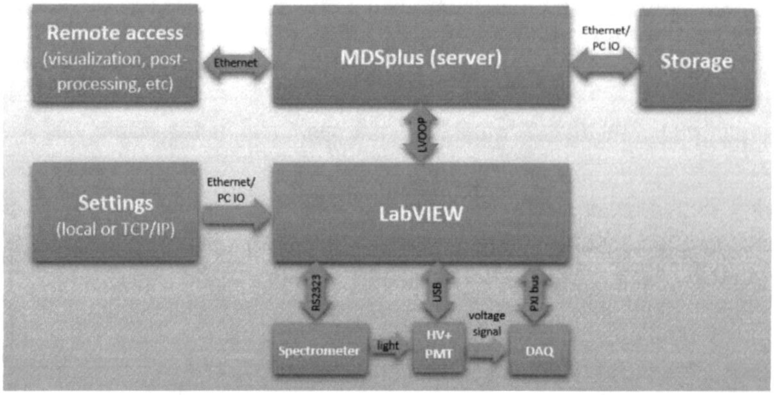

Figure 7. Flowchart of signals and parameters for the rotation diagnostic.

The LabVIEW virtual instrumentation, with the MDSplus, was programmed in a modular way, so the experiment diagnostic can run standalone of the main tokamak control system [9] (unless for the start trigger signal), and integrated by means of TCP/IP messages to control the acquisition. Furthermore, the data can easily merge with the main tokamak signals, since it can be set as a branch (i.e., an MDSplus subtree) of the main system, preserving its hierarchy.

4. Results for the TCABR

As an example, in the figures 8 and 9 we present the average toroidal and poloidal plasma rotation profile in the TCABR tokamak. The data was taken for several discharges and post-processed, with the calibration data, in the software MATLAB environment.

A characteristic feature of the plasma rotation in the TCABR tokamak is the change of direction of the toroidal velocity close to the edge. Besides it, the measured plasma rotation (except close to the edge) seems to agree with the neoclassical predictions in the collisional regime where the poloidal velocity is expected to be proportional to the gradient of the ion temperature [10, 11] as previously reported [2].

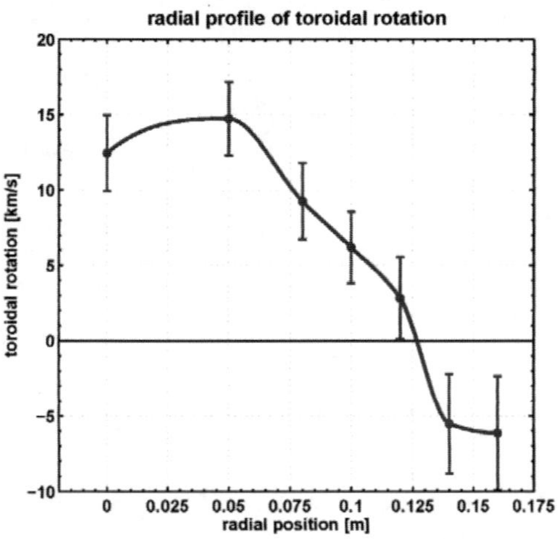

Figure 8. Radial profile for toroidal plasma rotation; positive means in the toroidal field direction.

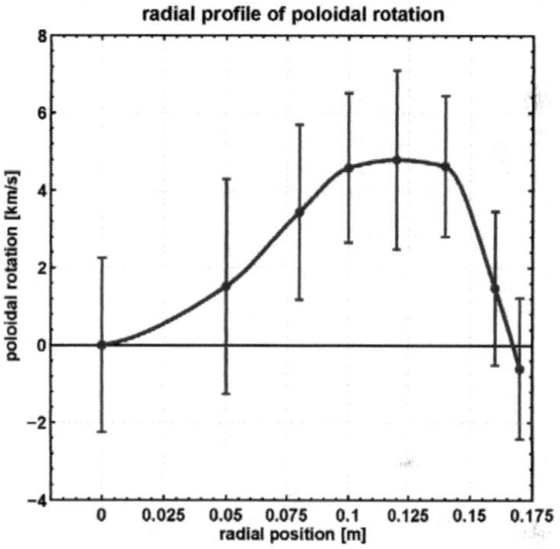

Figure 9. Radial profile for poloidal plasma rotation; positive means in the electron diamagnetic drift direction.

5. Conclusion

The plasma rotation diagnostic system was developed and tested using a single spectrometer and several optical chords to measure it at different positions along the plasma discharge. The automated system allows easy control of the instrumental setup as well as to record all relevant data parameters and instrumentation settings. All this process can be controlled locally or remotely by mean of the TCP/IP protocol.

Furthermore, the MDSplus server data allows remote access for data post-analysis, and makes it easier to retrieve specific information of any plasma tokamak discharge. Moreover, this time-resolved diagnostic allows both reconstruct the profile of plasma rotation and its temporal evolution as well.

Acknowledgments

This work was supported by the São Paulo Research Foundation (FAPESP Grant No. 10/50496-4 and 11/50773-0) and Capes, Brazil.

References

[1] Vries P and Zastrow K D 2012 *Phys. Rev. Lett.* **108**(9) 095001
[2] Severo J H F, Nascimento I C, Tsypin V and Galvão R M O 2003 *Nuclear Fusion* **43** 1047
[3] Severo J H F, Nascimento I C, Kuznetov Y K, Tsypin V S, Galvão R M O and Tendler M 2007 *Rev. Sci. Instrum.* **78** 043509
[4] Stryjewski W J 1991 *Rev. Sci. Instrum.* **62** 1921–1925
[5] Starn T K, Sesi N N, Horner J A and Hieftje G M 1995 *Spectrochimica Acta Part B: Atomic Spectroscopy* **50** 1147–1158
[6] Fredian T W and Stillerman J A 2002 *Fusion Engineering and Design* **60** 229–233
[7] Stillerman J and Fredian T W 1999 *Fusion Engineering and Design* **43** 301–308
[8] Manduchi G, Marchi E D and Mandelli A 2013 *Fusion Engineering and Design* 88 1196–1199
[9] de Sá W 2012 *Fusion Engineering and Design* **87** 2199–2202
[10] Rosenbluth M N, Hazeltine R D and Hinton F L 1972 *Physics of Fluids (1958-1988)* **15** 116–140
[11] Hazeltine R D 1974 *Physics of Fluids (1958-1988)* **17** 961–968

15th Latin American Workshop on Plasma Physics (LAWPP2014) IOP Publishing
Journal of Physics: Conference Series **591** (2015) 012008 doi:10.1088/1742-6596/591/1/012008

ISTTOK real-time control assisted by electric probes

I. S. Carvalho, H. Fernandes, P. Duarte, C. Silva, H. F. C. Figueiredo, T. Hekkert

Associação EURATOM/IST, Instituto de Plasmas e Fusão Nuclear, Instituto Superior Técnico, Universidade de Lisboa, P-1049-001 Lisboa, Portugal

Corresponding author: ivoc@ipfn.ist.utl.pt

Abstract. The ISTTOK tokamak (Ip = 7 kA, B_T = 0.5 T, R = 0.46 m, a = 0.085 m) was recently upgraded with a new multiple-input multiple-output control system, where the user can specify the combination of real-time diagnostics to produce the observed plasma quantities to control any actuator. The present observed quantities are: (i) plasma current, (ii) plasma position and (iii) plasma density, all of which are actively feeded back by the system.
Four-quadrant Langmuir probes were installed in ISTTOK on inner, outer, top and bottom parts of a poloidal section at a 75 mm radius. The floating potential signals measured by the four probes are fed to the control system to estimate the plasma position using a linear model. The model is based on a well known characteristic of the ISTTOK plasma, in that the floating potential varies linearly with respect to the radial position for an extension of about 10 mm (from ~70 mm to ~80 mm). By using only this set of electric probes for plasma position feedback, it was possible to obtain an alternate current (AC) discharge with 40 semi-cycles, surpassing 1 second without loss of ionization during plasma current reversal.

1. Introduction

The ISTTOK [1] tokamak (whose parameters are described in table 1) was upgraded [2] with a new control system for the automatic control of the alternate current discharges [3]. With this system it was possible to test several strategies for plasma control taking advantage of the Multi-Input Multi-Output (MIMO) [3] features installed. One of the strategies was the plasma position real-time control using electric probes. This type of control was successfully tested and demonstrated, thus providing an additional diagnostic to control the plasma position at the ISTTOK tokamak. This diagnostic can now be used as a standalone diagnostic for plasma position control or in conjunction with other diagnostics traditionally used for plasma position control.

In order to provide a real-time feedback control of the plasma parameters several diagnostics were integrated in the control system: (i) mirnov coils, (ii) tomography [4], (iii) interferometer, (iv) magnetic sine probe, (v) magnetic cosine probe, (vi) H_α radiation bolometer, (vii) loop voltage measurement and (viii) electric probes [5] which are this paper main subject.

To actuate on the plasma, several actuators were integrated in multiple a MIMO configuration; (i) two power supplies for the vertical and horizontal magnetic field based on a switching mode H-bridge with MOSFETs in parallel [6], (ii) a primary field power supply with the same characteristics as the other two but based on an IGBT H-bridge and (iii) gas injection piezoelectric valve.

The control system is based on the Telecommunications Computing Architecture (ATCA) standard [7] where two types of functional boards are used, the acquisition board and the control board. Each acquisition board includes 32 differential Analog-to-Digital Converter (ADC) channels which acquire data from the following real-time diagnostics/measurements: (i) tomography, (ii) Mirnov coils, (iii) interferometer, (vi) electric probes, (v) sine and cosine probes, (vi) bolometer, (vii) current delivered by the power supplies, (viii) loop voltage and (ix) plasma current. On the control board, the algorithms are executed with a control cycle of 100 μs on an Intel^{TM} Q8200 chip with 4 cores running at 2.33 GHz. The real-time control system was programmed in C++ on top of the Multi-Platform Real-Time Framework (MARTe) [8] which is also used in several other fusion research laboratories [9].

Table 6. ISTTOK parameters.

Parameter	Value
Plasma current	7 kA
Toroidal field	0.5 T
Major radius	0.46 m
Minor radius	0.085 m

2. ISTTOK electric probes

The ISTTOK electric probes are used to characterize the plasma poloidal asymmetries [10] and are also used for indirectly estimating the plasma position by measuring the floating potential in four poloidal angles. There are four sets of electric probes installed in ISTTOK similar to the probe depicted in fig. 1. These probes are poloidally separated by 90° and are located at the top, bottom, inner and outer positions of a poloidal cross-section. The distance from the pins to the vacuum vessel centre is 75 mm while the plasma limiter is located at 85 mm from the vacuum vessel centre.

The control system acquires the data (floating potential) from one pin of each set of electric probes depicted in fig. 1. The electric probe data collection is synchronized with the control cycle which runs at a fixed period of 100 μs.

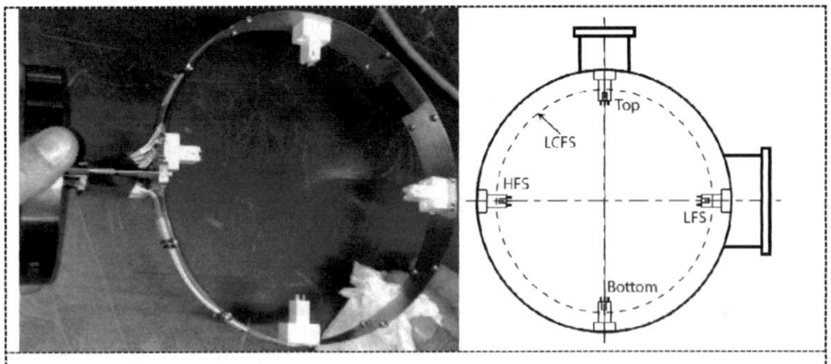

Figure 1. ISTTOK electric probes array and position inside the vacuum vessel

3. Position indirect determination using electric probes

The typical ISTTOK floating potential radial profile [11] is depicted in fig. 2. It can be observed that its value is monotonic on a region of about 15 mm inside the LCFS. Since the floating potential radial profiles are very similar for all poloidal positions, the plasma position on a given axis can be hinted from the difference between the floating potential of two opposed electric probes that define that axis. As such, the vertical position can be hinted from the floating potential measurements of the top and bottom electric probes, and similarly, a radial position estimation can be derived from the inner and outer probes

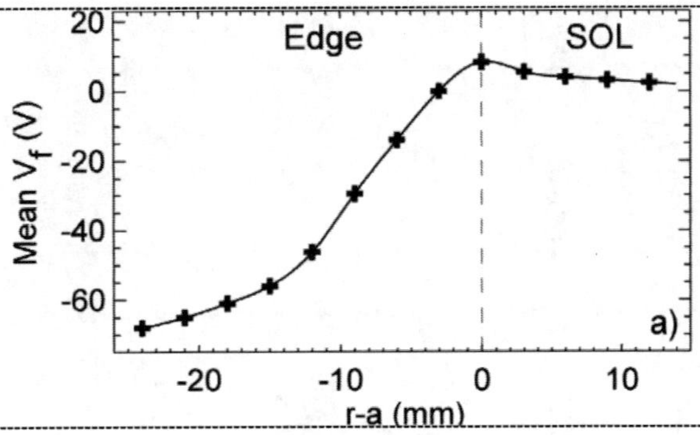

Figure 2. ISTTOK average floating potential versus distance to the LCFS

Taking into consideration the relation of the floating potential with the distance from the LCFS (fig. 2) and assuming circular plasma current only delimited by the ISTTOK inner carbon limiters, the approximate plasma position given by the electric probes is:

$$R_p = C_1 \text{ x } (Vf_{outer} - Vf_{inner}) + C_2$$
$$V_p = C_3 \text{ x } (Vf_{top} - Vf_{bottom}) + C_4$$

Where C_1-C_4 are constants estimated from the previous characheristic (fig. 2), R_p is the plasma radial position, V_p is the vertical position, Vf_{outer}, Vf_{inner}, Vf_{top}, and Vf_{bottom} are the floating potentials measured on the outer, inner, top and bottom probe positions respectively.

Figure 3 depicts a comparison between the radial position estimated with the electric probes and the same radial position measured by the magnetic probes. As it can be observed, apart from an offset during the negative plasma current cycles, the position obtained from the two diagnostics follows a similar pattern.

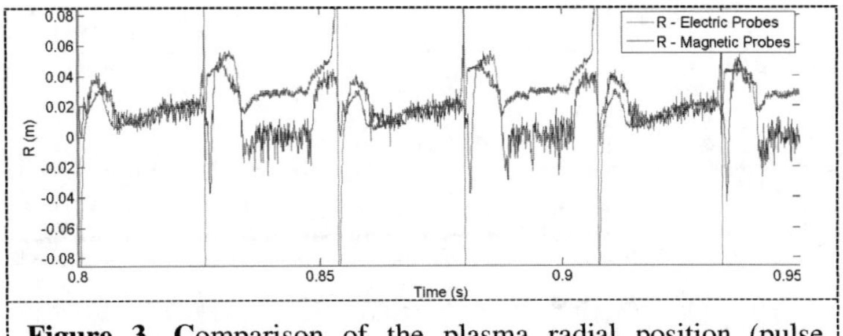

Figure 3. Comparison of the plasma radial position (pulse 34531) obtained from the magnetic probes diagnostic and from the electric probes diagnostic (without filtering).

The ISTTOK control can be programmed as a mixture of feedback control and pre-programmed currents during each time-window. The fig. 4 shows the typical behavior of the vertical field coils power supply during position feedback (color shaded areas) and pre-programmed current feed to the coils (unshaded areas) using the electric probes diagnostic in real-time.

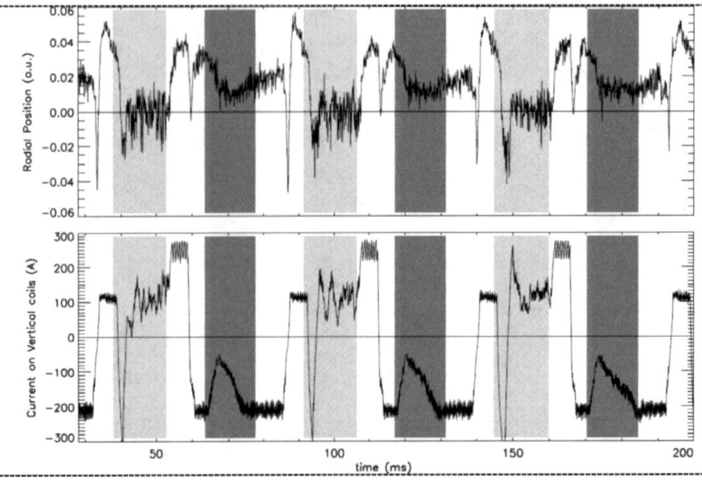

Figure 4. ISTTOK radial position feedback based on the plasma position estimation from the electric probes. The blue color represents the feedback time on the positive plasma current cycles and red color represents the feedback time on the negative plasma semi-cycles.

4. Results

The ISTTOK control system can be programmed to use weights for each diagnostic contribution for each plasma parameter. For the discharge depicted in fig. 4, the feedback control was exclusively done using the approximated plasma position given by the electric probes measurement.

As it can be observed in fig. 5 the feedback control of the plasma position using the electric probes enabled ISTTOK to obtain an AC discharge with more than one second of duration corresponding to 40 AC standard ISTTOK performance semi-cycles, while maintaining a finite density during plasma current reversals, as it can be observed in the line integrated density measurements from the ISTTOK interferometer.

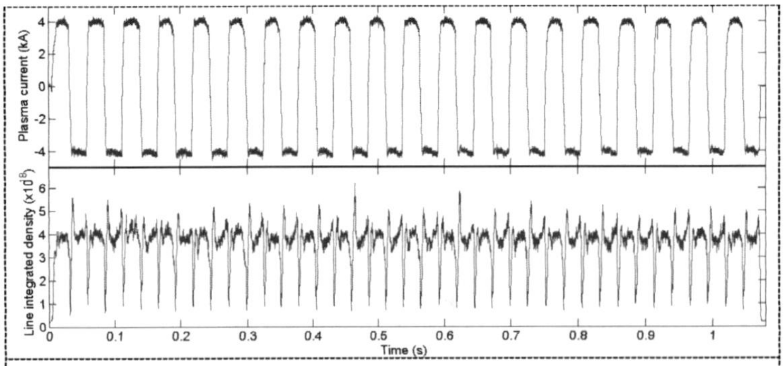

Figure 5. Pulse 34717, 40 AC cycles (1.07s) operation with electric probes position feedback and without loss of ionization during plasma current reversals.

5. Conclusions

The electric probes (as well as the other real-time diagnostics) were successfully integrated in the ISTTOK control system, providing an effective contribution for the ISTTOK real-time control, either by using it as a standalone feedback signal or integrated in a combination of diagnostics to provide an adequate estimation of the plasma position.

The floating potential linear region probe characteristic allows for a simple model suited for real-time control to be used. The method has proved its reliability by yielding the same results as the ones obtained from the magnetic probes, with the advantage of being able to output valid plasma position estimation during the low density plasma current reversal period. This last fact revealed to be crucial to the success of long AC discharges in ISTTOK tokamak, which enables it to be used as a plasma-material interaction testing facility.

Acknowledgments
This work was supported by EURATOM and carried out within the framework of the European Fusion Development Agreement. IST activities also received financial support from "Fundação para a Ciência e Tecnologia" through project Pest-OE/SADG/LA0010/2011. Additional support has been granted from IAEA technical contract CRP F1.30.14 on "Utilization of the Network of Small Magnetic Confinement Fusion Devices for Mainstream Fusion Research".
The views and opinions expressed herein do not necessarily reflect those of the European Commission or the IAEA.

References
[1] H. Fernandes, et al., "Engineering aspects of the ISTTOK operation in a multicycle alternating flat-top plasma current regime," Fusion Engineering and Design, 1998, vol.43, pp. 101-113.
[2] I. S. Carvalho, "ISTTOK control system upgrade", Fusion Engineering and Design, Volume 88, Issues 6–8, October 2013, Pages 1122-1126.
[3] I. S. Carvalho, "ISTTOK real-time architecture", (accepted for publishing in Fusion Engineering and Design, under publishing). DOI: 10.1016/j.fusengdes.2013.12.025
[4] Carvalho P.J., Thomsen H., Coelho R., Duarte P., Silva C., Fernandes H., "ISTTOK plasma control with the tomography diagnostic", 2010, Fusion Engineering and Design, 85 (2) , pp. 266-271.
[5] I. H. Hutchinson, "Ion collection by probes in strong magnetic fields with plasma flow", Phys. Rev. A 37, 4358–4366 (1988).
[6] I. S. Carvalho, et al., "Fast digital link for a tokamak current source control," Power Electronics, Electrical Drives, Automation and Motion, 2008. SPEEDAM 2008.
[7] A. Batista, et al., "ATCA control system hardware for the plasma vertical stabilization in the jet tokamak," IEEE Trans. Nucl. Sci., vol. 57, no. 2, pp. 583–588, Apr. 2010.
[8] A. Neto, et al., MARTe: A Multi-platform Real-time Framework, IEEE Transactions on Nuclear Science, Vol. 57, Issue 2, April 2010, pp. 479-486.
[9] A. C. Neto, et al., "A Survey of Recent MARTe Based Systems", IEEE Trans. Nucl. Sci., vol. 58, no. 4, 2011.
[10] C. Silva, "Characterization of the poloidal asymmetries in the ISTTOK edge plasma," Plasma Phys. Control. Fusion, vol. 53, 2011.
[11] C. Silva, "Interaction between mean and fluctuating E × B shear flows on the ISTTOK edge plasma," Plasma Phys. Control. Fusion, vol. 54, 2012.

Estimation of Electron Temperature on Glass Spherical Tokamak (GLAST)

S Hussain[1], M Sadiq[2], S I W Shah[3], GLAST Team[4]
NTFP, P. O. Box 3329, Islamabad Pakistan

E-mail: shahid92_pk@yahoo.com

Abstract. Glass Spherical Tokamak (GLAST) is a small spherical tokamak indigenously developed in Pakistan with an insulating vacuum vessel. A commercially available 2.45 GHz magnetron is used as pre-ionization source for plasma current startup. Different diagnostic systems like Rogowski coils, magnetic probes, flux loops, Langmuir probe, fast imaging and emission spectroscopy are installed on the device. The plasma temperature inside of GLAST, at the time of maxima of plasma current, is estimated by taking into account the Spitzer resistivity calculations with some experimentally determined plasma parameters. The plasma resistance is calculated by using Ohm's law with plasma current and loop voltage as experimentally determined inputs. The plasma resistivity is then determined by using length and area of the plasma column. Finally, the average plasma electron temperature is predicted to be 12.65eV for taking neon (Ne) as a working gas.

1. Introduction

Spherical tokamak is an evolutionary extension of conventional tokamak with additive advantages of low aspect ratio, low field, and tokamak confinement to achieve high beta. Glass Spherical Tokamak (GLAST) is a small spherical tokamak indigenously developed in Pakistan with an insulating vacuum vessel (Pyrex glass). The design parameters of GLAST are R= 15cm, a= 9 cm, κ =2.5, I_p=50kA, B_T= (0.1- 0.4) Tesla, τ_p=10ms, and T_e = (300-400) eV.

a b c d e

Figure 1. a) Vacuum vessel for GLAST-1 b) Vacuum vessel for GLAST-2 c) GLAST-1 d) GLAST-2 e) 2.45GHz waveguide system for pre-ionization

There are two tokamak devices (Fig.1c, 1d) currently operating at NTFP named GLAST-1 (having central tube of steel) and GLAST-2 (having central tube of glass). The toroidal field coil system consists of sixteen coils made from 1.5mm thick copper strip. In GLAST-1, a special ring (Fig.2) is used to join the coils in series, keeping inner legs straight while in GLAST-2, the inner legs are tilted at certain angle to join in series. This scheme provides an effect of additional pair of coils that helps to stabilize the plasma. The ohmic heating system consists of a central solenoid and two pairs of compensation coils connected in series. This system provides the necessary loop voltage for the startup and also deflects the unnecessary magnetic flux (due to the solenoid) out of the vacuum vessel. The size and the number of turns in these coils were determined experimentally with the help of double flux loops. The vertical field coil system, consisting of three pairs of coils, is currently being used to provide small vertical magnetic field to support the plasma current startup.

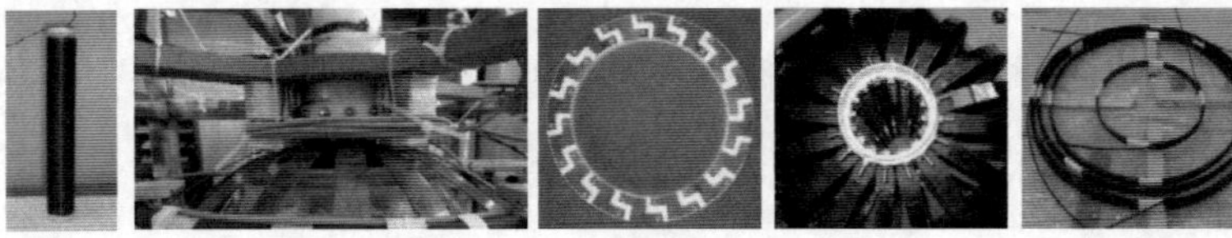

a b c d e

Figure 2. a) Central solenoid b) Compensation coils c) Support ring for TF coils on GLAST-1 d) TF coils of GLAST-2 e) Vertical field coils

The magnetic diagnostics such as induction coils, flux loops and Rogowski coils (Fig.3) are installed on the devices for the measurement of magnetic flux and the plasma current.

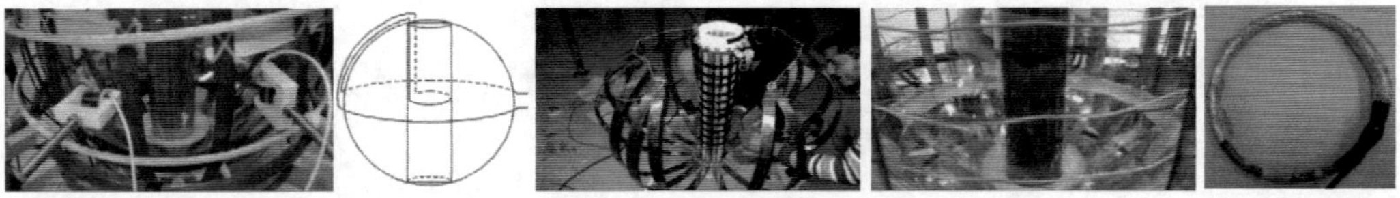

a b c d e

Figure 3. a) Magnetic induction coils b) Scheme of double flux loop c) Inner side of double flux loops d) Outer side of double flux loops e) Rogowski coil

The plasma diagnostics installed on GLAST (Fig.4) are two high speed cameras, fast phtodiode, Langmuir probe, Ocean Optics spectrometer HR4000 and a high resolution spectrometer JOBIN YVON THR1000.

a b c d e f

Figure 4. a) Fujifilm camera HS-10 b) Olympus high speed camera ispeed-2 c) Fast photodiode d) Langmuir probe e) Spectrometer THR1000 f) Spectrometer HR4000

A commercially available 2.45 GHz (0.8kW) magnetron source was modified for generating a microwave pulse of 4msec with enhanced output power of about 1.5kW. This pulsed waveguide microwave system is used as a source of pre-ionization for plasma startup in GLAST. In our initial studies neon was used as working gas with optimum pressure ranging from 1.0 mTorr to 0.5 mTorr. A plasma current of 5 kA for 0.5 msec have been generated in GLAST-2. The recorded experimental signals of microwave pulse, loop voltage, photodiode and of Rogowski coils are shown in Fig.5.

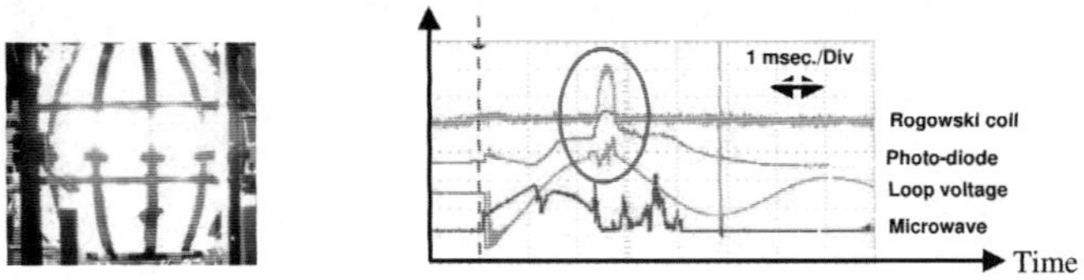

a b

Figure 5. a) Image at maximum plasma current recorded with high speed camera ispeed-2 b) Experimental signals

An image of the device recorded with high speed camera (5000fps) at the time of maxima of the plasma current is shown in Fig.5a. The intense glow has filled the whole space of the tokamak vessel giving effect of shining star. The temporal evolution of the plasma current, light intensity and the corresponding variations in the loop voltage and in the microwave pulse are clearly evident.

2. Estimation of Electron Temperature

In this work, Spitzer resistivity formula is exercised for estimating the plasma electron temperature using the measured values of plasma current and the variations in the loop voltage. The plasma inside GLAST-2 is assumed to be in Coulomb phase. The specific resistivity of such fully ionized plasma only depends on its electron temperature 'T_e' and effective charge number 'Z_{eff}' and this dependence is quantified by the Spitzer formula [1]. Through these considerations, an estimation of the average electron temperature 'T_e' can be obtained.

The plasma resistivity, using Spitzer resistivity formula, is given by

$$\eta_p = 5.23\times10^{-5} T_e^{-3/2} Z_{eff} \ln\Lambda \quad [\Omega m, eV] \tag{1}$$

where,
$$\ln\Lambda = \begin{cases} 16.34 + 1.5\ln T_e - 0.5\ln n_e \ ; & (T_e < 1.16\times10^5\,\mathrm{K}) \\ 22.81 + \quad \ln T_e - 0.5\ln n_e \ ; & (T_e > 1.16\times10^5\,\mathrm{K}) \end{cases}$$

and for tokamaks, $\ln\Lambda \approx 17$ [2].

The effective charge number 'Z_{eff}' is determined by the amount, composition and state of impurities in the plasma. For GLAST plasma the effective charge number is assumed to be $Z_{eff} = 1$.

Using these values in Eq. (1), the electron temperature 'T_e' is calculated as

$$T_e = \left(\frac{8.9\times10^{-4}}{\eta_p}\right)^{2/3} \quad [eV, \Omega m] \tag{2}$$

The plasma resistivity 'η_p' can be calculated by using experimentally determined plasma parameters as

$$\eta_p = R_p \frac{A_p}{l_p} \quad [\Omega m, \Omega, m^2, m] \tag{3}$$

where, 'l_p' and 'A_p' are the length and area of the plasma loop/column respectively and for GLAST, their values are determined as

$$\begin{aligned} l_p &= 0.97 \quad [m] \\ A_p &= 0.032 \quad [m^2] \end{aligned} \tag{4}$$

The plasma resistance 'R_p' is calculated as

$$R_p = \frac{U_{loop}}{I_p} \quad [\Omega, V, A] \tag{5}$$

where, 'U_{loop}' and 'I_p' are the loop voltage and plasma current respectively and for a typical shot of GLAST-2, their values are experimentally determined as

$$U_{loop} = 3 \quad [V]$$
$$I_p = 5\times10^3 \quad [A] \tag{6}$$

Using the values from Eq. (6) and Eq. (4), we get

$$R_p = 6\times10^{-4} \quad [\Omega]$$
$$\eta_p = 19.7\times10^{-6} \quad [\Omega m] \tag{7}$$

Resultantly, Eq. (2) gives the edge plasma electron temperature 'T_e' as

$$T_e = 12.65 \quad [eV]$$

It must be noted that this is a rough estimate of the edge plasma electron temperature basically for three reasons:

1. Plasma in the GLAST is not fully ionized, which makes us overestimate the plasma resistivity 'η_p' and resultantly underestimate the electron temperature 'T_e'.
2. Due to the presence of impurities, $Z_{eff} > 1$ but for GLAST, 'Z_{eff}' is assumed to be unity with large uncertainty, which again makes us overestimate the plasma resistivity 'η_p' and resultantly underestimate the electron temperature 'T_e'.
3. Adsorbed gases are released from the surface of plasma facing components during the discharge. These atoms enter the plasma and cool down it, thus making us underestimate the electron temperature 'T_e'.

The central plasma temperature is also estimated using the equilibrium temperature profile and found out to be 34.6 eV. This value is considerably higher than the estimated average value of the edge electron temperature because center of the plasma has lower resistivity with higher current density.

In our future work we are planning to estimate the electron temperature and the number density with better accuracy and also using other diagnostics like spectroscopy and interferometry.

[1] Spitzer L and Harm R 1953 *Phys. Rev.* **89** 977
[2] Woods L C 2006 *Theory of Tokamak Transport* (WILEY-VCH Verlag GmbH & Co. KGaA, Weinheim).

Journal of Physics: Conference Series **591** (2015) 012010 doi:10.1088/1742-6596/591/1/012010

Banana orbits in elliptic tokamaks with hole currents

P. Martin[1,2] , E. Castro[1] and J. Puerta[1]

[1] Departamento de Física, Universidad Simón Bolívar, Apdo: 8900, Caracas - Venezuela
[2] Departamento de Fisica, Universidad de Antofogasta, Chile

E-mail: jpuerta@usb.ve, pmartin@usb.ve, ecastro@usb.ve

Abstract. Ware Pinch is a consequence of breaking of up-down symmetry due to the inductive electric field. This symmetry breaking happens, though up-down symmetry for magnetic surface is assumed. In previous work Ware Pinch and banana orbits were studied for tokamak magnetic surface with ellipticity and triangularity, but up-down symmetry. Hole currents appear in large tokamaks and their influence in Ware Pinch and banana orbits are now considered here for tokamaks magnetic surfaces with ellipticity and triangularity.

1. Introduction

Banana orbit are very important for particles confinement in Tokamaks (1,2)[] The two most important parameters in banana orbits are the width and the amplitude. In this paper we are studying how this banana width depends on the geometry of the magnetic surface, as the elliptical, triangular, amplitude and other Tokamak characteristic, as the hole currents. The hole currents appear in large tokamaks and their influence in Ware Pinch and banana orbits have not been studied previously, which is the reason reason we have also considered how they could modify the banana width. The present treatment is performed using a general kind of coordinates previously developed (3,4).

2. Theoretical Treatment and Results

For trapped particles with parallel velocity $u \ll v$, the curvature drift will be a small correction to the ∇B drifts. Considering also the low β approximation, the drift velocity will be

$$\bar{v} = \frac{\mu}{e_i}\frac{\nabla \mathbf{B} \times \mathbf{B}}{B^2} \tag{1}$$

where μ is the adiabatic invariant $\mu = \frac{mv^2}{2B}$, e is the charge of the particle and $\mathbf{B}$ is the total magnetic field. The magnetic field has only two components toroidal and poloidal, that is, $\mathbf{B} = B_\varphi \hat{\varphi} + B_\gamma \hat{b}_\gamma$, and $B = \sqrt{B_\varphi + B_\gamma}$, where $\hat{\varphi}$ is the unitary vector along the toroidal angle and $\hat{b_\gamma}$ is the tangent unitary vector along the meridian magnetic curves, denoted as S-curves, in this work the tri-orthogonal frame of the curvilinear coordinates, will be denoted as $(\hat{N}, \hat{b_\gamma}, \hat{\varphi})$, where $\hat{N}$ is the unit vector normal to the surface, this vector is equal to the $-\mathbf{n}$, used in previous work. The unit vector along the magnetic lines is denoted here with $\hat{l}$.

$$\mathbf{v_d} = -\frac{\mu}{e_i B^2}\left(B_\phi \frac{\partial B}{\partial o}\hat{b}_p - B_\phi \frac{\partial B}{\partial o}\hat{N} - B_p \frac{\partial B}{\partial o}\hat{\phi}\right) \tag{2}$$

$$\Delta = -2\int_0^{\tau/4} \mathbf{v_d}\cdot\hat{N}dt = 2\frac{\mu}{e_i}\int_0^{\tau/4}\frac{B_\phi}{B^2}\frac{\partial B}{\partial S}dt \tag{3}$$

Where τ is the period of the banana orbit and $\mathbf{v_d}\cdot\hat{N}$ measures the displacement perpendicular to the magnetic surface, and $\tau/4$ denotes the time for the trapped particle to move from the mid plane to the reflection point with major radius R_l and R_b respectively. The time zero correspond to the mid plane position.

A long and cumbersome calculation leads to

$$\triangle = \sqrt{\frac{2\mu m_i}{e_i}}\left(\frac{R_b}{R_l(\lambda)B_{\rho_l}(\lambda)}\right)^{l/2} f(\lambda)\left\{\int_0^{s_b}\frac{\mu\kappa_0 dS}{G(\lambda,\theta)\sqrt{\mu^2+f(\lambda)^2}} - \int_0^{s_b}\frac{\sin(\theta)dS}{G(\lambda,\theta)R_\mu}\right) \tag{4}$$

where

$$f(\lambda) = \frac{2\pi q(\lambda)}{\oint \frac{ds}{R_\mu}} \tag{5}$$

and

$$G = G(\lambda,\theta) = \sqrt{\mu_b^2 + f(\lambda)^2} - \frac{R_b}{R}\sqrt{\mu^2+f(\lambda)^2} \tag{6}$$

Here $q(\lambda)$ is the safe factor.

These results can be combined with the equation for end presented in previous works

$$\begin{array}{ll} R = R(\lambda,\theta) = R_m + \lambda a\cos\theta - aT(1-\cos 2\theta) & \mu = \mu(\lambda,\theta) = \exp\left(\int_0^s \kappa_s ds\right) \\ R_l = R_m + \lambda a - \lambda^2\Delta_s(a) & R_b = R_m + \lambda a - \lambda^2(\Delta_s(a) + aT(a))(1-\cos\theta_b) \end{array} \tag{7}$$

Where R_m is the major radius of the minor axis, $E(a)$ and $T(a)$ are the ellipticity and triangularity distortions, $\Delta_s(a)$ and Shafranov shift, $2a$ is the width of the plasma along the mid-plane, λ is yhe parameter denoting the magnetic surface where the path of the particle is proyected and θ_b and θ_b' are the poloidal angle corresponding to the largest and smallest reflection points respectively. The coefficients of ellipticity and triangularity distortion, $E(a)$ and $T(a)$, are measured at the plasma on the 95% surface, and they are related to the ellipticity coefficient $\kappa(a)$ and triangularity $\delta(a)$ by expressions given in previous papers. In particular case in which the magnetic surfaces have a circular geometry the curvature $\kappa_0 = 0$ and $\mu(\lambda,\theta) = 1$. In this case the expression 3 and 4 have the form

$$\Delta = \sqrt{\frac{2\mu m_l}{e_l}}\left(\frac{R_b}{R_{lc}(\lambda)R_{\rho_{lc}(\lambda)}}\right)^{1/2} f(\lambda)\left\{-\int_0^{s_b}\frac{\sin\theta ds}{G(\lambda,\theta)_c R_c}\right\} \tag{8}$$

and

$$f_c(\lambda) = 2\pi q(\lambda)/\oint\frac{ds}{R_c},\quad G_c = 1 - \frac{R_{bc}}{R} = \sqrt{1+(f_c(\lambda))^2}, \tag{9}$$

where the sub index c is to denote circular surfaces.

Here we are interested on find the banana width effects due to current holes. In this case the Shafranov shift, triangularity and elliptic elongattion are taken from the work of Yavorky et al. They definitions are

$$\Delta(\lambda) = \Delta_0(1-\lambda^2),\quad \kappa(\lambda) = \kappa_a - (\kappa a - \kappa_0)(1-\lambda^2)^2 - \tfrac{1}{2}\kappa_a'\lambda^2(1-\lambda^2) \tag{10}$$

where Δ_0 denote the triangularity instead of the usual notation. The parameter $\delta(\epsilon\lambda)$ denoting as usual the magnetic surface, λ which is also defined as $\lambda = \tilde{o}/a = r/a$. In the Yavorsky paper a parameter x_m is introduced in order to adjust the analytic formto the experimental data. The calculation here will be carried out for the tokamak JT-60U and for this device a good value of x_m is 0.6. In this tokamak the toroidal current density has the analytic form

$$j_\phi = (1-\lambda^2)^2(1-(1-\lambda)^6)^{2m}, \quad m = \frac{3((1-x_m)^{-6}-1)}{1+x_m^{-1}} \tag{11}$$

The results of the application of these equations are shown in figures 4, 5, 6, 7 and 8. The two first figures can be considered as those with no hole currents and three last ones are for large Tokamaks, that is, those with hole currents. In figure 4 we see that the maximum width are almost for the same value of the ellipticity κ, which is different to one. This peculiarity also happens for the triangularity in the case of small Tokamaks, see figure 5, but it is not so clear for last Tokamaks as the JY-60U, see figure 6 and 7. The figure 8 is not related to the banana width, and it shows how triangularity, elasticity, toroidal current density and safety factors change with the location of the magnetic surfaces, that is, the value of λ. In the case of elasticity and triangularity both parameter increase with λ. In the case of toroidal current density, there is the maximum of some value of λ, and the safe factor increases rapidly to very large value nearby the magnetic axis.

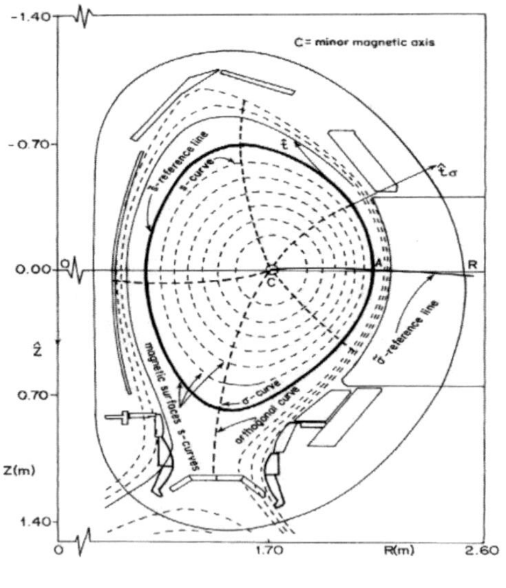

Figure 01: cross section od a Tokamak, showing the system of coordenates used in this paper.

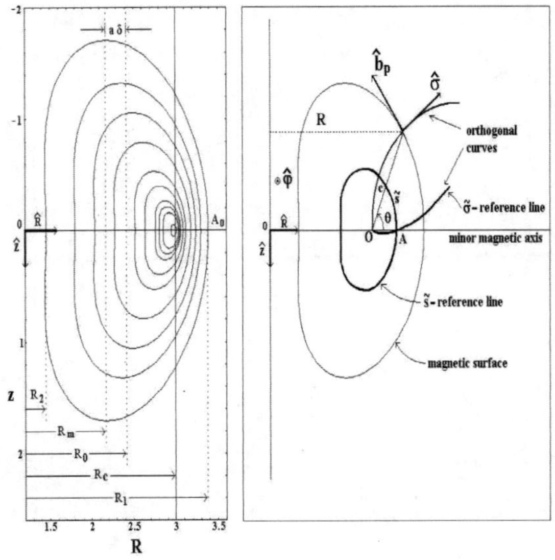

Figure 02: Cross section of a Tokamak showing the magnetic surfaces for different values of λ and all the important radius in this work.

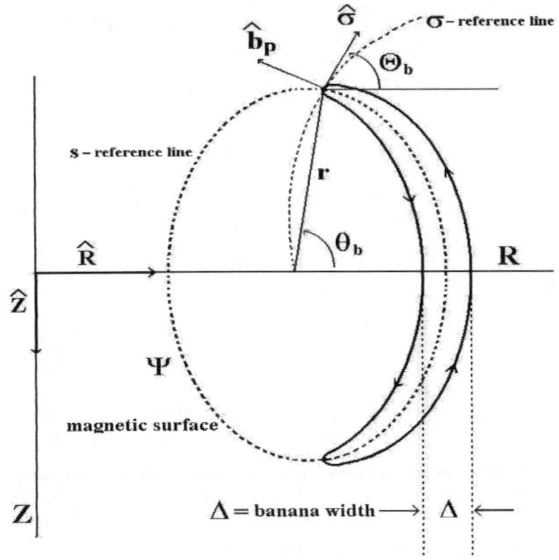

Figure 03: The banana orbit is shown in this figure with the most relevant parameters.

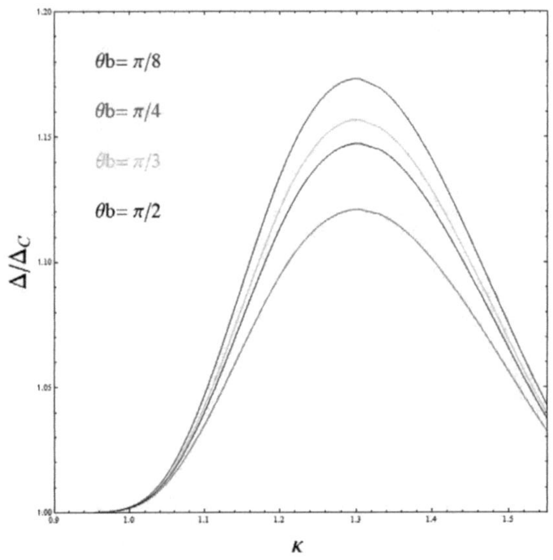

Figure 04: Normalized width of the banana orbit as a function of the ellipticity κ, for 4 different banana amplitudes. The widths are normalized with circular banana width in all the figures in this work.

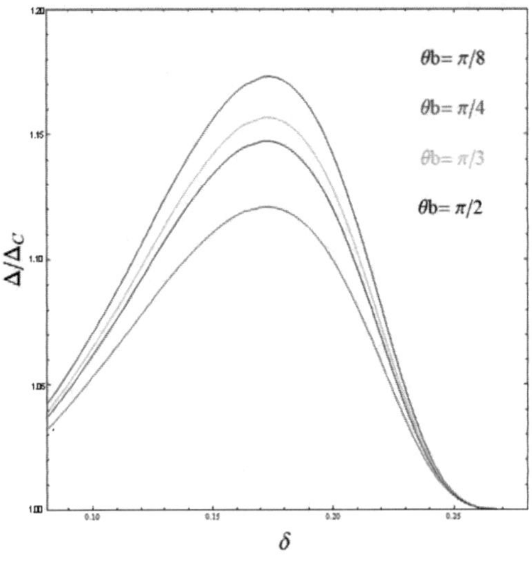

Figure 05: Normalized width of banana orbits as a function of the triangularity for 4 different banana amplitudes (no hole current).

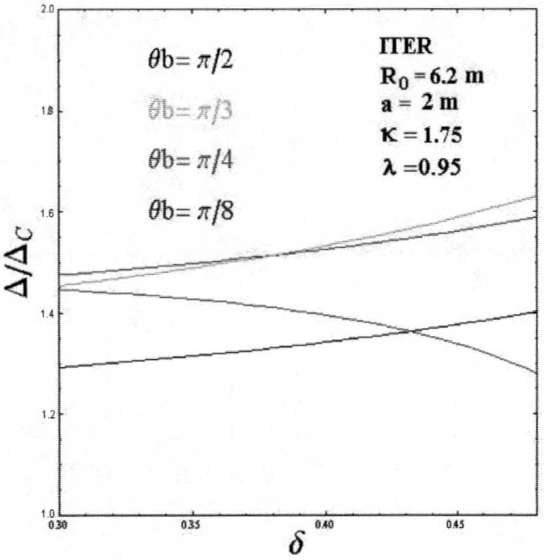

Figure 06: Normalized width of banana orbits as a function of the triangularity for ITER, in a range around the values of design.

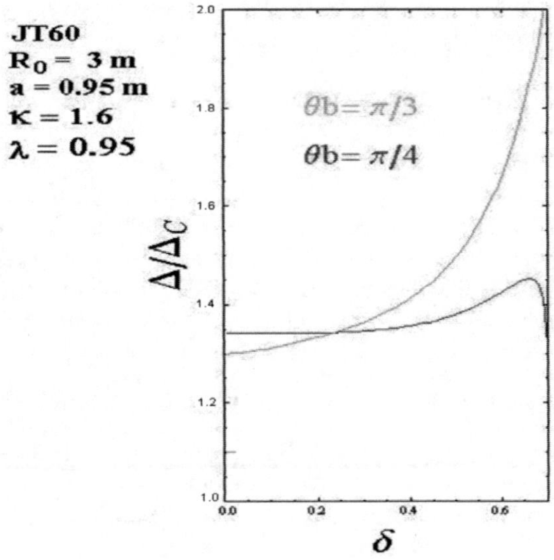

Figure 07: As in the preceding figure here the normalized width is considered as a function of triangularity for the Japanese Tokamaks JF-60, considered a Tokamak with hole currents.

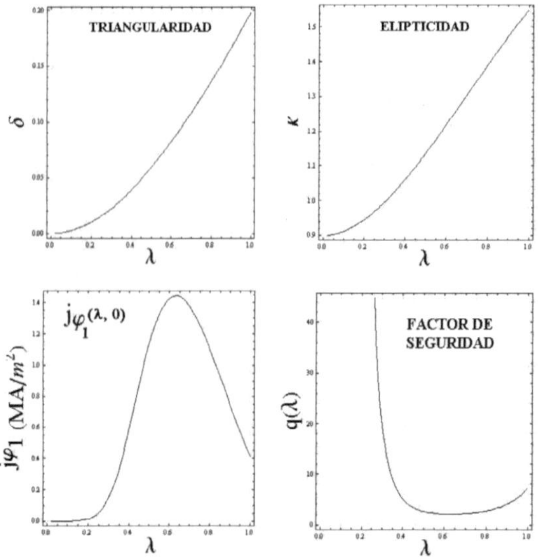

Figure 08: These four graph shows respectively how triangularity,ellipticity, toroidal current and safety factor change with the parameterλ, characteristic of the different magnetic surfaces in the JT-60 Tokamak

3. Conclusions

It is shown that the Ware Pinch effect can be obtained from the guided center first order theory , for tokamak magnetic surfaces with ellipticity and triangularity, and Current Hole. The Banana orbit width had been determined using general system of coordenates proposed by us in previous papers. It is also confirmed than the Ware Pinch is a consequence of a up-down simetry breaking due to the toroidal inductive electric field and dependent of the geometry properties of the magnetic surfaces. The results are in agreement with the clasical Ware Pinch. However the expressions of the new Ware Pinch are more precise and they depend of the aditional parameters of the surface as triangularity and ellipticity. There are some differences due to current hole , but the shape of the banana is similar and the differences seem to be no essentials.

4. Acknowledgements

The present work has been supported by grant DID-G22, Universidad Simón Bolívar Caracas-Venezuela
and by grant Proyecto ANT 128, Programa MECE 2, Universidad de Antofogasta, Chile.

5. References

[1] J.A. Wesson, "Tokamaks" (Clarendom, Oxford, second ed) pp. 164-166 (1997)
[2] M.G. Haines and P. Martin, Phys. Plasmas **3**, 4536 (1996)
[3] P. Martin, Phys. Plasmas, **7**, 2915 (2000)
[4] P. Martin, E. Castro and M. G. Haines, Phys. Plasmas **14**, 052502 (2007)
[5] G. O. Ludwig, plasma Phys. Controlled fusion **37**, 633 (1995)
[6] V. Yavorskij, V. Colaborod'ko, K Schoepf, S.E Sharapov, C.D Challis, S. Resnik and D. Stork. Nucl. Fusion **43**, 1077 (2003)

Neoclassical modeling of the radial electric field and comparison with measurements in the TJ-II stellarator

C. Gutiérrez-Tapia[1], J. J. Martinell[2], D. López-Bruna[3], A. V. Melnikov[4]

[1] Instituto Nacional de Investigaciones Nucleares, México
[2] Instituto de Ciencias Nucleares (UNAM), México D.F., México
[3] Asociación EURATOM-CIEMAT, Madrid, Spain
[4] Research Centre Kurchatov Institute, Nuclear Fusion Institute, Moscow, Russia

E-mail: `cesar.gutierrez@inin.gob.mx`

Abstract. The results of the radial electric field measurements by Heavy Ion Beam Probe (HIBP) in the TJ-II stellarator are compared with neoclassical transport computations. The role played by several plasma features is identified by studying a reduced analytical transport model, using both actual density and temperature profiles and representative model profiles for varying density. Additional electric field calculations are carried out numerically with the Astra code using three different expressions for the neoclassical transport coefficients in order to identify the common features characterizing E_r, particularly the sign reversal. For regimes with a wide variation of collisionality, a general qualitative agreement between modeling and experimental data is shown. The obtention of roots for E_r depends critically upon the temperature and density profiles as they determine the plasma collisional regime. It is found that the root transition (i.e. E_r sign reversal) occurs for a specific range of a collisionality parameter, consistent for all models employed. It is found that when density and temperature profiles give real roots of the electric field the pressure is almost constant. Discharges with high radial gradient of E_r are found to correlate well with high confinement regimes.

1. Introduction

In non-axisymmetric magnetic configurations like those of stellarators, the NC contribution is crucial in the analysis of phenomena related to E_r. Indeed, there have been joint efforts to proceed to a benchmark of numerically obtained NC electric fields and fluxes with experimental data in conditions suitable for a NC treatment [1]. In the particular case of the TJ-II device, an acceptable agreement between experiments and Montecarlo calculations has been obtained in low density regimes [2, 3]. Additionally, some works were dedicated to compare a particular formulation of the NC fluxes to describe the electric fields with experiment in restricted plasma conditions [4, 5].

Since the conditions and basic statements of NC theory are well grounded and provide a first approximation to transport, it is desirable to have practical implementations that evaluate this important contribution on transport and especially on the ambipolar radial electric field. The present work is also benchmark-minded, but from a more immediate and practical perspective: it is based on easy-to-implement formulations of the NC particle fluxes. The objective is to

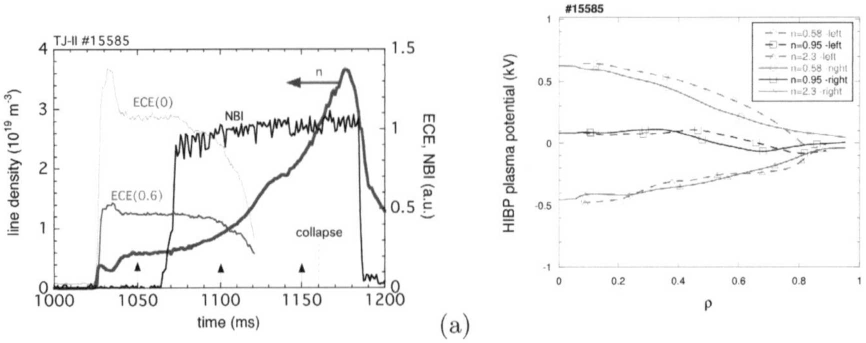

(a) (b)

Figure 1. (a) Time signals of Electron Cyclotron Emission (ECE) from the magnetic axis and from $\rho = 0.6$, line density $\bar{n}$ through the magnetic axis and NBI heating during an NBI discharge (TJ-II #15585) with continuous density rise. The ECE signals drop when the ECE cut-off is reached at $\bar{n} \approx 1.2 \times 10^{19}$ m^{-3} but the plasma remains hot until the radiative collapse at $t \approx 1160$ ms. (b) Plasma potential profiles corresponding to the times indicated with the arrows in (a).

provide a practical tool for the evaluation of the radial electric field under the assumptions that (i) the main non-ambipolar radial fluxes are of neoclassical origin and (ii) the functional dependencies of such fluxes for electrons and ions are enough to obtain E_r even if the fluxes themselves are not accurately described. If the results obtained from the NC formulations are found to be consistent with the experimental results, this model should be suited as a tool for the interpretation of experimental data, and also for predictive estimates. For this to be trustworthy, not one but different formulations of the NC problem should yield consistent and robust results with respect to the experimentally found trends of the radial electric field. We believe that this is especially significant when the exercise is done with plasmas operated in a complex magnetic geometry like that of a Heliac device, as the TJ-II is.

2. Experimental data

The TJ-II Heliac-type stellarator has a helical magnetic axis that winds around a circumference of radius $R_0 = 1.5$ m. The plasma has a bean-shaped cross section with average minor radius of $a \approx 0.2$ m and magnetic field at the axis $B_0 \approx 1.0$ T. The plasmas are always initiated with Electron Cyclotron resonance Heating (ECH); absorbed heating powers are normally $P_{\rm ECH} = (200 - 400)$ kW and in the absence of other heat sources the line averaged densities are $\overline{n} = (0.3 - 1.0) \times 10^{19}$ m^{-3} with central electron temperatures $T_e(0) \sim 1$ keV and ion temperatures of $T_i(0) \sim 0.1$ keV due to Coulomb heat exchange. Additional heating and fueling can be obtained with the help of one or two neutral beam injectors (NBI) delivering port-through powers $P_{\rm NBI} = (300 - 600)$ kW each. Above line densities of around 1.2×10^{19} m^{-3} the ECH is no longer effective and the plasmas are sustained with NBI heating alone up to densities $\overline{n} = (2.0 - 6.0) \times 10^{19}$ m^{-3} with lower temperatures: $T_e(0) \lesssim 0.3$ keV and $T_i(0) \sim 0.5 T_e(0)$.

The electron density and temperature profiles are measured using the Thomson Scattering diagnostic from the magnetic axis up to $\rho \approx 0.7$, where ρ is the normalized flux surface label proportional to the enclosed plasma volume. In order to obtain the density profiles, interferometry and reflectometry data are used to extend the profiles to the edge ensuring that their line integral matches the experimental value. The ion temperature is normally measured with the CX neutrals analyzer. $T_i(\rho)$ is quite homogeneous in the low density ECH plasmas and thus very different from T_e profiles due to the low collisional coupling in these cases. In what follows, we use T_i based on the central values provided by the diagnostic. The plasma potential in the bulk plasma is measured using the HIBP system described in [6].

Fig. 1(b) shows the plasma potential profiles $\phi(\rho)$ that correspond to the times indicated by vertical arrows in figure 1(a). Two profiles are shown at each density value because the diagnostic takes measurements at both sides of the magnetic axis (labelled 'left' and 'right'). This will be used as an indication of the uncertainty in the evaluation of the plasma potential. During the ECH phase $\phi(0)$ is positive with values in the 400–1000 V range that depend on $\bar{n}$. At high enough densities still in the ECH phase, the plasma potential becomes negative near the plasma edge and causes also the appearance of negative electric fields around the region of maximum density gradient [7]. In discharge #15585 this happens during the mixed ECH+NBI phase, when $\bar{n}$ increases, T_e decreases and T_i also experiences a small ($\sim$ 20%) rise correlated with the increased density. At still larger densities, like in the pure-NBI phase, $\phi(0)$ becomes negative as in the rest of the plasma. When $\bar{n} \approx 2 \times 10^{19}$ m^{-3}, the plasma potential is negative everywhere reaching values between -300 and -600 V near the magnetic axis. It should be noted that the change of sign of the plasma potential occurs in a rather continuous way starting near the edge where $T_e \sim 50$ eV and moving towards the center as n increases.

According to figure 1, three different types of plasma have been considered that represent different collisionality regimes and can be characterized by the corresponding average electron density:

LDHT (low density with high electron temperature): Low density ($\sim 0.6 \times 10^{19}$ m^{-3}), high electron temperature ($T_e(0) \sim 1$ keV) and low ion temperature ($T_0 \sim 0.1$ keV) as normally found in ECH plasmas. This regime is characterized by long mean free paths (LMFP) of the particles over most of the plasma, $\nu^* \ll 1$, giving rise to radial fluxes dominated by the contribution of bouncing trapped particles in the large magnetic ripple. Here, $\nu^* = \nu/\omega_t$ is the collision frequency normalized to a model transit frequency $\omega_t = v_{th}/(R_0 q) = \iota v_{th}/(2\pi R_0)$, involving the rotational transform in radians ι and the thermal speed v_{th}. In this regime the electrons, being much faster than the ions and having large magnetic moment due to the heating system (ECH), dominate the radial fluxes causing E_r to be positive and $\phi \sim T_e/e$ in the plasma core. The plasma potential is then positive over the entire plasma column.

IDT (intermediate density and electron temperature): Intermediate densities ($\sim 10^{19}$ m^{-3}) and electron temperatures ($T_e(0) \sim 0.5$ keV), typically found in mixed ECH +NBI plasmas, or in high density ECH plasmas. The electron temperatures are lower than in the typical ECH plasma, while T_i remains on the order of 0.1 keV. This regime is known from the first TJ-II experimental campaigns for inverting the sign of the plasma potential (inferred from floating potential measured with electric probes [8]), as well as the electric field (according to spectroscopic measurements of plasma rotation [9]) somewhere inside the plasma. In particular, E_r starts becoming more negative near the plasma edge [7]. With increasing average densities, the negative values of E_r cover also smaller radii.

HDLT (high density and low electron temperature): High density ($\gtrsim 2 \times 10^{19}$ m^{-3}) and low temperatures ($T_e(0) \lesssim 0.3$ keV, $T_i(0) \lesssim 0.14$ keV), corresponding to NBI plasmas. When the density is further increased from the IDT regime, the plasmas have closer electron and ion temperatures and high density giving rise to shorter mean free paths for electrons and ions, satisfying $\nu^* \lesssim 1$, ions being closer to one. The contribution from helically trapped particles is now much smaller and the radial transport in these conditions is probably dominated by passing and toroidally trapped particles over a large portion of the plasma column, which corresponds to the "plateau" regime. The larger tendency of ions to escape the plasma causes $E_r < 0$ in the entire plasma column.

Figure 2 shows density (a) and electron temperature (b) profiles taken from the same

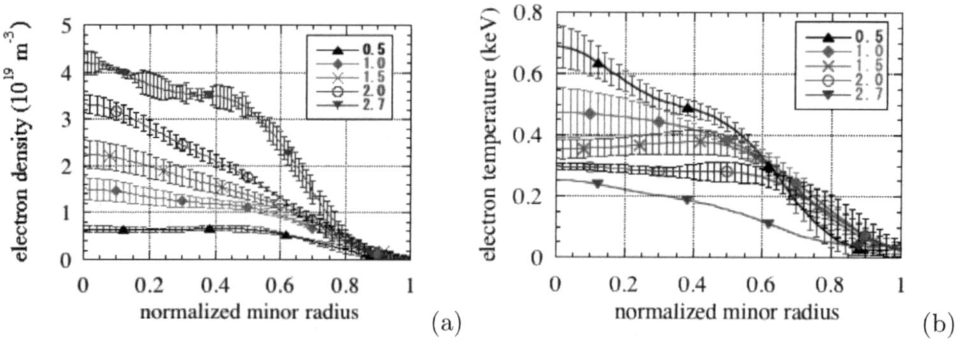

Figure 2. Averages of Thomson Scattering profiles at different line densities $\bar{n}$ (10^{19} m^{-3}) taken from N sample discharges according to the $\bar{n}(N)$ sequence 0.5(6), 1.0(6), 1.5(3), 2.0(5), 2.7(1). Large error bars in $T_e(\rho > 0.8)$ at $\bar{n} = 2.7$ have not been drawn for clarity.

experimental day of discharge #15585 (Fig. 1) at fixed line densities within a 10% range, with values (0.5, 1, 1.5, 2 and 2.5 in 10^{19} m^{-3} units). The profiles shown correspond to averages of Thomson Scattering profiles based on N discharges, with $N = (6, 6, 3, 5$ and $1)$ respectively — only one discharge was available for the highest density. Each individual profile is obtained after fitting a short expansion (typically 4th or 5th order) of Bessel functions in order to obtain $T_e(\rho)$ and $n_e(\rho)$ with reasonably smooth radial derivatives. Note that Thomson Scattering profiles are obtained from 256 points in the TJ-II diagnostic that show a fine structure [10], while in the present study we are only interested in the main macroscopic variations. The error bars in Fig. 2 are the standard deviation of the corresponding N profiles, except for the highest density case, where the errors obtained from Bayesian analysis [11] are shown for the only available discharge. Incidentally, this shows that these two errors are similar. The profiles and their radial derivatives are then mapped to a same calculation grid $[\rho_j]$. The ion density n_i is slaved to n_e through a prescribed effective charge $Z_{\text{eff}} = 1.2$.

Table 1. Constant values appearing in expressions (1) for different regimes of electron density and temperature ((a) LDTH regime, (b) IDT regime, and (c) HDLT regime) obtained from the fitting to an hyperbolic curve $T_{e0} = 1/(0.9226 + 0.7623n_{e0})$. Densities are expressed in 10^{19}m^{-3}, and electron and ion temperatures in KeV.

	n_a	n_0	a	b	T_{ae}	T_{0e}	a_e	b_e	T_{ai}	T_{0i}	a_i	b_i
LDHT	0.11	≤ 1	6.21	4.07	0.01	$(0.65, 0.7)$	1.9	1.66	0.01	0.077	40	10
IDT	0.08	$(1, 2)$	2.27	1.23	0.02	$(0.45, 0.5)$	2.91	1.64	0.01	0.09	35	10
IDT	0.09	$(2, 3)$	1.34	1.02	0.04	$(0.35, 0.4)$	6.64	5.43	0.01	0.09	35	10
IDT	0.11	$(3, 4)$	1.53	1.58	0.02	$(0.3, 0.35)$	5.23	2.12	0.01	0.09	35	10
HDLT	0.09	$(4, 5)$	4.78	5.11	0.02	$(0.25, 0.3)$	2.15	2.40	0.01	0.11	30	10

Finally, and since an algebraic model will be used later, we have prepared a set of analytic profiles that reproduce the main features of the experimental ones shown in Fig. (2), having the form

$$n(\rho) = n_a + n_0(1-\rho^a)^b; T_e(\rho) = T_{ae} + T_{0e}(1-\rho^{a_e})^{b_e}; T_i(\rho) = T_{ai} + T_{0i}(1-\rho^{a_i})^{b_i}, \quad (1)$$

where n_a, T_{ae}, and T_{ai} are the electron density, electron temperature and ion temperature at the boundary, while n_0, T_{0e} and T_{0i} give the corresponding values at the center. Since the corresponding central values of the experimental data shown in Fig. 2 can be well fitted by an inverse relation, $T_{e0} = 1/(0.9226 + 0.7623 n_{e0})$, as shown in Fig. 3(a), we have taken n_0, T_{0e} to fall on this curve. The parameters a and b in Eq. 1 selected for our calculations are given in Table 1. They produce the profiles shown in Figs. 3(b) and 4, where the model profiles are compared with the experimental ones.

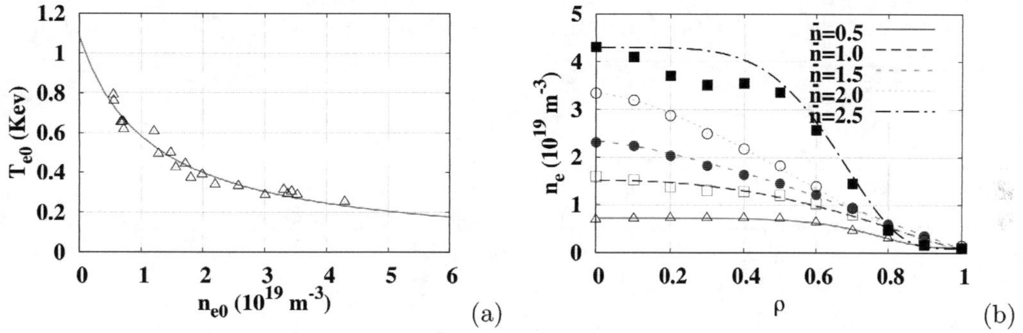

Figure 3. (a) Measured central electron temperature as function of central density showing hyperbolic- like dependence; (b) density profile: LDHT regime (solid line) and experimental (empty triangles), IDT regime (dotted lines) and experimental (empty squares, filled and empty circles), and HDLT regime (dash-dotted line) experimental (filled squares).

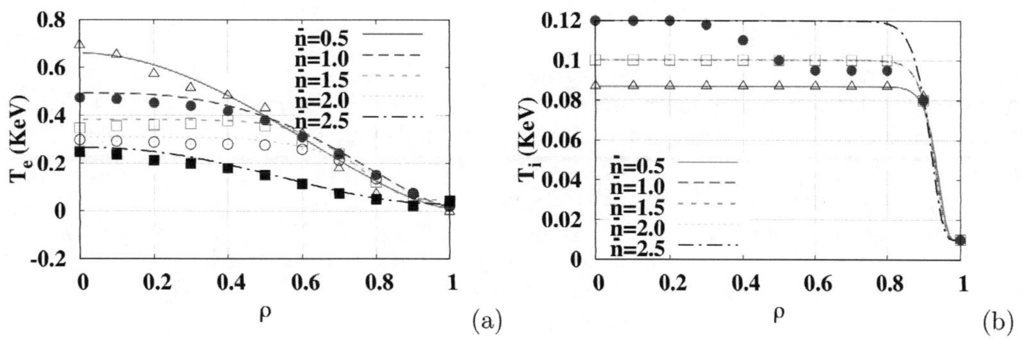

Figure 4. (a) Electron temperature: LDHT regime (solid line) and experimental (empty triangles), IDT regime (dotted lines) and experimental (empty squares, filled and empty circles), and HDLT regime (dash-dotted line) experimental (filled squares); (b) Ion temperature: LDHT regime (solid line) and experimental (empty triangles), IDT regime (dotted line) and experimental (empty squares), and HDLT regime (dash-dotted line) experimental (filled circles).

3. Neoclassical calculations of the radial electric field

The radial electric field in a stellarator can be computed from the neoclassical transport theory, starting from the diffusive fluxes for ions and electrons, Γ_j (with $j = e, i$). These fluxes have been calculated from kinetic theory or a two-fluid description to obtain closed forms in terms of the plasma parameters by several authors [12, 13, 14, 15]. From the dependence $\Gamma_j(E_r)$ it is possible to compute the radial electric field by applying the ambipolarity condition. Here we use three different formulations of the neoclassical fluxes to obtain E_r following certain procedures that numerically integrate the analytical formulas. For the algebraic model described below, the first procedure solves directly the ambipolarity equation

$$\Gamma_e(E_r) = Z_i\Gamma_i(E_r). \tag{2}$$

to determine E_r (additional conditions may be necessary if multiple roots exist for Eq. 2 which may produce a discontinuity of E_r as function of the radial coordinate). The other procedure, which is more convenient from the numerical point of view, follows the evolution of the electric field solving the equation

$$\frac{\partial E_r}{\partial t} \approx \frac{|e|}{\varepsilon_\perp}(\Gamma_e - Z_i\Gamma_i) \tag{3}$$

until the steady state is reached, which gives E_r. Here, $\varepsilon_\perp$ is the perpendicular dielectric constant and it is assumed that a term representing electric field diffusion is small [16]. Additionally, E_r is obtained from a simplified formulation due to Kovrizhnykh [17] which allows to follow an algebraic procedure to compute the electric field. For this model, E_r comes from the solution of a cubic equation which follows from Eq. 2.

The models used in this work simplify the complicated magnetic geometry in stellarators to obtain tractable NC transport fluxes, with the consequence that the resulting formulas do not give a full account of them, but they can be used to calculate E_r. No attempt is made to match the observed radial transport since it usually has an important anomalous component, particularly in regions far away from the plasma core. Since the non-ambipolar fluxes are mainly neoclassical, we expect that the E_r obtained from NC transport explain the measured values reasonably well. The ability to reproduce the electric field should depend on the way the NC fluxes react to this field, which is different in the various models considered.

3.1. Semi-analytical models for the neoclassical particle fluxes

Beidler's model. The first model provides mono-energetic transport coefficients for an idealized stellarator field with a single helical mode (see Eq. 4 below) obtained by Beidler [18, 19] from fits to DKES calculations. This model smoothly connects the three collisionality regimes that arise from the analytical theory of transport in a stellarator (ν, $\sqrt{\nu}$ and $1/\nu$), together with the axisymmetric transport, which becomes dominant at large collision frequencies. The monoenergetic transport coefficients have to be integrated over the thermal velocity distribution.

Kovrizhnykh's model. A second model reported by Kovrizhnykh [20], expresses particle fluxes as due to the contributions of an axisymmetric part Γ_j^S and an asymmetric one Γ_j^A (i.e. non-axisymmetric) for both the electrons ($j = e$) and ions ($j = i$), $\Gamma_j = \Gamma_j^S + \Gamma_j^{AS}$. The two parts have expressions that depend on the collisionality regime and Kovryzhnikh has derived interpolation formulas that are approximately valid for all the regimes. Formulas for the radial fluxes, were given in terms of the thermodynamic forces and the magnetic helical ripple ϵ_h entering the assumed representation for the standard stellarator magnetic field with a single helical harmonic,

$$B_T = B_0(R_0/R)[1 - \epsilon_h(r)\cos(l\theta - M\varphi)]; \quad \epsilon_h(r) = \epsilon_0 I_l(Mr/R_0). \tag{4}$$

In these expressions, ϵ_0 is a constant defining the amplitude of the stellarator field, related to $\iota(0)$, and $I_l(x)$ is the modified Bessel function of the first kind

The particle fluxes are based on a derivation that correctly describes the ambipolar field and the parallel (to $\mathbf{B}$) plasma velocity [14], and are given by expressions interpolated over the collisionality regimes for the symmetric and asymmetric components listed in [20].

Shaing's model. There is a third model presented in [12, 13]. In these reports authors argued that several values of the radial electric field can be possible that satisfy the ambipolar equation, but some of them are unstable. Thus, to find a stable solution for E_r from the thermodynamic point of view, this field must be at the minimum of the generalized heat production rate. Here, helical and toroidal ripples appear explicitly in the magnetic field taken of the form $B = B_0\left[1 - \epsilon_t \cos\theta - \epsilon_h \cos(l\theta - m\phi)\right]$, giving the asymmetric transport fluxes [13]. The symmetric part in this case is intrinsically ambipolar and is given by the usual expressions for axisymmetric devices [15]. The electric field is obtained solving Eq. 2.

The magnetic geometry in the models is taken from typical equilibrium computations and does not intend to capture the detailed geometry of the TJ-II device, but the main tendencies of collisional transport should be preserved. Profiles for the rotational transform ι and ripple amplitudes $\epsilon_{t,h}$ follow from those computations. For analytical uses the ripples are taken as

$$\epsilon_t = \alpha_t \rho^{\beta_t}; \quad \epsilon_h = \alpha_h \rho^{\beta_h}, \tag{5}$$

where the parameters $\alpha's, \beta's$ can be chosen to fit experimental data to account partially for geometrical effects. We will use $\alpha_t = 0.095, \beta_t = 0.55$ [21], but α_h and β_h will be used as free parameters to improve the fit of the E_r profiles.

In all these models the radial electric field enters only the non toroidally symmetric part since in a tokamak-like geometry steady state particle fluxes are ambipolar. The models are strictly valid for steady state conditions since no time dependent terms due to polarization drifts are included.

3.2. Algebraic formulation

Here we present an analytical model based on a reduced representation of Kovrizhnykh formulas, which allows to find the roots of the ambipolar equation (2) in closed form for given plasma profiles. The process followed is to give the model profiles of Eq. 1 for $n(\rho), T_e(\rho)$ and $T_i(\rho)$, compute the fluxes Γ_e and Γ_i and obtain E_r from Eq. (2). This simplified analytical model [17] assumes a simple magnetic geometry with a single helical harmonic. For the helical ripple we have taken ϵ_h with the values given after Eq. 5. The NC fluxes used are more appropriate in a low collisionality regime where the most important contribution to the transport coefficients comes from particles locally trapped in the helical ripple wells.

It is assumed that axisymmetric and anomalous fluxes are ambipolar and then only the asymmetric contribution is relevant: $\Gamma_j = \Gamma_j^{AS}$. In a quasi-stationary state with external particle sources Γ^{ext}, the particle balance equation

$$\Gamma_j = \Gamma^{ext} \tag{6}$$

has to be solved together with the ambipolarity equation (2). But in our analytical approach, only equation (2) is considered which gives an algebraic equation of third degree in the dimensionless electric field V. This can be solved once the equilibrium profiles $n(\rho)$ and $T_j(\rho)$ are given. However, not all profiles are physically possible, since the requirement of a single real root for $V = q_i E_r / T_e$ for all radial positions limits the choice of values of the profile parameters. This profile constriction can be understood, according to [17], in terms of the required particle sources obtained from Eq. (6) for the set of profiles $V(\rho), n(\rho)$ and $T_j(\rho)$. When V is not real and continuous the sources are not physically acceptable and then the profiles are not possible. The equation for the electric field V in steady state coming from Eq.(2) can be written as,

$$F(\rho, V) = V^3 + b(\rho)V^2 + c(\rho)V + d(\rho) = 0. \tag{7}$$

where the coefficients $b(\rho)$, $c(\rho)$, and $d(\rho)$ are given in [17]

Depending on the values of the coefficients $b(\rho)$, $c(\rho)$, and $d(\rho)$, equation (7) can have up to three roots, $V_1(\rho), V_2(\rho), V_3(\rho)$. In general, two roots are stable and one is unstable, but they have to be real in order to be acceptable. If the profiles are set arbitrarily, the solutions in certain regimes can present a jump in the E_r profile that corresponds to the transition from one root to another. As pointed out by Kovrizhnykh [17], this requires particle sources of the Dirac delta function type at the discontinuities, which is unphysical. Therefore, these solutions should be discarded arguing that the corresponding $T_j(\rho)$ and $n(\rho)$ profiles cannot be obtained. The profiles that have been verified to yield one real root solutions for E_r all over the plasma have central values of density and electron temperature that fall on the hyperbolic curve shown in Fig. 3(a). This assures that the ambipolarity condition is satisfied at each magnetic surface. The density and temperature profiles corresponding to the values given in Table 1, for each of the regimes defined in § 2 satisfy this criterion and are shown in Figs. (3(b), 4(a), 4(b)).

4. Comparison with experimental data

4.1. Semi-analytical models.

Here we present the results for the three models described in Sec. 3.1. The NC particle fluxes are obtained using the formulas for each model, and the radial electric field is obtained with a numerical code coupled to the ASTRA transport shell [22] that starts with an initial E_r, normally $E_r(r) = 0$, and evolves the equation (3) in every radial position r_j until a steady state is reached. When the experimental profiles of Fig. 2 are used the profiles obtained for the radial electric field are shown in Fig. 5. They are computed for three representative densities using each of the models, and the experimental electric field profiles are also shown for comparable densities computed from HIBP data of the electric potential. The experimental E_r-profiles can be matched better using $\epsilon_h(\rho)$ as a "fitting function". However, the purpose of this work does not require finding a "best fit" but a reasonably good behavior.

Since the potential is the quantity actually measured by HIBP we focus on the physical scalings of $\phi(\rho)$. The dependency with density is represented in a plot of the central potential which is the most sensible to density. The results of the three models can be seen in Fig. 6 when the experimental density and temperature profiles given in Fig. 2 are used. It shows that the behavior of all of the models is quite similar. We recall that the parameter ε_h has been adjusted to improve the fits in each model, thus allowing the good agreement among them. The values for the analytical computations are also shown for comparison. The shadowed region in the graph indicates the range of experimental values from HIBP and shows that the semi-analytical models are able to reproduce the measurements to some extent. At intermediate densities the agreement is good and a small departure is seen for low densities due to effects not included in the NC model such as the presence of suprathermal electrons. The values of the density for the potential sign change are almost the same for all models. Shaings's model gives somewhat smaller $\phi(0)$ values. It is worth mentioning that the smaller values of $|E_r|$ (and $|\phi|$) found in the HDLT case with the semi-analytical models coincides with the results found numerically based on Montecarlo calculations, which also underestimate the magnitude of E_r [1]. However it is noteworthy that the saturation of $|E_r|$ seen at high density is well reporduced by all models although the asymptotic value is model dependent.

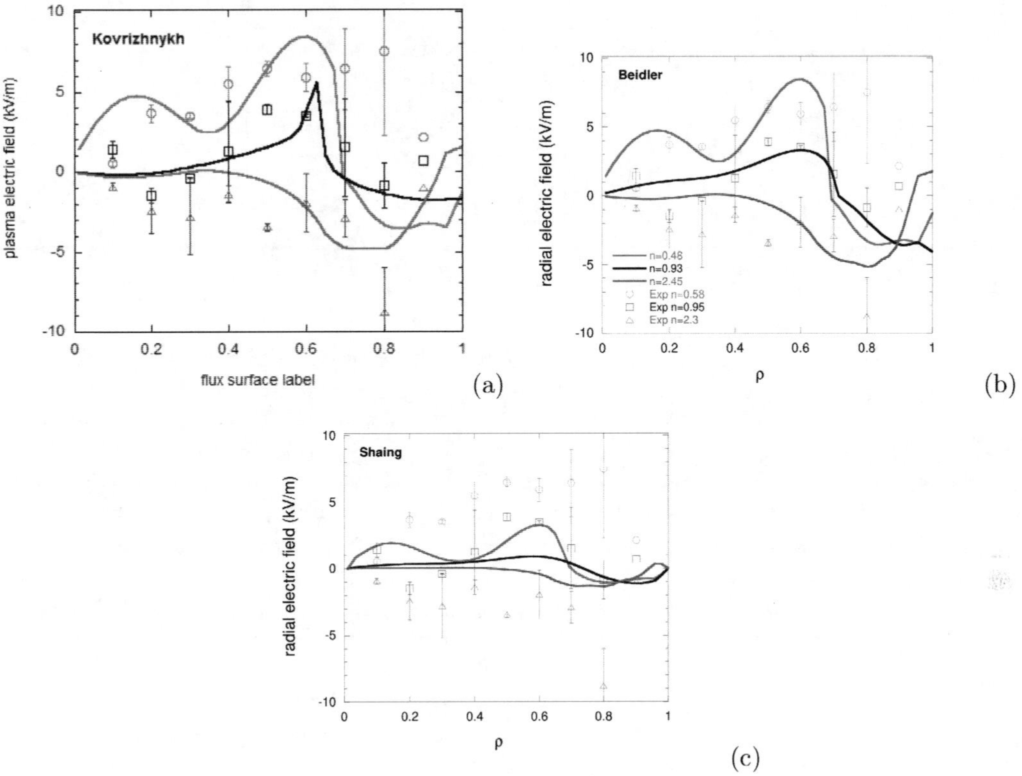

Figure 5. (a) E_r profiles for the models of: (a) Kovrizhnykh, (b) Beidler and (c) Shaing calculated for three experimental profiles from Fig. 2 for different line densities corresponding to the LDHT (cyan lines, $\overline{n} = 0.48 \times 10^19 m^{-3}$), IDT (black lines, $\overline{n} = 0.95$) and HDLT (blue lines, $\overline{n} = 2.45$) regimes; experimental profiles for line densities in the same regimes are shown with symbols and error bars.

4.2. Results of algebraic calculations.

The roots of Eq. (7) provide the radial electric field profile and the plasma potential profiles, $\phi(\rho)$, are obtained from $E_r(\rho)$ after radial integration from the edge to the magnetic surface at ρ, imposing the constraint $\phi(a) = 0$. The results for the E_r-profiles in each case are shown in Fig. 7(a), while the plasma potential profiles are shown in Fig. 7(b) (cf. Fig. 1), where they are compared with experimental results from each regime. It is seen that the same general features obtained with the semi-analytical models are also reproduced, approximately matching the experimental profiles, in particular, the sign change of E_r in the IDT regime at some radial position.

A common feature, already observed for TJ-II ECH plasmas [21], is that the plasma parameters for which the electric field is inverted in passing from LDHT to HDLT plasmas – the IDT regime discussed above – are found in the interval $0.5 < (n_0/10^{19}) < 2$ in agreement with the experiments, which corresponds to local collisionalities $\nu^* \sim 0.01$. It should be stressed then that the basic NC model, even for the complicated geometry of the TJ-II Heliac-type stellarator, provides a reasonable first approximation for the main experimental results shown. The correlation with experimental data shown in Fig. 7(b) indicates that the plasma potential has dependencies related mainly with the plasma parameters and the peculiarities of transport.

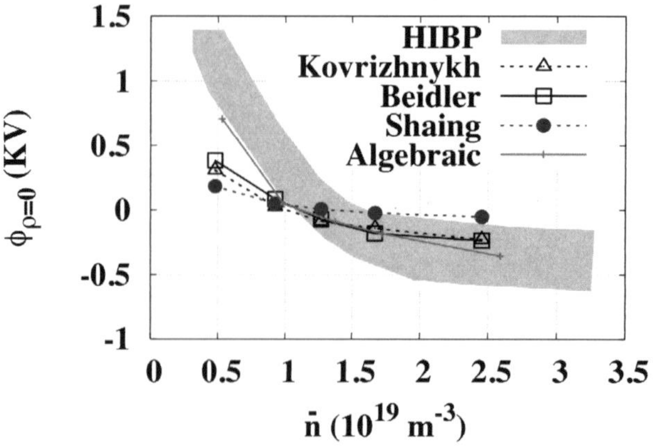

Figure 6. Plasma potential at the center obtained for Beidler, Kovrizhnikh, and Shaing models using the average experimental density and temperature profiles of Fig. 2. The analytical result is shown by the curve without symbols. Shaded area shows the range covered by the experimental values.

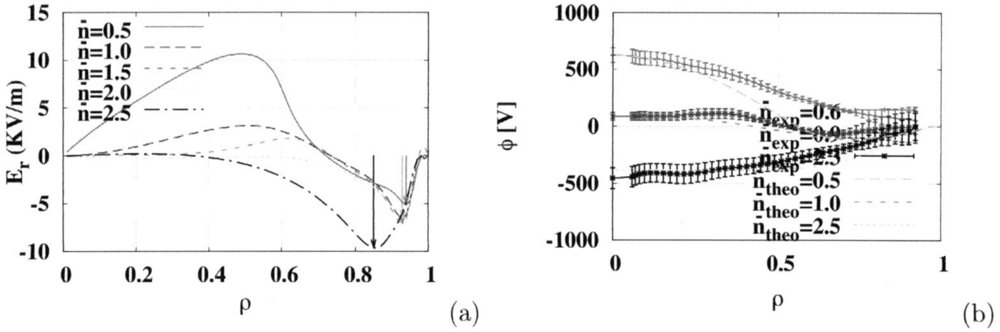

Figure 7. (a) E_r profiles from algebraic model using the five densities for model profiles in Figs. 3(b) and 4 going from LDHT to HDLT regimes; (b) ϕ profiles for the same model proflies, showing also the experimental profiles from HIBP for three densities for comparison, marked with symbols.

5. Discussion

In the light of the results obtained from our computations with the different approaches and models and compared with the HIBP measurements, we can make an appraisal of the role of NC transport regarding the radial electric fields. The main results of the comparison between experimental measurements and theoretical modeling are captured in figures 5 and 6. Figure 5 shows that both, experimental and theoretical E_r-profiles, approach to negative values as the density increases from the lowest values; in particular, a dip in E_r begins to develop near the edge ($\rho \sim 0.8$) when n rises, prior to becoming negative (also seen in Fig. 7). This is a systematic behavior in TJ-II plasmas that we find well described by the models. However, on the quantitative side, we find that the experimental data yield stronger fields than the models, in particular for the extreme LDHT and HDLT cases. This is apparent in figure 6, for the central potential: the crossing to negative values seems reasonably well represented by the models but the high (low) potentials at low (high) density are in general underestimated. On this respect

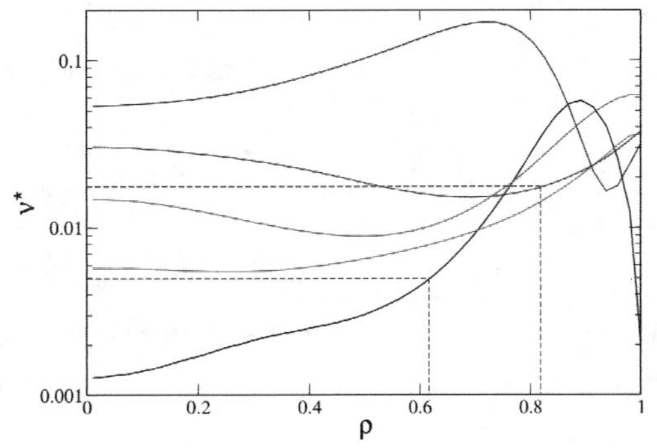

Figure 8. Collisionality profiles for the plasma profiles in Fig. 2 from discharge #15585, showing the collisionality range where E_r changes sign.

we must remember the simplifications of NC theory not accounted for in the models. In TJ-II, there are significant populations of supra-thermal electrons in LDHT conditions [23, 24, 25]. Furthermore, a low density transition to better particle confinement has been identified with the establishment of negative electric fields near the edge and the drastic drop of high energy ($>$ 20 keV) radiation from brehmsstrahlung [26]. Therefore, the presence of fast electrons and ECH pump-out effects are likely candidates to explain why the plasma electric potential presents higher $\phi(0)$ than the models, where these effects are not accounted for. In the case of the high density plasmas, it seems also the case that NC calculations considering the magnetic geometry in more detail yield lower electric fields than measured. It has been argued that this is due to non-local effects related with large-width banana orbits for the ions in TJ-II plasmas [27, 28]. The fact might be more general according to a recent benchmarking effort between numerical NC calculations and experimental data [1]. In general, then, the semi-analytical models behave quite like their numerical counterparts based on Montecarlo or other techniques.

Comparing the E_r profiles from the three models for experimental profiles we notice a general agreement among them, with Beidler and Kovrizhnykh models giving quite similar results—and reasonably close to the experimental data—and Shaing model yielding smaller values of the electric field and plasma potential in the extreme density cases. This consistency among models together with the fair comparison with the experiment indicates the robustness of the main assumptions of NC transport, which is remarkable in the complicated geometry of the TJ-II stellarator.

Once the models have been found acceptable to represent the radial electric field, we turn to the interpretation of some general properties of TJ-II plasmas. The transition from the LDHT regime to the IDT regime in TJ-II plasmas is characterized by a change in plasma rotation that gives rise to an $E \times B$-flow shearing layer near the edge; in particular, the rotation velocity reverses where the electric field changes sign. We analyze this fact in terms of the plasma collisionality. In Fig. 8 the collisionality profiles are plotted for the cases of experimental profiles (Fig. (2)) from the LDHT to the HDLT regimes and as one can see they have different behavior; near the center they are increasing with ρ for low $\overline{n}$ and decreasing for high $\overline{n}$. However, the radial position where the electric field changes sign, which is marked with the lines, always falls within a limited collisionality range. This means that, for all profile types, when the collisionality exceeds a certain value, trapped electrons become less important than circulating ions and there is a transition from the electron to the ion root.

The agreement of the algebraic model with the rest of the models seen in Fig. 6 points again to the robustness of the neoclassical particle fluxes in predicting the electric fields. There is also a qualitative agreement for the profiles as it is clear by comparing figures 7 and 5. Therefore, we can use the neoclassical formulation to study in more detail the passage from the LDHT regime to the IDT.

When a low density ECRH plasma in TJ-II evolves to a higher density via external gas puffing, the plasma rotation starts changing sign near the edge but inside the plasma [7]. To analyze this behavior, first we must note that, for a given E_r, the electron and ion fluxes change with collisionality at different rates. For low collisionalities Montecarlo calculations [29] show that $\partial_\nu \Gamma_i > \partial_\nu \Gamma_e$, *around the point where* $\Gamma_e = \Gamma_i$. This implies that, when ν is slightly increased (as when the density is increased by gas puffing) from its value for ambipolarity, $\Gamma_i > \Gamma_e$ which produces $\partial_t E_r < 0$, according to Eq. 3. The same relationship is satisfied for the fluxes with the algebraic model. Hence, for low collisionality (when $E_r > 0$), the electric field decreases when the density is raised. However, we may ask why there should be some $\rho < 1$ where the electric field drops faster and hits first the negative values. A typical observation is that the collisionality in low density plasmas shows a maximum around $\rho = 0.8 - 0.9$ and we hypothesize that this is related to the radius of the E_r inversion.

This hypothesis seems to be supported by simulations of TJ-II discharge #15585 (see figure 1(a)), shown in figure 9 giving the profile evolution of (a) the radial electric field and (b) the collisionality. For this simulation, ECE data have been used to build the evolving T_e-profile, while the density profiles have been constructed using a typical low density shape re-scaled to give the known line density. The collisionality shows a maximum near $\rho = 0.8$, which is the radial location where a dip in $E_r(\rho)$ develops and eventually becomes negative. Therefore, neoclassical transport suggests that the E_r inversion occurs at the maximum of collisionallity.

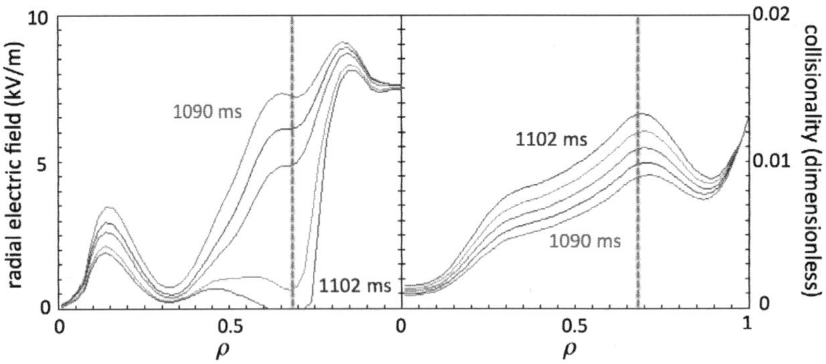

Figure 9. (a) Simulated evolution of a plasma from the LDHT to the IDL regimes where the line density evolves as is in TJ-II discharge #15585 (see Fig. 1(a)). The evolution of the radial electric field according to Kovrizhnykh semi-analytical model showing that it becomes zero at $\rho \approx 0.7$. (b) Corresponding collisionality has a maximum at the same position and increases with time since density increases and the electron temperature decreases.

It is interesting to analyze the equilibrium point for $E_r = 0$ from the point of view of rotation dynamics. In [30] it is shown that the transition from the electron to the ion root occurs in a poloidal rotation shear layer, where E_r can be obtained from a diffusion equation which may be interpreted as a balance between the non-ambipolar flux $e(Z_i\Gamma_i - \Gamma_e)$ and a viscous particle flux, driven by a poloidal viscous force,

$$\Gamma_{vis} \equiv \frac{2}{r^2}\frac{d}{dr}\left[r^2\hat{\eta}\left(E_r' - \frac{1}{r}E_r\right)\right] - \frac{\partial\hat{\eta}}{\partial E_r}\left(E_r' - \frac{1}{r}E_r\right)^2 = e(Z_i\Gamma_i - \Gamma_e), \tag{8}$$

where $\hat{\eta}$ is a viscosity coefficient. From here, we see that an ambipolar equilibrium ($e(Z_i\Gamma_i = \Gamma_e)$) can be reached at a point with $\rho < 1$ for vanishing E_r, when the $E_r(\rho)$ profile has a minimum (i.e. $E_r' = 0$) there, as in the left panel of Fig. 9. In that case $\Gamma_{vis} = 0$, which means that the viscous stress vanishes at the point where the poloidal rotation starts to change direction. This is in agreement with what was found in Ref.[31].

Finally, we turn to the problem of the formation of transport barriers in non-axisymmetric devices from the neoclassical viewpoint. In such devices the L-H transition happens at different rates, possibly due to the fact that the transport barriers can develop inside the plasma (instead of right at the plasma edge, as is normally the case in tokamaks) where neoclassical fluxes are not negligible in comparison with their anomalous counterpart. Assuming the paradigm of turbulence quenching due to the establishment of sheared electric drifts, it is in order asking whether the neoclassical electric field can provide a positive feedback loop for the establishment of a robust transport barrier. In other words: will the neoclassical E_r respond to the increasing gradients developing a stronger shearing rate in the $E \times B$ flows? This would give a positive feedback mechanism of the kind studied, e.g., in ref. [32], where the increasing pressure gradient intensifies the shearing rate that, in turn, eases a further increment of the pressure gradient. The models used in this paper provide such positive feedback indeed.

6. Conclusions

The results of calculations of the E_r field from the framework of neoclassical transport theory have been compared with experimental measurements of the plasma potential obtained with HIBP diagnostics in the TJ-II stellarator. Different collisionality regimes were analyzed with several theoretical models which in general terms yield E_r profiles in agreement with the experiment, indicating that neoclassical theory is a reasonable description for non-ambipolar fluxes [16]. However, the absolute values of the NC fluxes are smaller than the actual ones. From experimental observations, we can define three characteristic intervals for density and electron temperature. From the comparison of analytical results with results obtained from the three described models for the neoclassical fluxes, it can be noted that Kovrizhnykh and Beidler models are more adequate to model the non-axisymmetric NC transport in TJ-II plasmas (6), in the sense that solutions to the ambipolar equation (2) are closer to the experimental measurements and to those of the analytical model, which is entirely based on asymmetric NC transport. The qualitative agreement with experimental E_r profiles is remarkable considering that, for this model a simple magnetic geometry with a single helical harmonic (4), was assumed, with a simplified representation of helical ripple that is varied to improve the fits (Eq. 5). In the case of Beidler model, in a low collisionality regime we note it yields larger values than the other semi-analytical models (notice that the analytical fits of NC fluxes to the numerical DKES results, used in this model, deviate from them for small collision frequencies [18]).

Multiple solutions of the electric field that satisfy the ambipolar constraint (2) often arise when particle fluxes have a nonlinear dependence on electric field. This problem can be solved when the constants involved in density and temperature profiles are chosen to satisfy two criteria: the existence of a stable plasma equilibrium $\overline{P} = \overline{n_e T_e} \approx$ const over different density regimes, and the existence of a single real solution for the ambipolar equation (enforcing ambipolarity on each flux surface) [33]. The roots of E_r depend upon the plasma profiles for density and the two temperatures, thus allowing to conclude that the jumps between roots at two neighboring radii at steady-state, can be controlled by the plasma parameter profiles.

For the case LDHT we confirm the fact that with centrally peaked temperature profiles, a large positive "electron root" radial electric field is observed. In some cases, it can lead to the formation of an internal transport barrier which has been collectivelly named core electron root confinement (CERC); the large E_r has the effect of reducing the neoclassical transport by driving both plasma species out of the high-loss "$1/\nu*$" regime[33]. In the case of HDLT, we also obtain that the electric field becomes more negative near the plasma edge as was reported in other stellarators, with NBI heating. In general, the magnitude of the calculated E_r is smaller than HIBP measurements. Possible reasons are the presence of suprathermal particles since they would enhance the electric field, and non-local effects not accounted for in the theoretical models [34].

An important result is that the transition from the electron to the ion root is found to occur for a certain narrow range of collisionality which points to a threshold collisionality as the cause for the transition. The implication of this is that a right amount of collisions, which destroy the trapped particle orbits that produce large electron losses, is responsible for the appearance of the ion root. The determination of the change in the radial electric field sign as a function of collisions (see Fig. 8) could have a strong influence in the transport barrier formation [35].

The good performance of the analytical NC model in predicting E_r was exploited to explain the observation that E_r gets zero inside the plasma edge in going to from LDHT to IDT regimes. It was shown that this occurs where the collision frequency profile has a maximum. The maximum in turn would produce an accelerated reduction of E_r in a given low collisionality range. The NC model also provides support to the idea that an improved confinement mode can be maintained by the increased sheared $E \times B$ flows.

Acknowledgments: Partial support from projects PAPIIT-UNAM IN106911 and Conacyt 152905 is acknowledged.

[1] Dinklage A., Yokoyama M., Tanaka K., et al., 2013 *Nucl. Fusion* **53** 063022.
[2] Chmyga A.A., Dreval N.B., Khrebtov S.M., et al., 2002 *29th EPS Conf. Plasma Physics*, Montreaux, ECA Vol. 26B, O-1.09.
[3] Hidalgo C., et al., 2005 *Nucl. Fusion*, **45**, S266-S275.
[4] Garcia J., Dies J., Castejon F., et al., 2007 *Phys. Plasmas* **14** 102511.
[5] Vargas V.I, Lopez-Bruna D., Garcia J., et al., 2008 *35th EPS Conf. Plasma Physics*, Hersonissos, ECA Vol. 32D, P-5.018.
[6] Bondarenko I.S., et al., 2000 *Czech. J. Phys.* **50** 1397.
[7] Happel T., Estrada T. and Hidalgo C., 2008 *Europhys. Lett.* **84** 65001.
[8] Hidalgo C., Pedrosa M.A., Castellano J., et al., 2001 *Plasma Phys. Controll. Fusion* **43** A313.
[9] Baciero A., Zurro B., McCarthy K.J., et al., 2001 *Rev. Sci. Instrum.* **72** 971–974.
[10] Herranz J., et al., 2000 *Phys. Rev. Lett.* **85** 4715.
[11] van Milligen B.Ph., 2011 *Rev. Sci. Instrum.* **82** 073503.
[12] Shaing K. C., 1984 *Phys. Fluids* **27** 1567.
[13] Hastings D.E., Houlberg W.A., Shaing K.C., 1985 *Nucl. Fusion* **25** 445 .
[14] Kovrizhnykh L.M., 1999 *Plasma Phys. Rep.* **25** 760.
[15] Hinton F.L., Hazeline R.D., 1976 *Rev. Mod. Phys.* **48** 239.
[16] Turkin Yu., Beidler C.D., Maasberg H., et al., 2011 *Phys. Plasmas* **18** 022505.
[17] Kovrizhnykh L.M., 2005 *Plasma Phys. Rep.* **31** 14.
[18] Maasberg H., Beidler C.D., Simmet E.E., 1999 *Plasma Phys. Contr. Fusion* **41** 1135.
[19] Beidler C.D., 1996 *Neoclassical transport properties of HSR*, IPP-Report 2/331, p.194.
[20] Kovrizhnykh L.M., 2006 *Plasma Phys. Rep.* **32** 988 .
[21] López-Bruna D., Reynolds J.M., Cappa A., et al. 2010 *Programas Periféricos de ASTRA para el TJ-II.* Informes Técnicos Ciemat 1201.
[22] Pereverzev G.V., Yushmanov P.N., 2002 ASTRA *Automated System for TRansport Analysis*, Max-Plank-Institute für Plasmaphysik Rep IPP 5/98 Garching.
[23] Medina F., et al., 2001 *Rev. Sci. Instrum.* **72** 471.
[24] Ochando M.A., et al., 2003 *Plasma Phys. Control. Fusion* **45** 221.
[25] Ochando M.A., et al., 2006 *Plasma Phys. Control. Fusion* **48** 1573.
[26] Vargas V.I., et al., 2009 *Density dependence of particle transport in ECH plasmas of the TJ-II stellarator*, Informes Técnicos Ciemat No. 1162.
[27] López-Bruna D., Velasco J.L., Ochando M., et al., 2013 *Plasma Phys. Contrl. Fusion*, **55**, 015001.
[28] Velasco J.L., Castejon F., Tarancon A., 2009 *Physics of Plasmas*, **16**, 052303.
[29] Tribaldos V., 2001 *Phys. Plasmas* **8** 1229.
[30] Maassberg H et al., 1993 *Phys. Fluids B* **5** 3627.
[31] Velasco J.L. et al., 2012 *Phys. Rev. Lett.* **109** 135003.

[32] Carreras, B.A., et al., 1994 *Plasma Phys. Control. Fusion* **36** A93.
[33] Lore J., Guttenfelder W., Briesemeister A., et al., 2010 *Phys. Plasmas* **17** 056101.
[34] Tribaldos, V., Guasp, J., 2005 *Plasma Phys. Control. Fusion*,**47** 545 .
[35] Idei H., Ida K., Sanuki H., et al., 1993 *Phys. Rev. Lett.* **71** 2220.

Dynamics of magnetic islands and confinement transitions in TJ-II

J J Martinell[1], **D López-Bruna**[2] **and M A Ochando**[2]

[1] Instituto de Ciencias Nucleares (UNAM), México D.F., México
[2] Laboratorio Nacional de Fusión, Asociación EURATOM-CIEMAT, Madrid, Spain

E-mail: `martinel@nucleares.unam.mx`

Abstract. Bolometric observations over a large number of discharges under very different conditions in the TJ-II Heliac, together with magnetic and line-emission diagnostics have indicated that the plasma behavior near rational magnetic surfaces is rather peculiar, which involves the formation and destruction of internal transport barriers associated with magnetohydordynamic (MHD) activity [1]. The interpretation of the results is that around the magnetic islands that appear at resonant low-order rational surfaces there is a sheared flow that gives rise to the formation of a transport barrier, as the anomalous transport is decreased. This is evidenced by low recycling at the edge shown by low Hα emission. Then an MHD event arises, possibly due to magnetic reconnection, which destroys the barrier and releases high-energy particles. This sequence of events is observed as sawtooth-like oscillations in the bolometric emission with a well defined repetition time. In this work a model is presented where this ingredients are incorporated in a transport simulation that includes a turbulence model based on resistive ballooning modes [2]. The sheared flow is described by a neoclassical modeling of the ambipolar electric field given in [3]. A transition model [4] is included in the transport system of equations to verify that the appropriate signals are caused by the mechanisms described above. It is found that the model reproduces in an acceptable way the experimental results, supporting the assumed phenomenology: confinement regulated by rapid particle expulsions related to the break of magnetic islands locally and intermittently. The repetition rates are on the ms time-scale. The breaking of magnetic islands is described in terms of a magnetic reconnection model which is related to the presence of sheared flows just outside the islands. The unstable tearing mode includes the effect of the polarization drift in the outer region.

1. Introduction

Plasma confinement in stellarators has been found to be improved under certain circumstances leading to a transition in the confinement time with some features similar to the L-H transition in tokamaks [5]. The nature of the transport barriers in stellarators, however, has its own properties. For instance, in high-power electron cyclotron heating (ECH) an internal transport barrier (ITB) is created, characterized by peaked electron temperature profiles and a large positive radial electric field in the core plasma region, the so-called core electron root confinement (CERC) [6]. For neutral beam injection (NBI) heated plasmas, edge transport barriers (ETB) can be produced although the confinement improvement is more modest than in tokamaks [7]. The origin of the transport barriers is somewhat complex: it has been argued that the sheared flows associated with the radially varying radial electric field can suppress the turbulence-driven transport, following the usual paradigm for tokamaks; but there is evidence that MHD phenomena play an important role by modifying the confining magnetic field. A clear piece of evidence is the relationship that has been found between the presence of low-order rational magnetic surfaces and the formation of transport barriers [1, 6].

In the TJ-II flexible heliac stellarator [8] it has been possible to perform many studies about rational surfaces and related phenomena. Some facts emerged already in the first campaigns [9] and were confirmed over the years. A brief overview of the important results includes: close to the predicted locations of resonant layers in the bulk plasma the generation of enhanced electron confinement regimes has been observed [10], in addition to the de-trapping of barely trapped particles [11] and unexpected enhanced line impurity emissions [12]. Near the expected position of singular surfaces, radial electric field shearing at the time of L–H transitions and suprathermal electron accumulation have been detected [13]. These results are complemented with the knowledge of a systematic modification of plasma gradients that resembles the existence of (weak) transport barriers accompanying the main magnetic resonances throughout the plasma [14].

The specific characteristics of TJ-II, i.e. low magnetic shear and high magnetic configuration flexibility, allow controlling the position of low-order rational values within the rotational transform profile, which allows the study of how the magnetic topology affects the transport. It is known that rational surfaces are resonant with magnetic perturbations and thus tend to form magnetic islands. The presence of magnetic islands, in turn, contributes to the outward electron flux that creates a locally strong positive radial electric field [10]. This is due to the fact that the diffusion coefficient in magnetic islands depends on the parallel velocity of the species; consequently, the presence of the island will enhance the electron flux more than the ion flux, modifying the equilibrium ambipolar radial electric field. On the other hand, the transport reduction resulting from the strong electric field radial gradient may modify the behavior of the island and produce MHD activity. This interplay of phenomena is reflected in the various measurements using different diagnostics which has been described in [1], trying to identify a process that links MHD activity with transport barriers. Based on these data, we have started to develop a model that puts together all the relevant processes in order to reproduce the experimental observations. Here we present the basis of the model which describes transport through a turbulent process which is modulated by the presence of magnetic islands. To this end, we first give in Section 2 a description of the different experimental results on which we base our model. Then, in Section 3 we describe the transport model which assumes a resistive ballooning mode as the drive for turbulence, which incorporates the presence of magnetic islands in a phenomenological way. It will be shown that the main features of the time evolution of emission profiles are reproduced quite well. A physical study of the dynamics of the magnetic islands is presented in Section 4, where we incorporate the presence of sheared flows on the stability of tearing modes. Finally, in Section 5 preliminary conclusions of this work are given.

2. Experimental observations

The TJ-II device has a major radius of $R = 1.5$ m with bean-shaped cross-sectional magnetic surfaces of average minor radius $a \approx 0.2$ m and four toroidal periods. It operates at a mean magnetic field of ≈ 1 T with a low magnetic shear and high rotational transform $\iota(a)/2\pi \sim 3/2$. The plasmas considered are NBI heated with line-average densities $1.5 \times 10^{19} m^{-3} \leq \overline{n} \leq 4.0 \times 10^{19} m^{-3}$. However, features similar to those described next are also seen in high power ECH plasmas. The NBI power is around 0.5 MW, which produces a rather moderate plasma pressure (volume average $\langle\beta\rangle \leq 0.5\%$) and small net electric currents ($|I_p| \leq 2$ kA). In TJ-II plasmas the typical transition to the improved confinement manifests in a drop of electromagnetic activity and fluxes towards the wall, accompanied by a widening of the density profiles and a change in the radiation profiles from bell-shaped to dome-like. The formation of a sheared rotation layer is generally found at around $0.7 \leq \rho \leq 0.8$ [5] depending on the magnetic configuration, particularly

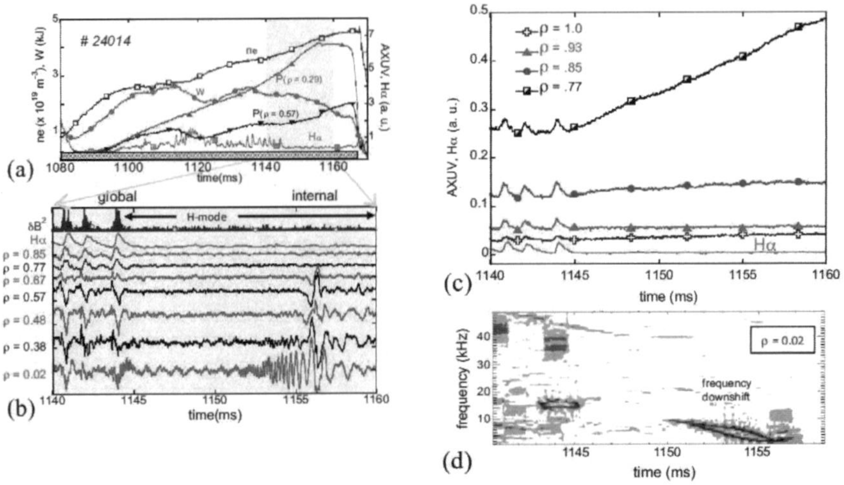

Figure 1. (a) Time evolution of discharge #24014 line-averaged electron density, diamagnetic energy, Hα emission and line-integrated emissivity, P, from two bolometer chords tangent to flux surfaces with $\rho \approx 0.29$ and $\rho \approx 0.57$. (b) Enlargement of Mirnov coil signal, Hα emission and multi-chord bolometry, showing sawtoothing activity, transition to H-mode and an internal crash. (c) Unfiltered traces of the outer bolometry chords and Hα. (d) Coherent modes in the spectrogram of a bolometer signal whose line of sight crosses the whole plasma.

due to the placement of different low-order rational values of the rotational transform. In Fig. 1 we show the time evolution of a discharge that presents three different phenomena: first sawtoothing activity, then a transition to high confinement and later on an internal crash.

After t=1135 ms bursts of activity develop and continue to t=1145 ms when an H-mode is established. This period is detailed in Figs. 1(b)-(d). A radius where the bolometry time traces invert growth (inversion point) is manifested during the pre-transition phase: there is a trace with no or little modification during the sawteeth, while the signals have opposite jumps inside and outside. During the bursty period of the Hα signal, the MHD events (Mirnov coil signal, δB^2) coincide with perturbations in all the viewing chords. Figure 1(c), shows details of unfiltered outer bolometry traces, $\rho \geq 0.77$, where the profile change near the plasma edge is apparent according to the different evolution of the $\rho = 0.77$ chord with respect to $\rho > 0.77$ chords. Figure 1(d) is the spectrogram of a bolometry chord passing very close to the magnetic axis, thus collecting information from practically all radii. A clear mode at 17 kHz becomes prominent just before H-mode onset. It is seen that the H-mode remains quiescent until another MHD event shows up in the core of the cooler, more collisional plasma at $t \approx 1155$ ms. However, the apparent avalanche gets stopped at a radial position $0.67 < \rho < 0.77$ without leaving a detectable trace further out.

The bolometric measurements of the three types of events that are usually observed are compared in Figure 2. These are the evolving profiles as a function of the effective minor radius. The off-axis sawteeth event (same discharge as in Fig. 1), is presented in Fig. 2 showing intermittent rapid radial redistributions of radiation with a pivot or inversion point near the plasma periphery and it is appreciated that the events cover the entire plasma despite their relative small magnitude. Coincident with sawtooth crashes there is the onset of a quasi-coherent mode of about 30 kHz detected in the edge bolometer and in the magnetic signals, which is resonant with the 8/5 magnetic rational surface. There is also an intensification of Hα activity, features that are in shown in Fig. 3. This points to a process started by and edge MHD mode that triggers the rise in edge radiation and the quasi-coherent mode, followed by a burst of particles reaching the wall. In Fig. 2(b) we present the profile evolution for an internal crash like the one in the later stage of Fig. 1 characterized by initial growth and subsequent damping of radiation oscillations suggesting the presence of an island with growing width. For this discharge (#24018) there are two internal crashes involving a small central volume accompanied by a modest increase beyond the inversion radius at around $\rho = 0.3$, but leaving the edge region unperturbed. Consequently, no edge magnetic signals are detected but the possible mechanism is that the pressure profiles increase until an internal MHD mode is triggered causing a momentary loss of confinement and a consequently rapid profile readjustment. The behavior of these two events can be compared with the profile evolution of the L-H transition presented in Fig. 2(c). Here the rise in emissivity produces a narrowing of the profiles indicating a cooler edge as a result of the presence of a transport barrier further in. It can be said that the same process involving transport barriers in Figs. 2(a) and (b) is present when an L-H transition happens, with the particularity that the resonant region in this case is more effective in creating a transport barrier.

The close correlation of rational surfaces with the inversion points in the emissivity profiles is supported by the fact that increasing the radial position of the vacuum rational over different discharges also increases the radial position of the inversion point. This link indicates that the resonances have tampered with transport in some way. Regardless of the position or the strength of the events, the plasma behavior is systematic in a way: particle flows are released and move radially leaving a magnetic signature detectable in Mirnov coils and radiation signals. The effect is clearly electromagnetic. In some cases, if the location of the resonances is appropriate the effectiveness of the transport barriers increases to the extent that they seem to be at the origin of the L–H transition, which occurs associated with low-order rationals located at about $\rho \sim 0.8$.

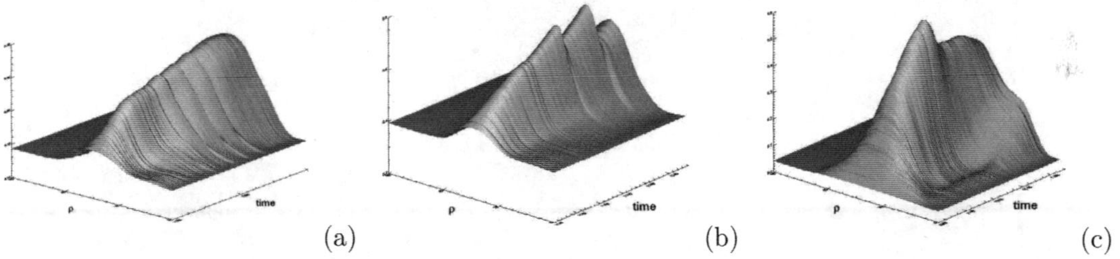

(a) (b) (c)

Figure 2. Time evolution of reconstructed emissivity profiles for (a) off-axis sawtoothing event in discharge #27222 (b) internal crash in discharge #24018 and (c) an improved confinement mode (H-mode) in discharge #21319.

The sequence of events just described has to be interpreted in terms of how the magnetic structures associated with rational surfaces (i.e. magnetic islands) can affect transport in the way observed.

3. Transport model in presence of magnetic islands

It is well known that neoclassical transport is strongly affected by the radial electric field which is important in a stellarator due to the non-ambipolar nature of fluxes. Therefore, any process that can potentially modify the electric field should affect plasma transport. Such is the case

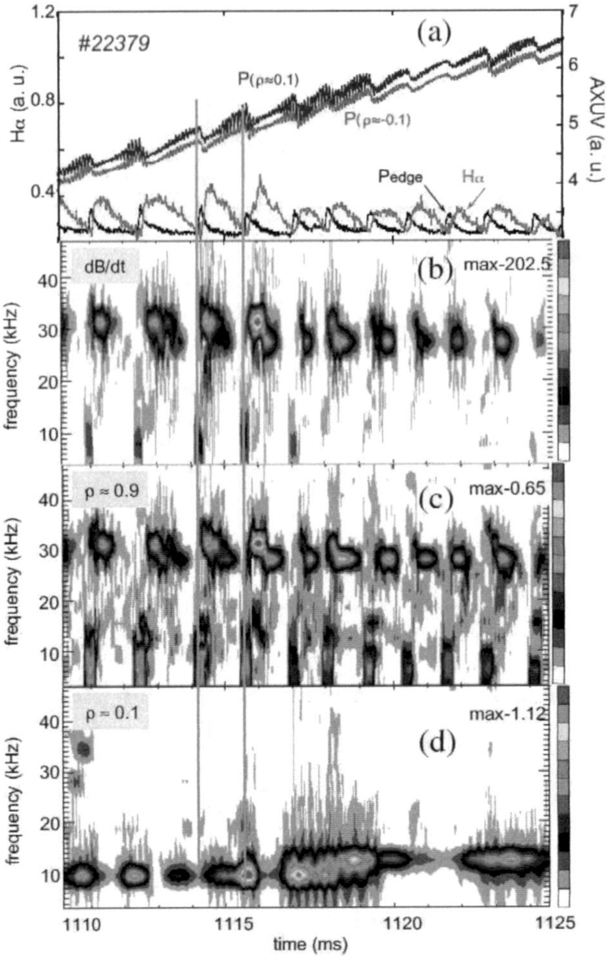

Figure 3. (a) Bolometer traces from central and edge chords and Hα emission for discharge #22379. Spectrograms of (b) a Mirnov coil signal, (c) an edge bolometer chord and (d) a central bolometer chord.

of a magnetic island since the presence of closed magnetic surfaces allows the particles to travel along the field lines efficiently producing fast radial transport, but more so for electrons. It is expected then that a radial electric field across the island be present, modifying in turn the confinement. In order to incorporate this effect in the overall plasma transport we used the Astra transport code [15] introducing a model for turbulent transport and the presence of the islands. We chose to adopt a model of transport due to resistive ballooning modes [2] since it seems to be a likely candidate for stellarators. In the model we have to include the effect of sheared-flow suppression of turbulence [4] in order to allow for the formation and destruction of transport barriers. In the Astra code the electric field is computed from the electron and ion

Journal of Physics: Conference Series **591** (2015) 012012 doi:10.1088/1742-6596/591/1/012012

fluxes $\Gamma_j(E_r)$ by solving the ambipolarity condition in the form

$$\frac{\partial E_r}{\partial t} = K[\Gamma_e(E_r) - Z_i\Gamma_i(E_r)]. \tag{1}$$

until the stationary state is established. In this relation, only the neoclassical fluxes are taken since the anomalous transport is usually intrinsically ambipolar. For the neoclassical coefficients we use the analytical expressions given by Kovrizhnykh [3]. In order to better model the TJ-II plasmas we have also included the presence of impurities which contribute to the radiation emission; three charge states of carbon are assumed. The equations considered in the model can be written as

$$\begin{aligned}
\frac{\partial n_e}{\partial t} &= -\nabla\cdot\Gamma_e + S_e \\
\frac{\partial}{\partial t}\left(\frac{3}{2}n_{e,i}T_{e,i}\right) &= -\nabla\cdot\mathbf{q}_{e,i} + Q_{e,i} \\
\frac{\partial n_{C2}}{\partial t} &= n_e(n_{C3}R_{C3} - n_{C2}(R_{C2}+I_{C2})) - \nabla\cdot\Gamma_{C2} \\
\frac{\partial n_{C3}}{\partial t} &= n_e(n_{C2}R_{C2} - n_{C3}(R_{C3}+I_{C3}) + n_{C4}R_{C4}) - \nabla\cdot\Gamma_{C3} \\
\frac{\partial n_{C4}}{\partial t} &= n_e(n_{C3}R_{C3} - n_{C4}(R_{C4}+I_{C4})) - \nabla\cdot\Gamma_{C4} \\
\Gamma_j &= -(D_{NC,j} + D_{tb,j}\epsilon)\nabla n_j \\
\mathbf{q}_j &= -n_j(\chi_{NC,j} + \chi_{tb,j}\epsilon)\nabla T_j + \frac{5}{2}\Gamma_j T_j \\
S &= S_{rec} + S_{gp} + S_{NBI} \\
Q_e &= -P_{ei} + P_{NBIe} - P_{rad} \\
Q_i &= P_{ei} + P_{NBIi} - P_{cx} \\
P_{rad} &= n_e(C_{eff}n_I + C_{C2}n_{C2} + C_{C3}n_{C3} + C_{C4}n_{C4}) \\
\frac{\partial\epsilon}{\partial t} &= (\gamma_{RB} - \alpha_1\epsilon + \alpha_2|\omega_s|)\epsilon - \nabla\cdot\Gamma_\epsilon
\end{aligned}$$

where all quantities have their usual meaning, with R_a, I_a the recombination and ionization rates, ϵ is the turbulence fluctuation level, whose evolution is determined by the linear growth rate of resistive ballooning modes (RBM)

$$\gamma_{RB} \propto \left(\frac{\eta\beta|p'|}{a\tau_A\rho\varepsilon t^2}\right)^{2/3}$$

the nonlinear saturation coefficient

$$\alpha_1 = \frac{|p'|\gamma_{RB}^{3/4}}{p\sqrt{a\hat{s}/\eta\sqrt{\tau_A}}}$$

and the shearing rate $\omega_s = \partial_r(cE\times B/B^2)$. Now, the presence of the island is included as an additional electron flux Γ_e^R at the island position in Eq. 1 in addition to the neoclassical contribution Γ_e^{NC}. Thus, the ambipolar state implies $\Gamma_e^{NC}(\rho) + \Gamma_e^R(\rho) = Z_i\Gamma_i^{NC}(\rho)$, with the island flux localized at the resonant surface ρ_R:

$$\Gamma_e^R(\rho) = \Gamma_0\exp[-((\rho-\rho_R)/\Delta_R)^2] \tag{2}$$

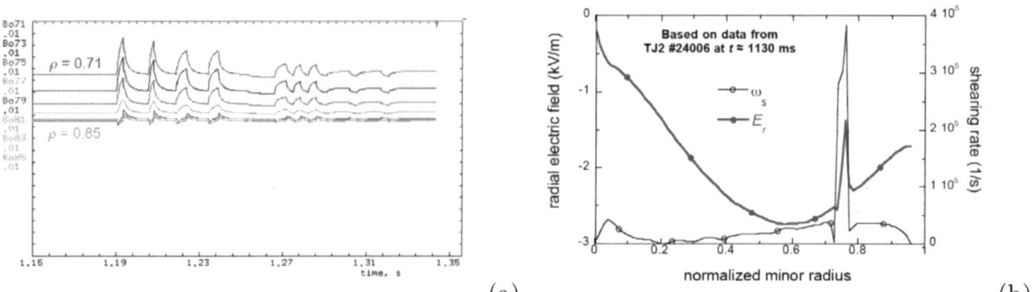

(a) (b)

Figure 4. (a) Simulated time evolution of a sawtoothing discharge for the emission at different radial chords. (b) Radial profiles of radial electric field and shearing rate showing large gradients at resonant surface position.

where Δ_R is a measure of the island width. This determines the radial electric field.

The effect of the resonant flux is to increase the radial electric field around $\rho = \rho_R$ and therefore the shear rate, $\omega_s = \partial_r E_r + (\hat{s} - 1)E_r/r$, increases substantially. This in turn, tends to stabilize the RBM turbulence, which reduces the transport, creating a barrier there. The transport reduction should affect the island too but this effect is not included in the present model; we force the feedback manually just to reproduce the cyclic evolution. The results of simulations with the model just described are shown in Figure 4, for parameters taken from discharge #24006. In panel (a) the time evolution of the total emission at different radial positions is presented for the case when an island is placed at $\rho_R = 0.76$ for a time of 1 ms and then removing it. After the barrier is formed the emission increases at radii smaller than ρ_R and decreases for $\rho > \rho_R$ in qualitative agreement with the observations. The different strength peaks are forced when the shearing parameter ω_s is varied keeping the electron-ion imbalance fixed. In panel (b) we have the radial electric field profile, exhibiting that the effect of the island is to increase the E_r values with the consequent high gradient of both E_r and shearing rate, which is responsible for the mode stabilization. The local perturbations in the simulation have maximum amplitudes consistent with the experimental values, namely, 10% for the density, 20% for the radiation emission, 4% for temperature and less than 2% for the potential.

The set of profiles of several parameters taken from the Astra simulation are shown in Figure 5 for a time when the barrier is not present. It shows that the carbon profiles for CIII and CIV peak near the edge and the E_r field is negative, as in all high density NBI discharges. The lower panel confirms that the instability dynamics, given by γ (`gamm`) is concentrated near the edge, as expected for resistive ballooning modes. Consequently, the electron thermal conductivity χ_e (`Xe`) is also quite large close to the edge.

4. Magnetic island evolution

In addition to the qualitative reproduction of the dynamics it is necessary to address the issue of what is the physical mechanism responsible for the flux enhancement assumed in the model. An ingredient from the experiment that has also to be considered is the expulsion of fast particles to the wall, which points to some mechanism that either accelerates particles or releases high-energy particles trapped in the islands when the MHD activity starts. In any case, the dynamics of the islands associated with the rational surfaces seems to be the important process to study. Regardless of whether the island is present in the vacuum magnetic field or it is formed during the

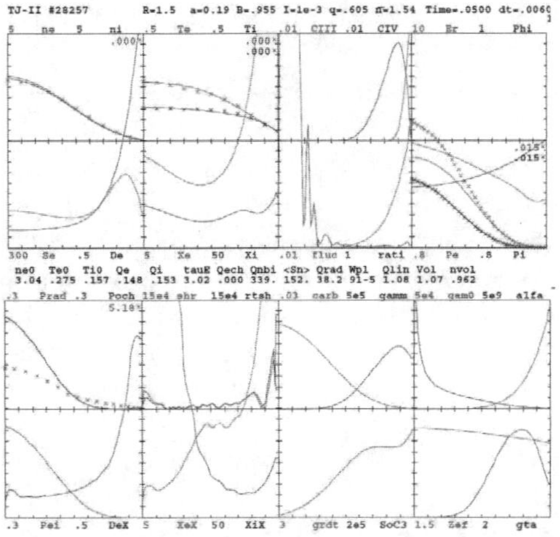

Figure 5. Astra profiles for a time when there is no transport barrier. Each panel contains two profiles in red and blue.

plasma discharge, the dynamics should be determined by tearing mode stability. The magnetic reconnection process resulting from the tearing instability can potentially accelerate the electrons detected by the Hα signal. Reconnection could also open island field lines and release trapped energetic particles.

To address the tearing mode stability in this case it is necessary to include the presence of shear flows, since they are always present when a transport barrier is formed. The presence of flows at the island location and their consideration in the nonlinear tearing mode dynamics should shed light on the interplay between barrier formation and island growth. An important effect to take into consideration when the island width is as small as the ion Larmor radius is the polarization drift. This effect is important when the island begins to grow and it has been variously described in the literature with some discrepancies as explained in [16]. The relevant effect that determines the contribution to the growth is the velocity profile across the island. It enters the evolution of the island width w through a contribution of the polarization drift Δ'_p to the tearing stability parameter Δ' as,

$$\frac{dw}{dt} \propto \Delta' + \Delta'_p = \Delta' + g(\omega - \omega_{*i})$$

where g depends on the velocity profile and its sign determines the stability. Depending on the frequency range, it can be stabilizing or destabilizing [16].

When there is shear flow the asymmetry in the velocity profiles modifies the stability; as shown in [17] the shear contribution is stabilizing for low and high shear but it destabilizes the mode for intermediate values of the shear. This may lead to oscillations of the island width.

An important issue to consider is the island rotation velocity. This is related to the frequency of the magnetic perturbations measured in the experiment because, in the plasma frame, it is determined by the electron or ion diamagnetic velocities as: $\omega = f\omega_{*e} + (1 - f)\omega_{*i}$, where f is the flattening parameter that measures the degree of temperature profile flattening inside the island. $f = 1$ means there is no flattening. Since in TJ-II NBI plasmas there is no flattening detected during off-axis sawteeth, one would expect $\omega \approx \omega_{*e}$. According to Connor et al. [16], for $w \leq \rho_i$, $g \sim \omega - \omega_{*e}$ giving for the polarization drift contribution

$$\frac{dw}{dt} = \frac{I}{k^2 w} 10^{-6} (\omega - \omega_{*e})(\omega - \omega_{*i}). \tag{3}$$

This would give $\Delta'_p \approx 0$ for for TJ-II when $\omega \approx \omega_{*e}$. Then there should be stable island rotation in the frame of reference where $E_r = 0$. Moving to a frame with $E_r = -3$ kV/m, Eq. 3 predicts there can be instability for frequencies outside the range $0 < \nu < 20$ kHz for a resonance 8/5 near the edge, for NBI plasma of moderate density. The corresponding growth rate is of order $10^5 s^{-1}$ when $I/(kW)^2 \sim 1$. This seems to be consistent with the experimental data.

Another ingredient one should consider is the fact that rational surfaces have magnetic islands even in the vacuum magnetic field. Therefore, the island dynamics analysis should start with these vacuum islands. This problem has been addressed recently in [18] in relation to island healing, considering the self-consistent torque balance. In our case it is relevant since an island can grow or shrink depending on the velocity profiles, so it is important to start with a finite island size. The program to follow is then to study the behavior of the polarization stability parameter from the initial vacuum island, providing the velocity profiles characteristic of the sheared flows observed around TJ-II rational surfaces. This should determine the island evolution.

5. Conclusions

The experimental results obtained over a large range of plasma conditions in TJ-II have evidenced the very important role magnetic rational surfaces play in the formation and destruction of transport barriers. We have presented here a summary of the measurements that associate the onset of MHD activity with the modification of plasma confinement at the position of a low-order rational surface. The picture that emerges is that a transport barrier can form at a rational resonant surface when an enhanced sheared flow is set there, due probably to the increased E_r-field that appears at the magnetic island produced by large parallel electron transport. The sheared flows can trigger a magnetic instability which eventually destroys the transport barrier releasing outward flows and fast particles that hit the wall.

This interpretation is incorporated in a transport model that includes the effect of the island through a localized radial electron flux centered at the resonant surface. The turbulent transport is produced by resistive ballooning modes that may be stabilized when the $\mathbf{E} \times \mathbf{B}$ shearing rate is large enough. The preliminary results of this model seem to reproduce in a satisfactory qualitative way the observations. Particularly, the increased radial electric field gradient is of the right size to stabilize the RBM turbulence and create a local small transport barrier, producing sawteeth in the total emission that have an inversion point at the island position. The model is imperfect in that it only describes the effect of the island on transport but the feedback effect of transport on the island is not included.

In order to study the later effect the island evolution is considered from the point of view of the tearing mode stability. The polarization drift contribution to the stability parameter is considered as a factor that modifies the island width evolution and this is critically determined by the velocity profiles across the island. Therefore, it is necessary to study the dynamics of an initial vacuum magnetic island that is subjected to the shear flows around the island typical of a transport barrier. This study is on-going and will be presented elsewhere. When this is completed, an improved transport model will be constructed that incorporates the ingredients for a self-consistent description.

Acknowledgments: Partial support from projects PAPIIT-UNAM IN106911 and Conacyt 152905 is greatly acknowledged.

[1] López-Bruna D, Ochando M A, López-Fraguas A, Medina F and Ascasíbar E 2013 *Nuclear Fusion* **53** 073051
[2] Carreras B A, García L and Diamond P H 1987 *Phys. Fluids* **30** 1388
[3] Kovrizhnykh L M 2006 *Plasma Phys. Reports* **32** 988
[4] Carreras B A, Newman D, Diamond P H and Liang Y-M 1994 *Phys. Plasmas* **1** 4014
[5] Estrada T, Happel T, Hidalgo C, Ascasíbar E and Blanco E 2010 *Europhys. Lett.* **92** 35001
[6] Estrada T, Medina F, López-Bruna D, Ascasíbar E, Balbín R, Cappa A, Castejón F, Eguilior S, Fernández A, Guasp J, Hidalgo C and Petrov S 2007 *Nucl. Fusion* **47** 305
[7] Toi K et al 2006 *Plasma Sci. Technol.* **8** 5
[8] Alejaldre C et al 1990 *Fusion Technol.* **17** 131
[9] Alejaldre C et al 2001 *Nucl. Fusion* **41** 1449
[10] Castejón F, López-Bruna D, Estrada T, Ascasíbar E, Zurro B and Baciero A 2004 *Nucl. Fusion* **44** 593
[11] Ochando M A and Medina F 2003 *Plasma Phys. Control. Fusion* **45** 221
[12] Baciero A, Zurro B, McCarthy K J, López-Fraguas A, Ascasíbar E, Estrada T, García-Cortés I, de la Luna E, Vega J 2001 *Plasma Phys. Control. Fusion* **43** 1039
[13] Ochando M A, Medina F, Zurro B, McCarthy K J, Pedrosa M A, Baciero A, Rapisarda D, Carmona J M and Jiménez D 2006 *Plasma Phys. Control. Fusion* **48** 1573
[14] López-Bruna D, Estrada T, Medina F, de la Luna E, Romero J A, Ascasíbar E, Castejón F and Vargas V I 2008 *Europhys. Lett.* **82** 65002
[15] Pereverzev G V and Yushmanov P N 2002 ASTRA *Automated System for TRansport Analysis*, Max-Plank-Institute für Plasmaphysik Rep IPP 5/98 (Garching)
[16] Connor J W, Waelbroeck F L and Wilson H R 2001 *Phys. Plasmas* **8** 2835
[17] Fitzpatrick R and Waelbroeck F L 2009 *Phys. Plasmas* **16** 052502
[18] Hegna C 2012 *Phys. Plasmas* **19** 056101

Are magnetic resonances practical transport controllers in fusion plasmas? The TJ-II experience

D López-Bruna[1], V I Vargas[2] and J A Romero[1]

[1] Laboratorio Nacional de Fusión, Ciemat, Madrid, Spain

[2] Plasma Laboratory for Fusion Energy and Applications, Cartago, Costa Rica

daniel.lopezbruna@ciemat.es

Abstract. The TJ-II stellarator is a mid-size Heliac design (high rotational transform, low magnetic shear) that allows for external control of the rotational transform profile. The long experience with locating diverse magnetic resonances at different plasma radii has assessed their reliability as transport controllers: in low density electron cyclotron heating plasmas, a "transport barrier effect" is found in most of the confinement zone accompanying magnetic resonances; in higher density and beta plasmas under neutral-beam heating operation, resonant layers show a clear incidence on the access to the H-mode of confinement or, likewise, on the back transition to L-mode. Moreover, confinement events similar to tokamak phenomenology have been also related with magneto-hydrodynamic activity around the magnetic resonances and are, consequently, amenable to external control. All in all, the TJ-II experience posits magnetic resonances as natural transport and stability controllers in toroidal plasmas. Further studies must either confirm or set operational limits to these findings.

1. Introduction

In toroidal magnetic confinement the geometry implies that some twisting (winding number) of the magnetic field lines is necessary to short-circuit charge separation due to curvature drifts. Magnetically resonant layers are regions where the magnetic field lines, owing to their ability to close on themselves after a few turns around the toroid, can break locally the ideal toroidal nesting of flux surfaces. This breaking of the topology of the magnetic field starts with resonant flux surfaces that are linearly unstable and give rise to volumes with cross sections that resemble islands; hence the close relationship between magnetic resonances and magnetic islands. Because of this, considerable effort was payed in early studies on stellarator confinement to the question of particle containment within the plasma volume. Using electron guns or other means to check for orbit containment, a fine structure of containment time was find in relation with the rotational transform: particles would get confined for a long time in some places, while being easily lost in very close locations supposedly associated with low order rational values of the rotational transform (manetic resonances); see, for instance, stellarator papers in [1]. Later, in the nineties, these magnetic resonances were confirmed to have a locally deletereous effect on confinement; however, and since lowest order rational values are "isolated" in terms of density of low order rational values, magnetic resonances in the confining volume are generally sided by robust flux surfaces with good particle containment properties. As a consequence, it was proposed that magnetic resonances must either be avoided working with very low magnetic shear (small radial variation of the rotational transform) or included with high enough magnetic shear so that the resonant regions are very narrow [2]. Results have accumulated in the early 2000's proving that transport can be also significantly reduced at, or beside, magnetically resonant locations. To cite a few examples, reductions of transport in the O-point of magnetic islands have been identified in LHD [3] and rational surfaces were found to possibly play an important role in the formation of internal barriers in high electron temperature stellarator/heliotron plasmas [4,5]. Further indication that magnetic resonances need not be detrimental for confinement is the well-known relation between the effect on L-H transition quality and the edge value of the rotational transform [6,7,8]. Finally, let us recall that the importance of magnetic resonances in the formation of transport barriers is also supported by many results from tokamaks (e.g. DIII-D [9], JET [10] to cite some). It looks like no theory of toroidal magnetic confinement for fusion applications would be complete without an assessment of the role of resonant regions inside the confining volume.

This work is based on the experience accumulated with the TJ-II stellarator, a machine with ~ 1 m^3 plasma volume inmersed in ≈ 1 T magnetic fields. It is a mid-size Heliac design (high rotational transform, low magnetic shear) with the notable property of allowing for external control on the rotational transform profile over a wide range of values and forms. This capability has been exploited to investigate the role of magnetic resonances on confinement properties. In rough numbers, the device produces plasmas with densities in the range 0.5–5.0 10^{19} m^{-3}, corresponding electron temperatures 1–0.2 keV and ion temperatures in the 0.1 keV range. Typical working gas is hydrogen and the heating systems are Electron Cyclotron Resonance (up to 0.6 MW) and Neutral Beam Injection (about 1.2 MW with two opposing tangential injectors). This work is a quick overview of results (Section 2) from the perspective of confinement control, which we summarize in Section 3.

2. TJ-II phenomenology

As pointed above, in terms of heating and fuelling there are two broad types of plasma produced with the TJ-II device. Low density (< 10^{19} m^{-3}) and high T_e (~1 keV) correspond to ECR plasmas. NBI plasmas have much more coupled electron and ion species and densities normally well above 10^{19} m^{-3}; the density is more difficult to maintain in steady state and, other conditions fixed, is very dependent on the wall conditioning. The response of most diagnostics is different depending on the plasma type, ECR or NBI. For these reasons the results are presented according to the heating/fuelling type.

2.1. Low density ECR heated plasmas

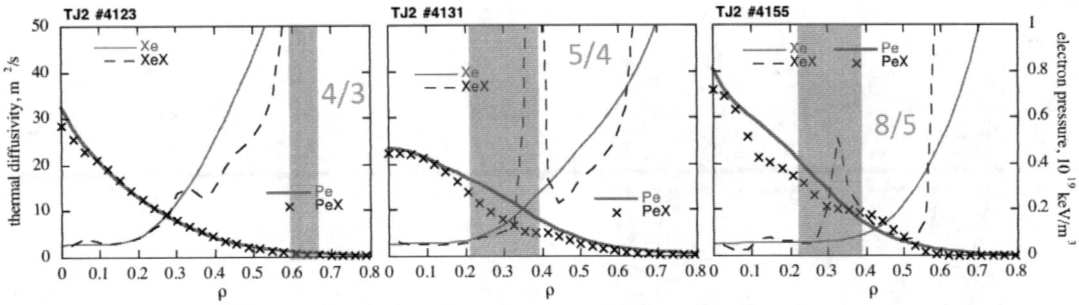

Figure 1. Profile of Thomson Scattering electron pressure (crosses) and χ_e from transport balance (dashes) for three ECR heated TJ-II discharges in magnetic configurations with resonances occupying approximately the shaded radial region. Corresponding smooth profiles from transport modelling are also shown (lines)

The effects of magnetic resonances in transport and stability have always been a matter of interest in TJ-II experiments. A systematic study about local effects due to the magnetic resonances was initiated in 2003 with the idea of clarifying the role of magnetic resonances in global transport scalings [11]. This motivated performing balance analysis based on Thomson Scattering data for the electron density and temperature profiles in different magnetic configurations. An example is shown in figure 1, where the experimental electron pressure is shown (crosses) along with the corresponding experimental electron heat diffusivity (χ_e, dashed line) for three discharges in configurations that have different magnetic resonances inside the plasma. χ_e is defined in a standard way considering the local net heat flux, electron density and electron temperature gradient scale-length. Shaded rectangles in the figures indicate the approximate location of the labelled resonances. A transport model was used to find steady states that approach the experimental electron pressure profiles (solid lines), although such features as a flattening in the electron temperature profile cannot be obtained unless some specific ingredient is included in the models.

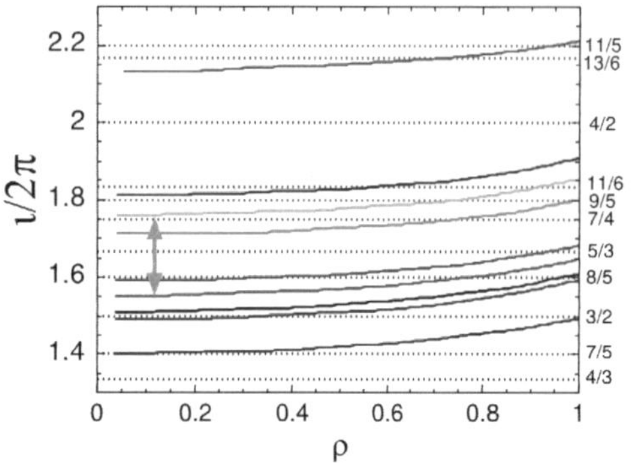

Figure 2. Vacuum rotational transform profiles calculated for different magnetic configurations of the TJ-II Heliac. Horizontal dashed lines with corresponding labels in the right axis indicate low-order rational values of $\iota/2\pi$. The crossings with the profiles mark the locations in plasma minor radius around which the magnetic resonances are expected to occur. A double arrow indicates the span of the magnetic configuration scan in figure 3.

From the results shown in figure 1 it would seem that occasional flattenings in the electron pressure profiles are due to the presence of magnetic resonances. Many more cases were analysed in this way [12] using data from configurations characterized by the vacuum $\iota/2\pi$ values shown in figure 2, but no systematic pattern was found: the eventual occurrence of flattenings may or may not be coincident with the expected location of the resonances, pointing to artifacts after fitting to the Thomson Scattering profiles. Bootstrap current density profiles were not expected to be important in these discharges with net plasma currents in the ~1 kA range. This called for a more systematic study where the magnetic configuration scan could be done in small steps, so as to place a certain resonance, say $\iota/2\pi = 8/5$, in nearby locations from one configuration to the next.

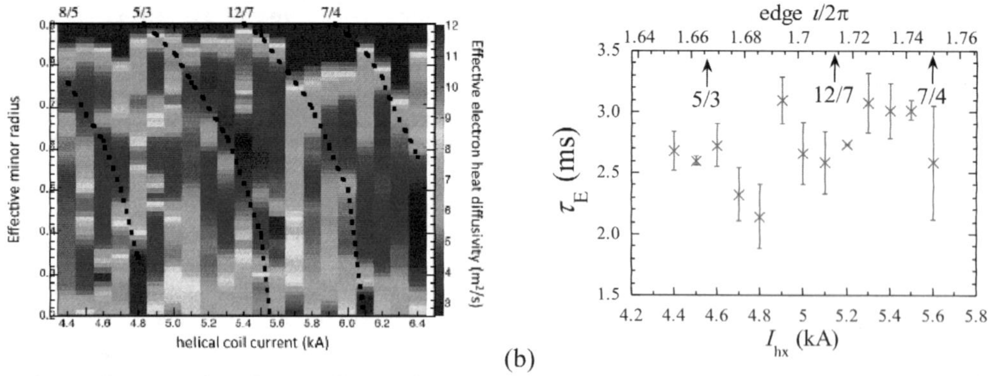

(a) (b)

Figure 3. (a) Contour plot of χ_e-profiles (color bar at right side) for different magnetic configurations characterized by their current in the helical conductor, I_{hx}. Dotted lines indicate the location of the main magnetic resonances (labelled on top). (b) Energy confinement times from balance calculations corresponding to some of the configurations shown in panel (a), as indicated with the I_{hx} values. Error bars are the standard deviation from repetition dicharges in each configuration.

Figure 3 (a) is a contour plot of effective electron thermal diffusivities (see values in color bar) obtained from a fine magnetic configuration scan covering, with 21 interleaved configurations, the range indicated with a double arrow in figure 2. Dotted lines indicate the "path" followed on minor radius (left axis) by the main vacuum magnetic resonances (labelled on top) along this scan. Abscissa values are the electric current in the helical conductor of the device, I_{hx}, which provides most of the rotational transform in TJ-II configurations and serves as an appropriate label for the scan.

The results in figure 3 (a) were obtained after averaging repetition discharges for every magnetic configuration. The analysis of such experiments [13] strongly suggested a different view of how magnetic resonances affect confinement: they can be placed essentially anywhere in the plasma despite the low magnetic shear of TJ-II plasmas and, systematically, larger temperature gradients accompany them as if a kind of barrier moved along. Of course, a precise location of the magnetic resonances is not known in these experiments. Thus, the path followed by the resonances in figure 3 (a) corresponds to vacuum locations and here we can only claim that lower elecron heat transport *follows* the resonances, but maybe beside them. Additionally, the experiments warned about the interpretation of rotational transform scalings for stellarators. Figure 3 (b) shows the average electron energy confinement time (τ_E, radiation not considered in the balance) for a part of the scan shown in figure 3 (a) as noted in the I_{hx} values. A fine structure, which somehow reminds one about old results in particle containment studies, seems to appear. The error bars in τ_E are standard deviations from repetition discharges, which unfortunately cannot account for all the uncertainties in the analysis. Despite this, the results of figure 3 (b) would be rather incomplete without the help of figure 3 (a).

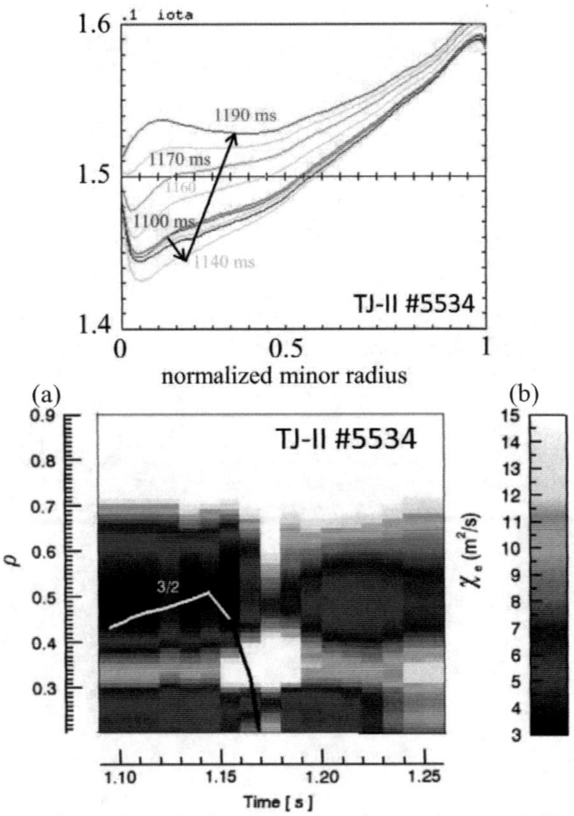

Figure 4. Time evolution of (a) the calculated rotational transform and (b) the electron diffusivity profiles obtained from heat balance, for a discharge with double ramp in the ohmic transformer.

Another motivation for these studies stemmed from the knowledge that, in dicharges with increased (negative) magnetic shear via ohmic induction, the simultaneous inclusion of several resonant layers within the plasma volume was compatible with improved confinement [11,14]. This result was well in line with W7-AS results [2], according to which the larger magnetic shear caused by the ohmic current would improve confinement. It was likewise possible to identify a so-called "signature", or characteristic evolution of the plasma signals due to the appearance and movement of the resonant layers throughout the plasma [15]. This systematic behavior suggested looking for the effect of resonant layers on transport by tracking them with the effective heat diffusivities —or, almost equivalently, the inverse gradient scale-lengths— during discharges with ohmic indiuction. Not surprisingly, the same "transport barrier effect" was found in most of the confinement zone [16] rather independently of the magnetic shear unless it was forced to be practically zero over a wide radial region [17] as figure 4 shows. There we see the calculated evolution of the rotational transform for a discharge with ohmic induction. Initially (1100 ms < t < 1140 ms) a negative current is induced, which increases the magnetic shear. Then the current ramp in the primary of the transformer is inverted provoking the induction of positive currents and an increment in $\iota/2\pi$, especially in the hotter plasma core. This process tends to force minima in the rotational transform profile. The resonance 3/2 initially present in the configuration disappears from the confining volume at approximately $t \approx 1180$ ms. The calculations of $\iota/2\pi$ cannot be accurate at this point because the flux surface coordinates are no longer a good geometrical description. However, the transport balance analysis (figure 4 (b)) clearly reflects a pronounced flattening of profiles, manifested in a very high χ_e. Interestingly, before that happens, a most prominent resonance like 3/2 is related with *lower* electron heat diffusivities (note the contour plot values around the resonant location for t < 1160 ms).

The results shown in figure 3 (a) were obtained on a shot-to-shot basis using Thomson Scattering as the main diagnostic tool, while dynamic results like figure 4 (b) were based on Electron Cyclotron Emission to allow for a practically continuous evolution of the T_e-profile. The TJ-II capability of working with variable configuration discharges [18,19,20] at close-to-vacuum magnetic shear was later used to reproduce the results of figure 3 (a) in one single discharge. The two discharges in figure 5 correspond to dynamic configuration scans where the offset of the rotational transform decreases (a) or increases (b), thus forcing respectively a movement outwards or inwards of the main resonances (see figure 1) through minor radius. These ECR plasmas are initiated around t = 1040 ms and the evolution of the net plasma currents can distort the vacuum paths (lines) of the resonances. However, the qualitative similarity with figure 3 (a) is clear: the presence of magnetic resonances does not necessarily deteriorate global confinement because they have local effects that are, as suggested in particular cases like the CERC [4], related with zones of reduced χ_e.

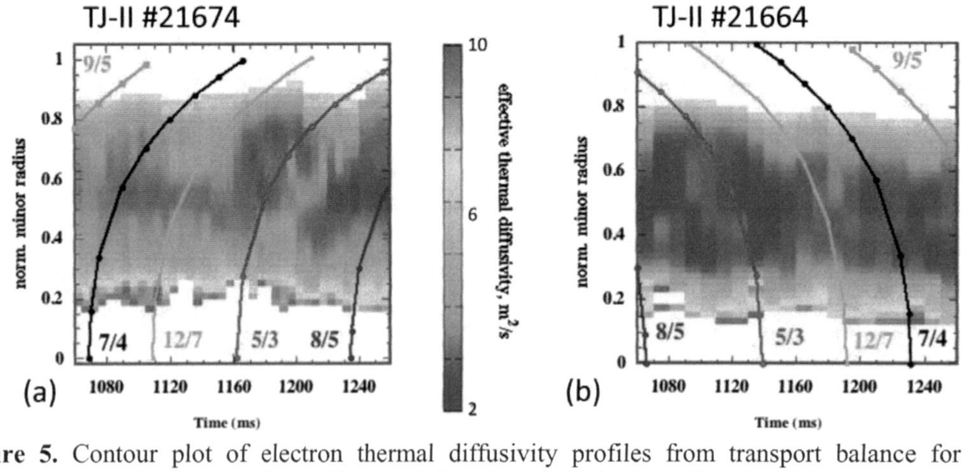

Figure 5. Contour plot of electron thermal diffusivity profiles from transport balance for two discharges with (a) downwards and (b) upwards sweepings of the rotational transform profile, which respectively cause outwards and inwards movements of the vacuum magnetic resonances through

plasma radius (lines).

The results shown up to here demanded a thorough investigation of possible causes leading to these local changes in confinement. In summary, associated with magnetic resonances, it was verified that there are local changes in plasma potential (edge electric probes) and rotation of density fluctuations (Doppler reflectometry) that convey considerable ~10^5 s^{-1} ExB flow shear. The reason why this happens is unclear despite the plausibility of some explanations. For example, the existence of the resonant regions can couple parallel and radial transport thus favoring the outwards movement of the fastest species (electrons in these cases), which would imply a tendency to electron-ion flow unbalance and the corresponding establishment of a locally strong radial electric field [21,22,23].

2.2. NBI plasmas

The studies on how magnetic resonances affect the plasma operation were extended to higher density and beta plasmas under NBI heating operation. Resonant layers were known to have a clear incidence on the access to the H-mode of confinement, thus defining "windows" in H-mode quality —or even existence— depending on the edge rotational transform [8], well in line with other stellarator devices. It was also found that, around the maximum (negative) pressure gradient region, the changes in perpendicular velocity shear depend on what magnetic resonance is placed in the region [24]. NBI plasmas do not allow for ECE measurements due to the large densities so we cannot track the presence of resonances in a way similar to the one presented for ECR plasmas. However, the technique of variable configuration discharge was once more used to provide cause-effect experiments. In this case the question was wether a magnetic resonance properly placed in the plasma could cause a transition; or, also importanly, wether it could also withdraw the plasma from the H-mode just by retiring from such appropriate plasma region. Effectively, by sweeping different magnetic resonances through NBI plasmas it was confirmed that magnetic resonances alter confinement in a sharp way (resembling typical L-H transitions) once they reach certain radial location in the main gradients zone; but such "transitioned" state disappears as the configuration change proceeds further [23]. This was proposed as a potentially powerful tool for plasma confinement control: if these results remain valid in reactor-grade plasmas, a reactor design able to move (just slightly) magnetic resonances along the plasma minor radius could have external control on the access to and from the H-mode of confinement. This is easy to conceive for a stellarator design, but also tokamaks would benefit just at the expense of using appropriate transformer control techniques [25] or other means of current drive.

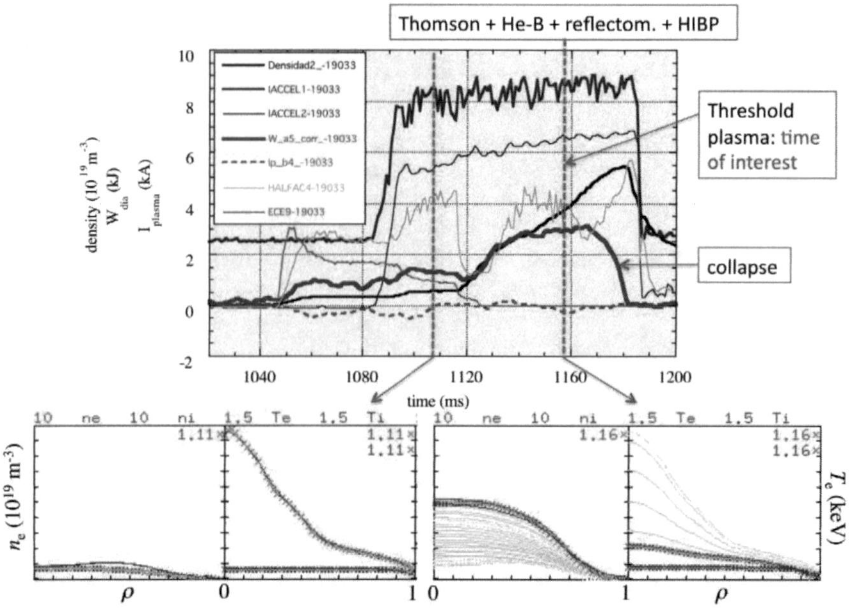

Figure 6. Time traces for balanced NBI discharge #19033. In arbitrary units: NBI currents (IACCEL), Hα light near a limiter (HALFAC4) and core electron cuclotron emission (ECE9). Also shown are the line density, diamagnetic energy and net plasma current. The discharge undergoes a transition at $t \approx 1157$ ms (vertical dashed line) soon before radiative collapse. The bottom figures show the Astra [27] outputs for the experimental (crossses) and model (lines) density and temperature profiles at t = 1115 ms (bottom left) and at the transition threshold time (bottom right). Thin light lines in the bottom right figure show the calculated evolution from t = 1115 ms according to a transport model that reproduces the evolution of the line density.

The strong effect that magnetic resonances have on the access to/from H-mode plasma seems to stem from their effect on confinement already in L-mode conditions. TJ-II plasmas heated and fuelled by NBI show a different evolution during the L-mode and have different density thresholds depending on the presence or not of magnetic resonances near the plasma edge [26]. The upper plot in figure 6 represents time traces of several TJ-II diagnostics, like beam currents (IACCEL, a.u.) from the co- and counter-NBI systems, Hα (HALFAC4, a.u.), and a central electron cyclotron emission signal (ECE9, a.u.) for TJ-II discharge #19033 with each NBI delivering around 0.45 MW (port-through power). In addition, the figure shows (left axis) line averaged density, diamagnetic energy and net plasma current. Note the very small values of the latter, from which we infer that the edge rotational transform must be very close to the vacuum one. Near t = 1160 ms there is an L-H transition, but it happens so close to the thermal collapse of the plasma that the increment in diamagnetic energy is barely notable.

Transport analysis is done in NBI discharges to study the evolution of the particle confinement time, τ_p, by means of a transport model in which the particle and energy sources are frequently updated. The bottom plots in figure 6 show the experimental profiles (crosses) and model calculations (lines) for the electron density. The temperatures for electron and proton species do not evolve from a model, but from diagnostic information (see [26] for details). The bottom right figure for the density also displays several lines in light color representing the evolution of the densities from the ECR phase to the L-H transition threshold. By reproducing the electron density evolution and checking for consistency with the diagnostics at the transition time (e.g. CX neutral particle fluxes), we obtain an estimate of the evolution of τ_p. The result is compared in figure 7 for discharges #19033 (figure 6) and #19065, which differ only on the edge rotational transform values: $\iota(a)/2\pi \approx 1.63$ and 1.61 respectively; i.e., the magnetic resonance 8/5 is placed around $\rho \approx 0.9$ in discharge #19033 but practically at the plasma edge in discharge #19065. According to figure 7 the particle confinement time is smaller during the L-mode phase for the discharge that has the resonance *inside* the plasma (#19033). This discharge reaches the transition threshold at a smaller line density. According to the small plasma current and the balanced NBI scheme, the magnetic shear must be quite small, i.e., close to the vacuum shear, in these discharges. A study of how the radial electric field is modified by the presence of a magnetic resonance in NBI plasmas is underway. For now, we have identified notable changes in the density gradient region that, as mentioned, affect transition dynamics

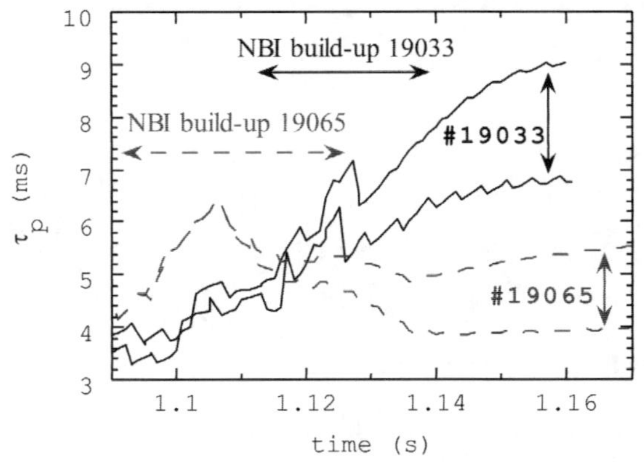

Figure 7. Calculated evolution of the electron confinement time for TJ-II discharges #19033 with $\iota(a) = 1.61$ (solid lines) and #19065 with $\iota(a) = 1.59$ (dashed lines). Two lines in each case delimit uncertainties in the calculations associated mainly with the edge temperatures.

We have mentioned a potential use of these findings for confinement control in a tokamak or stellarator reactor. Of course, our results cannot be extended beyond our plasma parameters. However, MHD dynamics in the TJ-II has been also studied keeping in mind a possible equal nature of tokamak and stellarator plasmas, particularly in matters concerning the effects of magnetic resonances on transport. At this respect, confinement events similar to tokamak phenomenology have been identified in TJ-II plasmas and related with MHD activity around the magnetic resonances [28]. In brief, and mostly based on the temporal and spatial resolution of bolometry diagnostics, magnetically resonant locations have been identified as well localized transport barriers (albeit weak) that can break suddenly causing sudden particle confinement losses. A repetitive breaking and restoration of confinement in the resonant regions provokes such events as sawteeth, internal crashes or ELM-like bursts. Up to now, an H-mode in the TJ-II looks like a state in which the barrier formed around the resonant region remains as such without subsequent failure.

3. Conclusions

In summary, the TJ-II experience is that the magnetic resonances provide another external knob because they—maybe their boundaries—modify significantly the radial fluxes and hence the plasma rotation in a local manner. This concept is quite old indeed, but we have provided a continued program to assess its practical implementation. Therefore, the notion of rarefaction of low order rational values of $\iota/2\pi$ may not imply that such magnetic resonances have to be avoided or allowed inside the plasma only with high magnetic shear. Good flux surfaces are desirable —this is not questioned at all— but isolated resonant locations, if "healed" or sufficiently narrow in minor radius extension, might become a necessary ingredient for transport and MHD control. At this respect, it is worth mentioning that a state of healed —not causing flattenings in plasma profiles— magnetic island associated to a major resonant surface is favored at *high plasma beta and low collisionality* in the LHD heliotron [29], the largest currentless magnetic confinement system to date. These are the conditions relevant for reactor plasmas.

Acknowledgments

All of the transport analyses presented here have been performed with the ASTRA transport shell [27]. This work is dedicated to the memory of G. V. Pereverzev.

References

[1] *Proc. 3rd Int. Conf. on Plasma Physics and Controlled Nuclear Fusion Research* held by the IAEA at Novosibirsk, 1–7 August 1968, Vol. 1
[2] Brakel R et al 1997 *Plasma Phys. Control. Fusion* **39** B273
[3] Inagaki S et al 2004 *Phys. Rev. Lett.* **92** 055002
[4] Castejón F et al 2004 *Nucl. Fusion* **44** 593
[5] Yokoyama M et al 2007 *Nucl. Fusion* **47** 1213
[6] Sano F et al 2005 *Nucl. Fusion* **45** 1557
[7] Wagner F et al 2005 *Phys. Plasmas* **12** 072509
[8] Estrada T et al 2010 *Contrib. Plasma Phys.* **50** 501
[9] Austin M E et al 2006 *Phys. Plasmas* **13** 082502
[10] Joffrin E et al 2002 *Plasma Phys. Control. Fusion* **44** 1739
[11] Ascasíbar E et al 2002 *Plasma Phys. Control. Fusion* **44** B307
[12] Vargas V I et al 2006 "Difusión térmica electrónica experimental en plasmas ECRH del TJ-II" Informes Técnicos Ciemat 1079, Ciemat, Madrid, Spain, July 2006 (in Spanish)
[13] Vargas V I et al 2007 *Nucl. Fusion* **47** 1367
[14] Romero J A et al 2003 *Nucl. Fusion* **43** 387
[15] López-Bruna D et al 2004 *Nucl. Fusion* **44** 645
[16] López-Bruna D et al 2008 *Europhys. Lett.* **82** 65002
[17] Ascasíbar E et al 2008 *Plasma Fusion Res.* **3** S1004
[18] Romero J A et al 36th EPS Conference on Plasma Phys. Sofia, June 29 - July 3, 2009 ECA Vol. 33E, P-4.183 (2009)
[19] López-Bruna D et al 2009 *Nucl. Fusion* **49** 085016
[20] López-Bruna D et al 2010 *Contrib. Plasma Phys.* **50** 600–604
[21] Bondarenko O et al 2010 *Contrib. Plasma Phys.* **50** 605–609
[22] Pedrosa M A et al 3rd EFDA Transport Topical Group Meeting & 15th EU-US Transport Task Force Meeting, Córdoba, Spain, September 7–10, 2010
[23] López-Bruna D et al 2011 *Plasma Phys. Control. Fusion* **53** 124022
[24] Estrada T et al 2009 *Plasma Phys. Control. Fusion* **51** 124015
[25] Romero J A et al 2012 *Nucl. Fusion* **52** 023019
[26] López-Bruna D et al 2013 *Plasma Phys. Control. Fusion* **55** 015001
[27] Pereverzev G V and Yushmanov P N 2002 "ASTRA: Automated System for TRansport Analysis," Tech. Rep. IPP 5/98, Max Plank Institut für Plasmaphysik, Garching, February 2002
[28] López-Bruna D et al 2013 *Nucl. Fusion* **53** 073051
[29] Narushima Y et al 2008 *Nucl. Fusion* **48** 075010

Identification of a Li III line relevant for TJ-II plasma diagnostics by passive spectroscopy

C. Quirós, A. Baciero, B. Zurro

Laboratorio Nacional de Fusión por Confinamiento Magnético, Asociación Euratom-CIEMAT, Madrid, Spain.

Abstract. The purpose of this work was to identify the Li III spectral line located at 4498 Å. First, the wavelength of an intense line lying close to 4498 Å was examined and compared with the information provided by different atomic databases. Some theoretical predictions were performed in order to simulate the behaviour of this line under different heating conditions. Afterwards, the spatial behaviour of this line was analysed and compared with other well-isolated spectral lines belonging to already-identified elements. Finally, an analysis of the temperature and intensity evolution from this line was performed. The results show a complex behaviour of the line, which is mainly dominated by Li.

1. Introduction

Spectroscopy is one of the most powerful techniques used to measure different plasma parameters. Passive spectroscopic methods rely on the line emission by impurities present in plasma. Most of these impurities come from the plasma facing components (PFC) of the reactor. In the TJ-II stellarator, the PFC have changed over the years. The stainless steel wall has been coated with a boron layer followed by a lithium one [1]. Shortly, the carbon tiles will be removed so there is an urgent need to identify new spectral lines in order to continue performing spectral diagnostics into the TJ-II plasma core. Spectral measurements show a prominent line close to 4498.4 Å whose identification is the main purpose of this work. This line could be from Li III located at 4498 Å [2,3] formed by charge exchange (CX) recombination with either cold neutrals whose population is maximum at the edge, or fast neutrals from the neutral beam injection (NBI) system, which penetrate into the plasma core [4]. This line is formed as follows: $Li^{3+} + H^0 \rightarrow (Li^{(2)+})^* + H^+ \rightarrow Li^{2+} + H^+ + h\nu$, where $(Li^{2+})^*$ is an excited state (n,l).

The main focus of this paper is to identify the line located at 4498.4 Å. To do this a wavelength analysis of the line was performed, as well as a space-time resolved study of its intensity and its intensity and temperature evolution over time.

TJ-II is a four period low magnetic shear medium size heliac type stellarator with a major radius of 1.5 m and an average minor radius of ≤ 0.22 m. A more detailed description can be found in [4]. To perform the wavelength analysis high-resolution mono and multi-channel spectrometers, described in [5,6], were used. The spectrometer explained in [7] was used for spatially resolved line emission. Finally, the system explained in [8,9] was used to perform the temporal evolution measurements.

2. Experimental Results

2.1. Wavelength analysis with spatial resolution

The first step followed in this work was to examine the wavelength of the line by performing a proper calibration and comparing its wavelength with the information found on different databases [2,3]. The atomic elements investigated were those that are most likely to be present in TJ-II plasmas. Elements with low relative line intensity, or whose strongest line is not present in TJ-II spectra, were not taken into account. The spectral line located at 4487.05 Å was successfully identified as B III. However, the line at 4498.4 Å could either be B III (4497.73 Å), C VI (4498.90 Å) or Li III (several wavelengths close to 4498 Å and one at 4500 Å due to excitation and recombination).

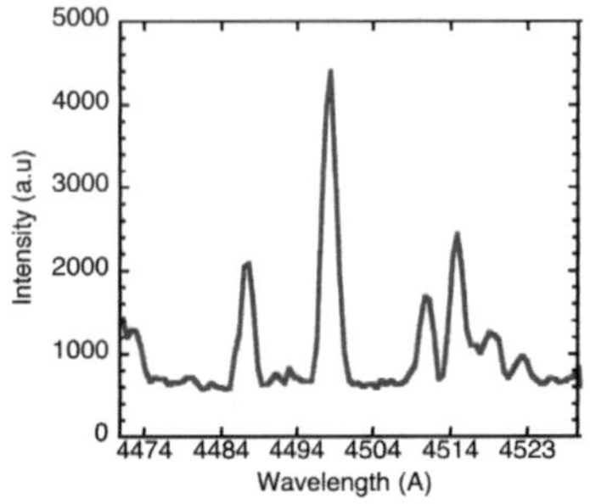

Figure 1: Spectrum obtained on a TJ-II plasma heated by NBI.

The next step consisted in doing a theoretical prediction of the behaviour of the Li line of interest during a NBI or ECRH plasma. We assumed solely a Doppler broadening with a Maxwellian velocity distribution as described in [10]. The intensity for all CX transitions with high-energy neutrals was assumed to be higher than for the excitation and recombination transitions, as detailed in [11].

2.1.1. NBI case

In this scenario there are fast energetic neutrals being injected into the plasma, which provoke CX reactions between stripped ions and fast energetic neutrals. Therefore, the CX processes mostly take place in the plasma core while the excitation and recombination processes should take place in the plasma periphery. The values of the temperatures are based on typical TJ-II parameters. The transitions for CX were summed and plotted along with one obtained for excitation and recombination (see Fig. 2). If all the transitions of Li III are occurring at an equal rate, the two peaks will not be easily distinguished in a NBI heated plasma.

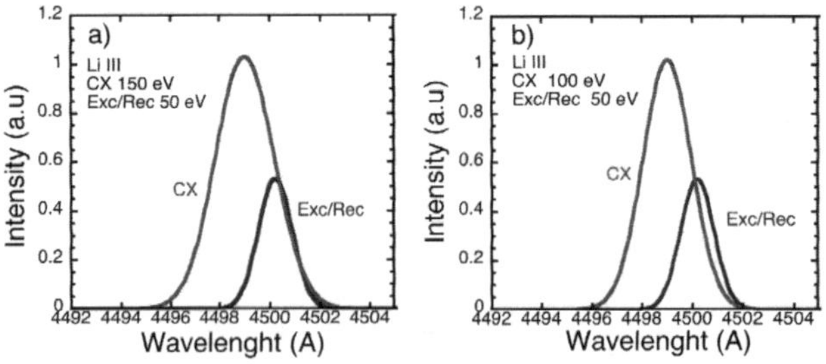

Figure 2: Gaussian calculated from the sum of all possible transitions (n=5 → n=4) for the Li III line formed by CX located around 4498.4 Å (red) and the one formed by the transition (n=5 → n=4) for excitation and recombination located at 4500 Å (blue). Theoretical temperatures are shown in the graph.

2.1.2 ECRH case

In this case the density of neutrals in the plasma core is negligible and the CX processes are assumed to be occurring mainly with the cold neutrals in the plasma periphery. Two cases are assumed: the first one consists on the excitation and recombination processes occurring closer to the plasma centre and the second one where they are assumed to take place in the plasma periphery. As in the previous case, the values of the temperatures are based on typical TJ-II parameters. The two cases are plotted in Fig. 3 (a) and 3 (b), respectively. If the excitation and recombination processes are taking place closer to the plasma centre, the line at 4500 Å can be easily distinguished from the 4498.4 Å one, whereas the lines blend if both processes are occurring in the plasma periphery.

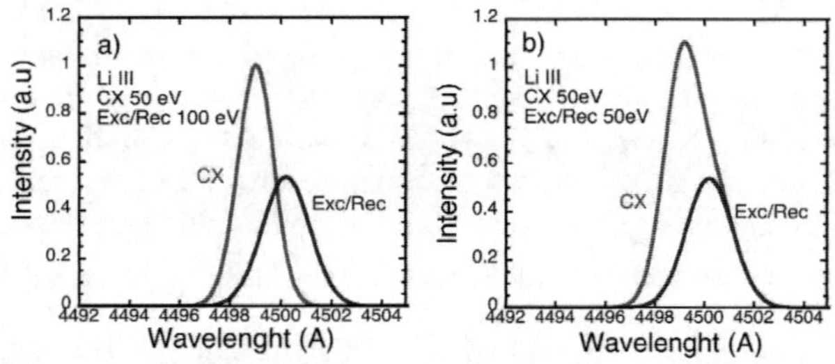

Figure 3: Gaussian calculated from the sum of all possible transitions (n=5 → n=4) for the Li III line formed by CX located around 4498.4 Å (red) and the one formed by the transition (n=5 → n=4) for excitation and recombination located at 4500 Å (blue). Theoretical temperatures are shown in the graph.

2.2. Spatial resolved study of the line intensity

2.2.1 ECRH case

Measurements were taken for two previously identified lines He I and B III and they were compared with the line of interest at three different phases of the discharge as shown in Fig. 4. The trace of B III shows a similar behaviour to He I; nevertheless its trace is narrower than the He I trace giving the clue that this element is peaking inner than neutral He I. To fully understand these traces, on has to keep in mind that the system maps the plasma from the "upper" edge of the plasma to the "lower" edge going through the plasma centre. In the He I case, the intensity valley in the central chords corresponds to a negligible concentration of this element in the plasma core. This is the typical shape for an ion that lies in the periphery. Fig. 4 (a) and (b) show how the emission corresponding to 4498.4 Å lies a bit closer to the plasma core than He I. In this case, the ion is not in the outermost part of the plasma periphery, but is not in the plasma core either. This fact also suggests that electron excitation and recombination rather than CX might be the working mechanisms in this case. This is consistent with the behaviour of a Li III peak formed by charged exchange with cold neutrals since these reactions should not take place in the plasma core.

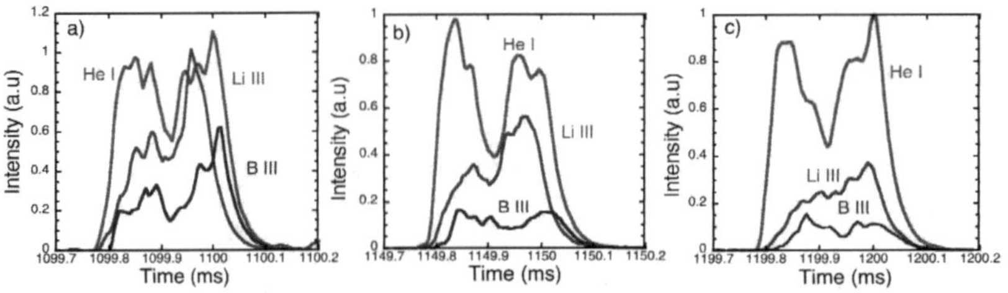

Figure 4: Spatial resolved study of He I (5486 Å) (red), B III (4487 Å) (blue) and the 4498.4 Å (green) spectral line at approximately (a) 1100 ms (b) 1150 ms (c) 1200 ms.

2.2.2 NBI case

Measurements were taken for two previously identified lines C VI, B III and its behaviour was compared with the line of interest at three different phases of the discharge: before, during and after the NBI injection. It is seen in Fig. 5 (a) that C VI does not emit entirely in the plasma core. After the NBI is injected, its emission mostly comes from the plasma centre as can be seen in Fig. 5 (c), which shows a typical shape for an ion emitting from the plasma centre. On the other hand, the B III intensity profile shows an irregular behaviour. Fig. 5 (a) also shows that at first the 4498.4 Å line emissions do not peak at the plasma core and after the NBI is injected (Fig. 5 (b) and (c)) it emits closer to the plasma core. This behaviour is expected for a Li III ion undergoing CX recombination with highly energetic neutrals.

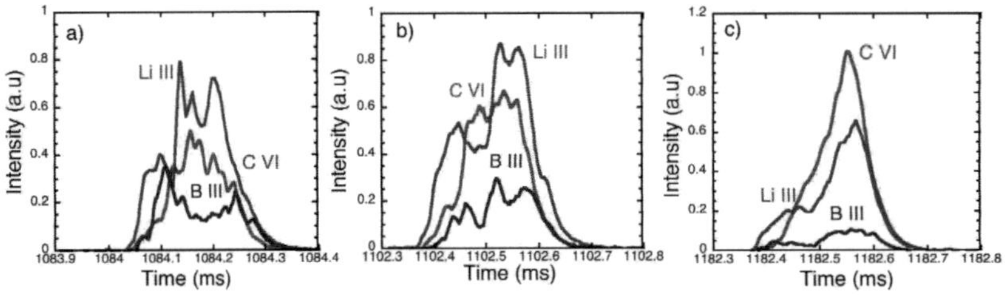

Figure 5: Spatial resolved study of C VI (5292 Å) (red), B III (4487 Å) (blue) and the 4498.4 Å spectral line (green) at approximately: (a) 1080 ms (b) 1102 ms (c) 1182 ms.

2.3. Measurement of the spatial evolution of ion temperatures

This section presents the analysis of the temperature and intensity evolution over time of the 4498.4 Å line. To perform this analysis the line was assumed to be either pure Li III, or a blend of different elements. The temperature and intensity evolution over time were calculated as explained in [7].

2.3.1 ECRH case

In this case, the line was analysed as being solely Li III. The temperature and intensity were plotted against time, as shown in Fig. 6 (b). The temperature, about 25 eV, is consistent with the peripheral ion temperatures found in TJ-II. Either charge exchange with cold neutrals or electron excitation could explain this emission.

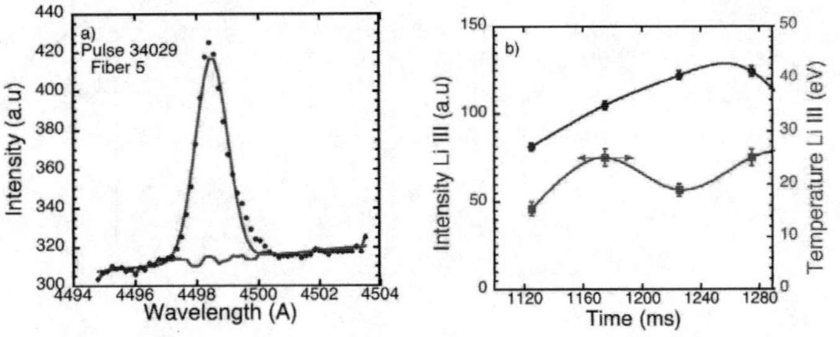

Figure 6: (a) The 4498.4 Å peak measured in first order (b) Intensity (blue) and temperature (red) evolution over time. The horizontal arrow represents the integration time, while the vertical bars represent the error percentage in the measurements.

2.3.2 NBI case

In this case, the line was analysed as being solely Li III. It can be seen, on Fig. 7 (c), how the line intensity starts to increase at 1150 ms. This increase coincides with the NBI injection. Notice that the maximum temperature reaches higher values than those seen in the previous case for ECRH.

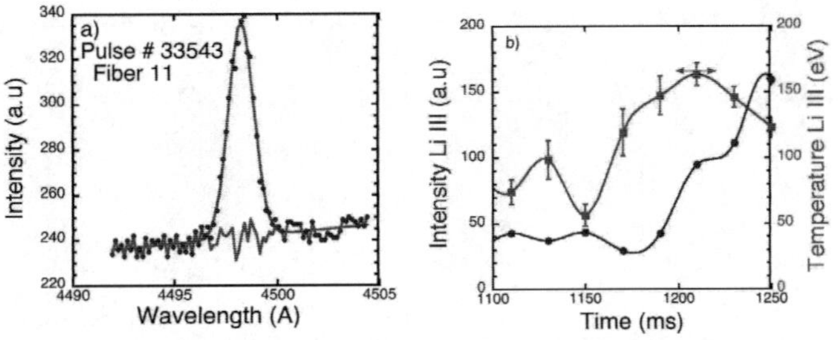

Figure 7: (a) The 4498.4 Å peak measured in first order. Assuming it was only Li III, it was analyzed using just one Gaussian. (b) Intensity (blue) and temperature (red) evolution over time when the peak is assumed to be only Li III.

It was intended to analyze these peaks using two Gaussians; however it could not be easily performed and the results obtained did not have a reasonable physical background. Therefore, the line was measured in second order with the purpose of examining the line with higher spectral resolution.

2.3.3 ECRH case in second order diffraction

These measurements show that the 4498.4 Å line is a combination of at least two peaks (Fig. 8). Hence, different assumptions were made concerning the nature of these peaks. First, it was assumed that there is a Li III peak at 4498.4 Å formed by CX and another one at 4500.2 Å due to excitation and recombination [3]. It can be seen from Fig. 9 that the temperature for the excitation and recombination processes is higher than for the CX processes, meaning that the former are occurring closer to the plasma core. Conversely, the intensity for the CX processes is higher. This result matches quite well with the results obtained with the theoretical predictions explained in section 2.1.

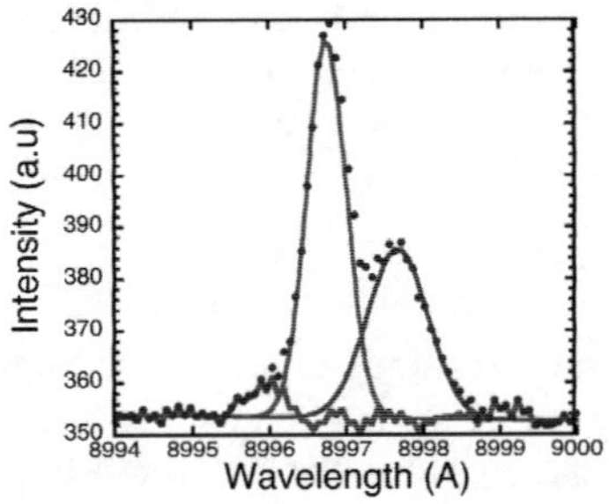

Figure 8: The 4498.4 Å peak measured in second order, showing two peaks. It was adjusted using two Gaussians.

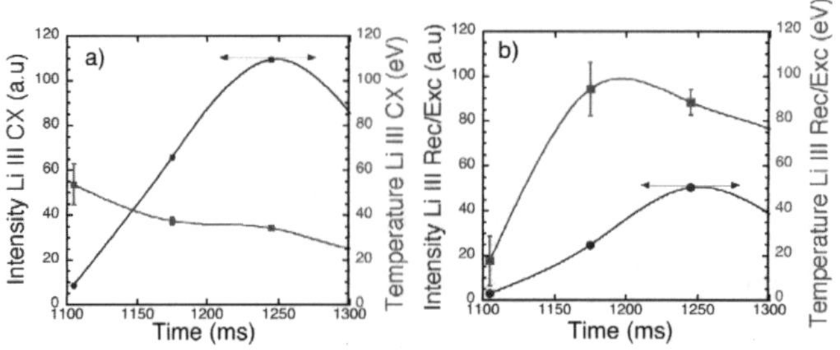

Figure 9: Results obtained under the assumption that the peaks correspond to two different Li III lines; the temperature (red) and intensity (blue) evolution of (a) Li III at 4498.4 Å undergoing CX (b) Li III at 4500.2 Å undergoing excitation and recombination. The blue horizontal arrow represents the integration time while the vertical bars the error percentage.

The second assumption was based on the fact that there is a B III line with a theoretical intensity of about 20% of the well isolated one located at 4487.6 Å [3]. Fig. 10 shows how a higher temperature for B III. The intensity of both lines shows a similar behaviour, Li III having a higher intensity.

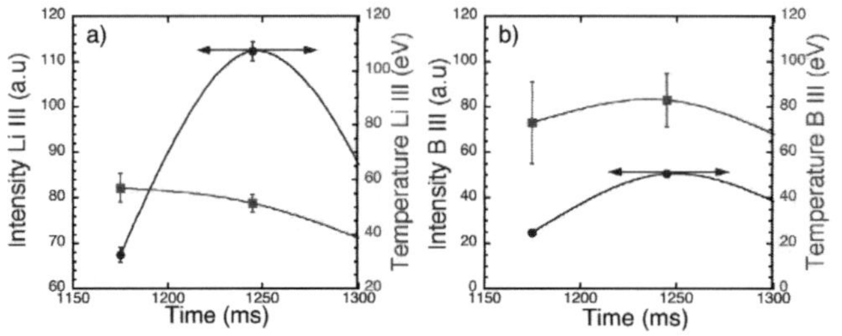

Figure 9: Results obtained under the assumption that the peaks correspond to Li III and B III; the temperature (red) and intensity (blue) evolution of (a) Li III at 4498.4 Å undergoing CX (b) B III at 4497 Å. The blue horizontal arrow represents the integration time while the vertical bars the error percentage.

Finally, the presence of C VI, which occurs at 4498.9 Å was analysed [2]. Fig 11 shows a higher temperature for Li III, however its temperature is lower. This means that C VI lies closer to the plasma core than Li III.

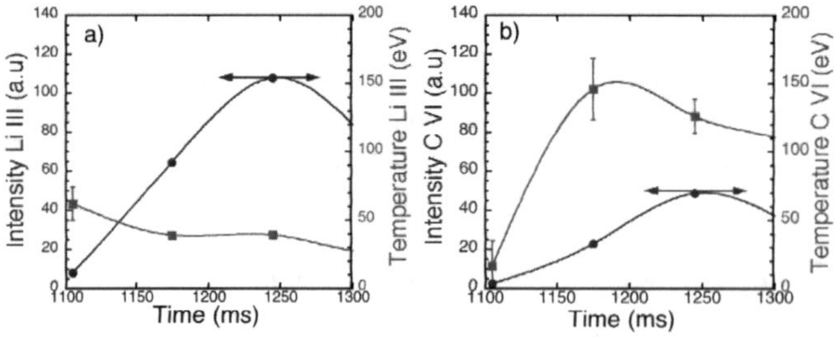

Figure 9: Results obtained under the assumption that the peaks correspond to Li III and C VI; the temperature (red) and intensity (blue) evolution of (a) Li III at 4498.4 Å undergoing CX (b) C VI at 4498.9 Å. The blue horizontal arrow represents the integration time while the vertical bars the error percentage.

3. Conclusions

The Gaussian fit performed for the different transitions of Li III shows how this line has a complex behaviour. This behaviour depends on the ion temperature and the location where the processes are taking place. If the excitation and recombination processes occur at a higher temperature, or closer to the plasma core than the CX processes, it would be easier to identify both lines. The results also show that the behaviour of this line is mainly dominated by Li III. However, a major source of uncertainty in the process to accurately identify this line is caused by the CV line located at a similar wavelength as Li III. Therefore, to fully understand the different atomic processes that may influence the behaviour of this line, it is recommended to perform measurements with an adequate charge exchange system.

References

[1] Tafalla D, Tabarés F, and Ferreira J 2010 *Fusion Eng. Des.* **85** 915
[2] http://physics.nist.gov
[3] http://open.adas.ac.uk
[4] Alejaldre C 2000 *Plasma Phys. Contr. Fusion* **41** 232
[5] Peláez R, Zurro B and Baciero A 2010 *J. Phys. B: At. Mol. Opt. Phys.* **43** 1440
[6] Baciero A, Zurro B and McCarthy K 2001 Rev. Sci. Instrum. **72** 971 (2001)
[7] Baciero A, Zurro B, *34th Conference on Plasma Physics* (ECA, Warsaw, 2007), Vol. 31, p.5.
[8] D. Rapisarda, B. Zurro, and V. Tribaldos. Plasma Phys. Control Fusion **49,** 309 (2007).
[9] D. Rapisarda, B Zurro and A. Baciero. Review of Scientific Instruments **77** 33506 (2006).
[10] Hutchinson I, *Principles of plasma diagnostics* (Cambrigde University Press, United States, 2002)
[11] Zou S, Kato T, and Murakami I 2001 *Nat. Inst. Fusion Sci.* **69** 50

Construction concepts and validation of the 3D printed UST_2 modular stellarator

V Queral

Laboratorio Nacional de Fusión, CIEMAT, 28040 Madrid, Spain

E-mail: vicentemanuel.queral@ciemat.es

Abstract. High accuracy, geometric complexity and thus high cost of stellarators tend to hinder the advance of stellarator research. Nowadays, new manufacturing methods might be developed for the production of small and middle-size stellarators. The methods should demonstrate advantages with respect common fabrication methods, like casting, cutting, forging and welding, for the construction of advanced highly convoluted modular stellarators. UST_2 is a small modular three period quasi-isodynamic stellarator of major radius 0.26 m and plasma volume 10 litres being currently built to validate additive manufacturing (3D printing) for stellarator construction. The modular coils are wound in grooves defined on six 3D printed half period frames designed as light truss structures filled by a strong filler. A geometrically simple assembling configuration has been concocted for UST_2 so as to try to lower the cost of the device while keeping the positioning accuracy of the different elements. The paper summarizes the construction and assembling concepts developed, the devised positioning methodology, the design of the coil frames and positioning elements and, an initial validation of the assembling of the components.

1. Introduction

Advanced stellarators are geometrically complex and simultaneously require high accuracy. Such requirements imply difficulties during fabrication and assembly, resulting high cost and long fabrication periods. The development of new enhanced manufacturing methods and assembling configurations may contribute to accelerate the advance of the stellarator research line.

Additive manufacturing is becoming cost-effective for direct final manufacturing of relatively small, geometrically complex parts when few identical items or single customized items are required. Fusion devices, particularly stellarators, are suited to such conditions, except for the size. Therefore, developing and validating 3D printing methods, particularly developed for fusion device manufacturing, may be useful. The use of additive manufacturing for stellarators has been previously proposed, [1].

Moreover, assembling and positioning the different stellarator components last long, contributing to the final cost and long construction period. The development of improved simpler positioning methods is essential for a faster assembling of advanced highly convoluted modular stellarators.

In this framework, the paper describes the construction concepts developed and the devised positioning methodology applied to the manufacturing of the UST_2 modular stellarator.

UST_2 is a small modular three period quasi-isodynamic stellarator based on the last closed flux surface (LCFS) of an optimised Quasi-isodynamic stellarator of 3 periods, [2]. The original LCFS is modified to obtain some enhanced engineering features. The torsion of the LCFS at the middle of the 3 straight mirror sections of the stellarator is eliminated and the torsion is increased in the remaining sections. A re-optimization process is carried out, starting from the original optimized magnetic configuration [2], to keep a reasonable neoclassical confinement, magnetic well in vacuum and lack of islands from low order rationals. Therefore, the coils at the middle of the straight mirror sections can be designed as non-circular planar coils and located far from the plasma, figure 1. Thus, the straight sections of the stellarator may contain wide ports for easy, fast and wide in-vessel access, and large volume for possible future divertors. The system is a combination of concepts somewhat similar to the ones proposed in Refs. [3].

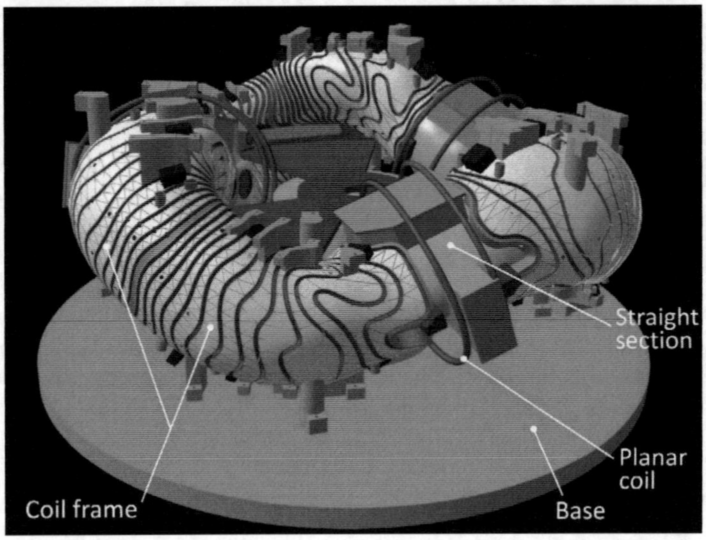

Figure 1. Concept and design of UST_2 stellarator.

A combination of 3D printing and moulding is used for the construction of the coil frames. The coil frames (figure 1 and figure 2) are the structures supporting the convoluted coils of the stellarator. The 3D printed piece for each coil frame is designed as a truss structure enclosed in a thin external surface. The internal volume of the 3D printed piece is filled with a material able to solidify, like resin or a low melting point metallic alloy. This method reduces 3D printed costs and increases the strength of the component. The method is currently applicable to small stellarators since large and accurate 3D printers are unavailable. However, it may be applicable to middle size or large stellarators in a next future.

The work is the continuation of the study and development of construction methods applied for UST_1 stellarator. UST_1 is a small stellarator designed and built by the author by a special toroidal milling machine developed on purpose, [4].

The work is funded by the author and private funds, and developed in his personal laboratory, with contribution of means and support from CIEMAT.

2. Coil frame construction concept

Each coil frame (figure 2) of the stellarator is devised as a 3D printed light hollow structure filled with a material able to solidify.

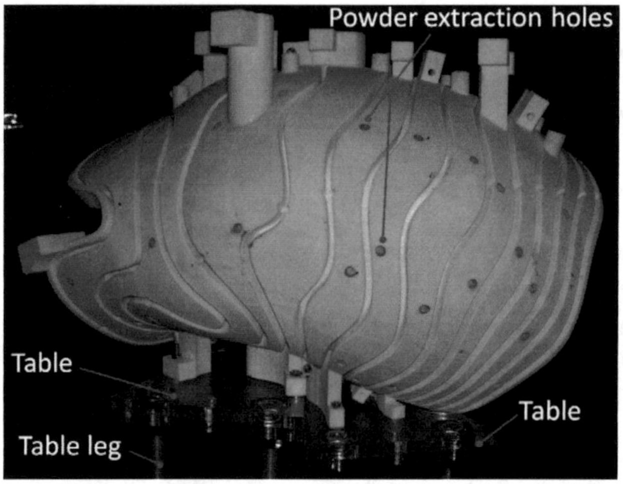

Figure 2. Photograph of the 3D printed coil frame filled with acrylic resin.

The 3D printed hollow structure is defined as a truss structure covered by a 3D printed thin closed surface equipped with upper filling openings (figure. 3) for fast and easy filling process. Four curved beams (figure. 3) define the edges of each winding pack and are parallel to the respective filamentary coil definition. The thin external surface envelops the curved beams to generate twisted grooves where the copper conductor is wound. Reinforcement bars are defined among the curved beams of the same coil and of consecutive coils. The result is a complex three-dimensional structure automatically generated by a VisualBasic Automation code for CATIA. The coil frame is divided in two halves to allow the introduction of the vacuum vessel in the coil frame. Legs are also part of the 3D printed piece in order to simplify the assembling process. Legs are designed to rest on horizontal surfaces located at different heights.

Diverse materials may be used as fillers depending on the design requirements, i.e. casting resins (epoxy, acrylic, polyurethane, etc.), fibre reinforced casting resins, low melting point alloys, type III or IV hard plasters, or fibre reinforced cements and plasters. The hollow volume is designed to allow satisfactory gravity flow of the filler as in conventional moulding or reinforced concrete buildings. The 3D printed hollow structure, acting as a mould, remains as part of the final component since the pieces are geometrically complex and only one or few identical items are produced of each piece. Producing and manipulating a mould that would allow extraction of such complex pieces would result more expensive. Consequently, the still costly 3D printing material and process is minimised and the strength of the component is supplied by the relatively low cost strong filler.

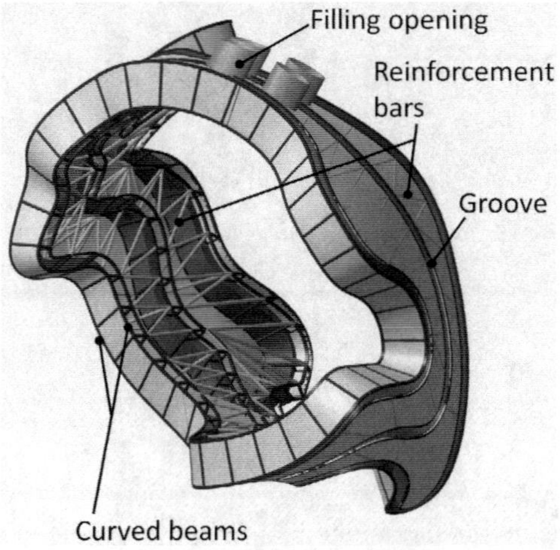

Figure 3. Design of three coils of the 3D printed coil frame for UST_2 (the internal surface is removed in the figure to show the internal bars and curved beams).

3. Assembling and positioning concepts

A series of coil frames are assembled contiguous (figure 1) and located on a mechanised circular base to form a full toroidal coil frame. All the coil frames of the torus are identical due to stellarator symmetry on condition that the legs are located both at the top and at the bottom of the coil frame, figure 2. One halfperiod coil frame of a stellarator period is rotated 180° with respect a horizontal axis and located on the respective legs to achieve stellarator symmetry, figure 6.

The positioning strategy for UST_2 coil frames is somewhat similar to the remote handling techniques used in JET and ITER. Thus, the assembling is fast but UST_2 is assembled hands-on. The inboard of the coil frame is equipped with two 3D printed positioning stops, figure 4. The stops contact on a central circular ring located in the central torus hole at the equatorial plane of the stellarator. Each coil frame is located on two tables (figure 2) of different height so as to accommodate the large vertical excursion of the magnetic axis of UST_2. Each table is formed by two flat trapezoidal horizontal plates located at the top and at the bottom of the table legs. Each table lays on a flat smooth circular base of slightly larger diameter than the external diameter of the torus, figure 1. The circular base is free of obstacles hindering the horizontal movement of the tables on the base. The coil frame, located on two contiguous tables, moves freely on the circular base until the positioning stops contact with the central circular ring. At this position, the coil frame has only one turning degree of freedom defined by the stops on the central circular ring. The position of the each coil frame is finally completely defined when all the coil frames are located around the central circular ring by successive horizontal movement of the coil frames. This assembling procedure is facilitated if straight mirror sections and planar coils between periods are designed, as in UST_2 stellarator.

Several particular features are implemented for UST_2. The central circular ring (figure 4) is formed by several bolted pieces to allow the vertical extraction of the central ring to facilitate inboard access on the coil frames. The central ring contains three positioning extensions since UST_2 is a 3 period stellarator, and it has two different diameters because the UST_2 magnetic axis viewed from the top is triangular, very different from a circle. One positioning stop contacts with the larger diameter circle of the circular ring and the other stop contacts with the smaller circle.

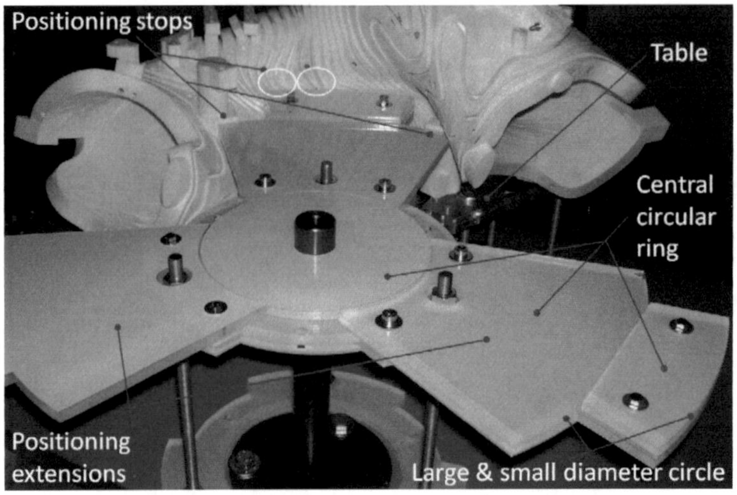

Figure 4. Central circular ring and positioning elements

4. Validation of the construction and assembling

Several components of the stellarator have been produced. Two halfperiod coil frames have been 3D printed and one has been filled with resin. The base and the central circular ring have been respectively milled and laser cut. The central circular ring has been installed on a provisional base. The vacuum vessel is still being designed and tested.

4.1. Construction and assembly of the coil frame

The first halfperiod coil frame has been produced in the company Shapeways by Selective Laser Sintering in Polyamide 12. Polyamide 12, a type of nylon, has reasonable mechanical properties and relatively low cost for Selective Laser Sintering. Tensile strength is 46 MPa and deflection temperature 177 ºC. Selective Laser Sintering has been selected against Fused Deposition Modeling (FDM). FDM is less appropriate for very hollow pieces since the abundant support material required must be printed and extracted from the interior of the piece. The cost in such cases is higher than for SLS. The SLS manufacturing cost of one UST_2 halfperiod was 700 €. Many holes of 6 and 3 mm diameter (figure 2) are designed on the external surface of the coil frame in order to allow the extraction of the nylon powder from the inside of the hollow piece. The laser melting process of the SLS manufacturing method tend to produce thermal warping. The two halves of each coil frame are 3D printed as a sole piece joined by small bars which are cut after filling both halves with the filler material. This strategy reduces the thermal warping.

Twenty small specimens 3x1.5x0.6 cm have been tested: hard plasters (Exaduro and Duro-6 plaster brands from Hëbor), cement, epoxy casting resin, polyurethane casting resin, glass fibre reinforced polyurethane resin, pita fibre reinforced hard plaster (Exaduro plaster). Viscosity, moulding operation simplicity and speed, fracture bending strength and effect of fibre reinforcement were compared. Casting acrylic resin moulds easier (less leaks, tools cleaned with water) than epoxy and polyurethane casting resins and it was selected to mould the first halfperiod coil frame. The first halfperiod of UST_2 stellarator was gravity filled with acrylic resin, brand Jesmonite, type AC-300. The tensile strength of this resin is 20 - 25 MPa and the expansion on set 0.15%. Expansion on set is low and it was favorable since the 3D printed piece resulted slightly smaller than designed. The small bars joining the two halves of the coil frame were cut after settling.

4.2. Results from the construction of the coil frame

- The deviation between the drawing dimensions and the real dimensions is lower than 0.3%. The deviation among different coil frames would be lower than 0.3% if the same 3D printer and position of the piece were used. 0.1-0.2% deviation among coil frames might be achieved. Therefore, Selective Laser Sintering with Polyamide 12 is preliminarily considered satisfactory for small stellarator construction.
- Nylon powder remained in the legs and other nooks inside the first 3D printed halfperiod. The powder was removed by drilling new holes on the surface of the halfperiod and using air compressed to extract the powder. Holes were added to the design where needed and the second 3D printed halfperiod did not contain internal nylon powders. Internal remaining nylon powders are undesirable since the casting resin does not fill the whole internal volume and strength is lower.
- Moulding the halfperiod coil frame was easy and fast. Filling the 3D printing piece lasted only 5 minutes. However, preparation of the 3D printed piece, tools and resin, and final cleaning of the components lasted about 2 hours. Preparation the 3D printed piece includes closing the powder extraction holes, leak testing of the closed volume to prevent resin leaks and, attaching the filler feeding basin to the filling openings, figure 5.

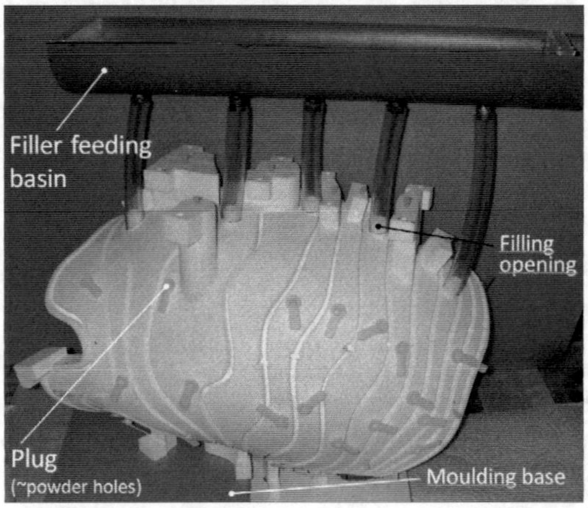

Figure 5. 3D printed coil frame prepared for moulding with resin

4.3. Results from the assembly

Figure 6 shows the current assembly status. The base is still the pre-assembling base though the definitive base has been mechanised and received. Collisions among components have not been found. Bolting standard bolts for all the coil frame legs was feasible and simple for most of the bolts. Bolting the bolts for the short table (figure 6) was more difficult than for the tall table, as expected, but the table design allowed bolting from the inside of the table. The distance between the positioning stops and the central circular ring (figure 4), at design position, has not been accurately measured since the base is not the definitive one. Coplanarity of the two or three legs corresponding to the same horizontal level is acceptable. Before moulding with resin, slight thermal warping of the coil frame legs was observed. Shim gauges of 0.2 mm can be inserted among certain coil frame legs and the table. This issue can be partially corrected if the leg bolts are attached to the moulding base (figure 5) while moulding. On the whole, the assembly of the central circular ring, tables and the two halfperiods have been satisfactory and the same design will be kept for the two remaining periods.

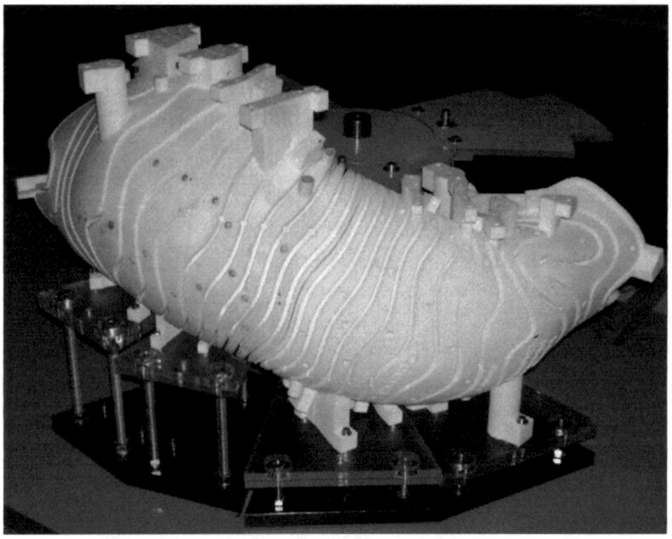

Figure 6. Construction status of UST_2 stellarator

5. Results and conclusions

Additive manufacturing combined with moulding for the construction of fusion components has been conceived and tested.

Coil frames for UST_2 stellarator are being produced to validate the construction methods developed. Each coil frame is devised as a 3D printed light hollow structure filled with a material able to solidify. The hollow structure is conceived as a truss structure covered by a 3D printed thin closed external envelop surface.

One coil frame has been moulded with acrylic resin. Filling the internal hollow volume of the coil frame with the filler was satisfactory. The issue of remaining nylon powder inside the hollow volume of the coil frame was solved for the second 3D printed halfperiod. Resin leaks did not occur.

The use of the still expensive 3D printing material is minimised, resulting a 3D printing cost for one UST_2 halfperiod of 700 €.

The assembly of UST_2 coil frames corresponding to one stellarator period has been feasible, simple and fast. The validation of the full stellarator assembly still remains.

Completion of the vacuum vessel and two periods remains for a future work.

The construction of small stellarators by 3D printing appears feasible. The feasibility will be determined by e-beam field mapping experiments after completion of the construction.

Acknowledgement

The author would like to thank the crowdfunding contributors to the project and to researchers from CIEMAT, Carlos III University of Madrid, IPP Max-Planck, ORNL and PPPL for their help with codes and for supplying LCFS definitions of different stellarators.

References

[1] Waganer L M et al 2008 ARIES-CS coil structure advanced fabrication approach *Fusion Sci. and Technol.* **54** Num. 3 878-889

[2] Mikhailov M I et al 2004 Comparison of the properties of Quasi-isodynamic configurations for different numbers of periods *Proc. 31st EPS Conf. on Plasma Phys. (London, 28 June - 2 July 2004)* ECA vol 28G P-4.166

[3] Kulygin V M, Arsenin V A, Zhiltsov V A, Zvonkov A V, Skovoroda A A and Timofeev A V 2006 Project EPSILON – a way to steady state high β fusion reactor *IAEA XXI Fusion*

Energy Conference (Chengdu, China, 16-21 October 2006) IC/P7-1

Spong D A and Harris J H 2010 New QP/QI Symmetric Stellarator Configurations *Plasma and Fusion Research: Regular Articles* **5** S2039

Imai T et al 2011 Status and plan of gamma 10 tandem mirror program *Fusion Sci. and Tech.* **59** Num. 1T 1-8

[4] Queral V 2013 Coil fabrication of the UST 1 modular stellarator and potential enhancements, *Fusion Engineering and Design* **88** Issues 6–8 683–686

Journal of Physics: Conference Series **591** (2015) 012016 doi:10.1088/1742-6596/591/1/012016

Constructing a small modular stellarator in Latin America

V.I. Vargas, J. Mora, J. Asenjo, E. Zamora, C. Otárola, L. Barillas, J. Carvajal-Godínez, J. González-Gómez, C. Soto-Soto and C. Piedras
Plasma Laboratory for Fusion Energy and Applications, Cartago, Costa Rica
Instituto Tecnológico de Costa Rica, Cartago, Costa Rica

E-mail: ivanvargasblanco@gmail.com.

Abstract. This paper aims at briefly describing the design and construction issues of the stellarator of Costa Rica 1 (SCR-1). The SCR-1 is a small modular stellarator for magnetic confinement of plasma developed by the Plasma Laboratory for Fusion Energy and Applications of the Instituto Tecnológico de Costa Rica (ITCR). SCR-1 will be a 2-field period small modular stellarator with an aspect ratio > 4.4; low shear configuration with core and edge rotational transform equal to 0.32 and 0.28; it will hold plasma in a 6061-T6 aluminum torus shaped vacuum vessel with an minor plasma radius 54.11 mm, a volume of 13.76 liters (0.01 m^3), and major radius R = 238 mm. Plasma will be confined in the volume by on axis magnetic field 43.8 mT generated by 12 modular coils with 6 turns each, carrying a current of 767.8 A per turn providing a total toroidal field (TF) current of 4.6 kA-turn per coil. The coils will be supplied by a bank of cell batteries of 120 V. Typical length of the plasma pulse will be between 4 s to 10 s. The SCR-1 plasmas will be heated by ECH second harmonic at 2.45 GHz with a plasma density cut-off value of 7.45×10^{16} m^{-3}. Two magnetrons with a maximum output power of 2 kW and 3 kW will be used.

Keywords. Magnetic confinement; Stellarator; low shear configuration; small modular stellarator

1. Introduction

Stellarators are toroidal devices where the required rotational transform of the magnetic field lines (needed to confine the plasma) is generated by external field coils and not via an induced net toroidal plasma current as in Tokamaks. This confinement scheme has the advantages that, in principle, steady-state plasma operation is possible and the machine does not have to brace itself against the strong impulses generated by short pulses of high current, such as the halo-currents generated by plasma disruptions in Tokamaks. At the cost of the increased complexity in toroidal asymmetry, the properties of the Stellarator's magnetic geometry can be tailored to suit reactor needs. Research focusing on the plasma confinement properties of different Stellarator fields and their suitability for reactor sized devices is an area of ongoing research.

In a toroidal device, the magnetic field lines need to be helically twisted in order to prevent polarization of the plasma due to the opposite vertical directions of ions and electrons curvature and gradient drifts that would otherwise prevent confinement [1]. In a Stellarator the net toroidal current that is required to be carried the plasma is zero and the confining magnetic field is generated solely by external field coils. This requires some of the coils to helically revolve around the plasma.

The goal of Stellarator research —as well as tokamak research— is to prove that the concept is suitable for a fusion reactor. Thus it is necessary to determine the magnetic field structure that can confine a plasma at sufficiently high density, n, and temperature, T, with sufficiently long energy confinement time τ_E, in order to meet the Lawson criterion [2]. Large Stellarator experiments have been developed to explore the attractiveness of the concept: for example, ATF in Oak Ridge, USA (R = 2.1 m); CHS (R = 1.0 m), LHD (R = 3.9 m) and CHS-qa (R = 1.5 m) in Tokio, Japan; Heliotron E (R = 2.2 m) and Heliotron J (R = 1.2 m) in Kyoto, Japan; TJ-II (R = 1.5 m) in Spain, W7-AS (R = 2.0 m) and W7-X (R= 5.5 m) in Germany, HSX (R = 1.2 m) and QPS (R = 0.9 m) in USA.

Recently, there has been an interest on developing small low cost Stellarators (major radius less than 0.9 m). A small Stellarator (R=119 mm) called Ultra Small Torus (UST_1) [3] was designed, built and operated by the Spanish engineer Vicente M. Queral on a budget of under $4000, demonstrating for the first time that low-cost techniques can be used to build small, functioning Stellarators.

UST_1 is a 2-field period modular Stellarator with an aspect ratio ≈ 6, formed by 12 partially optimized modular coils. Each coil is formed by 6 turns of flexible copper conductor wound in a groove machined in a circular toroidal plaster frame by a specially designed toroidal milling machine. Electron cyclotron frequency heating (ECH) at the second harmonic (B_0 = 46 mT) heats the plasma using a 0.8 kW, 2.45GHz commercial magnetron. Typical length of the plasma pulse is 2 s. Toroidal field (TF) current per coil is 2.3 kA-turn. Also the plasma volume is 1.1 liters, major radius R = 119.2 mm, average minor radius a ≈ 21 mm. Low shear configuration with core and edge rotational transform equal to 0.32 and 0.28, optimized to occupy a narrow range just below 1/3 in order to avoid high-order rationals and large magnetic islands. Additionally, UST_1 was optimized for other important plasma parameters, such as large plasma size, deep magnetic well, low ripple, and low variance of the minima of |B|. Optimization is modest because the coils are constrained to lie on a circular torus [3]. Plasma parameters for this small device deduced from ISS04v1 [4] with B_0 = 0.1 T, enhancement factor = 0.1, and P_{ECH} = 400 W are very modest, on the order of $T_e \sim 2$ eV, $n_e \sim 2\times10^{17}$ m^{-3}, and $\tau_E \sim 0.2$ μs, with $\beta \sim 0$ [3].

This paper presents the design of a small modular Stellarator that uses the same magnetic configuration as the UST_1. Stellarator of Costa Rica 1 (SCR-1) is built to a scale twice the size of UST_1 in length dimensions, stronger magnetic field (0.0438 T), increased heating power (5 kW), and has an minor plasma radius (54.11 mm), plasma volume of 13.76 liters (0.01 m^3), estimated electron temperature (13 eV) and electron density (5×10^{16} m^{-3}).

Figure 1. SCR-1 vacuum vessel drawing.

The main objectives of SCR-1 are to improve the engineering of the UST_1 device, focus on training human resources, identifying of problems related to the design and construction of small modular stellarators and also investigate the plasma physics in a small modular stellarator (R = 238 mm).

This article is organized as follows. The system description is described in detail in section 2, where each part of the SCR-1 will be presented, including its characteristics and parameters, and a description of engineering approach taken and problems overcome by the authors during the designing of the device, and their solutions. Expected plasma parameters are also presented. Conclusions are summarized in section 3.

2. System description

2.1. Organization of the SCR-1 Project (Heading 2)
The structure of the project involves about 4 engineering and physics undergraduate students, mainly from the ITCR, 3 supervisor engineers, under the supervision of two PhDs in plasma physic and nuclear fusion.
The project is divided into the following areas: Coil Systems and Layer Materials, Vacuum Systems, Power Supply Systems, Heating System, Diagnostics, Safety, Data Acquisition and Control Systems, the area of Magnetic Fields, Simulation and Modeling; and finally, the Administrative, Technical, and General Supervising areas.

2.2. Vacuum Systems
The torus-shaped vacuum vessel of the SCR-1 will be made of 6061-T6 aluminum. Although using austenitic 304L grade stainless steel was analyzed, it was discarded because of the difficulty to manufacture parts according to the device dimensions and the vacuum vessel price is higher.

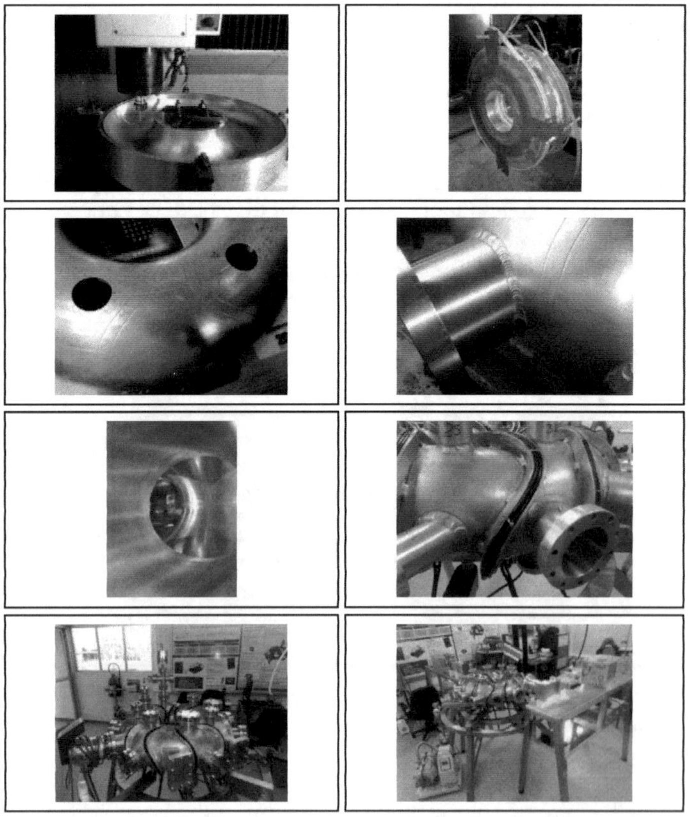

Figure 2. Stages of the vacuum vessel construction.

The vacuum vessel of 10 mm thickness will have a volume of 0.0434 m^3, with an external radius of 364.1 mm, internal radius of 112.1 mm and major radius R = 238.1 mm. Since the vacuum vessel must support a minimum pressure of 10^{-6} Torr (10^{-6} mbar), it will be used 6061-T6 aluminum in TIG welding Process.

The vessel will have 24 conflat ports with diametric dimensions of 6" CF, 4-1/2" CF, 3-3/8" CF and two rectangular shaped; available for different applications, as seen in Fig. 1 and Fig. 2.

The vacuum vessel design includes more ports than required by diagnostics, in order to support the future incorporation of more components if required.

For the construction of the torus shape two aluminum parts were melted subsequently they were polished on the inside by using a multi-axis CNC machine. Once the above was completed, the pieces were welded using the TIG welding process as shown in Figs. 2. Finally they were performed external polishing and liquid penetrant and porosity tests were done. CF Ports were also machined of aluminum as shown Fig 2. It has decided to use the same design the aluminum CF port of the vacuum vessel for a Cusp Confinement Plasma call MPDX (Madison Plasma Dynamo Experiment) from the University of Wisconsin, Madison [5].

The main vacuum system component is the vacuum pumps group, which is composed by one mechanical pump able to reach 10^{-4} Torr, and one turbo-molecular pump that can then achieve a further reduction in pressure to 10^{-10} Torr. An automated pump system was chosen, integrating the controllers of both pumps. This group has additional equipment such as RS485 communication, vacuum convectron, ion gauge sensors and RGA (Residual Gas Analyzers) for possible gas leak issues.

2.3. Coil Systems

The magnetic field that confines the plasma will be produced by 12 modular (irregularly shaped, non-planar) copper coils. Modular coils (see Fig. 1) in principle allow the best possible confinement of the plasma to be achieved. The geometry of the coils was obtained by engineer Vicente Queral [3], and has been optimized as discussed in section 1.

Each modular coil will have 6 turns, made of AWG#4 wire, and a current of 767.8 A per turn providing a total toroidal field (TF) current of 4.6 kA-turn per coil. It is not possible to add more turns due to the geometric constraints of the coils and vessel. Heat transfer simulation results from the modular coils are shown in Fig 3. The thermal behavior of copper wire by electrical current pass in the modular coil was simulated using COMSOL Multiphysics software and other methods. Temperature, resistance, voltage and power calculations as a function of time were performed for the electrical circuit under different wire configurations per modular coil to select the power supply taking into account the available budget. The wire configurations of 1, 2 and 4 turns per coil were discarded due to the high price of the power supply (USD > 50 k $). Also the wire configurations of 8 and 12 turns were also discarded because the clashes occurring between close coils when the height per coil increased. Finally it was decided 6 turns per coil (green color in the Fig. 3 and Fig. 4)

Active cooling systems were considered as a possible solution for the temperature rise of the coils but discarded due to the constraints imposed by the available space and complex geometry of the coils, though remain a possibility for future upgrades. Instead, temperature rises are limited to acceptable levels by increasing the number of turns, lowering the required current to maintain the same magnetic field strength; and increasing the cross section of the wire, thus lowering the resistance of the wire.

Having this in mind, authors realized that most of the difficulty of the coil system relies on the mechanical issues and not electrical. The number of extra turns and increased coil width is constrained by the available space for coils. A test bed model of one of the grooves was created and coils were wound exploring different AWG wire calibers, insulations and configurations of the turns, in order to gauge the ability of various coil configurations to be successfully implemented during the building of SCR-1. Once an optimal configuration based on these requirements was determined, the parameters of this configuration were applied to all coils on the Stellarator using a CAD program to check that the coils did not overlap, and calculations were performed to corroborate the temperature rise of the coils during the pulse are within acceptable levels. The results are shown in Figs. 3 y 4, where the current coils configuration provides 767.8 A.

For positioning of the coils in the vacuum vessel a CNC machine that generated little grooves in the vacuum vessel was used (see Fig. 2). A specialized automated machining device has been designed to cut the grooves into an acrylic material coating the torus in a pattern corresponding to the coil shapes [6], however, it was discarded because the supplier who constructed the vacuum vessel also has manufactured the coil guides using 3D printer and then casting technique. Subsequently the coil guides were welded to the vacuum vessel as shown in Fig. 2.

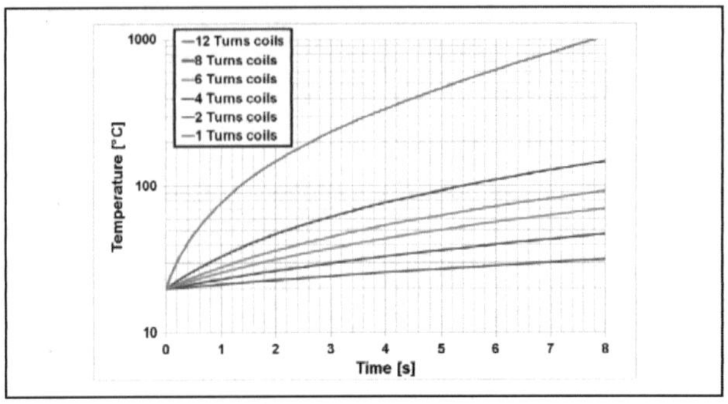

Figure 3. Temperature comparison between different coils configurations.

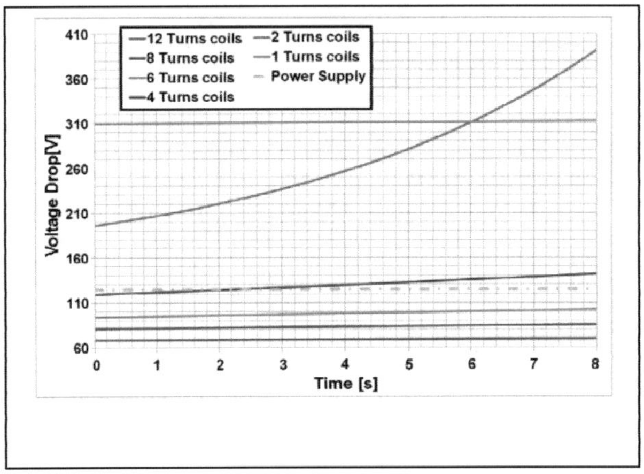

Figure 4. Voltage-Drop vs time for the battery bank of the SCR-1.

Before operation of the SCR-1, a magnetic field mapping system will be assembled to verify that the coils have been positioned correctly. This mapping system will consist of a movable thin rod impregnated with a substance (probably barium oxide or phosphor P15 or P24) that produces a fluorescent luminescence when electrons propelled by an e-gun hit with the rod. These impacts will be filmed by a CCD camera at approximately 1/10 – 1/30 frames per second; the number of frames per second chosen must be the same number of the frequency of the oscillation of the rod. Finally, the images will be superimposed to generate a final image of the vacuum magnetic surfaces, which will be compared to those calculated theoretically. This magnetic field mapping methodology was originally applied to UST_1 [3].

Journal of Physics: Conference Series **591** (2015) 012016 doi:10.1088/1742-6596/591/1/012016

2.4. Power Supplies systems

The electrical supply system consist of an array of 60 lead-acid electrochemical cells (battery bank); each cell will have a nominal voltage of 2 V and an electrical storage capacity of 150 Ah. The battery bank will have a voltage of 120 V and is capable to deliver a current of at least the required 767.8 A during one duty cycle of the SCR-1; as mentioned previously the duty cycle of SCR-1 is mainly limited by the ohmic heating experienced by the coils during operation (Fig. 3). The temperature rise leads to a rise in the coil resistance (Fig. 4), in order to maintain constant current over the estimated 4 to 10 seconds of a SCR-1 pulse, the system will use a switching current controller between the battery bank and coils system in order to allow the voltage across the coils to rise as the coils resistance increases [7]. The electrical current and electrical circuit for the SCR-1 is shown in Fig 5.

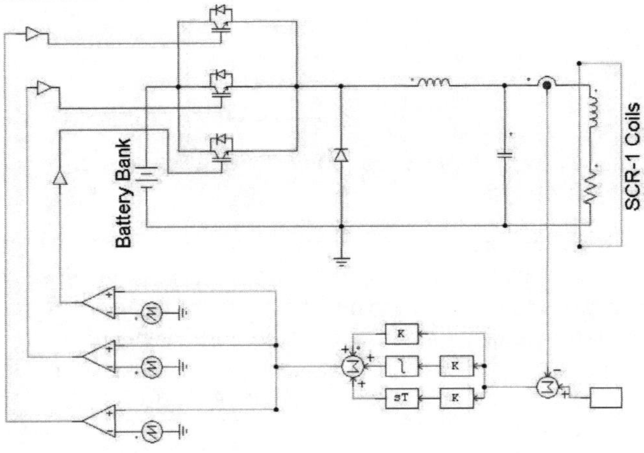

Figure 5. Electrical circuit for the SCR-1.

2.5. Heating Systems

The SCR-1 plasmas will be heated by ECH 2nd harmonic at 2.45 GHz with a plasma density cut-off value of 7.45×10^{16} m^{-3}. Two magnetrons with a maximum output power of 2 kW and 3 kW will be used, located in symmetrical positions. The heating system was specifically designed to meet the requirements of the characteristics of this small stellarator [8], with a focus on the efficiency of the energy transfer from the electromagnetic wave (EM) to the plasma bulk and reducing the reflection coefficient through impedance coupling, as well as safety and ease of measurement. Because the SCR-1 is much smaller than those commonly operated by other laboratories, the main engineering considerations have been: shot time, electron cyclotron frequency (2.45GHz) at a specific magnetic strength for the main heating frequency, and power consumption for the current supply affordable by the funding agency (ITCR).

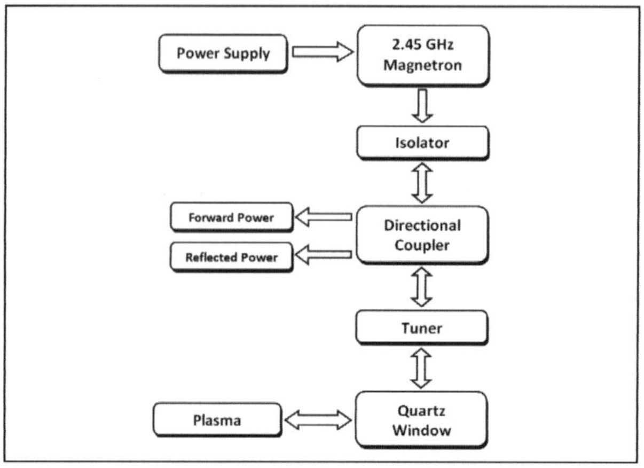

Figure 6. Simple block diagram of the Heating System of the SCR-1.

In terms of heating frequency, the wide selection and lower cost of components at 2.45GHz has defined several parameters of design such as the magnetron frequency and waveguide selection. In addition continuous wave (CW) magnetrons can be used and heating problems will not arise because of the short duration of shots (4 s to 10 s). Using the parameters discussed in previous sections, quantities such as skin depth, plasma density and others can be used to estimate the coupling of the plasma and heating power through selection of a tuner to match the impedance and avoid strong reflections, thus achieving maximum energy transfer and safety (see Fig. 6).

2.6. Heating Systems

The diagnostics consist of a Langmuir Probe, an iHR550 optical spectrometer and a Heterodyne Microwave Interferometer, specially designed for the requirements of this small scale stellarator design. A versatile remotely controlled reciprocating Langmuir probe presented in Fig. 7 (located in vertical position) has been developed for the SCR-1. The main components of the system are two removable heads, containing the measurement four tips each one, and a displacement system that enables the probe to be displaced in vacuum. Diagrams of tips and Langmuir Probe system especially designed for the SCR-1 is shown in Fig. 7.

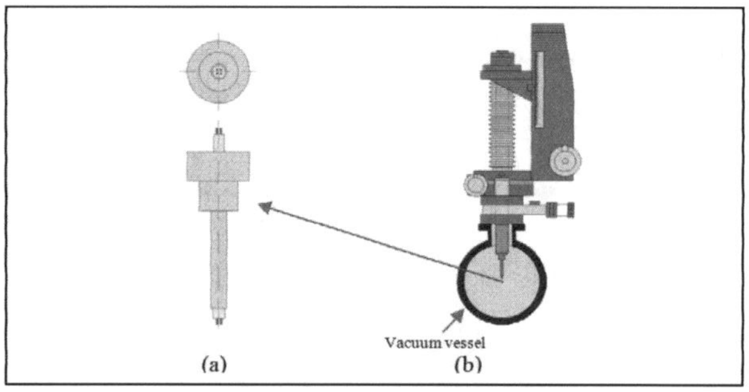

Figure 7. Diagrams of: (a) Close-up of the Langmuir probe tip. (b) Langmuir Probe system especially designed for the SCR-1.

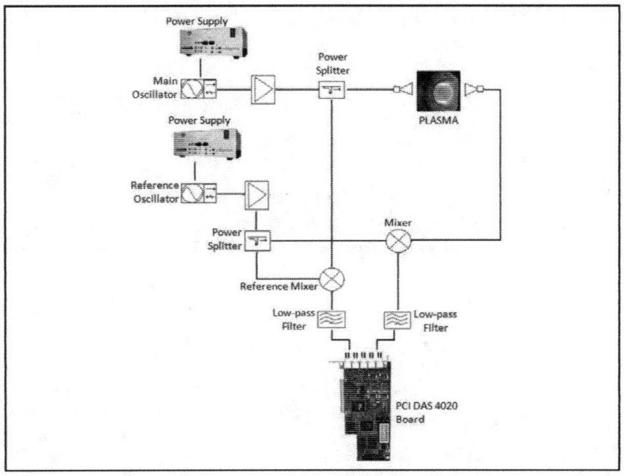

Figure 8. Diagram of the components of the Heterodyne Microwave Interferometer.

The probe heads that supports the Langmuir probes is easily removable. It consists of two boron nitride heads and an array of four tungsten tips per head. The probe will allow to investigate the local plasma density, temperature, plasma potential, and their fluctuations.
The Microwave Interferometer diagnostic is located vertical. The probing beam has a frequency of 28 GHz, corresponding to a wavelength of $\lambda = 10.71$ mm. The line-integrated electron density is deduced from the cumulative phase change of the probing beam and the theoretical length of the intersection of the probing beam with the plasma (obtained from the magnetic geometry in vacuum). It is important to state that the interferometer was developed along with a computer program to receive the data of the linear density of the plasma. Fig. 8 shows a diagram of the components of the interferometer.

2.7. Simulations, Modeling and Magnetic field calculations
A JAVA code named SimPIMF was developed by the creator of the UST_1, engineer Vicente Queral, to calculate three-dimensional (3D) magnetic fields. Later, the code evolved, and was able to calculate/simulate by field line tracing: Poincaré plots, rotational transform and magnetic well profile, plasma size, orbit simulation with drifts, particle losses, other 'plasma' parameters, minimum distance between coils, and optimization of such parameters by iterative generation of parametric 3D coils [3]. To obtain calculations of the SCR-1 magnetic structure and corroborate the UST_1 results, a computer code in MatLab was developed to estimate the 3D magnetic field, magnetic surfaces, rotational transform profile and magnetic well. Also particle tracking results and Poincaré maps were obtained using the COMSOL Multiphysics software. In Fig. 9 vacuum magnetic surfaces are presented on cero toroidal angle obtained on Poincaré simulations by MatLab code for SCR-1.
The magnetic resonant field for the SCR-1 will be 0.0438 T, and a period of 2 (m = 2) with a flat rotational transform profile with $\iota = 0.3$.
The magnetic field simulation has been approached by two different methods. In the first approach, the magnetic field is calculated using a Biot Savart solver implemented in JAVA and MatLab codes. The second approach on magnetic field simulation was done through a reduced model of the 12 modular coils set also using one turn per coil of the SCR-1 using the Magnetic and Electric Field module of COMSOL Multiphysics software. Using MatLab code and COMSOL Multiphysics software were obtained calculations of magnetic surfaces and where ECH resonance surface is located (Fig. 11) [9].

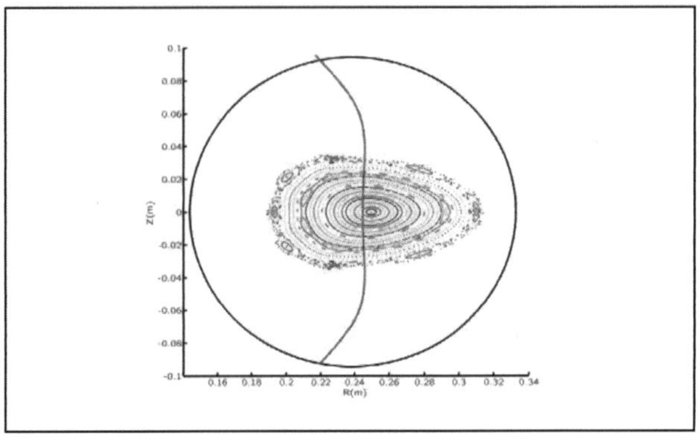

Figure 9. Magnetic surfaces at Φ = 0°.

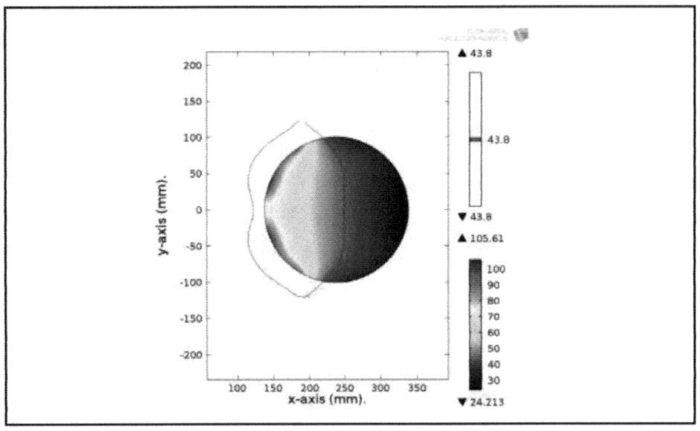

Figure 10. Contour map of magnetic field at Φ = 0° for I=4607 A.

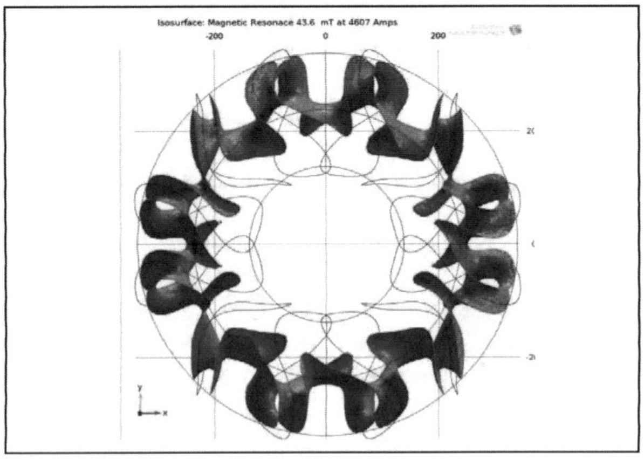

Figure 11. Resonant surface (2.45 GHz) for I=4607 A.

The magnetic field and the magnetic surfaces are important to define the best ECH heating system and diagnostics locations. Also, the magnetic surfaces and the rotational transform are both needed to evaluate the confinement of the device.

2.8. Data Acquisition and Control Systems

Currently, an application has been developed to generate and diagnose plasma using uniquely one click button. In order to make it possible it has been achieved primarily the identification and control of three threads associated with the plasma generation, which are: vacuum generation in the vessel, the automatic gas injection and heating system. For the gas injection process, a proportional – integrative - derivative (PID) compensator, for automatic control theory, was design using dynamic equations and physics of the system. The role of this compensator is to regulate the gas flow automatically to achieve a reference pressure within and specified time with a smooth behavior; this could be made by sending commands with the new set point, calculated by the compensator at every sampling time, to the Power Source and Readout device for the Mass Flow Controller.

2.9. Safety

Its main purpose is to achieve the goals of the project in a manner that guarantees the safety of the people and the equipment, as well as the correct development of the operations. The level of protection must be established as well as a risk reduction system on the development and operation of each stage of the SCR-1; coordinators in this aspect have knowledge on technical aspects of the SCR-1 and human issues. With this purpose a safety plan was developed and protocols are defined for the operation of stellarator

2.10. Plasma parameters

The following are the expected plasma parameters and other characteristics.

- Minor plasma radius: 54.11 mm.
- Electron temperature: 13 eV.
- Electron density: 5×10^{16} m^{-3}.
- Estimated confinement time: 5.70×10^{-4} ms (of ISS04 [10])
- Volume: 13.76 liters (0.01 m3).
- Aspect ratio: >4.4

The first SCR-1 plasma is expected to half of 2015.

3. Conclusions

The experience of designing and building this small Stellarator provides important opportunities for students, especially undergraduates, to develop the skills required for future opportunities in the field of experimental plasma physics and magnetic confinement fusion by working on professional research; engaging with the real engineering problems involved and finding their solutions; and contributing to the hands-on experience that is required before graduating.

As the future first Stellarator of Latin American, it is also very significant for the Plasma Physics Group from Instituto Tecnológico de Costa Rica (PlasmaTEC) to work on the design and construction of SCR-1; hoping that this can bring more plasma research opportunities to Latin American countries.

It is also hoped that the SCR-1 inspires more universities to develop similar devices that can work well for didactical purposes as well as achieving its main objective: contribute to the engineering and physics of small Stellarators [11,12,13].

Any kind of help, comment or suggestion is appreciated. Please contact us if any question.

4. Acknowledgment
The authors would like to thank the engineer Vicente Queral, CIEMAT and the Laboratorio Nacional de Fusión, in Madrid, Spain; for the support provided throughout this project.

5. References

[1] R.R. Weynants, "Fusion machines," Transactions of Fusion Science and Technology, Vol.49, pp. 36–42, February 2006
[2] E. Rebhan, and G. V. Oost, "Thermonuclear burn criteria for D-T plasmas," Transactions of Fusion Science and Technology, Vol.49, pp. 16–26, February 2006
[3] V. Queral, "UST_1 a small low cost stellarator," Stellarator News, Issue 118, pp. 1–6, December 2008. (references)
[4] H. Yamada, "Characterization of energy confinement in net-current free plasmas using the extended International Stellarator," Nucl. Fusion, Vol.45, pp. 1684–0000, May 2005
[5] M.M. Clark, A. Eckhart, K. Flanagan, J. Milhone, A. Rasmus, I. Reese, J.P. Wallace, D. Weisberg and C.B. Forest, "Fabrication of a Vacuum Vessel for a Cusp Confinement Plasma using Cast Al," Abs. Proc. 54th APS-DPP Meeting, Providence, US, November 2012
[6] L. Barillas, V.I. Vargas, et. al., An Engineering Approach to the Design and Construction of a Small Modular Stellarator for Magnetic Confinement of Plasma: SCR-1, in Proceedings of the19th International Conference on Nuclear Engineering, May 16-19, 2011, Chiba, Japan
[7] J. Mora. "Diseño del soporte y el control de corriente de las bobinas del Stellarator de Costa Rica 1. Informe de Proyecto de Graduación para optar por el título de Ingeniero en Mecatrónica con el grado académico de Licenciatura". Instituto Tecnológico de Costa Rica. Cartago, Noviembre de 2014, unpublished.
[8] H. Trimiño, V.I. Vargas, S. Tallents, L. Rivas, D. Chacon, J. Chacon, S. Fatjo, J. Jimenez, L.F. Villegas, J.Mora, J. Asenjo, J.M. Carranza, F. Cerdas, L. Barillas, J.I. Monge, J.Morera, H. Peraza, C. Rojas, G. Sánchez, M. Sandoval, J. Umaña, R. Gutiérrez, G. Solano. Design of an ECH system for a small modular stellarator, in Proceedings of the 38th International Conference on Plasma Science and 24th Symposium on Fusion Engineering, June 26-30, 2011, Chicago, Illinois, USA
[9] V.I.Vargas, J. Asenjo, A. Sura, S. García, F. Cerdas, D. Chacón, J. Chacón, S. Fatjó, J. Jiménez, and L. Rivas, Proc. 14th Latin American Workshop on Plasma Phys., Mar del Plata, Argentina, 2011
[10] Yamada H. et al., Nucl. Fusion 45, 1684 (2005).
[11] D.A. Hartmann, "Stellarators," Transactions of Fusion Science and Technology, Vol.49, pp. 43–55, February 2006.
[12] Stellarator News 2004-92.
[13] A. H. Boozer, "What is a stellarator?," Physics of Plasmas, Vol.5, pp. 1647–1655, May 1998.

Engineering of the Stellarator of Costa Rica: SCR-1

J. Mora[1], V. I. Vargas[1], C. Otarola[1], C. Piedra[1], W. Jimenez[1], L. Esquivel[1], R. Esquivel[1], K. Sanchez[1], J. Gonzalez[1], J. Asenjo[1] and L. Fonseca[2].
Plasma Laboratory for Fusion Energy and Applications[1]
Instituto Costarricense de Electricidad[2]
Instituto Tecnológico de Costa Rica-Cartago, Costa Rica
jaime.plasma.tec@gmail.com

Abstract. This Paper aims at briefly deSCRibing the challenge of the design and construction of the Stellarator of Costa Rica 1 (SCR-1) [1]. The SCR-1 is a small modular Stellarator for magnetic confinement of plasma (R_o=0.238 m, <a>=0.059 m, R_o/a>4.0, expected plasma volume ≈ 0.016 m^3, 10 mm thickness 6061-T6 aluminum vacuum vessel) developed by the Plasma Laboratory for Fusion Energy and Applications of the Instituto Tecnológico de Costa Rica (ITCR).

1. Introduction

The most challenging part of the design and construction is the vacuum chamber due to its size and its construction. The method of construction included a lot of testing and designing to determine the best way to build it. A second challenge of this construction is the support structure because the small size of the SCR-1 limited the places to fasten the support. During this process a coil support had to be developed, and, in this case, the challenge was obtaining accurately the coil shape. Limited resources to solve all these challenges translated into carefully designing each test.

Designing also needs to take into account the gas injection system. While making this design it is important to preserve the pureness of the gas and of the gas line. In addition, the vacuum pumps are important to the design because the same system will be used on the Spherical Tokamak Medusa [2]. As result, the challenge is to consider the needs of both machines.

Because of the dangerous equipment around the SCR-1, the automation of all of the systems is imperative. To accomplish this, it is necessary to test the equipment. In this case, the challenge is to design a test to make sure the lab staff is safe and that it includes the signals needed for automation; for example, ECH system test and the vacuum pump system test. Coils need a constant current and a constant voltage and the best way to achieve is with an automatic supply control.

2. Mechanical

2.1. Construction of the vacuum chamber.

The size of the vacuum chamber is the major issue because the regular methods to build vacuum chambers and the main recommendation cannot be followed either. The first question that we need to answer was regarding which material is the best for the construction. The first answer is stainless steel with 4 elbows at 90 degrees. But for the major radius (R_0=238.1 mm) does not fit with the standard elbows and manufacturing 4 customized elbows is too expensive. If two stainless steel plates were used to mill the shape it was more expensive than the elbows. Other option was to use aluminium cast; we could shape it as needed but the workshop couldn't do a good cast. Finally we decided to use aluminium plates of 6061-T6 and mill the shape of the vacuum chamber.

To make the vacuum chamber, we first mill the half interior shape of the vacuum chamber, as shown in figure 1.a. Then we repeated the process to obtain two halves. After that, the two halves were welded together from the outside as show figure 1.b. However, this have a problem because the torch used to weld the vacuum chamber from the outside didn't fit in the inside of the vacuum chamber. Then the outside of the vacuum chamber was milled, as show figure 1.c. After that, we made all the holes of the vacuum chamber to connect the CF-ports as shown figure 1.D. Finally, the support of the coils was placed and welded. On figure 2 it can be observed how the vacuum chamber is going to look when finished.

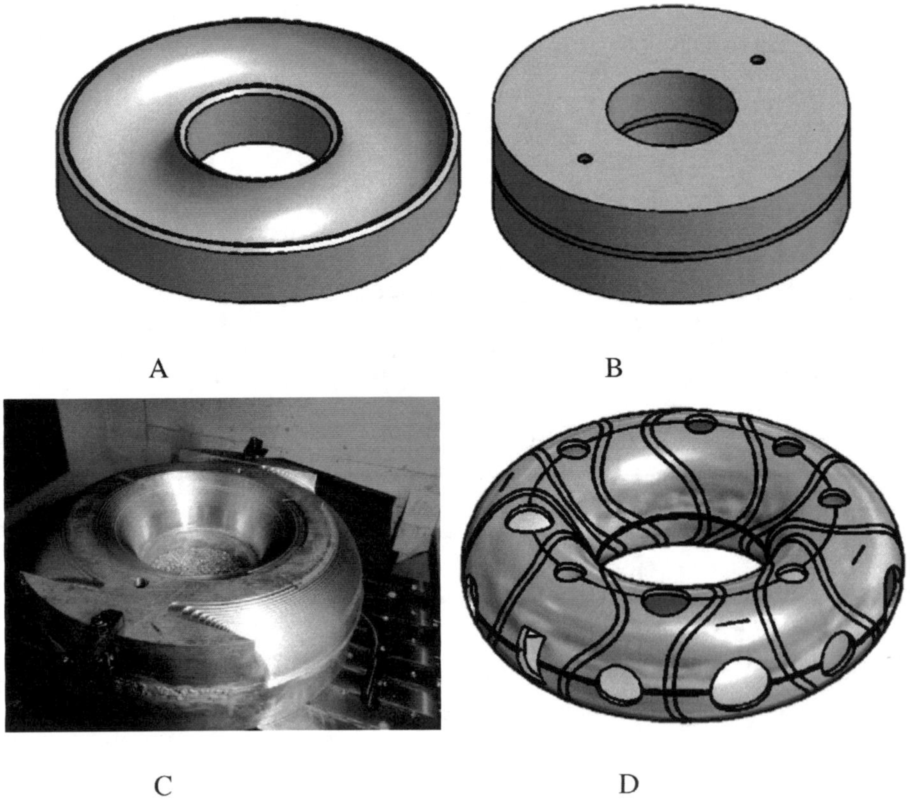

Journal of Physics: Conference Series **591** (2015) 012017 doi:10.1088/1742-6596/591/1/012017

Figure 1. A) It shows the interior shape. B) It shows how the two halves are welded with the interior shape. C) It shows how the exterior shape is milled. D) It shows the holes on the vacuum chamber.

Figure 2. Model in 3D cad of the final version of the vacuum chamber.

With the aluminium ports, it was necessary to find a way to make a CF that its knife does not damage the O-Ring. It was made a test in which in a vacuum chamber made of aluminium cast two ports without the Knife were welded, as shown in figure 3. The port on the right does not have the knife and it uses a fluorocarbon O-Ring, and the CF-port on the left is standard. This test proves we can use this to have a high vacuum. As a result it is better not to have a knife on the aluminium CF port.

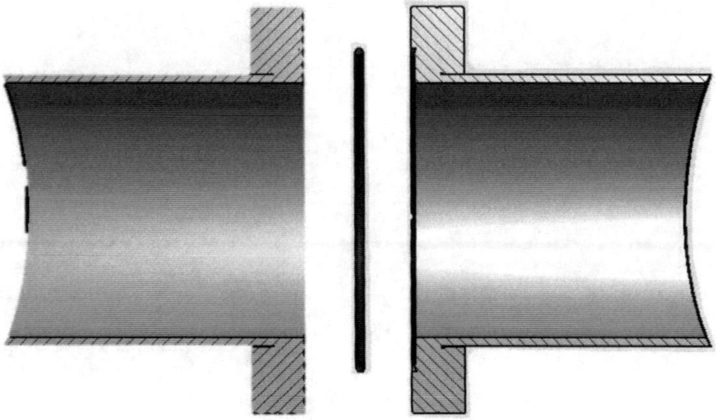

Figure 3. The CF-port without the knife.

Journal of Physics: Conference Series **591** (2015) 012017 doi:10.1088/1742-6596/591/1/012017

2.2. Developing the Coils

To build the coils the major issue was to create the shape and we had three possible ways to achieve this. The first one was to make a cover using a CNC milling machine to make the coil shape but it required a CNC milling machine with 6 axis not available in Costa Rica. Also, this method required a lot of material which in turn increases costs.

The second way was to make the shape using the embossed technique (see Figure 4) but it didn't achieve the expected accuracy, and that was enough to discard the method. The last method was the best and is the one we implemented: to use a 3D printer to develop a casting mold. Figure 5 shows the print version and figure 6 shows the cast and final version of the walls to make the grove. These pieces are going to be welded to the vacuum chamber, using the path marked by a CNC milling machine, that path look like figure 1.D.

With the last method we were able to achieve a precision of $\pm$ 0.5mm which is enough for this project.

2.3. Gas Injection System

This system has three lines because one of the lines is for injecting the gas for the main experiment. The second one is for making glow discharges to clean the vacuum chamber, and the last one is for ventilating the vacuum chamber. It is important to mention that it is necessary to keep clean the complete system to avoid the contamination of other gases. That is the reason why we have a vacuum system: to clean the tubes and inject only the gas that we want.

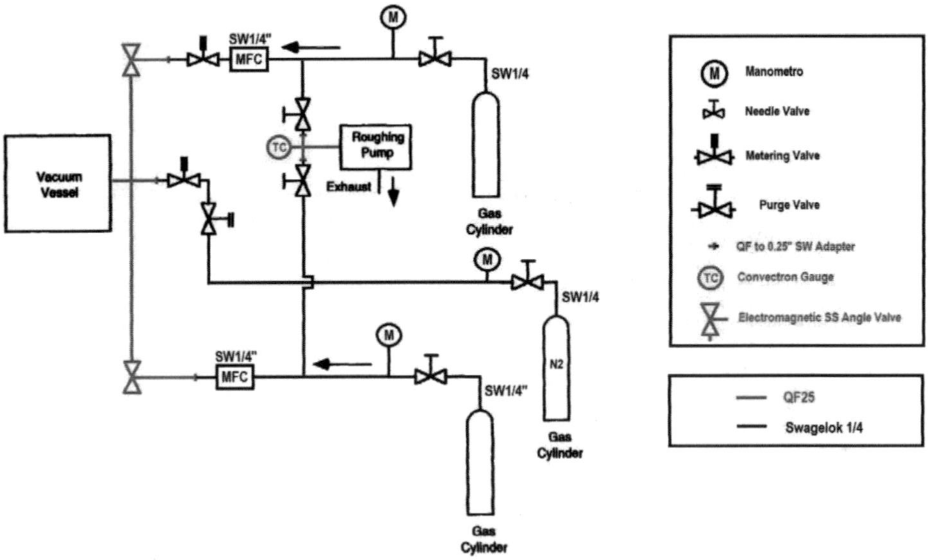

Figure 4. Costa Rica Stellarator 1 gas injection system diagram.

Journal of Physics: Conference Series **591** (2015) 012017 doi:10.1088/1742-6596/591/1/012017

2.4. Vacuum System

The real challenge of the vacuum system was to work on both machines: the SCR-1 and MEDUSA-CR. Figure 5 shows the system and the way it is connected to each machine. They have the same system because our laboratory does not have much space to fit both machines at the same time, and the laboratory has only one mechanical pump and one turbo pump. This system has a vent valve and a gate valve between the mechanical pump and the turbo pump because the turbo pump always needs to be in vacuum, but the mechanical pump can be stopped at any time. On the other side of the turbo pump we have the gauge to measure the pressure and RGA to find leaks on the system.

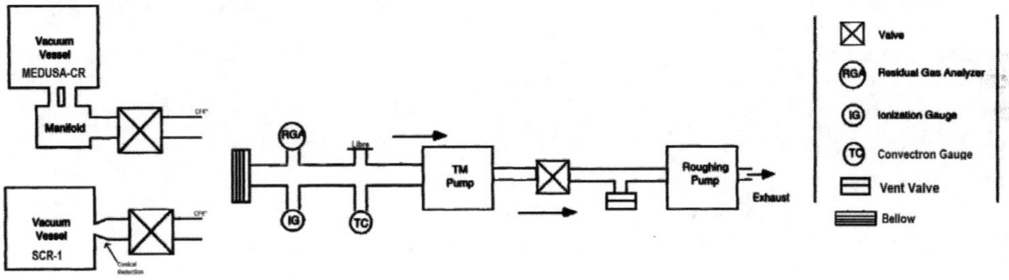

Figure 5. Vacuum system of the SCR-1 and MEDUSA -CR

3. Automation System

This aspect is one of the most important on our project because it directly relates to safety. At this moment, we have automatized the vacuum system, the ECH system and the current control for the coils

3.1. ECH Automation

We began the automation of the ECH system with tests. The first test was to connect a load and make some shots to determine how it worked. The second test was connecting the load in a controlled environment, and then make shots like the SCR-1 needs. Figure 6 shows some of the results that are necessary to automate the system; it shows the response time to deliver the power. The next step is to use a PIC to control the two magnetrons with one serial port. We will create a list of commands with a computer or other device like a PXI platform from national Instruments. The design took into count the possible ways to work: the first in which the two magnetrons works as one, and the second way in which each magnetron works separately. This allows the researchers to do different experiments.

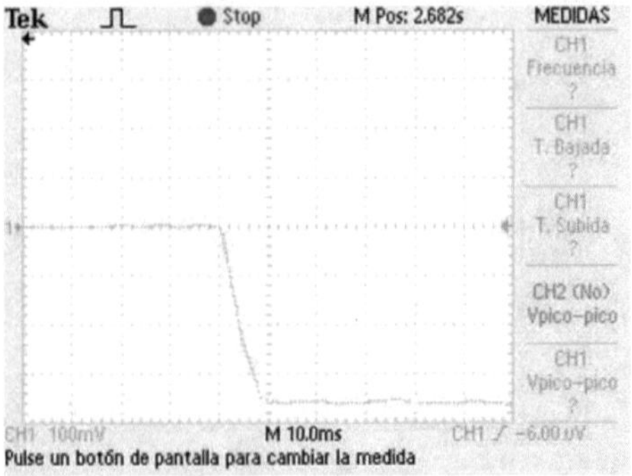

Figure 6. System's response time for delivering power

3.2. Vacuum system automation

Similarly to the ECH system, we are using a PIC to control the complete system with just one serial port. Thanks to a list of commands, we can control the complete system through the serial port with a computer or other device like a PXI or NI. The idea of one serial port with each system is to simplify the integration. On this system we will control the mechanical pump, the turbo pump, and the gauges of the system that are on figure 5.

3.3. Current control of the coils.

This system is critical for plasma stability, that is why it is important to have a constant current of 767 A $\pm$2%. In this system, MarteOs will be used to run the application to control an IGBT. This system will be running on a computer; for that reason, the other system will have a serial communication to be able to connect with a PC and in the future allow to integrate everything on one device.

4. Conclusion

The construction of the Stellarator of Costa Rica -1 comes with a lot of challenges, mainly regarding the size that its structures and systems need to have. All the manufacture processes are different from a large stellarator, and the limited space between the coils restricts the size and placement of the ports. Finally, the automation is very important in terms of security and to simplify the operation of the machine.

5. References

[1] V.I. Vargas, J. Mora, J. Asenjo, E. Zamora, C. Otárola, J. Carvajal-Godínez, J. González-Gómez, C. Soto-Soto, C. Piedras, L. Barillas and C. Ribeiro. ENGINEERING ISSUES TO THE STELLARATOR OF COSTA RICA 1 (SCR-1), in Proceedings of the 25th Symposium on Fusion Engineering, June 10-14, 2013, San Francisco, California, USA. (to be published)

[2]J. González et al., "Energy, Vacuum, Gas Fueling and Security Systems for the Spherical Tokamak MEDUSA-CR", 55th American Annual Meeting of the APS Division of Plasma Physics, Denver, Colorado, 2013

Planar geometry inertial electrostatic confinement fusion device

Daniel R. Knapp
Medical University of South Carolina and the Wilhelm Bratwurst Institute for Applied Physics Research, Charleston, SC 29425 USA

E-mail: knappdr@musc.edu

Abstract. In the classic gridded inertial electrostatic confinement (IEC) fusion reactor, ion bombardment of the grid leads to heating, thermionic electron emission, significant power loss, and ultimately melting of the grid. Gridless IEC devices have sought to overcome these limitations. Klein reported a gridless device in which ions are circulated as a linear beam in an electrostatic analogue of an optical resonator. To overcome limits of stored ions due to space charge effects at the turning regions, the device employed multiple overlapping traps. The work reported here seeks to further increase the turning region space in a gridless trap by employing a planar geometry. Ion trapping in the planar device was examined by simulating trajectories of $^{2}H^{+}$ ions with SIMION 8.1 software. Simulations were carried out using multiple potentials as in Klein's device and for a single potential trap as a planar analogue of the anharmonic ion trap. Scattering by background gas was simulated using a hard sphere collision model, and the results suggested the device will require operation at low pressure with a separate ion source.

1. Introduction

In the classic gridded inertial electrostatic confinement (IEC) fusion reactor, ion bombardment of the grid leads to heating, thermionic electron emission, and significant power loss; and the heating can ultimately melt the grid causing device failure [1]. Gridless IEC devices have sought to overcome these limitations. Klein reported a gridless device called the "Multiple Ambipolar Beam Line Experiment" ("MARBLE") [2] in which ions are circulated as a linear beam in an electrostatic analog of an optical resonator (figure 1). The MARBLE design was inspired by the linear electrostatic ion trap developed by Zajfmann, *et al.* (figure 2) [3] and the simpler "anharmonic ion trap" (figure 3) [4]. The number of ions stored in any electrostatic trap is limited by the space charge at the points of lowest velocity, i.e. the turning points of the recirculating ion beam, and the number of stored ions limits the fusion rate. The MARBLE device employed multiple overlapping traps in the same physical space to enable the trapping of more ions in an effort to increase fusion rate.

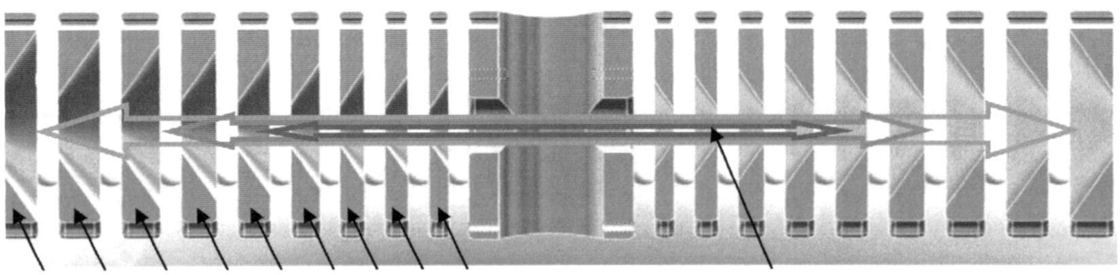

Figure 1. Cross section of the MARBLE device showing three of the five recirculating ion beams (adapted from [2]).

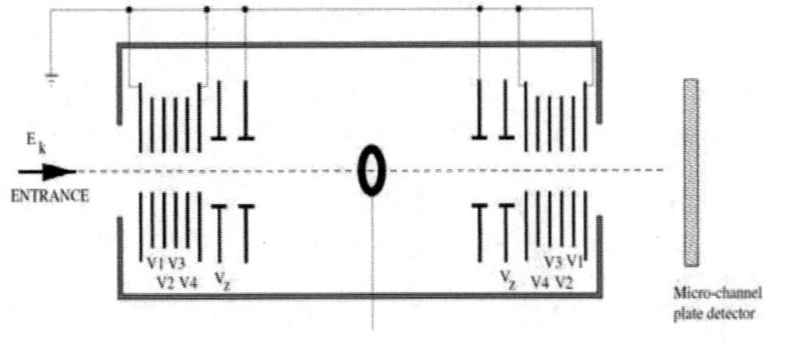

Figure 2. Linear electrostatic ion trap (from [3]).

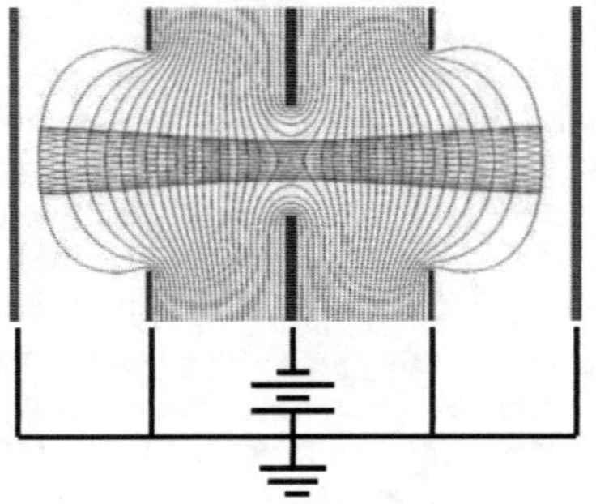

Figure 3. Anharmonic ion trap (adapted from [4]).

The spherical gridded fusor (shown in cross section in figure 4) operates by accelerating ions formed outside the grid toward the center of the device. Ions that pass through the grid continue toward the opposite wall where they are reflected and recirculate until undergoing a collision. In theory, the spherical fusor can trap recirculating ion beams throughout the full solid angle space of the sphere maximizing the area of the turning region. Thus, it should be able to accommodate more trapped ions than any configuration of linear ion traps. However, in the gridded spherical fusor the circulating ions self-organize into beams through the grid openings forming the "star mode" that seems to be essential for maximum fusion rate [5]. This cannot be overcome by increasing the number of openings in the grid, because practical limitations on wire size causes reduced transparency, and grid transparency is directly related to fusion rate. Thus, the spherical device is effectively a discrete beam device. The device described here has the potential for an improved gridless geometry by enabling a full circular angle of beam trajectories with the turning region spread around the full circle. Although a planar ion trap mass spectrometer was recently proposed [6], it does not appear that this geometry has been explored for an IEC fusion device.

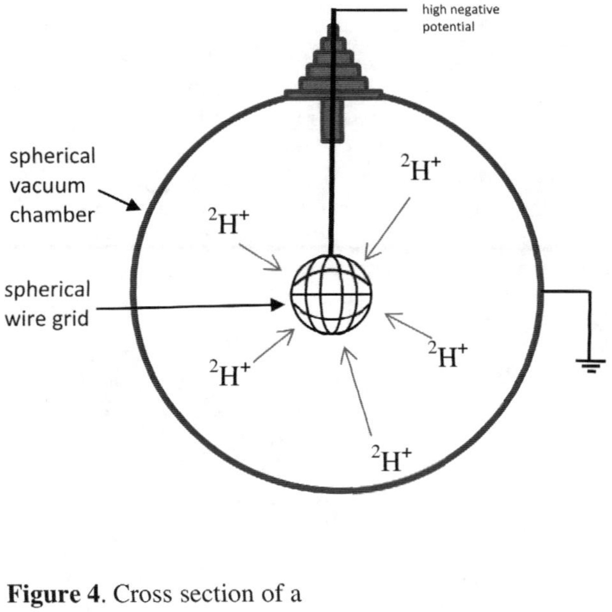

Figure 4. Cross section of a spherical gridded fusor.

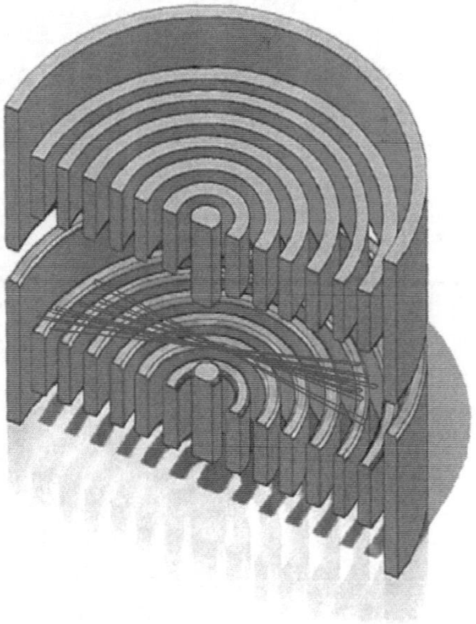

Figure 5. Cutaway view of a planar electrostatic ion trap.

2. Description of the Planar Geometry IEC Fusion Device

Linear electrostatic ion traps have cylindrical symmetry about the ion beam axis. If one envisions taking the cross section of a linear trap and rotating it about a central axis perpendicular to the ion motion, the result is a planar beam device with a series of concentric rings above and below the ion beams as shown in figure 5. If all of the ring electrodes are grounded and a negative potential is placed on the center pins, the device is a planar analogue of the trap shown in figure 3 (subsequently referred to as Case 1). Operating the device with a gradient of positive potentials on alternate rings, with the intervening rings and center electrodes grounded, yields a planar analogue of the MARBLE device (subsequently referred to Case 2). Both Case 1 and Case 2 were simulated using the electrode configuration shown in cross section in figure 6.

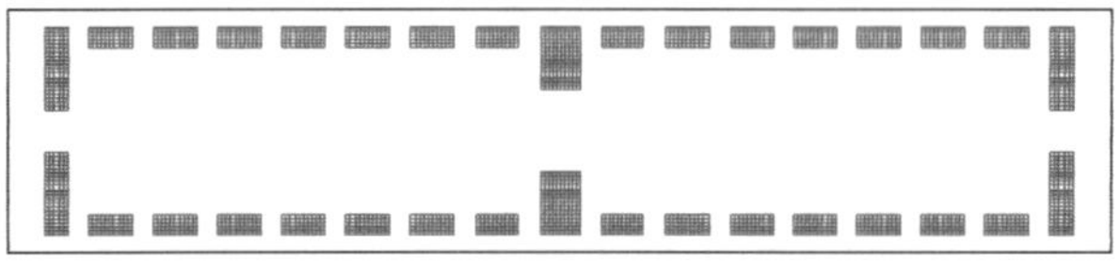

Figure 6. Cross section of the electrode configuration used for Case 1 and Case 2.

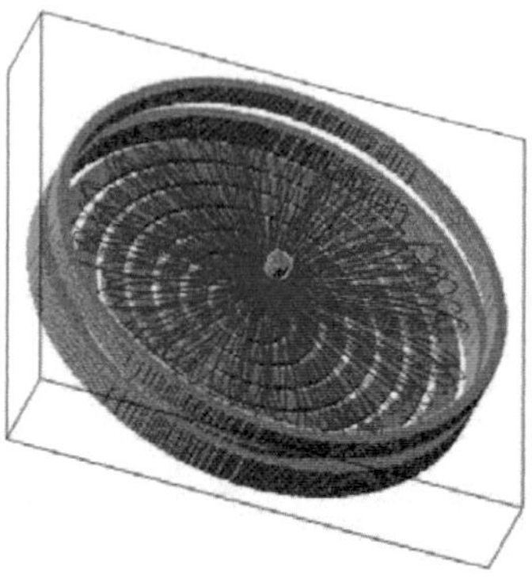

Figure 7. Simulated ion trajectories for Case 1 device; center electrode -10 kV; $^{2}H^{+}$ ions 0.1 eV tangential energy (cutaway view).

3. Simulation Studies

Simulations were carried out using SIMION 8.1 software to predict the trajectories of deuterium ions in a planar ion trap device. Case 1 and Case 2 were modelled under vacuum conditions. If the ions are generated in the model with even a small amount of energy (0.1 eV) in the tangential direction, they will quickly spread out all around the trap as shown in figure 7, suggesting that the planar trap will allow trapping of ions with trajectories throughout the full circular angle space.

Case 1 (figure 8) and Case 2 (figure 9) were modelled with 300°K isotropic ions generated in a single 2 mm radius spherical space near the periphery of the device. A series of different potentials were examined for Case 2, but none gave trapping as good as in Case 1. The ions had insufficient tangential velocity to spread around the full circular angle in the trapping time.

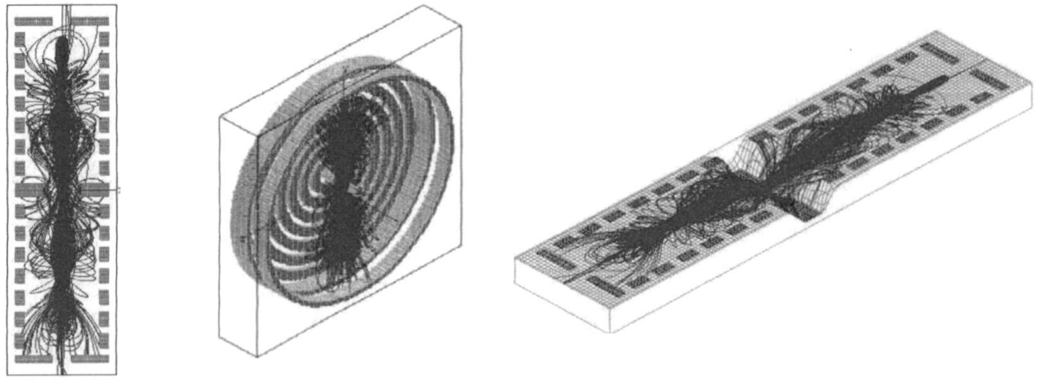

Figure 8. SIMION simulated ion trajectories for Case 1 device (-10 kV center electrode; 0 other electrodes; 300°K $^{2}H^{+}$ ions). Cross section view (left); cutaway view (center); potential surface view (right).

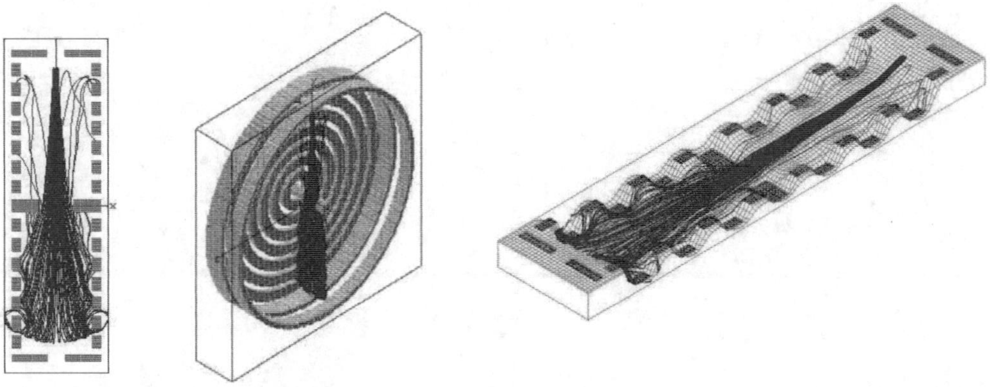

Figure 9. SIMION simulated ion trajectories for Case 2 device (+10, 0, 6.5, 0, 5.5, 0, 4.0, 0 kV; 300°K $^2H^+$ ions). Cross section view (left); cutaway view (center); potential surface view (right).

Simulations were also carried out for the device using bevelled edge electrodes as in the MARBLE device in figure 1 (Case 3). Figure 10 shows simulated ion trajectories for the Case 3 device. Using the same number of electrodes, the same cross section shapes, and the same potentials as in the reported device [2], $^2H^+$ ions could be trapped for as long as 16 seconds allowing the trajectories to spread around the full circular space.

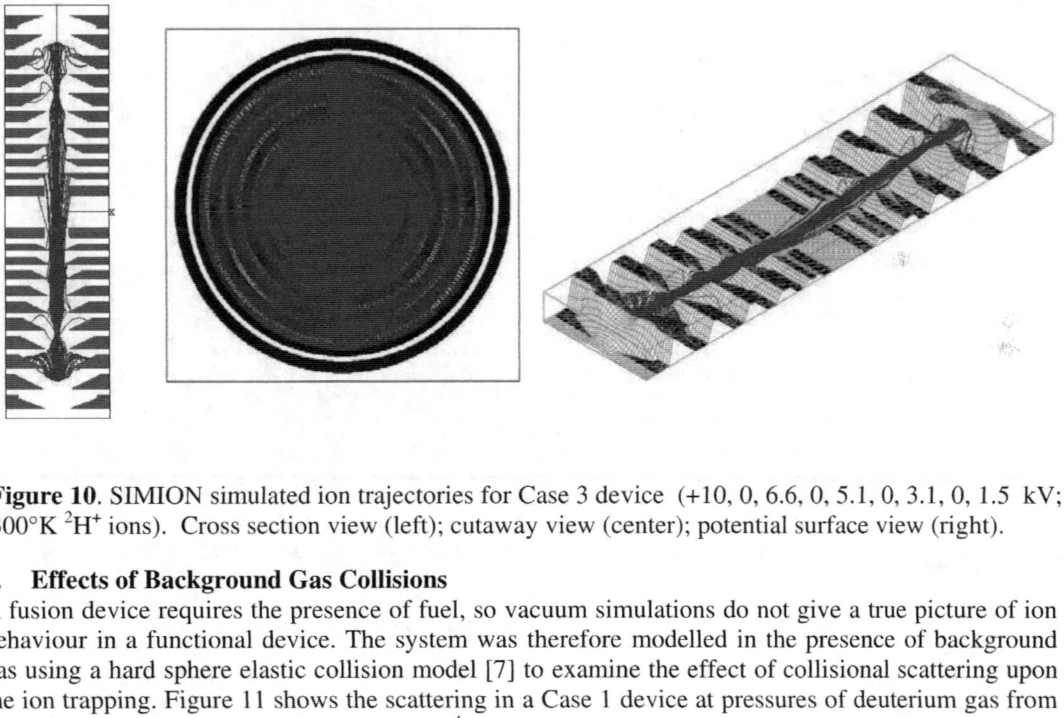

Figure 10. SIMION simulated ion trajectories for Case 3 device (+10, 0, 6.6, 0, 5.1, 0, 3.1, 0, 1.5 kV; 300°K $^2H^+$ ions). Cross section view (left); cutaway view (center); potential surface view (right).

4. Effects of Background Gas Collisions

A fusion device requires the presence of fuel, so vacuum simulations do not give a true picture of ion behaviour in a functional device. The system was therefore modelled in the presence of background gas using a hard sphere elastic collision model [7] to examine the effect of collisional scattering upon the ion trapping. Figure 11 shows the scattering in a Case 1 device at pressures of deuterium gas from 1 Pa (which totally destroys trapping) to 10^{-4} Pa. (where trapping is comparable to vacuum). Similar results were obtained for the Case 3 device (figure 12). These scattering simulations suggest that the planar trap may not be useful at pressures normally used in devices with ion formation from simple Paschen discharge, and that other means of ion generation will be required. A commercial residual gas analyser (Brooks Instruments, VQM) based upon the single potential linear trap shown in figure 3 generates ions within the trap using an electron beam injected from a filament outside the trap. The same method could be used in a planar trap fusion device to enable operation at lower pressures.

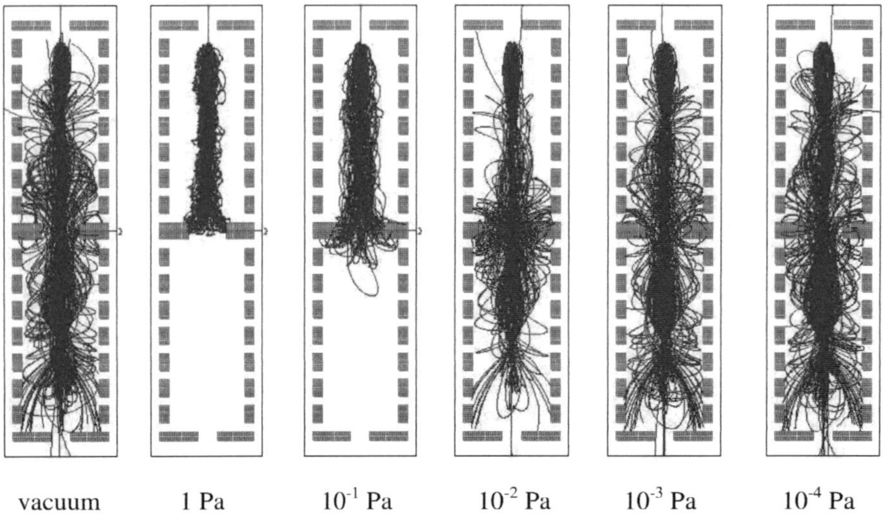

Figure 11. Simulated ion scattering results for the Case 1 device.

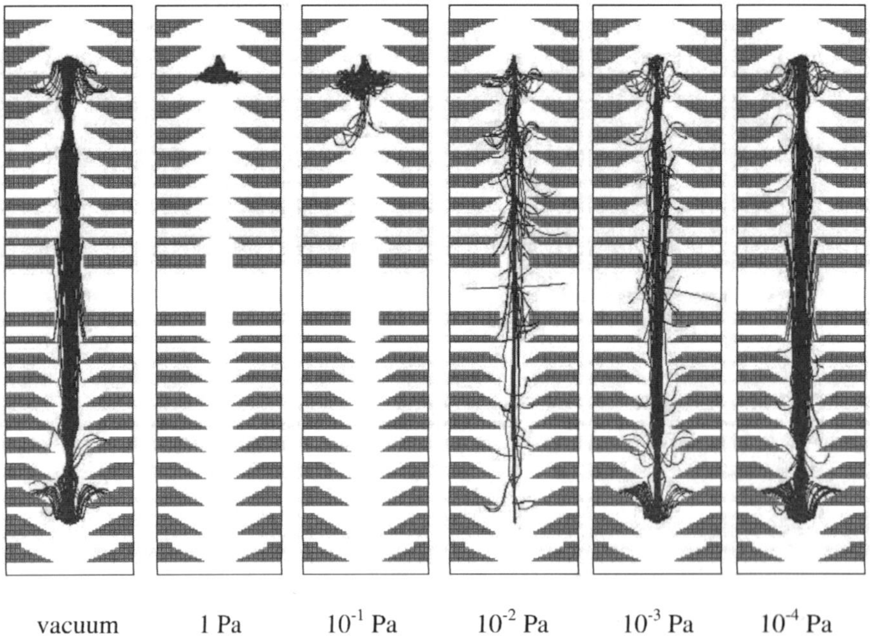

Figure 12. Simulated ion scattering results for the Case 3 device.

5. Conclusions

These simulations suggest that a planar geometry device could function as proposed and offer improved performance over previous geometries. The simulated trapping in the Case 3 device (figure 10) was sufficiently long to enable the recirculating ion trajectories to spread around the full circular space, thus supporting the basic premise that the planar geometry device should enable trapping of more ions and yield a higher fusion rate than previous geometries. However, the scattering simulations suggest that the device may require operation at lower pressure than normally used in gridded spherical devices. A prototype device is under construction.

References

[1] Miley G H and Murali S K *2014 Inertial Electrostatic Confinement (IEC) Fusion Fundamentals and Applications* (New York: Springer) p 115
[2] Klein A 2011 *The Multiple Ambipolar Beam Line Experiment (MARBLE)*, presented at the 13th U.S. - Japan Workshop on Inertial Electrostatic Confinement Fusion, Sydney, AU www.physics.usyd.edu.au/~khachan/IEC2011/Presentations/KleinMARBLEtalk.pdf
[3] Andersen L H Heber O and Zajfman D 2004 Physics with electrostatic rings and traps, *Journal of Physics B* 37, R57–R88
[4] Ermakov A V and Hinch B J 2010 An electrostatic autoresonant ion trap mass spectrometer, *Review of Scientific Instruments* **81**, 013107
[5] Miley G H, Gu Y, DeMora J M, Stubbers R A, Hochberg T A, Nadler J H, Anderl R A 1997 Discharge characteristics of the spherical electrostatic confinement (IEC) device, *IEEE Transactions on Plasma Science* 25 (4) 733 - 739, 1997.
[6] Ding L, Badheka R, Ding Z, and Nakanishi H, 2013 A simulation study of the planar electrostatic ion trap mass analyzer, *Journal of the American Society for Mass Spectrometry* 24: 356-364
[7] SIMION Collision Model HS1, www.simion.com

Analysis on the formation of a FRC by flux measurements and MHD simulations

M E Kayama[1], L C Nascimento[2], R P Mota[1], K G Kostov[1], R Y Honda[1], and M A Algatti[1]

[1]FEG-DFQ-UNESP, Av. Ariberto Pereira da Cunha 333, 12516-410 Guaratinguetá, SP, Brazil
[2]Universidade Federal de São João Del Rei, Campus de Santo Antonio, São João Del Rei, MG, Brazil

E-mail: mkayama@uol.com.br

Abstract. The evolution of the separatrix and excluded flux radius was investigated in a mirror-less theta pinch with external magnetic and loop probes and by numerical simulation. In the numerical hybrid code the anomalous effect was taken as evolution of lower hybrid drift instability. At mid-plane of the coil the experimental and theoretical values of lift-off time, implosion time and equilibrium plasma radius were very close. The implosion time is 5 ± 1 cm/μs and the separatrix radius at mid-plane is around one half of the discharge tube radius. Within the experimental uncertainty the excluded flux radius has the same value along the coil, suggesting an elongated configuration with closed magnetic field lines.

Keywords: excluded flux radius, separatrix radius, FRC

1. Introduction

The field reversed configuration (FRC) is a toroid compact with high β plasma. It is technologically simpler than the traditional toroidal system. For that reason it has been investigated as a alternative or complementary device for fusion reactor [1-4]. Traditionally this configuration is produced in theta-pinch devices. It starts with a bias field immersed in a resistive plasma and an intense and fast magnetic field to confine the plasma. This field is in the opposite direction of the main field. With the reconnection process driven by mirror coils at the ends of main coil these opposite fields reconnect forming the closed magnetic field line configuration.

The formation of the field reversed configuration can be investigated by measurements using internal magnetic probes [5,6]. That class of diagnostics uses small windings protected by quartz jacket. That causes two types of perturbation: firstly due to the presence of conductor in the plasma, and secondly, due to effects of plasma-solid interaction at surface of the jacket. The perturbations are notably high in high density and high temperature plasmas, as those produced in FRC's. In this case the usual technique is the excluded flux probe, a combination of the flux measurement by a single loop around the discharge tube and a local magnetic field to measure the magnetic field intensity. All diagnostic elements are outside the plasma region and can give the excluded flux ratio and topological structure of magnetic field lines in the configuration [7,8].

In the present work we investigate the formation of a field reversed configuration measuring the excluded flux radius using loop and magnetic probes. The evolution of the configuration was compared to the theoretical prediction based on a numerical code based on particle-in-cell technique. Experimental turbulence was theoretically assumed as evolution of lower hybrid drift instability. The actual plasma dynamics observed by the diagnostic was very close to the behavior predicted by this theoretical approach. It was also observed almost the same value of the excluded flux radius along the discharge tube suggesting a plasma column longer than the coil length.

The excluded flux probe uses the measurement of the loop or flux probe and the magnetic probe, respectively, with signal intensity *F* and B. The excluded flux ratio $r_{\Delta\phi}$ is given by:

$$r_{\Delta\phi} = r_L\left(1-\frac{B_V F_P}{B_P F_V}\right)^{1/2}$$

where r_L is the radius of the loop probe and the index *V* and *P* corresponds to discharges in vacuum and with plasma, respectively. At mid-plane of the coil where the magnetic field lines are straight lines this radius is equal to the separatrix radius (null of magnetic flux).

The present text is divided as follows: the experiment and simulation is explained in subsection 1.1. In section 2 are presented the results and the analysis of data. The Conclusion finishes the text.

1.1. Experiment and simulation

The theta-pinch coil is 21 cm long with inner radius of 4.5 cm. It is a pair of single turn coils separated by a 1 cm gap. The cylindrical Pyrex discharge tube placed in its interior has 4.0 cm of outer radius and *a*=3.8 cm inner radius. The tube's nominal length is one meter with two stainless steel vacuum chamber at its ends that are used for gas injection, vacuum system and diagnostic ports. In the present experiment we used hydrogen at 60 mTorr with continuous flow in the tube. The current in the coil is provided by the discharge of a capacitor bank (capacitance *C*=50 μF) by firing a pressurized spark-gap. The total inductance of the circuit is *L*=0.1 μH and the resistance is *R*=6 mΩ. The discharge produces a ringing waveform for the current with period *T*= 14.7 μs. We used the third-half cycle of the current for data analysis. According to previous study with magnetic probes this is the first of the half cycle where pinch occurs [9]. It starts with a magnetic field in the opposite direction to the main field with intensity around -1 kG. This is the natural field from the previous half cycle that remains frozen to the plasma during the transition of half-cycles. The reconnection of these opposite magnetic fields can give origin to the formation of a FRC.

The excluded flux probe is compounded by a single one turn loop around the discharge tube and a magnetic field next to the loop (figure 1). Three sets were assembled at different axial positions. The high voltage signal of the loop was measured using a compensated voltage divider made with high precision wire resistors and electrolytic solution. In spite of measurement repeatability three discharge signals were used to calculate the mean value and to estimate the uncertainty.

The theoretical approach takes the ion as particle and electrons as mass-less fluid. The equation of motion of each species is:

$$m_i\frac{d\mathbf{w}}{dt} = e\mathbf{E} + \frac{\mathbf{w}\times\mathbf{H}}{c} - \mathbf{P} \tag{1}$$

$$m_e\frac{d\mathbf{v}_e}{dt} = 0 = -en_e\left(\mathbf{E} + \frac{\mathbf{w}\times\mathbf{H}}{c}\right) - \nabla p_e - n_e\mathbf{P} \tag{2}$$

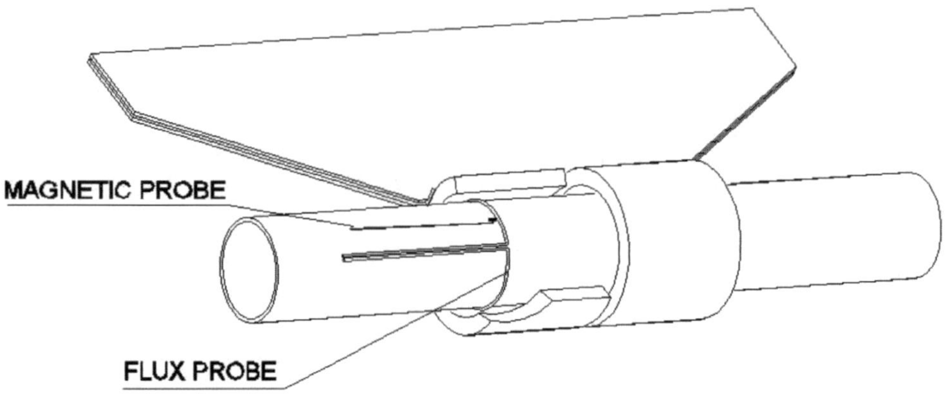

Figure 1. Schematic drawing of the experiment showing one set of the excluded flux probe.

where $f_i(r,w,t)$ is the distribution function, $n_i = \int f_i(r,w,t)d^3w$, $\mathbf{v}_i = (1/n_i)\int \mathbf{w}_i f_i(r,w,t)d^3w$, $\mathbf{P} = e\tilde{\eta}\mathbf{J}$ is the momentum transferred from ions to the electrons per ion per unit of time due to collisions, and $\tilde{\eta}$ is the resistive tensor calculated according to $\eta = m_e\nu / ne^2$, ν the anomalous collision frequency and $p_e = n_e kT_e$ is the pressure. Field equations are Maxwell's equations without displacement current term. The numerical code uses particle-in-cell technique in 1.1/2 dimension (space-time *r-t*) [10]. The anomalous collision frequency was calculated according to evolution of lower hybrid drift previously verified as the most important of the micro instability present in the plasma [11]. The spatial grid had 30 elements where particles at *t=0* were randomly distributed. For the present investigation we used 5000 particles and time steps of 5 ns. Due to the low *dB/dt* of the system the pinch evolves with a partially ionized gas and a broad current sheath profile in the discharge tube.

2. Results and discussion

The external loop and magnetic probes show variation on the waveform during the first 1.5 μs of the rising of the current in the coil. That is approximately the period of existence of the plasmas, including the current sheath formation next to the discharge tube wall, its implosion, equilibrium and extinction. The lift off of the current sheath from the wall is about 0.4 μs. From this instant the external magnetic pressure becomes higher than the total plasma pressure (magnetic plus kinetic) and the radial implosion begins take place. So the partially ionized gas in the tube is dragged toward the axis of the system leaving behind a small portion of neutral particles. Part of these particles is ionized by fast electrons at the border of the current sheath and them they are dragged to the axis. That creates a current sheath with electrons and ions concentrated respectively at the internal and external regions of the current sheath. An electrostatic field in radial direction is formed with intensity comparable to the electric field in the azimuthal direction. That creates a broad profile on particle densities and due to the low temperature, a diffuse magnetic field occurs as shown in figure 2. At the end of implosion the minimum plasma radius is around $r = a/2$, where a is the discharge tube radius.

The radial dynamic of the implosion is shown by the time evolution of experimental excluded flux radius at midplane of the coil (figure 3). In that region the magnetic field lines are straight and this radius is equal to the radius of the separatrix. The figure also shows the evolution of separatrix radius calculated from theoretical distribution of magnetic field in the discharge tube.

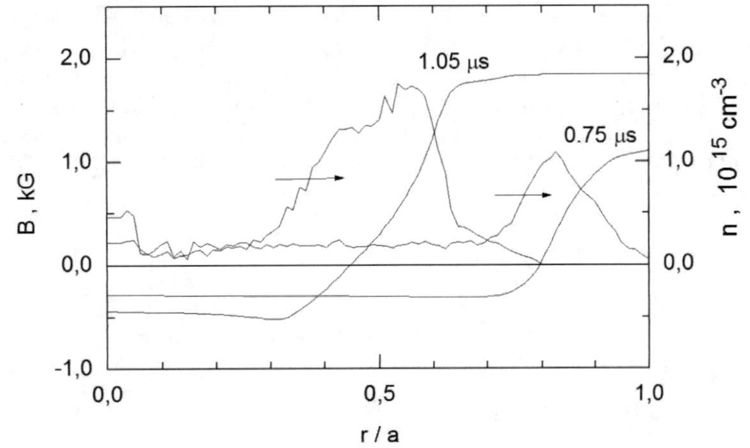

Figure 2. Radial profile of the magnetic field and electron density on early phase and at the end of the implosion.

The time lag between the instant of lift from the wall of the separatrix and the excluded radius is approximately within the estimated uncertainty on experimental time. After the lift off both radii reduce at same ratio. The mean speed of radial implosion is 5 ± 1 cm/μs and the experimental end of the implosion phase is around t= 1.1 μs. At this moment the configuration reaches the radius around one half of the inner radius of the discharge tube. The diffusion of external field starts to take place and the excluded flux radius reduces at speed 2 ± 1 cm/μs until t= 1.5 μs. The mechanism accelerates after this moment and the closed field line vanishes. Theoretically, the equilibrium phase shows a steady evolution with low amplitude radial magneto-acoustic. After the end of implosion the plasma the hot plasma expands radially. Since the external magnetic field is rising the external magnetic pressure promotes a second radial compression. On the simulation the process is interrupted due to high number of particles reflected by magnetic piston.

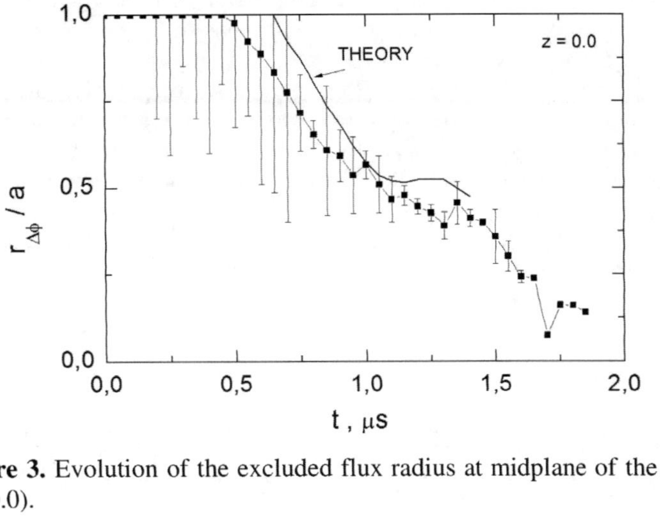

Figure 3. Evolution of the excluded flux radius at midplane of the coil (z= 0.0).

The time evolution of the excluded flux radius at end region of the coil has shown similar behavior of that at mid-plane of the coil. It is shown in figure 4, without the estimated error bars, for clarity. It shows that the displacement of the current sheath starts around the same time as at mid-plane. The final radius at the end of the implosion is about the same as well, around one half of the discharge tube radius. The excluded flux radius starts to reduce significantly after around t= 1.4 μs in the various axial positions inside the coil. That might indicate that the length of the plasma column extends beyond the limits of the coil. The applied diagnostic shows that the closed magnetic field line configuration is actually formed in the system. Therefore, the magnetic field reconnection process takes place outside the coil region. The configuration remains in equilibrium by the full length of the coil and is lost by radial diffusion or annihilation of the magnetic field.

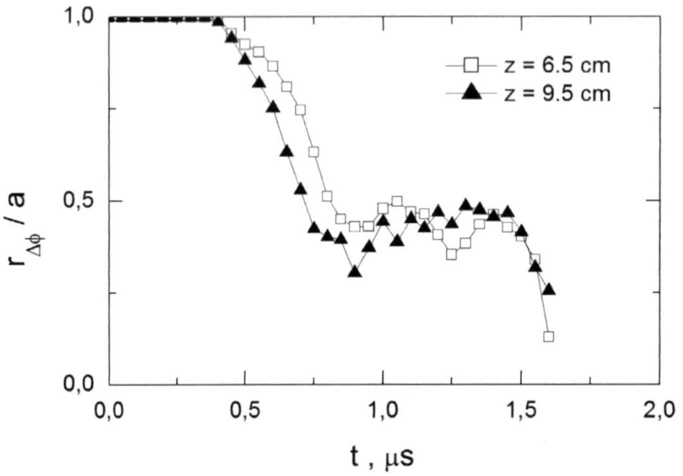

Figure 4. Evolution of the excluded flux radius at the end region of the coil. The border of the coil is at z= 10.5 cm.

3. Conclusion

The formation and dimension of a field reversed configuration was investigated with excluded flux probes and numerical simulation of a hybrid code. At mid-plane of the coil the experimental excluded flux ratio was compared to the theoretical separatrix radius where turbulence was taken as associated with evolution of lower hybrid drift instability. Both radii were comparable, showing close values of the lift off time and radial speed during the implosion phase. The final radii at the end of the implosion were close as well, around one half of discharge tube radius. The excluded flux radius along the axial direction shows the same variation during the confinement as that at mid-plane of the coil. That suggests the formation of a plasma column that extends beyond the coil length with magnetic field line reconnection process taking place outside the coil. Measurement of local magnetic field will be done to check the topology of magnetic field lines at the end of the coil. In spite of the absence of mirror coils, the excluded flux measurements shows that the formation of a reversed field line configuration might occur in this short theta-pinch.

Acknowledgments

We are grateful to J.B. Galhardo for technical help and to Sao Paulo Research Foundation (FAPESP) and National Council for Scientific and Technological Development (CNPq) for financial support.

References

[1] Steinhauer L C 2011 Review of field-reversed configurations. *Physics of Plasmas*, **18** 070501

[2] Tuszewski M, Smirnov A, Thompson M C, Korepanov S, Akhmetov T, Ivanov A, Voskoboynikov R, Schmitz L, Barnes D, Binderbauer M W, Brown R, Bui D Q, Clary R, Conroy K D, Deng B H, Dettrick S A, Douglass J D, Garate E, Glass F J, Gota H, Guo H Y, Gupta D, Gupta S, Kinley J S, Knapp K, Longman A, Hollins M, Li X L, Luo Y, Mendoza R, Mok Y, Necas A, Primavera S, Ruskov E, Schroeder J H, Sevier L, Sibley A, Song Y, Sun X, Trask E, Van Drie A D, Walters J K, and Wyman M D. 2012 Field reversed configuration confinement enhancement through edge biasing and neutral beam injection. *Phys. Rev. Lett.*, **108** 255008

[3] Ii T, Inomoto M, Gi K, Umezawa T, Ito T, Kadowaki K, Kaminou Y and Ono Y 2013 Stability and confinement improvement of an oblate field-reversed configuration by using neutral beam injection. *Nuclear Fusion*, **53** 073002

[4] Slough J, Votroubek G and Pihl C 2011 Creation of a high-temperature plasma through merging and compression of supersonic field reversed configuration plasmoids. *Nuclear Fusion*, **51** 053008

[5] Irby J H, Drake J F and Griem H R 1979 Observation and interpretation of magnetic-field-line reconnection and tearing in a theta pinch. *Phys. Rev. Lett.*, **42** 228

[6] Sevillano E and Ribe F L 1985 Reconnection studies in field-reversed configurations. *Physics of Fluids*, **28** 3142

[7] Tuszewski M 1981 Excluded flux analysis of a field-reversed plasma. *Phys. Fluids*, **24** 2126

[8] Gota H, Fujimoto K, Ohkuma Y, Takahashi T and Nogi Y 2003 Separatrix shapes and internal structures of a field-reversed configuration plasma. *Physics of Plasmas*, **10** 4763

[9] Kayama M E, Clemente R A, Honda R Y and Dobrowolsky M S 2009 Radial plasma dynamic in sequential pinches. *IEEE Trans. Plasma Sci.*, **37** 2186

[10] Sgro A G and Nielson C W 1976 Hybrid model studies of ions dynamics and magnetic field diffusion during pinch implosions. *Phys. Fluids*, **19** 126

[11] Kayama M E 2012 Resistivity in the dynamic current sheath of a field reversed configuration. *Physics of Plasmas*, **19** 032511

Dense Plasma Focus: physics and applications (radiation material science, single-shot disclosure of hidden illegal objects, radiation biology and medicine, etc.)

V.A. Gribkov[1,2,3,4], R. Miklaszewski[1], M. Paduch[1], E. Zielinska[1], M. Chernyshova[1], T. Pisarczyk[1], V.N. Pimenov[2], E.V. Demina[2], J. Niemela[3], M.-L. Crespo[3], A. Cicuttin[3], K. Tomaszewski[5], M.J. Sadowski[1,6], E. Skladnik-Sadowska[6], K. Pytel[6], A. Zawadka[6], G. Giannini[7], F. Longo[7], A. Talab[8], S.E. Ul'yanenko[9]

[1]Institute of Plasma Physics and Laser Microfusion, Warsaw, Poland,
[2]A.A. Baikov Institute of Metallurgy and Material Sciences, Russ. Ac. Sci., Moscow, R.F.
[3]The Abdus Salam International Centre for Theoretical Physics, Trieste, Italy
[4]A.I. Alikhanov Institute of Theoretical and Experimental Physics, Moscow, R.F.
[5]ACS Ltd., Warsaw, Poland
[6]National Centre for Nuclear Research, 05-400 Otwock, Poland
[7]Department of Physics, University of Trieste, Trieste, Italy
[8]Atomic Energy Authority, NRC, Plasma Physics and Nuclear Fusion Department, Egypt
[9] Medical Radiological Research Center, Russ. Ac. Med. Sci., Obninsk, Kaluga region, R.F.

E-mail: gribkovv@rambler.ru

Abstract. The paper presents some outcomes obtained during the year of 2013 of the activity in the frame of the International Atomic Energy Agency Co-ordinated research project *"Investigations of Materials under High Repetition and Intense Fusion-Relevant Pulses"*. The main results are related to the effects created at the interaction of powerful pulses of different types of radiation (soft and hard X-rays, hot plasma and fast ion streams, neutrons, etc. generated in Dense Plasma Focus (DPF) facilities) with various materials including those that are counted as perspective ones for their use in future thermonuclear reactors. Besides we discuss phenomena observed at the irradiation of biological test objects. We examine possible applications of nanosecond powerful pulses of neutrons to the aims of nuclear medicine and for disclosure of hidden illegal objects. Special attention is devoted to discussions of a possibility to create extremely large and enormously diminutive DPF devices and probabilities of their use in energetics, medicine and modern electronics.

1. Introduction

The most important effect ruled the physical processes in a DPF device [1] is the "current abruption" (CA) phenomenon [2, 3]. During this transient event (occurring in the ps time scale) a plasma diode is produced, which switches stored energy over to the beams of fast particles – electrons and ions. These fast particles produce extremely bright flashes of X-rays and neutrons. It gives us an opportunity to make *heavy* science by means of *light* technology.

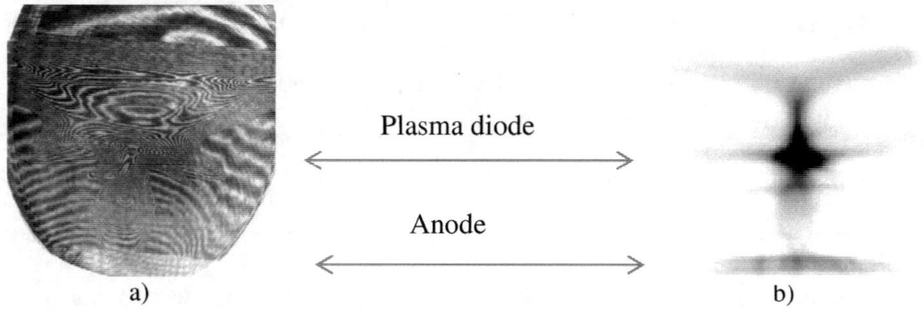

Figure 1. 1-ns laser interferometric (a) and self-luminescence (b) pictures of the pinch during the plasma-diode formation on the plasma column.

Value of current of fast electrons I_{fe} can be sharply inferred from X-ray dosimetry, i.e. by measuring an X-ray dose ($D = F \times \tau_x$) where F is energy flux with its spectral distribution (depends on U_{eff}), anode material (Z) and pulse duration (τ_x):

$$F \sim k_x Z U_{eff}^{\ 2} I_{fe} \text{ where } k_x = 10^{-9}\ [\mathrm{W/(V^2A)}]$$

Current of fast ions I_{fi} can be estimated by measuring neutron yield taking into consideration size (V_{pinch}) and density (n_{pl}) of the pinch plasma, time duration of the fast ions pulse (τ_{fi}) with the corresponding cross-section (σv) and magnetization degree of the ions k_m (5…10 [2, 3]):

$$Y_n \sim n_{pl} N_{fi} (\sigma v) V_{pinch} \tau_{fi} = n_{pl} (I_{fi} / e v_{fi} S_{fi}) (\sigma v) V_{pinch} \tau_{fi}$$

According to these measurements and formulae both types of currents (of fast electrons and high-energy ions) *sequentially in time* substitute almost the overall discharge current in DPF during the course of the CA phenomenon.

In particular one may see that the total number of fast deuterons generated in the PF-1000 facility [2, 3] is about 10^{18}. It means that a DPF is a good accelerator (efficiency of the bank energy conversion to the overall energy of the fast ion beam is $\eta \sim 10\%$) but a bad target (the product of density and dimensions is $n_{pl} \times l_{pinch} \ll 1/\sigma_{fusion}$, and the Lawson criterion is about $n_{pl}\tau \sim 10^{11}$ [$\mathrm{cm^{-3}s}$]).

2. Medium-sized DPF and their applications

At present time the following two types of energy transfer from the primary source to the load are mainly used: capacitive storage with the circuit closer and inductive storage with the circuit interrupter:

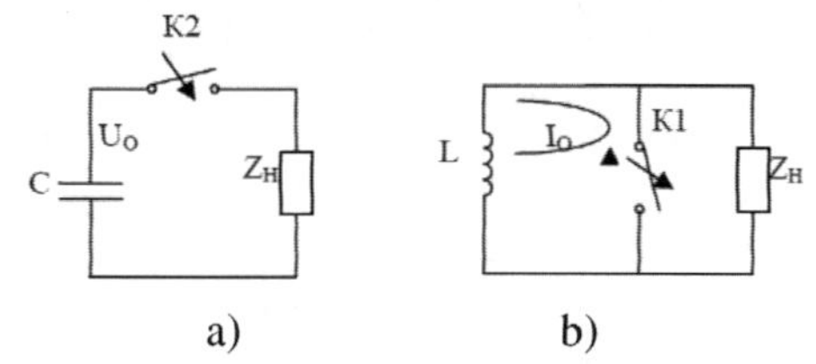

Figure 2. Capacitive and inductive storage systems with current closer and current interrupter correspondingly.

In the second case where L = const and when the load $Z_H \approx L_H$ in the most favorite case (when $L \approx L_H$) according to the Ampere's law the two portions by ~ ¼ part of the overall energy accumulated first in the magnetic storage system L will correspondingly be preserved in this primary storage L and another ¼ will be transferred to the load $Z_H \approx L_H$, whereas ½ part of it will be lost inside the breaker K_I.

In the medium-sizes DPF devices (with energies in the capacitor banks from 1 kJ till 100 kJ) one can reach the optimal (variable!) ratio of external (L_{ext}) and internal (L_{int}) inductances. Namely in the beginning of the discharge (at the breakdown phase) $L_{ext} >> L_{int}$, but during the pinching stage the ratio becomes just opposite $L_{ext} << L_{int}$ due to a variation of the dynamical internal inductance.

Moreover, during the maximal compression, when our pinch represents by itself a plasma inductive storage, the current breaker is situated *inside* the pinch (formation of a plasma diode). Energy that is usually lost in the external current interrupter in the classical scheme (*K1* in Fig. 2) is converted within the DPF pinch into beams of fast electrons and ions.

Because of the above two unique features (the dynamical change of the ratio L_{ext} / L_{int} and the active role of the current interrupter) it is possible to transfer from "plasma inductive storage" to the "load" (in the ideal theoretical approximation) up to 100% of the energy of azimuth magnetic field of the pinch, which is of the order of primary energy stored in the capacitor bank.

It is interesting to compare parameters of the discharge of the PF-6 device (IPPLM, Warsaw, Poland; C_{bank} = 32 μF, $E_{bank} \approx$ 6.4 kJ, $I_{max} \approx$ 760 kA, $\omega_{disch} \approx 10^6$ c^{-1}, $L_{pinch} \approx$ 30 nH): $\rho = \sqrt{L_{pinch}/C_{bank}} \approx Z_{pinch} = \omega L \approx R_{EMH} = 30uc \approx$ 30 mΩ (where u – current drift velocity, c – speed of light), i.e. the wave impedance ρ, the impedance of the pinch ($Z_{pinch} = \omega L$) and the electron magneto-hydrodynamic resistance ($R_{EMH} = 30uc$) [4] of the plasma diode are equal. Thus matching of all important circuit characteristics is excellent.

Such an apt combination of electrical parameters of the circuit facilitates a broad range of applications of these devices in various fields of science and technology with a very high efficiency of use of ionizing radiations generated by them.

Among these areas we shall present recent results obtained in radiation material science (testing of materials perspective for use in nuclear fusion devices), in unveiling of hidden fissile illegal items in a luggage, in dynamic quality control of machines and mechanisms in a course of their operation, in radiation biology (radio-enzymology), and in nuclear medicine (pulsed neutron treatment of cancer). The most important advantages in the applications of a DPF are demonstrated in those cases where nanosecond radiation pulses are used in pulsed radiation physics/chemistry/biology in their "perfect sense" [5], i.e. when micro-volumes of activity of secondary particles are overlapped during a time interval, which is short compared with the reciprocal physical/chemical/biological process. This time we used for irradiation 4 devices: PF-6 and PF-1000 (both IPPLM, Poland), the "Bora" facility (ICTP, Italy) and PF-5M (IMET, R.F.).

2.1. *Radiation material science*

In the past year we have irradiated targets made of different materials counted as the perspective ones for implementation of them in nuclear fusion reactors (Ti-based alloys, Mo, carbon-based composites CFC and SiC, low-activated stainless steels – all prepared in IMET, R.F., and in particular the tungsten-based material, identical samples of which were supplied to the participants of the IAEA CRP by the team of the FZJ, Juelich, Germany). Our main aim in these experiments was to test the above samples by fusion relevant pulses of hot plasma and fast ion streams generated in DPF. Results of their analytical examination after irradiation are presented in Fig. 3.

Fig. 3*a* shows a W-sample's photo after irradiation. One may clearly see a region (in the low-right corner) of the action of strong beam of fast deuterons (average energy is E_d ~ 100 keV). The rest part of the sample was irradiated mainly by a hot plasma stream (T_e ~ 300 eV, v_{pl} ~ 2×10^7 cm/s). In the border between these zones an optical microscopy (*b*) has shown pores and ridges in the plasma-action area (power flux density here is estimated as ~10^9 W/cm^2, penetration depth of plasma ions is about 1 nm). Remelted surface in the fast ion beam action part of the sample with cracks (fracturing pattern) on it is seen better in the SEM picture (*c*). Roughness of the surface was measured by atomic-force microscopy (Fig. 3*d*). In Fig. 3*e* we have presented a result obtained by an X-ray micro-tomographic computer technique using 150 kV HAMAMTSU micro-focus X-ray source L8121-03 and COBRA micro CT software. The picture is a cross-section of a Ti foil taken in a certain part of the sample after its irradiation. In the slice one may see some imperfections that are probably a result of a damage produced by a shock wave generated and penetrated through the foil after the action of the beam of fast deuterons.

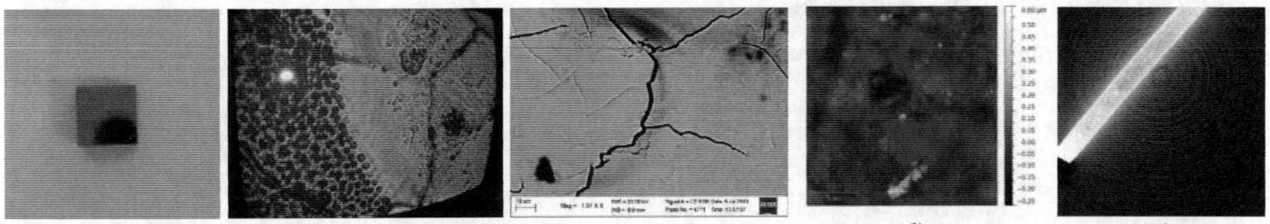

a) b) c) d) e)

Figure 3. Macroscopic image *(a)* of the tungsten specimen No. 40 after its irradiation in DPF, an optical microscopic picture of it *(b)* at the border between zones of irradiation by hot plasma and fast ion beam, a scanning electron microscopic view *(c)* of the part of the same sample with the strongest action of the ion beam, an atomic force microscopy *(d)* of the sample with its roughness, as well as one cross-section of the Ti foil made by the X-ray micro-tomography *(e)*.

It is important to note that we can investigate the samples not only *after* their irradiation by various analytical methods. In fact we are able *to monitor the irradiation process itself* with high temporal, spatial and spectral resolution by laser interferometry (Fig. 4 and 5) or by registering self-luminescence of plasma (Fig. 1*b* and Fig. 6).

Processing of an interferometric picture taken at the beginning of the formation of secondary plasmas produced on the anode by the self-focused beam of fast electrons (on the left side) and by the divergent stream of fast ions on the stainless steel plate of 2-mm thickness (on the right side) is shown in Fig. 5.

From processing of the interferometric pictures one may see that in secondary plasma produced by the electron beam near the anode surface the upper measured limit of density is higher than in secondary plasma produced by the ion beam. It results from the fact that the lateral dimension of the first type of plasma is small compared with the second one. In its turn it is a consequence of self-focusing of the electron beam in opposite to the divergent ions beam. Thus the upper limit is determined by crowding of interferometric fringes.

From Fig. 6 one may see that in the first period of time after the CB phenomenon our spectrometer registers mainly deuterium plasma lines (D_α, D_β...). This is our "primary" plasma generated in the pinch and colliding with the target. Then at the moments near the second microsecond after CA secondary plasma is generated near the target (in this case a sample of SiC).

One may see here mainly lines of the target's material (C and Si). The degree of ionization (e.g. an observation of C IV lines) confirms the high value of secondary plasma temperature that was deduced earlier from the velocity of its expansion. In the late stages the spectrometer registers a mixture of both substances.

In the very late stage (10μs after CA phenomenon) at using for an irradiation in the PF-1000 facility a thin (2 mm) sheet of stainless steel we were able to register a shock wave generated inside the target by the ion beam and escaping subsequently from the rear side of the blade (see Fig. 7).

What is of a particular interest in the picture – it is a "post pinch" that continues to exist during a very long period of time (much more than 10 microsecond after CA phenomenon).

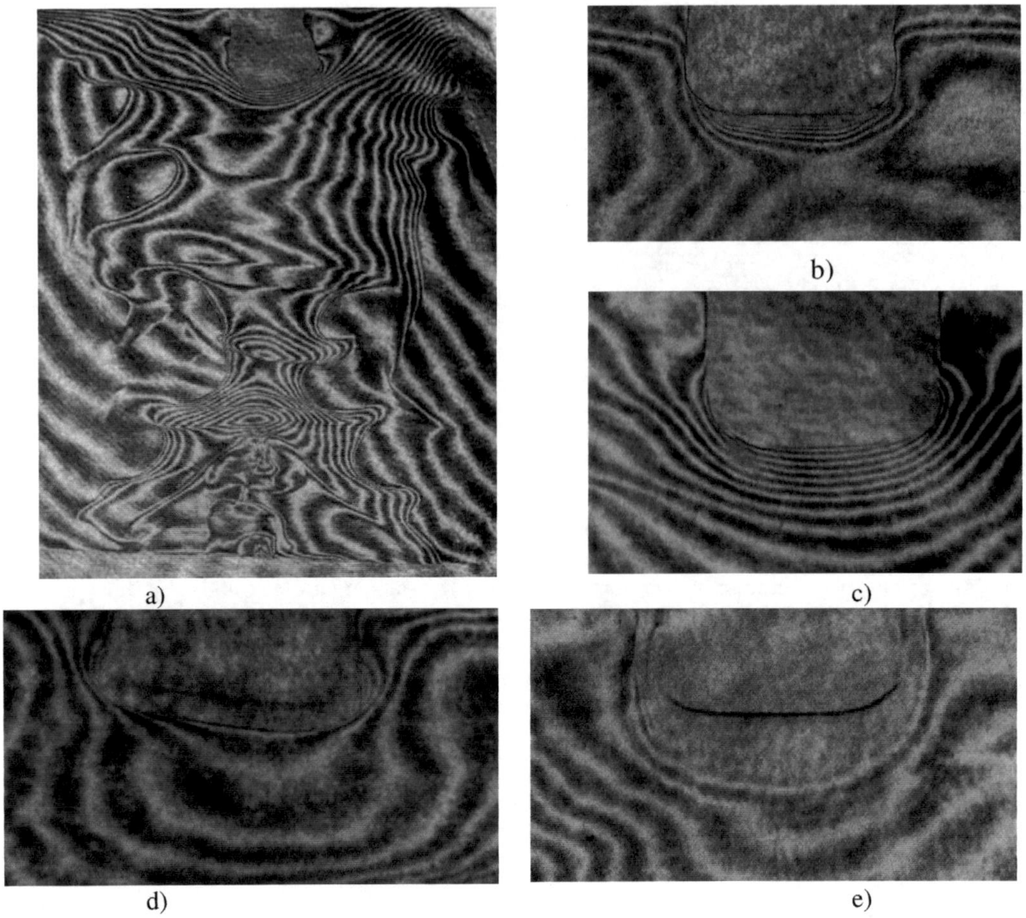

Figure 4. Laser interferometric images of target's plasma (specimen's material – SiC) taken at 4 consecutive time moments (temporal resolution – 1 ns): beginning of the secondary plasma formation (*a* and *b*), well-developed secondary plasma cloud (*c*), beginning of the formation of the unloading wave (*d*) and well-developed unloading wave (*e*).

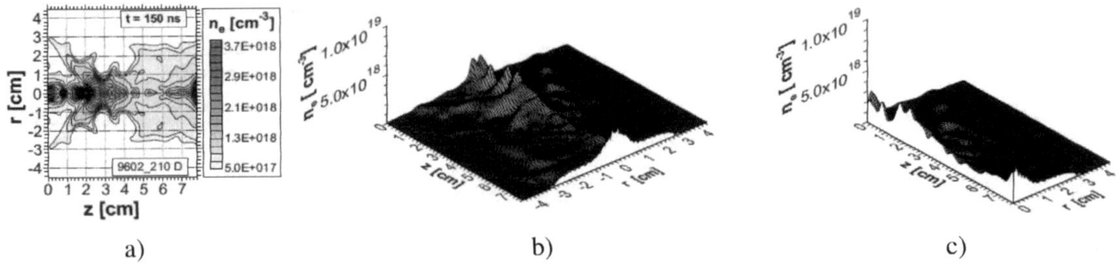

Figure 5. Electron density distribution (*a*) according to the processing of the interferometric image (Fig. 4 *a*) of the interaction of the powerful electron (left-hand side) and ion (right-hand side) streams with targets: *b*) – two-dimensional map, *c*) – cross-section along the Z-axis.

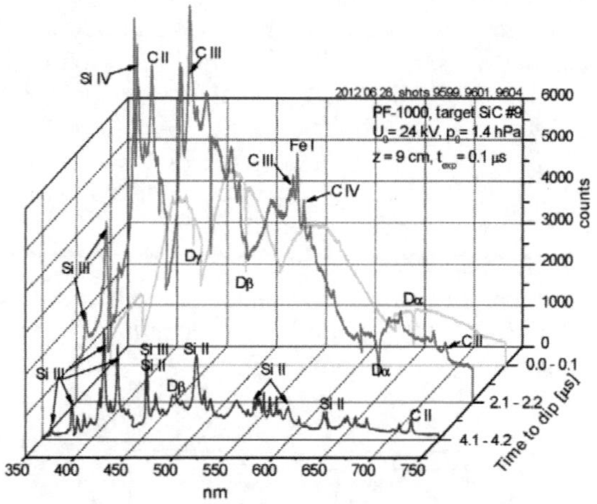

Figure 6. Time evolution of the secondary plasma spectrum taken in visual range with temporal resolution equal to 100 ns.

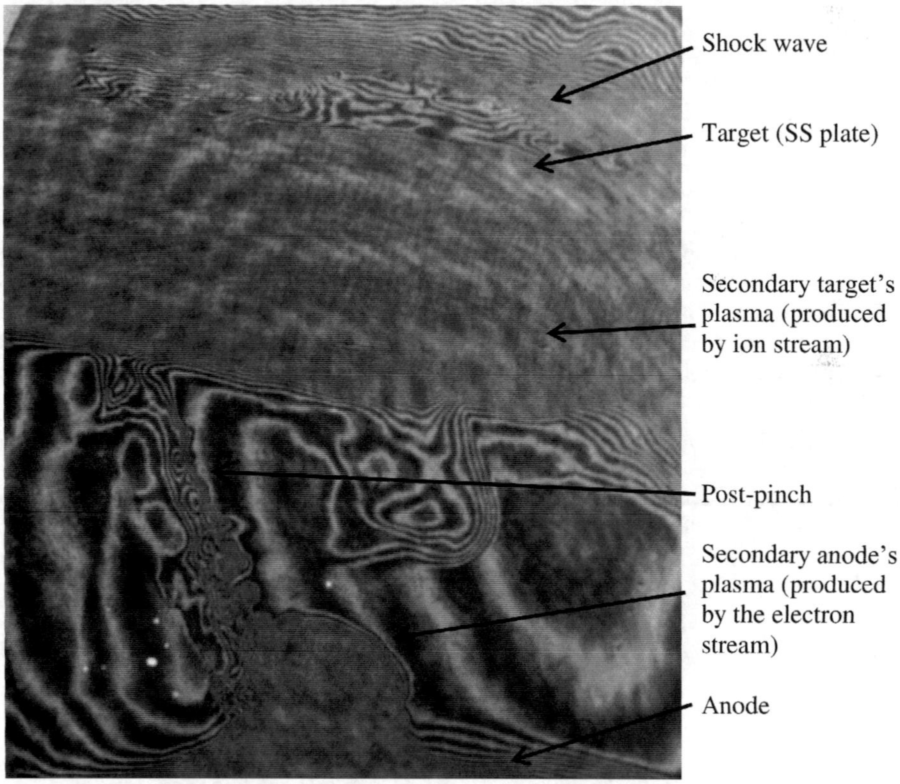

Figure 7. Late stage of the secondary plasmas/pinch development (10 µs after the current dip) taken at the irradiation of thin stainless plate with a post-pinch and a shock wave escaping from the rear side of the plate.

Together with the data of measurements provided in the collector part of the DPF chamber (in the PF-1000 facility) by a set of magnetic probes we have learned that at this configuration with the large-size metallic blade placed in a distance of ~ 10 cm from the anode the pinch is separated from the main electrical circuit and works as a torus-like self-organized magnetic storage dissipating its energy during period of time of the order: $\Delta t \sim L/R$ (~ 100 μs for the PF-1000), where L is the post-pinch inductance and R is its resistivity. This result is supported by the spectroscopic investigations (of the type of Fig. 6).

2.2. *Microlithography and micromachining*

DPF can be effectively used by means of irradiating of photoresists with soft X-rays for microlithography (with Ne-filled chamber) [6] and micromachining [7], i.e. in manufacturing of 3D structures (with Ar-filled chamber).

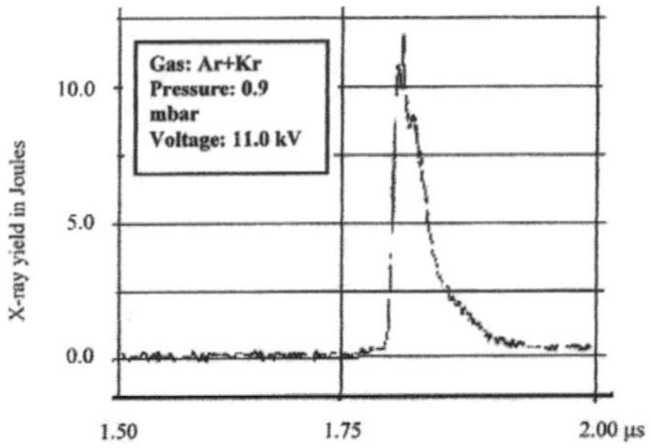

Figure 8. Soft X-ray pulse obtained at the DPF operation with Ar-Kr gas mixture.

In Fig. 8 one may see an X-ray pulse obtained in the second case whereas Fig. 9 is a picture of a structure obtained in such schemes at the National Institute of Education (Singapore) with a photoresist with chemical amplification.

As it will be seen from the subsequent part this micromachining technique can in principle be used for manufacturing of super-miniature DPF devices.

Figure 9. 3-dimentional structure obtained in the wave-length rang near 0.4 nm at the photoresist with chemical amplification.

Journal of Physics: Conference Series **591** (2015) 012020 doi:10.1088/1742-6596/591/1/012020

2.3. *Dynamic quality control of machines and mechanisms in a course of their operation*

DPF as a source of very bright nanosecond pulses X-rays and neutrons can be implemented for the dynamic quality control of machines and mechanisms in a course of their operation. In principle these two types of radiation are additive and can provide information on dissimilar features of the object under irradiation (e.g. X-rays can give us information about high-Z materials whereas neutrons – about low-Z – light – parts of it). At this stage we used hard X-ray flash generated by "Bora" device.

Spatial resolution of an image of a mechanism's detail (turbine's blade, car tire, piston of a car engine, etc.) taken during its operation by a flash of the hard X-Ray radiation from DPF determined by a number of factors:

- Pulse duration of X-Rays (ns)
- Size of the source of X-Rays and its remoteness from an object (< 100 μm, usually from 10 cm to a few meters)
- Diffraction (wavelength, distance)
- Contrast degree of an object's detail to be visualized (spectrum of hard X-Rays)

Theoretically for DPF it could be ~ 1 μm in a 10-cm distance at the rotation speed of a machine's part ~ 10,000 rotations per minute.

We have made several experiments in this field including the last one with a plastic fan during its operation. In the below Fig. 10 one may see the overview of the experiment (*a* and *b*) and two pictures taken with soft X-rays irradiating by the "Bora" device (ICTP) – one in a static position of the fan (*c* - 7 shots of DPF) and the other one during the fan's rotation (*d* – one shot of the "Bora" device). It is possible to see that the dynamic picture does not differ from the static one.

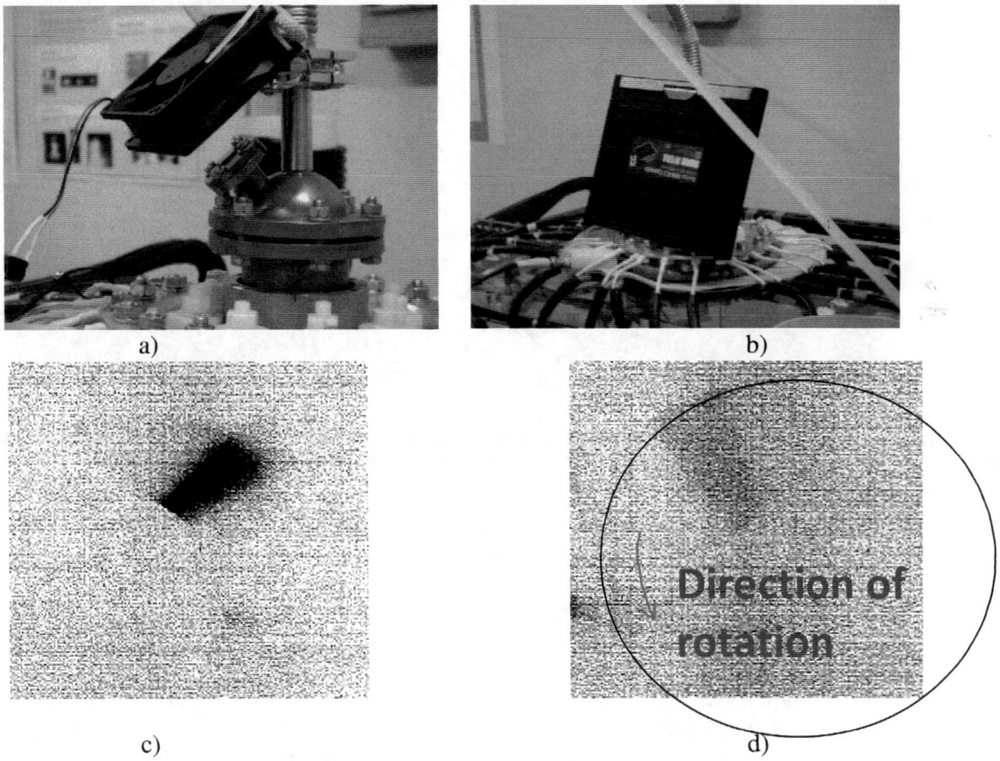

Figure 10. The arrangement of the experiment on the dynamic quality control (*a* and *b*) and two X-ray pictures of a fan in its static (*c*) and dynamic (*d*) states.

This technique opens opportunities in a dynamic quality control of machines during their operation. It can be important e.g. in mass production and defectoscopy of car tires moving in the conveyer, in a woodworking industry, in a quality control of turbine's blades when a "closed" defect might be visualized during the rotation because of a centrifugal force that may "open" this crack.

2.4. *Radiation chemistry/biology*

Formerly in these experiments enzymes were irradiated *in vitro* with various *doses, dose power and spectral range* by *X-Ray photons.*

We have found here a very large (4 orders (!) of magnitude) *difference in doses for the enzyme activation/inactivation* by their irradiation with X-Rays from DPF compared with the same procedure using an isotope source (Cs^{137}) [8].

However our attempts to find the same effect with *neutron irradiations* produced by DPF, neutron generators and pulsed fission reactor BARS were failed.

In these experiments the maximum of 14-MeV neutron tissue kerma obtained with DPF devices was 0.8 Gy, whereas for X-rays it was 0.5 Gy. ING-104 pulsed neutron generator is a Dense Plasma Focus (DPF) device, which serves as a source of nanosecond pulses of X-rays and neutron radiation. Pulses duration ranges from 20 to 30 ns. Below we present results of cytogenetic effects of mixed X-ray and 14-MeV neutron radiation of nanosecond pulses on human lymphocytes (Fig. 11 and 12).

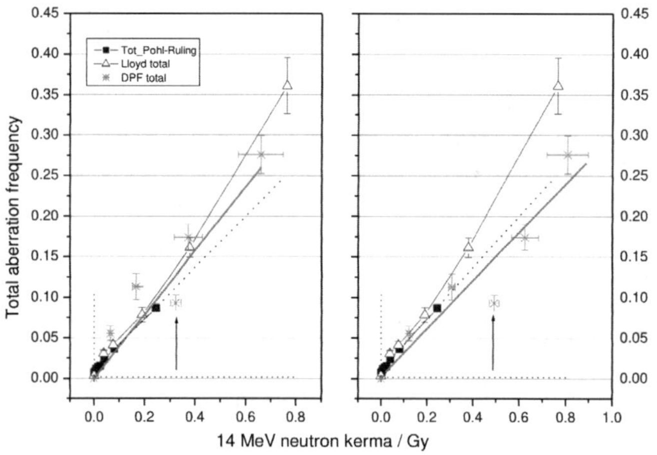

Figure 11. Dose curves of the total chromosomal aberration yield in human lymphocytes irradiated with 14 MeV neutrons. Left panel – neutron kerma was calculated using ^{237}Np fission detector; right panel – neutron kerma was calculated using SIVN 61 neutron yield monitor with activation silver detector. Linear regression equation are $Y_{tot} = (0.041 \pm 0.009) + (0.36 \pm 0.022)K$ and $Y_{tot} = (0.021 \pm 0.01) + (0.28 \pm 0.028)K$, respectively. DPF – Dense Plasma Focus.

5 plastic tubes with blood samples were placed in a cylinder plastic container using special holder. Diameter (inner) of the container was 55 mm, diameter of each tube was 10 mm, and wall thickness was 1 mm. Since DPF produced only 1 pulse every 10 minutes, to prevent damage reparation between pulses container was filled with melting ice and then it was mounted on the top of the plasma focus discharge chamber. Fission detectors with ^{237}Np were placed outside the container opposite to each tube with blood. Neutron kerma was calculated using two types of detectors – the above-mentioned ^{237}Np detector and the activation silver detector, which was the part of neutron yield monitor SIVN 61. In calculating kerma we supposed that neutron flux decreased according to $1/R^2$ law and the plasma focus neutron source was 25 mm below the top of the discharge chamber. To estimate the influence of scattered neutrons on detectors indications a Monte Carlo calculations were made with MCNP-4b code.

X-rays doses were measured with a 27012 clinical dosimeter (Dresden). The blood samples were received 16 pulses in total.

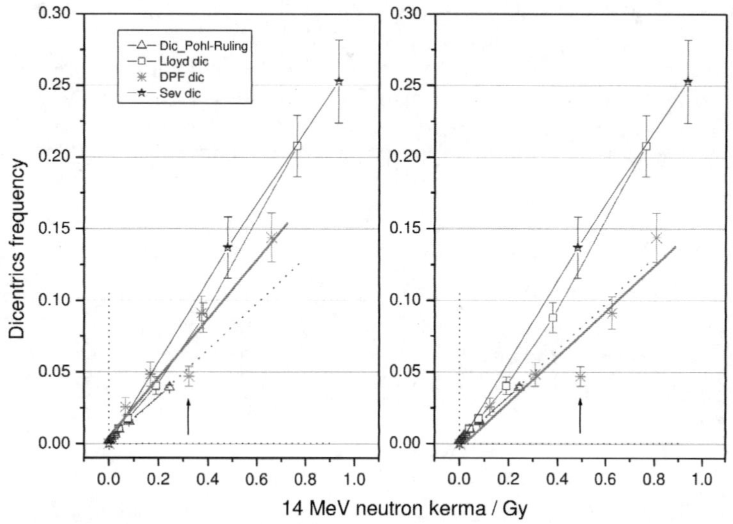

Figure 12. Dose curves of the dicentric chromosome yield in human lymphocytes irradiated with 14-MeV neutrons. Left panel – neutron kerma was calculated using ^{237}Np fission detector; right panel – neutron kerma was calculated using SIVN 61 neutron yield monitor with activation silver detector. Linear regression equation are $Y_{dic} = (0.012 \pm 0.006) + (0.21 \pm 0.026)K$ and $Y_{dic} = (0.005 \pm 0.01) + (0.15 \pm 0.026)K$, respectively.

At least three conclusions may be inferred from the above figures:

- First, the major contribution to the cytogenetic effect of the DPF neutron generator radiation is made by *neutrons.*

- Second, pulsed neutron radiation of *short, nanoseconds, duration is as effective as it is at continuous mode of neutron radiation, at least, within the mutual experimental uncertainties at its relatively high values (< 10^8 Gy/min)* for 14-MeV neutrons. It gives hope of applicability of DPF for the potential methods of neutron therapy *instead of cumbersome, expensive and dangerous nuclear reactors.*

- Third, our estimations made for boron-capture neutron therapy with Li ions and α-particles as products of nuclear reactions with their mean-free paths (6 and 9 μm respectively) have shown that for the above-mentioned overlapping of activity of the nuclear reaction products we need a flux of neutrons ~ 10^{12} n/cm^2.

2.5. *Nuclear medicine (pulsed neutron treatment of cancer)*

The main ideas to expect the same effects as with X-rays namely in malignant cells enriched with boron to higher degree compared with sound cells are as follows:

- Usage of *epithermal neutrons* (high cross-section, deep penetration)
- S*imultaneous destruction* of both spirals of DNA of malignant cells during the ns period of time by a high-density neutron flux (a "shock-like action" at a double-strand rupture)
- *Threshold-like behavior* of radiation damage of malignant cells due to the overlapping of micro-volumes of activity of free radicals and nuclear reaction products only in malignant cells.

All these opportunities might open ways for a *low-dose therapy* of cancer.

It is clear that to gain these opportunities we have to change spectrum of fast neutrons from DPF into spectrum with a large quantity of epithermal neutrons. However we must:

1) Not to lose too many neutrons

2) Not to increase neutron pulse duration too much (should be < 1 μs that is the time duration of biochemical reactions with free radicals).

In this work we focus on the development of a detailed simulation of interaction of short-pulse radiation generated by a DPF with tissue to estimate the absorbed dose by the cells for this dynamic case. The simulation was carried out by means of the Geant4 code [9], a toolkit to simulate the interaction of radiation with matter, originally developed for nuclear and particle physics. One of the results is presented in Fig. 13.

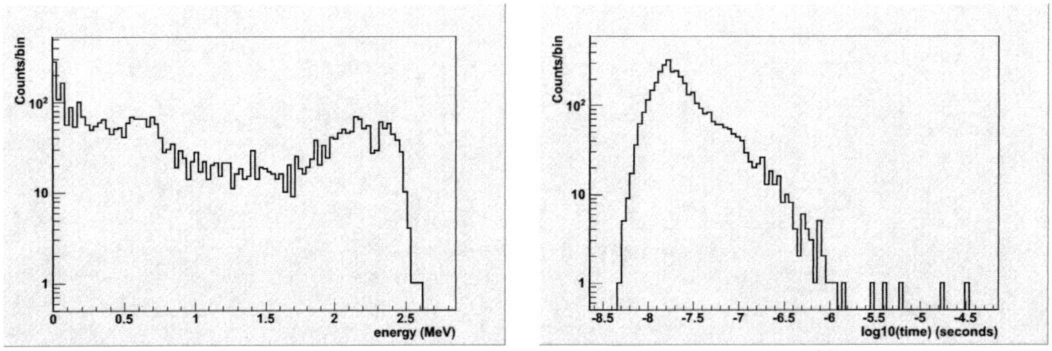

a) b)

Figure 13. Neutron energy spectrum (*a*) and time profile (*b*) obtained by simulating a monochromatic pulse of 10 ns of 2.45-MeV neutrons in the small chamber of the "Bora" device with a 15-cm Fluental TM moderator.

From this picture it is clearly seen that we have fulfilled the above-mentioned conditions. We investigated also some other moderators (e.g. tungsten) and we have also done such works for the 14-MeV neutrons that can be generated in DPF filled with D-T mixture.

2.6. *A single-shot nanosecond neutron pulsed technique for the detection of fissile materials*

In principle there are two methods that can be applied with DPF-generated neutrons due to the fact that spectrum of fission neutrons induced in fissile materials from an external source ("Watt spectrum") covers the range extended from 0.5 till 5.0 MeV with a maximum positioned near 1.5 MeV:

1) To use 2.5-MeV neutrons and look for a component of fission neutrons having energy *higher* than 2.5 MeV

2) To use 14-MeV neutrons and look for a component of fission neutrons having energy *lower* than 14 MeV.

Our MCNP modeling has shown that our neutron yield generated with pure deuterium (10^9 n/shot) is not enough for the first method; so we used 14-MeV neutrons.

Experimental results obtained with the PF-6 device working with D-T mixture on the irradiation of the EK-10 fuel element are shown in Fig. 14 and 15.

From these pictures one may clearly see neutrons elastically and inelastically scattered by elements ingressed into the fuel element. It is essential to note that these oscilloscope traces were obtained just in one shot of our DPF. This single-shot technique of collecting information during about 100 ns is in particular important for application in unveiling of fissile materials hidden in fast-moving objects (cars, trains, ships, etc.).

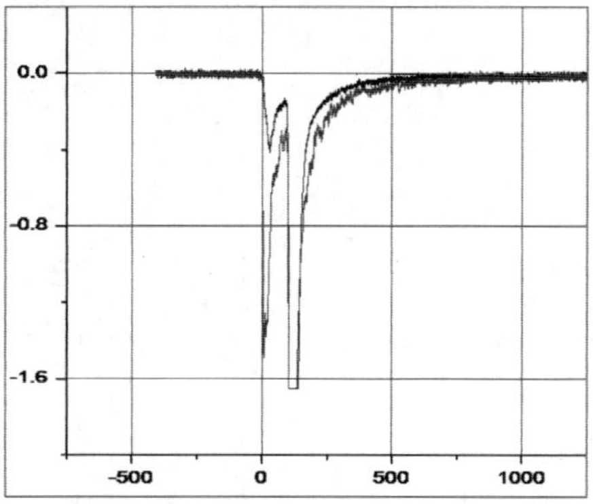

Figure 14. Two oscilloscope traces overlapping one another: one is taken without fuel element (the black one, smooth) and another one is taken with fuel element EK-10 (the blue one with multiple peaks).

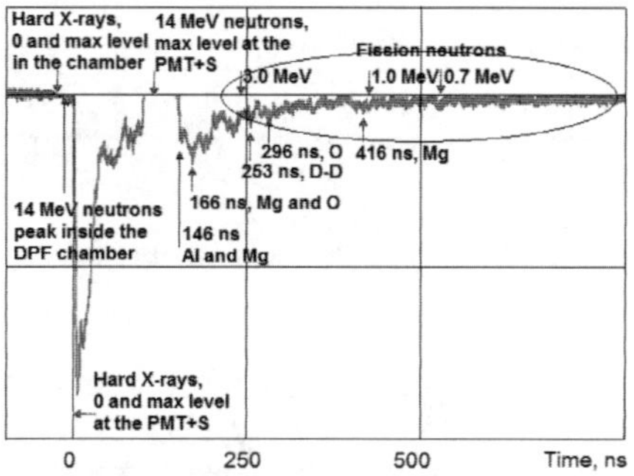

Figure 15. A result of the subtraction of oscilloscope traces with attribution of different peaks and comparison with the results of MCNP modelling calculations.

3. Contemporary scaling laws and drivers for DP foci

Fig. 16 presents a collection of experimental results on neutron and hard X-ray yields obtained to the moment with DPF devices of energy ranged from 1 J till few hundred kJ. We add some possible data using approximations to much higher and much lower currents/energies of the DPF discharge.

As it was found in the end of 80's of the last century for the medium-sized DPF neutron yield scaling may be roughly described by a formula:

$$Y_N \sim n_{pl} N_i (\sigma v) V_{pl} \Delta t \sim I_{pinch}^4 ,$$

where plasma density $n_{pl} = const$, *density of fast ions* $N_i \sim I_{pinch}$, $T_{pl} = const$, Energy Distribution Function of fast ions ~ *const, i.e.* $(\sigma v) \sim const$, $V_{pl} \sim I_{pinch}^2$, $\Delta t \sim I_{pinch}$ – Plasma Confinement Time.

For the hard X-ray yield the situation is a bit more complicated due to the facts that we have in DPF not just an electron beam but relativistic electron plasma ($v/\gamma >> 1$) and the absorption (deceleration) of fast electrons in the anode's material is accompanied by a collective plasma mechanisms and by a so-called "magnetic stopping" within the anode's plasma. But roughly:

$$Y_X \sim I_{pinch} U^2 Z_{anode} \tau_e \sim I_{pinch}^2,$$

where $U \sim E_i$ (more precisely speaking the distribution function of fast electrons) ~ *const*, $\tau_e \sim I_{pinch}$.

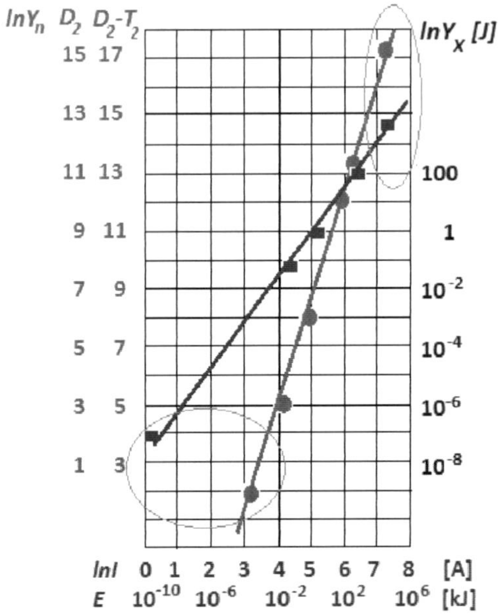

Figure 16. Experimental scaling for neutron and X-ray production in DPF. Regions in a circles has not been reached yet.

DPF is a high-current accelerator producing nanosecond pulses of fast particles without expensive and cumbersome intermediate elements of the type of PULSE FORMING LINES (e.g. the Blumlein line) that are not very reliable in exploitation, very expensive and have problems at their exploitation with a high repetition rate. However in the case of *large DPF devices* the usage of the in-between elements as in a classical scheme of high-current high-voltage generators (see e.g. in Fig. 17 a design map of GAMMA facility) might be productive from the point of view of matching all elements of it.

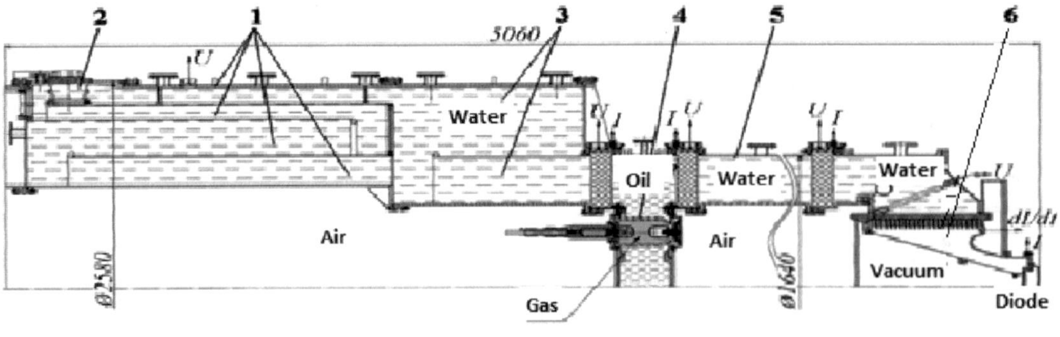

Figure 17. Design map of a high-voltage section of accelerator with a high-current vacuum diode (GAMMA facility, Sarov, R.F.): 1 – the first four double-stepped forming line – DSFL – cascades, 2 – DSFL multichannel switch, 3 – DSFL output cascade with a built-in duration converter, 4 – pre-pulse switch, 5 – WTL, 6 – vacuum insulator with MITL and vacuum diode.

Indeed after current abruption taking time of about an inverse increment of Langmuir plasma instability (in ps range) a DPF chamber of the Mather geometry is nothing more than a Transmission Line with Magnetic Insulation (MITL – see Fig. 18) where the overall e/m energy is transported as a e/m wave to the plasma diode as it is realized in the above Fig. 17. So in the case of a very large DPF it is impossible to expect a fulfillment of a good matching of current generator (primary energy storage) with the plasma diode.

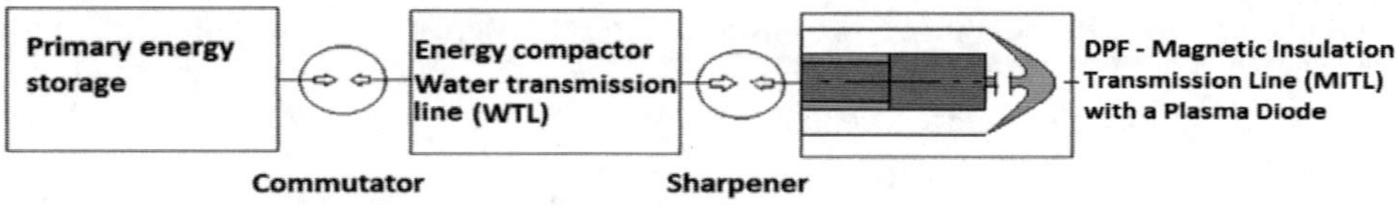

Figure 18. DPF with intermediate matching element.

In this case due to this intermediate element we can match the primary energy source with the load much easier. E.g. at the use of a double forming line (DFL – the so-called Blumlein line) the value of power released at the DPF will be (see e.g. [10]):

$$P_R = \frac{4U^2R}{(R+\rho)^2} = c\sqrt{\frac{\varepsilon}{\mu}}\frac{4E_m^{\ 2}R_1^2\ln^2(R_2/R_1)}{2(\ln(R_2/R_1)+\ln(R_3/R_2))}\frac{R\rho}{(R+\rho)^2} = c\sqrt{\frac{\varepsilon}{\mu}}\frac{2E_m^{\ 2}R_3^2\ln(X)}{X^2+X}e^{-\frac{2}{X}\ln X}\frac{R\rho}{(R+\rho)^2}$$

where E_m – maximal electric field in the line, R – load resistance (DPF's diode), R_i – radii of coaxial conductors of the line, $\rho = \rho_1 + \rho_2$ – summarized wave impedance of the line, $X = R_2/R_1$. By changing X it is possible to match DFL with DPF. Maximal power will be released in DPF at X = 1.52.

However at the numerical modeling of these schemes the full set of MHD equations must be used together with the *equations of telegraphy* at the final stage instead of simple electrotechnical equations of C-L-R circuitry to describe electro-magnetic wave travelling along the MITL.

Understanding the problems of matching of big devices with a large-scale DPF chamber in general and with its diode in particular as well as the problem of spending too much energy for the rundown phase in big DPF facilities we took out three patents (Fig. 19) in the beginning of 80's [11] with the following ideas:

1) Usual contemporary 1-MW electric power turbine has U = 30 kV and I = 30 kA. We may use transformer to change 30 kV into 10 V. In this case we shall obtain the output current I = 90 MA. We may use here an invertor (or convertor) to fit (match) operation frequency to the travelling time of the current sheath inside the DPF chamber, and/or -
2) To initiate the discharge by means of a laser pulse not to spend too much energy for the rundown phase (or to use one or several small DPF):

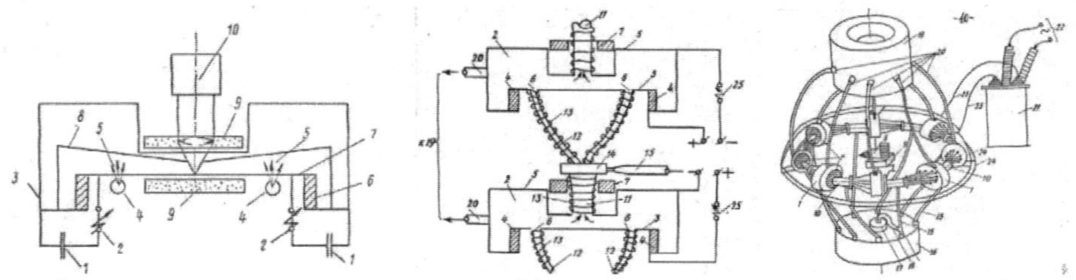

a) b) c)

Figure 19. 3 modifications of DPF or their chains to overcome problems with energy loss for a rundown phase.

Two important issues are raised in such schemes:

1) Whether recovering of electric insulation between anode and cathode of a DPF chamber after the discharge (or better to say a disconnection between the DPF chamber and primary source of energy) is possible?
2) How much power may we extract from the mains?

The answer to the first question has been obtained in our experiments. Namely, the interruption of current circulating in a DPF circuit may be ensured either in pseudo-spark switches (PSS) usually used by us [12] (see Fig. 20*a* and *b*) or directly in a gap between the anode and the cathode of the DPF chamber (as it is seen in Fig. 20*c* – during the pinch phase – and 20*d* – 10 μs later the CA phenomenon if the space above the anode is large enough).

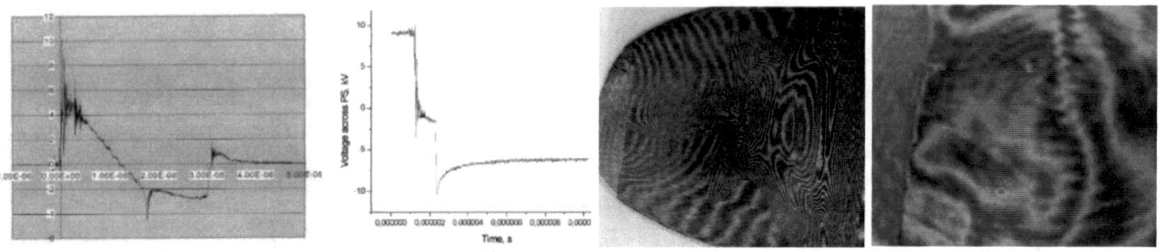

a) b) c) d)

Figure 20. PSS's disconnection: current (*a*) and voltage (*b*); blow-off of plasma in PF-1000: moment of the dip (*c*) and 10 μs later (*d*).

According to electric power generation laws an allowable unsinusoidality of the current generated may be not more than 5%. This demand implies the restriction upon our "dip" and consequently on the extracted power if we want to use the rest power to feed the common mains.

4. Possible applications of a very large DPF

4.1. *Medicine*

In previous paragraphs we have discussed a possibility of application of a DPF for the low-dose high intensity *single shot* neutron therapy of cancer. The characteristics of the radiation fields in this case of 'classical BNCT' by epithermal neutrons are as follows: neutron intensity on the tumor surface is 10^9 n cm^{-2} s^{-1}; time of irradiation – 40…60 min; dose – 60…70 Gy. It means that the overall neutron fluence is on the level of $3 \ldots 5 \times 10^{12}$ n cm^{-2} and the dose rate is equal to 1 $Gy \cdot min^{-1}$.

Unfortunately the present-day boron preparations are capable to increase concentration of boron atoms within the malignant cells by 3-5 times higher compared with normal cells only (instead of demanded figure equal to 20).

Large DPF (a few MJ) can ensure the dose power on the level exceeding 10^9 Gy/min at a neutron pulse of ~ 100-ns duration. For this we need a neutron flux ~ 10^{12} n/cm^2 acting during a time interval << the reaction time with Li, α and free radicals (i.e. < 1 μs). In this case we shall overcome the above-mentioned *threshold* in malignant cells *only*. Because of this we shall increase *selectivity*, and we shall be able to *decrease the overall fluence* by an order of magnitude, thus providing *treatment just in a single pulse*.

4.2. *Radiation material sciences*

One of the possible exploitations of DPF as a powerful source namely of *fusion* neutrons in material sciences is its use as the source for testing of materials perspective for the first wall components and construction elements in the magnetic confinement fusion and in the inertial confinement fusion reactors where the neutron interaction with materials will be of the "explosive-like" type.

Journal of Physics: Conference Series **591** (2015) 012020 doi:10.1088/1742-6596/591/1/012020

The IFMIF (~10^9 US$) as expected will be able to ensure during one year of irradiation *the neutron fluence on the level from 0.01 till 50 displacements per atom* (dpa) in various zones positioned near a target having the area 5×20 cm^2. The value of 1 dpa corresponds to the fluence of 10^{21} n/sm^2 for Be- and C-based materials.

If in the DPF facility the pinch current would be about 20 MA its neutron yield will exceed 10^{17} of the 14-MeV neutrons. Taking into consideration a possible geometry of the near-pinch anode part of the DPF of this energy level it is easy to estimate that 100 dpa can be succeeded during an operational year in a volume of about 1 liter with the irradiation area ~ 0.1 m^2. But for all that a DPF must work with a repetition rate 3-4 cps with the irradiating zone positioned 0.1 m apart from specimens. Such a device would be cost 100 times less compared with IFMIF in spite of the fact that during this experiment we shall have to change elements of the facility 10 time during the year.

4.3. *Energetics*

Let us assume a DPF device with energy consumption during the shot on the level 30 MJ (so the overall amount of energy circulated within the facility is about 100 MJ). Such DPF may produce about 5×10^{17} n/pulse (~ 1 MJ in neutrons). We have made calculations for the hybrid reactor with a fission blanket having 100 time coefficient of gain [13] (see Fig. 21).

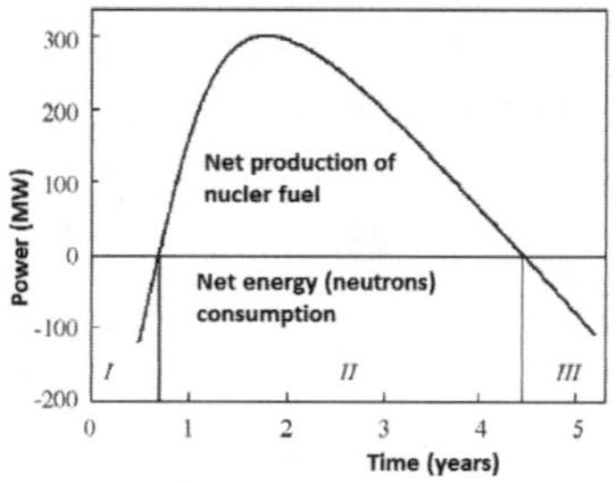

Figure 21. Evolution of energy in a hybrid reactor based on a DPF facility

We obtained the following parameters for the thermal power of the reactor equal to 0.5-1.0 GW and the operation period 2-5 year, efficiency of the fuel cycle 0.4, blanket thickness (for pure U^{238}) 12 cm and thickness of the SS wall 2 cm: uranium mass = 0.8-4 t, mean radius of energy zone is 0.33 m, mean operation rate of the DPF device is 13 cps, mean electric power consumed by the neutron source = 100-300 MW, mean useful electric power at the output = 270 MW. DPF size will be about 2.5 m.

However before reaching by the reactor of the positive value of the energy balance it will consume 6×10^8 kilowatt-hours of electric energy. This is an atonement for the low efficiency of the DPF-based neutron source. But we may use for this purpose the hydro-electric power generated in an estuary of a river in Siberia (see Fig. 22).

a) b) c)

Figure 22. Map of Siberia (*a*), Lena's estuary (60 km) (*b*) and Siberia by night (*c*)

Indeed, in these places because of a very large width of a river (*b*) a power dam may be very low (just 1-3 meters by height). It is good from ecological point of view. But before it was not used as in these regions of Russia are almost uninhabited (*c*). And transportation of electric energy from these territories to industrial regions or delivering of raw materials to these areas are economically ineffective. However with the above-described hybrid nuclear power stations we may interrupt the campaign at its summit and deliver a fission fuel to the industrial territories by a single ship per year.

5. How small a DPF can be?

From the above-presented scaling diagram one may see that at the values of the discharge current of DPF on the level of about 1 A we cannot expect any neutron yield. However the "hard X-ray" yield could be of the order 10^{-8} J that means up to 10^7 hard X-ray photons irradiated in full solid angle!

Pulse duration can be estimated using present day data on this parameter in the range of the bank energy 10^6 down to 1 J. It appears that this duration will be in the range 0.1 – 1 ps. This time is of the order of the increment of Langmuir turbulence presumably responsible for the current abruption phenomenon. So it is feasible.

Size of electrodes of such a device can be calculated using the Lee parameter. It appears that this "facility" will be in a sub-micrometer range of dimensions. With these dimensions the problem of ion saturation current becomes profound: in each successive shot we shall have or have not ion component. To support it we must change polarity and sputter a layer saturated with working gas. It means that such a source of the type of "current breaker" can be inserted in modern microchip structures. Manufacturing of these elements should be provided using micro-mechanics technology, e.g. with the help of medium-sized DPF working with Ar as a working gas.

6. Opportunities for a miniature DPF

In his 1971 paper, Chua extrapolated the conceptual symmetry between the resistor (voltage vs. current), capacitor (voltage vs. charge), and inductor (magnetic flux vs. current), and inferred the *memristor* [14] (see Fig. 23) as a similarly fundamental circuit element *linking magnetic flux and charge.* In contrast to a linear (or nonlinear) resistor the memristor has a dynamic relationship between current and voltage including a memory of past voltages or currents.

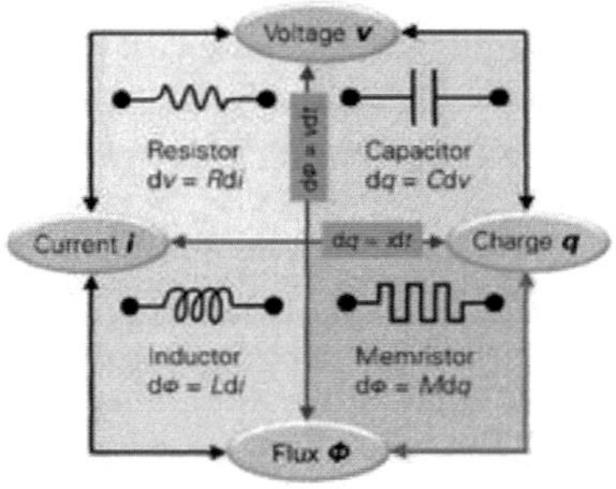

Figure 23. Zone occupied by a memristor in the symmetrical scheme of passive electrical elements.

Device	Characteristic property (units)	Differential equation
Resistor	Resistance (V per A, or Ohm, Ω)	$R = dV / dI$
Capacitor	Capacitance (C per V, or Farads)	$C = dq / dV$
Inductor	Inductance (Wb per A, or Henrys)	$L = d\Phi_m / dI$
Memristor	Memristance (Wb per C, or Ohm)	$M = d\Phi_m / dq$

Journal of Physics: Conference Series **591** (2015) 012020 doi:10.1088/1742-6596/591/1/012020

The resistance of a memristor depends on the *integral* of the input applied to the terminals, i.e. on the charge coming through it (rather than on the *instantaneous value of the input* as in a varistor). For some memristors, applied current or voltage will cause a *great change in resistance*. Such devices may be characterized as switches by investigating the time and energy that must be spent in order to achieve a desired change in resistance.

E.g. the HP memristor device is composed of a thin (50 nm) titanium dioxide film between two 5 nm thick electrodes, one Ti, the other Pt. Initially, there are two layers to the titanium dioxide film, one of which has a slight depletion of oxygen atoms. The oxygen vacancies act as charge carriers, meaning that the depleted layer has a much lower resistance than the non-depleted layer.

When an electric field is applied, the oxygen vacancies drift, changing the boundary between the high-resistance and low-resistance layers. Thus the resistance of the film as a whole is dependent on *how much charge* has been passed through it in a particular direction, *which is reversible* by changing the direction of current.

One of the resulting properties of memristors is the existence of a pinched hysteresis effect (see Fig. 24).

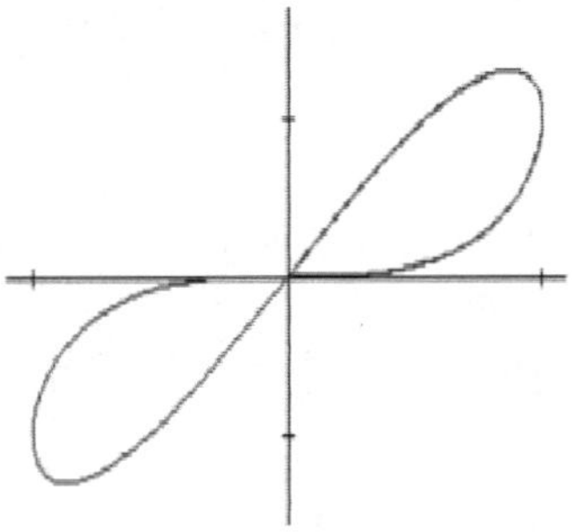

Figure 24. For a current-controlled memristive system, the input *x(t)* is the current *i(t)*, the output *y(t)* is the voltage *v(t)*, and the slope of the curve represents the electrical resistance.

The change in slope of the pinched hysteresis curves demonstrates a switching between different states of resistance, which is a phenomenon central to ReRAM and other forms of two-terminal resistance memory.

Following to these descriptions we may suppose that nano-sized foci may use the following opportunities to be counted (or hereafter used) as memristors:

1) DPF really is a circuit element *linking magnetic flux and charge:* $M = d\Phi_m / dq$ (the higher charge in a capacitor – the bigger values of current and reciprocal magnetic field)
2) Because of a very small size of the chamber we can control a discharge in time at any stage thus making *linkage between charge flowing through and magnetic flux*
3) We may use *ion saturation current effect* (in an analogy to the depletion of oxygen atoms in the titanium dioxide film of a HP© memristor that is under construction)
4) We may exploit a *great change in resistance* taking place twice in the DPF dynamics – in the breakdown phase and during CA phenomenon.

However additionally we may take *advantage from radiation properties* of this miniature DPF device. Indeed, because the overall number of hard X-ray photons is not small we can organize an electrical circuit in such a manner that this X-ray photon flux from DPF will open (or close) locally semiconductor elements (e.g. diodes or transistors) just through the space, i.e. without electrical connections. It can be done by interaction of X-rays with a semiconductor material and creation of uncombined carriers in the surrounding elements of a microchip.

In some cases it will be very important thus making such a memristor not simply a *passive element* of the circuitry but the *active one*.

7. Conclusions

With the above speculations we can see that in future there is a quite imaginable probability to gain prospects for:

1) Production of energy with *huge DPF facilities*

2) Implementation of these *large DPF facilities* in medicine for a *single-shot low-dose cancer therapy*

3) Use *this energy for feeding medium-sized DPF devices* that could be applied in different applications – radiation material sciences, biology, medicine, dynamic quality control, *single-shot* interrogation of illegal objects, but in particular for manufacturing of *very tiny elements of sophisticated computers* including super-miniature DPF as the parts of them

4) In the last case we may obtain IT technique not only with the super-miniature *passive* elements in its content but with the *active* ones.

8. Acknowledgments

This work was partially supported by the International Atomic Energy Agency CRP grants numbers RC-16932, RC-16954, RC-16955, RC-16956, RC-16960, RC-17167.

9. References

[1] Bernard A., H. Bruzzone *et al* 1998 *J. Moscow Phys. Soc.* **8** 1–93

[2] Gribkov V.A., Bienkowska B., Borowiecki M. *et al* 2007 *J. Phys. D: Appl. Phys.* **40** 1977–89

[3] Gribkov V.A., A. Banaszak, B. Bienkowska *et al* 2007 *J. Phys. D: Appl. Phys.* **40** 3592–607

[4] Kingsep A.S., Chukbar K.V. and Yan'kov V.V. 1987 Electron magnetic hydrodynamics *Voprosy Teorii Plazmy (Problems of Plasma Theory)* ed. by B.B. Kadomtsev **16** (Moscow: Energoatomizdat) 209 (in Russian)

[5] Gribkov V.A., Orlova M.A. 2004 *Radiat. Environ. Biophys.* **43** 303–309, DOI 10.1007/s00411-004-0259-2

[6] Gribkov, Lee S., Lee P. *et al* 2000 ISMA-2000: International Symposium on Microelectronics and Assembly, 27 November - 2 December 2000, Singapore

[7] Gribkov V.A., Lee S. *et al* 2000 Proc. of the ICPP-2000: International Congress on Plasma Physics, 42nd Annual Meeting of the Division of Plasma Physics of the American Physical Society, October 23 - 27, 2000, Quebec City, Canada

[8] Orlova M.A., Kost O.A., Gribkov V.A. *et al* 2000 *Applied Biochemistry and Biotechnology* **88** 243-55

[9] Giannini G., Gribkov V., Longo F., Ramos Aruca M. and Tuniz C. 2012 *Phys. Scr.* **86** 055801 (9pp) doi:10.1088/0031-8949/86/05/055801

[10] Arzhannikov A.V., Sinitskij S.L., Powerful pulsed beams, Teaching aid, Novosibirsk State University, 2007, 74 pp. (in Russian)

[11] Gribkov V.A., Tyagunov M.G., USSR patents Nos. 1012707 (26.01.1981), 1026439 (01.03.1981) and 1101838 (20.12.1982) (in Russian)

[12] V.A. Gribkov, M. Scholz, V.D. Bochkov *et al* 2004 *J. Phys. D: Appl. Phys.* **37** 2107-2111

[13] Gribkov V.A., Tyaganov M.C., Possible applications of a hybrid thermonuclear source based on the DPF device in modern energy complexes. – Nuclear Technologies in a Sustainable energy System, ed. by G.S. Bauer and A. McDonald, 1983, Springer-Verlag; B.,H., N.-Y.; 187-202

[14] http://en.wikipedia.org/wiki/memristor

Dense Plasma Focus – From Alternative Fusion Source to Versatile High Energy Density Plasma Source for Plasma Nanotechnology

R S Rawat[1]
Natural Sciences and Science Education, National Institute of Education
Nanyang Technological University, 1 Nanyang Walk, Singapore 637616

E-mail: rajdeep.rawat@nie.edu.sg

Abstract. The dense plasma focus (DPF), a coaxial plasma gun, utilizes pulsed high current electrical discharge to heat and compress the plasma to very high density and temperature with energy densities in the range of 1-10×10^{10} J/m^3. The DPF device has always been in the company of several alternative magnetic fusion devices as it produces intense fusion neutrons. Several experiments conducted on many different DPF devices ranging over several order of storage energy have demonstrated that at higher storage energy the neutron production does not follow I^4 scaling laws and deteriorate significantly raising concern about the device's capability and relevance for fusion energy. On the other hand, the high energy density pinch plasma in DPF device makes it a multiple radiation source of ions, electron, soft and hard x-rays, and neutrons, making it useful for several applications in many different fields such as lithography, radiography, imaging, activation analysis, radioisotopes production etc. Being a source of hot dense plasma, strong shockwave, intense energetic beams and radiation, etc, the DPF device, additionally, shows tremendous potential for applications in plasma nanoscience and plasma nanotechnology. In the present paper, the key features of plasma focus device are critically discussed to understand the novelties and opportunities that this device offers in processing and synthesis of nanophase materials using, both, the top-down and bottom-up approach. The results of recent key experimental investigations performed on (i) the processing and modification of bulk target substrates for phase change, surface reconstruction and nanostructurization, (ii) the nanostructurization of PLD grown magnetic thin films, and (iii) direct synthesis of nanostructured (nanowire, nanosheets and nanoflowers) materials using anode target material ablation, ablated plasma and background reactive gas based synthesis and purely gas phase synthesis of various different types of nanostructured materials using DPF device will discussed to establish this device as versatile tool for plasma nanotechnology.

1. Introduction

The dense plasma focus (DPF) device was independently developed in variant forms by Mather [1] in the U.S. and by Filippov *et al.* [2] in erstwhile the U.S.S.R. during the 1960s. The DPF device basically consists of a fast discharging energy storage capacitor bank connected to the coaxial electrode plasma focus load in a vacuum chamber through a low inductance fast switch. The typical DPF operation involves: (i) the initiation of a powerful electrical discharge across an insulator at the closed end of a coaxial electrode system in focus chamber filled with operating gas typically at few mbar pressure, (ii)

[1] To whom any correspondence should be addressed.

followed by a current sheath formation and acceleration in axial phase toward the open end of the electrode assembly, (iii) rapid compression of current sheath at the top of the central electrode in radial collapse phase to form dense (~10^{26} cm^{-3}) and hot (~1 keV) pinch plasma column and (iv) finally the disruption and decay of pinched plasma column. Almost immediately after its development in 1960s, it was realised that the DPF is an abundant source of multiple radiations like neutrons, relativistic electrons, soft/hard X-rays, and fast ions but much of the initial interest and progress in DPF devices was driven by its high neutron yield.

1.1 DPF for Mainstream Fusion Research

During 1960s, the observation of highly intense burst of neutrons from DPF led to the premature announcement of thermonuclear fusion being achieved. As a result of this, research on DPF started in many laboratories [3-8] all over the world leading to great progress in improving the associated technology and in understanding of physical phenomena taking place in this system. The scope of DPF in fusion research increased as technological development enabled successful operation of large DPF machines leading to better understanding and formulation of scaling laws of DPF performances. Some of the main investigations concerning the scope of DPF in fusion research program are presented next.

The neutron yield in plasma focus device has been reported to scale as $Y_n{\sim}E_0^2$ where E_0 is storage energy of the DPF capacitor bank [9-12]. This led to the construction of bigger DPF devices with larger storage energy for higher neutron yields. Rager [13] proposed that a 30 MJ plasma focus facility, working much below breakeven at a level at which α-particles do no influence as yet the scaling of neutron yield and the plasma dynamics, may act as a nuclear design test facility as the achievable neutron fluence in this facility will be sufficiently high. He predicted that a 300 MJ facility may act as breakeven plasma focus for thermonuclear fusion application with product $n\tau = 3 \times 10^{20}$ $m^{-3}s$, neutron yield of about $Y_n = 3 \times 10^{20}$ neutron/pulse and thermonuclear power of 1 GW for 1 Hz repetition rate operation. Korzhavin [14] concluded that DPF of 30 MJ storage energy with 30 kV capacitor charging voltage will be energetically close to the thermonuclear efficiency of 10%. He argued, if the efficiency is further increased then the DPF neutron source will be able to produce the fuel for several reactor-convertors. It is well known that the neutron production in DPF devices is by joint thermal and beam-target acceleration mechanisms and the beam-target type collisions associated with instabilities are more dominant in comparison to thermonuclear reactions. Korzhavin [14] proposed that if the discharge current in DPF is raised to ten mega ampere then neutron emission will be purely thermal in nature with negligible contribution from turbulent mechanism. Gratton *et al.* [15] studied the properties of the out flowing corona generated by the ablation of solid micro sphere within a hot, dense plasma and noticed that plasma columns such as those produced in DPF devices may drive the implosion of the solid microsphere. They showed that the pinch plasma column of kJ DPF device is potentially a driver comparable to a multi-beam kJ Nd-Yag laser facility concluding that the attention should be paid to DPF research in context of inertial fusion research.

Experiments conducted in various labs on DPFs with different storage energy invariably observed and reported strong anisotropy in neutrons measurements with number of neutrons along the electrode axis to be always much higher than in the perpendicular direction suggesting that neutrons are mainly produced by beam target mechanism and contribution of thermonuclear origin is rather low [3, 16-20]. This resulted in serious doubts about the originally envisioned potentiality of DPF as an alternative thermonuclear fusion reactor. Nevertheless, the scope of DPF of in mainstream fusion research and technology was recognized as early as in late seventies when Cloth and Conrads [21] and Zucker *et al.* [22] proposed a programme to develop DPF as neutron source for blanket and materials problems of controlled thermonuclear reactor. It is a well-established fact that one of the key issues still to be resolved in the quest for fusion energy production is the characterization, qualification and development of advanced plasma facing materials capable of withstanding the extreme radiation and heat loads expected in fusion reactors from high flux of neutrons in addition to a plasma containing energetic gammas, x-rays, ions and electrons. The DPF is an excellent plasma simulator for testing materials under extreme conditions as it provides a realistic fusion reactor environment including gamma rays, x-rays, ions, electrons and neutrons. The DPF is more energy efficient than most fusion plasma simulators because of the approximate E_0^2 scaling of neutron yield and the point source characteristics. It is accepted that large area reactor simulators would, no doubt, be necessary for screening of large number of materials, however, for specific material tests and to qualify the results of large scale fission reactor irradiations the DPF represents a good alternative. The advantages of DPF are high neutron yield with right neutron energy spectra, point source characteristics and possibility of higher repetition rate operations which allows high flux levels for faster tests or higher doses in reasonable time, relatively small size and cost and the lower power requirement. It is due to these reason the research efforts in this direction have continued and have recently gained more momentum through IAEA led coordinated research projects. IAEA has recognized and clearly stated in its report [23] that "*Dense magnetized plasma (DMP) devices serve as a first test bench for testing of fusion relevant plasma facing materials, diagnostic development and calibration, technologies and scaling to conceptual principles of larger devices while sophisticated testing facilities such as the International Fusion Materials Irradiation Facility (IFMIF) are being designed.*"

The concentrated and collaborative research efforts under the umbrella of IAEA's coordinated research project [24] is ongoing and expected to create (i) a data base of erosion rates of selected candidate materials (different tungsten grades and coatings, beryllium, CFC, SiC) under different heat and particle load conditions and different sample treatment scenarios and (ii) scientific base for improved understanding of mechanisms of plasma-material interaction and surface modification, possibly leading to the development of more advanced plasma-facing materials for fusion and other applications. While coming years will see the increased reporting of the research efforts and outcomes in this field this review is going to concentrate on another important development of the use and application of DPF devices for plasma nanotechnology.

1.2 DPF for Plasma Nanotechnology

The hot dense pinch plasma and plasma instabilities in DPF device makes it not only an intense source of fusion relevant neutron but also and an intrinsic, energetic and abundant source of multiple radiations like relativistic electrons [25-27], soft/hard X-rays [28-32], and fast ions [33-38]. These multiple radiations make it suitable for applications in wide variety of fields such as: pulsed activation analysis [39, 40], production of highly ionized species [41], X-ray lithography [42-48], X-ray radiography[49-57], and short lived radioisotope production [58-62]. In addition to the above mentioned wide ranging applications, another most regularly and routinely investigated application of DPF device has been in the areas of bulk and thin film material processing and synthesis. The first ever application of DPF device for processing of bulk material (AISI 304 stainless steel coupons) was reported by Feugeas *et al.* [63] while Rawat *et al.* [64] were first to report the processing of lead zirconate titanate thin film in DPF device. The first ever deposition of thin film, of carbon, was reported by Kant *et al.* [65]. After that more than hundred research papers have been written on various material processing and synthesis related work done by various plasma focus groups across the globe. Many of these papers highlight that in both cases, i.e. in case of processing of bulk substrates and thin films and also in case of direct synthesis of thin films, nanostructured materials are formed.

This review paper aims (i) to highlight several key and distinctive features of DPF device which sets it apart from routinely used low temperature plasma devices for nanostructured material synthesis, and (ii) to establish DPF as a novel high energy density plasma processing as well as deposition facility, in the fast emerging field of "Plasma Nanotechnology", for nanoscale material fabrications.

2. Plasmas for Nanoscience and Nanotechnology: Plasma Nanotechnology

The nanotechnology refers to the manipulation of matter with at least one dimension sized from 1 to 100 nanometers, through certain chemical and/or physical processes, to create materials, devices, and systems with fundamentally new properties and functions because of their small dimensions. The functionalization of nanoscale materials and their integration into nano-devices is leading to a revolution in wide ranging industries with numerous nanotechnology-enabled products fast becoming part of our daily life. The conceptualization, development and optimization of various possible routes and processes, for synthesis of nanoscale materials, are one of the most active areas of research in nanotechnology. There are numerous routes for nanoscale material synthesis, which include milling and vapor condensation routes, chemical routes and nature's biological routes. The plasmas based and plasma-assisted processes have become most versatile tools of nanoscale fabrication as plasmas provide a complex, reactive and far from equilibrium chemical factory. This has led to the development of new field which is popularly referred as "Plasma Nanoscience" and/or "Plasma Nanotechnology" [66, 67].

The plasmas used in plasma nanotechnology can be broadly classified into (i) low temperature (cold) plasmas which are used by industry and material research community, and (ii) high temperature (hot) high energy density plasmas from fusion, pinch and intense discharge sources with densities and temperatures many orders of magnitude higher than that of cold plasmas. The following subsections provide key features of low and high temperature plasmas used in plasma nanotechnology.

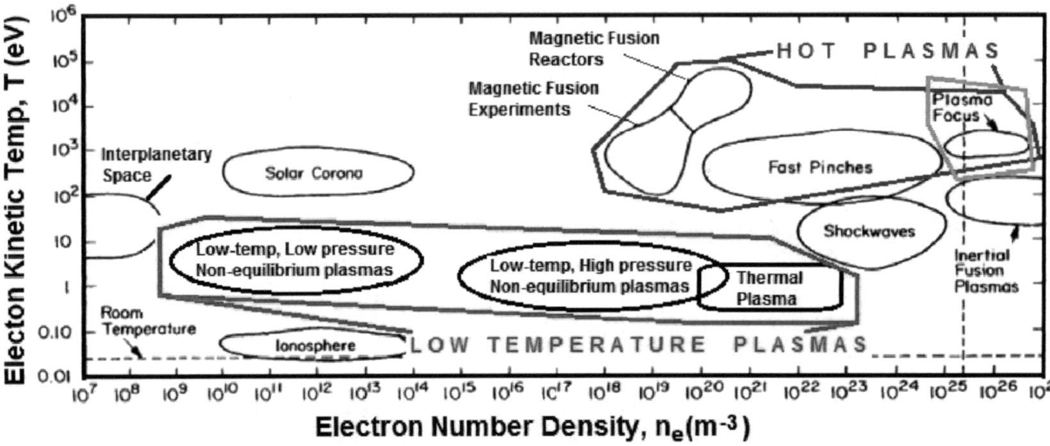

Figure 1. The plasma parameters of various types of plasmas. The low temperature plasmas routinely used in plasma nanotechnology are inside the red box while the high energy density DPF source is shown orange and blue box which belong to the category of hot plasmas.

2.1. Low Temperature (Cold) Plasmas for Plasma Nanotechnology

The low temperature plasmas are mainly produced by ac or dc electric gas discharge or by gas discharges initiated by RF or microwave electromagnetic fields and are characterized by low electron kinetic temperatures ranging from fractions to few tens of eV with low degree of gas ionization resulting in large fraction of the gas in neutral state. The low temperature plasmas used in plasma nanotechnology are grouped in red box in Figure 1 and can be further classified into non-equilibrium and equilibrium (thermal) plasmas. The non-equilibrium plasmas are formed when either the low operating pressure results in insufficient collisions between electrons and ions and neutrals and hence the equilibrium is not reached or high operating pressure having higher collision rates but the plasmas are short lived interrupting the equilibration process. They are described by relation $T_e >> T_i = T_g$, where T_e, T_i and T_g are temperatures of electrons, ions and background gas, respectively, and represent their mean kinetic energies. The thermal plasmas, on the other hand, are the plasmas in thermal equilibrium where the frequent collision between the electrons and ions and neutrals, for high pressure plasma discharges, results in equilibration of temperatures among the charged and neutral species i.e. $T_e \sim T_i \sim T_g$. The low temperature, equilibrium and non-equilibrium type, cold plasmas have been used extensively in plasma nanotechnology due to their interesting combination of electrical, thermal and chemical properties making them indispensable and versatile tool for plasma nanotechnology. The excellent reviews and reports on various types of low temperature plasma that are used in plasma nanotechnology can be found in references [67-69].

2.2. High Energy Density High Temperature (Hot) Plasmas for Plasma Nanotechnology

The hot plasmas, grouped in blue box in Figure 1, have temperatures from few to several keV making matter to be in almost fully-ionized state. In this group of hot plasmas one can find the DPF device, focus of this review paper, on almost the highest side of plasma density. The energy density of DPF devices, estimated by dividing the energy stored in the DPF capacitor bank by the volume of the final pinch plasma column, is reported to be in the range of $(1.2\text{-}9.5)\times10^{10}$ J/m^3 [70] making it a high energy density plasma facility. The high energy density plasmas, by definition, refers to the plasmas which are heated and compressed to extreme energy densities, exceeding 10^{11} J/m^3 (the energy density of a hydrogen molecule) [71]. The magnitude of physical parameters associated with high-energy-density physics is enormous: shock waves at hundreds of km/s (approaching a million km per hour), temperatures of millions of degrees, and pressures that exceeds 100 million atmospheres. The plasmas with energy densities in the range of $(1\text{-}10)\times10^{10}$ J/m^3 are also now classified as high energy density plasmas and hence the DPF devices fall in this category. The Figure 1 shows the clear differences, in terms of plasma parameters, between routinely used low temperature plasmas in plasma nanotechnology and high energy density DPF device plasmas which are emerging as novel plasma nanotechnology tool with both plasma densities and temperature to be about minimum of two to three orders of magnitude higher than low temperature plasmas. It is not only the plasma temperature and the plasma density that are different in DPF devices compared to low temperatures plasma devices but there are several other features that uniquely belong to DPF devices and set them significantly apart from conventionally used low temperature plasma nanotechnology devices. These features are discussed next.

3. DPF: Device Layout, Plasma Dynamics and Key Features

The overall experimental setup of DPF device, with its main subsystems, as a material processing and deposition facility is shown in Figure 2. The DPF processing/deposition facility consist of (i) the DPF vacuum chamber containing (a) the coaxial electrode assembly consisting of central electrode (anode) surrounded coaxially by multiple cathode rods, (b) an aperture assembly, (c) a shutter, and (d) an axially moveable substrate holder; (ii) the vacuum pump and gas inlet connections; (iii) a capacitor bank connected to high voltage power supply; (iv) low inductance high current fast discharging switch (or switches) and transmission lines between the capacitor bank and the DPF load; (v) trigger electronics to activate the high current switches; and (vi) various diagnostics as per the requirements of the experiment being conducted.

The capacitor bank is first charged to high voltages (typically in 10-30 kV range) using high voltage power supply. The electrical energy stored in the capacitor bank is then transferred across the electrode assembly by activating fast switches which initiates a discharge across the insulator sleeve at the closed end of the electrode assembly. This discharge then evolves into a well-defined sheath of plasma in about 100 to 500 ns for DPF devices having the quarter time period of about 500 and 3000 ns, respectively [32, 72]. The current sheath then moves in an inverse pinch manner forming an axially symmetric current sheath near the closed end of the electrode assembly. Driven by $\mathbf{J}\times\mathbf{B}$ force, the current sheath then accelerates along the electrode assembly in what is commonly referred as axial acceleration phase. The typical current sheath speed in axial acceleration phase is reported to be about 4-6 cm/μs. As the current sheath reaches the open end of the anode and starts to roll over anode it is said to enter into radial acceleration phase where it collapses on anode axis forming the hot and dense pinch plasma column. The radial speed of current sheath in radial compression phase is about 2 to 2.5 time the speed in axial phase which results in extremely fast moving shock. The fast moving shock front ahead of the collapsing current sheath can result into plasma temperature of about several hundred electron volts; while the reflected shock at the axis, together with the magnetic compression, can finally raise this temperature to around 1-2 keV. Soon after the formation of pinched plasma

Journal of Physics: Conference Series **591** (2015) 012021 doi:10.1088/1742-6596/591/1/012021

columns it breaks up due to $m = 0$ and $m = 1$ instabilities. The $m = 0$ mode instabilities accelerate the ions of the filling gas species to very high energies towards the top of the chamber and accelerate electrons to relativistic energies (100 keV and above) towards the positively charged anode. A fast moving axial ionization wave-front, produced by the ionization caused by energetic ions, later develops into a bubble-like structure [73]. The ionizing front coincides with the beginning of the hard X-ray emission due to the interaction of energetic electron with anode tip material and also the neutron pulses for deuterium filled DPF device. The pinched plasma column finally breaks up and disintegrates leading to the decay of hot and dense plasma. For more details about DPF devices readers are advised to refer to excellent reviews written by J. Tendys [8], A. Bernard *et al.* [74], V. A. Gribkov *et al.* [56], and M. Krishnan [12]. A detailed review highlighting several applications of DPF device in plasma nanotechnology can also be found in reference [75].

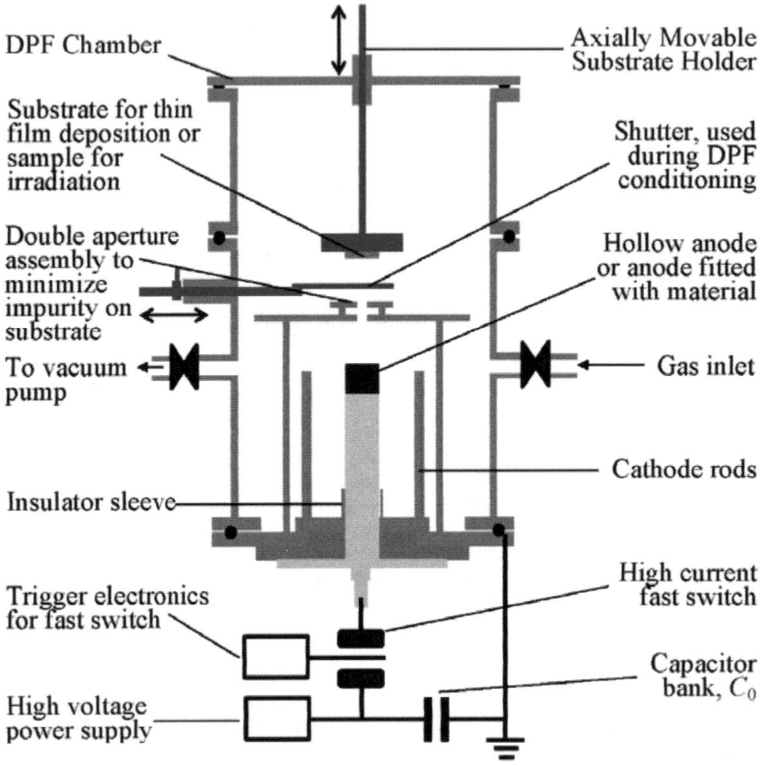

Figure 2. The schematic of DPF material processing/synthesis facility with its various subsystems.

The optimized DPF devices exhibit unique universality i.e. many typical parameters and characteristic features of pinch plasmas, current sheath dynamics, various radiation and energetic charged particles are very similar and are not affected by the dimensions of coaxial electrode assembly, capacitor bank energy, peak discharge current, and the operating gas type. The typical values of various key parameters of interest for optimized DPF devices that exhibit unique universality of DPF devices are given below.

(i) The current sheath speed in axial acceleration phase is typically in the range of 2 to 10 cm/μs [9, 76] which shock-heats the plasmas to electron and ion temperatures of about 100 and 300 eV, respectively, at the end of the axial acceleration phase [77].

(ii) The current sheath speed in radial compression phase is typically about 2 to 2.5 times that of the axial speed [76].
(iii) The electron/ion densities in pinch plasmas are in the range of $5\times10^{24}-10^{26}$ m^{-3} [37, 78].
(iv) The electron and ion temperatures of pinch plasmas are in the range of 200 eV – 2 keV [73] and 300 eV – 1.5 keV [78], respectively.
(v) The energies of instability accelerated electrons, which moves towards the anode, are in the range of few tens to few hundreds of keV [25, 79].
(vi) The energies of instability accelerated ions, which mostly move axial along the anode axis towards the top of the DPF chamber, are in the range of tens of keV to few MeV [33, 80]. The ions are mostly forward directed with most of the ions being emitted in narrow angel of 20° with respect to the anode axis.
(vii) The UV, soft and hard x-rays with photon energies ranging from hundreds to several hundred thousands of eV have been measured in DPF device [28-32].

The DPF devices are essentially transient discharge of a capacitive driver, C_0, into an inductive load, L comprising of fixed system inductance L_0 and dynamic plasma inductance L_p, with the characteristic transient discharge duration of the order of the $\sqrt{LC_0}$, which typically ranges from few hundred ns for low energy sub-kJ DPF devices to several μs for high energy hundreds of kJ or MJ large DPF devices. Hence the durations of pinch plasma, radiation and energetic particles in DPF devices, which are typically some fraction of characteristic discharge duration, are of the order of tens of ns to about hundred or several hundred ns [50 – 53]. *This makes all phenomena of interest in DPF device being highly transient in nature.* Hence, "*in DPF devices the very high densities and temperatures of pinch plasmas, very high flux of high energy instability accelerated ions and electrons, intense energetic radiations, fast moving shock and hot-dense decaying plasma combined with their transient nature offers a kind of plasma and radiation environment that is drastically different from the ambience of low temperature plasmas conventionally used in plasma nanotechnology for nanostructured material synthesis*".

4. Top-down and Bottom-up Nanoscale Fabrications using DPF devices

The nanostructured material synthesis methods involve two different strategies commonly referred as "bottom-up" and "top-down" approaches, shown in Figure 3. The "bottom-up" approach creates nanoscale materials from assembling atoms and molecules; while the "top-down" approach relies on the successive fragmentations or processing of macro-scale materials to smaller nano-sized objects. The radiation and charged particle emission characteristics of DPF allow the nanostructured material synthesis through both routes.

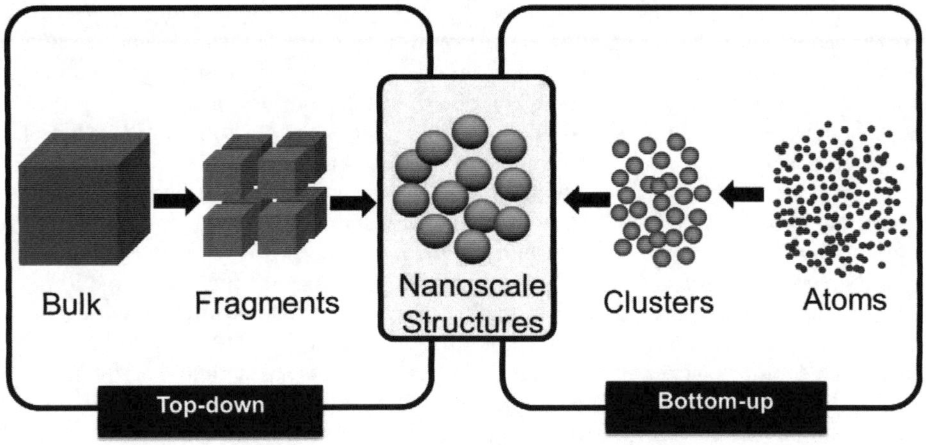

Figure 3. "Top-down" and "bottom-up" synthesis of nanofabrication.

4.1. Top-down Nanoscale Fabrication by Transient Processing of Bulk/Thin Films in DPF

This method involves the exposure of bulk or thin film samples to the different number of DPF shots. It may be pointed out that while the nanostructurization of the entire thickness of the thin film can be achieved by its exposure to DPF shots but for bulk material only the top surface layer of the bulk sample is processed and nanostructurized by its exposure to DPF shots. The setup for the processing of bulk or thin film sample in UNU-ICPT DPF facility, which is a 3.3 kJ device at 15 kV charging voltage with peak discharge current of about 170 kA, in our Plasma Radiation Source Lab in Nanyang Technological University, Singapore is shown in Figure 2. The sample, to be processed, can be placed either along the anode axis or different angular position with respect to the anode axis and at different distances using an axially moving sample holder

The sample exposed to DPF shot is processed by a complex mix of high energy high-flux photons, instability accelerated energetic ions, fast moving ionization wave-front, strong shock wave, and hot and dense decaying plasma. For efficient processing of samples in DPF device one should take note of following additional characteristics of DPF device and use the appropriate strategies:

(i) The DPF devices possess a point-source like characteristics and therefore it can be safely assumed that the above mentioned energetic ions, radiation and hot-dense decaying plasma are essentially emitted from point source (tip of pinched plasma column) and then spread out. Hence, with the increasing distance of exposure the ions/radiation/plasma flux will decrease or in words the energy flux delivered at different distances of exposure is different. This gives the possibility of different degree of processing of the sample by simply changing its exposure distance from the anode top and also by using different number of DPF exposure shots.

(ii) One of the drawbacks of DPF devices is the lack of exact reproducibility i.e. there is variation in device performance (and hence the radiation and charged particle yield) from one shot to another. To minimize the shot to shot variation in performance the device should be conditioned for consistent and efficient performance. Normally, few to few tens of conditioning shots are required, depending on device and its prior state of operation, to achieve efficient pinching in DPF device. During conditioning shots a mechanical shutter should be employed, refer Figure 2, to avoid any unwarranted exposure of sample to the less efficient conditioning shots. The conditioning of DPF ensures reliable shot to shot operation of DPF device. Once the good focusing efficiency, monitored by current and/or voltage probe, is achieved then the shutter is removed and the sample is exposed during the next DPF shot.

(iii) Another important area of concern for irradiation/processing experiments in DPF is the metal impurities from ablation of anode material by the instability accelerated electrons and hot dense pinch plasma. If substantial care is not taken then undesirable debris of ablated anode top material may deposit on the exposed sample surface. The ablation on undesirable anode material can be minimized, to a very large extent, by using deep hollow anode with relatively thin wall. A further reduction in anode wall material debris on the exposed sample surface is achieved with the use of aperture assembly, shown in Figure 2. The aperture assembly should preferably be made up of material which has higher ablation threshold such as brass.

(iv) Depending on the application, one should carefully choose the gas to be used as the operating media. Inert gases such as neon and argon have been used in many studies where the aim was to simply process the material to change its physical characteristics without affecting its composition. Reactive gases such nitrogen, oxygen, acetylene, methane etc. have been used for the formation of nitrides, oxides and carbides in the irradiated/processed materials.

(v) Due to delivery of high flux and high energy instability accelerated ions, energetic radiations, shock exposure and hot-dense decay plasmas in a very short time (with different events lasting from few tens of ns to several μs), the processing of material in DPF device is equivalent to ultra-fast annealing. Sanchez and Feugeas [81] estimated the generation of intense transient heating slopes and heating speeds as high as ~3600 K μm^{-1} and ~40 K ns^{-1} respectively, on various metal sample surface confirming the ultra-fast annealing rate in DPF device. The rapid surface temperature rise, beyond evaporation point, followed by the rapid cooling results in strong thermal effect on the exposed material which bring out the changes in their several physical properties and compositional characteristics. The extent of thermal excursion of the sample surface can be tailored easily either by changing the exposure distance or the angular position of the sample being exposed.

It may be important to highlight over here that though the processing of target material in DPF device is a very complex process due to the involvement of very complex radiation, shock and plasma wind but the main processing component undoubtedly is the forward-directed energetic ion beam. Typical characteristics of energetic ions have been mentioned earlier in section 3. The nano-structuring of thin film, through nanoparticle formation, using single DPF shot irradiation with hydrogen as the operating gas was firstly reported by Jiaji *et al.* [82]. To understand the mechanism of nanoparticle formation they used a Faraday cup and deduced H^+ ion energies to be in the range of about 35 keV to 1.5 MeV and ion flux to be about 2.63×10^{14} ions cm^{-2} at target placed at 5 cm from the anode top. As mentioned, for various processing and ion implantation experiments different gases such as argon, nitrogen, hydrogen, carbon etc. ions have been used in DPF devices. Recently, ion beam fluences and flux for various gases, at the exit point of the pinch, in a DPF device have been numerically estimated and reported by Lee and Saw [83, 84].

Following sub-subsections, using some selected examples, will illustrate the application of DPF device for nanoscale fabrication after being processed by transient complex mix of radiation, shock, plasma and charged particle.

4.1.1. Nano-structuring of Bulk Substrate Surface. Several experiments have conducted and reported on exposure of different bulk substrate materials, such as titanium [81, 85-90], stainless steel [63, 81, 91-93], silicon [94-100], aluminium [101-103], zirconium [104-110], zirconia [111], tungsten [112] etc, using plasma focus devices. Many of the SEM (scanning electron microscopy) images of the DPF irradiated sample surfaces shown in these research paper clearly show the surface reconstruction with nanostructures being formed on them. The formation of nanostructures on Ti and W samples exposed to different DPF devices for different number of shots, at different distances and angular positions, for different filling gas is shown in Figure 4. The surface of the unexposed Ti and W substrates are shown in Figure 4(a) and 4(g), respectively. The low-magnification image in Figure 4(a) shows linear scratch marks of mechanical polish on unexposed Ti sample surface by abrasive silicon carbide paper, while the high magnification image of unexposed W samples in Figure 4(g) shows variable sized and shaped grains and pits. The exposure of these bulk substrate samples to DPF shot with efficient pinching, as indicated by electrical probe signals, at lower distances of exposure causes extreme transient temperature rise of the top few micron thick layer of the sample surface which results in sputtering, melting and re-solidification of this top layer. This is clearly seen in Figure 4(b) which shows reconstruction with extensive smoothening of the exposed surface after the Ti sample, placed along anode axis at axial distance of 5 cm, is irradiated with 30 DPF shots with nitrogen as the operating gas. The SEM image at higher magnification, refer Figure 4(c), shows the formation of self-organized linear nanostructures on the sample surface. It may be mentioned that the dark grey islands observed in Figure 4(c) are not large grains but itself composed of very small nanoparticles (not visible at this magnification) within the boundaries of these linear nanostructures. The exposure of Ti substrate to DPF operation in nitrogen ambience led to the TiN layer formation on the substrate surface. The greater details about the structural, compositional and hardness of the surface TiN layer can be found in our paper [88].

Similarly, the exposure of Ti substrate to methane operated plasma focus devices led to formation of nancomposite titanium oxycarbide (TiC_xO_y) layer on the exposed Ti substrate surface where the carbon was provide by the dissociation of background reactive methane and oxygen probably came from the background impurity. The exposure of Ti substrate to methane operated DPF was conducted in two different DPF devices; low current 190 kA, 3.2 kJ single shot UNU/ICTP DPF and high performance 430 kA, 3.1 kJ NX2 DPF devices. The comparison of SEM images in Figure 4(d) and 4(f) shows the formation of titanium oxycarbide nanoparticles on the substrate surface for both DPF devices exhibiting the reproducibility of the results from two independent devices. The reproducibility of results is one big question mark on the credibility of DPF device for material related work that the critics would often raise due to the well-known shot to shot variation of DPF devices. But our experience has demonstrated that for a vigorously optimised DPF device this is not an issue as the running average performance of a well optimized device can be very consistent [113]. The size of nanoparticles in Figure 4(d) is about 35-50 nm whereas in Figure 4(f) is about 20 nm. The bigger size of nanoparticles in Figure 4(d) is attributed to the higher energy and material flux due to exposure at lower distance as well as along the anode axis. The off-axis irradiation at angular position of 10° with respect to anode axis as well as at higher distance of 9 cm results in lower material and energy flux at the irradiated substrate surface leading to smaller nanoparticle size. The SEM image of the Ti substrate exposed along anode axis in NX2 device, shown in Figure 4(e), exhibits features similar to the one observed in Figure 4(c) with linear nanostructures. The greater details about the structural, compositional and hardness of the nanostructured nanocomposite titanium oxycarbide layer formed on DPF irradiated Ti substrate surface can be found in our paper [87].

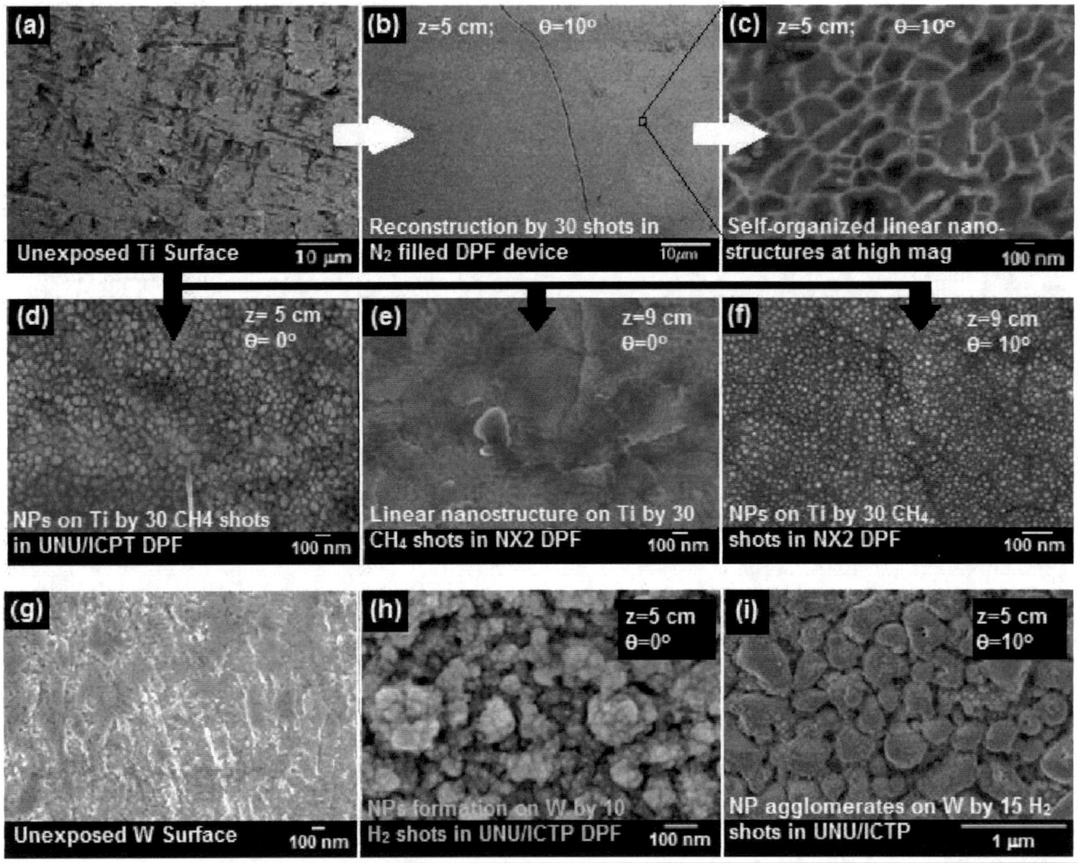

Figure 4. SEM images of unexposed and irradiated Ti and W samples in different DPF devices. The distance (z) and angular position (θ) of the samples, number of DPF irradiation shots and DPF operating gas for each of the exposure experiment are mentioned in SEM images.

The Figure 4(g-i) shows SEM images of the unexposed and irradiated tungsten samples. The unexposed mechanically polished sample in Figure 4(g) shows the rough surface but as such does not show any nanostructure formation. The SEM images of 10 and 15 shot irradiations in hydrogen operated UNU/ICTP DPF device, shown in Figure 4(h) and 4(i) respectively, exhibit the formation of nanoparticles and nanoparticle agglomerates on the irradiated tungsten surface. The size of nanoparticles is about 15-20 nm for 10 shot irradiation though the nanoparticles are also seen to agglomerate resulting in the formation of particle agglomerates with size varying from 100 to 300 nm. The tungsten sample irradiated with 15 shots shows mostly particle agglomerates with the further increase in their size to about 200-800 nm. It may however be noted that these big size particle agglomerates are made up of nanoparticles with size varying from about 40 to 60 nm. The increase in nanoparticle and particle agglomerate size with greater number of irradiation shots can be attributed enhanced transient thermal treatment of the sample surface.

4.1.2. Nano-structuring of Thin/Thick Films. The nano-structuring of pulsed laser ablation deposited FePt thin films in hydrogen operated DPF device [82, 114, 115] and polyaniline thick film in nitrogen operated DPF device [116] has been demonstrated. The FePt thin films were processed in main DPF chamber essentially by upwardly moving hydrogen ions as the aperture assembly was used whereas the polyaniline (PA) thick films were processed by backward moving relativistic electron beam after being extracted through the hollow anode in a separate chamber attached to bottom exit point of the anode as shown in Figure 5. The FePt thin films and PA thick films exposures were done in two different DPF devices and Figure 5 represents the combined schematic of the experimental setup of these two different experiments. The PLD grown 67 nm FePt samples were processed in 3 kJ hydrogen operated UNU-ICTP DPF device by Jiaji *et al.* [117] at a distance of the 5 cm from the anode top, while the few tens of micrometer thick PA films prepared by chemical oxidation method were exposed in 2.2 kJ Mather-type nitrogen operated DPF device [33] by Mohanty *et al.* [116].

The SEM images shown in Figure 5 show that the surface morphology of FePt thin films changed from smooth uniform film for as-deposited sample to the film with uniform and isolated nanoparticles after a single DPF shot exposure. The average size of the FePt nanoparticles formed upon single shot exposure was found to be about 9.1±2.3 nm. The increase in number of ion irradiation DPF shots to two resulted in agglomeration of FePt nanoparticles to bigger sized agglomerates of about 51.3±7.4 nm. Z.Y. Pan *et al.* repeated the experiments with 100 nm thick PLD grown FePt thin films and were able to reproduce nano-structuring of FePt thin films in hydrogen operated UNU-ICTP DPF device using different number of focus shots [114] as well as different exposure distances (5, 6 and 7 cm) [115]. The mechanism of nano-structuring of the DPF exposed thin film can be understood from the characteristics of instability accelerated ions of the filling gas species. Though the ion energy is found to vary over a very big range from few tens of keV to few MeV, but the mean energy of the bulk of the H^+ ions in UNU-ICTP DPF has been estimated to be 124 keV [82]. The projected range of the H^+ ions of this mean energy is about half-micrometer in FePt, as estimated from SRIM ®. Therefore, most of the H^+ ions stop and deposit bulk of their energy in silicon substrate at the Bragg peak position as the thickness of FePt thin films was only about 67 or 100 nm. This would result in heating of silicon substrate to very high temperature in a very short span of time. The thermal energy is then conducted to the FePt thin films and causes the diffusion of metal atoms either through the lattice or along grain boundaries. The diffusion releases the thermal expansion mismatch stresses between the silicon oxide layer of the silicon substrate surface and the PLD coated FePt thin film, leading to the formation of nanoparticles at the surface layer of FePt thin films. One of the biggest advantages of nano-structuring using a DPF device is that it can achieve nano-structuring of thin film in single shot exposure with ion pulse duration of the order of a few hundred ns compared with hours of irradiation time required for other reported continuous ion sources [118].

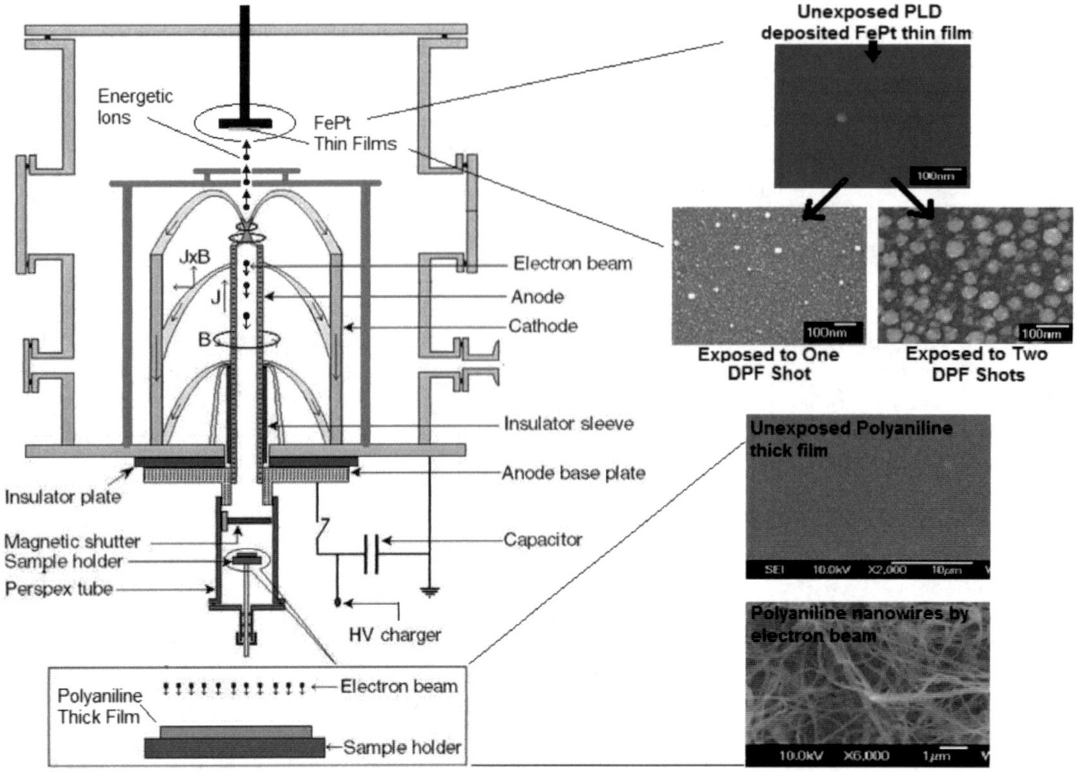

Figure 5. The combined schematic of DPF device showing the exposure of FePt thin film to energetic ions in UNU/ICTP device in Singapore [117] and PA thick film to energetic electrons in another DPF device in Guwahati, India [116]. The images on the right show the changes in surface morphology to nanoparticles and nanowires after DPF exposure.

The SEM images of the unexposed and energetic electron beam exposed PA films are also included in Figure 5. As-synthesized PA thick films are relatively smooth with some aggregates of polyaniline granules. Exposure of PA thick films to the electron beam in a single DPF shot resulted in the formation of network of interwoven nanowires, with diameter of about 50–80 nm and lengths extending up to 10 μm, at many places over the film sample surface, as shown in Figure 5. Liang *et al.* [119] took several hours for making PA nanowires using electrochemical deposition whereas Mohanty *et al.* [116] were able to grow network of PA nanowires in single DPF shot in ultrashort time-scale of about microsecond, the pulse duration of energetic electrons [120] in DPF device.

4.2. Bottom-up Nanoscale Fabrication in DPF

The DPF devices have been used for the deposition of different types of thin films [25, 65, 107, 110, 121-139]. It has invariably been observed that the surface morphologies of thin films deposited with DPF are essentially nanostructured with nanoparticles and nanoparticle agglomerates. Many different thin film deposition configurations, shown as setup C1 to setup C4 in Figure 6, have been used in DPF devices. Before moving on to discuss the bottom-up nanoscale fabrication in DPF device through some selected examples, the following subsection provides the details of different thin film deposition configurations that have been used in DPF devices.

4.2.1. Deposition Configurations of DPF Devices for Nanoscale Fabrication. The first ever configuration for thin film deposition in DPF device, used for carbon thin film synthesis by Kant *et al.*

[65], is shown as setup C1 in Figure 6. The setup C1 used the instability accelerated energetic ion beam and hot dense decaying plasma to ablate the target T placed down the anode stream in specially designed Perspex box in which the substrate was mounted in geometrical shadow region of the box to avoid its direct exposure to ions and plasma from the anode top. This setup used hollow anode to reduce impurities from anode rim to be deposited on the target. The method, however, has drawbacks of poor target ablation, deposition of anode material impurities on target, and ablation of Perspex box material by ions and hot-dense decay plasma resulting in more impurities.

One of the commonly used DPF based deposition configurations is the setup C2 where no substrate was used for thin film deposition rather the bulk target surface irradiated at lower distances gets sputtered, melted and re-solidified to form a top thin layer of different stoichiometry on the target itself. For example, when the targets T of materials such as Ti/Al/Zr/W etc. are exposed to DPF shots operated with reactive background gas such as nitrogen then the thin layer of TiN/AlN/ZrN/WN etc. are observed to form on target surface [86, 92, 105, 140, 141]. The operation of DPF with carbon containing gas such as methane or acetylene or combination of gases such methane and nitrogen has resulted in the formation of carbide [85, 97, 98, 142] or carbonitride [99, 108] thin films on the irradiated target surface. This method, though simple, has the disadvantage of thin film being not transferred to other substrate material and it is very difficult to find and control the thickness of the top layer formed. Moreover, these experiments were conducted mainly with hollow copper anode and would still have the probability of having ablated copper impurities to be deposited on the top layer.

The setup C3, shown in Figure 6 and first time used and reported by Rawat *et al.* [121], is the most flexible and versatile DPF based thin film deposition configuration whereby either the entire anode or anode tip is made of the target material (to save the cost) and the substrate (on which thin film is to be deposited) is placed at a certain fixed distance from the anode top either along the anode axis or at known angular position. A suitable (inert or reactive) operating gas and anode tip material combination is used for the deposition of various types of thin films on suitable substrate. For example, hard coating of TiN can be deposited on stainless steel (SS) substrate by using DPF fitted with Ti anode or Ti anode tip on copper anode and by operating the DPF in nitrogen ambience [122]. High pinch plasma temperature causes the complete ionization of the filling gas species (nitrogen) resulting formation of nitrogen ions, atoms and molecules. The interaction of backward moving relativistic electrons and energetic plasma cause the ablation of Ti anode material forming the Ti plasma which then reacts with ambient nitrogen plasma resulting in TiN formation on SS substrate placed down the anode axis. Similarly, for TiC coating the Ti anode fitted DPF just needs to be operated with gas containing carbon such as methane or acetylene. It has however been found that the DPF operation in these gases is non-reliable and inconsistent [121]. It is for this reason these gas are normally used as admixture with inert gases such as argon or neon. Thus, the deposition of thin films of carbides, nitrides or oxides of any metal can simply be achieved using suitable reactive background gases such nitrogen, acetylene, methane, and oxygen or their combinations. In case of synthesis of magnetic materials, such as Fe, FeCo, CoPt etc., reactive gases were not used rather the device was operated either in inert ambient or with hydrogen. The use of entire anode or the anode tip made up of the material which (or whose compound) is to be deposited on substrate removes anode impurity problem faced in the setup C1 and C2. The ablation rate of the target material is highest for the setup C3 resulting in higher deposition rates for this deposition configuration. As the depositions in DPF devices are conducted in multiple shots, the energetic ions of the filling gas species from next DPF shot process the material deposited in previous shot in a way similar to the one explained in setup C2. The setup C3, therefore, can be classified as *ion and plasma assisted pulsed electron deposition process* as the hot-dense decaying plasma and the energetic ions of gas species also process the thin film that is deposited on the substrate surface.

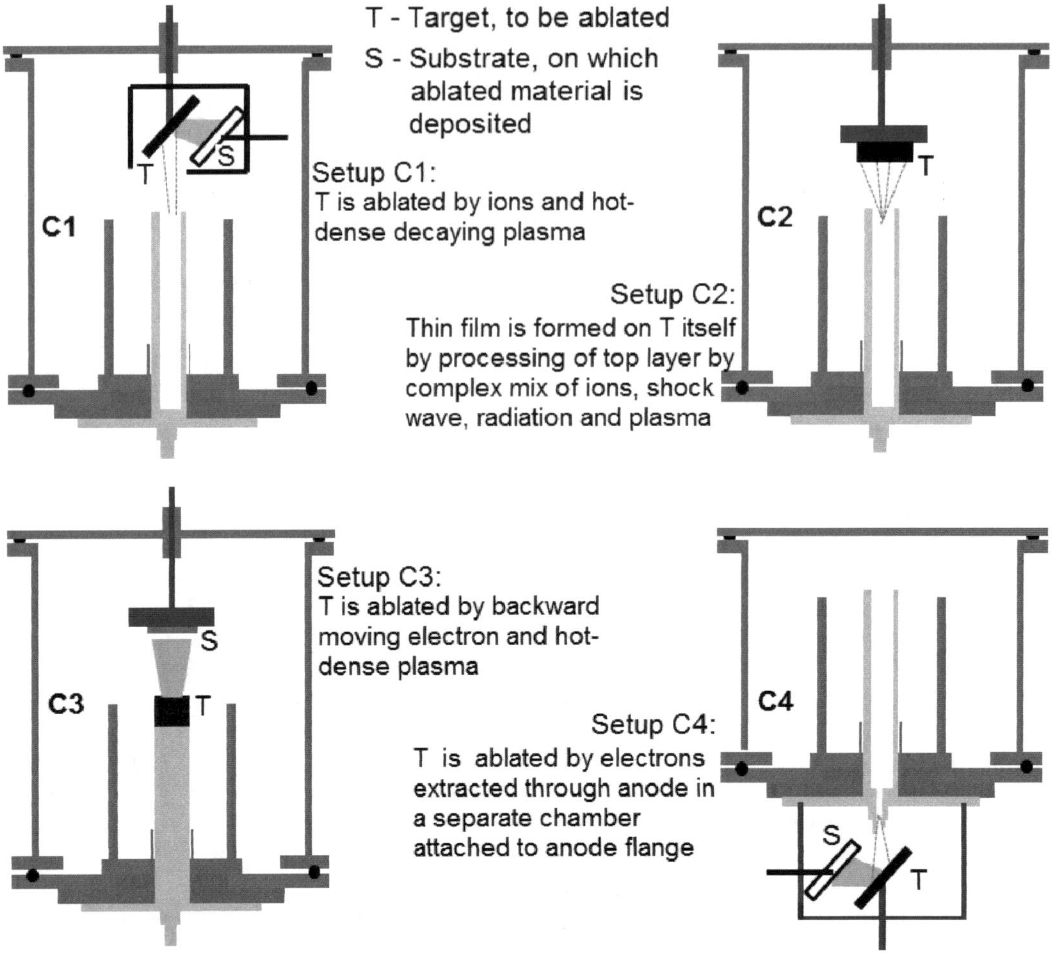

Figure 6. Various configurations of target (material to be ablated and deposited) and substrate in DPF devices used for thin film depositions. Different setups (C1 to C4) use different plasma/radiation components for target ablation.

The deposition configuration shown as setup C4 in Figure 6 was used Zhang *et al.* [25] for deposition of nanostructured thin film of FeCo. They referred their setup as PFPED (Plasma Focus assisted Pulsed Electron Deposition) facility in which an additional chamber was attached to the anode flange within which the ablation of the target material was essentially done by pulsed relativistic electron beam extracted through the hollow anode. The development of PFPED was motivated by the use of pulsed high power electron beams as novel tools for processing of material surfaces [143, 144] and their ability to exhibit stoichiometric deposition of complex multi-component materials [143, 145]. The routinely used deposition method of setup C3 has the advantage of efficient and effective ablation of anode material, however it also has disadvantages such as (i) the energetic ion bombardment of the deposited sample by the next deposition shot are inevitable, which may have a positive or a negative influence and also allows lesser control over the processing of the deposited material and (ii) the deposition chamber is always at high filling gas pressures (in the mbar range) compared with other plasma based deposition systems. The chamber pressure range in setups C1 to C3 is essentially controlled by mass number of filling gas species (bigger is the mass number and smaller is the pressure operation range of PF) etc. The PFPED of setup C4 provides much better control over the deposition process in the PF device as the unavoidable exposure to energetic ions and plasmas is not there. On the other hand, the setup C4 is more difficult to assemble and the target ablation rate (and hence the thin film deposition rate) is much lower compared to setup C3.

Each of the DPF based deposition processes shown in setup C1 to C4 has its own limitations and disadvantages but based on the extensive experience that our group has we strongly believe that the setup C3 is the most versatile and efficient for thin film deposition. The next subsection, using some selected examples will illustrate key features that DPF offers in nanoscale fabrication or nanostructured thin film formation.

4.2.2. Selected Examples and Features of Nanoscale Fabrication using DPF Devices. The Figure 7 shows the some selected SEM images of nano-structured material synthesized using DPF devices. It can be seen from SEM images that DPF devices is able to synthesize 0-, 1-, 2- and 3- dimensional (dim) nanostructures. The spherical nanoparticles (NPs) seen in Figure 7(a-d) are the examples of 0-dim nanostructures that are the most routinely synthesized in DPF devices. While NPs are visibly present over the entire substrate surface in Figure 7(a & b), they are also present as the background in Figure 7(c & d). It was interesting to note that once background carpet of 0-dim NPs is formed on the substrate surface then higher dimensional nanostructures, such as 1-dim linear network of beaded NPs (seen in Figure 7(c)) and 2- and 3-dim nanostructures such as graphene nanosheets (features in white circle marked as 1 in Figure 7(d)) and graphene nanoflowers (features in green circle marked as 3 or 3G in Figure 7(d)), can grow on this NP carpet. It may be important to point out that the graphene nanosheets and nanoflowers are grown in gas phase synthesis by using methane-nitrogen admixture discharge in UNU/ICTP device. These are very new and exciting results where the DPF device has been successfully used for synthesis of 2- and 3-dim nanostructures and will be presented in detail in near future.

It has invariably been observed that the surface morphologies of thin films deposited with DPF are essentially nanostructured with NPs formed over the entire substrate area particularly when deposition setup C1, C3 and C4 (shown in Figure 6) are used. The synthesis mechanism of NPs in DPF device can be understood by drawing analogy from PLD based synthesis of NPs. Happy *et al.* [146] investigate the surface morphology of FeCo thin grown by PLD method. They observed smooth and uniform FeCo thin films formation on substrate surface, without any nanofeatures, at low background pressure operation (about 10^{-6} mbar) of PLD chamber. The low background pressure lowers the interaction of laser ablated plasma with background gas resulting in higher kinetic energy of the plume species on the substrate surface leading to greater adatom mobility and hence the uniform thin film formation on the surface. The high pressure operation (mbar range), according to them and Jiaji *et al.* [147] on the other hand, results in lower kinetic energy of the ablated species on the substrate surface leading to less adatom mobility on substrate surface resulting in NP formation. The DPF deposition setups of C1, C3 and C4 are similar to PLD with the difference that the target ablation, instead of being done by intense laser light, is caused different radiation/charged particle/plasma component. The DPF operation however is always limited to higher pressure regime, typically in the range of about few mbar for all gases, for efficient plasma focus operation. This leads to 0-dim NPs formation on the substrate surface, as explained above, in all DPF based depositions.

Other features of nanostructured material synthesis in DPF devices, as deduced from several experiments conducted and reported by different groups, are as follows:

(i) Until certain number of DPF deposition shots, the size of NPs increases with the increasing number of focus shots. The increase in NP size with the increase in number of shots is seen in image sequence (a) to (d) in Figure 8. The average size of FeCo NP was found to increase linearly from about 45 nm for 10 shots deposition to about 75 nm for 30 shots deposition in UNU/ICTP DPF device for depositions done at substrates angular position of 18° with respect of the anode axis [126]. The increase in NPs size with the increasing number of deposition shots is also observed at higher angular position of 36°. This implies the NPs synthesized and deposited during the first DPF shot continue to act as the growth site resulting in increase in NP size with more shots. However, as mentioned earlier through the observation in Figure 7 that after certain number of DPF shots, depending on the nature of material being deposited, higher dimensional nanostructures may start to nucleate/grow on background NPs carpet layer.

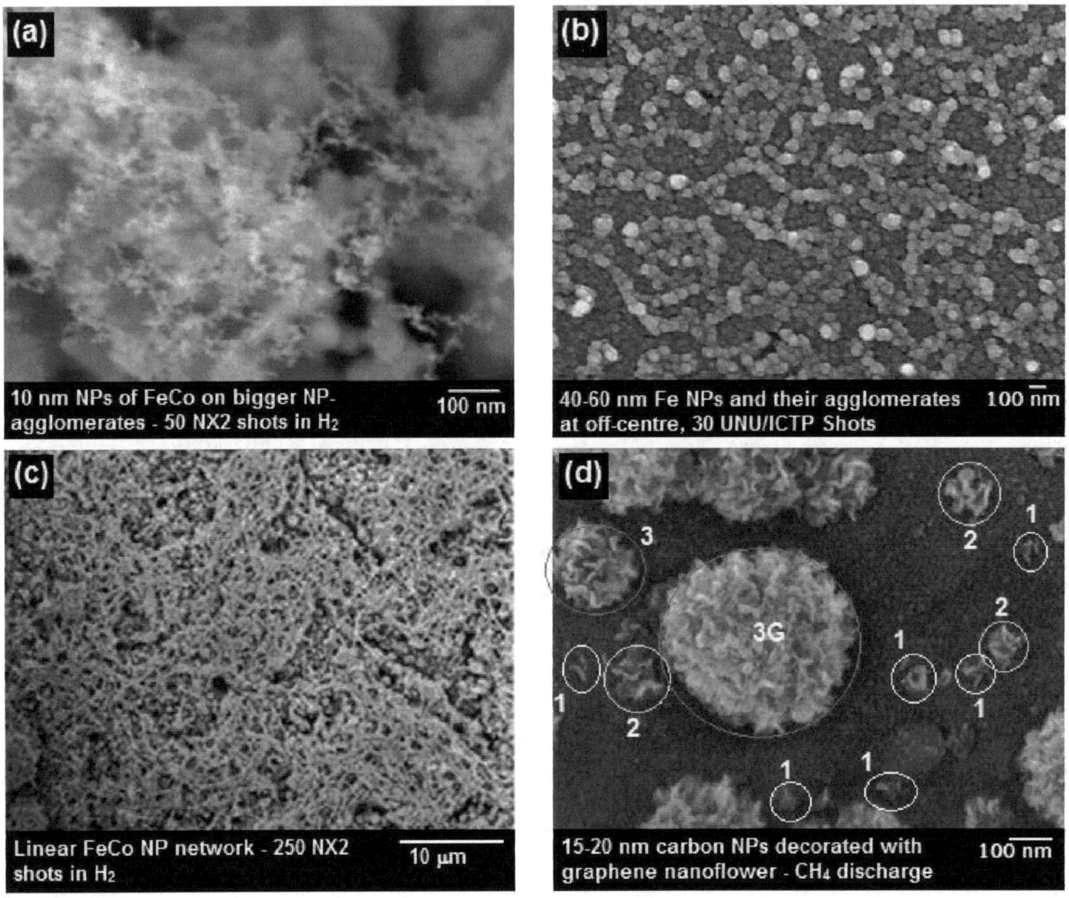

Figure 7. Synthesis of 0-, 1-, 2- and 3- dimensional nanostructures using DPF devices.

(ii) The comparison of FeCo NPs deposited at two different angular positions of 18 and 36 degrees with respect to anode axis shows that for similar number of deposition shots the NPs grown at higher angular position are smaller in size. This can be attributed to essentially forward-directed ablated material plume [123] from the anode top which results in decrease in ablated material flux with the increase in angle with respect to anode axis.

(iii) The average size of NPs decreases with the increase in ambient operating gas pressure. This is evident in the SEM image sequence shown in Figure 8(i-l) for CoPt NP deposition using NX2 DPF device using 25 deposition shots at deposition distance of 25 cm. The average CoPt NP size was found to decrease from about 35 nm for 4 mbar deposition to about 10 nm for deposition done at 8 mbar [128]. The average thickness of CoPt thin film was also found to decrease from about 58 nm for 4 mbar to about 22 nm for 8 mbar deposition. The decrease in NP size as well as film thickness with increasing filling gas pressure is due to two factors: (i) beyond certain operating gas pressure, the focusing efficiency decreases with the increase in filling gas pressure resulting in lesser ablated material from the anode target with lower kinetic energies, and (ii) with the increase in ambient gas pressure, the collision frequency between the ablation material plume and the ambient gas will increase, which causes the ablated material scatter more and also to lose more energy during their movement from the anode to the substrate.

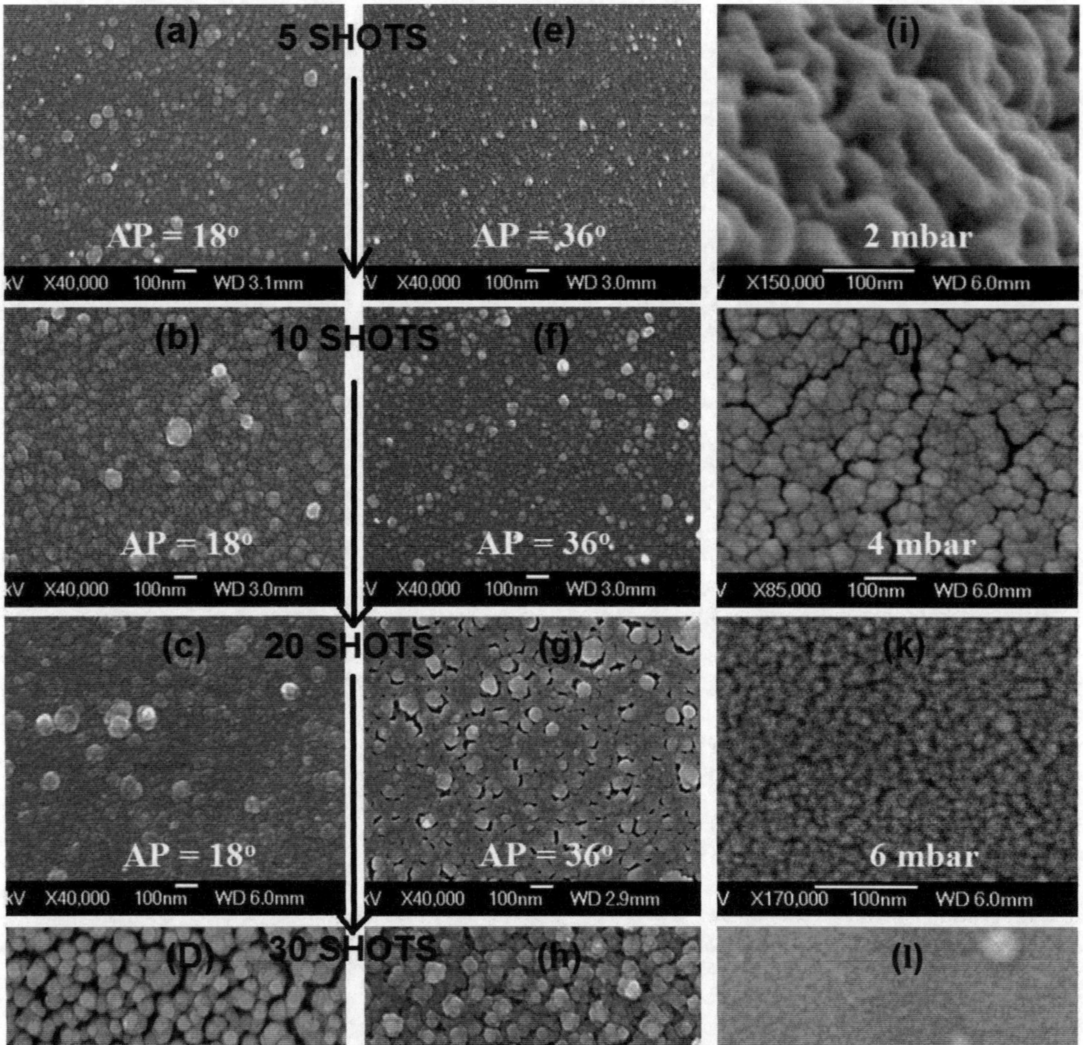

Figure 8. FeCo NPs using different number of DPF shots: (a-d) at angular position (AP) of 18°, and (e-h) at AP of 36° with respect to another. Other operating parameters of DPF devices such as charging voltage, ambient pressure, distance of deposition etc were kept fixed. (i-l) CoPt NPs synthesis with increase filling gas pressure while keeping other parameters fixed.

(iv) The nature of target material ablation, whose thin film is deposited using the setup C1, C3 and C4 of DPF shown in Figure 6, is quite similar to that of PLD as both are pulsed processes and use concentrated flux of energy to ablate the material. The only difference is that instead of using the concentrated flux of laser light energy used in PLD, a complex mix of pulsed energetic electrons/ions and hot-dense plasma is used in DPF device for material ablation. In both cases, PLD and DPF based ablation, the ablated material plume is also forward directed (normal to the target surface) with material flux being highest along the plume centre gradually decreasing radially outward. The DPF device, therefore, suffers from the criticism of non-uniform depositions at the substrate. However, with the availability of 10 Hz or higher repetition rate DPF systems [148-150] one can simply implement the off-axis rotation substrate strategy normally used in PLD device to achieve uniform thin film depositions.

(v) The DPF devices, compared to PLD system, have shown to have significantly higher material ablation rates and hence have thin film deposition rates. For example, the average deposition rate of CoPt thin film was about 1.78 nm/shot for NX2 plasma focus device operating at 880 J at 6 mbar hydrogen ambience at the substrate distance of 25 cm from the anode top [128] which is more than 30 times higher as compared to that of conventional PLD for similar material such as FePt. Similarly, for ZnO nanostructured thin film deposition rates were found to be about 75 nm per shot giving rise to probability of achieving outstanding deposition rates of about 45 μm/min at 10 Hz repetition rate operation of NX2 DPF device [75]. The higher ablation rate of anode top material actually provide significant amount of material even at large distances of deposition. We have routinely observed that in NX2 plasma focus device we can achieve uniform thin film depositions over one inch substrate disc at the larger deposition distance of 25 cm without using any substrate rotation.

(vi) Most of the depositions conducted in DPF devices are essentially multiple-shot depositions which results in transient processing of the thin film, already been deposited previously, by the instability-accelerated energetic ions and hot-dense decay plasma during the next DPF shot. This results in high packing density in the deposited films as well as simultaneous energetic processing (by transient heating) of thin films during deposition. The extent of transient annealing of the deposited material depends on the distance of deposition as well as on the efficient of DPF shot. This, however, most of the time results in direct crystalline phase formation in thin films deposited using DPF device [121, 122, 125, 127, 130, 132-135] without any need for post deposition annealing.

5. Conclusions

The DPF device has successfully been used by many different research groups across the globe for nanostructured material synthesis using both "bottom-up" and "top-down" approaches. Several thin film deposition configurations, in term of target-substrate orientation and placement and use of different ablative components, have been used in DPF devices for nanostructured material syntheses. Each of the DPF based deposition configurations has its own limitations and disadvantages but based on our extensive experience and survey of results from different groups we conclude that most versatile and efficient for thin film deposition setup is the where the target material (to be ablated) is used either as anode insert or as anode tip, background reactive gas is used in the DPF chamber and the substrate (on which thin film is deposited) is placed downstream either along or at some angle with respect to the anode axis. This setup allows deposition of nanostructured metallic, bi-metallic, carbide, nitride, oxide and composite thin films. The DPF device can also be used synthesis for 0-, 1-, 2- and 3-dim nanostructures proving its immense potential in plasma nanoscience and nanotechnology. The DPF device offers a complex mixture of high energy ions of the filling gas species, immensely hot and dense decaying plasma, fast moving ionization wavefront and a strong shockwave that provides a unique plasma and physical/chemical environment that is completely unheard of in any other conventional plasma based deposition or processing facility making it a novel and versatile high energy density pulsed plasma device with immense potential for Plasma Nanotechnology.

Acknowledgement
The author wish to acknowledge research collaborations and scientific discussions with many of his colleagues, collaborators and past PhD students. Financial support from NIE and NTU for various research projects and IAEA research contract is also gratefully acknowledged.

References

1. Mather, J.W., *Formation of a high-density deuterium plasma focus.* Physics of Fluids, 1965. **8**(2): p. 366-&.
2. Filippov, N.V., T.I. Filippova, and V.P. Vinogradov, *Dense, high-temperature plasma in a noncylindrical Z-pinch.* Nuclear Fusion, 1962: p. 577-587.
3. Lee, J.H., et al., *Neutron production mechanism in a plasma focus.* Physics of Fluids, 1971. **14**(10): p. 2210.
4. Schonbach, K.H., L. Michel, and H. Fischer, *Correlation of soft x-ray with hard radiation and neutron emission in 1 kJ plasma focus device.* Applied Physics Letters, 1974. **25**(10): p. 547-549.
5. Oppenländer, T., et al., *The Plasma Focus Current in the Compresion Phase.* Plasma Physics, 1977. **19**: p. 1075.
6. Gullickson, R.L. and H.L. Sahlin, *Measurements of high-energy deuterons in the plasma-focus device.* Journal of Applied Physics, 1978. **49**(3): p. 1099-1105.
7. Bernard, A., et al., Nuclear Instruments & Methods, 1977: p. 191-218.
8. Tendys, J., *Internal report of Australian Atomic Energy Commission*, L.H. Research Establishment, Australia, Editor 1976.
9. Krompholz, H., et al., *A scaling law for plasma-focus devices.* Physics Letters A, 1981. **82**(2): p. 82-84.
10. Lee, S. and A. Serban, *Dimensions and lifetime of the plasma focus pinch.* IEEE Transactions on Plasma Science, 1996. **24**(3): p. 1101-1105.
11. Lee, S. and S.H. Saw, *Neutron scaling laws from numerical experiments.* Journal of Fusion Energy, 2008. **27**: p. 292-295.
12. Krishnan, M., *The Dense Plasma Focus: A Versatile Dense Pinch for Diverse Applications.* Ieee Transactions on Plasma Science, 2012. **40**(12): p. 3189-3221.
13. Rager, J.P., ed. *Uncoventional Approach to Fusion*. ed. B. Brunelli and G. Leotta1982, Plenum: New York. 157.
14. Korzhavin, V.M., ed. *Unconventional approach to fusion*. ed. B. Brunelli and G. Leotta1982, Plenum: New York.
15. Gratton, R., A.R. Piriz, and J.O. Pouzo, Nuclear Fusion, 1986. **26**: p. 483.
16. Dyachenk.Vp and Imshenni.Vs, *Plasma focus and neutron emission mechanism in a Z-pinch.* Soviet Physics Jetp-Ussr, 1969. **29**(5): p. 947-&.
17. Shomo, L.P., J.H. Lee, and K.H. Kim, *Dependence of neutron fluence anisotropy on neutron yied of a plasma focus.* Bulletin of the American Physical Society, 1970. **15**(11): p. 1463-&.
18. Bernstein, M. and F. Hai, *Evidence for nonthermonuclear neutron production in a plasma focus discharge.* Physics Letters A, 1970. **A 31**(6): p. 317-&.
19. Lee, J.H., L.P. Shomo, and K.H. Kim, *Anisotropy of neutron fluence from a plasma focus.* Physics of Fluids, 1972. **15**(12): p. 2433-2438.
20. Pecorella, F., et al., *Time and space resolved neutron measurements on a dense-plasma focus.* Physics of Fluids, 1977. **20**(4): p. 675-682.
21. Cloth, P. and H. Conrads, *Neutronics of a dense-plasma focus - Investigation of a fusion plasma* Nuclear Science and Engineering, 1977. **62**(4): p. 591-600.
22. Zucker, O., et al., *Plasma focus as a large fluence neutron source.* Nuclear Instruments & Methods, 1977. **145**(1): p. 185-190.

23. *Intergrated approach to dense magnetized plasma applications in nuclear fusion technology* in *Report of Coordinated Research Project 2007-2011*2013, International Atomic Energy Agency: Vienna.
24. *Investigations on Materials under High Repetition and Intense Fusion Pulses.* http://cra.iaea.org/crp/project/ProjectDetail?projectId=1754].
25. Zhang, T., et al., *Optimization of a plasma focus device as an electron beam source for thin film deposition.* Plasma Sources Science & Technology, 2007. **16**(2): p. 250-256.
26. Patran, A., et al., *A magnetic electron analyzer for plasma focus electron energy distribution studies.* Journal of Fusion Energy, 2006. **25**(1-2): p. 57-66.
27. Zakaullah, M., et al., *Correlation study of ion, electron and X-ray emission from argon focus plasma.* Physica Scripta, 1998. **57**(1): p. 136-141.
28. Raspa, V., et al., *Plasma focus as a powerful hard x-ray source for ultrafast imaging of moving metallic objects.* Brazilian Journal of Physics, 2004. **34**(4B): p. 1696-1699.
29. Zakaullah, M., et al., *Characteristics of x-rays from a plasma focus operated with neon gas.* Plasma Sources Science & Technology, 2002. **11**(4): p. 377-382.
30. Bhuyan, H., et al., *Comparative study of soft x-ray emission characteristics in a low energy dense plasma focus device.* Journal of Applied Physics, 2004. **95**(6): p. 2975-2981.
31. Barbaglia, M., et al., *Experimental study of the hard x-ray emissions in a plasma focus of hundreds of Joules.* Plasma Physics and Controlled Fusion, 2009. **51**(4): p. 045001.
32. Zhang, T., et al., *Current sheath curvature correlation with the neon soft x-ray emission from plasma focus device.* Plasma Sources Science & Technology, 2005. **14**(2): p. 368-374.
33. Bhuyan, H., et al., *Analysis of nitrogen ion beam produced in dense plasma focus device using Faraday Cup.* Indian Journal of Pure & Applied Physics, 2001. **39**(11): p. 698-703.
34. Kelly, H., et al., *Analysis of the nitrogen ion beam generated in a low-energy plasma focus device by a Faraday Cup operating in the secondary electron emission mode.* Ieee Transactions on Plasma Science, 1998. **26**(1): p. 113-117.
35. Heo, H. and D.K. Park, *Measurement of argon ion beam and X-ray energies in a plasma focus discharge.* Physica Scripta, 2002. **65**(4): p. 350-355.
36. Szydlowski, A., et al., *Measurements of fast ions and neutrons emitted from PF-1000 plasma focus device.* Vacuum, 2004. **76**(2-3): p. 357-360.
37. Gribkov, V.A., et al., *Plasma dynamics in the PF-1000 device under full-scale energy storage: II. Fast electron and ion characteristics versus neutron emission parameters and gun optimization perspectives.* Journal of Physics D-Applied Physics, 2007. **40**(12): p. 3592-3607.
38. Moreno, J., et al., *Preliminary Studies of Ions Emission in a Small Plasma Focus Device of Hundreds of Joules*, in *Dense Z-Pinches*, D.A. Hammer and B.R. Kusse, Editors. 2009. p. 215-218.
39. Rapezzi, L., et al., *Development of a mobile and repetitive plasma focus.* Plasma Sources Science & Technology, 2004. **13**(2): p. 272-277.
40. Tartari, A., et al., *Improvement of calibration assessment for gold fast-neutron activation analysis using plasma focus device.* Measurement Science & Technology, 2002. **13**: p. 939-945.
41. Beckner, E.H., *Production and diagnostic measurements of kilovolt high-density deuterium, helium and neon plasmas.* Journal of Applied Physics, 1966. **37**(13): p. 4944-&.
42. Tan, T.L., et al., *Characterization of chemically amplified resist for X-ray lithography by Fourier transform infrared spectroscopy.* Thin Solid Films, 2006. **504**(1-2): p. 113-116.
43. Kato, Y. and S.H. Be, *Generation of soft x-ray using a rare gas-hydrogen plasma focus and its application to x-ray lithography.* Applied Physics Letters, 1986. **48**(11): p. 686-688.
44. Gribkov, V.A., et al., *Operation of NX2 dense plasma focus device with argon filling as a possible radiation source for micro-machining.* IEEE Transactions on Plasma Science, 2002. **30**(3): p. 1331-1338.

45. Kalaiselvi, S.M.P., et al., *Optimization of neon soft X-rays emission from 200 J fast miniature dense plasma focus device: A potential source for soft X-ray lithography.* Physics Letters A, 2013. **377**(18): p. 1290-1296.
46. Petr, R., et al., *Performance summary on a high power dense plasma focus x-ray lithography point source producing 70 nm line features in AlGaAs microcircuits.* Review of Scientific Instruments, 2004. **75**(8): p. 2551-2559.
47. Kato, Y., et al., *Plasma focus x-ray source for lithography.* Journal of Vacuum Science & Technology B, 1988. **6**(1): p. 195-198.
48. Bogolyubov, E.P., et al., *A powerful soft X-ray source for X-ray lithography based on plasma focusing.* Physica Scripta, 1998. **57**(4): p. 488-494.
49. Rawat, R.S., et al., *Soft X-ray imaging using a neon filled plasma focus X-ray source.* Journal of Fusion Energy, 2004. **23**(1): p. 49-53.
50. Hussain, S., et al., *Low energy plasma focus as an intense X-ray source for radiography.* Plasma Science & Technology, 2004. **6**(3): p. 2296-2300.
51. Hussain, S., et al., *Plasma focus as a possible x-ray source for radiography.* Plasma Sources Science & Technology, 2005. **14**(1): p. 61-69.
52. Moreno, C., et al., *Plasma-focus-based tabletop hard x-ray source for 50 ns resolution introspective imaging of metallic objects through metallic walls.* Applied Physics Letters, 2006. **89**(9).
53. Raspa, V., et al., *Effective hard x-ray spectrum of a tabletop Mather-type plasma focus optimized for flash radiography of metallic objects.* Journal of Applied Physics, 2007. **102**(12).
54. Di Lorenzo, F., et al., *Hard x-ray source for flash radiography based on a 2.5 kJ plasma focus.* Journal of Applied Physics, 2007. **102**(3).
55. Castillo, F., et al., *High contrast radiography using a small dense plasma focus.* Applied Physics Letters, 2008. **92**: p. 051502.
56. Gribkov, V.A., *Current and perspective applications of Dense Plasma Focus devices*, in *Plasma and Fusion Science*, C. Varandas and C. Silva, Editors. 2008. p. 51-64.
57. Verma, R., et al., *Miniature Plasma Focus Device as a Compact Hard X-Ray Source for Fast Radiography Applications.* IEEE Transactions on Plasma Science, 2010. **38**(4): p. 652-657.
58. Roshan, M.V., et al., *Short-Lived PET Radioisotope Production in a Small Plasma Focus Device.* IEEE Transactions on Plasma Science, 2010. **38**(12): p. 3393-3397.
59. Brzosko, J.S., et al., *Breeding 10(10)/s radioactive nuclei in a compact plasma focus device*, in *Application of Accelerators in Research and Industry*, J.L. Duggan and I.L. Morgan, Editors. 2001, Amer Inst Physics: Melville. p. 277-280.
60. Sumini, M., et al., *Preliminary design of a 150kJ repetitive plasma focus for the production of 18-F.* Nuclear Instruments & Methods in Physics Research Section a-Accelerators Spectrometers Detectors and Associated Equipment, 2006. **562**(2): p. 1068-1071.
61. Talaei, A., S.M.S. Kiai, and A.A. Zaeem, *Effects of admixture gas on the production of F-18 radioisotope in plasma focus devices.* Applied Radiation and Isotopes, 2010. **68**(12): p. 2218-2222.
62. Shirani, B., F. Abbasi, and M. Nikbakht, *Production of N-13 by C-12(d,n)N-13 reaction in a medium energy plasma focus.* Applied Radiation and Isotopes, 2013. **74**: p. 86-90.
63. Feugeas, J.N., et al., *Nitrogen implantation of AISI 304 stainless steel with a coaxial plasma gun.* Journal of Applied Physics, 1988. **64**(5): p. 2648-2651.
64. Rawat, R.S., et al., *Crystallization of an amorphous lead zirconate titanate thin film with a dense plasma focus device.* Physical Review B, 1993. **47**(9): p. 4858-4862.
65. Kant, C.R., M.P. Srivastava, and R.S. Rawat, *Thin carbon film deposition using energetic ions of a dense plasma focus.* Physics Letters A, 1997. **226**(3-4): p. 212-216.
66. Ostrikov, K. and A.B. Murphy, *Plasma-aided nanofabrication: where is the cutting edge?* Journal of Physics D-Applied Physics, 2007. **40**(8): p. 2223-2241.

67. Ostrikov, K., U. Cvelbar, and A.B. Murphy, *Plasma nanoscience: setting directions, tackling grand challenges.* Journal of Physics D-Applied Physics, 2011. **44**(17).
68. Meyyappan, M., *Plasma nanotechnology: past, present and future.* Journal of Physics D-Applied Physics, 2011. **44**(17).
69. Shiratani, M., et al., *Nano-factories in plasma: present status and outlook.* Journal of Physics D-Applied Physics, 2011. **44**(17).
70. Soto, L., *New trends and future perspectives on plasma focus research.* Plasma Physics and Controlled Fusion, 2005. **47**: p. A361-A381.
71. Committee on High Energy Density Plasma Physics, P.S.C., National Research Council *Frontiers in High Energy Density Physics: The X-Games of Contemporary Science*2003: National Academies Press. 176.
72. Hassan, S.M., et al., *Pinching evidences in a miniature plasma focus with fast pseudospark switch.* Plasma Sources Science & Technology, 2006. **15**(4): p. 614-619.
73. Bernard, A., et al., *Experimental studies of plasma focus and evidence for non-thermal processes.* Physics of Fluids, 1975. **18**(2): p. 180-194.
74. Bernard, A., et al., *Dense plasma focus - high intensity neutron source.* Nuclear Instruments & Methods, 1977. **145**(1): p. 191-218.
75. Rawat, R.S., *High-Energy-Density Pinch Plasma: A Unique Nonconventional Tool for Plasma Nanotechnology.* Ieee Transactions on Plasma Science, 2013. **41**(4): p. 701-715.
76. Lee, S., et al., *Numerical experiments on plasma focus neon soft x-ray scaling.* Plasma Physics and Controlled Fusion, 2009. **51**(10): p. 105013.
77. Toepfer, A.J., D.R. Smith, and E.H. Beckner, *Ion heating in dense plasma focus.* Bulletin of the American Physical Society, 1969. **14**(11): p. 1013-&.
78. Qi, N., et al., *Space and time resolved electron density and current measurements in a dense plasma focus z-pinch.* Ieee Transactions on Plasma Science, 1998. **26**(4): p. 1127-1137.
79. Choi, P., et al., *Characterization of self-generated intense electron beams in a plasma focus.* Laser and Particle Beams, 1990. **8**(3): p. 469-476.
80. Roshan, M.V., et al., *Magnetic spectrometry of high energy deuteron beams from pulsed plasma system.* Plasma Physics and Controlled Fusion, 2010. **52**(8): p. 085007.
81. Sanchez, G. and J. Feugeas, *The thermal evolution of targets under plasma focus pulsed ion implantation.* Journal of Physics D-Applied Physics, 1997. **30**(6): p. 927-936.
82. Lin, J.J., et al., *FePt nanoparticle formation with lower phase transition temperature by single shot plasma focus ion irradiation.* Journal of Physics D-Applied Physics, 2008. **41**(13).
83. Lee, S. and S.H. Saw, *Plasma focus ion beam fluence and flux-Scaling with stored energy.* Physics of Plasmas, 2012. **19**(11).
84. Lee, S. and S.H. Saw, *Plasma focus ion beam fluence and flux-For various gases.* Physics of Plasmas, 2013. **20**(6).
85. Gupta, R. and M.P. Srivastava, *Carbon ion implantation on titanium for TiC formation using a dense plasma focus device.* Plasma Sources Science & Technology, 2004. **13**(3): p. 371-374.
86. Hassan, M., et al., *Nitriding of titanium by using an ion beam delivered by a plasma focus.* Journal of Physics D-Applied Physics, 2007. **40**(3): p. 769-777.
87. Hassan, M., et al., *Synthesis of nanocrystalline multiphase titanium oxycarbide (TiCxOy) thin films by UNU/ICTP and NX2 plasma focus devices.* Applied Physics a-Materials Science & Processing, 2008. **90**(4): p. 669-677.
88. Hassan, M., et al., *Dense plasma focus ion-based titanium nitride coating on titanium.* Nuclear Instruments & Methods in Physics Research Section B-Beam Interactions with Materials and Atoms, 2009. **267**(11): p. 1911-1917.
89. Bhuyan, H., et al., *High energy ion beam irradiation on titanium substrate in a pulsed plasma device operating with methane.* Journal of Physics D-Applied Physics, 2009. **42**(20): p. 205207.

90. Valipour, M., et al., *Increasing of Hardness of Titanium Using Energetic Nitrogen Ions from Sahand as a Filippov Type Plasma Focus Facility.* Journal of Fusion Energy, 2012. **31**(1): p. 65-72.
91. Gribkov, V.A., et al., *Interaction of high temperature deuterium plasma streams and fast ion beams with stainless steels in dense plasma focus device.* Journal of Physics D-Applied Physics, 2003. **36**(15): p. 1817-1825.
92. Shafiq, M., et al., *Dense plasma focus-assisted nitriding of AISI-304.* Radiation Effects and Defects in Solids, 2008. **163**(9): p. 729-736.
93. Feugeas, J., et al., *Austenite modification of AISI 316L SS by pulsed nitrogen ion beams generated in dense plasma focus discharges.* Surface & Coatings Technology, 2010. **204**(8): p. 1193-1199.
94. Sadiq, M., et al., *Amorphization of silicon by ion irradiation in dense plasma focus.* Physics Letters A, 2006. **352**(1-2): p. 150-154.
95. Sadiq, M., et al., *Nitrogen ion implantation of silicon in dense plasma focus.* Nuclear Instruments & Methods in Physics Research Section B-Beam Interactions with Materials and Atoms, 2006. **252**(2): p. 219-224.
96. Bhuyan, H., et al., *Effect of high energy ion irradiation on silicon substrate in a pulsed plasma device.* Applied Surface Science, 2007. **254**(1): p. 197-200.
97. Bhuyan, H., et al., *Formation of hexagonal silicon carbide by high energy ion beam irradiation on Si(100) substrate.* Journal of Physics D-Applied Physics, 2007. **40**(1): p. 127-131.
98. Wang, Z.R., et al., *Preparation of silicon carbide film by a plasma focus device.* Physics Letters A, 2008. **372**(48): p. 7179-7182.
99. Jabbar, S., et al., *Carbonitriding of silicon using plasma focus device.* Journal of Vacuum Science & Technology A, 2009. **27**(2): p. 381-387.
100. Ahmad, M., S. Al-Hawat, and M. Akel, *Porous Structure Formation on Silicon Surface Treated by Plasma Focus Device.* Journal of Fusion Energy, 2013. **32**(4): p. 471-478.
101. Roshan, M.V., et al., *High energy ions and energetic plasma irradiation effects on aluminum in a Filippov-type plasma focus.* Applied Surface Science, 2008. **255**(5): p. 2461-2465.
102. Rad, Z.S., M. Shahriari, and F.A. Davani, *Investigation of Spatial Distribution of Hydrogen and Argon Ions and Effects of them on Aluminum Samples in a 2.5 kJ Mater Type Plasma Focus Device.* Journal of Fusion Energy, 2011. **30**(5): p. 358-366.
103. Afrashteh, M. and M. Habibi, *Study of Dense Nitrogen Plasma Irradiation of Aluminum Targets by APF Plasma Focus Device.* Journal of Fusion Energy, 2012. **31**(3): p. 223-226.
104. Rico, L., et al., *Crystallization of amorphous zirconium thin film using ion implantation by a plasma focus of 1 kJ.* Applied Surface Science, 2007. **254**(1): p. 193-196.
105. Khan, I.A., et al., *Nitridation of zirconium using energetic ions from plasma focus device.* Thin Solid Films, 2008. **516**(23): p. 8255-8263.
106. Khan, I.A., et al., *Synthesis of zirconium oxynitride (ZrON) nanocomposite films on zirconium substrate by dense plasma focus device.* International Journal of Modern Physics B, 2008. **22**(23): p. 3941-3955.
107. Khan, I.A., et al., *Synthesis of nano-crystalline zirconium aluminium oxynitride (ZrAlON) composite films by dense plasma Focus device.* Applied Surface Science, 2009. **255**(12): p. 6132-6140.
108. Khan, I.A., et al., *Deposition of zirconium carbonitride composite films using ion and electron beams emitted from plasma focus device.* Nuclear Instruments & Methods in Physics Research Section B-Beam Interactions with Materials and Atoms, 2010. **268**(13): p. 2228-2234.
109. Murtaza, G., et al., *Carburizing of zirconium using a low energy Mather type plasma focus.* Surface & Coatings Technology, 2011. **205**(8-9): p. 3012-3019.
110. Khan, I.A., et al., *Deposition of alumina stabilized zirconia at room temperature by plasma focus device.* Applied Surface Science, 2014. **288**: p. 304-312.

111. Mohanty, S.R., et al., *Energetic ion irradiation of American diamond in a plasma focus device and characterization of irradiated material.* Nuclear Instruments & Methods in Physics Research Section B-Beam Interactions with Materials and Atoms, 2006. **243**(1): p. 113-118.
112. Bhuyan, M., et al., *Plasma focus assisted damage studies on tungsten.* Applied Surface Science, 2013. **264**: p. 674-680.
113. Wong, D., et al., *Soft X-ray optimization studies on a dense plasma focus device operated in neon and argon in repetitive mode.* IEEE Transactions on Plasma Science, 2004. **32**(6): p. 2227-2235.
114. Pan, Z.Y., et al., *Nanostructuring of FePt thin films by plasma focus device: pulsed ion irradiation dependent phase transition and magnetic properties.* Applied Physics a-Materials Science & Processing, 2009. **96**(4): p. 1027-1033.
115. Pan, Z.Y., et al., *Lowering of L1(0) phase transition temperature of FePt thin films by single shot H+ ion exposure using plasma focus device.* Thin Solid Films, 2009. **517**(8): p. 2753-2757.
116. Mohanty, S.R., et al., *Self-organized transformation to polyaniline nanowires by pulsed energetic electron irradiation in a plasma focus device.* Physics Letters A, 2009. **373**(22): p. 1962-1966.
117. Lee, S., et al., *A simple facility for the teaching of plasma dynamics and plasma nuclear fusion.* American Journal of Physics, 1988. **56**(1): p. 62-68.
118. Alexander, D.E., et al., Applied Physics Letters, 1993. **62**: p. 1597.
119. Liang, L., et al., *Direct assembly of large arrays of oriented conducting polymer nanowires.* Angewandte Chemie-International Edition, 2002. **41**(19): p. 3665-3668.
120. Patran, A., et al., *Spectral study of the electron beam emitted from a 3 kJ plasma focus.* Plasma Sources Science & Technology, 2005. **14**(3): p. 549-560.
121. Rawat, R.S., et al., *Room temperature deposition of titanium carbide thin films using dense plasma focus device.* Surface & Coatings Technology, 2001. **138**(2-3): p. 159-165.
122. Rawat, R.S., et al., *Deposition of titanium nitride thin films on stainless steel - AISI 304 substrates using a plasma focus device.* Surface & Coatings Technology, 2003. **173**(2-3): p. 276-284.
123. Soh, L.Y., et al., *Shadowgraphic studies of DLC film deposition process in dense plasma focus device.* IEEE Transactions on Plasma Science, 2004. **32**(2): p. 448-455.
124. Gupta, R., et al., *Deposition of nanosized grains of ferroelectric lead zirconate titanate on thin films using dense plasma focus.* Journal of Physics D-Applied Physics, 2004. **37**(7): p. 1091-1094.
125. Rawat, R.S., et al., *Nano-structured Fe thin film deposition using plasma focus device.* Applied Surface Science, 2006. **253**(3): p. 1611-1615.
126. Zhang, T., et al., *Characteristics of FeCo nano-particles synthesized using plasma focus.* Journal of Physics D-Applied Physics, 2006. **39**(10): p. 2212-2219.
127. Rawat, R.S., et al., *Nano-phase titanium dioxide thin film deposited by repetitive plasma focus: Ion irradiation and annealing based phase transformation and agglomeration.* Applied Surface Science, 2008. **255**(5): p. 2932-2941.
128. Pan, Z.Y., et al., *Nanostructured magnetic CoPt thin films synthesis using dense plasma focus device operating at sub-kilojoule range.* Journal of Physics D-Applied Physics, 2009. **42**(17): p. 175001.
129. Malhotra, Y., et al., *Extremely non-equilibrium synthesis of luminescent zinc oxide nanoparticles through energetic ion condensation in a dense plasma focus device.* Journal of Physics D-Applied Physics, 2009. **42**(15): p. 155202.
130. Ghareshabani, E., et al., *Synthesis of nanostructured multiphase Ti(C,N)/a-C films by a plasma focus device.* Nuclear Instruments & Methods in Physics Research Section B-Beam Interactions with Materials and Atoms, 2010. **268**(17-18): p. 2777-2784.

131. Ghareshabani, E., et al., *Low energy repetitive miniature plasma focus device as high deposition rate facility for synthesis of DLC thin films.* Applied Surface Science, 2010. **256**(16): p. 4977-4983.
132. Pan, Z.Y., et al., *Miniature plasma focus as a novel device for synthesis of soft magnetic FeCo thin films.* Physics Letters A, 2010. **374**(8): p. 1043-1048.
133. Macharaga, G., et al., *TiO2 Nano-cluster Thin Films by Dense Plasma Focus and Ion Implantation Effect on its Photocatalytic Activity.* Journal of Advanced Oxidation Technologies, 2011. **14**(2): p. 308-313.
134. Etaati, G.R., et al., *Deposition of tungsten nitride on stainless steel substrates using plasma focus device.* Nuclear Instruments & Methods in Physics Research Section B-Beam Interactions with Materials and Atoms, 2011. **269**(10): p. 1058-1062.
135. Hosseinnejad, M.T., et al., *Deposition of tungsten nitride thin films by plasma focus device at different axial and angular positions.* Applied Surface Science, 2011. **257**(17): p. 7653-7658.
136. Javadi, S., et al., *Deposition of Chromium Thin Films on Stainless Steel-304 Substrates Using a Low Energy Plasma Focus Device.* Journal of Fusion Energy, 2012. **31**(3): p. 242-248.
137. Ngoi, S.K., et al., *Formation of Nano-Crystalline Phase in Hydrogenated Amorphous Silicon Thin Film by Plasma Focus Ion Beam Irradiation.* Journal of Fusion Energy, 2012. **31**(1): p. 96-103.
138. Hosseinnejad, M.T., et al., *Using Mather-Type Plasma Focus Device for Fabrication of Tungsten Thin Films.* Journal of Fusion Energy, 2012. **31**(5): p. 426-431.
139. Umar, Z.A., et al., *Hard TiCx/SiC/a-C:H nanocomposite thin films using pulsed high energy density plasma focus device.* Nuclear Instruments & Methods in Physics Research Section B-Beam Interactions with Materials and Atoms, 2013. **301**: p. 53-61.
140. Borthakur, T.K., et al., *Surface hardening of high carbon steel by plasma focus nitriding.* Surface Engineering, 1999. **15**(1): p. 55-58.
141. Sadiq, M., et al., *The nitriding of aluminium by dense plasma focus.* Plasma Sources Science & Technology, 2006. **15**(3): p. 295-301.
142. Shafiq, M., et al., *Pulsed ion beam-assisted carburizing of titanium in methane discharge.* Chinese Physics B, 2010. **19**(1): p. 012801.
143. Kovaleski, S.D., et al., *Electron beam ablation of materials.* Journal of Applied Physics, 1999. **86**(12): p. 7129-7138.
144. Stark, R., et al., *Pseudospark produced pulsed electron beam for material processing.* Ieee Transactions on Plasma Science, 1995. **23**(3): p. 258-264.
145. Strikovski, M. and K.S. Harshavardhan, *Parameters that control pulsed electron beam ablation of materials and film deposition processes.* Applied Physics Letters, 2003. **82**(6): p. 853-855.
146. Happy, et al., *Effect of deposition parameters on morphology and size of FeCo nanoparticles synthesized by pulsed laser ablation deposition.* Applied Surface Science, 2006. **252**(8): p. 2806-2816.
147. Lin, J.J., et al., *Effects of target-substrate geometry and ambient gas pressure on FePt nanoparticles synthesized by pulsed laser deposition.* Applied Surface Science, 2009. **255**(8): p. 4372-4377.
148. Lee, S., et al., *High rep rate high performance plasma focus as a powerful radiation source.* IEEE Transactions on Plasma Science, 1998. **26**(4): p. 1119-1126.
149. Soto, L., et al., *Nanofocus of tenth of joules and a portable plasma focus of few joules for field applications*, in *Dense Z-Pinches*, D.A. Hammer and B.R. Kusse, Editors. 2009. p. 219-222.
150. Verma, R., et al., *High Performance High Repetition Rate Miniature Plasma Focus Device: Record Time Averaged Neutron Yield at 200 J with Enhanced Reproducibility.* Journal of Fusion Energy, 2013. **32**(1): p. 2-10.

Multi-scaling of the dense plasma focus

S H Saw[1,2,4] and S Lee[1,2,3]
[1]INTI International University, 71800 Nilai, Negeri Sembilan, Malaysia
[2]Institute of Plasma Focus Studies, Melbourne, Australia
[3]University Malaya, Kuala Lumpur, Malaysia

E-mail: sorheoh.saw@newinti.edu.my

Abstract. The dense plasma focus is a copious source of multi-radiations with many potential new applications of special interest such as in advanced SXR lithography, materials synthesizing and testing, medical isotopes and imaging. This paper reviews the series of numerical experiments conducted using the Lee model code to obtain the scaling laws of the multi-radiations.

1. Introduction

Plasma focus machines of various energies are increasingly being studied as sources of neutrons, soft x-rays and ions. An exciting prospect is for scaling the plasma focus up to regimes relevant for fusion energy studies. However, even a simple machine such as the UNU/ICTP PFF 3 kJ machine consistently produces 10^8 neutrons in deuterium [1]. Plasma focus machines operated in neon have also been studied as intense sources of soft x-rays [2-4]. Whilst many recent experiments have concentrated efforts on low energy repetitive devices [2-4], other experiments have looked at larger plasma focus devices [5, 6] extending to MJ regime. Numerical experiments are also gaining interest [7, 8] with the Lee model code [9] demonstrating that it computes realistic focus pinch parameters and absolute values of neutron yield Y_n and soft x-ray yield Y_{sxr} which are consistent with those measured experimentally. A comparison was made for the case of the NX2 machine [4], showing good agreement between computed and measured Y_{sxr} [8-10]. This gives confidence that the Lee model code gives realistic results in the computation of Y_n and Y_{sxr}.

In recent years, we see increasing investigations on the ion beams and plasma streams emission from PF devices. The motivation for these studies is the potential applications for materials synthesis, signatures and damage studies of candidate wall materials of fusion reactors. Hence we have extended our model code to enable numerical experiments to be carried out on defining properties of beam ions in various gases.

In this review, we show the comprehensive range of numerical experiments conducted to derive scaling laws on neutron yield Y_n [11, 12] and neon Y_{sxr} [8, 10], in terms of storage energy E_0, peak discharge current I_{peak} and peak focus pinch current I_{pinch} obtained from studies [13-15] carried out over E_0 varying from 0.2 kJ to 25 MJ for optimised machine parameters and operating parameters. We also present as yet unpublished results of the scaling of some beam ion defining properties.

2. The Lee Model Code

The Lee model code couples the electrical circuit with plasma focus dynamics, thermodynamics and radiation, enabling realistic simulation of all gross focus properties. The basic model, described in 1984 [16] was successfully used to assist several projects [17-19]. Radiation-coupled dynamics was included in the five-phase code leading to numerical experiments on radiation cooling [20]. The vital role of a finite small disturbance speed discussed by Potter in a Z-pinch situation [21] was incorporated together with real gas thermodynamics and radiation-yield terms. Before this 'communication delay effect' was incorporated, the model consistently over-estimated the radial speeds. This is serious from the point of view of neutron yields. A factor of two in shock speeds gives a factor of four in temperatures leading to a difference in fusion cross-sections of ~1000 at the range of temperatures we are dealing with. This version of the code assisted other research projects [22-27] and was web-published in 2000 [28] and 2005 [29]. Plasma self-absorption was included in 2007 [27] improving SXR yield simulation. The code has been used extensively in several machines including UNU/ICTP PFF [1, 17, 22, 23, 25-27, 30, 31], NX2 [24, 27, 32], NX1 [3, 32] and adapted for the Filippov-type plasma focus DENA [33]. Neutron yield Y_n using a beam–target mechanism [11, 12, 14, 15, 34], is incorporated the code [9] (versions later than RADPFV5.13), resulting in realistic Y_n scaling with I_{pinch}[11, 12]. The versatility and utility of the model are demonstrated in its clear distinction of I_{pinch} from I_{peak}[13] and insights of current and neutron yield limitations [14, 15], neutron saturation [12, 35], radiative collapse[36] and current-stepped PF[37-39] and extraction of diagnostic data [13,40-45] and anomalous resistance data [46,47] from current signals. The description theory, code and a broad range of results of this 'Universal Plasma Focus Laboratory Facility' are available for download from [9]. We summarise the five phases used in the model code.

i) Axial Phase: Described by a snowplow model with an equation of motion coupled to a circuit equation. The equation of motion incorporates the axial phase model parameters: mass and current factors f_m and f_c [9,45,48-50] respectively; f_m accounting for the porosity of the current sheet, the inclination of the moving current sheet-shock front structure and all other unspecified effects which have effects equivalent to increasing or reducing mass in the moving structure; f_c accounting for the fraction of current effectively flowing in the moving structure (due to all effects including current shedding at or near the back-wall and current sheet inclination).

ii) Radial Inward Shock Phase: Described by four coupled equations using an elongating slug model. The first equation computes the radial inward shock speed from the driving magnetic pressure; the second the axial elongation speed of the column; the third the speed of the current sheath (aka the magnetic piston), allowing the current sheath to separate from the shock front by an adiabatic approximation. The fourth is the circuit equation. Thermodynamic effects due to ionization and excitation are incorporated, being important for gases other than hydrogen and deuterium. A communication delay between shock front and current sheath due to the finite small disturbance speed is crucially implemented. The model parameters, radial phase mass swept-up and current factors f_{mr} and f_{cr} respectively are incorporated in all three radial phases.

iii) Radial Reflected Shock (RS) Phase: When the shock front hits the axis, because the focus plasma is collisional, a RS develops which moves radially outwards, whilst the radial current sheath piston continues to move inwards. Four coupled equations are used, these being for the RS moving radially outwards, the piston moving radially inwards, the elongation of the annular column and the circuit. The plasma temperature behind the RS undergoes a jump by a factor of approximately two.

iv) Slow Compression (Quiescent) or Pinch Phase: When the out-going RS hits the in-coming piston the compression enters a radiative phase, with inclusion of energy loss/gain terms from Joule heating and radiation losses into the piston equation of motion; so that for gases such as krypton, radiation emission may enhance the compression. Three coupled equations describe this phase; the piston radial motion, the pinch column elongation and the circuit equations.

v) Expanded Column Phase: To simulate the current trace beyond this point, we allow the column to suddenly attain the radius of the anode. Two coupled equations are used; similar to the axial phase above.

2.1 Computation of Neutron Yield

The neutron yield is computed using a phenomenological beam-target neutron generating mechanism described recently by Gribkov et al [34]. A beam of fast deuteron ions is produced by diode action in a thin layer close to the anode, with plasma disruptions generating the necessary high voltages. The beam interacts with the hot dense plasma of the focus pinch column to produce the fusion neutrons. The beam-target yield is derived [11, 12, 14, 28] as:

$$Y_{b\text{-}t} = C_n n_i I_{pinch}^2 z_p^2 (\ln(b/r_p))\, \sigma / U^{0.5} \quad (1)$$

where n_i = ion density, b = cathode radius, r_p = radius of the plasma pinch with length z_p, σ =cross-section of the D-D fusion reaction, n-branch [51] and U= beam energy. C_n is treated as a calibration constant combining various constants in the derivation process.
The D-D cross-section is sensitive to the beam energy in the range 15-150 kV. The code computes induced voltages (due to current motion inductive effects) V_{max} of the order of only 15-50 kV. However it is known, from experiments that the ion energy responsible for the beam-target neutrons is in the range 50-150 keV [34], and for smaller lower-voltage machines the relevant energy could be lower at 30-60 keV [31]. Fitting with extensive experimental observations of machines from sub-kJ to near MJ, the D-D cross section σ is reasonably obtained by using $U=3V_{max}$. A value of C_n=2.7x10^7 was obtained by calibrating the yield [9], [13]-[14] at an experimental point of 0.5 MA.
The thermonuclear component is also computed in every case and it is found that this component is negligible when compared with the beam-target component.

2.2 *Computation of Neon SXR Yield*
In the code [9, 32, 52], neon line radiation Q_L is calculated as follows:

$$dQ_L/dt = -4.6 \times 10^{-31} n_i^2 Z Z_n^4 (\pi r_p^2) z_f / T \quad (2)$$

where for the temperatures of interest in our experiments we take the SXR yield $Y_{sxr} = Q_L$. Z_n is the atomic number.

This generated energy is then reduced by the plasma self-absorption which is included by computing volumetric plasma self-absorption factor A derived from the photonic excitation number M which is a function of Z_n, n_i, Z and T. For SXR scaling there is an optimum small range of temperatures (around 200-500 eV T-window necessary to produce the He-like, H-like neon ions) [23, 24] to operate.

2.3 *Computation of Beam ion properties*
In the latest (2013 version) the Lee code computes the flux of the ion beams $J_b=n_b v_b$ where n_b=number of beam ions N_b divided by volume of plasma traversed is derived from pinch inductive energy considerations; and v_b=effective speed of the beam ions is derived from the accelerating voltage taken as diode voltage U. All quantities are expressed in SI units, except where otherwise stated. The resulting equation is given below:

$$\text{Flux} = J_b = 2.75\text{x}10^{15}\ (f_e/[MZ_{eff}]^{1/2})\{(\ln[b/r_p])/(r_p^2)\}(I_{pinch}^2)/U^{1/2} \quad \text{ions m}^{-2}\text{s}^{-1} \quad (3)$$

where M=ion mass, Z_{eff}=average effective charge of the ion in the pinch, b=cathode radius, r_p=pinch radius and I_{pinch}=pinch current. The parameter f_e = fraction of energy converted into beam energy from the inductive energy of the pinch. Analyzing neutron yield data [11, 14, 53, 54] and pinch dimensional-temporal relationships [22] we estimate f_e=0.14 and use the approximate scaling [22]: $\tau=10^{-6} z_p$. This condition f_e=0.14 is equivalent to ion beam energy of 3%-6% E_0 in the case when the pinch inductive energy holds 20%-40% of E_0. Our extensive study of high performance low inductance PF classified [46] as Type 1 shows that this estimate of f_e is consistent with data. We summarise the assumptions:

(i) Ion beam flux J_b is $n_b v_b$ with units of ions m^{-2}s^{-1},
(ii) Ion beam is produced by diode mechanism [34],

(iii) The beam is produced uniformly across the whole cross-section of the pinch,
(iv) The beam speed is characterized by an average value v_b,
(v) The beam energy is a fraction f_e of the pinch inductive energy, taken as 0.14 in the first instance; to be adjusted as numerical experiments indicate,
(vi) The beam ion energy is derived from the diode voltage U,
(vii) The diode voltage U is $U=3V_{max}$; a relationship obtained from data fitting in extensive earlier numerical experiments [11, 14].

The value of the ion flux is deduced in each situation by computing Z_{eff}, r_p, I_{pinch} and U.

3. Numerical Experiments and Results

The Lee code is configured to work as any plasma focus by inputting the bank parameters, L_0, C_0 and stray circuit resistance r_0; the tube parameters b, a and z_0 and operational parameters V_0 and P_0 and the fill gas. The computed total current waveform is fitted to an experimentally measured total current waveform [11, 13-15, 28-29] using the four model parameters f_m, f_c for the axial phase and f_{mr} and f_{cr} for the radial phases.

3.1 Scaling laws for neutrons from numerical experiments over a range of energies from 10kJ to 25 MJ

We apply the Lee model code to the MJ machine PF1000 over a range of C_0 to study the neutrons emitted by PF1000-like bank energies from 10kJ to 25 MJ.

First, we fitted a measured current trace to obtain the model parameters. A measured current trace of the PF1000 with C_0=1332 μF, operated at 27 kV, 3.5 torr deuterium, has been published [34], with cathode/anode radii b=16 cm, a=11.55 cm and anode length z_0=60 cm. In the numerical experiments we fitted external (or static) inductance L_0= 33.5 nH and stray resistance r_0 = 6.1 mΩ (damping factor $RESF= r_0/(L_0/C_0)^{0.5}$ = 1.22). The fitted model parameters are: f_m=0.13, f_c=0.7, f_{mr}=0.35 and f_{cr}=0.65. The computed current trace [11, 15] agrees very well with the measured trace through all the phases; axial and radial, right down to the bottom of the current dip indicating the end of the pinch phase as shown in figure 1.

This agreement confirms the model parameters for PF1000. Once the model parameters have been fitted to a machine for a given gas, these model parameters may be used with some degree of confidence when operating parameters such as the voltage are varied [9].

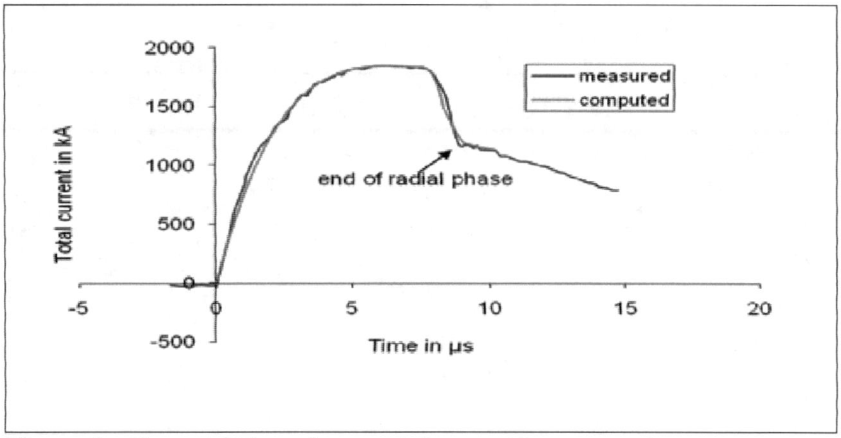

Figure 1. Current fitting of computed current to measured current traces to obtain fitted parameters f_m= 0.13, f_c= 0.7, f_{mr} = 0.35 and f_{cr}= 0.65.

This series is carried out at 35 kV, 10 Torr D, L_0=33.5 nH, r_0=6.1 mΩ (damping factor $RESF=r_0/(L_0/C_0)^{0.5}$ =1.22). The ratio $c=b/a$ is kept at 1.39; C_0 ranges from 14 μF to 39960 μF corresponding E_0 of 8.5 kJ to 24 MJ [12]. For each C_0, anode length z_0 is varied to find the optimum. For each z_0, a is varied so that end axial speed is 10 cm/μs.

We find that the Y_n scaling changes from $Y_n \sim E_0^{2.0}$ at tens of kJ to $Y_n \sim E_0^{0.84}$ at the highest energies (up to 25MJ) investigated (figure 2).

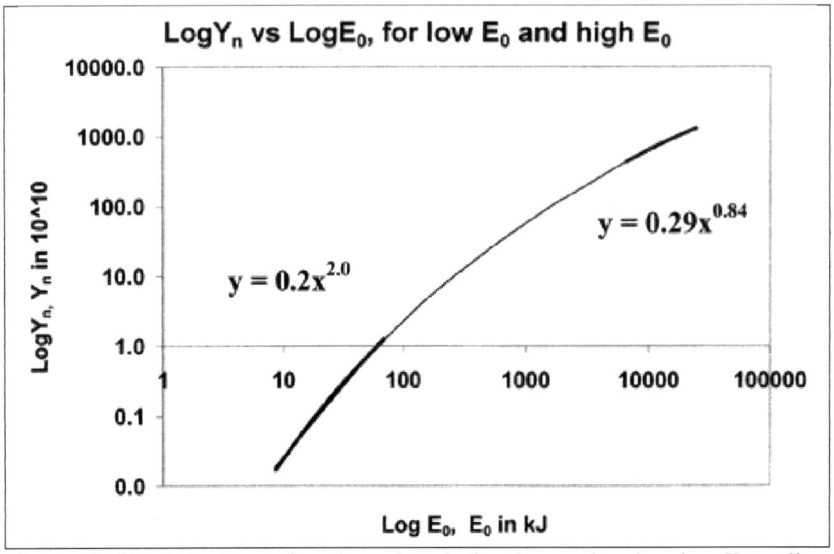

Figure 2. Y_n plotted as a function of E_0 in log-log scale, showing Y_n scaling changes from $Y_n \sim E_0^{2.0}$ at tens of kJ to $Y_n \sim E_0^{0.84}$ at the highest energies (up to 25MJ). The scaling deterioration observed in this figure is discussed in the Conclusion section.

The scaling of Y_n with I_{peak} and I_{pinch} over the energy range up to 25 MJ (figure 3) is: $Y_n = 3.2\times10^{11} I_{pinch}^{4.5}$ and $Y_n = 1.8\times10^{10} I_{peak}^{3.8}$ where I_{peak} ranges from 0.3 to 5.7 MA and I_{pinch} ranges from 0.2 to 2.4 MA.

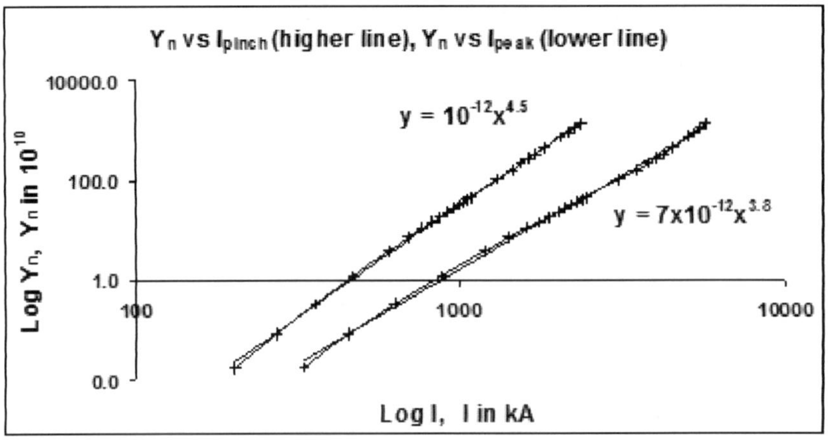

Figure 3. Log(Y_n) scaling with log(I_{peak}) and log(I_{pinch}) for the range of energies up to 25 MJ.

This scaling result confirms an earlier study carried out on several machines with published current traces with Y_n yield measurements, operating conditions and machine parameters including the PF400, UNU/ICTP PFF, the NX2 and Poseidon [11].

4. Scaling laws for neon SXR from numerical experiments over a range of energies from 0.2 kJ to 1 MJ

We next use the code to carry out numerical experiments for bank energies from 0.2 kJ to 1 MJ [52] using a fast PF machine with optimised values for c and typical low L_0.
The following parameters are kept constant: (i) $c=b/a$ (kept at 1.5, which is practically optimum; (ii) the operating voltage V_0 (kept at 20 kV); (iii) static inductance L_0 (kept at 30 nH, which is low enough to reach the I_{pinch} limitation regime [13,14] and (iv) ratio of stray resistance to surge impedance *RESF* (kept at 0.1). The model parameters [8-14] f_m, f_c, f_{mr}, f_{cr} are also kept at fixed values 0.06, 0.7, 0.16 and 0.7 representing average values of large range of machines we have studied. A typical current waveform is shown in figure 4.

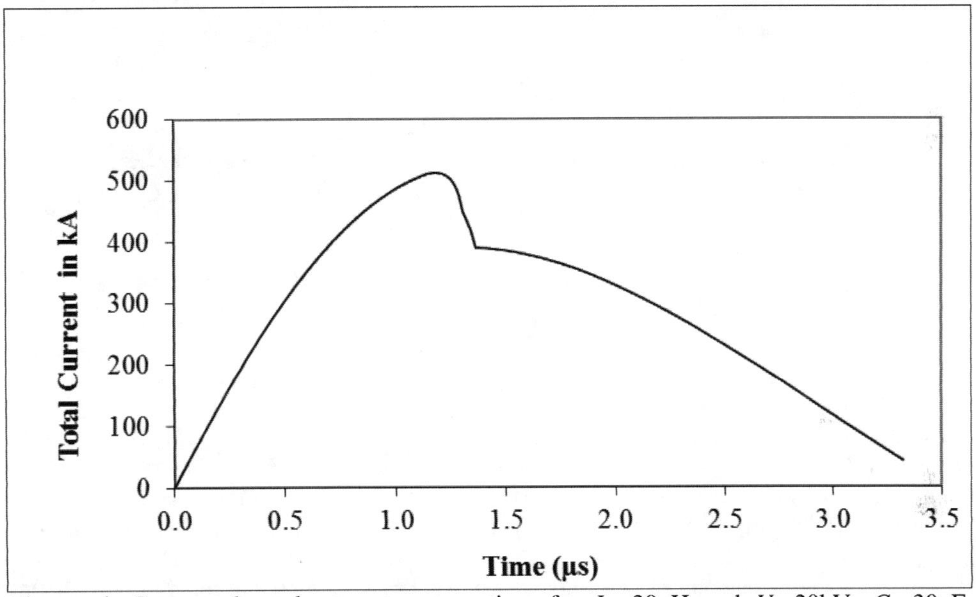

Figure 4. Computed total current versus time for L_0=30nH and V_0=20kV, C_0=30uF, *RESF*=0.1, c=1.5 and model parameters f_m, f_c, f_{mr}, f_{cr} are fixed at 0.06, 0.7, 0.16 and 0.7 for optimised a=2.285 cm and z_0=5.2 cm.

The storage energy E_0 is varied by changing the capacitance C_0. Parameters that are varied are operating pressure P_0, anode length z_0 and anode radius a. Parametric variation at each E_0 follows the order; P_0, z_0 and a until all realistic combinations of P_0, z_0 and a are investigated; the number of runs totalling some 2000. A plot of Y_{sxr} against E_0 is shown in figure 5.

We then plot Y_{sxr} against I_{peak} and I_{pinch} and obtain SXR yield scales as $Y_{sxr}\sim I_{pinch}^{3.6}$ and $Y_{sxr}\sim I_{peak}^{3.2}$. The I_{pinch} scaling has less scatter than the I_{peak} scaling. We next subject the scaling to further test when the fixed parameters *RESF*, c, L_0 and V_0 and model parameters f_m, f_c, f_{mr}, f_{cr} are varied. We add in the results of some numerical experiments using the parameters of several existing plasma focus devices including the UNU/ICTP PFF [17, 26] (*RESF*=0.2, c=3.4, L_0=110nH and V_0=14kV with fitted model parameters f_m = 0.05, f_c = 0.7, f_{mr} = 0.2, f_{cr} = 0.8) [7-9, 23], the NX [10, 24, 27] (*RESF* = 0.1, c = 2.2, L_0 = 20nH and V_0 =11 kV with fitted model parameters f_m = 0.06, f_c = 0.7, f_{mr} = 0.16, f_{cr} = 0.7) [7-10, 24] and PF1000 (*RESF*=0.1, c=1.39, L_0=33nH and V_0=27kV with fitted model parameters f_m=0.1, f_c=0.7,

f_{mr}=0.15, f_{cr}=0.7) [7-9,14]. These new data points (un-blackened data points in figure 6) contain wide ranges of c, V_0, L_0 and model parameters. The resulting Y_{sxr} versus I_{pinch} log-log curve remains a straight line, with scaling index 3.6 unchanged and with no increased scatter. However the resulting Y_{sxr} versus I_{peak} curve now exhibits larger scatter and the scaling index has changed.

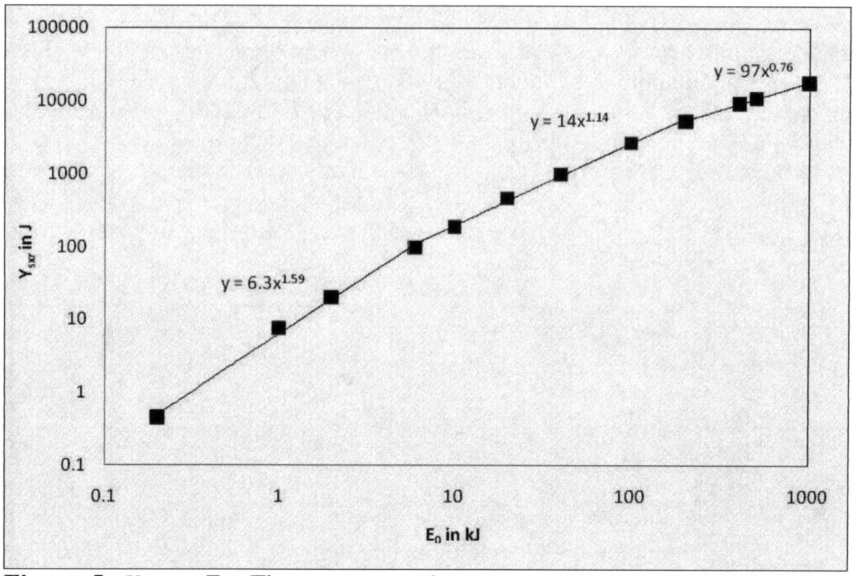

Figure 5. Y_{sxr} vs E_0. The parameters kept constants are: *RESF*=0.1, c=1.5, L_0=30nH and V_0=20 kV and model parameters f_m, f_c, f_{mr}, f_{cr} at 0.06, 0.7, 0.16 and 0.7 respectively. The scaling deterioration observed in this figure is discussed in the Conclusion section.

We highlight that the consistent behaviour of I_{pinch} in maintaining the scaling of $Y_{sxr} \sim I_{pinch}^{3.6}$ with less scatter than the $Y_{sxr} \sim I_{peak}^{3.2}$ scaling particularly when mixed-parameters cases are included, strongly support the conclusion that I_{pinch} scaling is the more universal and robust one. Similarly conclusions on the importance of I_{pinch} in plasma focus performance and scaling laws have been reported [11-15].

It is remarkable that our I_{pinch} scaling index of 3.6, obtained through a set of comprehensive numerical experiments over a range of energies 0.2 kJ to 1 MJ, on Mather-type devices is within the range of 3.5 to 4 postulated on the basis of sparse experimental data, (basically just two machines one at 5 kJ and the other at 0.9 MJ), by Filippov [6], for Filippov configurations in the range of energies 5 kJ to 1 MJ.

We point out that the results represent scaling for comparison with baseline PF devices that have been optimized in terms of electrode dimensions. It must also be emphasized that the scaling with I_{pinch} works well even when there are variations in device from L_0 =30 nH, V_0 =20 kV and c=1.5. However there may be many other parameters which can change and could lead to a further enhancement of x-ray yield.

Journal of Physics: Conference Series **591** (2015) 012022 doi:10.1088/1742-6596/591/1/012022

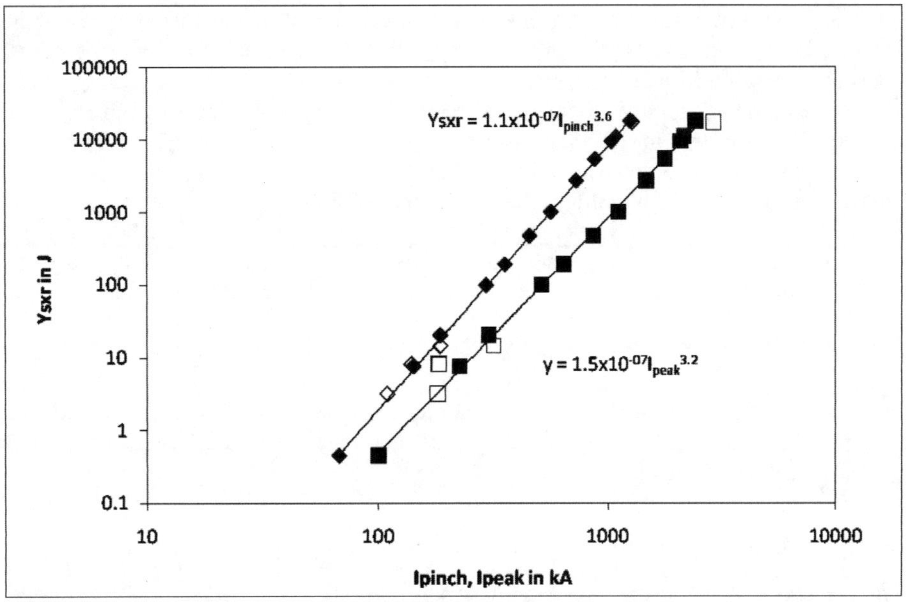

Figure 6. Y_{sxr} is plotted as a function of I_{pinch} and I_{peak}. The parameters kept constant for the black data points are: *RESF* = 0.1, *c* = 1.5, L_0 = 30nH and V_0 = 20 kV and model parameters f_m, f_c, f_{mr}, f_{cr} at 0.06, 0.7, 0.16 and 0.7 respectively. The unblackened data points are for specific machines which have different values for the parameters *c*, L_0, V_0 and *RESF*.

5. Scaling Laws for Beam Ions

Another set of series of numerical experiments on machines from sub-kJ to 1 MJ [50,51] were carried out to obtain relevant scaling laws for ions beam emitted from the pinch plasma. The results for PF deuteron beams (beam energy versus E_0) is presented in figure 7.

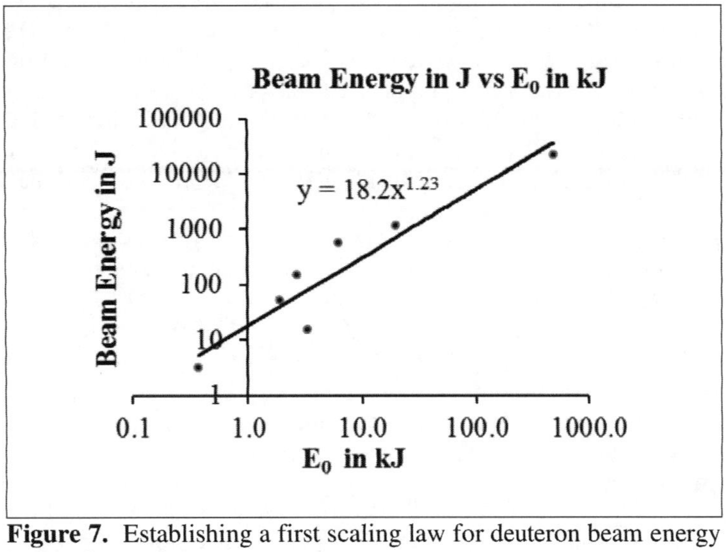

Figure 7. Establishing a first scaling law for deuteron beam energy in the plasma focus.

The beam ions (in J) at exit of a deuterium plasma pinch dependent of the currents (in kA) have the following scaling: Y_{beamions}=2.8x10$^{-7}I_{\text{pinch}}^{3.7}$; Y_{beamions}=8.4x10$^{-7}I_{\text{peak}}^{3.16}$; $Y_{\text{beamions}}=18.2E_0^{1.23}$; where Y_{beamions} is in J and E_0 is in kJ ranging from 1 kJ to 1MJ.

We note the considerable scatter with square of residual 'R^2' of 0.89, deviating considerably from the perfect value of 1.00 in this initial attempt to present quantitative ideas of ion beams to provide reference data for laboratory measurements. More numerical experiments and laboratory measurements are needed to put the scaling laws on a firmer footing.

The PF operating in any gas and over a wide range of pressure emits ion beams and plasma streams. These ion beams are being studied to synthesize nano-materials and damage candidate materials of reactor walls. The Lee model code computes the ion beam yields emitted from the pinch column.

6. Conclusion

Numerical experiments carried out using the universal plasma focus laboratory facility based on the Lee model code gives reliable scaling laws for neutrons production and neon SXR yields for plasma focus machines. The scaling laws obtained:

For neutron yield: (yield in number of neutrons per shot)
$Y_n = 3.2\text{x}10^{11} I_{\text{pinch}}^{4.5}$; $Y_n = 1.8\text{x}10^{10} I_{\text{peak}}^{3.8}$; I_{peak}(0.3 to 5.7) and I_{pinch}(0.2 to 2.4) in MA.
$Y_n \sim E_0^{2.0}$ at tens of kJ to $Y_n \sim E_0^{0.84}$ at MJ level (up to 25MJ).

For neon soft x-rays: (yield in J per shot)
$Y_{\text{sxr}} = 8.3\text{x}10^3 I_{\text{pinch}}^{3.6}$; $Y_{\text{sxr}} = 6\text{x}10^2 I_{\text{peak}}^{3.2}$; I_{peak}(0.1 to 2.4) and I_{pinch}(0.07 to1.3) in MA.
$Y_{\text{sxr}} \sim E_0^{1.6}$ (kJ range) to $Y_{\text{sxr}} \sim E_0^{0.8}$ (towards MJ).

For beam ions at exit of a deuterium plasma pinch: (yield in J per shot)
Y_{beamions}=2.8x10$^{-7}I_{\text{pinch}}^{3.7}$; Y_{beamions}= 8.4x10$^{-7}I_{\text{peak}}^{3.16}$; and currents in kA.
$Y_{\text{beamions}} = 18.2E_0^{1.23}$; where E_0 is in kJ; averaged over 1 kJ to 1MJ

These laws provide useful references and facilitate the understanding of present plasma focus machines. More importantly, these scaling laws are also useful for design considerations of new plasma focus machines particularly if they are intended to operate as optimized neutron or neon SXR sources. More recently, the scaling of Y_n versus E_0 as shown above has been placed in the context of a global scaling law [38] with the inclusion of available experimental data. From that analysis, the cause of scaling deterioration for neutron yield versus energy as shown in figure 2 (which has also been given the misnomer 'neutron saturation') has been uncovered as due to a current scaling deterioration caused by an almost constant axial phase 'dynamic resistance' interacting with a reducing bank impedance as energy storage is increased at essentially constant voltage. The deterioration of soft x-ray yield with storage energy as shown in figure 5 could also be ascribed to the same axial phase 'dynamic resistance' effect as described in reference [38]. This deterioration of scaling will also appear in the scaling trends (with stored energy) of beam ions.

We emphasis here that the scaling laws with I_{pinch} is the more fundamental and robust one compared to I_{peak}. This is because although the PF is reasonably consistent in its operations, there will be occasions when even the best optimized machines may not focus or poorly focused although having a high I_{peak} with no neutrons. However, I_{pinch} being the current actually flowing in the pinch is more consistent in all situations.

The numerical experiments gives robust scaling laws for PFs covering a wide range of energies from sub kJ to tens of MJ. It supplements the limited (non-existent in the case of beam ions) scaling laws available to predict PF radiations yields. Now, we have on stronger footing the useful scaling laws for neutron, SXR and ion yields from PF machines.

7. References

[1] Lee S 1998 *Twelve years of UNU/ICTP PFF—A Review* IC/ 98/ 231 Abdus Salam ICTP, Miramare, Trieste ICTP Open Access Archive http://eprints.ictp.it/31/ 5-34
[2] Kato Y and Be S H 1986 *Appl. Phys. Lett.* **48** 686
[3] Bogolyubov E P et al 1998 *Phys. Scripta.* **57** 488-494
[4] Lee S et al 1998 *IEEE Trans. Plasma Sci.* **26** 1119
[5] Filippov N V et al 1996 *IEEE Trans Plasma Sci. 24* 1215-1223
[6] Filippov N V et al 1996 *Phys Lett A* **211**(3) 168-171
[7] Lee S *Institute for Plasma Focus Studies* http://www.plasmafocus.net
[8] Lee S *Internet Workshop on Plasma Focus Numerical Experiments(IPFS-IBC1)14April-19May 2008* http://www.plasmafocus.net/IPFS/Papers/IWPCAkeynote2ResultsofInternet-basedWorkshop.doc
[9] Lee S *Radiative Dense Plasma Focus Computation Package: RADPF* http://www.intimal.edu.my/school/fas/UFLF/File1RADPF.htm http://www.plasmafocus.net/IPFS/modelpackage/File1RADPF.htm
[10] Lee S, Rawat R S, Lee P and Saw S H 2009 *J. Appl. Phys.* **106** 023309
[11] Lee S and Saw S H 2008 *J Fusion Energy* **27** 292-295
[12] Lee S 2008 *Plasma Phys. Control. Fusion* **50** 105005
[13] Lee S, Saw S H, Lee P C K, Rawat R S and Schmidt H 2008 *Appl. Phys. Lett.* **92** 111501
[14] Lee S and Saw S H 2008 *Appl. Phys. Lett.* **92** 021503
[15] Lee S, Lee P, Saw S H and Rawat R S 2008 *Plasma Phys. Control. Fusion* **50** 065012
[16] Lee S, 1984 Plasma focus model yielding trajectory and structure in *Radiations in Plasmas*, ed B McNamara (Singapore: World Sci Pub Co, ISBN 9971-966-37-9) vol. II 978–987
[17] Lee S et al 1988 *Am. J. Phys.* **56** 62-68
[18] Tou T Y, Lee S and Kwek K H 1989 *IEEE Trans. Plasma Sci.* **17** 311-315
[19] Lee S 1991 *IEEE Trans. Plasma Sci.* **19** 912-919
[20] Jalil bin Ali 1990 *Development and Studies of a small Plasma Focus* PhD thesis, Universiti Teknologi Malaysia, Malaysia
[21] Potter D E 1978 *Nucl. Fus.* **18** 813-823
[22] Lee S and Serban A 1996 *IEEE Trans. Plasma Sci.* **24**(3) 1101-1105.
[23] Liu Mahe 2006 *Soft X-rays from compact plasma focus* PhD thesis, NIE, Nanyang Technological University, Singapore. ICTP Open Access Archive: http://eprints.ictp.it/327/
[24] Bing S 2000 *Plasma dynamics and x-ray emission of the plasma focus* PhD Thesis, NIE, Nanyang Technological University, Singapore. ICTP Open Access Archive: http://eprints.ictp.it/99/.
[25] Serban A and Lee S 1998 *J Plasma Physics* **60**(1) 3-15.
[26] Liu M H, Feng X P, Springham S V and Lee S 1998 *IEEE Trans. Plasma Sci.* **26** 135–140
[27] Wong D, Lee P, Zhang T, Patran A, Tan T L, Rawat R S and Lee S *2007 Plasma Sources Sci. Technol.* **16** 116-123
[28] Lee S 2000–2007 http://ckplee.myplace.nie.edu.sg/plasmaphysics/
[29] Lee S 2005 ICTP Open Access Archive: http://eprints.ictp.it/85/
[30] Mohammadi M A, Sobhanian S, Wong C S, Lee S, Lee P and Rawat R S 2009 *J. Phys. D: Appl.Phys.* **42** 045203
[31] Springham S V, Lee S and Rafique M S 2000 *Plasma Phys. Control. Fusion* **42** 1023-1032
[32] Lee S, Lee P, Zhang G, Feng X, Gribkov V A, Liu M, Serban A and Wong T 1998 *IEEE Trans. Plasma Sci.* **26**, no. 4 1119-1126
[33] Siahpoush V, Tafreshi M A, Sobhanian S and Khorram S 2005 *Plasma Phys. Control. Fusion* **47** 1065-1072
[34] Gribkov V A, Banaszak A, Bienkowska B, Dubrovsky A V, Ivanova-Stanik I, Jakubowski L, Karpinski L, Miklaszewski R A, Paduch M, Sadowski M J, M Scholz M, A Szydlowski and Tomaszewski K 2007 *J. Phys. D: Appl. Phys.* **40** 3592-3607.

[35] Lee S 2009 *Appl. Phys. Lett.* **95** 151503
[36] Lee S, Saw S H and Jalil Ali 2013 *J Fusion Energy* **32** 42-49
[37] Lee S and S H Saw 2012, *J Fusion Energy* **31** 603-610
[38] Saw S H 1991 *Experimental studies of a current-stepped pinch* PhD Thesis Universiti Malaya, Malaysia
[39] Lee S 1984 J *Phys D: Appl Phys* **17**,733-739
[40] Lee S et al 2011 *IEEE Trans Plasma Sci* **39** 3196-3202
[41] Saw S H et al 2010 Rev Sci Instruments, **81** 053505
[42] Lee S et al 2012 *J Fusion Energy* **31** 198–204
[43] Lee S, Saw S H, Soto L, Springham S V, Moo S P 2009 *Plasma Phys Contr Fusion* **51** 075006
[44] Saw S H, Rawat R S, Lee P, Talebitaher Alireza, Abdou A E, Chong P L, Roy F Jr, Ali J and Lee S 2013 *IEEE Trans Plasma Sci* **41** (11) 3166-3172
[45] Lee S, Saw S H, Hegazy H, Ali J, Damideh V, Fatis N, Kariri H, Khubrani A, Mahasi A *J Fusion Energy* published online 5 January 2014
[46] Lee S, Saw S H, Abdou A E and Torreblanca H 2011 *J Fusion Energy* **30**, 277-282
[47] Aghamir F M and Behbahani R A 2012 *J. Plasma Physics* **78**(5) 585-588
[48] M. Akel, Sh Al-Hawat, S H Saw and S Lee 2010 *J Fusion Energy* **29**(3) 223-231
[49] Sh. Al-Hawat, M. Akel, S H Saw, S Lee 2012 J Fusion Energy **31** 13-20
[50] Akel M, Lee S, Saw S H, *IEEE Trans Plasma Sci* **40** 3290-3297
[51] Huba J D 2006 *Plasma Formulary* 44
[52] Lee S, Saw S H, Lee P and Rawat R S 2009 *Plasma Phys and Controlled Fusion* **51** 105013
[53] Lee S and Saw S H 2012 *Phys. Plasmas* **19** 112703
[54] Lee S and Saw S H 2013 *Phys. Plasmas* **20** 062702

Acknowledgments
This work is partially supported within IAEA Research Contract: 16934 of CRP F1.30.13 (Investigations of Materials under High Repetition and Intense Fusion-relevant Pulses). The author also acknowledges research grants INT-CPR-01-02-2012 and FRGS/2/2013/SG02/INTI/01/1.

Ion emission study using visible spectroscopy and ToF method in a plasma focus device of two kilojoules

J. Moreno[1,2], D. Morales[3], G. Avaria[1,2], O. Cuadrado[4], and L. Soto[1,2]
[1]Comisión Chilena de Energía Nuclear, Chile
[2]Center for Research and Applications in Plasma Physics and Pulsed Power, P4, Chile
[3]Universidad Andrés Bello, Chile
[4]Universidad Austral de Chile, Chile

E-mail: jmoreno@cchen.cl

Abstract. Different studies have been developed in order to understand the dynamics and processes involved in the particle emission from plasma focus devices operating in the kilo-joule range. The use of chemical compound gasses and noble gas mixtures has proven to produce different charged particles, as well as increase the neutron yield from deuterium plasma discharges. Nevertheless, the processes and parameters involved in these discharges are not fully understood. In this work we will present results of visible spectroscopy and "time-of-flight" observations of the different ion species and ionization levels obtained in a 2kJ plasma focus device, when using deuterium or hydrogen with small percentage impurities.

1. Introduction

It is an interesting phenomenon that high energy ions with energy of more than 1 MeV are produced in plasma focus devices [1]. The energy of these ions is sometimes larger than hundred times the charging voltage of the capacitor bank. These energetic ions are considered to play an important role in the production of the intense neutron flux in the plasma focus device when using deuterium gas [2, 3]. The determination of ion beam characteristics is very important not only in understanding the mechanism of production of high-energy ion or neutron, but also for their potential applications.
In general, the studies of ion beams have been performed in plasma focus devices that operate with energies from 1kJ to 1MJ [4]. In this work, preliminary results of some characteristics of the emitted ions from a plasma focus device of 1.6 kilojoules, PEPUTNIK, will be presented. In particular, the ion energy and flux is measured with a biased graphite ion collector. Furthermore, to characterize the ionization degree evolution of the gas impurities present in the plasma we used a visible spectrometer, ANDOR Shamrock-500i-B1.

2. Experimental setup

The experiments presented here were conducted in a plasma focus device operating at a kilojoule energy level (PEPUTNIK: 8000nF, 20-30kV, 1600-3600J, ~960ns time to peak current, dI/dt ~ 2.7×10^{11} A/s) (see figure 1), which has been designed and constructed at the Chilean Nuclear Energy Commission, CCHEN. In table 1 the dimensions and parameters of the device are shown. Discharges in different gasses like hydrogen, deuterium and a mixture of 98 % hydrogen and 2% neon at pressure of 9mbar to 10mbar were done. A charging voltage of 20 kV was used.

Table 1. Electrical parameters of the PEPUTNIK device.

Parameter	Value
C_0[nF]	8000
L_0[nH]	47
E[J]	1600
V_{op}[kV]	20
$I_{máx}$[kA]	260
T/4[ns]	960
r_a[mm]	12
z_{eff}[mm]	27

Figure 1. Image of the experimental setup of the 2kJ plasma focus device, PEPUTNIK.

3. Electrical and spectral diagnostics

Regular voltage divider probes and current derivative probes, as well as a Faraday Cup (FC) for ion collection, were used. The ion collector is a conical probe designed to maintain an impedance of 50 Ω [5, 6]. It is composed of a graphite conical inner electrode of 15mm in length with a base diameter of 7mm to reduce the secondary electron emission (SEE) and a brass external electrode separated by a Teflon insulator (figure 2). The graphite collector inside the drift tube was used to detect the axial ion beams emitted from the PF and it was placed 66 cm away from the anode tip. The collector is negatively biased to a dc voltage between -150 to -200V. The negative voltage also removes the secondary electrons from the surface of the collector that are generated at the collector surface by the incoming ion beams. Thus, a suitable biasing voltage on the collector is essential for proper interpretation of the collector signal. Due to the negative biasing, the ion probes also act as X-Ray Diode (XRD).

The characterization of the ionization stages of the gas impurities present in the plasma is being done by means of the visible spectrometer: ANDOR Shamrock-500i-B1. With the use of an ICCD available in the laboratory, time resolved images of the spectra emitted by the discharge are acquired and processed. To be able to have a spatial resolution we use a system involving two plano-convex lenses that allow focusing a small volume of the plasma sheath into the fiber optic (FO) that connects to the spectrometer. A 300 groove/mm diffraction grating was used due to the broad spectral region available and good response to the observed region, which in this report spans from ~ 620 nm to ~ 705 nm. Two plano-convex lenses were used to focus the plasma emission into a 19 circular-to-linear fiber optic bundle connected to the entrance of the spectrometer (see figure 3). The ICCD-Spark gap system was synchronized externally, allowing the acquisition of time resolved measurements with integration times of 20 ns.

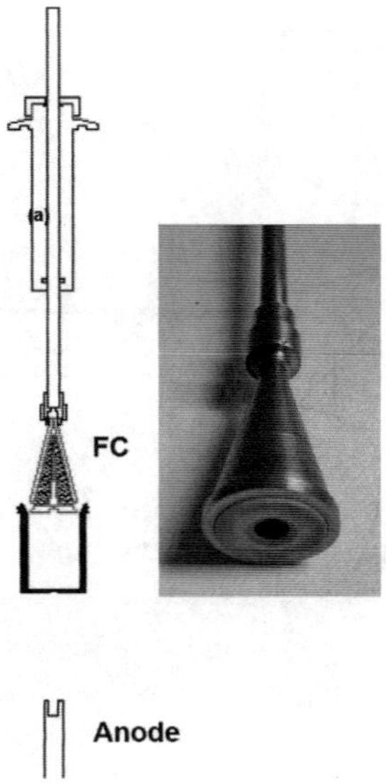

Figure 2. Picture and scheme of Faraday cup.

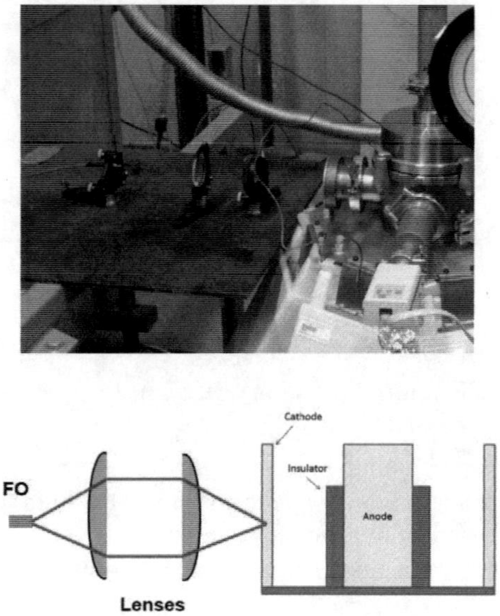

Figure 3. Picture and scheme of the experimental setup for the spectroscopy measurements.

4. Results and discussion

In general it is possible to observe of the voltage and current derivative signals that the dip is seen to coincide with both the x-ray pulse in the FC. This kind of peak on these detectors has been reported earlier as x-rays emitted during pinching [5, 6]. Time of flight (TOF) measurements performed using two ion probes indicate that the time delay between probes for the different beam components with simultaneous emission at a time which, within the time resolution, coincides with the emission of the x-ray pulse. These facts allow single ion probe signals to be used for the ion energy spectrum measurements [6]. The time correlation between x-ray and ion beam emission indicates that in PF discharges ion acceleration takes place at some time close to the maximum radial compression phase. Therefore, it is reasonable to assume that ion acceleration takes place in existing plasma, with a multi-charged ion composition. Under these conditions, all ions are accelerated by the same electric field and the energy gained in the acceleration process is proportional to the ion charge [7, 8].

Figure 4 shows the distribution of the ions energies for all data obtained in hydrogen (H_2) at 9 mbar. It is clearly observed, in this case, that the ions energies are distributed in two regions with a maximum energy around 370keV. The region of higher energy corresponds to the ion species with the highest ionization stages, Al^{+5}, Al^{+6} (from the alumina insulator), and O^{+5}. On the other hand, the region of lower energy corresponds to the ion species with lower ionization stages, H^{+}, Al^{+} and Al^{+2}, N^{+}, Cu^{+} and Cu^{+2} (from the anode), and O^{+2}.

Spectral measurements were done in order to identify the elements present and their ionization degree. Figure 5 shows the spectra acquired during the whole current pulse at the pinch volume. As seen from the image, only some ionic species are detected, as Al^{+} and Al^{+2}, Cu^{+2} and Cu^{+3}, and O^{+}. Similar ionization stages and flux level have been observed in devices of the same energy range operated with methane and nitrogen [6, 9, 10].

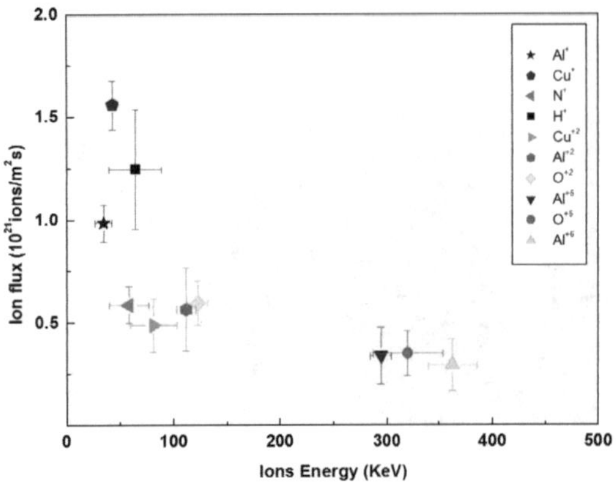

Figure 4. Graph of ion flux vs ion energy at 9 mbar in hydrogen

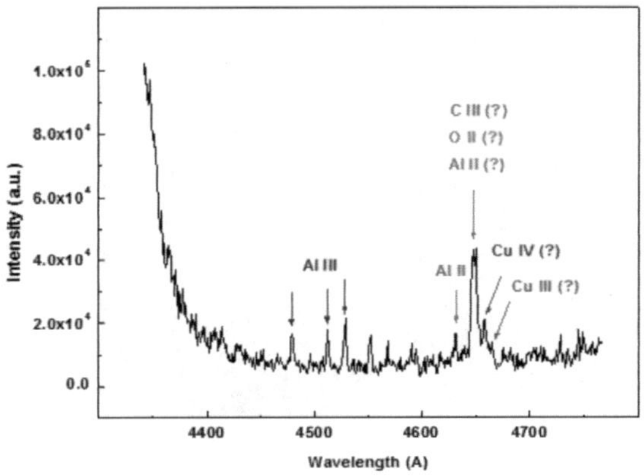

Figure 5. Spectral evidence of the presence of some ion species obtained at 9 mbar in hydrogen.

These results show some agreement between both diagnostics, ToF and spectrometer, but it is necessary to extend the range of measurements to other pressure using mixture of gases and improving the method of time of flight to determine the ionization degree.

5. Conclusion and future work

The results obtained with ions ToF method using a Faraday Cup show the presence of several ion species with energies between 40 keV and 370 keV. The comparison of the results of both diagnostics shows a small agreement between the ionization degree and type of species present in the discharge. It is necessary to increase the number of shots for each experimental condition. It is necessary to measure at the axial position with the spectroscopic diagnostic to directly compare the results and enhance the measuring conditions in case some radiation absorption is present.

Acknowledgments
The authors would like to acknowledge financial support by FONDECYT grant 1120801, FONDECYT Iniciación 11121587, CONICYT PAI grant 791100020, and financial support partial CONICYT-PIA Anillo ACT1115.

References

[1] K. Takao, Y. Doi, S. Hirata, M. Shiotani, I. Kitamura, T. Takahashi, and K. Masugata, Jpn. J. Appl. Phys. **40**, 1013-1015 (2001).
[2] W. H. Bostik, H. Kilic, V. Nardi, and C. W. Powell, Nucl. Fusion. **33**, 413-420 (1993).
[3] M. Sadowski, J. Zebrowski, E. Rydygier, and J. Kucinski, Plasma Phys. Control. Fusion **30**, 763-769 (1988).
[4] S. R. Mohanty, T. Sakamoto, Y. Kobayashi, I. Song, M. Watanabe, T. Kawamura, A. Okino, K. Horioka and E. Hotta, Rev. Sci. Instrum. **77**, 04506 (2006).
[5] S. R. Mohanty, H. Bhuyan, N. K. Neog, R. K. Rout, y E. Hotta, Jpn J. of Appl. Phys. **44**, 5199-5205 (2005).
[6] H. Bhuyan, H. Chuaqui, M. Favre, I. Mitchell, and E. Wyndham, J. Phys. D: Appl. Phys. **38**, 1164-1169 (2005).
[7] M. J. Bernstein. Phys. Fluids, **13**, p. 2858 (1970).
[8] S. P. Gary. Phys. Fluids **17**, p.2135 (1974).
[9] H. Kelly, A. Lepone and A. Márquez, IEEE Trans. Plasma Sci. **25** 455 (1997).
[10] H. Kelly, A. Lepone, A. Márquez, M. J. Sadowski, J. Baranowski and E. Skladnik-Sadowska, IEEE Trans. Plasma Sci. **26** 113 (1998).

Time resolved visible spectroscopy studies of the plasma sheath evolution in a low energy plasma focus device

G. Avaria[1,2], O. Cuadrado[3], J. Moreno[1,2] and L. Soto[1,2]

[1]Comisión Chilena de Energía Nuclear, Casilla 188-D, Santiago, Chile
[2]Center for Research and Application in Plasma Physics and Pulsed Power, P4, Chile
[3]Universidad Austral de Chile, Valdivia, Chile

E-mail: gavaria@cchen.cl

Abstract. We present the study of the evolution of the ionization degree of the plasma sheath in a low energy plasma focus device with time and spatial resolution, by means of visible spectroscopy. The measurements were developed in the low energy plasma focus device PF-400J (176-539 J, 880 nF, 20-35 kV, quarter period ~300 ns) [1], using an ANDOR Shamrock 500i visible spectrometer with an ICCD that enabled the acquisition of spectra with time resolution in the order of tens of nanosecond. The use of a lens system improved spatial resolution, allowing the study of the plasma sheath in different stages of the discharge dynamic. In this work we present results from the pinch volume to study the evolution of the ionization degree of the plasma. The results include discharges in pure Hydrogen and a mixture of noble gases and Hydrogen, such as Neon and Helium, which were introduced in the discharge by small percentages ranging from 2 to 5%.

1. Introduction

New interest on small plasma focus devices has been developing over the years for its capacity as a pulsed neutron source [2–4] and a pulsed X-ray source [5]. Several reports indicate an increase in the neutron emission depending on the gas impurity used, yield that can improve in an order of magnitude [6; 7]. Although this is an important finding, studies about the ionization degree evolution for those impurities in the plasma sheath are none existent. Plasma sheath kinematics on a low energy plasma focus device have been studied in detail by Veloso et al.[8; 9]. Few studies using visible spectroscopy have shown the ionization degree at the pinch for discharges in Argon [10] and the evolution of the plasma when it interacts with a target [11]. The following report shows preliminary results for the measurements of the ionization degree of impurities at the pinch volume of low energy plasma focus devices.

2. Experimental details

The experimental observations were developed at the low energy plasma focus device PF-400J (176-539 J, 880 nF, 20-35 kV, quarter period ~300 ns) [1] and at the plasma focus device PF-2kJ "Peputnik" (1.6-3.6 kJ, 8000nF, 20-30 kV, ~ 960 ns time to peak current, dI/dt ~ $2.7 \times 10^{11} A/s$). For the observations presented here, the PF-400J device was operated with Hydrogen at 9 mbar or with a mixture of Neon (2% - 5%) and Hydrogen (98 % - 95%) at an equivalent pressure of 9 mbar in Hydrogen. The device was operated at a repetitive scheme with a charging voltage of 27 kV. The spectra was acquired by means of a ANDOR Shamrock 500i spectrometer which had a Princeton Instruments ICCD-576/EMG camera attached to the exit port. A 300 groove/mm diffraction grating was used due to the broad spectral region available and good response to the observed region, which in this report spans from $\sim$ 620 nm to $\sim$ 705 nm. The plasma emission from the pinch volume was focused into a 19 circular-to-linear fiber optic bundle connected to the entrance of the spectrometer (see figure 2).The ICCD-Spark gap system was synchronized externally, allowing the acquisition of time resolved measurements with integration times of 20 ns. The measurements taken at the PF-2kJ device were done with a mixture of Neon (2%) and Hydrogen (98%) with the same optical setup as the one used with the PF-400J.

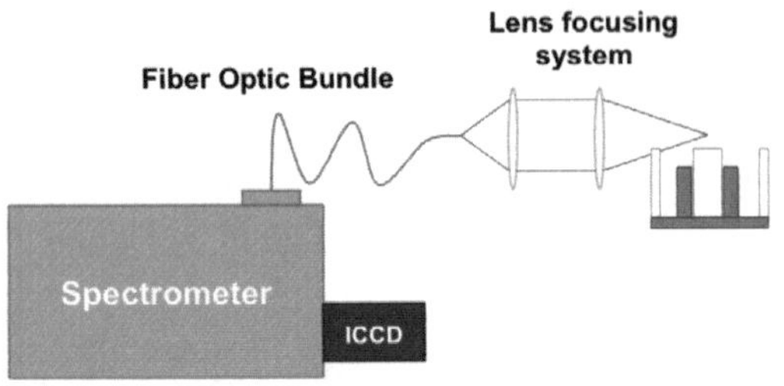

Figure 1. Scheme of the experimental setup for the spectroscopy measurements in both devices.

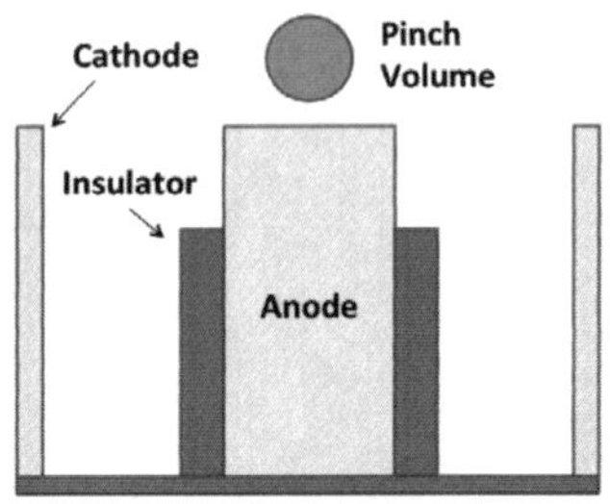

Figure 2. Plasma emission from the pinch volume (green circle) was focused into the lens-to-fiber optic system.

3. Experimental results

Figure 3 shows the evolution of the plasma emission for the PF-400J discharge at 9 mbar of Hydrogen. Times are taken with respect to the pinch moment: a negative time corresponds to a spectra taken before the pinch and a positive time corresponds to a spectra acquired after the pinch. In the sequence it is possible to observe the appearance of a very broad H_α line with a self-absorption dip. The broadest profile is observed at the pinch time which is related to a high density plasma[12]. For later times the line becomes narrower and the self absorption dip disappears. The wings and central dip observed in the Balmer-α line have been reported before [13]: the wings are related with high velocities of the excited hydrogen emitters and electric fields of the discharge [14] and the dip is associated with ion dynamic effects [15]. A more detailed study of this phenomena is going to be treated in a future report.

The evolution of the plasma emission for the PF-400J with a mixture of Neon (2%) and Hydrogen (98%) is shown in figure 4. The image shows the spectra emitted from the pinch volume at different times during the evolution of the discharge where a similar behavior as figure 3 is seen in the sequence. Notice that at a time close to the pinch, the intensity and width of the H_α line does not resemble the same characteristics seen in the pure hydrogen discharge. In this case, it looks like the plasma is taking longer to reach the anode center which could be related with the mass on the sheath. Even though the gas mixture is set up to achieve an equivalent mass, it appears from the experimental observations that the plasma sheath could be heavier than calculated. Future measurements will include a different calculation of the gas proportion which takes into account sheath velocity parameters [16]. At a time around +50 ns, the line profile is similar to the one seen in pure hydrogen, nevertheless, the evolution of the line at times larger than +100 ns is different from the pure hydrogen case. It can be seen that for

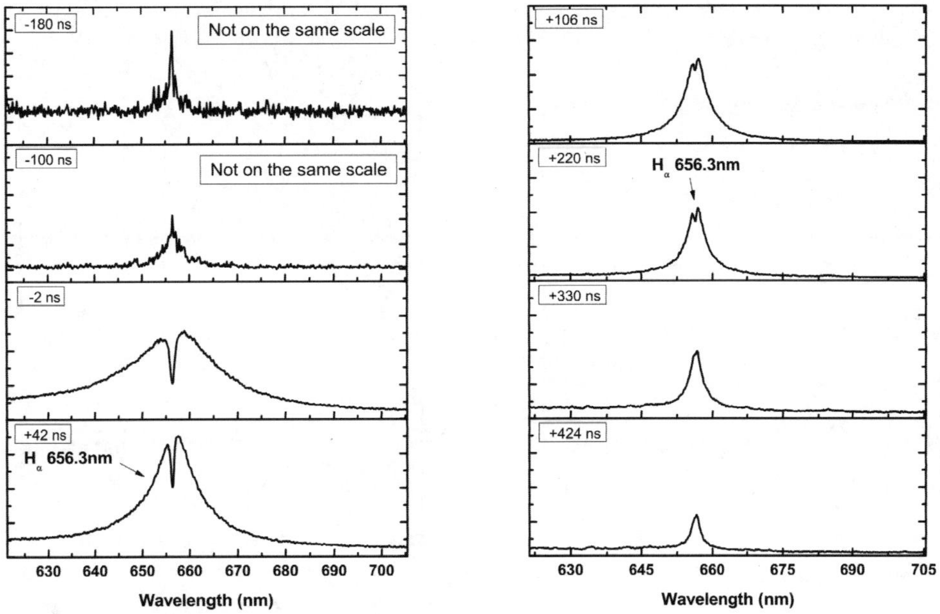

Figure 3. Evolution of the spectra taken at the pinch volume for a discharge in pure Hydrogen on the PF-400J device.

times larger than +100 ns, the balmer-α line stays broader and it starts to become narrower at times over 400 ns after the pinch. The spectra taken at +112 ns shows the very dim appearance of the Ne I emission in the region of 630-645 nm, emission that is evident at later times.

In figure 5 a detailed spectra sequence for times larger than +100 ns can be seen. From the image it is clear that a very small emission from Ne I is present and its intensity grows from the background level at times larger than 200 ns after the pinch. Spectral regions where emission of Ne II-IV ions is expected were probed, but no line emission was detected.

In order to check if the stored energy of the discharge is the reason for the results obtained, the optical setup was moved to the PF-2kJ discharge. The gas mixture was kept on the same proportion, but the equivalent pressure was increased to 12 mbar in order to follow the scaling rules determined by L. Soto et al [17] for devices with this stored energy. Time evolution of the spectra showed identical results as the ones seen in the PF-400J: only excited Neon and no ionization of the impurity. To observe if the emission was appearing at a different time during the discharge evolution, time integrated shots were acquired. Figure 6 shows a spectra integrated during the whole current pulse at the pinch volume (figure 2). As seen from the image, only excited Neon is detected and some ions of Aluminium appear in the profile. The latter emission comes from the Alumina (Al_2O_3) insulator used to cover the anode of the discharge. Nevertheless no emission from the electrode material was detected, these results are in agreement to the ones presented by Skladnik-Sadowska in PF-1000 [11]. Similar observations have been made by Feugeas [10] were emission from the insulator material was measured. The reason for the observed behavior in both devices (PF-400J and PF-2kJ) is not understood and further research is needed to fully comprehend these results.

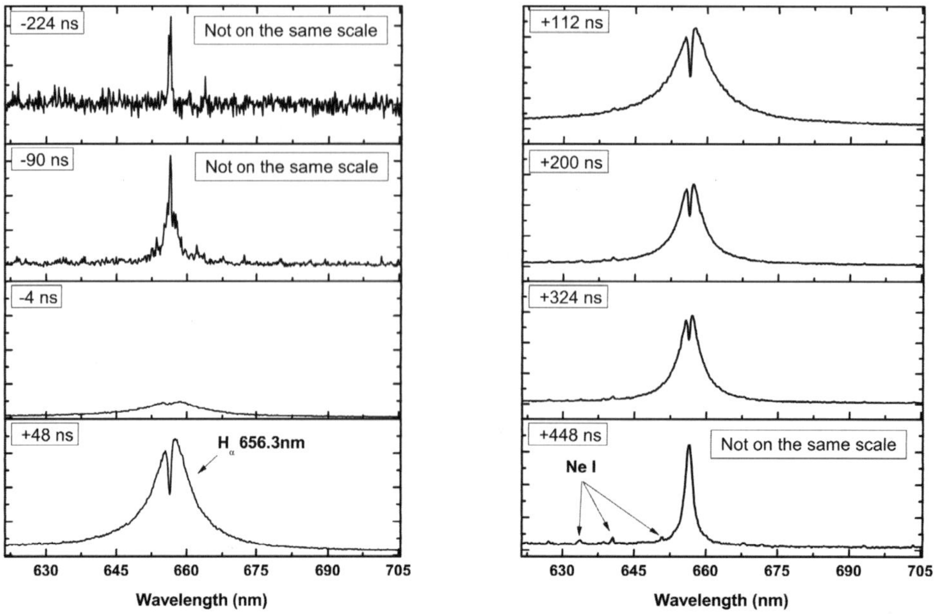

Figure 4. Evolution of the plasma sheath in time at the pinch volume. The discharge was operated at 9 mbar with a mixture of Neon (2%) and Hydrogen (98%) in the PF-400J discharge.

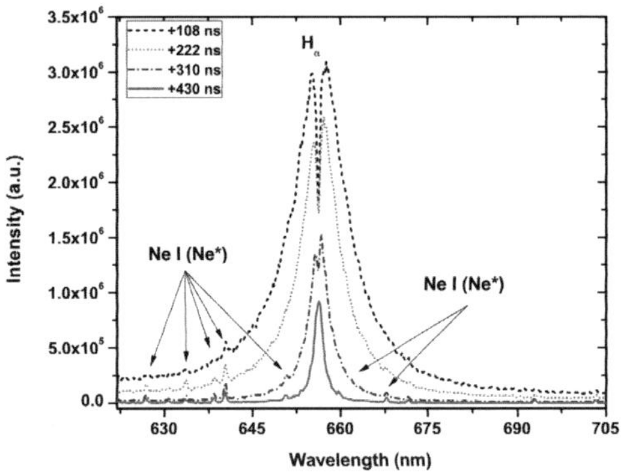

Figure 5. Detail of the appearance of Ne I line emission for the PF-400J discharge. Notice that the line emission appears at times that are hundreds of nanoseconds after the pinch.

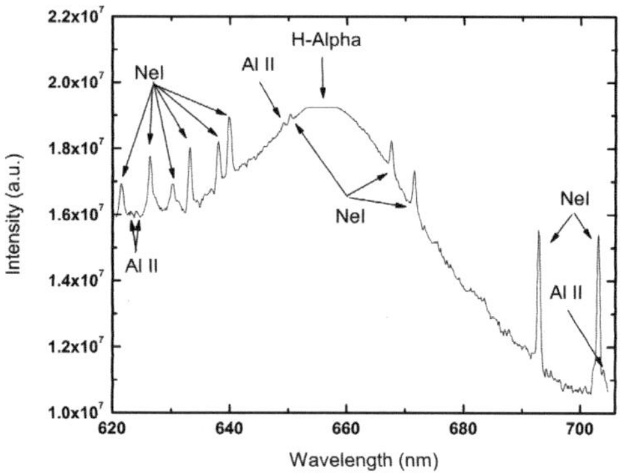

Figure 6. Time integrated spectra from the PF-2kJ discharge. From the image is evident that only excited Neon (Ne I) is detected, as well as some aluminium that comes from the insulating sleeve.

4. Conclusion

The results presented above show the behaviour of the plasma sheath for different conditions on different plasma devices. From the measurements done in pure Hydrogen it is possible to observe the presence of a dense plasma at the time of the pinch, which relaxes in times in the order of hundreds of nanoseconds. When the discharge is done in a mixture of Neon and Hydrogen, a change on the evolution of the sheath is evident: the broadening of the H_α line occurs later than the pinch time and this broadening continues until times larger than the ones observed in pure Hydrogen. The appearance of excited Neon at later times than the pinch is still under study and not understood at this time.

Acknowledgments

The authors would like to acknowledge the financial support from CONICYT-PAI Inserción 791100020, FONDECYT Iniciación 11121587, CONICYT-PIA Anillo ACT1115 and FONDECYT Regular 1120801.

References

[1] Silva P, Moreno J, Soto L, Birstein L, Mayer R E and Kies W 2003 *Applied Physics Letters* **83** 3269–3271

[2] Soto L 2005 *Plasma Physics and Controlled Fusion* **47** A361

[3] Klir D and Soto L 2012 *Plasma Science, IEEE Transactions on* **40** 3273–3279 ISSN 0093-3813

[4] Veloso F, Pavez C, Moreno J, Galaz V, Zambra M and Soto L 2012 *Journal of Fusion Energy* **31** 30–37

[5] Pavez C, Pedreros J, Zambra M, Veloso F, Moreno J, Ariel T S and Soto L 2012 *Plasma Physics and Controlled Fusion* **54** 105018

[6] Verma R, Lee P, Lee S, Springham S, Tan T, Rawat R and Krishnan M 2008 *Applied Physics Letters* **93** 101501–101501

[7] Bures B L, Krishnan M, Madden R E and Blobner F 2010 *Plasma Science, IEEE Transactions on* **38** 667–671

[8] Veloso F, Tarifeño-Saldivia A, Pavez C, Moreno J, Zambra M and Soto L 2012 *Plasma Physics and Controlled Fusion* **54** 095007

[9] Veloso F, Moreno J, Tarifeño-Saldivia A, Pavez C, Zambra M and Soto L 2012 *Measurement Science and Technology* **23** 087002

[10] Feugeas J and Grigioni G 1997 *Journal of Physics D: Applied Physics* **30** 2026

[11] Skladnik-Sadowska E, Malinowski K, Sadowski M, Kubkowska M, Jakubowska K, Paduch M, Scholz M, Garkusha I, Ladygina M and Tereshin V 2011 *Contributions to Plasma Physics* **51** 288–292

[12] Parigger C, Plemmons D H and Lewis J W L 1995 *Appl. Opt.* **34** 3325–3330

[13] González M A, González V R and Mar S 1989 *Journal of Quantitative Spectroscopy and Radiative Transfer* **42** 247–251

[14] Ayers E L and Benesch W 1988 *Phys. Rev. A* **37**(1) 194–199

[15] Wiese W L, Kelleher D E and Helbig V 1975 *Phys. Rev. A* **11**(6) 1854–1864

[16] Knoblauch P 2011 *Caracterización de la emisión neutrónica y del espectro de radiación x de alta energía generados en una descarga plasma focus* Ph.D. thesis Universidad de Buenos Aires

[17] Soto L, Pavez C, Tarifeno A, Moreno J and Veloso F 2010 *Plasma Sources Science and Technology* **19** 055017

Calibration methodology for proportional counters applied to yield measurements of a neutron burst

Ariel Tarifeño-Saldivia[1,2], Roberto E. Mayer[3,4], Cristian Pavez[1,2] and Leopoldo Soto[1,2]

1) Comisión Chilena de Energía Nuclear, Casilla 188-D, Santiago, Chile.
2) Center for Research and Applications in Plasma Physics and Pulsed Power, P[4], Chile.
3) Comisión Nacional de Energía Atómica, Univ. N. de Cuyo, Argentina.
4) PLADEMA, Argentina.

E-mail: atarifeno@cchen.cl

Abstract. This work introduces a methodology for the yield measurement of a neutron burst using neutron proportional counters. The methodology is based on the calibration of the counter in pulse mode, and the use of a statistical model to estimate the number of detected events from the accumulated charge resulting from detection of the burst of neutrons. An improvement of more than one order of magnitude in the accuracy of a paraffin wax moderated ^{3}He-filled tube is obtained by using this methodology with respect to previous calibration methods.

1. Introduction

The measurement of the neutron yield of a single neutron burst is required in pulsed fusion experiments like dense z-pinches [1], plasma focus [2], laser fusion experiments [3], among others. A methodology for the yield measurement of a neutron burst using neutron proportional counters has been developed. This methodology is to be applied when a continuous current can not be measured at the output of the counter, or when single neutron events can not be resolved in time by nuclear standard electronics. The methodology is based on the calibration of the counter in pulse mode, and the use of a statistical model to estimate the number of detected events from the accumulated charge resulting from detection of the burst of neutrons. The calibration of the counter can be achieved by using a radioactive neutron source with a spectrum similar to that of the neutron burst. In that condition, it is assumed that after moderation the two spectra are not significantly different to produce a bias in the detector efficiency. For a D2 plasma focus neutron source, a ^{252}Cf neutron source satisfies this requirement. For a ^{3}He-filled counter moderated by paraffin wax, an improvement of more than one order of magnitude in the accuracy of the detection system is obtained by using this methodology with respect to previous calibration methods. More details about the methodology, including theoretical framework, experimental implementation for measurements in D_2 plasma focus and experimental verification of the accuracy of the methodology can be found in reference [4].

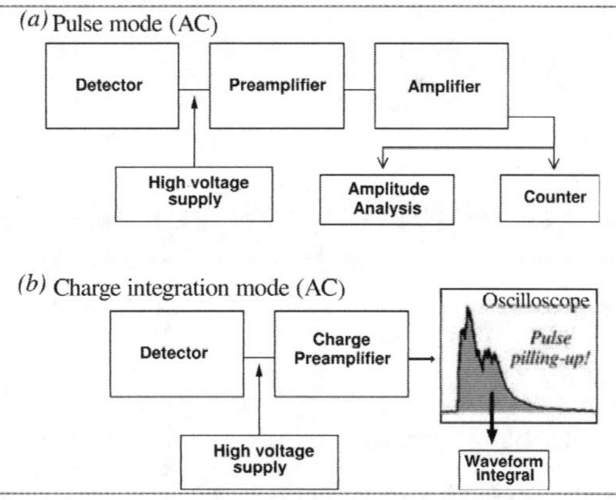

Figure 1. Operation modes and acquisition schemes for neutron proportional counters.

2. Step by step implementation of the methodology

Flowcharts of the acquisition system for the implementation of the methodology are shown in fig. 1.a (pulse mode) and 1.b (charge integration mode). The methodology is implemented carrying out the following steps:

1. Experimental characterization of the injection probability (e_D) and calibration factor (j_c) for different positions from the neutron source. This characterization can be made with a radioactive neutron source and the acquisition system in pulse mode.
2. Characterization of the waveform area distribution for a single detected event: Measurement of estimators for the expected value (μ_{PA}) and variance (σ^2_{PA}). The waveform area distribution for a single detected event can be obtained from a detector pulse height spectrum in pulsed mode and the characterization of single neutron signals in charge integration mode.
3. Calculation of the parameters $a = 1/\mu_{PA}$ and $f_0 = 1 - e_D + a^2\sigma^2_{PA}$.
4. On the experimental run, when no detection is registered, sampling the EBN distribution for measurements of estimators for expected value (μ_B) and variance (σ^2_B) at the specific oscilloscope setup.
5. Given the total waveform area (X_{gross}), the measurement of the neutron yield for a single shot is obtained from the following calculations,

$$X_{net}= X_{gross} - \mu_B,$$

$$G(X_{net})=a\ X_{net},$$

$$Y_\Omega(X_{net})= j_c\ G(X_{net}) \pm [f_0/G(X_{net}) + (a\ \sigma_B/\ G(X_{net}))^2 + (\Delta j_c/j_c)^2]^{1/2}$$

Acknowledgments

Supported by FONDECYT grant 1110940.

References

[1] M. G. Haines, Plasma Phys. Control. Fusion **53**, 093001 (2011).
[2] L. Soto, C. Pavez, A. Tarifeño, J. Moreno and F. Veloso, Plasma Sources Sci. and Technol. **19**, 055017 (2010).
[3] O. A. Hurricane et al. , Nature **506**, 343–348 (2014)
[4] Ariel Tarifeño-Saldivia, Roberto E. Mayer, Cristian Pavez, and Leopoldo Soto. *Rev. Sci. Instrum.* **85**, 013502 (2014).

Wire array experiments in a low impedance and low current generator

Nibaldo Cabrini[3], Cristian Pavez[1,2,3], Gonzalo Avaria[1,2], Patricio San Martin[1,2], Felipe Veloso[5], Barbara Zúñiga[4], Adolfo Sepúlveda[4], and Leopoldo Soto[1,2,3]

[1]Comisión Chilena de Energía Nuclear – Chile
[2]Center for Research and Applications in Plasma Physics and Pulsed Power – P4 Chile
[3]Departamento de Ciencias Físicas, Universidad Andrés Bello, Chile
[4]Facultad de Ingeniería, Universidad Mayor, Santiago, Chile
[5]Instituto de Física, Pontificia Universidad Católica de Chile, Santiago, Chile

E-mail: cpavez@cchen.cl

Abstract. In this work, a preliminary study about the behavior of a low impedance generator on different wire array configurations is reported. The experimental measurements were carried out on a small multi-purpose generator (1.2μF, 345J, 47.5nH, T/4=375 ns and Z=0.2Ω in short circuit) which produces currents up to 122 kA with 500 ns quarter period, when a charging voltage of 24kV and a wire load are used. Two types of configurations were tested: parallel wires (two and four) and X-pinch configurations. The experiments were carried out on W, Al, and Cu wires with different diameters. The discharge was characterized by means of a set of diagnostics which included: Rogowski coil; filtered PCD detector; filtered PIN diode; gated VUV/soft X-ray pinhole camera, Shadow diagnostic and dark field Schlieren technique. From the set of experimental results, the following observations can be established: (i) The generator is highly sensitive to the changes of load impedance due to its low impedance design. (ii) Every shot shows a dip in the current derivative signal shortly after the discharge onset time (from 6 to 40 ns), which is inversely related to the load resistance. (iii) Both configurations show a similar dynamic to those observed in experiments of higher current and shorter quarter period. (iv) At the X-pinch experiments, two or more hard X-ray bursts are detected, around 200 ns from the current onset time. These X-ray bursts are correlated with the dips observed in the current derivative signal.

1. Introduction

Nowadays is widely known that the Z-pinches are the most powerful X-ray radiation source produced at the laboratories, particularly those configurations based on wire array [1]. Of these, two types capture the attention by the characteristics of the emission source: the cylindrical wire array and the X-pinch configurations. In the first, several metallic wires cylindrically arranged connect the electrodes. In the X-pinch configuration, two diagonal wires are assembled between the electrodes, touching only at the cross point, forming an "X". In both configurations, either the skin-depth effect or the low inductance path favors a current flow in the wire surface. This current flow ablates mass by Joule effect creating a hot, low-density coronal plasma that surround a cold and dense structure. On this coronal plasma flows the bulk of the current. The subsequent dynamic is governed by the characteristics of the precursor plasma and its interaction with the global magnetic field. In the wire array experiments, the migration of the precursor plasma to the symmetry axis is accompanied in the final phase of the discharge by a violent collapse of the residual plasma on the discharge axis. In the stagnation, the thermalized hot plasma produces intense soft X-ray pulses, which have been used in the Z-machine to produce inertial confinement fusion [1, 2]. At the X-pinch configuration a significant effect of the global magnetic field is obtained in the cross point, where a hot, dense micro-pinch is originated. In these conditions, the plasma yields an intense and small soft X-ray source of short duration, ideal for applications. At the Cornell University facilities, X-pinch experiments driven in a large generator, size sources of less than $1\mu m$ and time-durations of less than $1ns$ have been reported [3]. In all experiments of wire arrays Z-pinch above mentioned, the experiments are driven in expensive and complex generators, characterized by high peak currents and current rates faster than 1kA/ns.

The interest for intense and short duration X-ray pulses has been extended to wire array experiments of lower power, whose rise times are between 200 ns-800 ns [4, 5]. In this direction, a new generation of X-pinches is being developed with lower energy generators and current rise-rates lower than 1kA/ns (low-current and long-time), as those driven in capacitive storage systems [6-8], or in a compact device based in transmission line for applications [9-12]. Motivated by recent developments, we decided to study the behavior of a low impedance capacitive generator [13], in the range of low currents (~$120kA$), using metallic wire arrays in two configurations: two and four parallel wire configuration and X-pinch configuration. In both, a wide variety of different diagnostics was used.

2. Experimental setup

2.1 The Multipurpose generator

The experiments reported here were carried out on a compact capacitive generator (Multipurpose generator) made at the Chilean Nuclear Energy Commission (CCHEN). These experiments correspond to the first time the generator operated with a plasma load. The generator consist of a capacitive bank formed by four capacitors of 0.3μF connected in parallel, a triggerable switch (spark-gap) and a discharge chamber with nine windows for diagnostics and vacuum system. Figure 1 show a picture of the generator assembled on a surface of $1.1\times1.1m^2$, with the discharge chamber at the center. For the wire array experiments, compact electrode configuration and isolation systems were built. The assembly of electrodes is shown in Figure 1b. A summary of electrical parameters of the generator are listed in Table 1. The generator inductance with the electrodes corresponds to the minimum value in order to see the wire array through the chamber windows.

2.2 Diagnostics

In order to define the scope of the wire array experiments using the multipurpose generator, an ample set of diagnostics were mounted, both time-resolved and time integrated. Of the first type, a Rogowski coil; filtered PCD detector; filtered PIN diode; gated VUV/soft X-ray pinhole camera, Shadow diagnostic and dark field Schlieren diagnostic were used. To image the plasma for photon energies higher than 3keV, a time-integrated filtered pinhole system was implemented. On the other hand, the series of experiments developed here produced the ideal environment, which allowed us to test many of the above mentioned diagnostics.

(a)

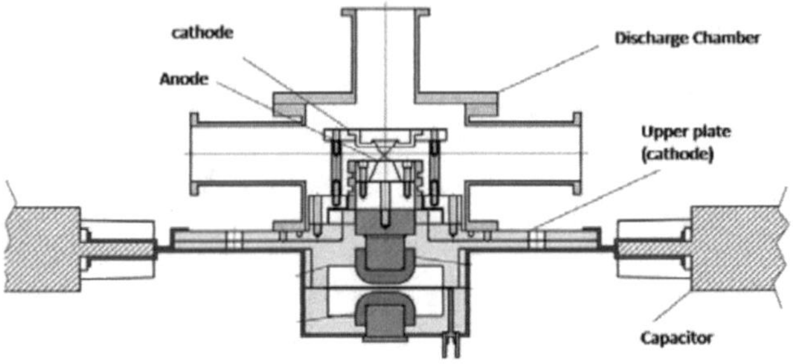

(b)

Figure 1. (a) Panoramic view of generator, (b) Cross section the generator and electrode configuration.

Table 1. Electrical parameters of the generator.

	C(μF)	L(nH)	V_{max}(kV)	E_{load}(J)	I_{peak}(kA)	T/4(ns)	Z(Ω)
Short circuit without electrodes	1.2	33	30	540	180	313	0.17
Short circuit with electrodes	1.2	47.5	30	540	150	375	0.2

2.3 Used wire array configurations

Figure 2 depicts the wire assemblies for each configuration. The side-on view gives account the effective spacing between wires that the imaging system records. Both the electron gap and the wire array diameter are $1cm$.

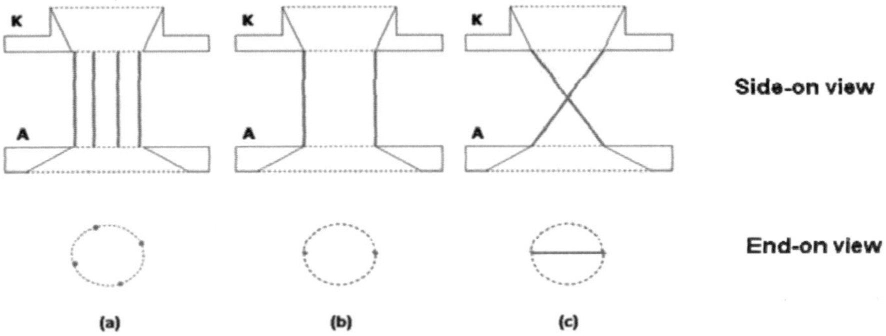

Figure 2. Schematic representation of the assembly of wire array.

Journal of Physics: Conference Series **591** (2015) 012026 doi:10.1088/1742-6596/591/1/012026

3. Experimental results

3.1 Parallel wire array. General behavior

From figure 3, it is possible to recognize the three typical phases observed in wire array experiments of higher current and rise rate. (a) Modulation of the plasma density along the axis. Appearance of coronal plasma and instabilities. (b) Ablation and precursor plasma migration. (c) Plasma column formation (stagnation). The pictures correspond to pin-hole images recorded on a VUV-MCP camera with an exposure time of 4ns. In wire array experiments x-rays were not detected either on the PIN diode or on the PCD detector.

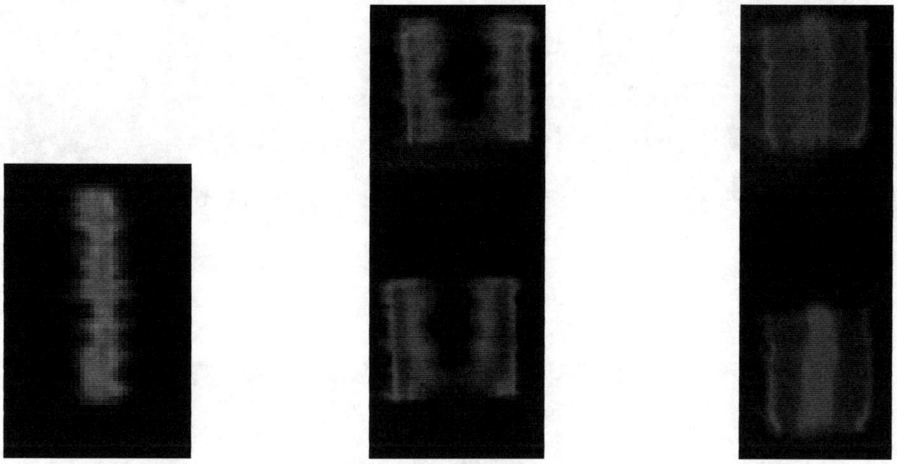

(a) (b) (c)

Figure 3. Formation and dynamics of plasma. (b) and (c) correspond to two aluminum wire of 50 μm.

3.2 X-pinch

In this subsection we report the main characteristics observed on the X-pinch discharge by means of the diagnostics above mentioned. In this configuration all radiation diagnostics recorded X-rays (from soft-X-ray to harder radiation): both the filtered PIN diode and the PCD detector detected X-rays since the beginning of the fast compression of the micro-pinch until the subsequent rupture. Figure 4 shows the structure and dynamics of the X-pinches on different systems of image recording. Figure 4(a) shows a sequence of pin-hole images in the VUV spectral range. Figure 4(b) and 4(c) show a Shadow and Schlieren record respectively, for which it was used a Nd:YAG power laser at 532 nm and 8-ns pulsewidth.. In each diagnostics the time is taken with respect to the discharge onset time.
Figure 5 shows a set of electrical signals for a typical shot on Tungsten X-pinch (W of 7.5 μm). In this shot, multiple singularities (dip) appear in the current derivative signal (about 50 kA and 75 kA) thus, the current signal shows a decrease in the rise rate of current, followed by multiple bursts of X-rays. This is evident from the consecutive peaks observed in both the PCD detector and PIN diode.

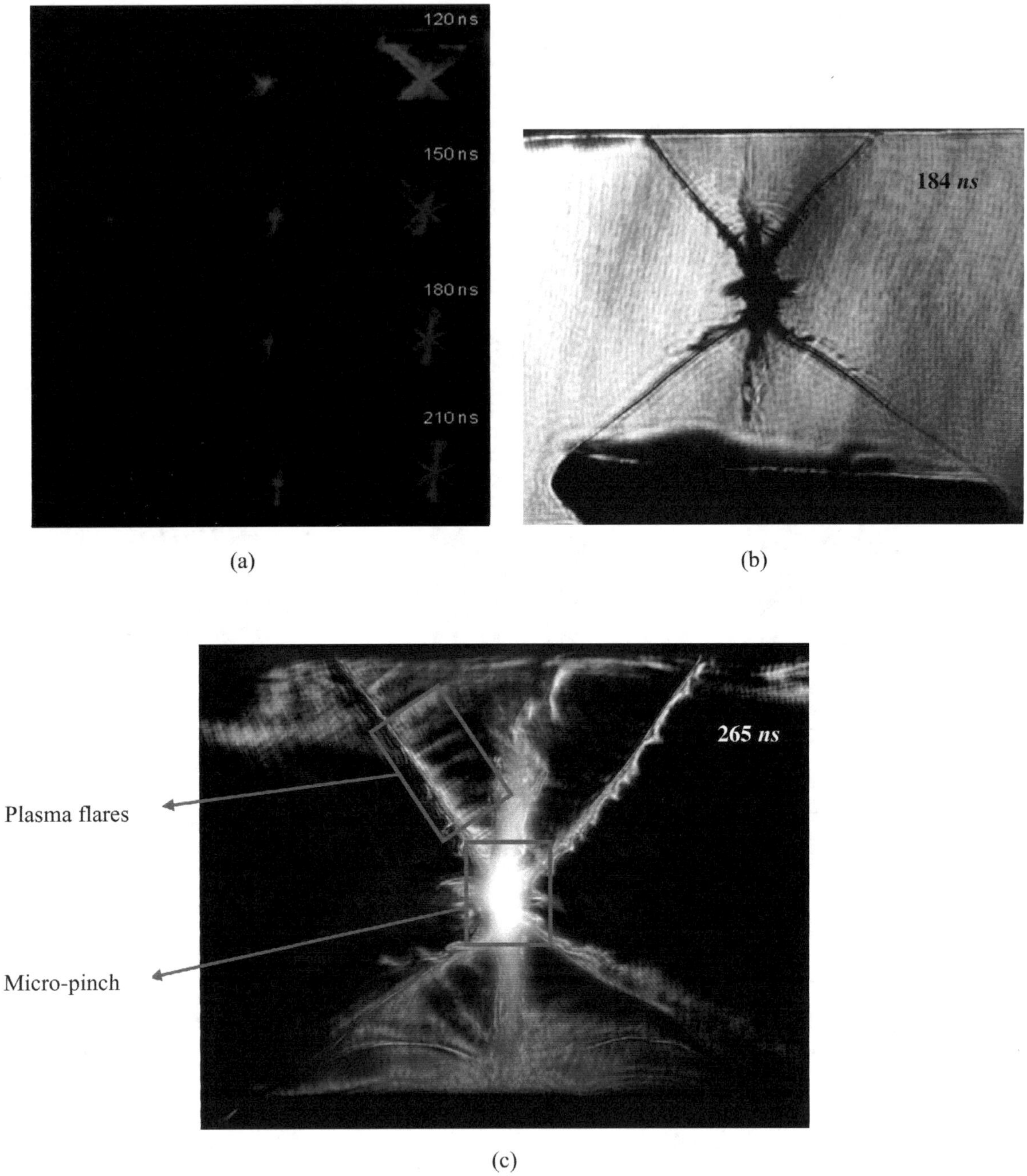

Figure 4. (a) VUV pinhole image of aluminum X-pinch of 25 um. (b) Image-plane Shadowgraph of Tungsten X-pinch of 13 μm. (c) Dark field Schlieren of Tungsten X-pinch of 13 μm

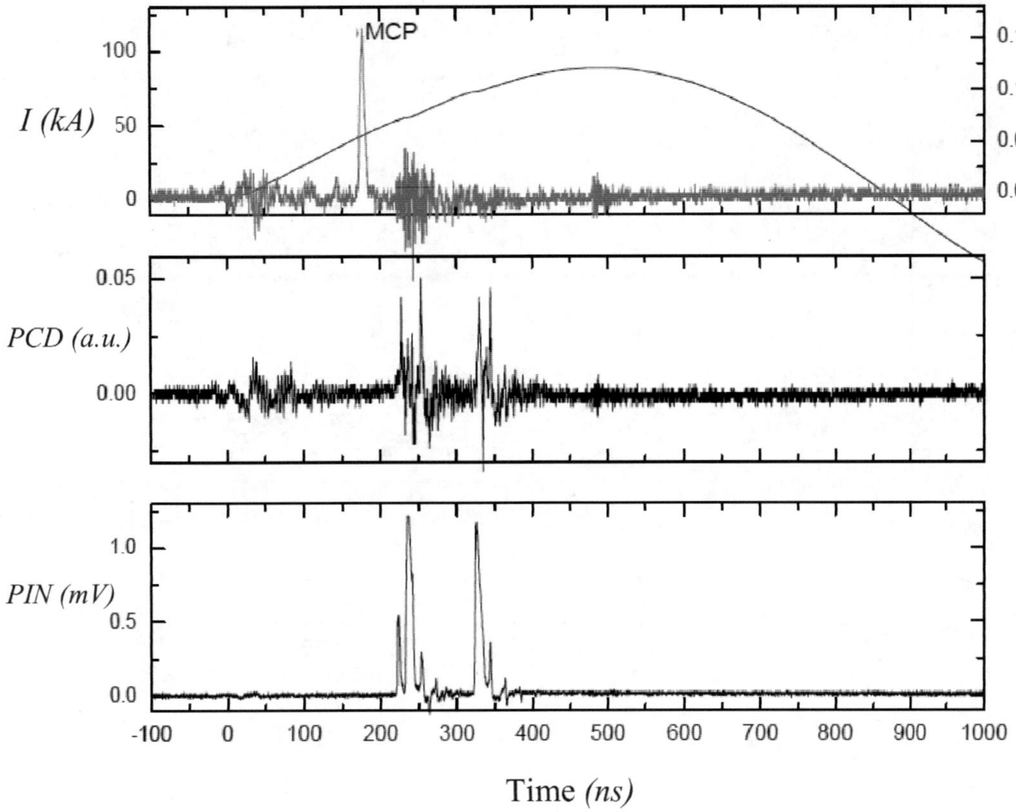

Figure 5: Electrical signals for a typical shot on Tungsten X-pinch (W of 7.5 μm).

3.2 Initial load behavior

In this subsection, we report the electrical behavior observed in the current derivative signals (I_{dot}) shortly after the discharge onset time (from 6 to 40 ns). In all wire array signals a dip was observed whose onset time is inversely related with the initial resistance of the wire array. Here, we calculate the initial resistance by means of the relation $R=\rho L/A$, where ρ is the resistivity of the material, L the length of the wire and A the cross section area. This reproducible behavior could be giving account of the solid-liquid-vapour transition in the wire, dominated by an increase of the resistivity, which is easily detected for a low impedance generator as the Multipurpose generator. The relation between the load resistance and the dip onset time is shown in Fig. 6.

3.2 Applications

In order to illustrate the potentiality of the X-ray sources based in X-pinch, furthermore to make a complementary qualitative diagnostic, an insect radiograph was made. The X-pinch was made of Tungsten of 7.5 μm and the radiography was recorded on a dental film filtered with an Al foil of 15 μm, without the original packaging. The radiography is shown in Fig. 7.

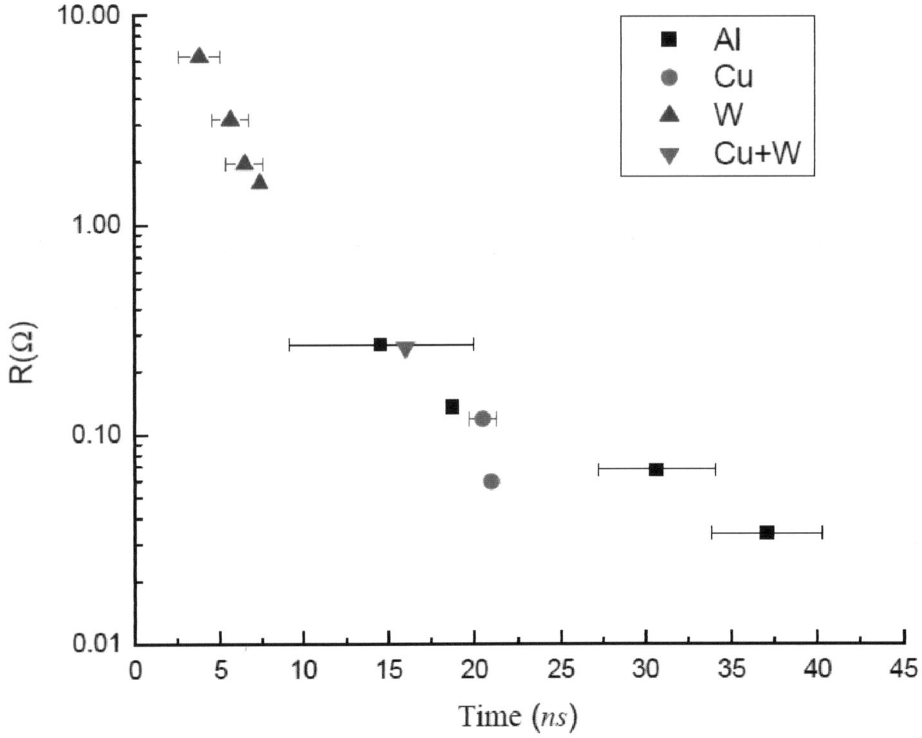

Figure 6: Experimental relation between the wire array resistance and the dip onset time.

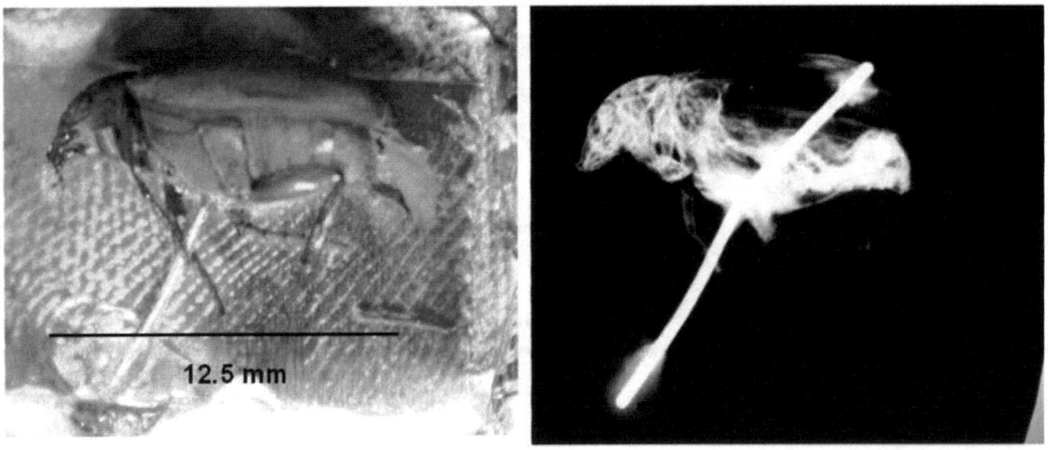

Figure 7: Insect radiography using Tungsten X-pinch of 7.5 μm.

4. Remarks and conclusion

From the set of diagnostics and experimental results, a first approach to the dynamic and emission of the plasma, under wire array experiments can be summarized as following:

- The generator is highly sensitive to the changes of load impedance due to its low design impedance.
- Experiments with metallic wire conducted on the multipurpose generator showed similar characteristics to those observed in higher currents and high current rate generators ($\geq 10^{12}$ A/s), namely: density modulation along the wire and migration of precursor plasma because of the global magnetic field (similar dynamics).
- Every shot shows a dip in the current derivative signal shortly after the discharge onset time (from 6 to 40 ns), which is inversely related to the load resistance.
- In wire array experiments, a central plasma column at the final stage of the ablation is observed. For X-pinch experiments, a plasma jet and a micro-pinch with high emission of X-ray were observed in the crossing point of the wires.
- Up to now, there is no experimental evidence that a significant X-ray production is taking place. The presence of soft X-ray gives account of a high temperature plasma, which is observed in the micro-pinch region (crossing point).
- Two or more hard X-ray bursts were detected around 200 ns from the current onset time. These X-ray bursts are correlated with the dips observed in the current derivative signal.
- From the refractive optical diagnostics, the x-pinch dynamic is similar to that observed in other x-pinch experiments, characterized by the formation of plasma flares feeding a plasma jet in the axis of symmetry of the discharge in addition to the formation of a bright micro-pinch in the cross point.

New experiments are being developed in order to quantify and characterize each observation here described.

The authors would like to acknowledge financial support by PIA CONICYT grant ACT 1115. G. Avaria acknowledges the financial support from PAI CONICYT 791100020 and FONDECYT Iniciación 11121587.

References

[1] M. Haines, "A review of the dense Z-pinch", Plasma Phys. Control. Fusion **53** (2011).
[2] C. L. Ruiz, G. W. Cooper, S. A. Slutz, J. E. Bailey, G. A. Chandler, T. J. Nash, T. A. Mehlhorn, R. J. Leeper, D. Fehl, A. J. Nelson, J. Franklin, and L. Ziegler. *Phys. Rev. Lett.* Vol. **93,** 1 (2004).
[3] S. A. Pikuz, et al, "High energy density z-pinch plasma conditions with picosecond time resolution". *Phys. Rev. Lett.,* **89**, 035003 (2002).
[4] C. Deeney, C. A. Coverdale, and M. R. Douglas, *Laser and Particle Beams*, **19**, 497-506, (2001).
[5] H. Calamy, F. Hamann, F. Lassalle, J. Chittenden, S. V. Lebedev, F. Bayol, Ch. Mangeant, A. Morell, D. Huet, and J-P. Bedoch, *IEEE Trans. Plas.Sci.*, **34**, 2279 (2006).
[6] Christou C, Dangor A E and Hammer D A 2000 *J Appl Phys* **87** 8295-8303.
[7] Aranchuk L E, Chuvatin A S and Larour J 2004 *Rev Sci Instrum* **75** 69-74.
[8] Aranchuk L E, Larour J and Chuvatin A S 2005 *IEEE Trans Plasma Sci* ***33*** *990-996.*
[9] G. W. Collins IV, Ms. Of Science in physic, "Study of X-pinch dynamics using a low current

(25kA) and long-rising (400ns) pulse". San Diego State University, 2013.
[10] G. W. Collins IV, et al "Study of X-pinch dynamics using a low current (25kA) and slower current (400ns) pulse". *Phys. Plasmas.* **20**, (2013).
[11] G. W. Collins IV, et al "Effect of the global to local magnetic field ratio on the ablation modulations on X-pinches driven by 80kA peak current". *New Journal of Physics.* **14**, (2012).
[12] D. M. Haas, et al "Supersonic jet formation and propagation in X-pinches". *Astrophys. Space Sci* (2013) 336: 33-40.
[13] A. Tarifeño, C. Pavez, M. Cardenas, L. Soto, *Phys. Scr.* **T131** (2008) 014028 (5pp).

Journal of Physics: Conference Series **591** (2015) 012027 doi:10.1088/1742-6596/591/1/012027

Effects of uneven mass distribution on plasma dynamics in cylindrical wire array Z-pinches.

Felipe Veloso[1], Luis Donoso[1], George Swadling[2], Jeremy Chittenden[2], Gonzalo Muñoz[1], Vicente Valenzuela[1], Francisco Suzuki-Vidal[2], Mario Favre[1] and Edmund Wyndham[1]

[1] Instituto de Física, Pontificia Universidad Católica de Chile, Chile

[2] Blackett Laboratory, Imperial College London, United Kingdom

E-mail: fveloso@fis.puc.cl

Abstract. The formation and dynamics of cylindrical wire array Z-pinch plasmas using uneven mass distribution is studied through experiments and computer simulations. In this study, the initial mass distribution is modified using different wire sizes within the same array, in contrast to the most standard wire arrays. Since the ablation rate of a particular wire material is related to current driver characteristics but not to the wire size, the use of different wires within an array produces time differences for complete ablation of each size. This changes the plasma dynamics and precursor plasma formation as compared to standard arrays. The experiments have been carried out on the Llampudken pulsed power generator (~350kA in ~300ns) using a 1:6 mass ratio among different wires of a single array. Plasma dynamics are studied using time-resolved laser interferometry (532nm, 5ns FWHM) and XUV imaging (5ns exposure time) in both side-on and end-on directions respect to the array. Experimental results show the formation of a dense, precursor plasma column on the array axis at early times, which shifts its position toward the thicker wires at later times at velocities of the order of 10^4 m/s. Numerical simulations using the 3-D MHD code GORGON are able to reproduce the experimental observations. They show that the larger mass of thicker wires induces modifications in the global magnetic field topology, producing the shifting in the precursor position as observed in the experiments. Further details on the changes of ablation dynamics and precursor formation are presented and discussed.

1. Introduction

Wire array z-pinches have been widely studied as efficient and powerful soft x-ray sources, producing peak powers up to 280TW [1]. In addition, very rich and complex physical phenomena can be studied with z-pinches [2] e.g. with applications to inertial confinement fusion, radiation hydrodynamics, high energy density physics, astrophysical objects among others. Standard wire array z-pinches typically use thin metal wires evenly distributed in a cylindrical configuration as a load of a pulsed power generator. At very early times of the high current pulsed discharge, each wire transforms into a structure made of a solid core surrounded by ablated coronal plasma that carries most of the electrical current [3]. The cores act as stationary mass reservoirs for the coronal plasma which is continuosly driven toward the array axis by the global Lorentz force surrounding the cylindrical array. This ablation process dominates the mass redistribution and the conditions for the final implosion of the array which occurs when the wires becomes fully depleted of mass. The implosion of the wires produces the final strong x-ray emission at stagnation.

Journal of Physics: Conference Series **591** (2015) 012027 doi:10.1088/1742-6596/591/1/012027

A combination of interesting physical aspects occur during precursor formation, such as collisional and collisionless regimes, effects in the plasma by radiation cooling, among others[4, 5]. Thus being able to control the precursor properties and dynamics is an interesting subject by itself. Since precursor formation depends on wire ablation, one way of controlling its properties is by varying the wire parameters as they directly affect their ablation rate. It is known that the ablation rate of the wires does not depend on the wire material, but it can depend on the rate of the current provided by the pulsed power driver [6, 7, 8]. In this work, the dynamics of the precursor plasma are intentionally modified by changing the mass distribution within the array using different wire diameters of the same material. Hence, the precursor is initially located on the array axis until thinner wires are mostly fully ablated. This will lead to the precursor to shift its initial position toward thicker wires in which the core-corona structure is still present. Preliminary results using both experiments and numerical simulations on controlling precursor position are presented next.

2. Experimental set-up

The experiments are driven on the Llampudken pulsed power generator [9] which delivers a peak current of ~350kA in ~350ns, measured using an integrated Rogowski coil. The wire arrays are mounted in a cylindrical configuration (23mm height and 16mm diameter) of eight equally spaced tungsten wires which act as the load for the pulsed power driver. For the experiments described here, the array loads are overmassed meaning that the stagnation phase will not be observed. In order to produce an uneven mass distribution on the array, two consecutive wires on the array have larger diameter in comparison to the other six. For the experiments shown here, a 1:6 mass ratio has been used by using 10μm and 25μm diameter wires in the array

The plasma is diagnosed using time-resolved laser probing and time-resolved extreme ultraviolet (XUV) imaging. Mach-Zehnder interferometry is performed using a frequency doubled Nd:YAG laser (5ns FWHM, 532nm) probing the plasma along the axial direction of the array (i.e., end-on direction). These time-resolved laser images are recorded using open-shutter SLR digital cameras filtered to the laser light frequency. Besides this, microchannel plates (MCPs) are used to record the self-emission from the plasma in the XUV using small pinholes. These MCP cameras are located perpendicular to the array axis (i.e., side-on direction) so that thicker wires correspond to edge wires on each image. Temporal resolution for the MCPs are ~5ns and images are recorded on HP5 photographic paper. An schematic diagram of the experimental setup is shown in Figure 1

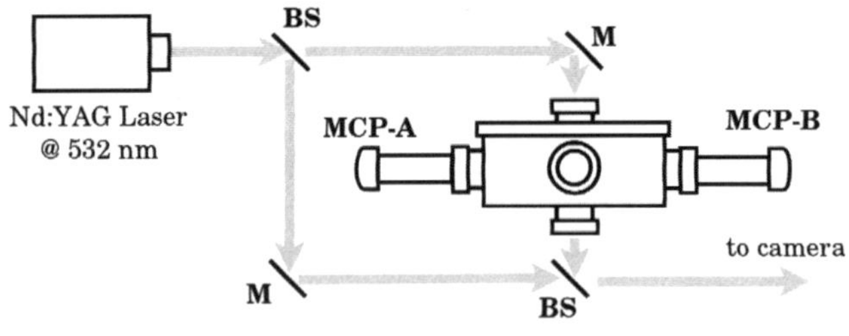

a)

b)

Figure 1. (a) Schematic diagram of the imaging diagnostics on the discharge chamber (imaging diagnostics omitted), (b) Actual photograph of the wire array load, in the line-of-sight of the side-on XUV images.

3. Discussion

During the ablation phase, the ablated plasma is driven toward the region where the B-field is zero (i.e., no Lorentz force). Later in the discharge, the thinner wires will lose their material producing wire breakages along the axis, which implies small gaps for the current path along those wires. Hence, the current will be redistributed toward the thicker wires, where the breakages have not occured yet, moving the null B-field region toward them. In consequence, the precursor plasma moves toward the thicker wires as well. Figure 2 shows the temporal evolution of the discharge in the side-on direction. On these images, the edge wires correspond to two consecutive wires overlapped on the image; where the right hand side wires correspond to thicker wires. At times $\leq$ 210ns the precursor is still close to the array axis, and ablation plasma streams can be seen flowing to it. However, as the time evolves, the precursor begins its movement toward the thicker wires position at an average speed $\sim 10^4$m/s. Figure 3 shows an end-on interferogram at t = 220ns, which is close to the time when noticiable movement of the precursor can be seen in side-on XUV images. It is clear that larger fringe shift (i.e., larger electron density) corresponds to an off-axis position. Even though the end-on measurement correspond to axially average measurements, the presence of larger density located off-axis is a clear indication that variations in the ablation process from an standard wire array have occurred. This larger fringe shift corresponds to average axial densities $\sim 10^{18}$cm^{-3} located roughly 1mm apart and twice the density of the plasma at the array axis.

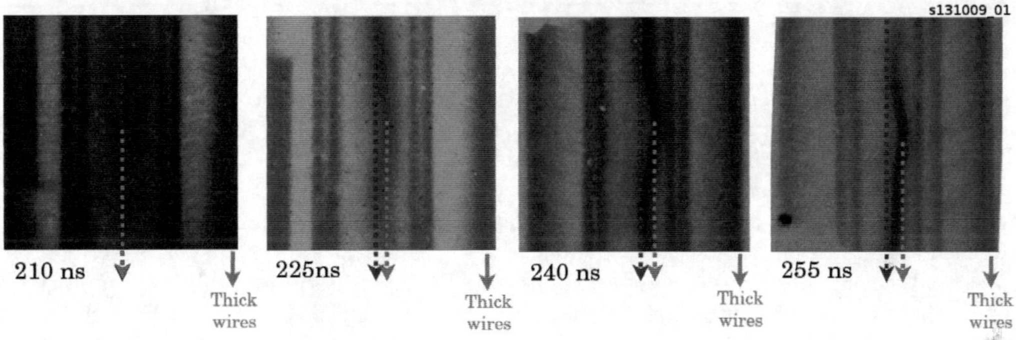

Figure 2. Side-on XUV images showing the movement of the precursor plasma. Green arrows indicate thicker wires position, blue arrows indicate array axis and red arrow indicate average precursor position

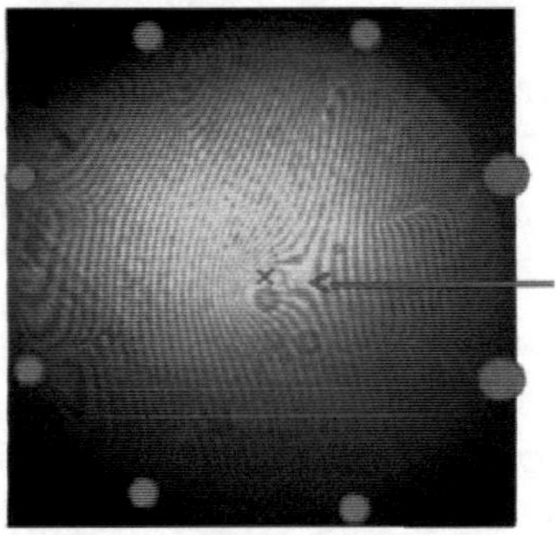

Figure 3. End-on interferogram imaging of the wire array at 220ns. The cross at the center indicates the array axis. Colored dots indicate the original position of the wires (blue=thin wires, red=thicker wires) The red arrow indicates a higher density region formed off-axis

These experimental results can be compared with MHD computer simulations using the

GORGON code. GORGON is a three-dimensional numerical code designed to integrate the system of magnetohydrodynamics equations in a single fluid approximation on a Cartesian grid. It has proved to be a robust and accurate tool for Z-pinch modeling [10]. Figure 4 shows the temporal evolution of the corresponding simulations of the experiments as mass density plots in 2D projections simulating end-on probing of the plasma. It can be seen that precursor is uniformly formed as a round object at the array axis at times $\leq$200ns. At later times, its shape changes to an ellipse-like object, moving later toward the position of the thicker wires. The velocity of the denser region of the precursor is 10^4m/s, which are in agreement with the experimental results.

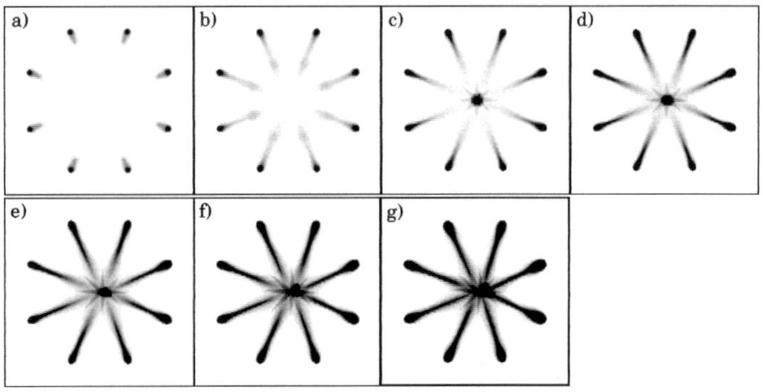

Figure 4. GORGON simulations of tungsten arrays. The secuence shows end-on density plots in grayscale at different times from (a) 100ns to (g) 400ns in 50ns intervals.

The effect of using thicker wires within the array produced modifications on plasma dynamics and precursor position when compared to standard wire arrays. When thinner wires are mostly ablated, both numerical simulations and experimental results show the shifting of the precursor position toward the position of thicker wires with velocities of the order of 10^4 m/s. Since plasma accumulates at the zero B-field region, the use of thicker wires induces modifications on the global magnetic field topology. As far as the authors are aware, this is the first report of an intentional movement of the precursor column which can be useful for studying properties of stagnated plasma under non-stationary conditions. Future experiments will include the use of aluminium wires in smaller diameter arrays, where precursor movement and collisional effects on the plasma will be much more relevant.

This work has been funded by project FONDECYT/Iniciacion No. 11121621. G. Muñoz is funded by a postgraduate studies scholarship awarded by CONICYT.

References

[1] Deeney C, *et al* 1998 *Phys Plasmas* **5** 2431
[2] Haines M G 2011 *Plasma Phys Controlled Fusion* **53** 093001
[3] Lebedev S V, Beg F N, Bland S N, Chittenden J P, Dangor A E, Haines M G, Pikuz S A and Shelkovenko T A 2000 *Phys Rev Lett* **85** 98
[4] Swadling G F, *et al* 2013 *Phys Plasmas* **20** 062706
[5] Bott S C, *et al* 2006 *Phys Rev E* **74** 046403
[6] Bott S C, *et al* 2009 *Phys Plasmas* **16** 072701
[7] Knapp P F, Greenly J B, Gourdain P A, Hoyt C L, Martin M R, Pikuz S A, Seyler C E, Shelkovenko T A and Hammer D A 2010 *Phys Plasmas* **17** 012704
[8] Veloso F, Suzuki-Vidal F, Molina F, Mitchell I H, Chuaqui H, Favre M and Wyndham E 2012 *IEEE Trans Plasma Sci* **40** 3319
[9] Chuaqui H, Wyndham E, Friedli C and Favre M 1997 *Laser and Particle Beams* **15** 241
[10] Chittenden J P, Lebedev S V, Jennings C A, Bland S N and Ciardi A 2004 *Plasma Phys Control Fusion* **46** B457

Journal of Physics: Conference Series **591** (2015) 012028 doi:10.1088/1742-6596/591/1/012028

Study of the Rogowski coil response in a plasma focus device

S. Perusquía[1], F. Castillo[2], P. G. Reyes[1], H. Martínez[2].
[1] Facultad de Ciencias, Laboratorio de Física Avanzada, Universidad Autónoma del Estado de México, C.P. 50000, Toluca, Estado de México, México.
[2] Instituto de Ciencias Físicas, Laboratorio de Espectroscopia, Universidad Nacional Autónoma de México, A.P. 48 3, C.P. 62251, Cuernavaca, Morelos, México.
ciro@fis.unam.mx

Abstract. In plasma physics is very important understand the behaviour and amplitude of high intensity currents. The high intensity current pulses are measured by resistive and inductive sensors. Coil configurations are employed in the same form as inductive sensors. Rogowski coil is a representative example; that coil consists of a toroidal-shaped winding. In this work we built a Rogowski coil. The coil operation was realized in a plasma focus device. The results show a good performance. We also used the coil to measure the current and voltage changes with respect to time, during the normal operation of the plasma focus device. The maximum value of the current (348kA) was obtained. As a complementary work, we analysed the results by changing the central electrode of the device (needle electrode, gap electrode and plane electrode). It founds that the pinch time is independent of the configuration of the central electrode.

1. –Introduction

In power system applications as well as in other scientific fields, it is necessary to determine the characteristics of high intensity currents. High current pulses have been studied in lightning, electric arcs, and in plasma physics experiments. The current amplitude can vary from several few hundred amperes to kilo – amperes, in a few nanoseconds [1, 2].
These current pulses in plasma physics experiments are measured by resistive and inductive sensors. Unlike resistive sensors, inductive transducers are a non-intrusive measurement and provide good insulation between the circuit under test and measuring equipment. Different types of configurations of magnet coils can be used as inductive sensors.
A representative example of such sensors is a Rogowski coil. Walter Rogowski (1881 - 1947), best known for his research in electrical engineering, electro-technology, telecommunications and electronics primarily physical. In 1912 proposed the use of Rogowski coil, with toroidal shaped for polling the current changes with respect to time.
There are several forms to generate plasma in the laboratory, discharges in gases, and plasmas by magnetic confinement, inertial confinement plasmas, among others. However, the magnetic confinement systems can be classified into two groups:
• Equilibrium conditions, where the plasma pressure is balanced by external magnetic fields. Usually these systems are also characterized by using closed or quasi-closed geometry of the device: Tokamak, Stellerator, magnetic mirrors, etc. [3, 4].

• Non-equilibrium condition, where the plasma is accelerated by its own magnetic forces generated by the same link and self-confined, the geometry is opened in these devices: Z- Pinch, θ -Pinch Plasma Focus, etc. [5, 6].

The magnetic confinement consists in generating electrical pulses in a gaseous medium, producing plasma, without allowing inhomogeneity between the center and the walls of the device.

Rogowski coils are employed to measure the electrical currents of the system [7]. The principal objective of this work, it is build a Rogowski coil to measurement the current variations (*dI/dt*) in a pulsed discharge device. It is why in this section will explain the Rogowski coil and the pulsed discharge device briefly. The Rogowski coil (RC) consists in a helical winding around a toroidal-shaped, a representative form of this coil is shows in Figure 1 [8 – 15]. His principal operation is based in the measurement of the current induced by the magnetic field produced by the device. In that case, the total induced voltage is given by:

$$\boldsymbol{V_{coil} = -\mu_0 AN\frac{dI}{dt}} \tag{1}$$

The equation (1) represents the relationship between the current variation (*dI/dt)* and the total potential difference induced ($\boldsymbol{V_{coil}}$) in the coil, where A is the coil cross section, N is the turns number and μ_0 is the magnetic permeability.

In general, it can be said that the magnetic field variation produced by the current flowing through the coil conductor induces a voltage proportional to the change of the current with respect to time. Being proportional with the mutual inductance (*M*), then:

$$\boldsymbol{V_{coil} = -M\frac{dI}{dt}} \tag{2}$$

Where:

$$\boldsymbol{M = \mu_0 AN} \tag{3}$$

Interest in devices capable of generating high voltage and current in a very short time, of the order of nanoseconds, they have been grown considerably recently. Then, for that purpose several machines have been designed and constructed. In particular, pulsed discharges are formed to generate high voltage and current, specifically in this work we use a plasma focus device [16, 17]. In that device the variations of high voltage and current are critical, hence the use of a Rogowski coil for measuring the current variation can be done [18].

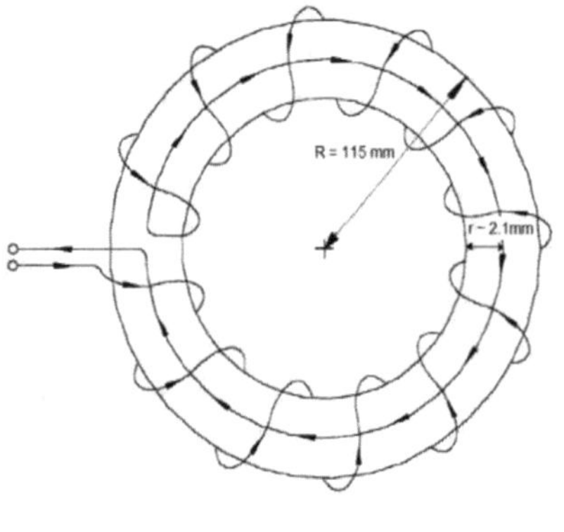

Figure 1.*Dimensions of the Rogowski coil.*

II. - Experimental Setup

This work was performed using a small plasma focus device (PFD) [16]; its capacity is 5kJ of energy at 38 kV. The device has an oxygen-free copper anode, with 40 mm long, and 50 mm diameter. The coaxial cathode is formed by ten copper rods arranged in a squirrel cage configuration at a radius of 50 mm. The insulator is an annular *Pyres®* tube located at the base of the anode, Figure 2. Throughout the present work the plasma focus was operated in its neutron optimised regime, corresponding to a 2.75 torr of deuterium gas pressure.

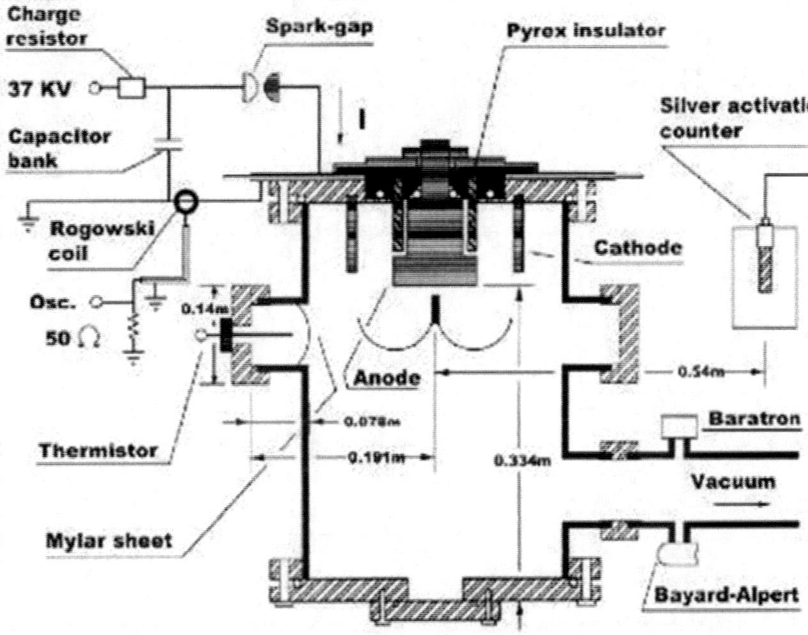

Figure 2. Pulsed Discharge device experimental setup [16].

The Rogowski coil [9 - 12] consists of a winding of 95 turns, toroidal-shaped of 21 cm in diameter and 4.2 mm internal diameter. Then, the mutual inductance provided by the Rogowski coil is:

$$M = \frac{\mu_0}{2\pi r} N^2 A = 317.6 \text{ nH} \tag{4}$$

It is noteworthy that the coil was connected to an oscilloscope for recording the signal. Figure 1 shows the dimensions of the coil built.

III. –Results

Typical Rogowski coil signals during normal operation are shown in Figure 3.

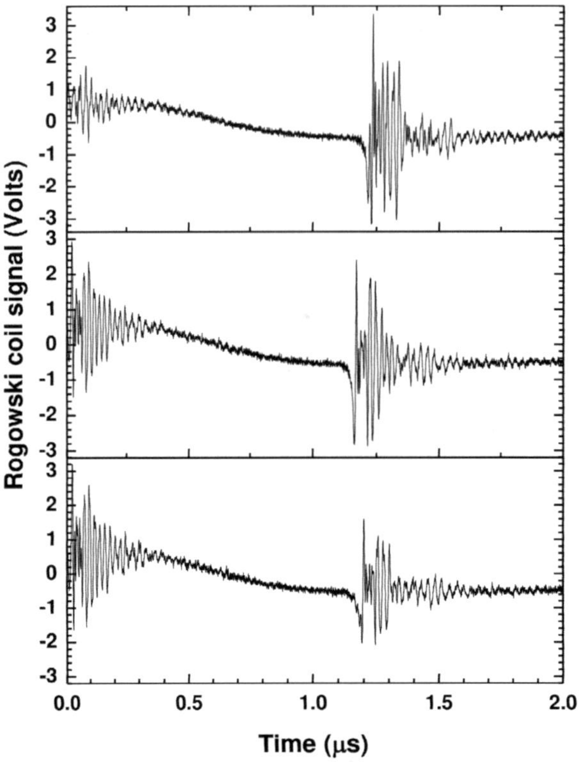

Figure 3. Rogowski coil signals during normal operation.

In a period time of 1.16 μs for pulse, it produced 35.2 ns of pinch time. Taking these values, the current was obtained by:

$$I(t) = I_0 \sin(\omega t) \approx (348 \pm 3\ 5)\ \mathrm{kA} \tag{5}$$

Different configurations of the central electrode were performed (Figure 4): with gap, planar electrodes and needle-shaped electrodes. The change in the pinch time with these configurations will be measure in each case.

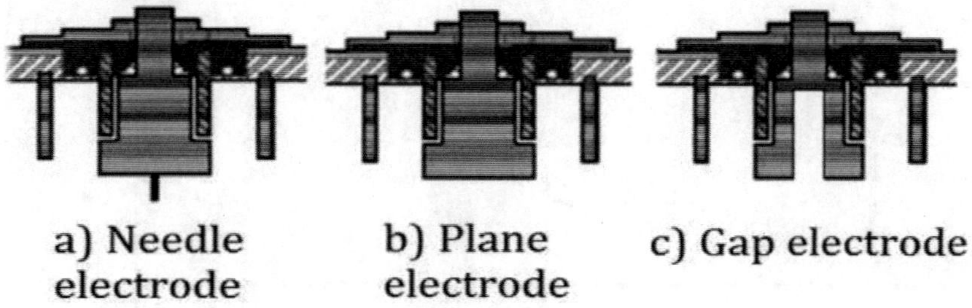

Figure 4. Central electrodes arranged on the PFD.

The Rogowski coil response produced in each configuration is displays in Figure 5. It can be that the pinch time is approximately the same for the three configurations. That implies that the pinch time does not depend on configuration of the central electrode.

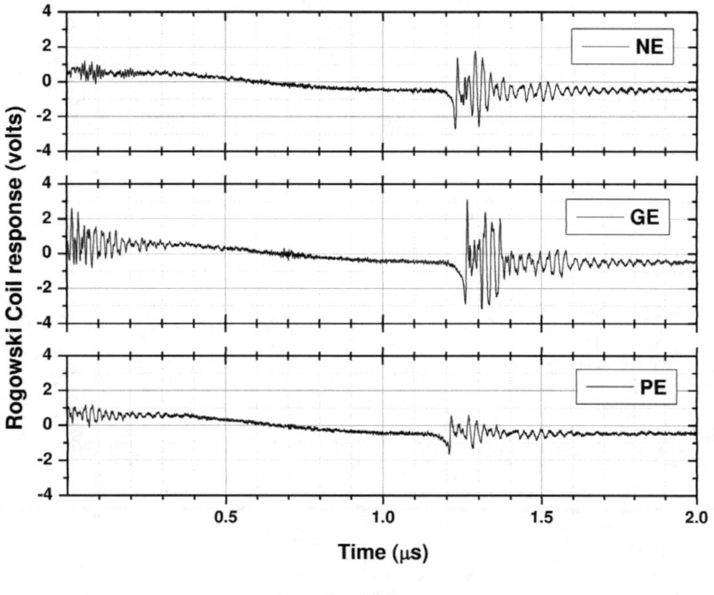

Figure 5. Rogowski Response. NE - Needle electrode, GE – Gap electrode and PE - Plane electrode.

Individual results for each configuration are shown in Figures 6, 7 and 8.

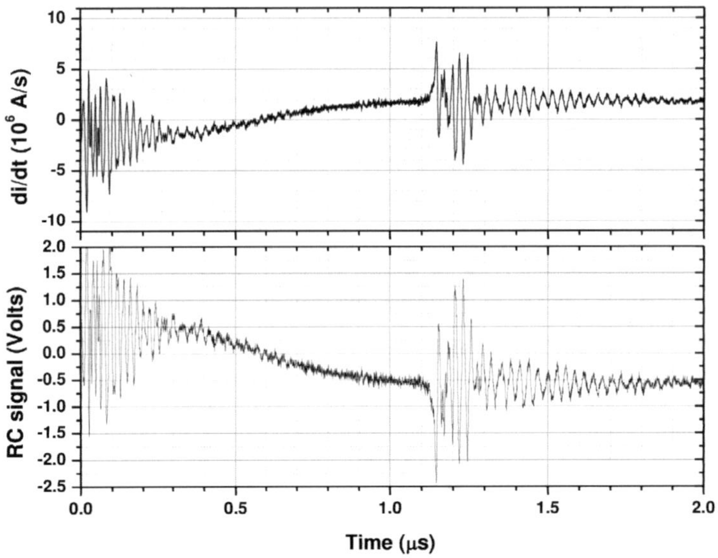

Figure 6. Rogowski coil signals using Needle electrode.

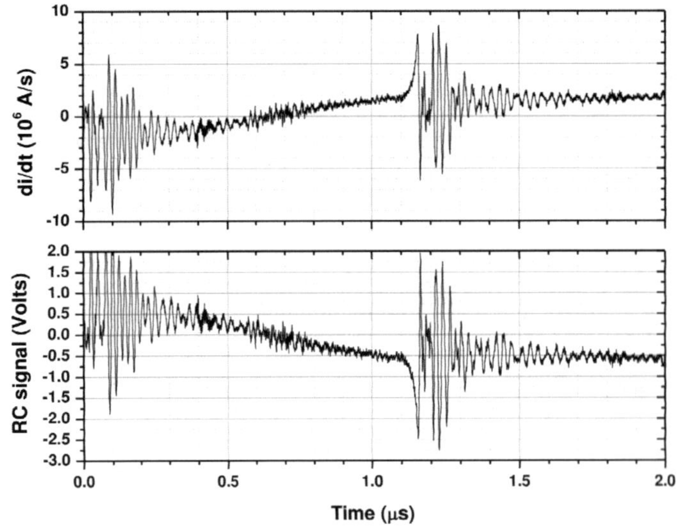

Figure 7. Rogowski coil signals using Gap electrode.

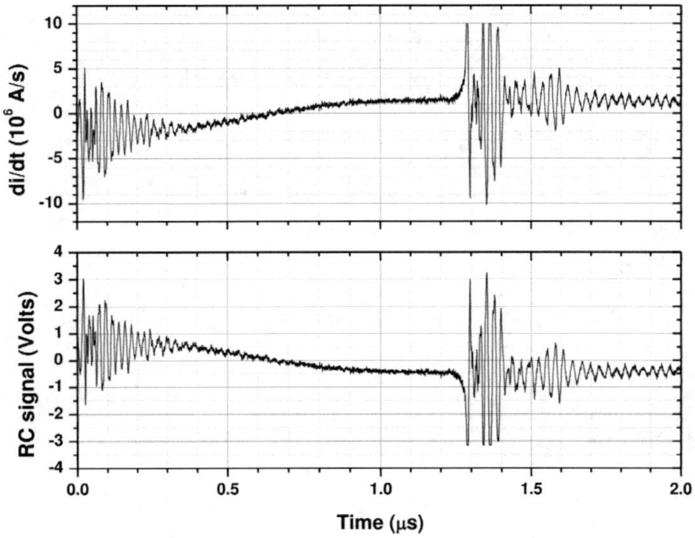

Figure 8. Rogowski coil signals using Plane electrode.

From Figures 6, 7, and 8, it observes that the maximum values of de dI/dt are between $5x10^6$ and $10x10^6$ A/s, in a pinch time of approximately 37 ns. The pinch time is approximately the same for the three types of electrodes studied in this work (see Figures 5, 6, 7 and 8), which is consistent with existing theories [19, 20, 21]. That pinch time depends on the dimensions of electrodes and not from the configuration of the central electrode.

These changes in the configurations of the central electrode are present in the final stage (compression) and its subsequent evolution. For example, in PE configuration the hard x-rays are the radiation main production; in GE configuration, the ions and electrons become more important and in NE configuration, the production of neutrons and soft x-rays generation emerge as the principal source of emission. These results verified that the changes in the central electrode configuration do not modify or change the evolution of the current sheet, only modifies the compression stage. Likewise the current calculated for each case shown. It known that in plasma focus, the signal changes from shot to shot.

IV. -Conclusions

The conclusions of this work are:

- A Rogowski coil was built.
- The Rogowski coil operation in short circuit testing was proof.
- Maximum values of current variation (*dI/dt*) were found in the order of 10^6 A/s.
- Time pinch were approximately of 37 ns.
- The pinch time is independent of the configuration of the central electrode.
- The maximum value of the pinch current was (348 ± 35) kA.
- The Rogowski coil is a very good instrument to measure the current variation in pulsed devices.

Acknowledgements

The authors are thankful to J. Herrera and J. Rangel for technical assistance. This research was partially sponsored by DGAPA 1IN101613 and CONACYT 128714.

Referencies:

[1]Sato E, et. al., 1994 *Med.Biol. Eng. Comput.* **32** (3), 295-301.
[2]Hurricane O. A., et. al., 2014 *Nature* **506,** 343 – 348.
[3]Harris, J.H., et al. (1985) Nuclear Fusion, 25, 623.
[4]Shultis, J.K., and Faw, R.E. (2002) Fundamentals of nuclear science and engineering, CRC Press ISBN 0-8247-0834-2.
[5]Castell, R., et al. (2010) Revista Mexicana de Física, 56 (1), 9-13.
[6]Milanese, M., et al. (2012) Anales AFA, Malargüe 22, 64-70.
[7]Pellinen, D. G., et al., 1980 *Rev. Sci. Instrum.* **51**(11), 1535 – 1540.
[8]Cooper, J. 1963 *Plasma Physics (Journal of Nuclear Energy. Part C)* **5,** 285 – 289.
[9]Ramboz, J. D., 1996 *IEE Transactions on Instrument and Measurement*, **45** (2), 511 – 515.
[10]Ray, W. 1993 *EPE Journal 3* (1), 51 – 59.
[11]Ray, W. 1993 *EPE Journal* **3** (2), 116 – 122.
[12]Rogowski, W. and Steinhaus, W., 1912 *Arch Electrotech* **1,** 141 – 150.
[13]Smirnov, V.P., 2010 *Nucl. Fusion* **500**, 14003.
[14]Habibi, M., et. al., 2008 *Brazilian Journal of Physics* **38,** (2).
[15]Milanese, M., et. al., 2012 *Anales AFA, Malargüe* **22,** 64 – 70.
[16]Castillo, F. et. al., 2002 *Brazilian Journal of Physics* **32,** (1).
[17]Castillo, F., et. al., 2007 *J. Phys. D: Appl. Phys.* **40,** 5902–5906.
[18]Podlesak M., 1990 *Rev. Sci. Instrum.* **61,** (2) 892 – 896.
[19]Soto, L., et. al., 2004 *12th International Congress on Plasma Physics ICPP2004,* Niza France
[20]Lee S. and Saw S.H. (2008) Applied Physics Letters, 92 (2), 021503.
[21]Lee S., Saw S.H., Lee P., Rawat R.S. and Schmidt H. Applied Physics Letters, 92 (11), 111501.

Experimental studies of the interactions between a hydrogen plasma and a carbon or tungsten wall

K. Ouaras, L. Colina Delacqua, C. Quirós, G. Lombardi, M. Redolfi, D. Vrel, K. Hassouni and X. Bonnin

LSPM-CNRS Université Paris 13, Sorbonne Paris Cité, F-93490 Villetaneuse, France

Abstract. We present work done at LSPM (Laboratory of Sciences of Processes and Material Sciences), using the CASIMIR ECR plasma reactor device, aimed at answering questions about hydrogen isotope fuel retention and dust production in the context of the plasma-facing components (PFCs) of the International Thermonuclear Experimental Reactor (ITER). The plasma is characterized by means of optical spectroscopy, mass spectrometry and electrostatic probe; furthermore the dust density and size distribution will be measured by a laser diagnostic system. We present some early results obtained from hydrogen plasma exposure of pure tungsten samples, as well as samples of ITER-relevant tungsten-rich powders, produced in-house by the ball-milling technique, which are likely to be a by-product of material erosion and migration during tokamak operation. In particular, we have performed measurements of the specific surface area of these powders as a proxy to their capacity to absorb hydrogen.

1. Introduction

At LSPM, we have developed a low pressure / high density ECR plasma reactor named CASIMIR II, along with a smaller prototype reactor, so-called 'monosource'. In these devices, we consider the behaviour of nanometric ternary powders from the tungsten-rich corner of the W-C-Mg system, as representative of the mixed materials likely to form as a consequence of the plasma-wall interactions and impurity transport phenomena in ITER. We diagnose the plasma phase by means of Optical Emission Spectroscopy (OES), Langmuir Probe, and Mass Spectrometry, as well as with two laser diagnostic systems. The first is an IR (Infra-Red) quantum cascade laser (QCL) devoted to the detection and absolute measurement of light hydrocarbon dust precursors. The second system is a continuous wave (cw) visible laser, in the course of installation, that will be used in extinction and scattering modes to measure the dust density in our plasmas as well as determine the dust size distribution.

2. Short presentation of the CASIMIR reactor

Our reactor CASIMIR II (for Chemical Ablation, Sputtering, Ionization, Multi-wall Interaction and Redeposition) relies on the Electron Cyclotron Resonance (ECR) principle. Its aim is to produce low pressure / high density plasmas, using an innovative dipolar plasma source [1, 2], to simulate some plasma/surface processes occurring in parasitic discharges under the divertor dome and in the far Scrape-off Layer (SOL) regions of tokamaks. A full description of the device is given elsewhere [3]. The main idea of this paper is to study the formation under plasma exposure of mixed materials with compositions similar to those foreseen in fusion devices (C/W).

The plasma is generated at the electron cyclotron resonance location, in our case 875 G for a 2.45 GHz excitation frequency. Langmuir probe and optical spectroscopy measurements show the plasma to be homogeneous within the volume enclosed by the 16 ECR sources arranged in a circle. Typical electron densities and temperatures are of the order of 10^{11} cm^{-3} and 2-3 eV, respectively, for working pressures between 10^{-3} and 10^{-2} mbar and an injected microwave power of 3 kW. An energy balance of the reactor with a single ECR source was performed in order to estimate the power absorbed by the plasma to the power supplied to the system ratio $P_{absorbed}/P_{injected}$. It was shown that around 1/3 of the injected power is in fact coupled to the plasma [4]. These electromagnetic parameters are important to understand the homogeneous growth mechanism of dusts.

To understand the mechanism of dust formation, we installed an Infrared Absorption Spectroscopy system, using a cw Quantum Cascade Laser as light source, emitting around 1347 cm^{-1} [5, 6]. Its aim is to perform quantitative measurements of carbon dust precursors: CH_4 and C_2H_2, before going further with mixed materials. This system provides finely-tuned IR laser radiation that is absorbed by specific molecular species within the plasma, before being detected by a very high resolution dedicated spectrometer. The absorption length was increased to 5 m using a custom-made optical cell that makes the laser beam do some 24 passes inside the plasma volume.

In order to detect the apparition of soot in the discharge, we are presently implementing an additional laser setup relying on a cw visible solid-state laser emitting at 532 nm (up to 1 W power). Two sets of measurements will be attempted: (i) Extinction (which corresponds to the decrease of the laser signal after crossing the plasma volume, when the soots start to appear), and (ii) light scattering, in order to gain information on the size distribution of these soots. A photograph of the first step of setup assembly is shown as Figure 1.

Figure 1. The cw visible laser assembly during installation and alignment.

3. Carbon dust formation under ECR plasma exposure

Carbon dusts were synthetized at low pressure (10^{-3} mbar) plasma with feed gas composed of (50:50) H_2:C_2H_2 or H_2:CH_4. SEM (Scanning Electron Microscopy) pictures were performed on Si substrates, showing spherical particles of tens of nm, obtained after 30s. TEM (Transmission Electron Microscopy) measurements exhibit graphitic structures [7]. These observations are quite unusual at such low temperatures, due to the very small residence time of neutral particles. Therefore, ions (positive and negative) were tracked since they are sensitive to the permanent magnetic field of the ECR sources.

Ions detection was performed by means of in-situ mass spectrometry. It is seen that positive ions are by far much more prevalent in the gas phase than negative ions (Figure 2), which corresponds to a different behaviour with respect to higher pressure (1 mbar) RF reactors, in which particle growth is due to trapping of negative ions. In our case, positive ions, confined by the device's magnetic field, seem to be responsible for the dust synthesis.

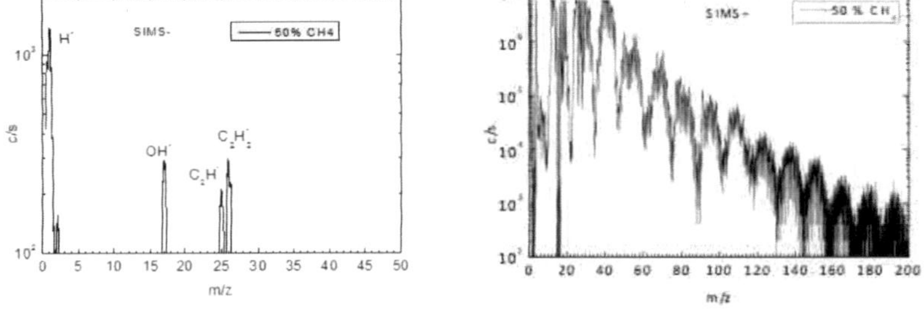

Figure 2. Mass spectra of positive ions (left) and negative ions (right) obtained in the single source ECR reactor. Each broad peak in the left spectrum corresponds to a different number x of carbon atoms in the C_xH_y hydrocarbons collected.

Journal of Physics: Conference Series **591** (2015) 012029 doi:10.1088/1742-6596/591/1/012029

4. Plasma exposure of prepared mixed materials dust samples

We can also synthesize ternary (W/C/Mg) powders of arbitrary but controlled stoichiometry by the ball milling method [8], also known as mechano-synthesis, as examples of the material mixes that will occur in ITER following erosion of the plasma-facing components [9]. Here we use Mg as a stand-in for Be, which is too dangerous to handle in our facility. The advantage of working with a composite material is to assess the influence of carbon and magnesium on the incorporation of hydrogen by studying the effect of their concentration on the hydrogen amount implanted near the surface and the kinetics of hydrogen diffusion into the bulk of the material. To assess the amounts of diffused and implanted hydrogen, techniques such as thermal desorption are expected to be used.

Measurements of specific surface areas (SSA), done prior to plasma exposure, have shown a value of up to 2 m^2g^{-1} for pure W powder, and increase further to more than 7 m^2g^{-1} when up to 10at% C is added; on the contrary, adding Mg has a clear negative influence on the measured SSA, probably due to an incomplete insertion of Mg into the W lattice, where Mg then has a tendency to bind together the W particles, thereby reducing the SSA. Measurements of the pore size distributions of binary and ternary materials are presented below (Figure 3). The majority of the pores (80%) have a diameter greater than 20 nm. These diameters correspond to mesopores (20-50 nm) and macropores (above 50 nm diameter). Large pore sizes mean that species can be inserted more easily into the material, in particular for hydrogen to get incorporated into the material during plasma implantation. This is beneficial for our future studies on the hydrogen incorporation into the material by plasma implantation.

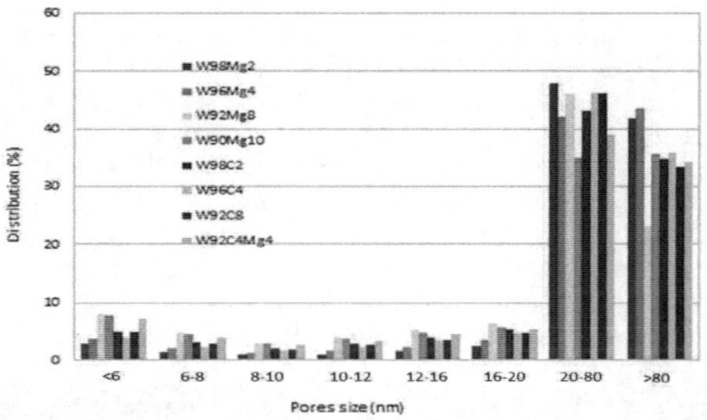

Figure 3: Distribution of pore sizes in binary W-C and W-Mg as well as ternary W-C-Mg materials.

The presence of oxygen is a problem because it changes the material properties and thus distorts the results. To enable the study of hydrogen incorporation in powders, it is necessary to reduce the oxygen concentration. For this we treat our powders with argon plasma. We exposed a sample of milled tungsten powder in ethanol. We performed the experiments in an argon plasma reactor for 1 hour. After treatment, the oxygen concentration decreases to 10%. The treatment is effective for oxygen content reduction on the material surface. It must, however, be renewed after each air exposure event.

The hydrogen plasma exposure experiments were carried out in our reactor with a single dipolar ECR source, which can deliver a power of 180 W and generate high density deuterium plasmas at low pressure. The neutral species flux is 10^{22} $m^{-2}.s^{-1}$ and ion flux is 10^{20} $m^{-2}.s^{-1}$.The samples were placed on a ceramic support for electric insulation, at floating potential, and situated 1 cm away from the ECR resonance location for 6 hours of plasma exposure.

Three compounds (pure W, binary W/8%atMg and ternary W/4%atC/4%atMg) were chosen and exposed to a deuterium plasma. To detect structural changes in the tungsten due to plasma interaction, a series of X ray diffraction patterns were obtained after various exposure times. We can see in Figure 4, the appearance of new peaks after 1 hour which indicates a phase change in the tungsten structure, the exact nature of which is still under study.

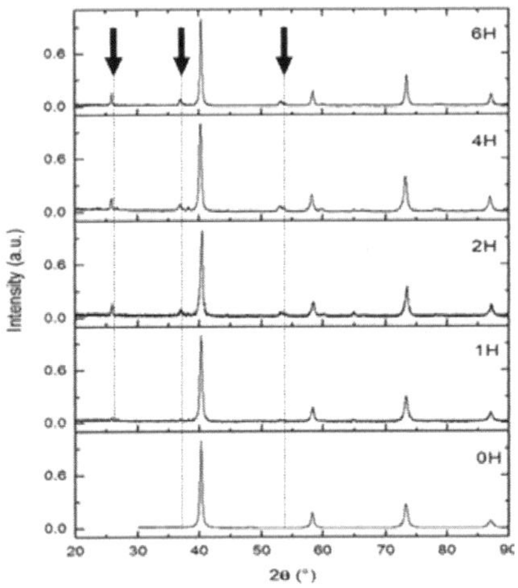

Figure 4 : X-ray diffraction spectra of the pure-W sample as a function of the plasma exposure time. The arrows indicate the positions of new peaks not present before plasma exposure.

The 3 mixed materials samples were imaged by SEM before and after deuterium exposure (Figure 5). The first (a) is the pure tungsten sample; it presents a real surface change, with disappearance of the roughness and a molten aspect that could have a link with phase changes that we can see in Figure 4. The second (b) is the binary compound (W/8%atMg) sample. Before exposure, the surface is flat and has few cracks. Thermal cracking appears after exposure. This is because of deuterium loading leading to large strains and aging of the samples. The black spots appearing are due to SEM chemical contrast between W and Mg, and could be interpreted as the resurgence of the latter to the surface. The third sample (c) is the ternary compound (W/4%atMg/4%atC) sample. Interpretation is the same as for sample (b). The presence of carbon manifests itself through a much larger porosity than for the sample containing only magnesium.

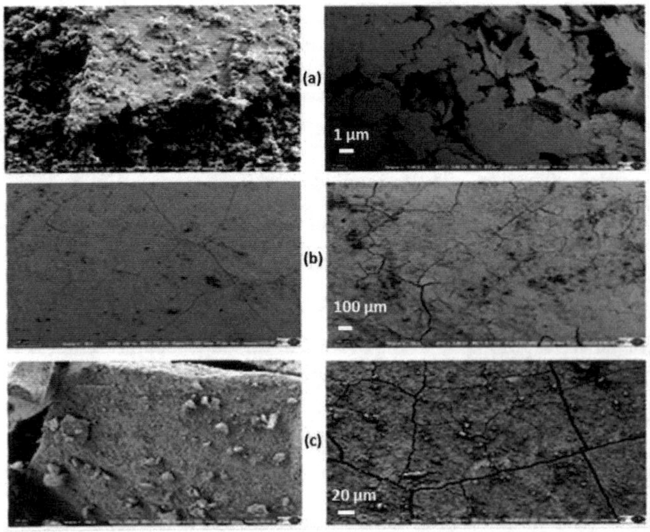

Figure 5: SEM micrographs showing samples before (left) and after plasma exposure (right). Each panel pair is shown at the same magnification. Sample (a): pure W; sample (b): W/Mg8%at; sample (c): W/C4%at/Mg4%at.

5. Conclusion

In this paper, we have presented briefly the work performed at LSPM using the CASIMIR II ECR plasma device, and a companion monosource prototype, aimed at plasma-wall interaction studies of fusion-relevant material mixes, concerning dust formation, hydrogen retention, and material changes due to plasma exposure. To date we have synthetized, by mechano-synthesis, a wide variety of tungsten-rich mixtures (with small atomic percentages of C and Mg, as a stand-in for Be) and exposed them to a deuterium plasma in the CASIMIR device and in our single source reactor. Characterization of these samples showed the appearance of new phases of W and the resurgence of the Mg from the W matrix after plasma exposure. In the single source reactor, we also witnessed growth of carbon nanoparticles, in a plasma dominated by positive ions, which is rather unusual and will require further investigation. We have also investigated the changes in specific surface area as function of carbon and magnesium concentrations included in the tungsten matrix. The pore size distribution of the powders showed that the majority of the pores (80%) are meso- and macro-pores, promoting surface adsorption processes.

References

[1] G. Lombardi, *et al.,* J.Nucl.Mater. **390-391** (2009) 196.
[2] S. Béchu *et al.*, Physics of Plasmas **20** (2013) 8
[3] Ligia Colina Delacqua, "Modeling and diagnostic of dust production in an H_2 plasma in contact with a C/W target. Contribution to the study of plasma/surface interactions in thermonuclear fusion devices", PhD Thesis, Université Paris 13, 2012.
[4] L. Colina Delacqua *et al.,* submitted to Plasma Sources Sciences and Technology.
[5] K. Ouaras *et al.*, to be published in J. of Plasma Physics (2014).
[6] J. Röpcke *et al.*, J Applied Physics B **92** (2008) 335
[7] C. Arnas *et al.*, Plasma Phys. Controlled Fusion **52** (12), 124007 (2010).
[8] C. Suryanarayana, "Mechanical Alloying and Milling," Prog. Mater. Sci., **46**, 1 (2001).
[9] F.R.A. Onofri *et al.*, Fusion Sci. & Technol. **62** (2012) 39.

High power plasma interaction with tungsten grades in ITER relevant conditions

I E Garkusha[1], V A Makhlaj[1], N N Aksenov[1], O V Byrka[1], S V Malykhin[2], A T Pugachov[2], B Bazylev[3], I Landman[3], G Pinsuk[4], J Linke[4], M Wirtz[4], M J Sadowski[5] and E Skladnik-Sadowska[5]

[1] Institute of Plasma Physics, NSC Kharkov Institute of Physics and Technology, Kharkov, Ukraine;
[2] National Technical University "Kharkov Polytechnical Institute", Kharkov, Ukraine;
[3] Karlsruhe Institute of Technology, IHM, P.O. Box 3640, 76021 Karlsruhe, Germany;
[4] Forschungszentrum Juelich GmbH, EURATOM Association, 52425 Juelich, Germany;
[5] National Centre for Nuclear Research (NCBJ), 05-400 Otwock, Poland

E-mail: garkusha@ipp.kharkov.ua

Abstract. Experimental simulations of ITER-like transient events with relevant surface heat load parameters (energy density up to 1.1 MJ/m^2 and the pulse duration of 0.25 ms) as well as particle loads (varied in wide range from 10^{23} to 10^{27} ion/m^2 s) were carried out with a quasi-stationary plasma accelerator QSPA Kh-50. Particular attention was paid to elaboration of damage of tungsten as a main candidate material for ITER divertor surfaces and also as prospective material for DEMO design. Erosion features, cracks evolution and changes in their thickness with increasing exposition dose are studied for different W grades, including deformed material with elongated grains, as well as WTa5 and sintered tungsten.

1. Introduction

Tungsten is considered now as plasma facing material (PFM) for ITER divertor and also it is primary choice for DEMO design [1]. Key advantages of W material are low sputtering yield and tritium retention, high sputtering threshold energy and melting temperature, good thermal conductivity etc. However its brittleness, high ductile to brittle transition temperature (DBTT), ability to melt during the transient events and large Z are among critical issues for tungsten application in fusion reactor [2,3]. Under the high heat loads macroscopic mechanisms of W erosion would be dominating rather than microscopic ones. Macroscopic erosion due to the brittle destruction (cracks, debris and dust particles), tungsten melt losses caused by melt motion and droplets splashing, material modification and changed properties after the repetitive plasma pulses, possible synergetic effects from heat and particle loads require comprehensive studies in ITER relevant conditions [4,5].

Database of tungsten behavior under the high heat loads from the existing machines is insufficient. Only few tokamaks have experience with W. The expected ITER loads (especially for disruption) will considerably exceed the available energy loads in the experimental devices. Resulting damage effects from million ELMs in ITER also remain to be among key topics of predictive studies. Thus fundamental understanding of plasma-surface interaction (PSI) processes in magnetic confinement devices requires dedicated R&D activity in plasma simulators used in close connection with material characterization facilities as well as with theory and modeling activities [5-7]. For such investigations different simulators of transient loads are now involved (quasi-stationary plasma accelerators, e-beams, dense plasma foci and plasma guns), that are cost effective, flexible, able to provide faster results and important comparison of damage features from various machines [8-13].

While a number of unresolved issues in plasma edge physics (e.g. scrape-off layer, plasma flows and turbulent transport) can only be addressed in tokamaks, issues related to fuel retention, dust formation, impact of transient heat loads on materials, erosion and re-deposition mechanisms, as well as new material concepts can be explored using dedicated PSI devices. Besides a good access to the plasma-material interaction zone, allowing specific diagnostics to be operated and an easy exchange of material samples, the reasons to explore PSI issues using this approach include the ability of plasma simulators to achieve ITER relevant particle and energy fluxes, provide well diagnosed test cases for plasma wall interaction codes benchmarking, as well as test beds for PSI diagnostics.

Thus, using the pool of existing plasma accelerators, guns, e-beam facilities allows contribute to the knowledge-based understanding of the performance and adequacy of candidate plasma facing materials of next step fusion devices under extreme energy and particle load conditions. Such activity is in progress now within IAEA CRP "Investigations of Materials under High Repetition and Intense Fusion-Relevant Pulses" which includes, in particular, round-robin tests of candidate materials in available simulators as well as investigation of combined plasma treatments of samples in different devices.

This paper presents the results of tungsten samples exposures in ITER relevant conditions that have been performed with powerful plasma streams generated by quasi-stationary plasma accelerator QSPA Kh-50. The targets were exposed to a varied number of pulses (up to 400) of different energy load Q. Also we compare the behavior of samples of different grades of tungsten subjected to plasma loads, including those with deformed elongated grains that is chosen as ITER reference material.

2. Experimental setup

Experimental simulations of ITER ELMs with relevant surface heat load parameters (energy density and the pulse duration) as well as particle loads were carried out with a quasi-stationary plasma accelerator QSPA Kh-50 described elsewhere [14-16]. Hydrogen plasma streams, generated by QSPA Kh-50, are injected into the magnetic system of 1.6m in length and 0.44m in inner diameter with a magnetic field of up to 0.54T in diagnostic chamber (2.3m from accelerator) where the target has been installed [14-16]. General view of the experimental installation and scheme of the targets exposure are presented in Fig.1. Plasma parameters were varied both by changing dynamics and the amount of gas filled in the accelerator channel and by the variation of the working voltage of capacitor battery of the discharge in accelerating channel.

Main plasma diagnostics used in the present experiments were as follows: The plasma stream velocity was measured by the time-of-flight of the plasma stream between two magnetic probes and/or photodiodes, the electron density in the plasma stream was evaluated on the basis of Stark broadening of the H_β and H_α spectral lines, radial distributions of the plasma stream energy density were monitored with a movable thermocouple calorimeters. Plasma pressure was measured with piezoelectric detectors. The power density was calculated on the basis of measurements of the time dependencies of the plasma pressure, plasma stream density and its velocity.

For ELM simulations, the main parameters of the plasma stream were as follows: ion impact energy 0.4-0.6 keV, plasma stream density mean plasma density $(5\text{-}7)\text{x}10^{15}$ cm^{-3}, maximum plasma pressure ~0.32 MPa and the plasma stream diameter 0.18 m. It allows achievement of very small pressure gradient values in the target region, where plasma pressure is practically constant. The plasma pulse shape is triangular with pulse duration of 0.25 ms. Experiments were performed with several fixed surface energy loads chosen in the range of 0.45-1.1 MJ/m^2 (measured precisely with calorimetry) i.e. either below the melting threshold, or resulted in strong melting of exposed surface [16].

Deformed W grade (Plansee AG), including both single (F) and double forged (DF) samples, was used for the plasma loads tests. Cylindrical shaped specimens with a diameter of 12 mm and a height of 5 mm were prepared from a 1 m long rod (Ø 12 mm, the deformation axis along the rod) [17]. The grain orientation was parallel to the heat transfer direction, which corresponds to the ITER specification. Sintered tungsten samples and WTa5 tungsten that consist of pure tungsten alloyed with 5 wt% tantalum were investigated also.

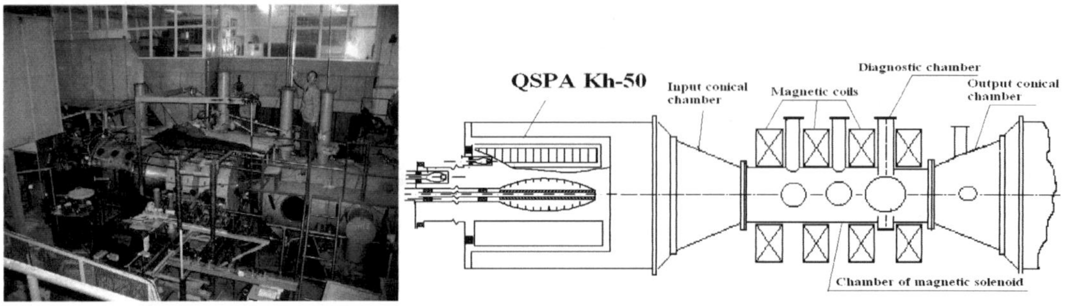

Figure 1. General view of QSPA Kh-50 and scheme of the samples exposures

Before each plasma pulse the sample was kept either under the room temperature (RT) or preheated at T_0 = 200 ^{0}C. The Ohmic heater was installed at the back side of the target. It provided the target heating in vacuum conditions (to avoid oxidation) prior to the exposure. A molybdenum diaphragm with the hole diameter of 1 cm was applied to protect the target edges from plasma impact providing a reference line for profilometry measurements and also to simulate the case when the target size exceeds the diameter of the plasma stream.

Observations of plasma interactions with exposed surfaces have been performed with a high-speed 10 bit CMOS pco.1200 s digital camera PCO AG (exposure time from 1μs to 1s, spectral range from 290 to 1100 nm). Information from consecutive camera frames with traces of particles flying from the tungsten surface after plasma shot allowed calculation of velocities of emitted particles and the time moments of their start-up from the target surface.

Surface analysis was carried out with an optical microscopy, SEM, AFM and profilometry. Measurements of weight losses and microhardness of the surface were performed also. XRD method was applied for structure analysis and measurements of stresses in thin surface layer (X-ray tensometry).

3. Results of experiments

For studies of tungsten response to the plasma exposures, which are not resulted in surface melting, the energy load Q=0.45 MJ/m^2 has been applied. All tungsten grades that have initially room temperature suffer from the major crack meshes which are developed on the surface. Cracking leads to increasing surface roughness as it is measured with profilometry. Evolution of surface roughness for DF samples in the course of plasma exposures is shown in Fig.2. Fluctuation peaks on profilograms correspond to the average distance between developed major cracks. With increasing number of exposures depth and width of the profile fluctuations grow.

Plasma exposures of tungsten in conditions of pronounced melting of the surface layer have been performed with energy load of 0.75 MJ/m^2. Corresponding changes in the profile of the surface after 10 plasma pulses are presented in Fig.3. Smoothing crack edges due to the melting is observed by profilometry. However surface roughness after plasma exposures with surface melting is essentially higher in comparison with similar number of pulses with heat load below the melting threshold.

Besides major cracks, micro-cracks network along the grain boundaries is always detected in experiments. It is attributed to the surface melting and following resolidification. Typical cell sizes of intergranular micro-cracks network are within 10–40 μm, which corresponds to the grain size of the W grade.

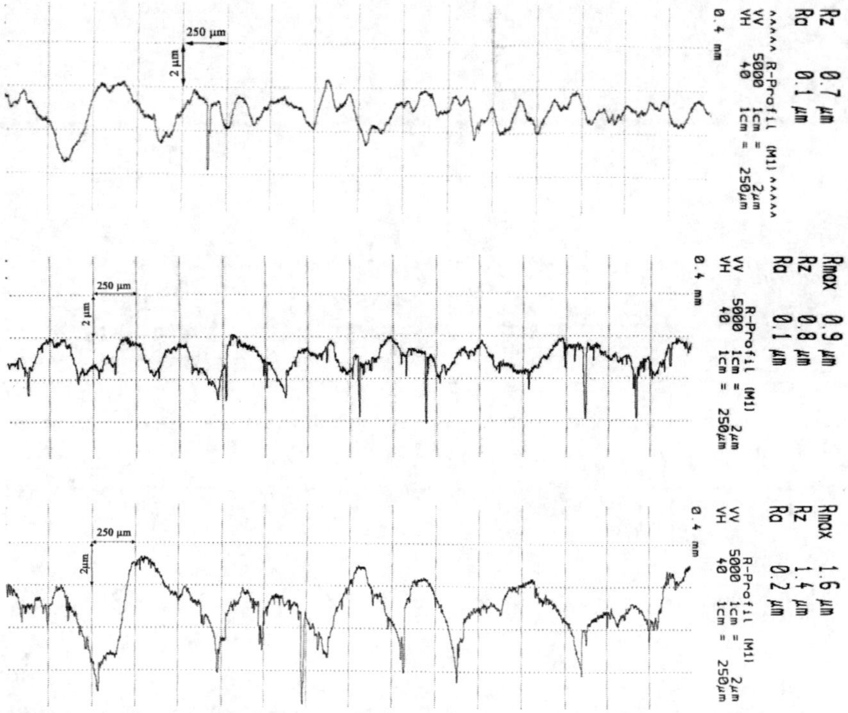

Figure 2. Surface profile development in the course of plasma exposures, Q=0.45 MJ/m^2.

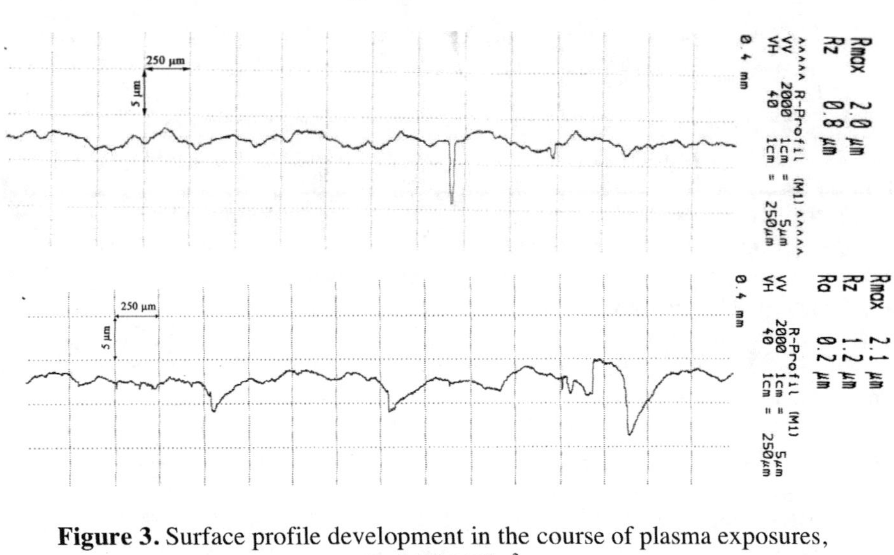

Figure 3. Surface profile development in the course of plasma exposures, Q=0.75 MJ/m^2.

Roughening of the surface caused by plasma exposures are demonstrated also by dedicated measurements with AFM (Fig.4). Specific surface morphology is developed due to the crack meshes arisen on the surface and also it is influenced by melt layer appearance and surface tension effects.

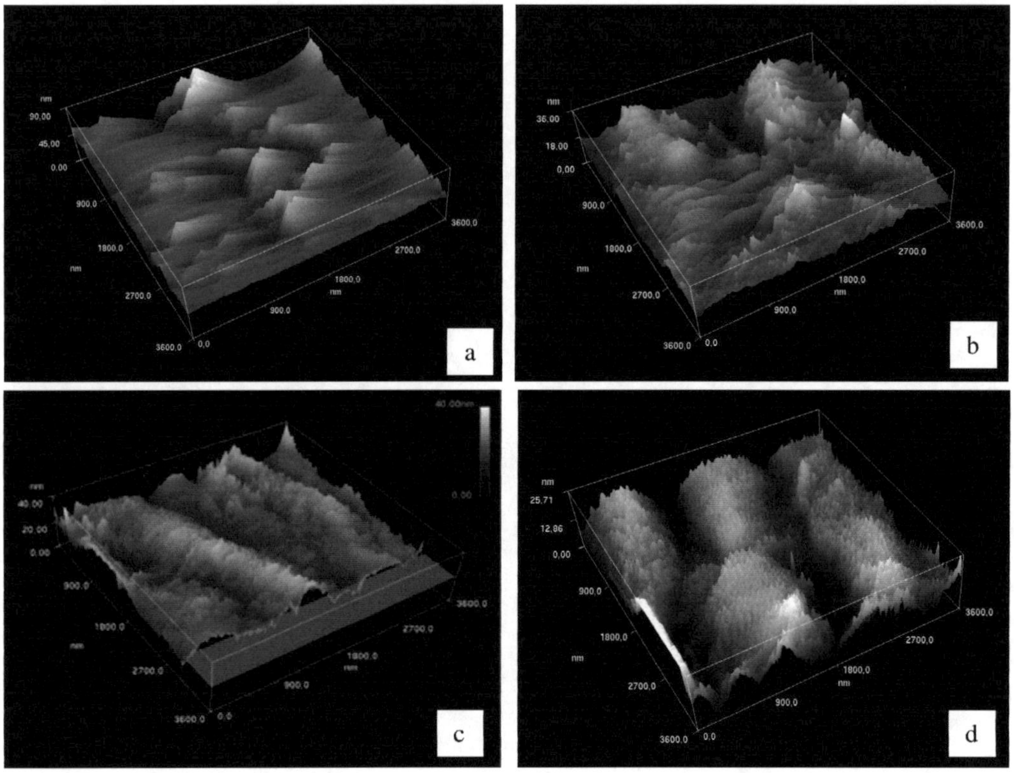

Figure 4. AFM images of single forged tungsten sample; a-initial surface, b- exposed with 10 pulses of 0.75 MJ/m^2, c,d- exposed with 20 pulses of 0.75 MJ/m^2.

Tungsten cracking leads to separation of material peaces with further ejection of erosion products from the surface. Depending of the heat load value the eroded material could be in the form of solid dust or, for exposures with melting, as splashed droplets. However, the droplets appearance is registered also in regimes of plasma exposures with the energy loads below the melting threshold. Formation of dust particles in the course of cracking and development of sharp edges of the cracks are further resulted in their melting under the plasma energy deposited because of decreased heat transfer to the bulk. Being initially solid they are subjected to melting during the later stage of plasma pulse or with next plasma impact in spite of other surface remains non melted.

Formation of dust particles in the course of surface cracking with their subsequent melting under the action of plasma pulses is shown in Fig. 5. Melting of the crack edges is also clearly observed. Performed analyses confirms that registered droplets consist of tungsten material and its appearance do not caused by other effects (e.g. from electrodes erosion, diaphragm, holder etc.).

Target preheating to 200 °C do not resulted in suppression of tungsten cracking. Nevertheless it influences on the cracking dynamics. Cracks are starting to appear only after several pulses, they grow slower, their depth and width remain to be smaller in comparison with RT targets exposures. Some examples of surface cracks after 10 and 100 plasma pulses of 0.45 MJ/m^2 target are presented in Fig.6

for preheated to 200 °C. It is seeing that surface damage by cracking for 100 plasma pulses became more pronounced as compare to initial 10 exposures.

Detailed analysis of the crack patterns for deformed tungsten material has been performed and results of characterization of the crack meshes are summarized in Table 1. Average crack depth is grown practically twice with simultaneous increase of crack width also in 2 times. Average size of crack mesh, i.e. distance between cracks, is practically unchanged. It influenced only by some new cracks appearing on the surface and completing the crack pattern.

Element	Weight %	Atomic%
Fe K	1.12	3.56
Cu K	0.65	1.82
W M	98.22	94.62
Total	100.00	

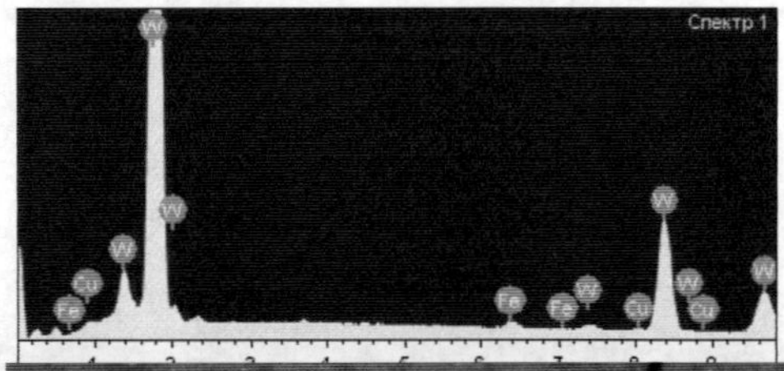

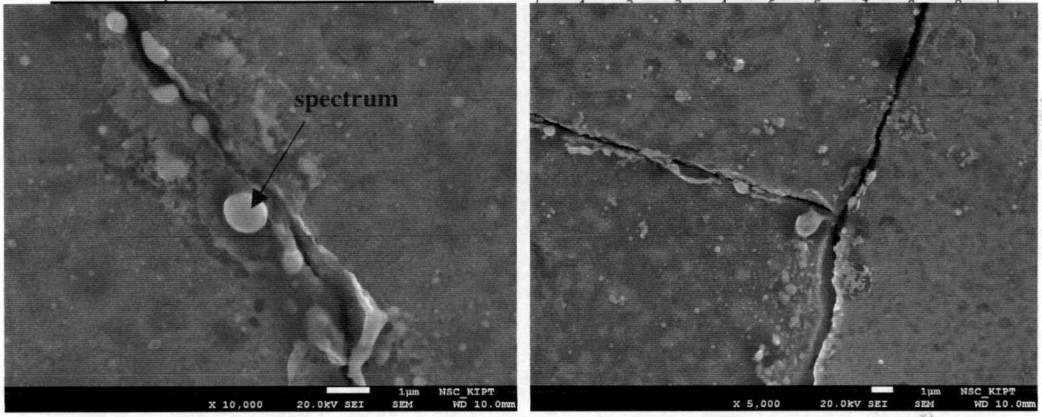

Figure 5. Formation of dust particles on DF tungsten sample in the course of surface cracking with their subsequent melting by plasma pulses. Exposition dose 400 pulses, Q=0.45 MJ/m^2.

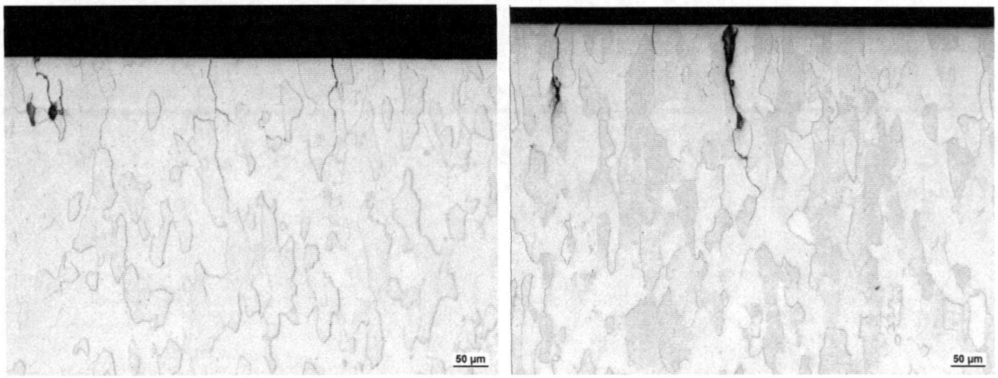

Figure 6. Comparison of surface cracks for 10 and 100 plasma pulses of 0.45 MJ/m^2.
Initial temperature is 200 ^{0}C.

It should be mentioned that size of the major crack mesh is about half millimeter and it is similar for all investigated tungsten grades being characteristic of brittle destruction of W surface in the course of plasma pulses. However, major cracks that are arisen in sintered tungsten for exposed RT targets have average width of 5μm after 100 pulses and it achieved 7 microns after 300 plasma exposures.

Table 1. Characterization of crack patterns on the single forged W material exposed with 10 and 100 pulses of 0.45 MJ/m^2, T=200 ^{0}C

	Crack depth		**Crack width**		**Distance between cracks**	
	10 pulses	100 pulses	10 pulses	100 pulses	10 pulses	100 pulses
average [μm]	92.78	171.72	1.11	2.12	546.32	432.55
minimum [μm]	20.62	68.04	0.30	0.78	22.33	39.08
maximum [μm]	119.59	329.9	1.70	2.91	2171.59	1278.39
standard deviation [μm]	26.72	55.94	0.30	0.54	358.77	223.55
median [μm]	103.09	154.64	1.12	2.22	474.51	418.69

Thus, it is obtained that deformed W material is found to be more resistive against cracking and grain losses as compare to other pure W samples. No major cracks are detected after single pulse for the target preheated at 200°C. However, with further pulses, once a major crack develops for deformed W, it grows quickly and propagates transversely to the surface along the grain boundary to the depth of ≈200 μm. In result, after 100 plasma pulses of 0.45 MJ/m^2 the tungsten surface is essentially damaged by the cracks, as it is demonstrated by cross-section image in Fig.7. After 400 plasma pulses crack depth achieves ~400 μm.

Cross-section metallography shows that crack thickness can be maximal not on the surface, but in some depth below. It could be attributed to the fact, that DBTT transition under the target cooling after impacting plasma pulse occurs firstly in deeper layers, while the surface is still remained to be above DBTT. Then, cooling front achieves the surface. Therefore the crack could be originated in some depth below the surface, and the maximal width of major cracks sometimes is registered in the depth of 10-20 microns and higher, confirming this conclusion. In very deep layers the temperature changes are rather small to initiate the cracking process. Nevertheless the major crack, if arrisen, propagates to the bulk regions which were not heat affected by plasma pulses.

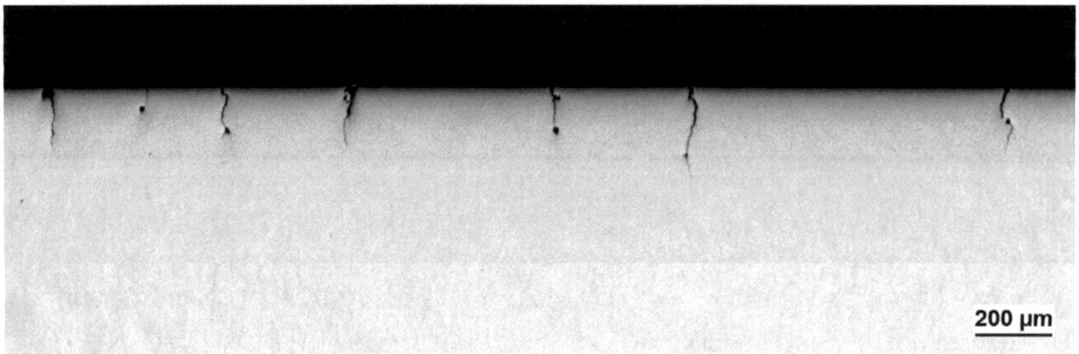

Figure 7. Cross-section of single forged W exposed with 100 pulses of 0.45 MJ/m^2 T=200 ^{o}C.

Cracking process is originated by the stresses which induced the plasma pulses. Residual stresses in surface layer in the course of plasma pulses were analyzed with XRD using the $\sin^2\psi$-method. Plasma irradiation results in a symmetrical tensile stress in thin subsurface layer. Maximal residual stress in the plasma affected layer is reached after first plasma pulses, and further pulses change the stress not substantially. It is observed as a rule that increase of number of exposures decreases the residual stress.

It is obtained that the residual stress does not practically depend on initial target temperature and significantly grows with increasing thermal loads up to achievement of melting threshold for exposed material. Surface melting with repetitive exposures leads to fast relaxation of stresses in the surface layer.

Results of XRD analysis of residual stresses in surface layer for DR targets exposed in different regimes are summarized in Table2. As follows from the performed measurements, the lattice parameter does not changed significantly in the course of plasma exposures. For surface heat load of 0,45 MJ/m^2 the maximum stress level in surface layer is ~400 MPa and it does not depend on the initial state (as compare DF 73 and 74). With increased number of pulses up to 100, stresses are relaxed to 270 - 284 MPa. Decreasing line width (B) as observed in result of exposures is caused by changing dislocation density in the plasma affected layer of the material. It should be mentioned that asymmetry δS (the difference between left and right side of the line profile) become negative, i.e. increasing the right-hand side is detected. This may be explained by corresponding increase in the number of interstitial complexes, which is consistent with a slight grows in the lattice parameter. For plasma exposures with Q=0,75 MJ/m^2 which are always accompanied by strong melting of the surface layer the trends are similar to those obtained for exposures with Q=0,45 MJ/m^2.

It was found that under similar loading conditions the magnitude of residual stresses for deformed tungsten material is smaller practically in 2 times in comparison with those observed for sintered samples and rolled W plates in similar conditions. The obtained result can be explained by combination of two factors: initial compressive stresses in deformed W grade and its improved structure in comparison with sintered and rolled W grades. Threshold residual stresses for cracks formation are 300-350 MPa.

Table 2. XRD analysis of stresses in surface layer for DR targets

targets	Heat load	Number of pulses	a_0, å	σ, MPa	B, degree	δS, %
DF 73	0 (initial)	0	3.1640	-33	0.56	+8
	0.45 MJ/m^2	10	3.1649	440	0.52	-3
DF 74	0 (initial)	0	3.1640	-360	0.55	+7
	0.45 MJ/m^2	10	3.1645	390	0.53	-12
DF 75	0 (initial)	0	3.1641	-380	0.56	+6
	0.45 MJ/m^2	100	3.1644	270	0.44	-7
DF 76	0 (initial)	0	3.1640	-365	0.58	+7
	0.45 MJ/m^2	100	3.1646	284	0.48	-13
DF 21	0 (initial)	0	3.1642	-47	0.5	+2
	0.75 MJ/m^2	10	3.1647	301	0.47	-1.5

Plasma exposures of WTa5 tungsten grade, that enriched be 5% of tantalum demonstrate considerably suppressed brittle destruction effects for this material with no essential influence on melting threshold energy. Repetitive plasma pulses with energy load of 0.3 MJ/m^2 do not result in cracking development. It means that threshold load for the cracking initiation for this material is above 0.3 MJ/m^2. Due to improved ductility, no major cracks are found on the surface even for RT targets exposures with several tens pulses of 0.45 MJ/m^2 which lead to crack meshes for all other grades. For energy load of 0.45 MJ/m^2 isolated thin cracks are appeared on the surface with increasing the exposition dose up to 100 plasma pulses. Thus WTa5 shows better performance in terms of cracks due to its ductility, only single cracks which do not form the complete mesh on the surface are registered after QSPA exposures of this W grade. Fig.8 shows exposed surface after 100 plasma pulses of 0.45 MJ/m^2 with examples of droplet and arisen crack. Resulting formation of micro-sized debris separated by cracking development is also seen. Droplet is W material. However, tantalum is difficult to distinguish in the spectrum on the background of W.

Besides the erosion products in form of solid particles and droplets, the important feature of plasma-surface interaction is surface modification with formation of fine cellular structures on the surface exposed with repetitive pulses. The SEM shows that the size of cellular structure globules (up to 200 nm) is essentially less than initial size of tungsten grains. Such "foam-plastic" ordered structures of nano-size on the tungsten surface have been detected also for other W grades [18]. It is obvious, that the observed surface modification effects with formation of submicron or nano-cellular structures may play an important role in the material behaviour under the transient plasma heat loads.

Element	Weight %	Atomic%
Fe K	0.25	0.78
Cu K	2.99	8.15
W M	96.76	91.07
Total	100.00	

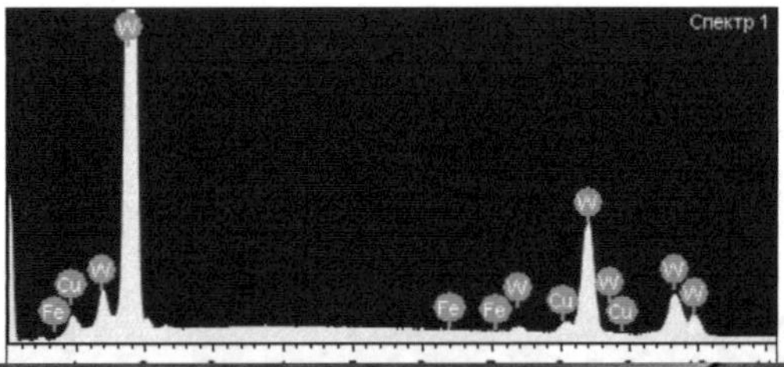

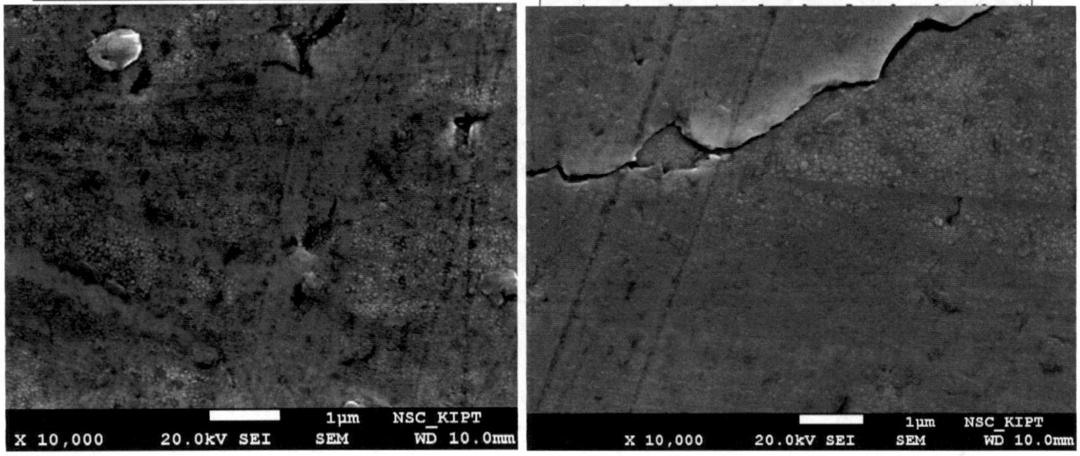

Figure 8. Surface modification of WTa5 with formation of nanosized structures and single crack, Exposition dose is 100 pulses, Q=0.45 MJ/m^2. Droplet is W material.

4. Summary

Experimental simulations of ITER-like transient events with relevant surface heat load parameters (energy density up to 1.1 MJ/m^2 and the pulse duration of 0.25 ms) as well as particle loads (varied in wide range from 10^{23} to 10^{27} ion/m^2s) were carried out with a quasi-stationary plasma accelerator

QSPA Kh-50. Particular attention was paid to elaboration of damage of tungsten as a main PFM for ITER divertor surfaces and also as prospective material foreseen for DEMO design.

Tungsten damage during the transient events is dominated by influence of macroscopic erosion mechanisms related with cracking development and surface melting effects.

Erosion products in the form of solid dust and splashed droplets are analyzed. Performed studies demonstrate that main sources of W dust are cracks development, resolidification process and material surface modification resulted in formation of cellular nanostructures in the course of repetitive plasma pulses.

Besides the melt motion and splashing effects for exposures with pronounced melting, the tungsten droplets appearance is registered also in regimes of plasma exposures with the energy loads below the melting threshold. It was found that main source of the droplets in this case is melting of separated material pieces and sharp crack edges. Due to decreased heat transfer to the bulk their melting occurs during later stage of plasma pulse or with next plasma impacts in spite of other surface remains solid.

For investigated pure W grades the damage threshold energy as well as the observed major cracking and roughening of the samples that exposed above these thresholds are qualitatively similar Performed experiments show just quantitative differences for the measured crack parameters. Deformed W material is found to be more resistive against cracking and grain losses as compare to other pure W samples.

The most pronounced differences of material performance are observed for W material alloyed with tantalum. Among investigated W grades WTa5 shows better performance in terms of cracks due to improved ductility. Only single isolated cracks which do not form the complete mesh on the surface are registered after QSPA exposures threshold load for the cracking initiation for this material is above 0.3-0.35 MJ/m^2 for single pulse. Threshold behaviour with multipulsed plasma exposures and possible degradation of the material properties could be a subject for further studies. It should be mentioned that even preheating up to 200 C does not result in suppressing of cracking, thus DBTT for this grade is still higher.

Within further round-robin tests, some of exposed samples could be subjected to irradiation with other involved simulators. In particular, synergetic effects would be important. Also, combination of steady state and pulsed plasma exposures is of especial interest aiming at damage character and erosion products, hydrogen retention etc. Basing on this planned activity comprehensive database on material surface characterization after exposures with different devices should be created including roughness values, crack distance, crack depth and width etc.

Acknowledgments

This work has been performed in part within IAEA CRP F1.30.13 and also partially supported by the Targeted Program of NAS of Ukraine on Plasma Physics. The authors would like to acknowledge N.V. Kulik, V.V. Staltsov, S.I. Lebedev, and P.V. Shevchuk for their assistance in the experiments.

References

[1] Philipps V 2011 *J. Nucl. Mater.* **415**, 1, Suppl 1, S2

[2] Rieth M, Armstrong D, Dafferner B, Heger S, Hoffmann A, Hoffmann M D, Jäntsch U, Kübel C, Materna-Morris E, Reiser J, Rohde M, Scherer T, Widak V and Zimmermann H 2010 *Advances in Science and Technology* **73** 11.

[3] Pintsuk G 2012 *Comprehensive Nuclear Materials* **4** 551

[4] Pitts R A, Carpentier S, Escourbiac F, Hirai T, Komarov V, Kukushkin A S, Lisgo S, Loarte A, Merola M, Mitteau R, Raffray A R, Shimada M and Stangeby P C 2011 *J. Nucl. Mater.* **415** 1, Suppl 1, S957.

[5] Landman I S, Pestchanyi S E, Safronov V M, Bazylev B N and Garkusha I E 2004 *Physica*

Scripta **T111** 206

[6] Miloshevsky G and Hassanein A 2011 *J. Nucl. Mater.* **415** 1 S74

[7] Pestchanyi S, Garkusha I and Landman I 2010 *Fusion Engineering and Design* **85** 1697

[8] Tereshin V I, Bandura A N, Bovda A M, Brown I G, Byrka O V, Chebotarev VV, Garkusha I E and Tortika A S 2002 *Review of Scientific Instruments* **73** (2) 831

[9] Loewenhoff Th, Bürger A, Linke J, Pintsuk G, Schmidt A, Singheiser L and Thomser C 2011 *Physica Scripta* **T145** 014057

[10] Skladnik-Sadowska E, Czaus K, Malinowski K, Sadowski M J, Nowakowska-Langier K, Ladygina M S and Garkusha I E 2012 *Nukleonika* **57(2)** 193

[11] Poznyak I M Arkhipov N I, Karelov S V Safronov V M and Toporkov D A 2012 *Problems of Atomic Science and Technology, Series: Plasma Physics* **6** 52

[12] Makhlaj V A, Bandura A N, Byrka O V, Garkusha I E, Chebotarev V V and Landman I S 2007 *Physica Scripta* **T128** 239

[13] De Temmerman G, Daniels J, Bystrov K, van den Berg M A and Zielinski J. 2013 *Nucl. Fusion* **53** 023008

[14] Makhlai VA, Garkusha I E , Malykhin S V, Pugachov A T, Landman I, Linke J, Pestchanyi S, Chebotarev V V and Tereshin V I 2009 *Physica Scripta* **T138** 014060

[15] Tereshin V I 1995 *Plasma Phys. Control. Fusion* **37** A177

[16] Garkusha I E, Bandura A N, Byrka O V, Chebotarev V V, Landman I, Makhlaj V A, Pestchanyi S and Tereshin V I 2009 *J.Nucl. Mater.* **386-388** 127

[17] Wirtz M, Linke J, Pintsuk G, Singheiser L and Uytdenhouwen I. 2011 Comparison of the thermal shock performance of different tungsten grades and the influence of microstructure on the damage behaviour *Physica Scripta T145* 014058

[18] Makhlaj V A, Garkusha I E, Aksenov N N, Chuvilo A A, Chebotarev V V, Landman I. Malykhin S.V. Pestchanyi S and Pugachov A T 2013 *J. Nucl. Mater.* **438** S233

The plasma focus as a tool for plasma-wall-interaction studies

G Ramos [1], M Martinez[2], J J E Herrera[2] and F Castillo[3]
[1] *Instituto Politecnico Nacional, CICATA Unidad Queretaro, 76090 Queretaro, Mexico*
[2] *Instituto de Ciencias Nucleares, Universidad Nacional Autónoma de México, México*
[3] *Instituto de Ciencias Físicas, Universidad Nacional Autónoma de México, México*

E-mail: gramos@ipn.mx

Abstract. The study of the interaction of magnetized plasmas with candidate materials for fusion reactors, as for example tungsten, is a main topic in fusion research. Many studies simulate the plasma wall interaction using ion beams, while only a few use plasma simulators. Plasma foci can produce dense magnetized plasmas of deuterium and helium among other species. We used the plasma focus Fuego-Nuevo II, to expose tungsten samples to deuterium and helium plasmas. The samples were analysed by means of SEM, RBS and NRA, evidencing surface erosion, surface melting and retention of deuterium in a shallow surface layer of 250 nm amounting $6.5{\bullet}10^{16}$ D/cm^2. The plasma temperature has been measured at the position of the samples using a triple Langmuir probe and compared to calculations of a snowplow model. The modelling of the electrode to reach desired plasma parameters is discussed.

1. Introduction

The interaction of plasmas with wall surfaces is important for the development of magnetic confinement fusion devices such as tokamaks, as they can release impurities into the plasma, erode the surfaces or produce retention of fuel in the wall, among other effects [1]. This effects are enhanced in the event of edge localized modes (ELMs) when large loads are deposited onto the wall. In a divertor-type Tokamak the particles reaching the first wall have usually relatively low energy and low density. The ion temperature averaged over an ELM is of the order of 20-200 eV outside the separatrix [2]. The study of the interaction of plasmas (and particles) with candidate first-wall materials is carried out mainly with ion beams, while only a few use plasma devices. Plasma simulators have usually very low energies of the order of tens of eV [3], while energies up to several hundreds of eV are desired.

The plasma focus [4,5] is a simple device that can produce a hot and dense plasma. During the initial axial phase energies up to several hundreds eV can be easily obtained This phase is characterised for producing a large plasma volume of low energy. In the radial or compression phase a short living hot dense plasma is produced (small volume, high energy). Many of the early studies

[1] To whom any correspondence should be addressed.

were concentrated on the study of the production of neutrons [6], while in the last time the research is more focused on applications in many fields including materials sciences [7].

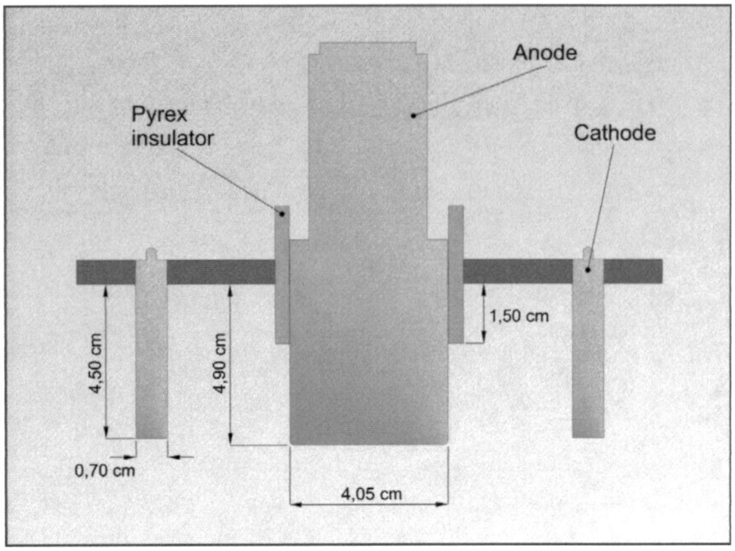

Figure 1. Schematic diagram of the FN II electrode showing the cathode, anode and insulator together with their dimensions.

2. Fuego Nuevo II results

For plasma wall interaction studies the device has to show the capability of producing surface modification (erosion, surface melting), have the flexibility of having different plasma compositions, and it is also desirable to have the capability of studying retention of plasma species and intermixing.

Fuego Nuevo II (FN II) is a Mather-type Plasma Focus device with 4.8 kJ of stored energy described in more detail elsewhere [8,9]. The coaxial electrode is formed by an inner anode made out of solid OFHC copper and an outer cathode formed by twelve brass bars. The dimensions of the electrodes are shown in figure 1. For our experiments the inner electrode had a dip in the center to minimize erosion.

Samples were prepared from a tungsten foil 0.1 mm thick, with 99.95% purity, from Goodfellow [10], handled carefully, in order to avoid contamination, cut to proper dimensions and mounted on a copper massive holder in order to provide a proper heat sink during experiments. No further treatment was performed on the samples prior to plasma exposure. They were placed at 11 from the end of the electrode, as illustrated in figure 2, and exposed at room temperature to 20-30 shots each using deuterium or helium as filling gas, typically around 2.5 Torr, operating the device at 4.8 kJ (38 kV 7.4 μF). Additional Ti (0.127 mm thick, 99.7% purity[11]) and Cu samples were prepared, avoiding contamination but without further treatment, and placed either at 11 cm or 30 cm from the electrode.

After exposure the samples were analysed using different techniques: electron microscopy, cross section analysis using focused ion beam milling, Rutherford backscattering and nuclear reaction analysis.

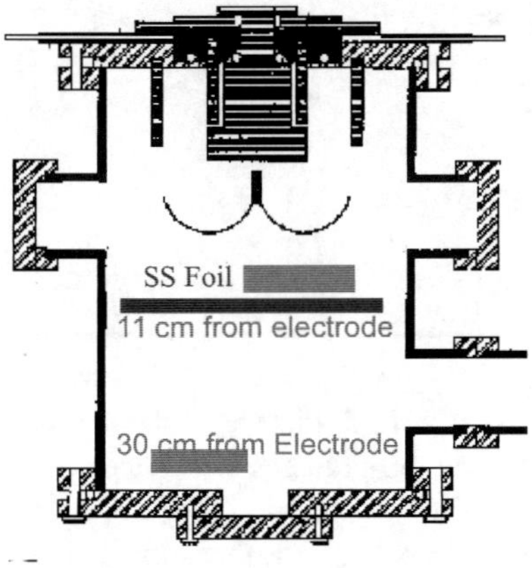

Figure 2. Schematic cross section of the chamber showing the electrode on the top, upside down, and the position for the stainless steel foil and the targets at 11 and 30 cm from electrode end.

2.1 Characteristics of the FN II discharge

A typical 4.8 kJ discharge with deuterium (~ 2 Torr) has a duration of a few microseconds. During the axial phase of the discharge the current increases steadily reaching its maximum of 240 kA at approx. 1 μS, while the plasma sheet reaches the end of the electrode shortly after, the radial phase starts, and the pinch occurs at 1.2 μS. The total duration of the discharge is about 9 μS.

Due to the interaction of the deuterium plasma with the insulator and the copper / brass electrodes material is eroded and projected into the chamber, as can be observed by the deposits on a stainless steel sheet placed 11 cm form electrode end (figure 3). In the center of impact a relatively large amount of copper can be observed in distances up about 1.3 cm from the axis. For larger distances from the axis the amount of coper drops, and for distances larger than 1.8 cm copper is not visible on the sheet. In the different studies we performed looking for impurities coming from the electrode only glass and copper could be observed [12] .

In order to minimize the influence of the impurities coming from the electrode all our samples were placed off-axis and at distances larger than abourt1.8 cm from the center of impact. This also possible reduces the influence of the ion beam on the samples.

The evolution of the plasma after the collapse of the plasma column at the pinch has not been extensively studied, since the attention in this device was focused on the neutron production. An intense ion beam together with the formation of a plasma dome and an additional "bubble" structure above the dome are observed [13]. The ion beam can reach energies in excess of 1 MeV according to different measurements under different filling gases and pressures [14,15], having a broad energy distribution. The angular distribution is also broad reaching angles of 80 degree [15]. There is apparently no information of the ions produced compared to the number of particles in the plasma or a model that could help us estimate the ion flux expected a target position. Specific measurements of the ion distribution or energy have not been performed in the FN II.

Besides the ion energy and flux it is also important to characterize the plasma that arrives at target position. To evaluate this a series of measurements using a triple Langmuir probe (TLP) were performed [13] giving plasma electron temperatures of the order of 65 eV .

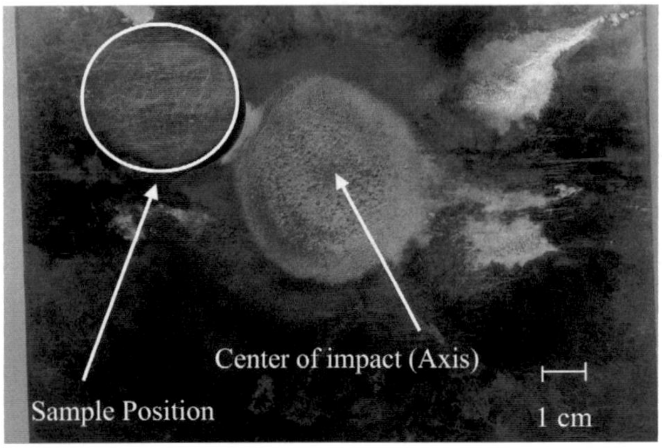

Figure 3. Picture of the deposits on the stainless steel sheet, showing the center of impact and the position of the samples.

2.2 Erosion and melting of tungsten samples

After exposure to 25 shots with 4.8 kJ each with deuterium the surface shows regions with partial melting together with region with erosion. Figure 4 shows SEM micrographs of tungsten samples before and after exposure. The images have both a magnification of 4000X. The image before exposure shows a clear structure of laminated crystallites on the left and a rough surface structure on the right. The image on the right (b) shows a very smooth surface, evidencing erosion. Some of these areas show a black deposit, whose composition according to Energy Dispersive Xray-Spectroscopy (EDS) is essentially iron and chromium, which may come from the clamps used to fix the sample. Higher magnifications of the eroded areas show a smooth surface without voids or cracks even for the highest magnification used: 60000X (not shown here). The exposure to helium plasmas shows a similar picture.

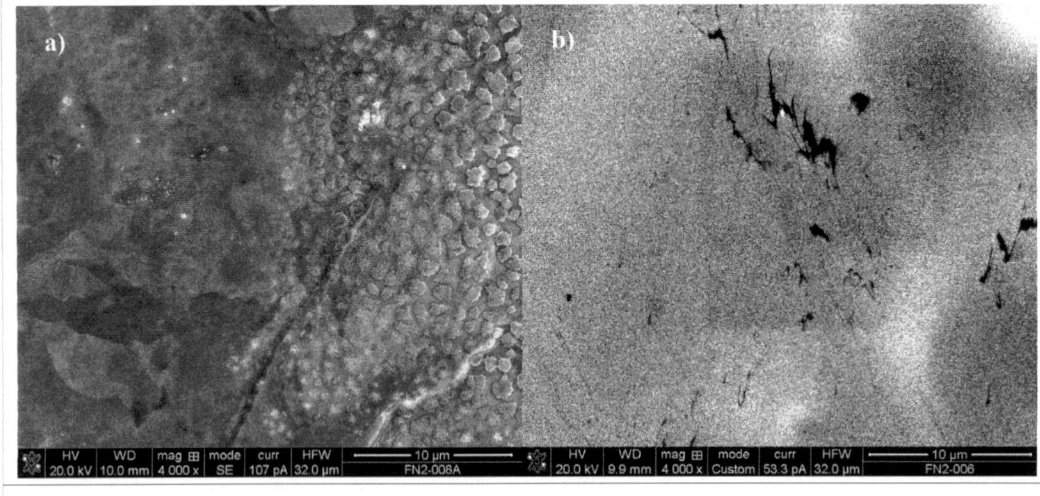

Figure 5. SEM micrographs of tungsten unexposed (a) and exposed (b) to 25 shots deuterium. Magnification 4000X.

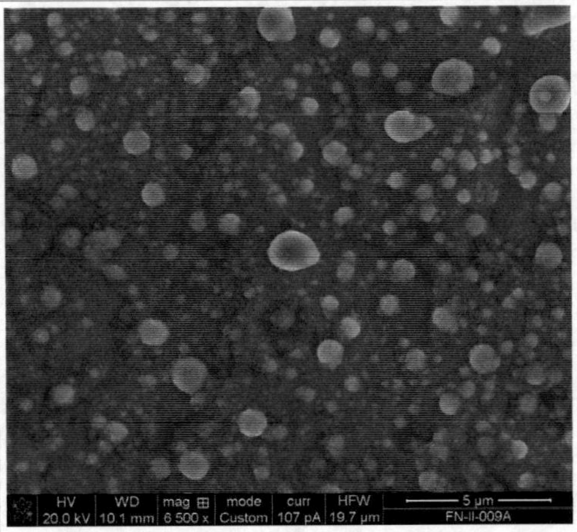

Figure 6. SEM micrographs of a region of the tungsten sample showing the formation of droplets. Magnification 6500 X.

Figure 6 shows a SEM micrograph of a region with the formation of small tungsten droplets. This micrograph has a higher magnification than the previous images. An EDS study (not shown here) demonstrates that the composition of the droplets is essentially tungsten. The droplets have sizes of the order of 1 μm with a broad distribution. Formation of droplets is observed also in the case of helium plasmas.

2.3 Deuterium retention in tungsten

In order to study the retention of deuterium in the tungsten samples, some of them were analysed by means of nuclear reaction analysis (NRA) in an ion beam facility. In this technique ^3He ions ions with energies above 690 keV produce protons and helium ions by means of the nuclear reaction $D(^3He,^4He)p$. As well as ^4He ions and protons are detected in a certain solid angle and account for the amount of deuterium in the sample. The nuclear reaction has a narrow energy window allowing the measurement of a very shallow surface layer of about 250 nm. Increasing the ion energy the penetration is larger and the maximum of the reaction is shifted towards larger depths allowing probing at higher depths, the scale conversion, however, is non trivial [16].

Figure 7 shows the result of the NRA analysis of a tungsten and a titanium sample exposed at the same time to 25 shots of deuterium with 4.8 kJ each. The energy of the probing ions was varied from 690 keV to 2400 keV, to probe into the depth. In tungsten a high amount of deuterium is observed in the near surface region ($6.5*10^{16}$ D/cm^2), falling rapidly into the depth. With a probing energy of 1200 keV the deuterium signal 1/7 of that in the surface. This indicates, that deuterium is retained in a near-surface layer of about 250 nm. Extensive EDS studies have been performed to look for co-deposited species that could be responsible for the deuterium retention. The impurities found [12] mainly iron cannot explain the large amount of retained deuterium. 250 nm agree well with the range of 20 keV deuterium ions in tungsten [12].

In titanium we observe, that a significant amount of deuterium is detected in the near surface region, increasing for probing energies around 1800 keV and falling for higher probing energies. Titanium is a material that forms hydride with deuterium, thus retains well deuterium. The reduced amount in the near surface region evidences that the sample was heated during exposure, thus allowing some of the retained deuterium to de-trap.

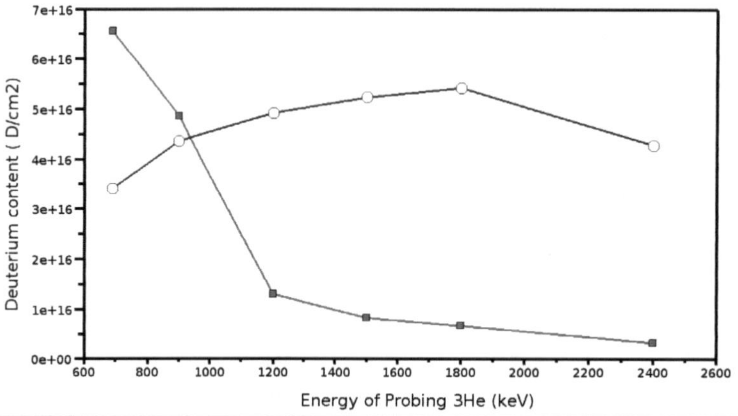

Figure 7. Deuterium content in tungsten (solid squares) and in titanium (open circles) exposed to 25 D shots in FN II according to NRA. The energy of the probing 3He ions was varied from 690 keV to 2400 keV.

3. Modelling of the electrode

Many models have been developed to simulate Plasma Focus discharges [15]. So called "snowplow" models have shown to predict reasonable well current-sheet velocities, however, these models require of a "sweeping" efficiency to fit experimental data. These models as well as other MHD models, coupled with an external circuit equation, can describe the plasma parameter of the run-down phase [15]. We use the "snowplow" model with external circuit equation presented by Lee and Saw [17] since it was applied in the past to the FN II finding good agreement with experiments. This model describes the evolution of the plasma during the run-down phase and the compression phase, but does not consider any dome or bubble structures formed after collapse of the pinch.

Table 1 shows the results of the calculations for selected electrode and pressure parameters. T_{axial} is the plasma temperature in eV reached at the end of the electrode, T_{pinch} the temperature reached in the pinch, C the capacity of the capacitor bank, V its charging voltage, and P the deuterium pressure in Torr. The electrodes dimensions of the first 3 row correspond to the FN II electrode. The details of the calculations are presented elsewhere [18].

In the first row, which reproduces the parameters of our experiments, the axial temperature reaches 126 eV. Assuming, that only this plasma component reaches the target, this is the highest expected temperature at target position. The TLP measurement gave 65 eV (section 2.1). Since the plasma has to travel form the end of the electrode to the target position a distance two times larger than the length of the electrode, it is expected that it gives up some energy to the neutral gas, thus the measured value seems in reasonable agreement with the model calculations. The TLP measurement can have contributions of the plasma of the run-down phase as well as contributions from the dome and bubble structures, that we cannot separate in this moment; we have to assume, that the experimental results are a combination of the three contributions. To confirm the accuracy of the model calculations, accurate measurements of the plasma temperatures in both positions are required, specifically the study of the dome and bubble structures would be needed.

Table 1. Plasma temperatures of the axial (T_{axial}) and radial (T_{radial}) phases according to the snowplow model for discharges using an electrode the dimension given by Inner Radius, Outer Radius and Length, and operated with a capacitor of capacitance C at voltage with a deuterium pressure P in the chamber.

T_{axial} (eV)	T_{pinch} (eV)	Inner Radius (cm)	Outer Radius (cm)	Length (cm)	C (uF)	V (kV)	P (Torr)
126	1300	2.02	5.5	4.8	7.4	38	2
200	2100	2.02	5.5	4.8	7.4	38	1
29	259	2.02	5.5	4.8	7.4	38	10
11	97	3	5	2.5	7.4	10	2

Nevertheless, the measured plasma temperatures at target position is in the desired range of temperatures for PWI studies. The penetration depth of these energies is far less than the penetration depth of deuterium as observed by NRA. This plasma component cannot explained the observed deuterium retention. The temperatures expected at pinch are of the order of 1.3 keV, much higher than the temperatures of the axial phase, but still to low to explain the penetration depths of D in W. The energy of this plasma component is to high to be observed with a TLP, and did not appear in or measurements. A penetration depth of 250 nm can only be caused by deuterium ions in the range of 20 keV. We have to assume that these ions are part of the ion beam produced after the collapse of the pinch. Thus, the main component of the retained deuterium has to come from the ion beam, rather than the plasma projected into the chamber. To take advantage of the axial plasma component, either the pinch has to be inhibited, or the ion beam has to be blocked.

The second row shows the expected plasma temperatures for the same experimental parameters, but with a lower pressure (1 Torr). In his case the axial final temperature is significantly higher than for 1 Torr, the same happens with the pinch temperature. Increasing significantly the pressure (third row) would produce lower plasma temperatures. Another way to obtain lower temperatures would be selecting shorter electrodes with a narrower gap and operating them at lower voltages (fourth row). This parametric study shows, that there is room to select energies suitable for plasma wall interaction varying the operation parameter of the plasma focus or selecting a more suitable geometry. For a fixed geometry, tailoring the operation parameters, a wide range of plasma temperatures, suitable for PWI studies can be produced.

4. Conclusions

The plasma focus has been evaluated as a device for the study of plasma wall interactions. The plasmas of this device, operated with deuterium, produce erosion and retention in fusion relevant materials like tungsten. The plasma at target position has been measured with a triple Langmuir probe and found to be in the range suitable for plasma wall interaction studies. The retention of deuterium is attributed mainly to the ion beam produced after the pinch. For a fixed electrode geometry the temperature of the plasma of the axial phase can be adjusted changing the operation parameters like capacitance and voltage of the voltage source and the pressure in the chamber.

Acknowledgments
This work was partially supported by CONACYT under project 84176 and by IPN under projects SIP20130132 and SIP20140523. The FN-II work was supported by PAPIIT-DGAPA-UNAM IN120409.

5. References

[1] Federici G et al 2003 *Journal of Nuclear Materials* **313–316** 11
[2] Kočan M et al. 2012 *Nucl. Fusion* **52** 023016
[3] van den Berg M A, Brons S, Kruijt O G, Scholten J, Pasquet R, Smeets P H M, Schweer B and De Temmerman G 2011 *Fusion Engineering and Design* **86** 1745
[4] Mather J W 1964 *Phys. Fluids* **7** S28
[5] Filippov N V et al. 1971 *Plasma Phys. Contr. Nuc. Fusion Res.* 573
[6] Imshennik V S, Filippov N V and Filippova T I 1973 *Nucl. Fusion* **13** 929
[7] Pouzo J O and Milanese M M 2003 *IEEE Transactions on Plasma Science* **31** 1237
[8] Castillo F, Herrera J J E, Rangel J, Milanese M, Moroso R, Pouzo J, Golzarri J I and Espinosa G 2003 *Plasma Phys. Control. Fusion* **45** 289
[9] Castillo F, Herrera J J E, Rangel J, Alfaro M, Maza R and Sakaguchi 2003 *Braz. J. of Phy.* **32** 3
[10] Goodfellow Cambridge Ltd, Huntington U.K., Tungsten LS369668
[11] Sigma-Aldrich, Inc. 3050 Spruce Street, St- Louis MO 63103 USA, Titanium 348791-13G
[12] Ramos G 2013 *14th Plas. Facing Materials Conf.* Aachen Germany
[13] Martínez-Fuentes M, Castillo-Mejía F, Herrera-Velázquez J J E, Nieto-Pérez M and Ramos-López G 2012 39 th EPS Conference & 16 th Int. Congress on Plasma Physics P5.160
[14] Bhuyan M, Neog N K, Mohanty S R, Rao C V S and Raole P M 2010 *J Fusion Energ* **29** 177
[15] Bernard A et. al. 1998 *J. Moscow Physical Society* **8** 93
[16] Schmid K and von Toussaint U 2012 *Nuclear Instr. and Meth. in Physics Research B* **281** 64
[17] Lee S and Saw S H 2008 *Journal Fusion Energy* **27** 292
[18] Ramos G 2013 *Superficies y Vacio* submitted

Journal of Physics: Conference Series **591** (2015) 012032 doi:10.1088/1742-6596/591/1/012032

Analysis of swarm coefficients in a gas for bi-modal electron energy distribution model

Gorur Govinda-Raju

Department of Electrical and Computer Engineering
University of Windsor, Windsor, On., N9B 3P4, Canada

Email: raju@uwindsor.ca

Abstract

Cross sections for collision between electrons and neutrals in a gas discharge are essential for theoretical and computational developments. They are also required to interpret and analyze the results of experimental studies on swarm parameters namely drift velocity, characteristic energy, and ionization and attachment coefficients. The cross sections and swarm coefficients are interconnected through the most important electron energy distribution function. The traditional method of solving the Boltzmann equation numerically yields the required distribution (EEDF). However there are many situations where a simpler approach is desirable for deriving the energy distribution analytically. Energy distribution in non-uniform electric fields, in crossed electric and magnetic fields, breakdown in mixtures of gases for electrical power or plasma applications, calculation of longitudinal diffusion coefficients are examples.

In other studies the swarm parameters are employed to derive the cross sections in an unfolding procedure that also involves the energy distribution function. Application of Boltzmann solution method, though more rigorous, consumes enormous efforts in time and technical expertise. In an attempt to provide a simpler method the present author has previously suggested a bimodal electron energy distribution in gases. In this paper the author has generalized the idea of bi-modal energy distribution by considering a model gas with representative cross sections and adopted numerical methods for greater accuracy. The parameters considered are the nature of the two distributions, their relative ratio, and the dependence of cross sections on electron energy. A new method for determining the combination of distributions has been shown to be adequate for calculation of swarm parameters. The results for argon are shown to yield very good agreement with available experimental and theoretical values.

Introduction

Interaction between electrons and neutrals in a gas discharge is studied by swarm methods in which the average properties such as drift velocity, diffusion coefficient, ionization and attachment coefficient are measured experimentally. The controlling parameter in these experiments is the reduced electric field, E/N, in which E is the electric field and N is the gas number density. Alternately, various collision cross sections are measured using mono-energetic electrons and the controlling parameter is the electron energy. The beam experiments and the swarm experiments are interlinked by the electron distribution function. Given the collision cross sections as a function of electron energy the swarm coefficients may be calculated at various values of E/N using the electron energy distribution function.

In other studies, the results of beam measurements are employed to calculate the swarm parameters by an unfolding procedure which involves calculation of electron energy distribution functions. Historically, this approach was developed by Frost and Phelps et al. [1] to obtain momentum transfer cross sections at low energies in H_2.

It is appropriate to mention here that swarm properties at low values of E/N ($\leq$ 10 Td, 1 Td = 10^{-21} Vm^2) are relatively easier to measure than cross sections at low electron energies ($\leq$ 1 eV) though the experimental techniques for cross section measurements employing low energy electrons have been mastered since the pioneering work of Phelps and his group [2]. The interrelation ship between the beam measurements, swarm measurements and electron energy distribution is well known. Substitution of the determined electron energy distribution into integral equations, given below, yields the respective swarm quantity. A large number of publications in the literature show very good correspondence between the theoretically calculated swarm parameters, starting from cross sections and experimentally determined values [3]. However, the unfolding procedures for deriving the cross sections from measured swarm parameters are not quite satisfactory.

The motivation for the present work is to examine whether a simpler approach, to the elaborate Boltzmann equation method that requires a specialization on its own, can be devised for the calculation of the four quantities shown in equations (2)-(5). Such an approach would be beneficial where one wishes to use the results of simpler calculations rather than examine detailed collision processes. The present author has previously suggested a bimodal electron energy distribution and demonstrated that accurate values of swarm parameters could be calculated by this method.

Theory

The electron distribution function, F(ε) is calculated using the Monte Carlo procedure or numerically solving the Boltzmann equation.

The swarm parameters are defined by the following relationships:

$$W = -\frac{E}{3N}\left(\frac{2e}{m}\right)^{1/2}\int_0^\infty \frac{\varepsilon}{Q_M(\varepsilon)}\frac{d}{d\varepsilon}\left[\frac{F(\varepsilon}{\varepsilon^{1/2}}\right]d\varepsilon \tag{1}$$

$$D_T = \frac{1}{3N}\int_0^\infty \frac{\varepsilon}{Q_M(\varepsilon)}F(\varepsilon)d\varepsilon \tag{2}$$

$$\frac{\alpha}{N} = \frac{1}{W}\int_{\varepsilon_i}^\infty \varepsilon^{1/2}Q_i(\varepsilon)F(\varepsilon)d\varepsilon \tag{3}$$

And in general [2],

$$k_j = \left(\frac{2}{m}\right)^{1/2}\int_0^\infty \varepsilon Q_{M(j)}(\varepsilon)F(\varepsilon)d\varepsilon \tag{4}$$

In equations (1)-(4) the symbols are defined as follows:

D_T = radial diffusion coefficient (m^2/s); e/m = electron charge to mass ratio; j = specific inelastic process, k_j = rate constant for j^{th} process; (m^3/s); Q_i = ionization cross section (m^2); Q_M = momentum transfer cross section (m^2); W = drift velocity (m/s); ε = electron energy (eV), ε_i = ionization threshold energy; ε_j = threshold energy for j^{th} process. Substitution of determined electron energy distribution into equations (1) to (4) yields the respective quantity.

Assuming that the cross section is independent of energy Morse, Allis and Lamar [4] derived an analytical equation to the electron energy distribution according to:

$$F(\varepsilon) = G\frac{\varepsilon^{1/2}}{\bar{\varepsilon}^{3/2}}\exp\left[-H\left(\frac{\varepsilon}{\bar{\varepsilon}}\right)^{2(p+1)}\right] \tag{5}$$

where $\bar{\varepsilon}$ is the mean energy, p is a constant having value of $-1 < p$ and no upper limit seems to exist though at large values the distribution tends to be so small that it ceases to have any practical significance. G and H are functions given by:

$$G = 2(p+1)\left[\Gamma\frac{5}{4(p+1)}\right]^{3/2} \times \left[\Gamma\frac{3}{4(p+1)}\right]^{-5/2} \tag{6}$$

$$H = \left[\Gamma\frac{5}{4(p+1)} \times \frac{1}{\Gamma\frac{3}{4(p+1)}}\right]^{2(p+1)} \tag{7}$$

There is now a renewed interest in the problem because of the fact that gas discharges are being used in a number of different areas involving specialists from various disciplines, for example plasma medicine. Towards this end a bimodal energy distribution is proposed, expressed as,

$$G(\varepsilon) = (1-\gamma)F_1(\varepsilon) + \gamma F_2(\varepsilon) \tag{8}$$

Where γ is a constant, $0 \le \gamma \le 1$, F_1 and F_2 are individual distributions respectively and G is the combined distribution, which is normalized according to

$$\int_0^\infty F(\varepsilon)\,d\varepsilon = 1 \tag{9}$$

In this paper the author has generalized the idea of bi-modal energy distribution by considering a model gas with representative cross sections and adopted numerical methods to calculate the swarm parameters using equations (1)–(3) for greater accuracy. The parameters considered are the nature of the two distributions, their relative ratio, and the dependence of cross section on electron energy. A new method for determining the combination of distributions has been shown. The results for argon are shown to yield excellent agreement without the need for scaling factors for cross sections.

Results for argon and discussion

Application of the theory to argon gives extremely satisfactory results for as shown in Figures 1–2 for W, D (Not shown) and the first ionization coefficient α/N respectively. Comparisons with published data for W, D and E/N are very satisfactory.

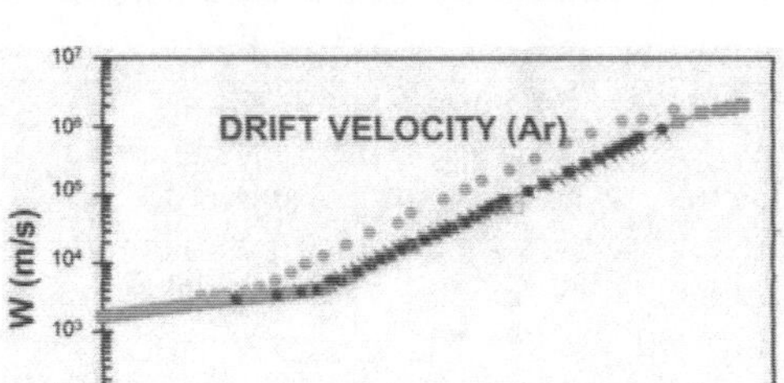

Figure 1. Drift velocity as a function of E/N. (Δ) Kücükarpaci and Lucas [3]; (□) Nakamura and Kurachi [3]; both these sets of symbols are obscured by other points; (•) Lisovskiy et al. [3]; (■) Raju [2, 3]; (×) Boltzmann method; (•) This paper (2014).

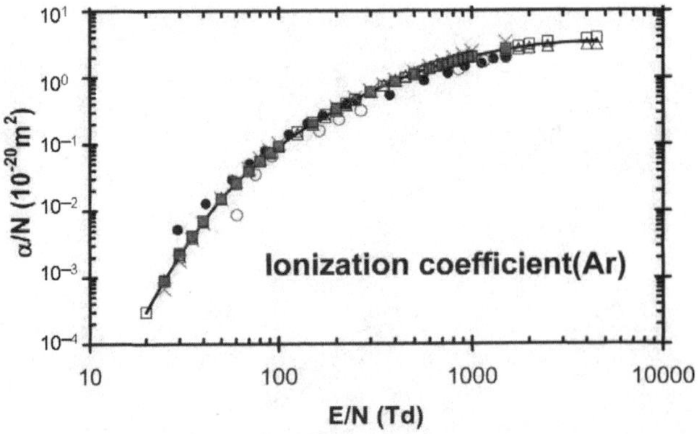

Figure 2. Density reduced ionization coefficients as a function of E/N for argon. (□) Kruithoff [3]-Experimental; (○) Davies and Milne [3]-Experimental; (•) Lakshminarasimha and Lucas [3]-Experimental; (Δ) Phelps and Petrović [3]-analytical expression; (——) Raju [3]; (×) Boltzmann code [5]; (■) This paper (2014)-Theory.

Conclusions

A bi-modal distribution method for calculating the electron energy distribution has been developed and shown to yield extremely satisfactory results for argon. The method is much simpler and does not require detailed knowledge of collision processes. The semi-empirical energy distribution function is independent of the momentum transfer cross section though its influence is reflected through the mean energy. An interactive spread sheet gives the results with no need for an iterative procedure and it is expected that the method will be tested for other gases to verify its general applicability. The EEDF can be expressed analytically which is a major advantage in many applications.

Acknowledgement

The results in the paper for Boltzmann distribution are obtained by using the freeware www.Bolsig. Com.

References

[1] Frost, L. S. and A. V. Phelps, "Rotational Excitation and Momentum Transfer Cross Sections for Electrons in H_2 and N_2 from Transport Coefficients," Phys. Rev., 127, 1621, 1962.
[2] Raju, G. G. "Gaseous Electronics-Theory and Practice" Taylor and Francis Group LLC., Boca Raton (Florida), 2006.
[3] Raju, G. G. "Gaseous Electronics-Tables, Atoms, and Molecules," Taylor and Francis Group LLC., Boca Raton (Florida), 2012. Data for over 500 target particles are compiled.
[4] Morse B. M., W. P. Allis and E. S. Lamar, "Velocity Distributions for Elastically Colliding Electrons," Phys. Rev. 48, 412, 1935.
[5]www. bolsig.com, Also see G. J. M. Hagelaar, and L. C. Pitchford, Plasma Sources Sci. Technol., Vol. 14, 722, 2005.

Conjugate influence of current relaxation and of current-vortex sheet formation on the magnetorotational instability

F E M Silveira[1] and R M O Galvão[2]

[1] Centro de Ciências Naturais e Humanas, Universidade Federal do ABC,
Rua Santa Adélia, 166, CEP 09210-170, Bairro Bangu, Santo André, SP, Brazil
[2] Instituto de Física, Universidade de São Paulo,
Rua do Matão, Travessa R, 187, CEP 05508-090, Cidade Universitária, São Paulo, SP, Brazil

E-mail: francisco.silveira@ufabc.edu.br

Abstract. The conjugate influence of current relaxation and of current-vortex sheet formation on the magnetorotational instability is reviewed as firstly derived in [13]. It is shown that the relative amplification of the magnetic viscosity from marginal stability to the instability determined by the maximum growth rate is around 924% when resistive effects dominate, while the corresponding quantity is around 220% in the ideal limit. This means that the conjugate influence is much more efficient to amplify the magnetic viscosity than just the effect due to the standard magnetic tension. The results presented here may contribute to the understanding of the various processes that play a significant role in the mechanism of anomalous viscosity observed in Keplerian disks. It is argued that the new effect shall be most relevant in thin accretion disks.

1. Introduction

The magnetorotational instability was first dealt with by Chandrasekhar [1, 2] and Velikhov [3] about half a century ago. It did not attract much attention from the community of plasma physics as a whole in the immediately following decades, except from a few interested in its more basic aspects [4, 5, 6]. However, since the work of Balbus and Hawley [7] about twenty years ago, the magnetorotational instability became a popular starting point for several investigations, mainly in astrophysical plasmas. Actually, thin disks of conducting gases are frequently observed to rotate about forming stars or in binary systems, where they are subjected to stelar fields. They may be found at the center of galaxies as well, where they can be very luminous due to the action of galactic fields. For instance, a conducting disk rotating about a compact object (a neutron star or a black hole) may originate a quasar (an intense source of X-ray) [8]. Such systems are known as accretion disks. Early investigations of the magnetorotational instability were pursued on the assumption of a perfectly conducting fluid (neglecting resistivity). The studies evolved rapidly to include finite resistive effects since they can contribute to the effective viscosity of a conducting fluid. Given the difficulty to deal analytically with the resulting dispersion relation, the first approaches to the problem were conducted numerically [9]. Another important motivation to take the influence of a finite resistivity into account is that the magnetorotational instability can be observable in laboratory plasmas (liquid metals) whose unperturbed motion is determined by a Couette flow [10]. Actually, there have been some interesting attempts to tackle the problem analytically, although the essence of the physics remains somewhat obscured [11, 12]. This work reviews how the assumption of the formation of a current-vortex sheet modifies the description of the magnetorotational instability in a conducting fluid as firstly derived in [13]. Particularly, since we describe the dynamics of thin accretion disks and the natural length scale of the problem is given by the characteristic wavelength of the perturbative fields along the width of the system (negleting compressive effects), we include more rich physical processes. Actually, we require the influence of the inertia of charged species to be also as important. In this case, the current density flowing in the plasma relaxes typically in a finite time scale. Such a situation implies a time-dependent correction to Ohm's law. As a result, we may state that this work presents the conjugate effect of current relaxation and of current-vortex sheet formation on the magnetorotational instability in a conducting fluid.

2. Dispersion relation

Let us start by considering a fluid element describing a circular orbit with angular speed Ω. We allow Ω to be a function of the radius R of the circumference. The centripetal acceleration of the fluid element is $-R\Omega^2$, the minus sign indicating its direction toward the center of the circle. For instance, if this acceleration is of gravitational origin, due to a particle of mass M located at the center, then it must be identified with $-GM/R^2$, where G is Newton's constant. We regard now small departures from the circular orbit due to some disturbing process. Let us attach the origin of Cartesian coordinates (x, y) to the unperturbed orbit determined by the radius, say, R_0. At the initial instant, the actual orbit is determined by the distance $R = R_0 + x$, from the center of the unperturbed orbit, and the fluid element moves with an angular speed $\Omega(R)$. We assume that x increases in the same direction of the radius of the unperturbed orbit and that y increases in the same direction of the unperturbed angular speed $\Omega(R_0) \equiv \Omega_0$. Needless to say, the conditions $x \ll R_0$ and $y \ll R_0$ must be satisfied. As time evolves, x and y are determined from the coupled equations $\ddot{x} - 2\Omega_0\dot{y} + R\Omega^2 - R\Omega_0^2 = -\omega_0^2 x - \beta\dot{x}$ and $\ddot{y} + 2\Omega_0\dot{x} = -\omega_0^2 y - \beta\dot{y}$, where a dot denotes total time derivative. The terms $-2\Omega_0\dot{y}$ and $2\Omega_0\dot{x}$ are the x and y components, respectively, of the Coriolis acceleration, $R\Omega^2$ is the centrifugal acceleration, and ω_0 and β denote time rates. By Taylor expanding Ω^2 about R_0, we get $\Omega^2(R_0 + x) \simeq \Omega^2(R_0) + x\,[\Omega^2]'(R_0)$, where the prime denotes a total derivative with respect to R. Then, by assuming that x and y depend on time as $\sim e^{\gamma t}$, we obtain the dispersion relation

$$\gamma^4 + \left[2\left(\omega_0^2 + \beta\gamma\right) + \kappa^2\right]\gamma^2 + \left[\left(\omega_0^2 + \beta\gamma\right) - 2q\Omega_0^2\right]\left(\omega_0^2 + \beta\gamma\right) = 0, \qquad (1)$$

where we have introduced the usual epicyclic frequency κ and shear parameter q, which are defined through the formulae $\kappa^2 = 4\Omega_0^2 + R_0\,[\Omega^2]'(R_0)$ and $-2q\Omega_0^2 = R_0\,[\Omega^2]'(R_0)$.

3. Current relaxation and current-vortex sheet

In the presence of a magnetic field $\vec{B}$, a (electrically neutral) perfectly conducting fluid (neglecting resistivity), moving with velocity $\vec{v}$, shields itself against the Lorentz force per unit charge of carriers, $\vec{v} \times \vec{B}$, by producing the internal electric field $\vec{E} = -\vec{v} \times \vec{B}$. But, if heating processes cannot be neglected, this ideal limit does not provide a satisfactory description of the plasma dynamics. Then, the usual approach to circumvent this issue is to simply add the term $\eta\vec{J}$ to $-\vec{v} \times \vec{B}$ on the right hand side of the equation above, where η denotes the resistivity and $\vec{J}$, the current density. However, even this standard procedure fails to provide a satisfactory description of dissipative phenomena occurring at sufficiently fast rates in the plasma. The reason is that it completely ignores inertial effects due to charged species. Actually, when inertial effects are taken into account, the balance of field inside the plasma shall be determined appropriately through [14, 15]

$$\vec{E} - \eta\tau\partial_t\vec{J} = -\vec{v}\times\vec{B} + \eta\vec{J}, \tag{2}$$

where τ is a time constant. Note that Eq. (2) contains the two opposite limiting situations mentioned above. On one hand, in the perfect conductor approximation (the resistivity η becomes vanishingly small), the ideal law $\vec{E} = -\vec{v}\times\vec{B}$ is recovered. On the other hand, if $\vec{B}$ and $\vec{E}$ are suddenly removed from the presence of the conducting fluid, (of course, provided that η is finite) then $\vec{J} = \vec{J}_0 e^{-t/\tau}$. This means that any initial current $\vec{J}_0$ dissipates in a time scale of the order of τ inside the plasma. Thus, τ must be interpreted properly as the relaxation time of the current density. If τ were ignored, then $\vec{J}_0$ would quench instantaneously inside the plasma. Clearly, such a situation could be justifiable only if the other rates that are relevant to the problem were slow enough. Thus, inertial effects due to charged species should be negligible. As it appears, Eq. (2) provides an extension of Ohm's law for a resistive plasma, $\vec{E} = -\vec{v}\times\vec{B} + \eta\vec{J}$, when fast enough rates are involved. Additionally, let us assume that the current density $\vec{J}$ can be expressed in terms of the usual flow vorticity $\nabla\times\vec{v}$ through

$$\vec{J} = -\imath a\left(\frac{\rho_0}{\mu_0}\right)^{1/2}\nabla\times\vec{v} + \nabla f, \tag{3}$$

where μ_0 denotes the vacuum magnetic permeability and ρ_0, the (constant and uniform) mass density, and f is a (generally complex) function and $a > 0$, a constant. Note that if f is uniform, then $\vec{J}$ and $\nabla\times\vec{v}$ become aligned. In any case, f is a solution of Laplace's equation. Notice also that a measures the strength of $\vec{B}$ with respect to that of $\vec{v}$. Finally, we remark that $\vec{v}$ lags $\vec{B}$ in time by $\pi/2$. An expression such as Eq. (3) is known as a (local) current-vortex sheet.

4. Magnetorotational instability

By assuming that during an infinitesimal interval of time δt the resulting infinitesimal displacement $(\vec{v})\,\delta t$ and magnetic field $\left(\partial_t\vec{B}\right)\delta t$ both vary harmonically in space as $\sim e^{\imath kz}$, the combination of Eqs. (2) and (3) with Maxwell's equations imply the term between parentheses in Eq. (1) may be expressed as

$$\omega_0^2 + \beta\gamma = k^2 v_A^2\left[\frac{1 + (kv_A T)(a)}{1 + (kv_A T)(kv_A\tau)}\right], \tag{4}$$

where we have introduced the usual Alfvén speed $v_A = B_0/\sqrt{\mu_0\rho_0}$ and time $T = \eta\rho_0/B_0^2$ of decay of the flow on the plane (x,y) perpendicular to the equilibrium magnetic field $\hat{z}B_0$ [16]. In astrophysics, one is often concerned with a conducting gas, which is supported by a differential rotation of angular speed Ω_0 against the gravitational attraction to a central object of mass M. The identification of the centripetal acceleration as being of gravitational origin leads to $\Omega_0^2 = GM/R_0^3$, at a distance R_0 from the object, where G is Newton's constant. Such a result is nothing but Kepler's third law. The system is known as a Keplerian disk. According to Sec. 2, the shear parameter $q = 3/2$. Since $q > 0$, if the rotating gas is subjected to a perpendicular constant and uniform magnetic field, then Eq. (1) shows that $\gamma > 0$ and the disk becomes magnetorotationally unstable, provided that the condition $\omega_0^2 + \beta\gamma < 3\Omega_0^2$ is satisfied in accordance with Eq. (4). An useful notion in the study of plasma instabilities is that of marginal stability, the (still stable) state of the fluid which is determined by the minimum value attained by the growth rate (actually, by requiring that $\gamma = 0$). According to Eq. (1), a Keplerian disk is marginally stable if the condition $\omega_0^2 + \beta\gamma = 3\Omega_0^2$ is satisfied. In this case, such a condition implies the wavelength $\lambda = 2\pi/k$ is given by

$$\lambda_{\text{id}}(\gamma = 0) = \frac{2\pi v_A}{\Omega_0\sqrt{3}}, \qquad \lambda_{\text{res}}(\gamma = 0) = \frac{2\pi a v_A}{3\Omega_0^2\tau}, \tag{5}$$

where the suffix id (res) stands for ideal (resistive). Needless to say, the opposite limiting situation for which the growth rate acquires its possible maximum value should be of interest as well. In accordance with Eq. (1), such a value is $\gamma = 3\Omega_0/4$, which occurs if the condition $\omega_0^2 + \beta\gamma = 15\Omega_0^2/16$ is satisfied. In this case, such a condition leads to

$$\lambda_{\text{id}}\left(\gamma = 3\Omega_0/4\right) = \frac{8\pi v_{\text{A}}}{\Omega_0\sqrt{15}}, \qquad \lambda_{\text{res}}\left(\gamma = 3\Omega_0/4\right) = \frac{32\pi a v_{\text{A}}}{15\Omega_0^2\tau}. \tag{6}$$

We arrive now at the main result of this work. As mentioned in [16], if L is a characteristic length scale of the disk, then $L^2B_0^2/\eta$ may be interpreted as the magnetic viscosity, where B_0 denotes the strength of the magnetic field in the state of equilibrium of the gas. Thus, by assuming that $L \sim \lambda$, Eqs. (5) and (6) show that the relative amplification of the magnetic viscosity from marginal stability to the instability determined by the maximum growth rate is given by

$$\frac{\lambda_{\text{id}}^2\left(\gamma = 3\Omega_0/4\right) - \lambda_{\text{id}}^2\left(\gamma = 0\right)}{\lambda_{\text{id}}^2\left(\gamma = 0\right)} = 220\%, \qquad \frac{\lambda_{\text{res}}^2\left(\gamma = 3\Omega_0/4\right) - \lambda_{\text{res}}^2\left(\gamma = 0\right)}{\lambda_{\text{res}}^2\left(\gamma = 0\right)} = 924\%. \tag{7}$$

Quite interestingly, we see that the conjugate influence of current relaxation and of current-vortex sheet formation is much more efficient to amplify the magnetic viscosity than just the effect due to the standard magnetic tension in the ideal limit.

5. Conclusion

The main result of this work is expressed by the contrast of Eqs. (7). The former provides the relative amplification of the magnetic viscosity from marginal stability to the instability determined by the maximum growth rate in the ideal limit and the latter, the corresponding quantity when resistive effects dominate. We have found that the conjugate influence of current relaxation and of current-vortex sheet formation is much more efficient to amplify the magnetic viscosity than just the effect due to the standard magnetic tension in the ideal limit. The results presented here may contribute to the understanding of the various processes that play a significant role in the mechanism of anomalous viscosity observed in Keplerian disks. Since the wavelength λ is in the direction perpendicular to the plane of rotation of the gas and inertial effects are important at short wavelengths, the new effect shall be appreciable in thin accretion disks.

References

[1] Chandrasekhar S 1953 *Proc. Roy. Soc. (London) A* **216** 293
[2] Chandrasekhar S 1981 *Hydrodynamic and Hydromagnetic Stability* (New York: Dover)
[3] Velikhov E P 1959 *J. Exp. Theor. Phys. (USSR)* **36** 1398
[4] Fricke K 1969 *Astron. Astrophys.* **1** 388
[5] Acheson D J and Hide R 1973 *Rep. Prog. Phys.* **36** 159
[6] Acheson D J and Gibbons M P 1978 *Phil. Trans. Roy. Soc. London Ser.: A* **289** 459
[7] Balbus S A and Hawley J F 1991 *Astrophys. J.* **376** 214
[8] Prendergast K and Burbidge G R 1968 *Astrophys. J. Lett.* **151** L83
[9] Balbus S A and Hawley J F 1998 *Rev. Mod. Phys.* **70** 1
[10] Ji H, Goodman J and Kageyama A 2001 *Mon. Not. R. Astron. Soc.* **325** L1
[11] Lominadze J, Pashitskii E, Mikhailovskii A, Kharshiladze O, Churikov A and Avsarkisov V 2008 *Bull. Georg. Natl. Acad. Sci.* **2** 48
[12] Pessah M E and Chan C 2008 *Astrophys. J.* **684** 498
[13] Silveira F E M and Galvão R M O 2013 *Phys. Plasmas* **20** 082126
[14] Goedbloed H and Poedts S 2004 *Principles of Magnetohydrodynamics with Applications to Laboratory and Astrophysical Plasmas* (Cambridge: Cambridge)
[15] Spitzer L 2006 *Physics of Fully Ionized Gases* 2nd ed. (New York: Dover)
[16] Jackson J D 1975 *Classical Electrodynamics* 2nd ed. (New York: Wiley)

Journal of Physics: Conference Series **591** (2015) 012034 doi:10.1088/1742-6596/591/1/012034

Kinetic plasma instabilities due to charge exchange and elastic collisions

M. Horký[1,2], W. J. Miloch[3]

[1] Department of Physics, Faculty of Electrical Engineering, Czech Technical University in Prague, Technická 2, 166 27 Prague, Czech Republic

[2] Astronomical Institute, Czech Academy of Sciences, Boční II 1401/1a, 141 00 Prague, Czech Republic

[3] Department of Physics, Univeristy of Oslo, P. O. Box 1048 Blindern, N-0316 Oslo, Norway

E-mail: horkymi1@fel.cvut.cz; w.j.miloch@fys.uio.no

Abstract.
With numerical particle-in-cell simulations, we study kinetic plasma instabilities induced by collisions between plasma particles and neutral background in magnetized drifting plasmas. We consider the role of charge exchange as well as elastic collisions in the evolution of the system. Charge exchange collisions can give rise to velocity distributions in the form of loss-cone or ring shaped distributions that can become linearly unstable. Elastic collisions also lead to the instability, but in this case the principal mechanism may be attributed to the generalized two-stream instability. We investigate the growth rates and saturation levels for instabilities associated with these collisional processes, and find higher saturation levels and stronger fluctuations for the case with charge exchange collisions. Characteristics of the studied system are similar to the E- and F-regions of the Earth's ionosphere. Our results are relevant for explaining some of the low frequency oscillations observed in the lower parts of the Earth's ionosphere, and are also relevant for some laboratory experiments.

1. Introduction
In the lower parts of the Earth's ionosphere, the plasma collisional processes can dominate the plasma dynamics. In those regions, the collision rates between the charged particles and neutrals can surpass the rates of collisions between charged particles [1]. In general, the frequency of collisions with neutrals strongly depends on the altitude, and this process is mostly relevant for the ionospheric E-region, which extends above ca. 100 km. In the lower E-region, the ions can become effectively unmagnetized, which can give rise to the two-stream like instabilities [1, 2]. In the upper E-region and lower F-region, all charged particles are magnetized, and the plasma is weakly collisional, and is in general drifting due to external electric and magnetic fields in the $\mathbf{E} \times \mathbf{B}$ direction. In those regions, the collisions may significantly modify the ion velocity distribution functions. Similar conditions can be found in dedicated laboratory experiments [3, 4].

Distorted ion velocity distribution functions can lead to different kinetic microinstabilities. For example, charge-exchange collisions can give rise to loss-cone or ring shaped velocity distribution functions that can become linearly unstable. These distributions will generally not be cylindrically symmetric [5, 6]. Elastic collisions can, on the other hand, distort the $\mathbf{E} \times \mathbf{B}$

drift of the ions, and in a limiting case, give rise to a generalized two-stream instability [7]. Since the instability origin is often due to distorted velocity distribution functions, one should use a kinetic approach or first-principle particle simulations to address this problem.

Different ion-neutral collision types can thus lead to distinct instability regimes. In this paper, we focus on two limiting cases, where we consider charge exchange or alternatively elastic scattering for ions in two-component plasmas. We address the problem with a self-consistent particle-in-cell code that accounts for collisions. By simulating weakly collisional plasmas, we investigate nonlinear plasma dynamics with the emphasis on the question on how the collision type affects instability characteristics such as the instability growth, level of the electrostatic potential fluctuations, and the potential distribution.

2. Numerical Model

Our analysis is carried out with a three-dimensional, electrostatic particle-in-cell (PIC) code. We simulate the dynamics of electrons and ions in self-consistent fields in a periodic system, and account also for external electric and magnetic fields, as well as collisions with a neutral background gas. The code is based on our previous PIC codes, with details on its numerical implementation given in earlier works [5, 8]. Collisions between plasma particles and neutral background are implemented with the null-collision method, which allows for arbitrary, energy dependent collision cross-sections. In the present work, however, we choose to use a constant collision frequency ν in order to focus on the basic physical mechanisms associated with a given collision type. We account for charge exchange collisions or alternatively elastic collisions for ions. In both cases the electrons experience elastic collisions.

The simulated system is a box of size of 0.5 m in each direction, and spatial grid resolution of 0.7 cm. Initially, the system is spatially homogeneous. The plasma density is $n = 10^{12}$ m^{-3}, with the electron temperature $T_e = 8.6$ eV, and the electron to ion temperature ratio $T_e/T_i = 4$. The external magnetic and electric fields are $B_0 = 0.005$ T and $E_0 = 550$ Vm^{-1}. The $\mathbf{E} \times \mathbf{B}/B^2$ drift v_d is supersonic, with $v_d = 2\sqrt{T_e/M}$, where M is the ion mass. The ion collision frequency is $\nu_{in} = 3.52 \cdot 10^5$ s^{-1}, which is lower than the ion gyrofrequency Ω_{ci}, $\nu_{in} = \Omega_{ci}/5$, while the electron collision frequency is given by $\nu_{en} = \nu_{in}\sqrt{M/m}$, where m is the electron mass. For computational reasons, we use the reduced mass ratio $M/m = 500$, while keeping the electron mass realistic, and the mass of cold neutral species $m_n = M$. This reduced mass ratio speeds up the simulations, while it still gives credible results [9]. We typically run large-scale simulations for times up to $t = 95000\Delta t$, which corresponds to 19 ion gyroperiods with a timestep Δt being a fraction of the electron gyroperiod.

The simulated plasma parameters can be related to scaled conditions in the upper parts of the ionospheric E-region, where the plasma is only weakly collisional [1, 2, 5]. However, while the respective ratios of the parameters are within the relevant range, the one-to-one correspondence is not maintained; rather the scaled system is simulated. Thus, while we expect that the main phenomena will be present in the simulations, some of the ionospheric processes might be not well represented.

3. Results and Discussion

In the simulated system, the charge exchange collisions lead to an asymmetric ring distribution function for ion velocities, which after one ion gyroperiod, when the full ring shape velocity is formed, becomes unstable and triggers the instability [5]. This instability mechanism is equivalent to the loss-cone instability discussed in the context of fusion research [2, 10]. The instability grows during a few ion gyroperiods until the ring in the distribution function is filled-in in the velocity phase-space. It is characterized by an enhanced level of fluctuations in the electrostatic potential distribution, and the growth of harmonics in the wave spectra. A partially filled ion velocity distribution at a later stage is shown in Figure 1(a). In Figure 2 we

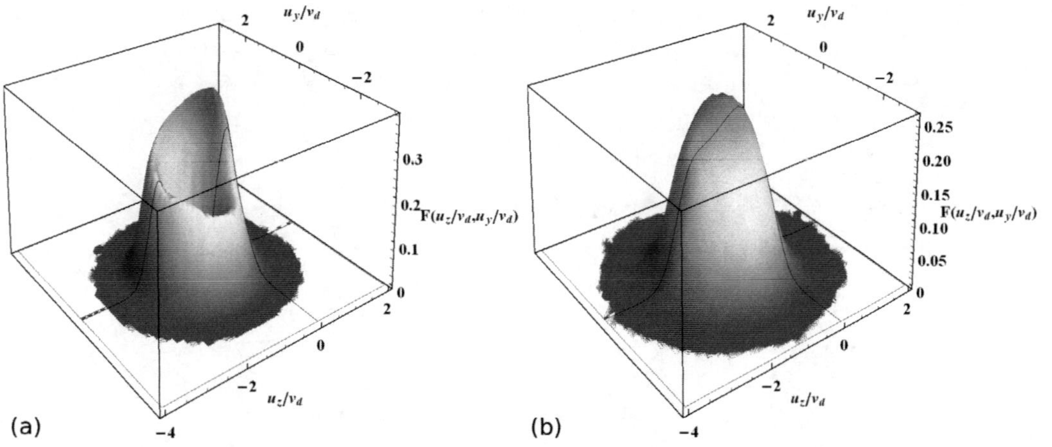

Figure 1. Ion velocity distribution function for the charge exchange (a) and elastic (b) collisions at $t = 4 \cdot 2\pi/\Omega_i$.

show the absolute value of potential fluctuations as a function of time for different collision types averaged over a chosen subset of grid points in the simulated system. The average amplitude of potential fluctuations in the case of charge exchange collisions increases by a factor of three during the simulation, and is characterized by strong fluctuations in the saturated stage with amplitudes up to seven times the initial value. For elastic scattering the growth rate is similar as in the charge-exchange case, while the saturation level is lower, being characterized by much smaller fluctuations. We observe that the elastic scattering modifies the ion-velocity distribution function by strongly broadening the distribution, see Figure 1(b). Thus, the mechanism for the instability will be different than in the case of charge exchange collisions, where the typical loss-cone instability due to non-Maxwellian distribution can be expected [6]. Moreover, in this nonlinear regime there may also be a competition between several other instabilities, thus the instability due to collisions might act to only modulate other instabilities present in the system [1, 11].

The potential structures in the saturated stage of instability are strongly aligned with the **B**-field. The long wavelengths along **B** can be crucial for the dynamics of the system [1, 12]. However, to study waves at small wavelengths, it suffices to consider only the plane perpendicular to the **B**-field direction. In Figure 3, the potential distribution in the plane perpendicular to the direction of magnetic field is shown for charge exchange collisions, together with the corresponding potential distribution at the beginning of the simulations and at the onset and saturated stage of instability. Initially, the potential distribution is normal, with the maximum values of ±5 V, and there are no coherent structures observed, see Figure 3(a). Coherent structures start forming during the onset of the instability, and in the saturated stage large, coherent structures propagating in the $\mathbf{E} \times \mathbf{B}$ direction are observed (i.e., in the negative $\hat{z}$ direction, see Figures 3b-c). The angle of propagation of these structures with respect to the $\mathbf{E} \times \mathbf{B}$ direction can be related to the supersonic drift regime. As the amplitudes of the fluctuations increase, the potential distribution becomes more shallow, but can still be approximated by a normal distribution.

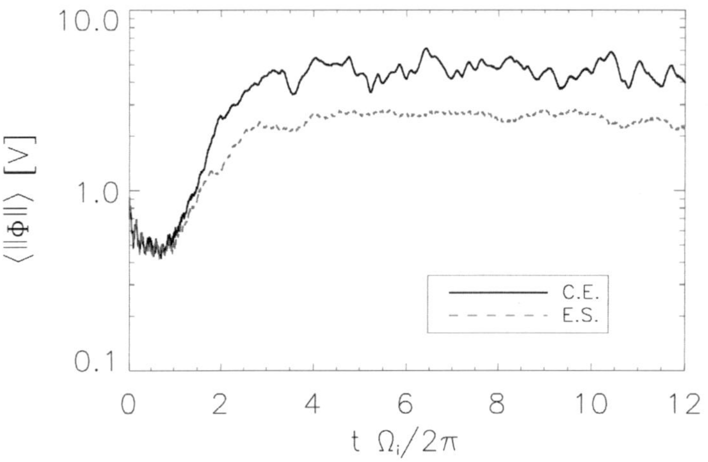

Figure 2. The amplitude of the potential fluctuations $||\Phi||$ averaged over a part of the simulation box as a function of time for elastic (dashed line), and for charge exchange (full line) ion-neutral collisions. Time is normalized to the ion gyroperiod.

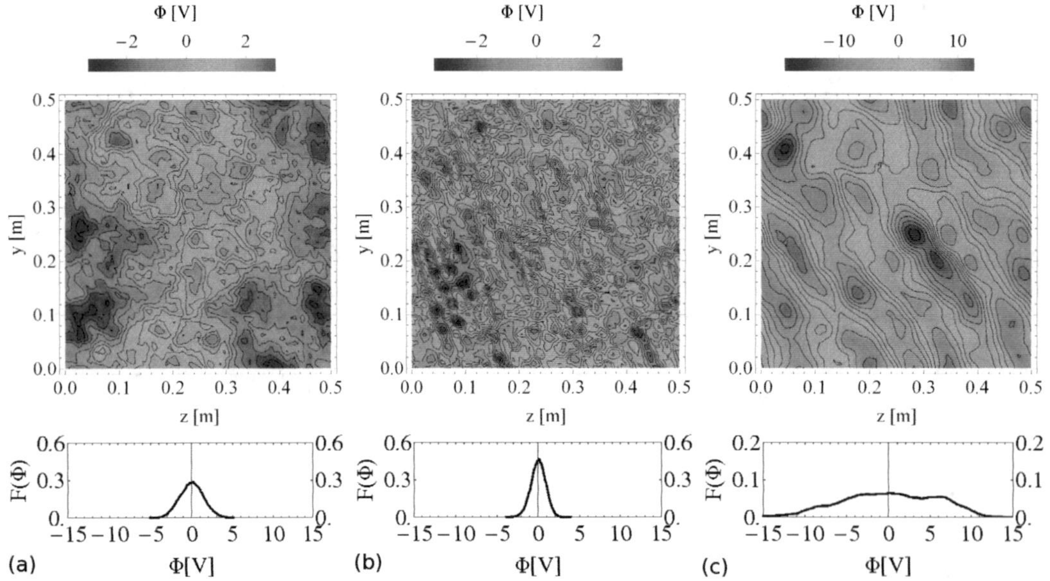

Figure 3. The potential distribution in the plane perpendicular to the direction of the magnetic field in the middle of the box together with the corresponding distribution of potential fluctuations in the whole system. The $\mathbf{E} \times \mathbf{B}$-drift is in the negative $\hat{z}$-direction. The results shown are for charge exchange collisions in the beginning of the simulations (a), at the onset of instability (b), and for the developed stage of the instability (c). Note that the range of the $F(\Phi)$-axis in (c) is reduced.

4. Conclusions

With numerical PIC simulations, we have studied the kinetic plasma instabilities in the $\mathbf{E} \times \mathbf{B}/B^2$-drifting weakly collisional plasma, with the parameter regime that can be relevant for ionospheric conditions. Two limiting cases have been considered: charge exchange and elastic collisions. For the case with charge exchange collisions, a significant saturation level for the amplitudes in the potential fluctuations is reached after a few ion gyroperiods. Elastic collisions give rise to much smaller potential fluctuations and a lower saturation level. While the ion velocity distributions are different for these two collision types, the instability growth rate is the same for both cases. It is expected that in this nonlinear regime there is a competition between several instabilities, and thus further detailed analytical and numerical studies are required to identify different competing and dominant processes in the evolution of this system.

Acknowledgments

M.H. acknowledges support by the CTU grants SGS12/181/OHK3/3T/13 and SGS13/194/OHK3/3T/13 that allowed him to carry out this research during his stay at the University of Oslo. This work is also a part of the 4DSpace Strategic Research Initiative at the University of Oslo.

References

[1] Kelley M C 2009 *The Earth's Ionosphere: Plasma Physics and Electrodynamics* (Amsterdam: Elsevier)
[2] Treumann R A and Baumjohann W 2001 *Advanced Space Plasma Physics* (London: Imperial College Press)
[3] Saito S, Sato N and Hatta Y 1964 *Appl. Phys. Lett.* **5** 467
[4] Mikkelsen T, Pécseli H L 1980 *Phys. Lett. A* **77** 15962
[5] Miloch W J, Pécseli H L and Trulsen J K 2013 *Plasma Phys. Control. Fusion* **55** 124006
[6] St-Maurice J-P and Schunk R W 1979 *Rev. Geophys.* **17** 99134
[7] Farley D T 1963 *J. Geophys. Res.* **68** 60836097
[8] Miloch W J, Kroll M and Block D 2012 *Phys. Plasmas* **17** 103703
[9] Miloch W J, Pécseli H L and Trulsen J 2007 *Nonlin. Processes Geophys.* **14** 575
[10] Teller E 1981 *Fusion Vol 1, Magnetic Confinement Part A* (New York: Academic Press)
[11] Hasegawa A 1975 *Plasma Instabilities and Nonlinear Effects* (Berlin: Springer-Verlag)
[12] Dimant Y S and Milikh G M 2003 *J. Geophys. Res.* **108** 1350

Simulation of ultra-relativistic dynamics of electrons in the interaction of a short-intense laser pulse with plasma

N O Ramirez-Soto[1] and R Ondarza-Rovira[2]

[1]Universidad de Sonora, Depto. Física, C.P. 83000, Hermosillo, Sonora, Mexico
[2]Instituto Nacional de Investigaciones Nucleares, A.P. 18-1027, México 11801, Distrito Federal, Mexico

E-mail: [1]naharai.ramirez@gmail.com , [2]ricardo.ondarza@inin.gob.mx

Abstract. This paper presents numerical simulations of the interaction of an ultra-intense laser field with a single electron and studies the nonlinear plasma effects produced when an ambient harmonic restoring force is considered during the relativistic acceleration of the particle. The motion of a charged particle in a strong laser pulse of varying wave form and polarization is explored. Mainly, the objective is to study how linear polarization leads to different energy distributions for electrons stripped from atoms by the laser field. The numerical solutions that make use of the variable-step Runge-Kutta technique are presented and describe the relativistic dynamics of a single charged particle in plasma, through the relativistic equations of motion. A harmonic restoring force is used as a simple model to simulate charge separation effects and the generation of electrostatic forces during the laser field interaction in a plasma. The motion of the relativistic electron is discussed at length considering different initial conditions for time-varying pulse intensities. The dynamics of particles at rest are studied before the interaction with the laser pulse. In addition, the time dependent longitudinal momentum for various pulse intensities is compared after the interaction with the laser for a range of optical cycles. Specifically, we focus on the time dependence of the longitudinal momentum with linear polarization for different laser intensities. In addition to the longitudinal position and momentum, the effect of the residual momentum and energy of the electrons in the plasma is carefully studied. It was found a strong dependence of the residual momentum on plasma density and laser intensity for a number of ultra-relativistic laser-plasma scenarios considered

1. Introduction

A highly intense laser pulse that interacts with matter can stimulate the atoms leading to ionization [1-2]. Those electrons –the so-called photoelectrons- are due to a number of light ionization mechanisms. The incident light can provide the photoelectrons with a high energy that can be measured in the wake of a laser field [3]. It is possible to analyze the energy spectrum of such particles that after being detached from the atom attain different energy states and interact with the laser [4]. Depending on the polarization, the photoelectrons will have a wide range of energies [5]. Also, the polarization leads to preferential ionization in which the electrons in the atoms are separated from the atom at points in which the electromagnetic radiation is highly intense [6]. For circular polarization it is found no preferential points of ionization in the optical cycles of the light [7]. The different polarizations of light lead to differing dynamics for the charged particles in the wake of the plasma. Another factor of consideration is the intensity of the light pulse which clearly has an effect on the dynamics of photoelectrons. Although the relativistic dynamics of charged particles in plane-wave laser fields can be solved exactly (as demonstrated by Sarachik and Schappert [8]) for particles at rest prior to the arrival of the pulse, in this paper, the general solution is found by means of numerical simulation. In this paper, the electrons are considered to be at rest before the arrival of the pulse, but studies have been conducted in which the velocity is non-zero [9]. Thus, it would be possible to study the relativistic dynamics of photoelectrons after being stripped from the atoms by a strong laser pulse.

2. Exact solution for a single electron in a strong plane wave

The relativistic equations of motion can be solved exactly for a vector field of the form [7]

$$\mathbf{A}(\eta = \eta_0) = \mathbf{A_0}. \tag{1}$$

Specifically, the form of the pulse is given by

$$\mathbf{A}(\eta) = A_p P(\eta) \left[x\delta cos\, \eta + y\sqrt{1-\delta^2} sin\, \eta\right]. \tag{2}$$

Here, the phase is given by

$$\eta = \omega t - kz, \tag{3}$$

and δ is the polarization factor and values $\delta = 0, \pm 1$ stand for linear polarization. The kinetic momentum of a charged particle in this case is given by

$$p = p_0 = p_{OT} + p_{OL}\hat{z}. \tag{4}$$

The canonical momentum on the other hand is

$$\Pi = \Pi_0 = \left(p_{OT} - \frac{e}{c}A_0\right) + p_{OL}\hat{z}. \tag{5}$$

with the initial energy given by

$$E_0 = \sqrt{p^2{}_{0T} + p^2{}_{0L} + m^2c^4} \tag{6}$$

Here, the subscripts T and L refer to the transverse and longitudinal components, respectively. We assume the propagation of the laser wave to be along the z direction. Initially at rest at the origin, the electron is strongly accelerated by the action of an incident laser field. The laser pulse is assumed to be transverse, plane and arbitrarily polarized. Throughout the simulations the laser frequency is taken as $\omega = 0.18$, value that corresponds to a Kr:F laser. A pulse-shape factor $P(\eta)$ emulating a Gaussian shape given by a sine-squared envelope is used. For solving the relativistic equations of motion we consider the Hamilton-Jacobi equation

$$[\nabla S(\mathbf{r},t) - (^e/_c)A(\eta)P(\eta)]^2 - \left(^1/_{c^2}\right)\left[\frac{\partial S(\mathbf{r},t)}{\partial t}\right]^2 + mc^2 = 0 \tag{7}$$

and assume a solution form for the Hamilton principal function $S(\mathbf{r},\mathrm{t})$ of the form

$$S(\mathbf{r},t) = \boldsymbol{\alpha}\cdot\mathbf{r} + \beta ct + \Phi(\eta), \tag{8}$$

here $\boldsymbol{\alpha}$ and β are constants determined by the initial conditions, and $\Phi(\eta)$ is determined by the Hamilton-Jacobi equation. In our particular case, $\boldsymbol{\alpha} = \boldsymbol{\Pi}_0$ and $\beta = {}^{E_0}/_c - p_{0L}$. By imposing these boundary conditions on the particle and differentiating the principal function with respect to the constant $\boldsymbol{\alpha}$ and equating the result to the initial coordinate, the complete solutions can be obtained. Therefore, the kinetic momentum of the electron is

$$\mathbf{p} - \mathbf{p_0} = \boldsymbol{\sigma} + \hat{z}\frac{\boldsymbol{\sigma}\cdot(\boldsymbol{\sigma} + 2\mathbf{p}_{0T})}{2\beta} \tag{9}$$

where

$$\boldsymbol{\sigma} = \frac{e}{c}[\mathbf{A}(\eta) - \mathbf{A}_0]. \tag{10}$$

Using this the energy of the electron attains the form

$$E^2 = m^2c^4 + p^2c^2. \tag{11}$$

The momentum of the electron after the laser interaction is obtained by setting $\mathbf{A}= 0$ in the kinetic momentum, equation (9). Thus

$$\mathbf{p}_f = \left(\mathbf{p}_0 - \frac{e}{c}\mathbf{A}_0\right) - \hat{z}\frac{e\mathbf{A}_0}{2\beta c}\cdot\left(2\mathbf{p}_{0T} - \frac{e}{c}\mathbf{A}_0\right). \tag{12}$$

3. Numerical simulations

To study the dynamics of a single electron, in these calculations space charge forces were modeled by adding a harmonic restoring force of the form

$$\mathbf{F} = -\mathrm{m}{\omega_p}^2\mathbf{z}\ . \tag{13}$$

Here ω_p is the plasma frequency. We solved the relativistic Newton-Lorentz equation,

$$\frac{d}{dt}\mathbf{P} = -\mathrm{e}\mathbf{E} - \frac{e}{c}\mathbf{v}\times\mathbf{B} - \mathrm{m}{\omega_p}^2\mathbf{z}\ , \tag{14}$$

using a Runge-Kutta variable-step method. Furthermore, the time-varying laser intensity is defined by the sine-squared envelope

$$E_x(\eta) = E_0 \sin\eta\, \sin^2\left(^\eta/_N\right). \tag{15}$$

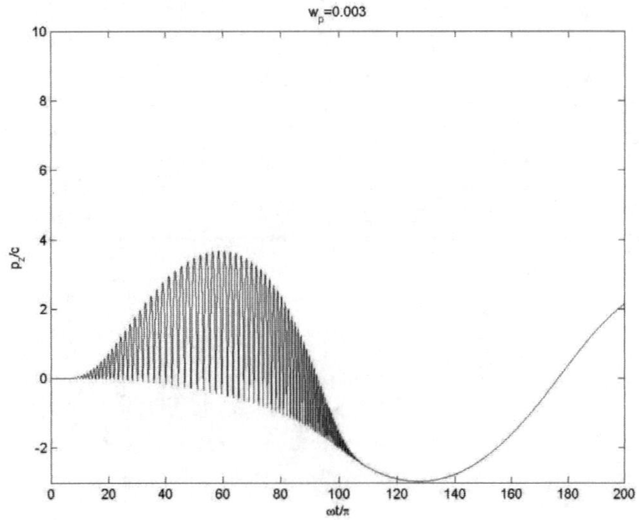

Figure 1. Time dependence of the longitudinal momentum for $\omega_p = 0.003\ a.u.$ and pulse length $100\ \pi$.

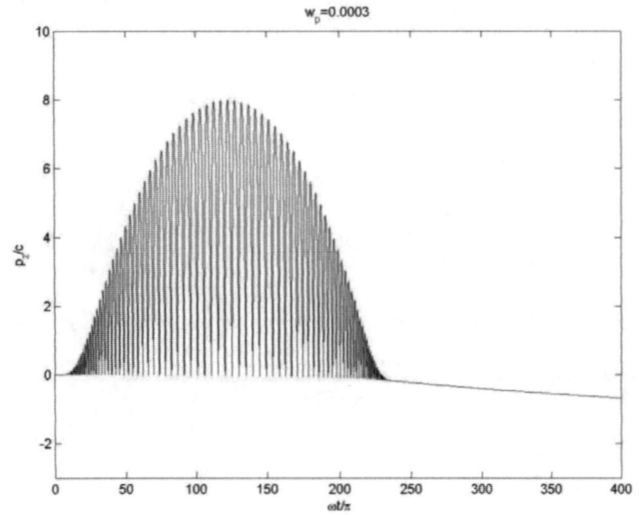

Figure 2. Time dependence of the longitudinal momentum for $\omega_p = 0.0003\ a.u.$ and pulse length $100\ \pi$.

In figure 1, the residual longitudinal momentum of the electron is relatively large when compared to the amplitude of the longitudinal momentum during the interaction with the pulse. In figure 2, the plasma frequency is one-tenth the size of the plasma frequency in figure 1. Here, we see that the residual longitudinal momentum is much less than in the previous case.

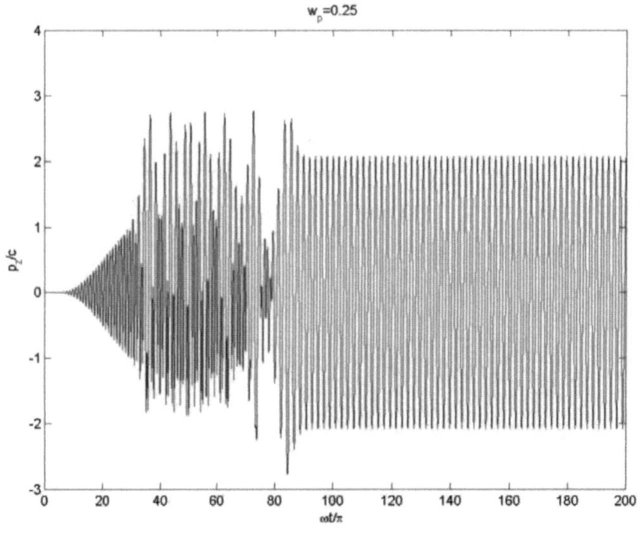

Figure 3. Time dependence of the longitudinal momentum for $\omega_p = 0.25\ a.u.$ and pulse length $100\ \pi$.

In figure 3, we observe another possible resonance region for the specified plasma frequency, which is close to the natural frequency of the laser. We note that the longitudinal residual momentum is comparable to the pulse interaction momentum and on average larger after the passage of the pulse when compared to figure 1.

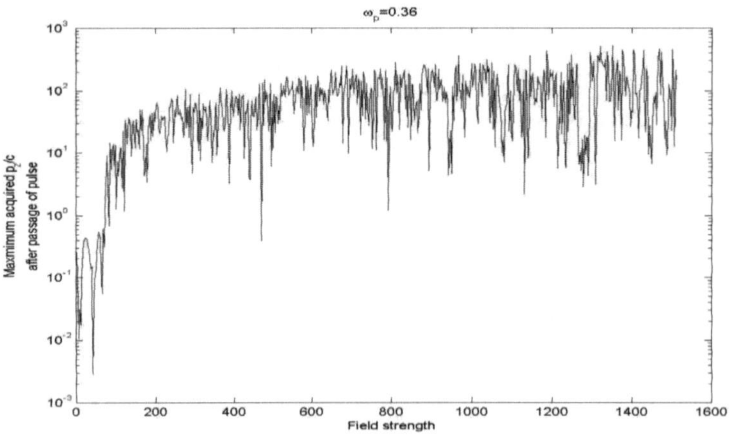

Figure 4. Maximum value of the residual momentum for a 100 cycle pulse as a function of field strength (in atomic units).

In figure 4 we observe the dependence of the maximum residual momentum (momentum obtained by the electron after the passage of the pulse) for a fixed value of the plasma frequency and pulse length. For small values of the field strength, the maximum residual longitudinal momentum increases greatly. As the field strength increases, however, saturation seems to be reached for the maximum residual momentum.

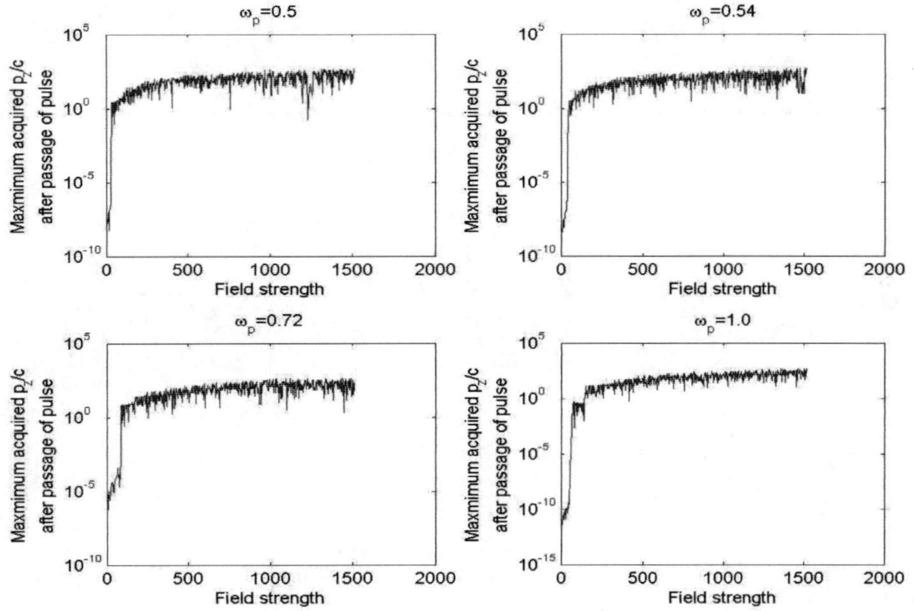

Figure 5. A comparison of various values of the maximum residual momentum for a fixed frequency. As for other graphs of this form saturation seems to be reached for small values of the field strength.

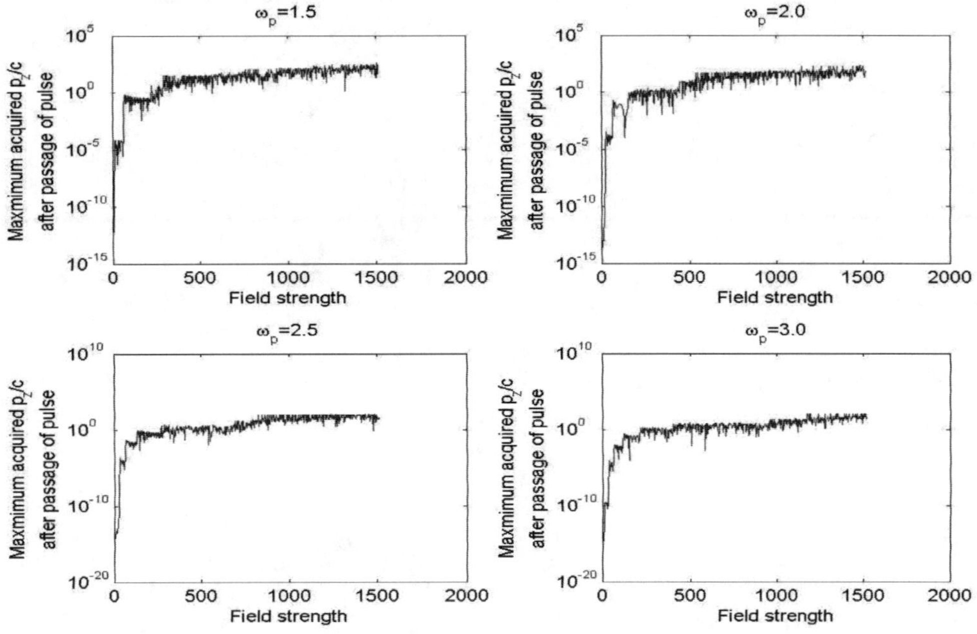

Figure 6. Same as in figure 5, except the plasma frequency is significantly higher. The maximum residual momentum for higher values of the plasma frequency appears to decrease slightly.

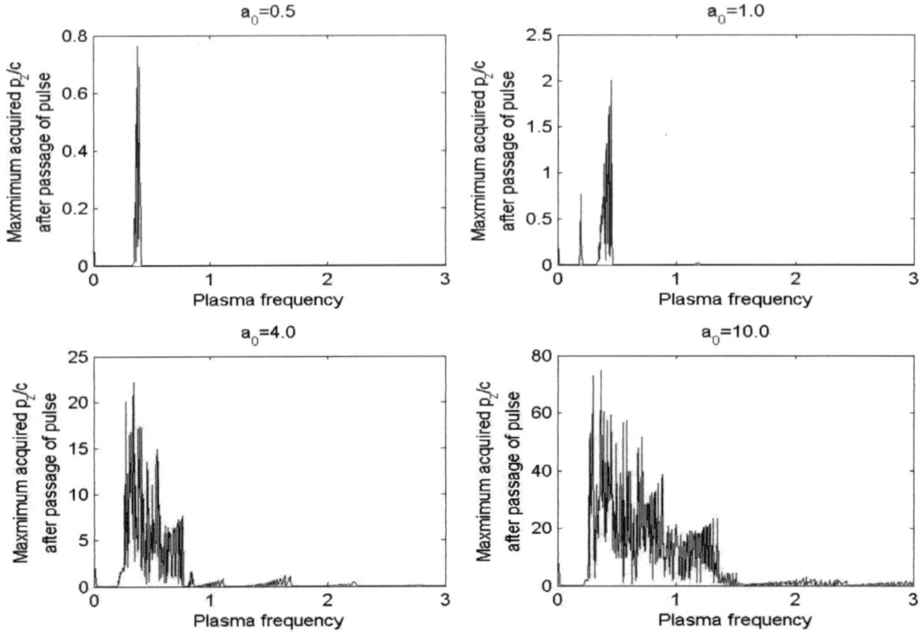

Figure 7. Residual momentum of the electron for a 100 pulse length laser as a function of the plasma frequency. Here the intensity of the field is held constant at various values.

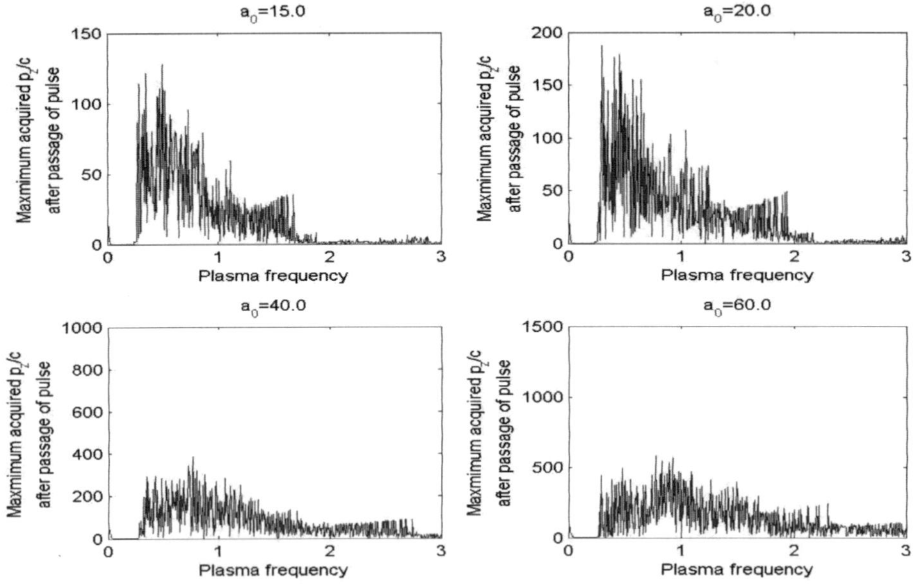

Figure 8. Residual momentum of the electron for a 100 pulse length laser as a function of the plasma frequency. The intensity of the field is held constant at various values.

In figure 7 and figure 8 we track the dependence of the maximum residual momentum on the plasma frequency itself while keeping the laser intensity fixed. We notice that for higher values of the field intensity, the maximum residual momentum increases. The behavior of the electron as the plasma frequency increases has no discernible pattern. It is evident that for low values of the field strength, the resonance regions are far more pronounced than that for fields in the ultra-relativistic regime.

4. Conclusions

Although the analytic methods for solving the electron motion are important for the understanding of the dynamics of charged particles in intense laser fields, it is necessary to use detailed numerical calculations to find solutions to more complex and thereby realistic interactions. In order to consider the effect of the plasma, we considered a simple model that allowed us to simulate the effects of different laser and plasma parameters in the highly irregular and nonlinear behavior of the plasma electrons. We found that in the regime of relativistic dynamics the pattern of electron residual momentum gain completely decayed into unpredictable electron motion. At higher field intensities the irregular behavior of the electron becomes conspicuous, independent of the plasma density. The evident increase in the maximum and average residual momentum of electrons for certain parameter conditions suggests that there should exist alternative methods for heating plasmas. It is possible that the dynamics for different residual momenta when an electron emerges from an ion corresponds when the laser amplitude is close to its maximum, leading to a mismatch between the phase of the electromagnetic field and the subsequent oscillations of the photoelectrons.

References

[1] T S Luk, T Graber, H Jara, V Johann, and C K Rhodes 1987 J. *Opt. Soc. Am. B* **4**, 847
[2] J A Cobble, G A Kyrala, A A Hauer, A J Taylor, C C Gomez, N D Delameter, and G T Schappert 1989 *Phys. Rev. A* **39**, 4
[3] M M Murnane, H C Kapteyn, and R W Falcone 1989 *Phys. Rev. Lett.* **62**, 155
[4] L D Landau and E M Lifshitz 1965 *Mechanics* (Pergamon, Oxford, England, p. 276)
[5] P B Corkum, N H Burnett, and F Bruenel 1989 *Phys. Rev. Lett.* **62**, 1259
[6] K Boyer, T S Luk, and C K Rhodes 1988 *Phys. Rev. Lett.* **60**, 557
[7] J N Bardsley, B M Penetrante, and M H Mittleman 1989 *Phys. Rev. A* **40**, 3823
[8] E S Sarachik and G T Schappert *1970 Phys. Rev. D* **1**, 2738
[9] J Peyraud and N Peyraud 1972 *J. Appl. Phys.* **43**, 2993
[10] N H Burnett and P B Corkum 1989 *J. Opt. Soc. Am. B* **6**, 1195
[11] G A Kyrala 1987 *J. Opt. Soc. Am. B* **4**, 731
[12] J Kruger and M Bovyn 1976 *J. Phys. A* **9**, 1841

Plasma emission and radiation from ultra-relativistic Brunel electrons in femtosecond laser-plasma interactions

R Ondarza-Rovira[1] **and TJM Boyd**[2]

[1] Instituto Nacional de Investigaciones Nucleares, A.P. 18-1027, México 11801, Distrito Federal, Mexico
[2] Centre for Theoretical Physics, University of Essex, Wivenhoe Park, Colchester, CO4 3SQ, United Kingdom

E-mail: ricardo.ondarza@inin.gob.mx

Abstract. A highly intense femtosecond laser pulse incident on a plasma target of supercritical density, gives rise to reflected high-order harmonics of the laser frequency. The radiation model adopted here considers Brunel electrons -those reinjected into the plasma after performing a vacuum excursion- perturbed by a localized turbulent region of an electrostatic field that is generated during the interaction, and characterized by a soliton-like structure. The observed power spectrum is characterized by a power-law decay scaled as $P_m \sim m^{-p}$, where m denotes the harmonic order. In this work an appeal is made to a radiation mechanism from a single particle model that shows harmonic power decays described -as previously reported from particle-in-cell simulations- within the range $2/3 \leq p \leq 5/3$. Plasma emission is strongest for values of the similarity parameter $S=n_e/(n_c a_0)$ in the range $1 \leq S \leq 5$, and where n_e and n_c are the ambient electron plasma density and the critical density, respectively, and a_0 is the normalized parameter of the electric field of the incident light. It was found that the radiation spectra obtained from the single particle model here presented is consistent with previously reported power-law 5/3 decays from particle-in-cell simulations.

1. Introduction

High harmonic generation in relativistic interactions from femtosecond laser pulses on solid density targets has stimulated interest in a number of potential applications in plasma diagnostics and technological developments in ultrafast optics for studying dynamical processes in matter, and more recently as sources for generating ultrashort-intense pulses in the attosecond range [1].

It has long been known from experiments and particle simulations that laser-generated harmonics from the interaction of a highly intense light pulse with overcritical density plasmas are characterized by a power-law decay that can extend over hundreds or even thousands of harmonic orders. The spectral decay has been characterized in different approximations from the laser-plasma parameters [2,3]. In the mildly relativistic regime Gibbon [2] first proposed an empirical relation from PIC simulations for the harmonic efficiency η, and showed that this scaled as $\eta \sim 17.2\, a_0^4\, m^{-p}$, where the decay index p=5, and $a_0 = 8.5\,(I_{20}\,\lambda_L^2)^{1/2}$ is the normalized quiver momentum, with I_{20} denoting the light intensity in units of 10^{20} Wcm^{-2} and λ_L the laser wavelength measured in microns. As values of a_0 increased above a_0=1, relativistic effects arise in the interaction physics. In picosecond-pulse experiments, the harmonic spectra was observed to span a range of values of p for increasing intensities, from p=5.5 at $a_0 \sim 0.64$ to p=3.38 at $a_0 \sim 2.85$ [3]. More recently, Baeva *et al.* [4] proposed a model based on a similarity model, that the underlying physics is similar when the similarity parameter S remains constant, with an associated so-called "universal" value 8/3 to the harmonic power decay. Furthermore, the S-similarity analysis predicts a cut-off in the harmonic spectra at $m = \sqrt{(8\,\alpha)}\,\gamma_s^3$, where γ_s is the relativistic factor of the plasma surface and α is of order one. No clear-cut evidence has been found for the spectral cut-off experimentally, though the spectral power decay was shown to be fitted by indexes in the range $2.3 \leq p \leq 2.7$, with the upper limit corresponding to the predicted "universal" value.

2. Plasma effects

In mildly relativistic interactions, Boyd and Ondarza-Rovira (BOR) found evidence of emission at the plasma line and at its harmonics [5]. In the UR regime, the ponderomotive force of the Langmuir field is strong enough to govern plasma behaviour over time scales $t > \omega_p^{-1}$ and highly localized electrostatic fields are generated in correspondence with density structures inside the plasma. These structures were identified as sources of emission at the plasma frequency and at its harmonics. Through PIC simulations BOR explored the harmonic power decay in UR laser-plasma interactions [6] and spectral decays characterized by an index p=5/3 were found and, at higher intensities, by p=4/3 for p-polarized light. When the incident laser pulse is normal to the plasma surface, the spectrum is invariably characterized by an 8/3 power law decay, regardless of the combination (a_0, n_e/n_c). Contrasting this, plasma effects are evident in the power harmonic decay for oblique incidence, since the excitation of plasma oscillations enable the enhancement of emission with lower harmonic power decay indexes, resulting in much flatter spectra. As an example, the plasma oscillations for the interaction a_0 =10, n_e/n_c=40 and a pulse length τ_p=17 fs are shown in figure 1a around the strong electrostatic field generated, as obtained from a PIC simulation. Figure 1b shows the corresponding density contours where the high plasma density spot represents a region of strong electrostatic field generation. Also Brunel trajectories of electrons can be discerned travelling deep inside the plasma.

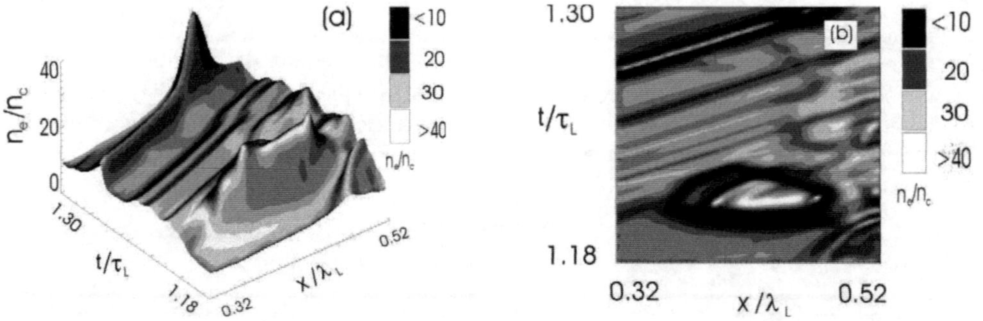

Figure 1: Regions in the plasma where the emission source is correlated: (a) plasma electron density oscillations and (b) electron density contours at sites of high electrostatic field generation inside the plasma.

The Brunel electron trajectories were also observed to be correlated with the electrostatic fields in places of beam crossings, as shown in figure 2a. For p-polarization field amplification is also observed (figure 2b), being an order of magnitude greater than the s-polarized case. These reflected attosecond spikes correlate with the electron bunching where strong electrostatic fields are found, and identified as sources of emission [6].

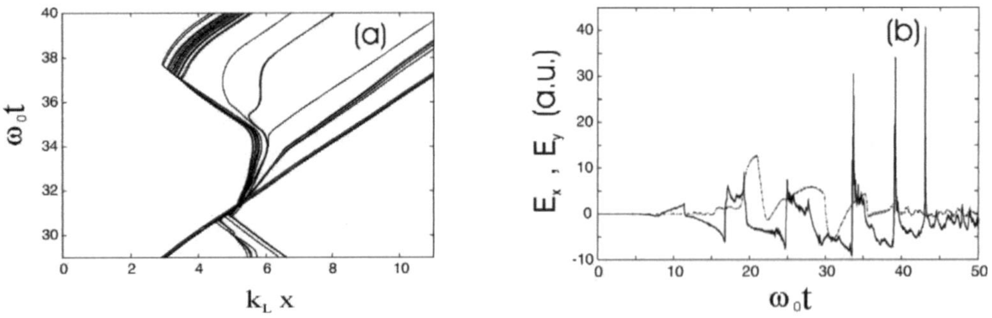

Figure 2: Correlation of the electrostatic field generated in the plasma (E_x, dotted line) with: (a) Brunel electron trajectories and beam crossings, and (b) the reflected attosecond electric spikes (E_y, solid line).

3. Plasma and XUV harmonic emission from ultra-relativistically laser-accelerated electrons perturbed by soliton-like electrostatic plasma fields

In the following we make use of a simple dynamical model to explore the plasma effects on harmonic radiation emission from a charged particle relativistically accelerated by a strong electromagnetic pulse, and perturbed by a soliton-like electrostatic field generated in the plasma during the laser-plasma interaction -as discussed in the previous section.

Radiation from electron bunching in large-amplitude Langmuir waves has been considered previously in cases where these have been excited in beam-plasma interactions. In particular, Weatherall and Hobbs [7] showed that electrostatic bunching was an effective means of generating plasma harmonics. Subsequently, Weatherall and Benford (WB) [8] modelled these fields as soliton-like structures generated inside the plasma and showed that these sources gave rise to strong nonlinear perturbations when high energy electron beams traverse such structures resulting in broadband bremsstrahlung emission. This model describes the radiation effects produced by the scattering of the electron beams that propagate inside the plasma through regions of highly intense localized electrostatic fields. These localized electrostatic structures cause the beam electrons to be scattered and to emit distinctive radiation. The WB model considers a specific shape and size of the soliton in which the characteristic turbulent density oscillation has a dipole structure within a localized region

$$\rho = \rho_0 \frac{\hat{\mathbf{p}} \cdot \mathbf{r}}{D} \exp\left(- \frac{r^2}{D^2} \right) \exp\left(- \mathrm{i}\, \omega_p t \right) , \tag{1}$$

here ρ_0 denotes the amplitude of the density oscillation localized by a Gaussian envelope of scale length D, oscillating at the plasma frequency ω_p, with a dipole moment oriented in the direction of $\hat{p}$. It can be showed that the electric field strength near the centre of the soliton is

$$\mathbf{E}_s = - \frac{2}{3} D \rho_0 \exp\left(- \mathrm{i}\, \omega_p t \right) \hat{\mathbf{p}} . \tag{2}$$

The acceleration of a charged particle through a soliton field generates electromagnetic wave packets containing a broad spectrum of frequencies. The electrons in a uniform electron beam would radiate with no phase coherence, resulting in weak scattered emission. In contrast, an electron beam with nonuniform density can maintain a degree of phase coherence over a range of wavelengths comparable with the scale length of density fluctuations. Therefore, the superposition of coherent radiation from plasma electrons can result in strong emission if the beam is modulated or bunched. Furthermore, when the beam is ultrarelativistic the radiation could be greatly enhanced by relativistic beaming along the direction of propagation of the laser wave. Beam radiation would be expected to dominate plasma emission when the beam density is high and the electrons highly relativistic.

Thus electron bunching results in a state of strong turbulence within the scale of the localized electrostatic field and a broad spectrum of density fluctuations develop. Beam density fluctuations produce regions of high E_s fields which would represent centres of electromagnetic radiation producing a plasmon condensate spectrum with a power-law index of p=5/3. It can be demonstrated that the spectrum of beam density correlations leads to a structure function that gives the total energy radiated by the soliton, in the form

$$\frac{\mathrm{d}E(\omega)}{\mathrm{d}\omega} \approx V\left(\frac{\omega-\omega_p}{v_0}\right), \tag{3}$$

where v_0 is the electron beam velocity and V the beam density correlation, given by [9]

$$V(k) \approx \frac{\Gamma(\nu+1/2)}{\sqrt{\pi}\,\Gamma(\nu)} \frac{r_0}{\left(1+k^2 r_0^2\right)^{\nu+1/2}}, \tag{4}$$

where Γ is the gamma function, r_0 the perturbation scale length and the wave number $k=(\omega-\omega_p)/v_0$. The turbulent spectrum of density fluctuations gives coherence to the emission spectra when the electron beam is perturbed by scattering density centres.

By integrating (4) we found that the power-law index of p=5/3 that governs the spectrum decay is retrieved when the parameter ν=1/3. Figure 3a shows the function $V(k)$ representing the beam radiation power characterized by a prominent line at the plasma frequency, a power spectral decay 5/3 over high harmonic orders and an sudden cut-off representing an upper bound at the highest frequencies emitted by the electrons that are scattered by the solitonic structure.

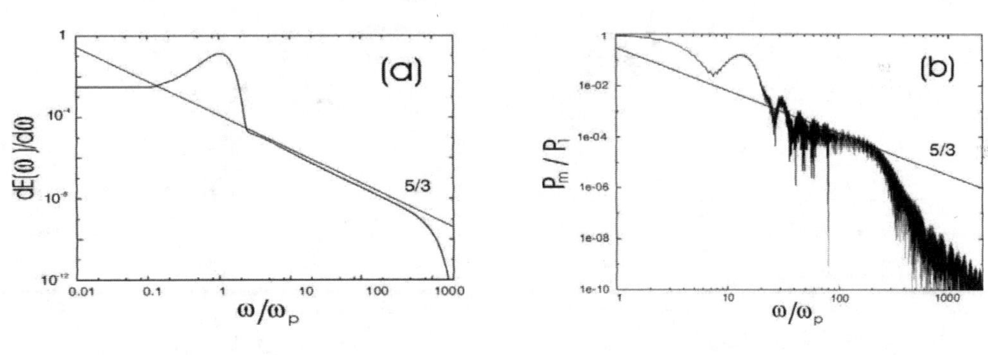

Figure 3: (a) Beam radiation power spectrum from density correlations $V(k)$ as given by equation (4), and (b) single particle radiation from a laser-accelerated electron perturbed by a soliton field (equation (2)), for a_0=40, n_e/n_c=167, both having power-law index decays p~5/3. The electron dynamics was integrated from equation (5).

Thus, the XUV spectral decay is described by a power law $P(\omega) \sim \omega^{-5/3}$, below an abrupt cut-off at a critical frequency $\omega \leq \omega_c = 2\,\gamma_b^2 c/D$, where γ_b is the relativistic factor for the electron beam [8].

By integrating the Lorentz equation we determine the motion of an electron plasma relativistically accelerated by a short highly intense field of a laser pulse. The dynamics was obtained from the relativistic equations for the energy and momentum [10]

$$mc\gamma = \boldsymbol{\beta}\cdot\mathbf{p},$$
$$\dot{\boldsymbol{\beta}} = \frac{e}{mc\gamma}\left[\mathbf{E}-\boldsymbol{\beta}\left(\mathbf{E}\cdot\boldsymbol{\beta}\right)+\boldsymbol{\beta}\times\mathbf{B}\right]+\mathbf{E}_s, \tag{5}$$

where $\gamma = (1 - \beta^2)^{-1/2}$ is the relativistic factor, with $\boldsymbol{\beta} = \boldsymbol{v}/c$ the normalized electron velocity and c the speed of light in vacuum. The electric and magnetic fields of the monochromatic plane-polarized wave are given by $\mathbf{E}=-\partial\mathbf{A}/\partial t$ and $\mathbf{B}=\nabla\times\mathbf{A}$, respectively, where $\boldsymbol{A}(\eta) = A_p\, P(\eta)[\,\hat{\mathbf{x}}\,\delta\cos\eta + \hat{\mathbf{y}}\,(1-\delta^2)^{1/2}\sin\eta]$ is the electromagnetic potential and $\eta=\omega_L\, t - k_L\, x$, the phase of the wave with propagation along the x direction. A_p, $P(\eta)$ and δ are the amplitude, the shape pulse and the polarization parameter, respectively. The relativistic dynamics of an electron is characterized by strong nonlinearities that transform the periodic motion of a driven harmonic oscillator into irregular motion, giving rise to very high order harmonics in the radiation spectra. A typical spectrum is shown in figure 3b for the combination (a_0, n_e/n_c)=(40, 167). The spectrum is characterized not only by a prominent emission at the plasma line but also by a spectral 5/3-decay.

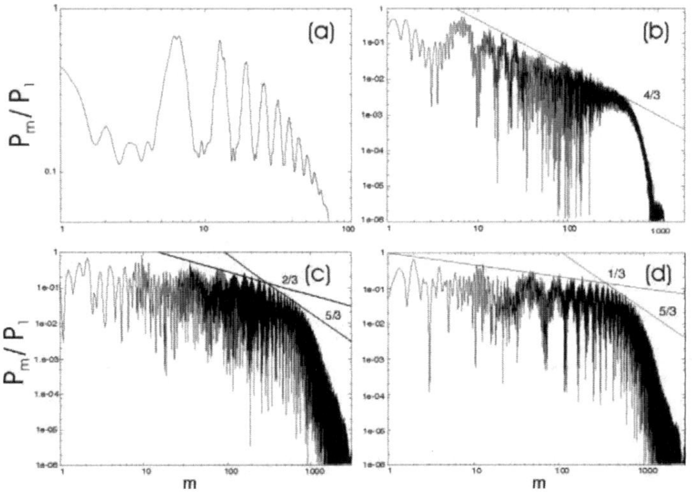

Figure 4: Plasma harmonic excitation for (a) a_0=10, n_e/n_c=40, t_p/τ_L=15, (b) a_0=30, n_e/n_c=40, t_p/τ_L=10, (c) a_0=40, n_e/n_c=80, t_p/τ_L=10, and (d) a_0=80, n_e/n_c=120, t_p/τ_L=10.

Figures 4 a-d show spectra obtained from different combinations of the parameters (a_0, n_e/n_c), and where emission at the plasma line, and its harmonics, are clearly observed together with a harmonic power decay index in the range 2/3-5/3 at high orders. Some modulation around the plasma emission is also discerned in these simulations with a cut-off at a high harmonic order as predicted by the parameter ω_c.

4. Conclusions

The formulated single particle radiation model is not inconsistent with our previous results obtained from PIC simulations, with the observed highly localized electrostatic fields generated inside the plasma during the interaction, that correlate with the electron trajectory crossings produced when strongly accelerated beam-plasma electrons -the so-called Brunel electrons- are re-injected into the target by the laser electric field after performing vacuum excursions outside the plasma. The radiation from perturbed plasma electrons by a soliton field representing the electrostatic fields in the plasma is reminiscent of the 5/3 UR power law decay observed in PIC emission spectra.

Acknowledgments
One of us (R.O.R.) acknowledges support from CONACyT under Contract No. 167185.

References

[1] Teubner U and Gibbon P 2009 *Rev. Mod. Phys.* **81** 445
[2] Norreys PA *et al.* 1996 *Phys. Rev. Lett.* **76** 1832, Gibbon P 1996 *Phys. Rev. Lett.* **76** 50
[3] Dromey B *et al.* 2006 *Nat. Phys.* **2** 456
[4] Baeva T, Gordienko S and Pukhov A 2006 *Phys. Rev. E* **74** 046404
[5] Boyd TJM and Ondarza-Rovira R 2008 *Phys. Rev. Lett.* **101** 125004; 2008 *Laser-Driven Relativistic Plasmas, AIP Conf. Proc.* **1024** 233
[6] Boyd TJM and Ondarza-Rovira R 2000 *Phys. Rev. Lett.* **85** 1440, *idem* 2010 *Phys. Lett. A* **374** 1517, *idem* 2012 *Phys. Rev. E* **86** 026407
[7] Weatherall JC and Hobbs WE 1986 *Phys. Fluids* **29** 2292
[8] Weatherall JC and Benford G 1991 *Astrophys. J.* **378** 543
[9] Tatarski VI 1967 *Wave Propagation in a Turbulent Media* (New York: Dover)
[10] Boyd TJM and Sanderson JJ 2003 *The Physics of Plasmas* (UK: Cambridge University Press)

Adaptive Finite Volume numerical method

L Bilbao

INFIP-CONICET, and Physics Department (FCEN-UBA),
Ciudad Universitaria, Pab. I, 1428 Buenos Aires, Argentina

E-mail: bilbao@df.uba.ar

Abstract. This work describes a Finite Volume computational method for the parametric study of phenomena in plasmas, i.e. in situations where numerous runs of three-dimensional simulations are required (in many cases, thousands or tens of thousands). Such problems require simple codes, robust, modular (to add or remove physical processes) and do not require high precision. The code is based on a complex multi-component species program with transport and radiation terms. The integration domain is represented with a structured irregular mesh, with fixed connectivity. A new algorithm for the hydrodynamics was implemented in order to improve the computational efficiency plus the improved capability of adapting the mesh to the solution. The improved hydrodynamics method worked well in an ample range of Mach number from subsonic (10^{-3}) to supersonic. After each calculation cycle, mesh vertices are moved arbitrary over the fluid. This is done in order to dynamically adapt the mesh to the solution. The adaptive method consists of shifting mesh vertices over the fluid in order to keep a reasonable mesh structure and increase the spatial resolution where the physical solution demands. As an example, we show the results of the development of the Kelvin-Helmholtz instability in local plane slab models of the magnetopause, showing the development and saturation of the instability in an initially unperturbed structure, i.e., the temporal response approach; and the response of a background equilibrium to the excitation by finite amplitude perturbations generated upstream, i.e., the spatial response of the system.

1. Introduction

The scope for developing the present numerical method was to perform parametric studies for optimization of several problems in plasma physics. Nowadays there exist several efficient numerical codes in the subject. However the construction of own computational codes brings the following important advantages: (a) to get a deeper knowledge of the physical processes involved and the numerical methods used to simulate them; and, (b) more flexibility to adapt the code to particular situations in a more efficient way than a closed general code would.

On those premises a new Arbitrary Lagrangian-Eulerian (ALE) Finite Volume algorithm was developed based on a previously reported code [1]. As in the previous version, Three-Dimensional (3D), time depend, multi-component, two-temperature code was used.

The code includes ion viscosity, thermal conduction (electrons and ions), magnetic diffusion, thermonuclear or chemical reaction, Bremsstrahlung radiation, EOS (from the ideal gas to the degenerate electron gas). In the next paragraph the main improvements will be highlighted.

2. Numerical resolution

We have used an ALE Finite Volume method with hexahedral cells that move at arbitrary velocity. The notation and definition of the geometry is described elsewhere [1]. The main advantage of this method is that the geometry is imposed to the cells and not to the operators.

This method consists of integrating conserved quantities over a cell. For a general fluid property Ψ, the following identity holds

$$\frac{d}{dt}\int_{V(t)} \Psi dV = \frac{\partial}{\partial t}\int_{V(t)} \Psi dV + \oint_{S(t)} \Psi\, \mathbf{w}\cdot d\mathbf{S} \tag{1}$$

where integration is performed over a cell of arbitrary volume *V(t)* with boundary *S(t)* moving at an arbitrary speed **w**. By choosing **w** = 0, that is both *V* and *S* remain fixed, a full Eulerian formulation is retrieved. On the other hand, setting **w** = **v**(**x**,*t*) –fluid velocity- leads to the standard Lagrangian formulation. The adaptive method consists of choosing an appropriated **w**(**x**,*t*) over the fluid in order to keep a reasonable mesh structure and increase the spatial resolution where the physical solution demands.

For the ideal MHD, the equations for mass (*M*), momentum (**P**), internal energy (*E*), and, magnetic flux (Φ), are

$$\frac{dM}{dt} = \frac{d}{dt}\int_{V(t)} \rho dV = -\oint_{S(t)} \rho(\mathbf{v}-\mathbf{w})\cdot d\mathbf{S} \tag{2}$$

$$M\frac{dY_k}{dt} = -\oint_{S(t)} \rho(\mathbf{v}-\mathbf{w})\cdot d\mathbf{S} - \oint_{S(t)} d\mathbf{S}\cdot \mathbf{q}_k^D + \int_{V(t)} dV\,\omega_k \tag{3}$$

$$\frac{d\mathbf{P}}{dt} = \frac{d}{dt}\int_{V(t)} \mathbf{v}\rho dV = -\oint_{S(t)} \rho\mathbf{v}(\mathbf{v}-\mathbf{w})\cdot d\mathbf{S} - \int_{S(t)} p d\mathbf{S} + \int_{V(t)} \mathbf{j}\times\mathbf{B}\, dV \tag{4}$$

$$\frac{dE}{dt} = \frac{d}{dt}\int_{V(t)} \varepsilon\rho dV = -\oint_{S(t)} \rho\varepsilon(\mathbf{v}-\mathbf{w})\cdot d\mathbf{S} + \int_{S(t)} \left(\frac{c_p kT}{e}\mathbf{j}\right)\cdot d\mathbf{S} - \int_{V(t)} p\nabla\cdot\mathbf{v}\, dV \tag{5}$$

$$\frac{d\Phi}{dt} = \frac{d}{dt}\int_{\Sigma(t)} \mathbf{B}\cdot d\mathbf{S} = -\oint_{C(t)} d\mathbf{l}\cdot\mathbf{E}' + \oint_{C(t)} [(\mathbf{v}-\mathbf{w})\times\mathbf{B}]\cdot d\mathbf{l} \tag{6}$$

where **j** is the electrical current density, $\varepsilon = p/\rho(\gamma-1)$ the internal energy density (c_p is the specific heat at constant pressure) and *e* the charge of the electron. In the last equation Σ is an arbitrary open surface bounded by a closed path *C*, which moves according to **w**(**x**,*t*).

The integration domain is represented with a structured irregular mesh, with fixed connectivity made of hexahedral cells. Each cell is surrounded by 6 faces and 8 vertices. Fluid variables are assigned to staggered locations in the mesh. Pressure, internal energy, density, species concentration, cell volume and mass are all assigned to cell centres. Coordinates (x,y,z) and velocities (u_x,u_y,u_z) are assigned to cell vertices.

For example, using **w** = 0 the evolution of the cell mass $M \approx \langle\rho\rangle V$, $\langle\rho\rangle$ being a mean value of the density over de cell, and V the volume, is written as

$$\frac{dM}{dt} = -\sum_i \rho_i \mathrm{v}_i \cdot \mathrm{S}_i \tag{6}$$

where $\mathbf{S}_i$, $\mathbf{v}_i$, ρ_i, are the surface, velocity and mass density of the i-th face of the cell. Analogously, the momentum of a cell, $\mathbf{P} \approx <\rho><\mathrm{v}>\ V$, evolves according to

$$\frac{d\mathbf{P}}{dt} = -\sum_i \rho_i \mathbf{v}_i (\mathbf{S}_i \cdot \mathbf{v}_i) - \sum_i p_i \mathbf{S}_i + \frac{1}{4\pi}(\sum_i \mathbf{S}_i \cdot \mathbf{B}_i) \times <\mathbf{B}> \tag{7}$$

where <**B**> is the mean value of the magnetic field on a cell.

Using the same ideas all integral equations are discretized. In order to set the values on the faces, appropriate mean values over adjacent cells are used.

Global conservation of the above magnitudes is checked during calculation. Also, condition $div(\mathbf{B})=0$ is controlled over the whole integration domain.

The same method may be applied for more complex modelling of plasmas that includes ion viscosity, thermal conduction (electrons and ions), magnetic diffusion, chemical or thermonuclear reactions, Bremsstrahlung radiation, and realistic equation of state (EOS) from the ideal gas to the degenerate electron gas. The integration in time is sequential. This means that each process is integrated with a different uncoupled method during a time step. The overall truncation error is $O(\Delta x, \Delta y, \Delta z, \Delta t)$. Solution does not dependent on the integration order of the terms.

Calculation proceeds as follows:

- Hydrodynamics is integrated using an improved Predictor-Corrector method [2].
- Diffusion processes (of species concentration and energy) are integrated with ADI methods (Alternating Direction Implicit) for the 2D/3D case, and with full implicit methods for the 1D case.
- Source terms from chemical or nuclear reactions are explicitly integrated in time.
- Mesh vertices are moved arbitrary over the fluid as described below.

Lagrangian cell methods are not adequate for describing flows undergoing large distortions, because cells may undergo severe deformations.

Nuclear burning, combustion structures, or instabilities may present steep gradient of temperature, species concentration, and heat release. The adaptive method consists of shifting mesh vertices over the fluid in order to keep a reasonable mesh structure and increase the spatial resolution where the physical solution demands.

The rezone velocity **U** with which vertices are moved over the fluid is calculated in order to increase spatial resolution where steep temperature gradients and high chemical/nuclear heat release are produced. The new coordinates for each vertex are calculated in such a way that the Lagrangian distance remains uniformly distributed. A possibility is to define **U** in order to restore an Eulerian grid. This calculation is performed following one computational direction. The procedure is repeated over several paths where corrections are needed.

Vertex i moves from i to i' according to

$$\mathbf{x}_i' = \mathbf{x}_i + \mathbf{U}\Delta t \tag{8}$$

where **U** is the rezone velocity. During this process there is an exchange of mass, species, energy and momentum among neighbourhood cells.

Then the following steps must be applied

- Calculate the new coordinates using (8), then calculate the volume exchange among neighbourhood cells
- Associated with the volume exchange there will also be a mass, species concentration and total energy exchange. The mass, concentration or energy per unit mass assigned to this volume are computed using a donor cell flux. Finally, fluid conserved properties are updated.

- Once updated the conserved cell and vertex properties, cell volume, density and species concentration are calculated from the new vertex coordinates.

3. Development of the KH instability

A physical problem studied by means of the numerical code is related to the turbulent mixing layers of plasma at the near equatorial flanks of the terrestrial magnetopause. The numerical 3D code was used to simulate the large amplitude perturbations and waves at the dusk side low latitude boundary layer of the magnetopause generated by an interplanetary tangential discontinuity.

We focused on the motion and structure of the magnetopause/boundary layer observed in response to a joint tangential discontinuity/vortex sheet (TD/VS) observed by the ACE spacecraft on December 7, 2000. Sharp polarity reversals in the east-west components of the field and flow occurred at the discontinuity. These rotations were followed by a period of strongly north-ward IMF. These two factors elicited a two-stage response at the magnetopause, as observed by Cluster spacecraft situated in the boundary layer at the dusk side terminator.

These studies were aimed to clarify some particular physical properties of the instability of velocity shear flows and mixing layers, such as the 3D vorticity amplification mechanism and non conservation of vorticity flux.

The most common of these is the Kelvin-Helmholtz instability feeding on the flow shears which exist across the magnetosheath/magnetosphere interface (see figure 1). Ever since Dungey [3] pointed out the possibility of the magnetopause going Kelvin-Helmholtz unstable, many studies have been devoted to this instability, the surface waves it causes, their coupling to the geomagnetic field giving rise to field line resonances, and how this instability might mediate the entry of magnetosheath plasma into the magnetosphere [4-13].

Attention has been drawn to two other interplanetary parameters whose changes may occasion considerable deformation and motion of the magnetopause. One is a variable orientation of the interplanetary magnetic field (IMF). Fairfield et al. [14] showed that, even when the dynamic pressure is constant, the magnetopause and inner-magnetosphere boundaries may be in motion. They attributed these motions to pressure changes generated in the foreshock region by the IMF changes. These pressure changes subsequently convect through the bow shock and impinge on the magnetopause. The authors thus concluded that even in the presence of a solar wind that is absolutely steady in velocity and density but which carries an embedded interplanetary magnetic field of variable orientation; there will be a variation of pressure exerted on the magnetopause.

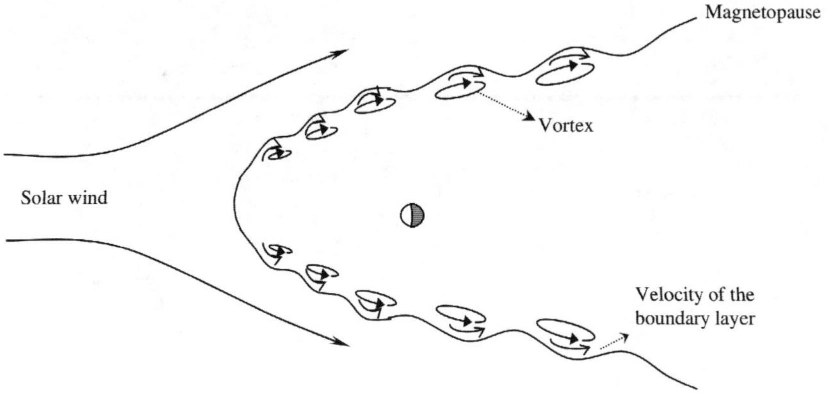

Figure 1. Sketch of vortex formation due to the Kelvin Helmholtz instability in the magnetopause

For the observations at the magnetopause we focused on Cluster 3 data. The Cluster spacecrafts were on an outbound trajectory just tail ward of the dusk terminator and at northern latitudes. (At this early stage of the mission, the spacecraft separation was of order a few hundred km and all spacecraft see the same features.). For some hours prior to our observations, the spacecraft stayed in the vicinity of the magnetopause because the magnetopause boundary was expanding slowly outward due to a slow decrease in dynamic pressure. From 14:00 to 14:30 UT the spacecraft stays within a distance of 0.4 RE from the magnetopause. We may reasonably assume that Cluster 3 is inside the boundary layer according to the model, in agreement with instrument readings.

We carried out a stability analysis of a limited segment of the magnetopause, where the boundary layer may be considered to lie in a local tangent plane. The boundary layer is then modelled as a planar plasma slab with stratified density, flow velocity, and magnetic field. The physical quantities on either side of the magnetosheath-magnetosphere transition are obtained from the Cluster 3 readings at the extremes of the large oscillations recorded after 13:58 UT. Specifically, we selected the second oscillation, where the lowest speed was recorded at maximum T. That is, we assume that at the extremes of the oscillation, Cluster 3 samples alternately the high density-low temperature magnetosheath plasma, and the low density-high temperature magnetosphere plasma. In this way the local instability is examined by parameters measured locally and in temporal proximity.

The full description of the problem and main results are presented in [15]. Anyhow, some sample outputs showing the 3D, complex character of the problem are presented. Figure 2 shows streamlines complexity and formation of a swirling flow in the large eddy at $t = 84$ s. Figure 3 shows the development of kinetic helicity, $t = 132$ s, contours of kinetic helicity density are displayed. Figure 2 illustrates the progression from the simple, parallel shear flow, where the kinetic helicity density is initially zero, to a complex stage with a large vortex core and deformations of large magnitude of the vorticity layer. Kinetic helicities of both sign arise from the initial velocity field perturbation. Note that the KH waves appear already after a few theoretical e-folding times, due to the great oscillations generated by the tangential discontinuity stress that has reduced the amplification time. In order to get an appropriated resolution, up to 1,000,000 cells were used.

Some interesting features of the phenomenon are obtained from the numerical results. For example, high temperature rises in a compact vortex core correlated with local density depletion while gas, and magnetic pressure (roughly constant across the structure) are near to surrounding values. Also detected a high density patch in a vorticity structure correlated with a local increase of gas pressure, whereas temperature stays near to surrounding values.

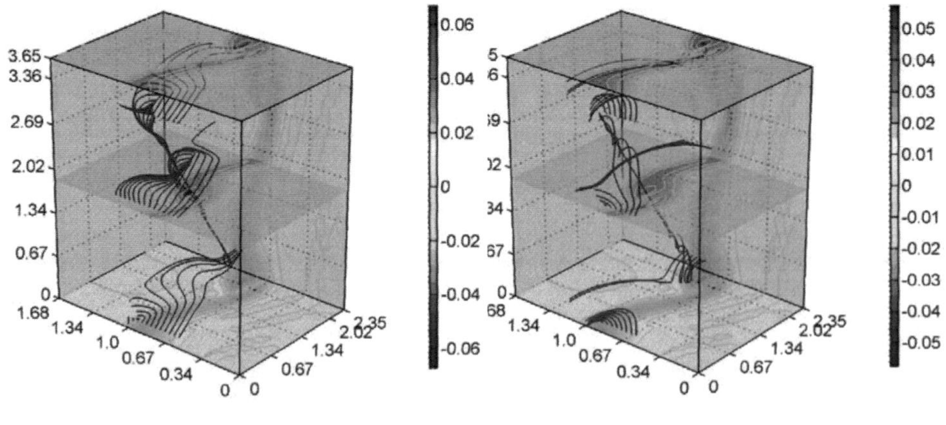

Figure 2. Mixing layer, $t = 84$ s, vorticity contours and streamlines. Comparing streamlines views: left, fixed (Earth) frame, and right, moving frame with average velocity ($U = 78.5$ km/s). In the moving frame magnetosheath and magnetopause plasmas are seen counter streaming.

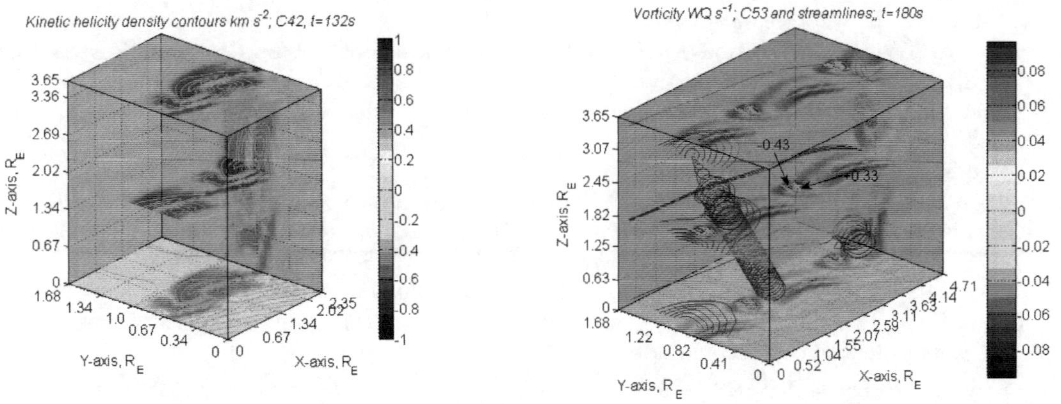

Figure 3. Development of kinetic helicity, t = 132 s. The plot show contours of kinetic helicity density (left). Details of vorticity evolution at t = 180 s with a view of the distortions of shape of the BL flow and some streamlines of swirling flows.

Concomitant with 3-D vortex stretching, kinetic helicity density rises at the vorticity cores. Finally, non conservation of vorticity becomes manifest at about (or even before) one rollover time, where in addition to vortices with positive rotation (the same sign of the original vorticity sheet) other coherent structures with strong negative vorticity also arise.

The rise of these coherent structures has important consequences for the properties of the mixing layer. There is a broadening of the boundary layer's width, and a substantial lengthening of the transit time of plasma, compared with that of a stable laminar boundary layer.

4. Conclusions

The numerical simulation was a powerful tool for the parametric analysis of the development of Kelvin-Helmholtz instability. The sequential implementation resulted of practical use, because each physical process was treated with a different numerical method. The improved hydrodynamics method worked well in an ample range of Mach number from subsonic (10^{-3}) to supersonic.

The time step was restricted by efficiency considerations rather than stability conditions. When time step becomes larger, the calculation progress faster because fluid variables, transport properties, etc., are updated with less frequency. But the implicit stage of the hydrodynamic advance requires more iterations to perform. An optimum time step has been used for which the minimum computational effort is obtained.

We reported here a study of the MHD development of vortices in compressible, non-homogeneous, mixing layers due to the velocity gradient instability. The non-linear evolution was computed with a new 3D+t FV code of temporal mixing layers tailored to represent distinctive conditions of the terrestrial magnetopause. The boundary layer is characterized by the growth of large-scale vortices, and becomes a site of mass mixing favouring enhanced plasma diffusion. The rise of counter-rotating vortices from the evolution of the original vortex sheet is one of the noteworthy processes revealed by the simulation. The adaptive grid process was able to keep a reasonable mesh structure and resolution inside the vortices.

5. Acknowledgments

Work partially supported by grant UBA 20020100100866, University of Buenos Aires.

6. References

[1] Bilbao L 1990 *J Comp Physics* **91** 361.
[2] Bilbao L 2009 “Computational Plasmadynamics applied to parametric studies” *Journal of*

Physics: Conference Series **166**, 012020.

[3] Dungey J W 1955 "Electrodynamics of the outer atmospheres" Report 69, Ions Res Lab. Pa State Univ, University Park.

[4] Southwood D J 1979 *Proceedings of Magnetospheric Boundary Layers Conference* ESA SP-148 Eur Space Agency Paris 359

[5] Miura A 1984 *J Geophys Res* **89** 801

[6] Chen S H, Kivelson M G, Gosling J T, Walker R J, Lazarus A J 1993 *J Geophys Res* **98** 5727.

[7] Kivelson M G, Chen S H 1995 in *Physics of the Magnetopause* (edited by P Song, B U O Sonnerup and M F Thomsen) *Geophys Monogr Ser* **90** AGU Washington DC 257.

[8] Seon J L, Frank A, Lazarus A J, Lepping R P 1995 *J Geophys Res* **100** (11) 907.

[9] Fairfield D H, Otto A, Mukai T, Kokubun S, Lepping R P, Steinberg J T, Lazarus A J, Yamamoto T 2000 *J Geophys Res* **105** (22),159.

[10] Otto A, Fairfield D H 2000 *J Geophys Res* **105** (A9) 21175-21190.

[11] Farrugia C J, Gratton F T, Contin J, Cocheci J C C, Arnoldy R L, Ogilvie K W, Lepping R P, Zastenker G N, Nozdrachev M N, Fedorov A, Sauvaud J A, Steinberg J T, Rostoker G 2000 *J Geophys Res* **105** (A4) 7639-7667.

[12] Gratton F T, Bender L, Farrugia C J, Gnavi G 2004 *J Geophys Res* **109** (A04211) 1-13.

[13] Farrugia C J, Gratton F T, Torbert R 2001 *Space Sci Rev* **95** (1/2) 443-456

[14] Fairfield D H, Baumjohann W, Paschmann G, Luehr H, Sibeck D G 1990 *J Geophys Res* **95** 3773-3786.

[15] Gratton F T, Bilbao L, Farrugia C J, Gnavi G 2009 "Large eddy simulations in MHD: the rise of counter-rotating vortices at the magnetopause" *Journal of Physics: JPCS* **166**, 012023.

Journal of Physics: Conference Series **591** (2015) 012038 doi:10.1088/1742-6596/591/1/012038

Some numerical simulation results of swirling flow in d.c. plasma torch

C L Felipini and M M Pimenta

Polytechnic School, University of São Paulo, São Paulo, Brazil

E-mail: celsofelipini@gmail.com

Abstract. We present and discuss some results of numerical simulation of swirling flow in d.c. plasma torch, obtained with a two-dimensional mathematical model (MHD model) which was developed to simulate the phenomena related to the interaction between the swirling flow and the electric arc in a non-transferred arc plasma torch. The model was implemented in a computer code based on the Finite Volume Method (FVM) to enable the numerical solution of the governing equations. For the study, cases were simulated with different operating conditions (gas flow rate; swirl number). Some obtained results were compared to the literature and have proved themselves to be in good agreement in most part of computational domain regions. The numerical simulations performed with the computer code enabled the study of the behaviour of the flow in the plasma torch and also study the effects of different swirl numbers on temperature and axial velocity of the plasma flow. The results demonstrated that the developed model is suitable to obtain a better understanding of the involved phenomena and also for the development and optimization of plasma torches.

1. Introduction

Thermal plasma torch designs using electric arc (usually called *plasma torches*) should be specifically developed according to their applications [1]. Therefore, the optimization of electrothermic processes involving thermal plasmas is closely related to the proper conception of plasma torches, mainly concerning to geometry/dimensions of electrodes (anodes and cathodes) and to operating parameters: working gas (plasmogenic gas), electric current intensity, electric voltage, gas flow rate and temperature range [2]. Besides the experimental procedures used in the development of plasma torches, the Computational Fluid Dynamics (*CFD*) is an interesting tool for torch design and optimization because it allows reducing the development costs and it helps to understand the related phenomena [3].

The model of this work approaches non-transferred arc torches (torches where the electric circuit closes among electrodes, cathodes and anodes, inside the torches themselves and the plasma jet does not conduct the current to their outer part), specifically the ones destined to industrial applications (thermal spraying, advanced material production, etc.) and the solutions for garbage, contaminated residues, toxic material and pollutants which affect the environment (figure 1).

We should highlight that in many non-transferred arc plasma torch, the gas is introduced with swirl in order to reduce the anode erosion rate through the arc rotation [4, 5]. It also shows that the swirling in plasma jet is useful for the synthesis of some materials [4]. This way, the effect of swirling flow in the behavior of plasma arc and plasma jet is an important investigation aspect which is a little divulged in literature motivating its insertion into the goals of this work.

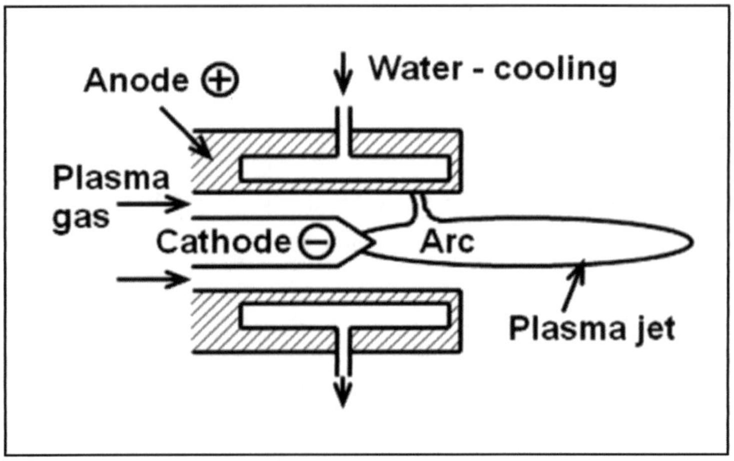

Figure 1. Schematic representation of a non-transferred arc plasma torch.

This article presents a mathematical model for the numerical simulation of swirling flow in thermal plasma torches and, besides that, in some results obtained with initial simulations.

2. The model

A magnetohydrodynamic model (*MHD* model) was adapted to thermal plasma torches based on [3-17]. Its main characteristics are: two dimensional mathematical model (axisymmetric) of swirling flow in non-transferred arc thermal plasma torches which operate in direct current (*d.c.*), involving all regions of study interest and influence in phenomena (gas inlet; inside torch– where there is arc/flow interaction; plasma jet free in the environment).

2.1. Assumptions

The following assumptions have been adopted for this model: (1) plasma (electrons; heavy particles) has a *Newtonian* and single-species fluid behavior; (2) flow is approached in a *continuum*, two dimensional (axisymmetric), laminar, with incompressible behavior and steady state; (3) the local thermodynamic equilibrium (*LTE*) prevails in the study domain; (4) plasma is optically thin (black body radiation emitted by plasma is not absorbed by it); (5) heat dissipation due to viscous tensions may not be taken into consideration (constituted by second order terms); (6) *MHD approach* is applicable: displacement current is not considered when compared to the conduction current; the electrostatic field strength is not considered; (7) gravity effects are not considered; (8) the external environment is formed by the same gas as the work one.

2.2. Model equations

By taking into consideration the assumptions previously adopted, the governing equations for the axisymmetric (*2D*), laminar flow in stationary status with plasma torch swirl may be presented in cylindrical coordinates as follows:

Mass conservation equation:

$$\frac{1}{r}\frac{\partial}{\partial r}(r\rho v)+\frac{\partial}{\partial z}(\rho u)=0 \tag{1}$$

Momentum conservation equations in z (axial), r (radial) and θ (azimuthal) directions, respectively:

$$\frac{1}{r}\frac{\partial}{\partial r}(r\rho uv)+\frac{\partial}{\partial z}\left(\rho u^2\right)=\frac{1}{r}\frac{\partial}{\partial r}\left[r\mu\left(\frac{\partial u}{\partial r}+\frac{\partial v}{\partial z}\right)\right]+\frac{\partial}{\partial z}\left(2\mu\frac{\partial u}{\partial z}\right)-\frac{\partial p}{\partial z}+j_rB_\theta \quad (2)$$

$$\frac{1}{r}\frac{\partial}{\partial r}\left(r\rho v^2\right)+\frac{\partial}{\partial z}(\rho uv)=\frac{1}{r}\frac{\partial}{\partial r}\left(2r\mu\frac{\partial v}{\partial r}\right)+\frac{\partial}{\partial z}\left[\mu\left(\frac{\partial u}{\partial r}+\frac{\partial v}{\partial z}\right)\right]+\rho\frac{w^2}{r}-2\mu\frac{v}{r^2}-\frac{\partial p}{\partial r}-j_zB_\theta \quad (3)$$

$$\frac{1}{r}\frac{\partial}{\partial r}(r\rho vw)+\frac{\partial}{\partial z}(\rho uw)=\frac{1}{r}\frac{\partial}{\partial r}\left(r\mu\frac{\partial w}{\partial r}\right)+\frac{\partial}{\partial z}\left(\mu\frac{\partial w}{\partial z}\right)-\frac{1}{r^2}\mu w-\rho\frac{vw}{r} \quad (4)$$

where u, v and w are the velocity components in z (axial), r (radial) and θ (azimuthal) directions, respectively; μ, ρ and p are the following gas properties: dynamic viscosity, specific mass and pressure, respectively; j_z and j_r are the axial and radial components of the electric current density vector $\vec{j}$, respectively and B_θ is the azimuthal component of the of magnetic induction intensity vector (self-induced) $\vec{B}$ (note: "magnetic field inducted by the arc electric field"). Observe that j_rB_θ and j_zB_θ terms are components of *Lorentz* force vector: $\vec{F}$ ($\vec{F}=\vec{j}\times\vec{B}$).

Energy conservation equation ("written" in function of the gas specific enthalpy, h: $h=\int C_p dT$):

$$\frac{1}{r}\frac{\partial}{\partial r}(r\rho vh)+\frac{\partial}{\partial z}(\rho uh)=\frac{1}{r}\frac{\partial}{\partial r}\left(r\frac{k}{C_p}\frac{\partial h}{\partial r}\right)+\frac{\partial}{\partial z}\left(\frac{k}{C_p}\frac{\partial h}{\partial z}\right)+\frac{j_r^2+j_z^2}{\sigma}-S_r+\frac{5}{2}\frac{k_b}{e}\left(\frac{j_r}{C_p}\frac{\partial h}{\partial r}+\frac{j_z}{C_p}\frac{\partial h}{\partial z}\right)$$
$$+u\frac{\partial p}{\partial z}+v\frac{\partial p}{\partial r} \quad (5)$$

where C_p is the gas specific heat at constant pressure; k is the gas thermal conductivity and σ is the gas electric conductivity. One should observe that besides the terms which represent the enthalpy transport through convection and diffusion, the energy conservation equation is constituted by the following source terms: $\frac{j_r^2+j_z^2}{\sigma}$, heating through Joule effect; S_r, heat dissipation through radiation per unit of plasma volume ($S_r=4\pi\varepsilon_N$, where ε_N is NEC emissivity NEC - plasma is considered optically thin – experimental values [12]); $\frac{5}{2}\frac{k_b}{e}\left(\frac{j_r}{C_p}\frac{\partial h}{\partial r}+\frac{j_z}{C_p}\frac{\partial h}{\partial z}\right)$ (*electron drift*) is the enthalpy transport through electron current which is directed to the anode (k_b is Boltzmann constant; e is the elementary electric charge); $u\frac{\partial p}{\partial z}+v\frac{\partial p}{\partial r}$, energy variation due to pressure variation.

The reference [2] presents the thermodynamic and plasma transport properties for some gases (air; argon; others).

The current densities and the azimuthal component of the magnetic field are obtained through the electric current conservation equation solution, and through the third *Maxwell* equation, with the help of the "adapted" *Richardson-Dushmann equation* (applicable for thermionic emission).

Electric current conservation equation (in the form of electric potential):

$$\frac{1}{r}\frac{\partial}{\partial r}\left(r\sigma\frac{\partial \Phi}{\partial r}\right)+\frac{\partial}{\partial z}\left(\sigma\frac{\partial \Phi}{\partial z}\right)=0 \qquad (6)$$

where Φ is the electric potential.

Third *Maxwell* equation (in differential form):

$$\frac{1}{r}\frac{\partial}{\partial r}(rB_\theta)=\mu_0 j_z \qquad (7)$$

where μ_0 is the magnetic permeability in free space (vacuum).

One should observe that this equation is applicable only for symmetric current distributions (in this specific case, symmetry related to axis: axisymmetry).

The current density vector $\vec{j}$ is related to the electric field intensity vector $\vec{E}$, or to Φ electric potential, through $\vec{j}=\sigma\vec{E}=-\sigma\nabla\phi$. So, for such case (2D model): $j_z=-\sigma\frac{\partial \Phi}{\partial z}$ and $j_r=-\sigma\frac{\partial \Phi}{\partial r}$.

2.3. Computational domain and boundary conditions

The computational domain used for the simulations of this work is schematized in figure 2 and the employed boundary conditions are summarized in table 1.

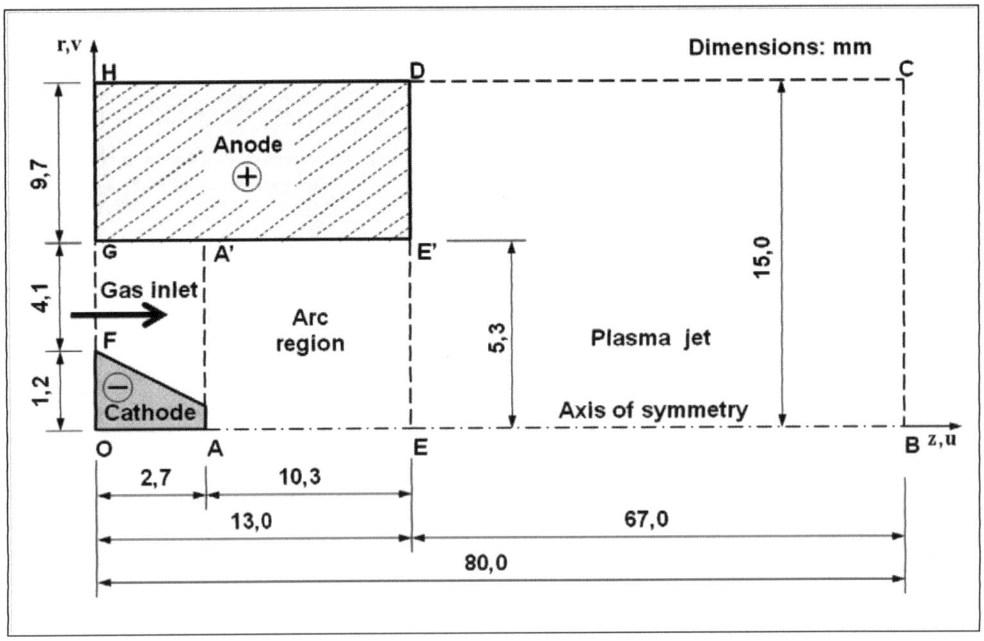

Figure 2. Computational domain.

Table 1. Summary of the boundary conditions.

	u	v	w	h	Φ
AB	$\partial u/\partial r=0$	0	0	$\partial h/\partial r=0$	$\partial j_z/\partial r=j_r=0$
BC	$\partial(\rho u)/\partial z=0$	$\partial v/\partial z=0$	$\partial w/\partial z=0$	$\partial h/\partial z=0$	
CD	0	$\partial(r\rho v)/\partial r=0$	0	$300K$	
FG	$u=u(r)$	0	$w=w(r)$	$500K$	
AA'					$j_z(r)$ (specified)
EE'					$j_z=j_r=0$
Anode	0	0	0	$1000K$	0
Cathode	0	0	0	$3000K$	

The description of the boundary conditions is presented as follows:

Inlet (**FG** line): in the inlet section, the axial velocity presents a parabolic profile and depends on the specification of the gas flow rate. Radial velocity is zero at the inlet. Azimuthal velocity presents a profile partially formed by the behavior of free vortex and forced vortex. The azimuthal velocity depends on the swirl number in torch inlet. The definition of the swirl number used in this work is presented by [4]: ratio of the axial flux of the azimuthal momentum to the axial flux of the axial momentum, normalized by an appropriate/characteristic radius (R_0). In this work, R_0 radius is the torch nozzle radius. Therefore, the swirl number is given by:

$$Sw=G_\theta / G_z R_0 \tag{8}$$

where the axial flux of the azimuthal momentum is

$$G_\theta=\int_0^\infty \rho u w r^2 dr \tag{9}$$

and the axial flux of the axial momentum is

$$G_z=\int_0^\infty \rho u^2 r dr \tag{10}$$

Other conditions: on the surfaces of electrodes (anode and cathode) the *non-slip* conditions are assumed $(u=v=w=0)$. In the symmetry axis (line **AB**): $\partial u/\partial r=0$ and $v=w=0$. Conditions of *free boundary* are assumed for **BC** and **CD**. For **BC**: $\partial(\rho u)/\partial z=0$, $\partial v/\partial z=0$ e $\partial w/\partial z=0$. And for **CD**: $u=w=0$ and $\partial(r\rho v)/\partial r=0$. The boundary conditions for enthalpy are provided in temperature and converted to gas enthalpy. For the torch inlet, the temperature is $500K$ and for **CD** line the temperature is $300K$. The temperatures of cathode and anode surfaces are assumed $3000K$ and $1000K$, respectively. In the symmetry axis (line **AB**): $\partial h/\partial r=0$; and in line **BC**, $\partial h/\partial z=0$. Lines **AA'** and **EE'** are internal boundary conditions and are known as "fictitious" or "porous" boundary conditions (in these lines the boundary conditions for the electric potential (Φ) are assumed). In line **AA'**, the electric potential is determined by assuming a axial current density profile [7, 18], adapted from *Richardson-Dushmann* equation (applicable for thermionic emission): $j_z(\mathrm{r}) = j_0 \mathrm{e}^{-r/r_c}$, where j_0 and r_c are constants which depend on the electric current and are experimentally obtained. In line **AB**: $\partial j_z / \partial r = 0$ and $j_r = 0$. In **ED**: $j_r = j_z = 0$. In anode: $\Phi = 0$.

2.4. Numerical method

The model was implemented in the Computer Code for Simulation of the Swirling Flow in Thermal Plasma Torches (CTP), developed in order to enable the numerical solution of governing equations. CTP code uses the Finite Volume Method (FVM) and it is based on [9, 10, 19]. CTP was elaborated in *FORTRAN 90* language (*Microsoft FORTRAN Powerstation 4.0*) and the simulations were executed in a microcomputer with *INTEL CORE i7* processor and 6GB of *RAM* memory. The average computing time for each simulation is 15min, for meshes of 120x80 nodes and approximately 20000 iterations.

3. Initial results and comments

This article presents some results of initial simulations executed with *CTP* code. The results are categorized according to their goals: comparison with experimental results described in literature; study of the behavior of swirling flow in plasma torches.

3.1. Comparison with literature experimental results

In order to verify the quality of results of simulations with *CTP*, there were researches in literature to find experimental results obtained in plasma torches similar to the ones of this work and, also, in comparable operating conditions. In fact, there are a few experimental results in literature due to technical difficulties to execute the measurements of temperature and, mainly, velocity. Besides that, in general, there are considerable differences between the torches experimentally researched and the one used in this work.

The results of the performed experiences in Idaho National Engineering Laboratory (*INEL*) were chosen for the comparisons due to the similarity between its torch and the torch simulated in this work and, also, because the swirl was generated in it, enabling a proper comparison for the interest of this work in the study of the effects of swirl in the behavior of arc and plasma jet. The results obtained in *INEL* were published by [20].

The experimental apparatus used in *INEL* consisted of a non-transferred arc plasma torch and a chamber for pure argon atmosphere control. The torch operated in laminar mode, so that the flow in its interior part and in the jet, near its nozzle outlet, may be assumed as laminar (once the number of *Reynolds* at outlet is in the range of 100-200) [20]. An emission spectroscopy technique was used to measure the temperature of the plasma jet coming outside the torch. Investigators estimated the following measurements errors: ±0,2% for 12000K and ±1,5% for 9000K [20].

The numerical simulations were executed for two operating conditions of *INEL* torch and they are presented in table 2.

The comparisons between the experimental results and the temperature calculated results for both operating conditions are showed in figures 3, 4, 5 and 6. In these figures, it is possible to observe that discrete figures denote experimental results and the lines indicated the calculated results.

Figure 3 refers to BES23-INEL case and presents the temperature values of plasma jet coming out of the torch. The values were experimentally obtained (measured) in function of the axial distance (origin of the graph coincides with the origin (0) of the computational domain) for the radial position: torch symmetry axis (r=0). The temperature profile which was numerically obtained (calculated) through computer code (*CTP*), for the same operating conditions, is also presented.

The comparison presents the following differences: the biggest difference between the measured temperature values (experimental ones) and the calculated values (simulation with *CTP* code) is -5,4% (Texp=11196 K; Tcalc=11800 K), in the axial distance z=33 mm. From the distance z=41 mm, the

Table 2. Operating conditions of INEL torch.

INEL Code	Plasma Gas	Gas Flow (l/min)	Gas Flow (scmh)	Electric Current (A)	Electric Voltage (V)	Input Power (kW)	Swirl Number
BES23	Argon (Ar)	9,83	0,59	250	19,16	4,79	Sw=5
BES24	Argon (Ar)	13,83	0,83	250	19,44	4,86	Sw=5

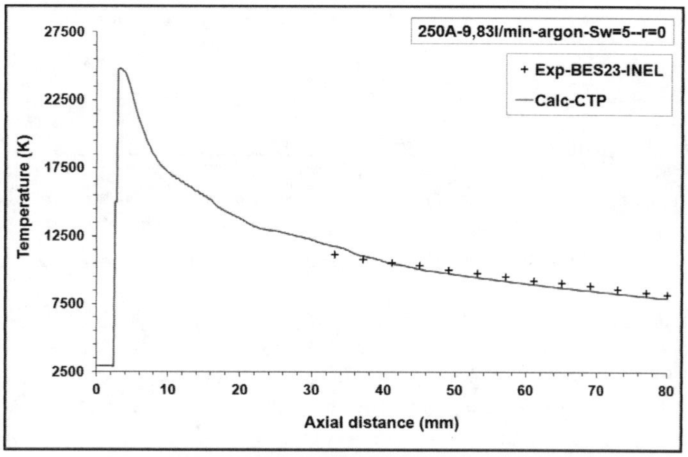

Figure 3. Temperature profiles in function of axial distance in the radial position: symmetry axis (r=0). Comparison between experimental results (BES23-INELcase) and calculated results (simulation-*CTP*).

measured values are bigger than the calculated values and the biggest difference between them is +4,3%, in the axial distance z=69 mm. In other axial distances, the differences are between +2,8% and +3,4%.

The comparison between the radial temperature values that were measured and the temperature radial profiles that were obtained with *CTP* for BES23-INEL case is shown in figure 4, for axial positions: z=33 mm, z=45 mm, z=57 mm, z=69 mm and z=80 mm.

The differences between the experimental and calculated values are the following ones:

Axial position z=33 mm: in the torch symmetry axis, r=0, the difference between the experimental and the calculated value is -5,4%, already showed in figure 3 as well. The biggest difference, -5,5%, occurs in r=0,5 mm. Between r=0,5 mm and r=2,3 mm, the difference varies between -3,7% and -2%, and between r=2,7 mm and r=4,4 mm the differences vary from -1,3% to +1,3%.

Axial position z=45 mm: the differences vary from +1,0% to 3,1%.

Axial position z=57 mm: between r=0 and r=3,4 mm, the differences are between +3,4% and +4,3%. In r=3,8 mm the difference decreases to 1,3%.

Axial position z=69 mm: between r=0 and r=2,7 mm, the differences are between +4,3% and +1,5%. In r=3,0 mm the difference is -1,5%. Axial position z=80 mm: in the torch symmetry axis (r=0) the difference between the experimental and the calculated value is +3,0% (already showed in figure 3 as well). Between r=0,2 mm and r=0,9 mm, the differences vary from +2,2% to 1,4%. Between r=1,1 mm and r=1,4 mm, the differences vary from +1,1% to 0,5%. In r=1,5 mm the difference is -1,0%.

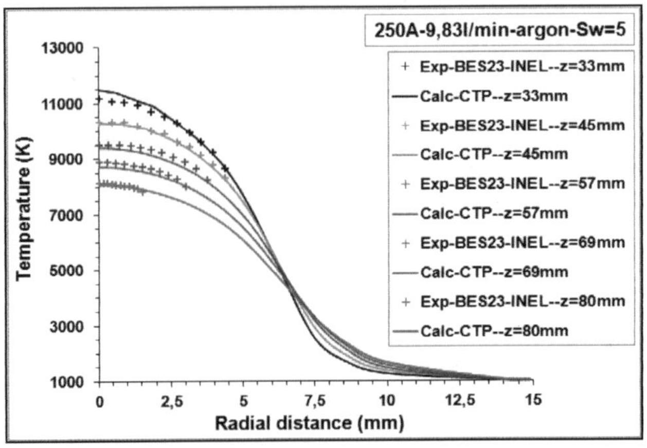

Figure 4. Temperature profiles in function of the radial distance in axial positions (plasma jet): 33 mm; 45 mm; 57 mm; 69 mm; 80 mm. Comparison between experimental results (BES23-INEL case) and calculated results (simulations-*CTP*).

The comparisons between experimental results and calculated ones for BES24-INEL case are showed in figures 5 and 6.

Figure 5 presents the temperature in function of the axial distance for the radial position r=0 (torch symmetry axis). The comparison between the measured and the calculated values shows the following differences: in z=33 mm the difference is -2,7%. From the axial distance z=37 mm, the measured values are bigger than the calculated ones and the biggest difference between them is +4,8% (Texp=10390 K; Tcalc=9890 K), in axial distance z=45 mm. In other axial distances, the differences are between +1,5% and +3,5%.

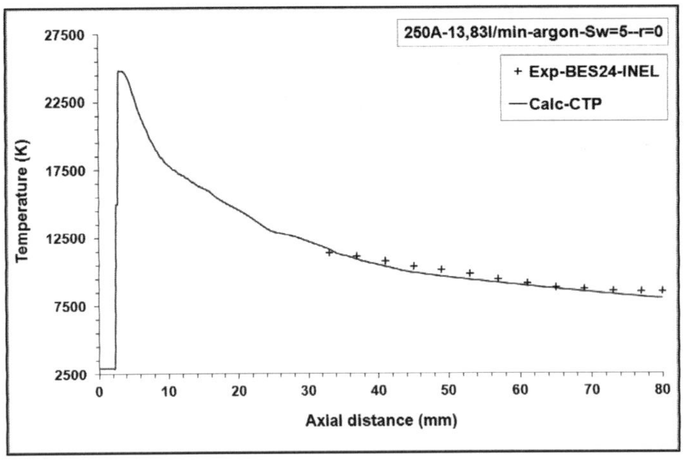

Figure 5. Temperature profiles in functions of the axial distance in radial position: symmetry axis (r=0). Comparison between the experimental results (BES24-INEL case) and the calculated results (simulation-*CTP*).

Figure 6 presents the temperature in function of the radial distance for the axial positions: z=33 mm, z=45 mm, z=57 mm and z=69 mm.

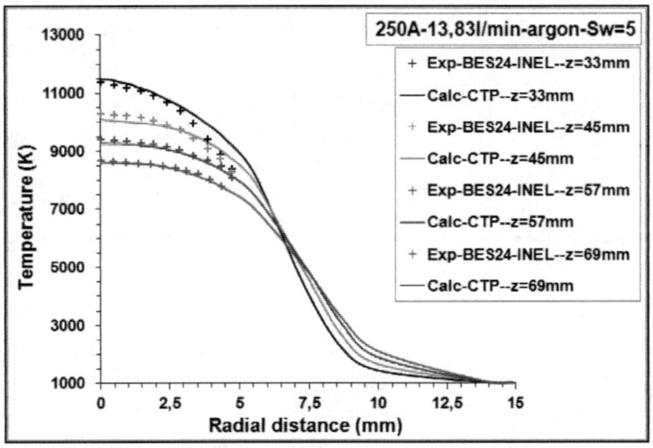

Figure 6. Temperature profiles in function of the radial distance in axial positions (plasma jet): 33 mm; 45 mm; 57 mm; 69 mm. Comparison between the experimental results (BES24-INEL case) and the calculated results (simulations-*CTP*).

The differences between the experimental values and the calculated values are the following ones:

Axial position z=33 mm: in the torch symmetry axis (r=0) the difference between the experimental and the calculated value is -2,7% (showed in figure 5 as well). The biggest difference, -10,0%, occurs in r=4,8 mm. Between r=0,5 mm and r=2,9 mm, the difference is around -2,5%. Between r=3,4 mm and r=4,4 mm the differences vary from -4,5% to -7,2%.

Axial position z=45 mm: between r=0 and r=1,9 mm the differences vary from +4,8% to +3,0%. From r=3,4 mm, the measured values are smaller than the calculated values and the biggest difference between them is -6,0%, in r=4,7 mm. Axial position z=57 mm: the differences vary from +2,4% to +1,9% between r=0 and r=4,4 mm. In r=4,8 mm, the difference is -1,6%. Axial position z=69 mm: between r=0 and r=4,4 mm the difference are between +2,4% and +1,3%.

Although the biggest differences between the experimental values and the calculated values are negative (-5,5% for BES23 case and -10,0% for BES24 case), that is, experimental values are smaller than calculated values, a systematic tendency is not characterized in this sense because there are also several regions where the experimental values are bigger than the calculated values (positive differences).

The predominance of the biggest differences in one of the two investigated cases in relation to each other is not characterized either. Considering that the average of differences is approximately ±3,2% and also considering the simplifying assumptions of the developed mathematical model, the concordance of the results of the numerical simulations with the results of measurements is good and, therefore, the model is adequate to the intended investigations in this work and in future ones as well.

3.2. Results of simulations of swirling flow in plasma torches

The numerical simulations carried out until the moment with *CTP* code for different plasma torches and operating conditions have enabled the study of behavior of flow in torches, as well as the study of effects of different swirl intensities in temperature and axial velocity in plasma flow. This paper presents (figure7 to figure 12) some results of simulations of an academic torch operating with the conditions of BES23-INEL case (table 2).

Figure 7(a) presents the temperature axial profiles in the radial position r=0 (torch symmetry axis), obtained with different swirls numbers and figure 7(b) presents the temperature radial profiles in the axial position z=33 mm, also obtained with different swirl numbers.

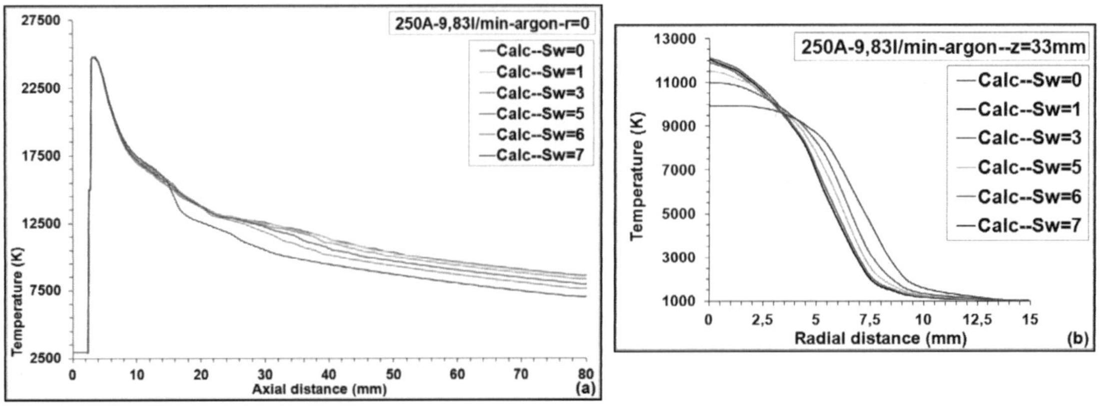

Figure 7. Temperature profile and swirl effect: (a) temperature axial profiles (radial position: symmetry axis (r=0)); (b) temperature radial profiles (axial position: 33 mm).

Figure 8 shows the temperature radial profile, for each swirl number, in the axial positions z=45 mm (a) and z=57 mm (b) and figure 9 in the axial positions z=69 mm (a) and z=80 mm (b).

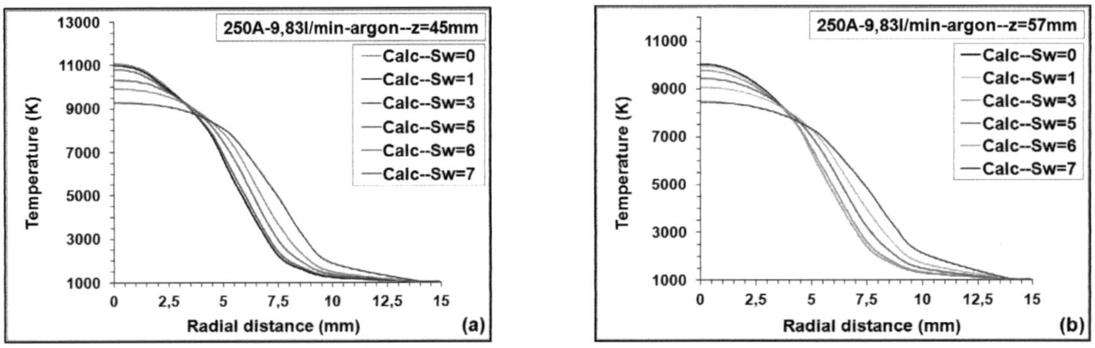

Figure 8. Temperature radial profiles and swirl effect: (a) axial position: 45 mm; (b) axial position: 57 mm.

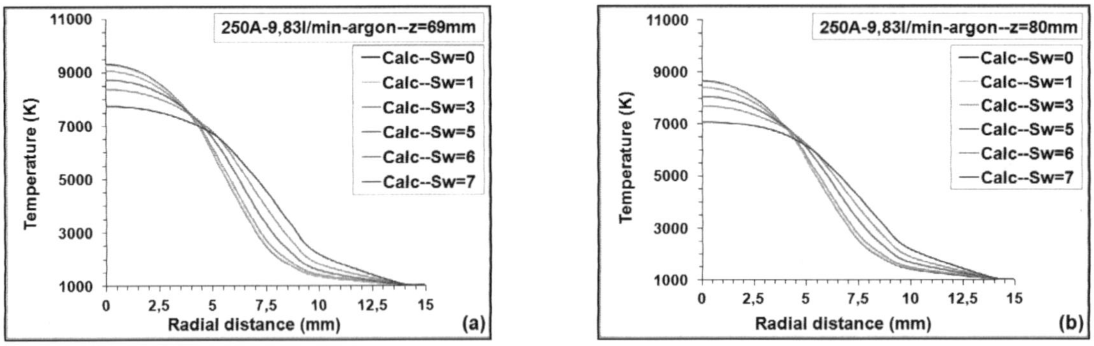

Figure 9. Temperature radial profiles and swirl effect: (a) axial position: 69 mm; (b) axial position: 80 mm.

Figure 10(a) presents the axial velocity profiles in function of axial distance, in the radial position r=0 (torch symmetry axis), obtained with different swirl numbers and figure 10(b) presents the axial velocity profiles in function of the radial distance, in the axial position z=33 mm, also obtained with different swirl numbers.

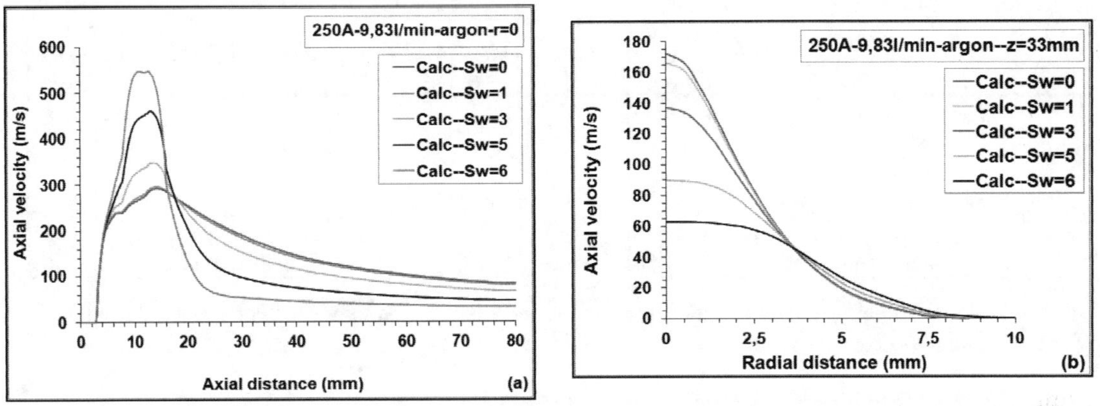

Figure 10. Axial velocity profile and swirl effect: (a) in function of the axial distance (radial position: symmetry axis (r=0)); (b) in function of the radial distance (axial position: 33 mm).

Figure 11 shows the axial velocity profile in function of the radial distance, for each swirl number, in the axial positions z=45 mm (a) and z=57 mm (b) and figure 12 in the axial positions z=69mm (a) and z= 80 mm.

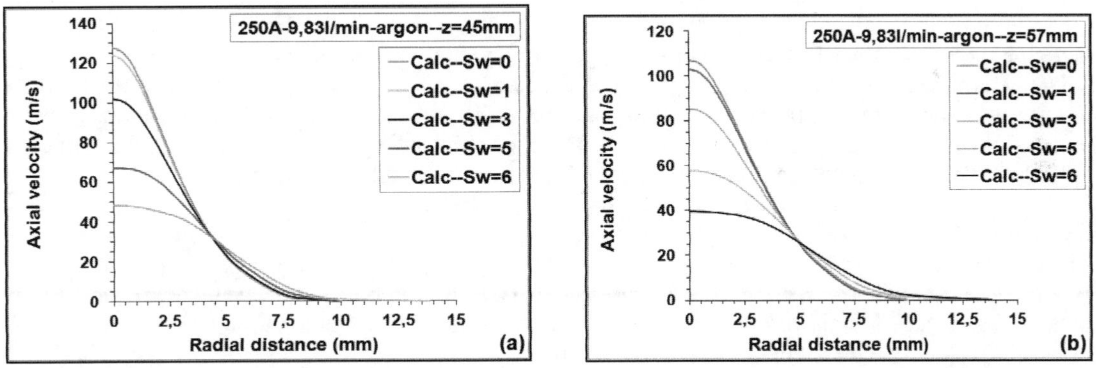

Figure 11. Axial velocity profile in function of the radial distance and swirl effect: (a) axial position: 45 mm; (b) axial position: 57 mm.

Through the analysis of figures it is possible to verify the following phenomelogical characteristics:

Due to gas ionization and, consequently, to its increased electric conductivity, in the region next to the cathode (where there is the maximum current density), there is a big increase of the gas temperature due to its heating through *Joule* effect.

Still inside the torch (up to the distance z=13 mm), there is a big decrease of the plasma temperature due to the convection with the anode cooled surface. When the plasma jet comes out of the torch, it presents a less intense reduction of temperature, in function of the axial distance, due to the convection with the environmental gas. Next to the cathode there is plasma acceleration due to its expansion (caused by the temperature increase) and due to *Lorentz* force effect (resulting from the interaction between the arc current and the inducted magnetic field). Then, there is a continuous deceleration of the velocity of plasma jet when in contact with the environmental gas.

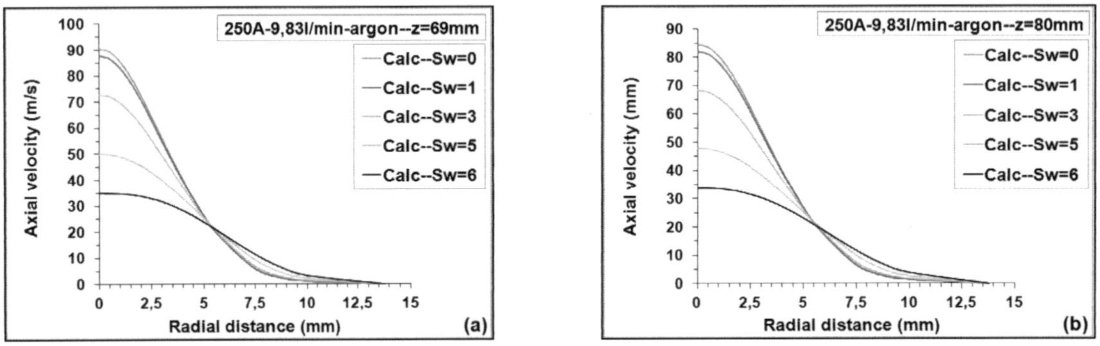

Figure 12. Axial velocity profile in function of the radial distance and swirl effect: (a) axial position: 69 mm; (b) axial position: 80 mm.

It is also possible to observe the "flattening" of the radial profile of the axial velocity of plasma jet and of the temperature radial profile, in the axial direction, due to the dragging with the environmental gas. Such behavior is also reported by [4, 6, 7].

The temperature profiles (figure 7 to figure 9) with Sw=0 and Sw=1 practically coincide, which indicates that a bigger swirl number is required to modify the arc behavior. Only for Sw>3, the swirl significantly affects the temperature profiles (axial and radial ones). The increase in the azimuthal momentum of plasma jet due to an increase in swirl intensifies the jet heat exchange with the environmental gas and, consequently, it causes a more intense decrease in temperature (that can be confirmed in the axial and radial profile figures).

Concerning the axial velocity (figure 10 to figure 12), its decrease is significantly intense for Sw≥3 due to the increase in the jet energy dissipation interacting with the environmental gas.

The determination of the relationship between the swirl number which is appropriate to rotate the arc and significantly reduce the anode erosion rate and its effects on the profiles of temperature and velocity of plasma jet, which are important for the efficient work of torches in determined applications, requires both the experimental surveys and the numerical simulations.

4. Conclusion

This work presented and discussed the results of some numerical simulations of thermal plasma torch flow carried out with the developed computer code (*CTP*), which was concisely presented. The mathematical model was also presented including the swirl in order to enable the study of its effect on the arc behavior inside the torch and in the free plasma jet.

With the purpose of verifying the quality of the simulation results, part of them was compared to the experimental results and was observed that -10% is the biggest difference obtained between the experimental temperature values and the calculated ones, and ±3,2% is the average obtained from the differences between the values.

Simulations were made for this study with the following characteristics /operating conditions: conic cathodes; current intensities of 100 and 250 A; argon at atmospheric pressure; gas flow of 9,83 and 13,83 l/min; flow with and without swirl (swirl numbers, Sw= 0 to 7). Through the analysis of the results, it is possible to point out the following remarks: the flow temperature curves with Sw=0 and Sw=1 practically coincide, which indicates that a bigger swirl number is required to modify the arc behavior. Only for Sw≥ 3, the swirl significantly affects the temperature profile. The increase in the azimuthal momentum of plasma jet due to an increase in swirl intensifies the jet heat exchange with the environmental gas and, consequently, it causes a more intense decrease in temperature. Concerning the axial velocity, its decrease is significantly intense for Sw≥3 due to the increase in the jet energy dissipation interacting with the environmental gas. Therefore, it is possible to conclude that the model is able to carry out numerical investigations of flow in plasma torches and swirl effects in the arc/flow interaction.

Acknowledgements
The authors acknowledge the support from Escola Politécnica da Universidade de São Paulo and Centro de Pesquisa da Universidade São Judas Tadeu.

References

[1] Feinman J 1987 *Plasma technology in metallurgical processing* (Feinman and Associates, USA)
[2] Boulos M I, Fauchais P and Pfender E 1994 *Thermal plasma: Fundamentals and applications* vol 1 (New York: Plenum Press)
[3] Gleizes A, Gonzalez J J and Freton P 2005 *J. Phys. D: Appl. Phys.* **38** 153
[4] Westhoff R and Szekely J 1991 *J. Appl. Phys.* **70** 3455
[5] Sun X and Heberlein J 2005 *Journal of Thermal Spray Technology* **14**(1) 39
[6] Scott D A, Kovitya P and Haddad G N 1989 *J. Appl. Phys.* **66** 5232
[7] Murphy A B and Kovitya P 1993 *J. Appl. Phys.* **73** 4759
[8] Bauchire J M, Gonzalez J J and Gleizes A 1997 *Plasma Chem. Plasma Process.* **17** 409
[9] Favalli R C 1997 *Master Dissertation*, Universidade de São Paulo
[10] Bianchini R C 2000 *Master Dissertation*, Universidade de São Paulo
[11] Klinger L 2002 *PhD Thesis*, L'école Polytechnique de Lausanne
[12] Baudry C 2003 *PhD Thesis*, Universite de Limoges
[13] Li H P, Pfender E and Chen X 2003 *J. Phys. D: Appl. Phys.* **36** 1084
[14] Yuan X Q, Li H, Zhao Z, Wang F, Guo K and Xu P 2004 *Plasma Chem. Plasma Process.* **24**(4) 585
[15] Li H P and Pfender E 2007 *Journal of Thermal Spray Technology* **6**(2) 245
[16] Trelles J P, Chazelas C, Vardelle A and Heberlein J V R 2009 *Journal of Thermal Spray Technology* **18**(5-6) 728
[17] Huang R, Fukanuma H, Uesugi Y and Tanaka Y 2013 J*ournal of Thermal Spray Technology* **22** 183
[18] Hsu K C, Etemadi K.and Pfender E 1983 *J. Appl. Phys.* **54** 1293
[19] Vatavuk P 1996 *PhD Thesis*, Universidade de São Paulo
[20] Dilawari A H, Szekely J, Batdorf, J, Detering R and Shaw C B 1990 *Plasma Chem. Plasma Process.* **10** 321

Helium-3 Generation from the Interaction of Deuterium Plasma inside a Hydrogenated Lattice: Red Fusion[1]

Edbertho Leal-Quiros, David A. Leal-Escalante
Neopowertech, LLC. 2224 E. Oakmont Ave, Fresno, CA 93730 (US)

E-mail: edbertho@neopowertech.us; david@neopowertech.us

Abstract. Helium-3 has been created in a nuclear fusion reaction by fusing deuterium ions from deuterium plasma with hydrogen ions in a "RED" (the Spanish word for net) or crystal lattice, a method we called red fusion ("*Fusion en la red cristalina*"), because is a new method to make nuclear fusion reaction. In this paper, it will be show the experimental results where the helium-3 has been generated for the first time in this kind of new method to confine deuterium and hydrogen inside the RED or lattice of the hydrogenated crystal and that confinement inside the RED facilitated overcoming the Coulomb barrier between them and helium-3 and phonons are produced in this fusion reaction. The results of a long time research in which helium-3, has been created in a fusion reaction inside the lattice or RED of the crystal that contained hydrogen after adequate interaction of deuterium plasma at appropriate high temperature and magnetic confinement of the Mirror/Cusp Plasma Machine at Polytechnic University of Puerto Rico, designed by the authors. Several mass spectra and visible light spectrum where the presence of helium-3 was detected are shown. The experiment was repeated more than 200 times showing always the generation of helium-3. In this experiment no gamma rays were detected. For this experiment several diagnostic instruments were used. The data collection with these control instrumentation are shown. Thus, it is an important new way to generate Helium-3. © 2013 NEOPOWERTECH, LLC. All rights reserved.

1. Introduction

This experiment consists of exploring several iterations to find a viable process for the creation of Helium-3 with Nuclear Fusion by using Julian Schwinger's theories. Helium-3 (also known as He-3 or ${}_2He^3$) is a very rare stable isotope of Helium that has two protons and one neutron in the nucleus. He-3 is an odourless, colorless, and tasteless high pressure gas. It has a mass of 3.016029310 (U), equivalent to 5.0086715x10-27 kg, and the percentage of abundance is only 0.000137 in a naturally occurring mixture of Helium isotopes. Its atomic number is Z=2 (it has two protons) and its atomic mass is A=3 (it has three nucleons, comprising two protons and only one neutron). He-3 has essential uses in cryogenics physics because in a mix alone at 2.2 K, it presents superfluidity. He-3 is also utilized in Medicine, mainly in nuclear medicine in cardiology procedures and in Hyperpolarized MRI (magnetic resonance imaging) of the lungs and respiratory system. It is possible to see more amplified details of the lungs when the patient breaths He-3 for the hyperpolarized MRI than is possible to view in conventional MRI lung images. Furthermore, He-3 has been used in radiation detectors, mostly of the proton recoil type for neutron detectors, which are of big importance in accelerators, nuclear

[1] **United States Patent AP. Pub. No.: US 2013/0329844A1, Pub. Date: Dec. 12, 2013.**
Inventors: Edbertho Leal-Quiros and David Leal-Escalante. Neopowertech LLC, Fresno, CA (US)

reactors, non-proliferation, and national Homeland Security, among others. At the present time, there is large deficit of He-3 in all countries of the world, but the greatest needs are mainly in USA and Europe. It is known in astrophysics, mainly in the stellar structure and stellar evolution theories, that in the center of a star like the Sun, the fusion reaction: ${}_1P^1 + {}_1D^2 = \hbar\omega + {}_2He^3$, may happen. Where ${}_1P^1$=Proton, ${}_1D^2$=Deuterium, $\hbar\omega$=phonon and ${}_2He^3$=Helium-3. Moreover, the stellar interior is at very high pressure, about 10^{12} atmospheres, and very high temperatures, about $15X10^6$ K. This fusion reaction has not been widely emphasized in the normal Fusion Research programs. Julian Schwinger (the writer's former professor) presented theoretical calculations that show evidence for the assertion that an H-ion in a deuterided lattice encounters a relatively narrow Coulomb barrier before fusing to form He-3 according to the reaction presented below;

$$ {}_1P^1 + {}_1D^2 = \hbar\omega \text{ (phonon)} + {}_2He^3 \qquad (1) $$

Julian Schwinger in his late days suggested that the final product of the fusion reaction (1), [3-4] could be ${}_2He^3$ and Phonons instead the Photons, [3-4] if the reaction were to occur under special conditions, such as inside a particular lattice of a crystal, in which the phonon produced in the reaction could be absorbed by the crystal itself. This experiment use proprietary information that includes Plasma and targets of appropriate crystals to obtain Helium-3. From the authors' previous experiences using plasmas, and Ion sources design, there was very high probability of obtaining positive results. This meant that using Mass Spectrometry techniques, it was possible to identify the generation of He-3 and therefore to proving the Schwinger suggestion of this particular reaction. This idea pushed us to believe that the: ${}_1P^1 + {}_1D^2 = \hbar\omega$ (phonon) + ${}_2He^3$, **could be possible.**

For obvious reasons, more specifics with regards to the description of the experiment cannot be provided. The needs of Helium-3 are real and vital for our human civilization.

2. Experiment and location

The experiment was performed at the Polytechnic University of Puerto Rico's Plasma Laboratory on the Mirror/Cusp Plasma Machine [5-8]. According to Schwinger's Theoretical reaction, the PUPR Plasma Machine was used to create the necessary conditions for the reaction to take place inside the crystal lattice of several hydrogenated lattice materials.

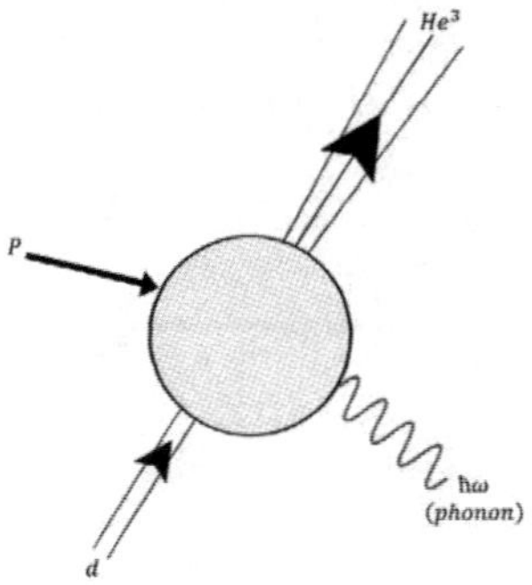

Figure 1. Feynman Diagram of the Reaction ${}_1P^1 + {}_1D^2 = \hbar\omega$ (phonon) + ${}_2He^3$ to get Helium-3.

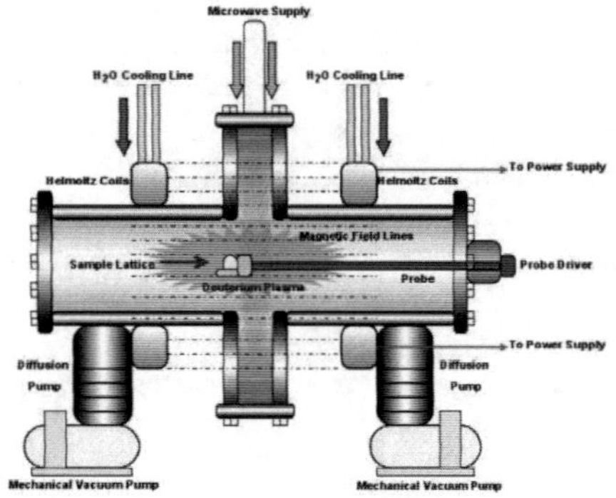

Figure 2. Helium-3 experimental diagram using the PUPR plasma machine.

For this experiment several Diagnostic instruments were used such as: Langmuir Probes (Single and Double), Mass Spectrometer, Visible Spectrometer, NaI Detector, Geiger Mueller, Hall Probe, Vacuum Pressure Gauges, Laser Thermometers, among others. Using a Sodium Iodine detector (NaI Detector), no gamma or beta radiation was detected during this experiment. By using Mass

Journal of Physics: Conference Series **591** (2015) 012039 doi:10.1088/1742-6596/591/1/012039

Spectrometry techniques available at the PUPR Plasma Laboratory, such as the Quadrupole mass analyzer, the detection of Helium-3 was possible. During every step of the experiment, data was recorded to constantly monitor all the experimental plasma parameters. The experiment was repeated more than 200 times showing always the generation of Helium-3.

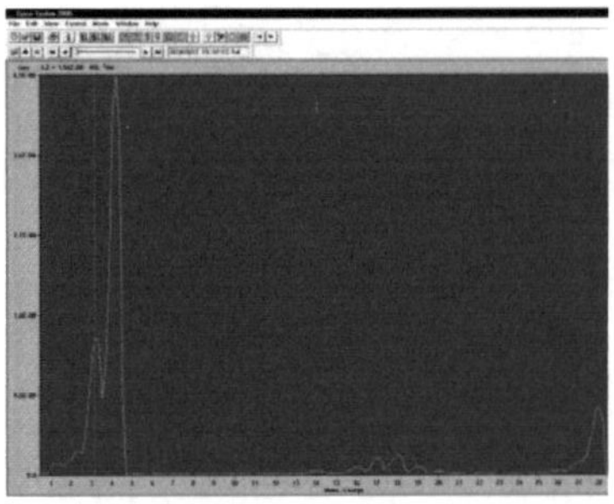

Figure 3. Mass Spectrum of a hydrogenated sample 1. He-3 is clearly identified.

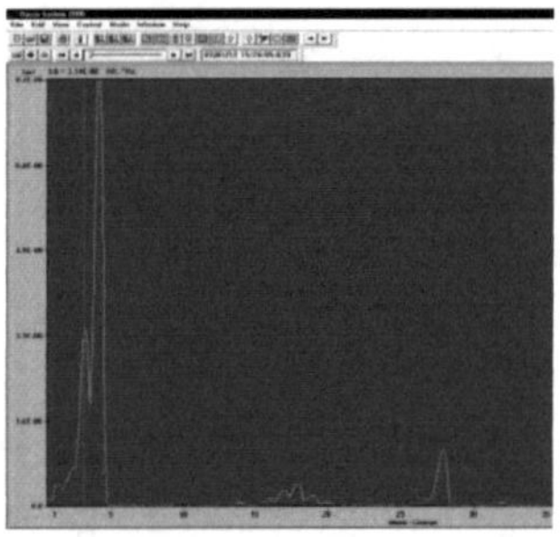

Figure 4. Mass Spectrum of a hydrogenated sample 2. He-3 is clearly identified.

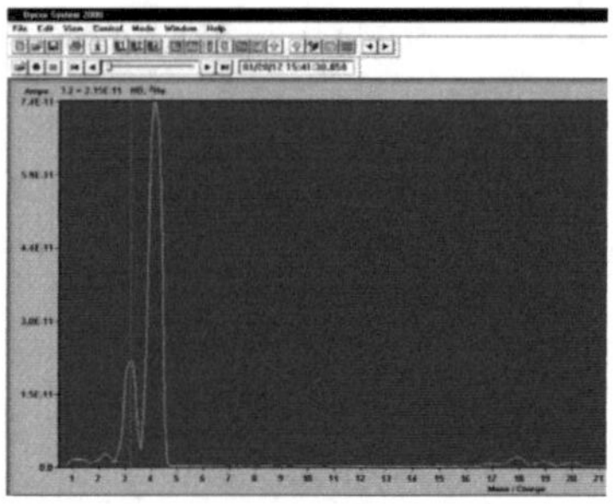

Figure 5. Mass Spectrum of a hydrogenated sample 3. He-3 is clearly identified.

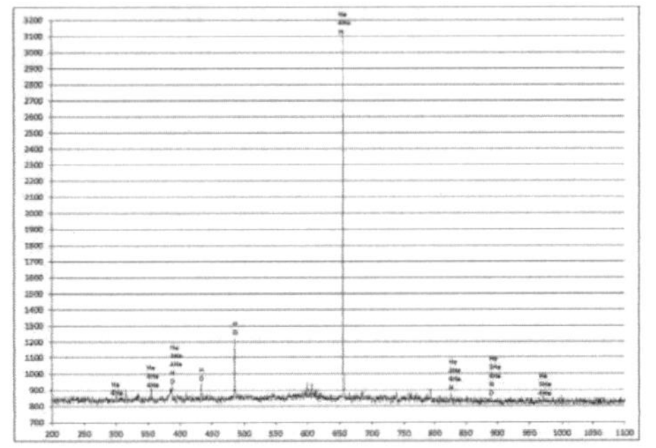

Figure 6. Optical Spectra. Intensity vs. Wavelength. He-3 peaks are clearly identified.

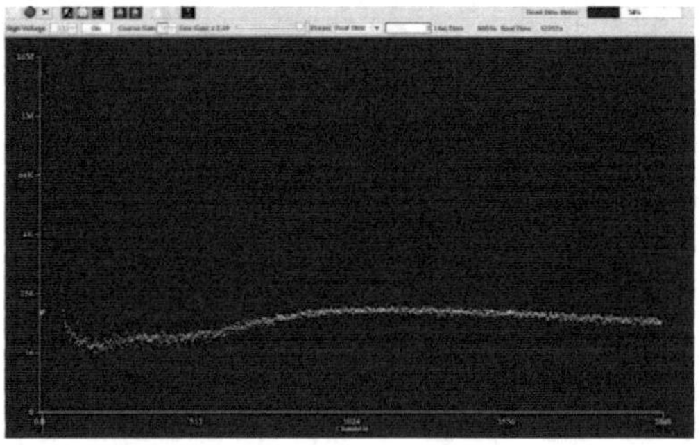

Figure 7. The Sodium Iodine Detector graph showed ***no spikes of Gamma radiation*** during the experiments.

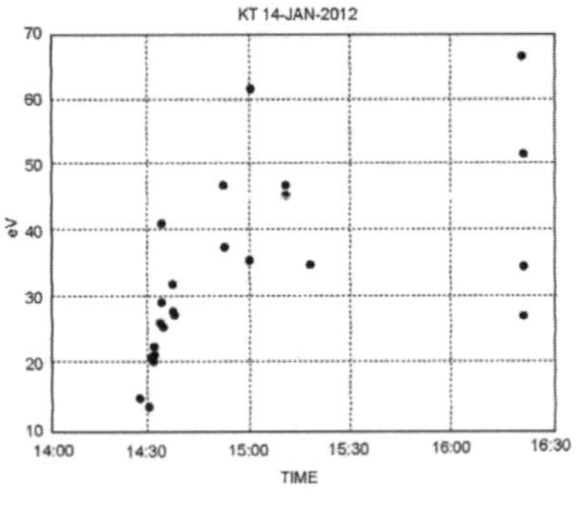

Figure 8. Temperatures measured during the deuterium plasma experiment with the Single Langmuir probe. None of the temperatures are lower than 10 eV.

3. The PUPR plasma machine advantages

The PUPR Plasma Machine can generate a very stable plasma (several days) and it is possible to obtain a range of plasma densities from $n=10^4$ to 10^{12} ions per cubic centimeter and temperatures from $2x10^3$ to $30x10^6$ Kelvin degrees in the hot electron beams region.

Figure 9. Dr. Edbertho Leal-Quiros (right) and David A. Leal-Escalante with the PUPR Mirror Cusp Plasma Machine, San Juan, Puerto Rico

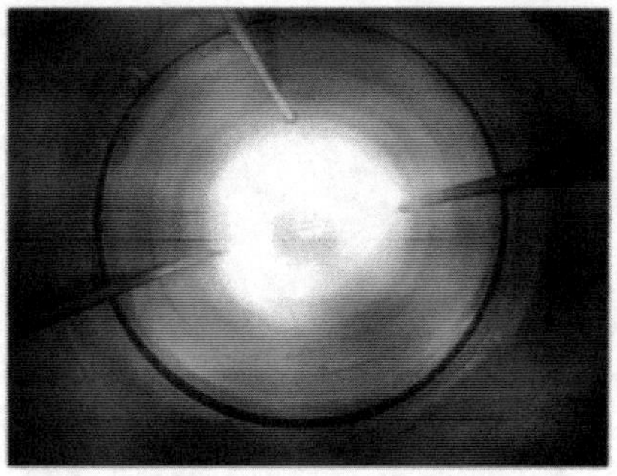

Figure 10. Stable plasma at the PUPR Plasma Lab.

4. Conclusions

It was to demonstrated that Helium-3 could be produced using Deuterium-Ions interacting with Hydrogen-Ions already inside our Hydrogenated Lattice or “Red”, according to Julian Schwinger's Hypothesis. The results obtained with the Quadrupole Mass Analyzer show masses of three (3) in the mass spectra. Because the average kinetic energy of the plasma was over 10 eV, which is larger than the bonding energy of the molecule of H-D (4.6 eV), H-D molecules are dissociated in the Deuterium Plasma. The Optical Spectra show the distinct peaks that are the visual fingerprint of Helium-3. Therefore, the Atomic Mass three (3) corresponds to Helium-3, as Schwinger predicted.

5. Acknowledge

This work has been supported by Omar Leal-Quiroz and Neopowertech, LLC.

6. References

[1] Private communication in Los Alamos National laboratory (LANL), spring 1985. Wheatley J was the first Physicist to obtain experimentally less than 1 K degree of temperature; I had the fortune to be in his cryogenic research group in LANL since June to Dec 1986 working with Ecke R and Swift G, Post Docs of Prof John Wheatley and also my mentors

[2] Private communication of Prof. Doulas Osheroff (Nobel Laureate 1996) to E.Leal-Quiros in Cuzco, Peru; Sept 1998. He discovered the Superfluidity of He-3, for which he was awarded the Nobel Prize in Physics

[3] Private communication at UCLA 1986. Prof. Julian Schwinger (Nobel Laureate 1965) was professor of E.Leal-Quiros of three courses in Advanced Quantum Mechanics at UCLA in 1984-1985, and Leal-Quiros translated his Advanced Quantum Mechanics book to Spanish.

[4] Schwinger J collection papers 1920-1994, UCLA Library Dept. of Special Collections, Collection 371. Box 25, Folder 8, Physics 221A Mecanica Cuantica Avanzada, traducida por Edbertho Leal, manuscript in spanish, fall 1984.

[5] Leal-Quiros E 1981 “Some Experimental Facts that Indicate the Elimination of Astigmatism in Ion Beams with Separators EXB.” Journal of Applied Physics **52** 1152

[6] Leal-Quiros E, Prelas M 1990 “Focused High Intensity Proton Beam from Lithium Source by Using EXB Stigmatic Selector.” Rev. Sci. Instrum **61** (1) 636

[7] Leal-Quiros E 1999 “Advanced Analyzers and Probes for Fusion-Plasma Diagnostics,” Trends in International Fusion Research. Second Symposium, edited by E. Panarella, NRC Research Press, National Research Council of Canada, Ottawa, ONK1A ,0R6 Canada, pp 1-23

[8] Leal-Quiros E 2006 Solar Probe Thermal Protection System Risk Mitigation Study, submitted to APL-NASA and NASA Goddard Space Flight Center, Living With a Star Program office. Sec 4.6 pp 4-23 to 4-27 http://solarprobe.gsfc.nasa.gov/SP_RMStudy_HiRes.pdf

p-ZnO Thin Films Deposited by RF-Magnetron Sputtering

L S Zambom[1] and R D Mansano[2]

1DSE -Faculdade de Tecnologia de São Paulo – CEETEPS, Brasil
2DEE-PSI-LSI-Escola Politécnica da Universidade de São Paulo, Brasil

Email: zambom@fatecsp.br, mansano@lsi.usp.br

Abstract. In this work, we study ZnO thin films deposited onto silicon substrate by RF-magnetron sputtering for semiconductor applications. For this, we analyses resistivity, photoconduction, composition and semiconducting characteristics, in these films. The results indicate that the films are nearly stoichiometric, p-type, with relative high resistivity and some samples shown photocurrent when exposed to white light.

1. Introduction

Undoped ZnO thin films are a promising optoelectronic and gas sensing material because of its band gap (3.37 eV) and have n-type conductivity. ZnO can be obtained by various techniques such as spray pyrolysis [1], chemical vapor deposition [2], pulsed laser deposition [3], sputtering [4,5] and sol-gel process [6,7]. Its semiconducting property is function of the native defects oxygen vacancies and zinc interstitials [7]. On the other hand, is difficult to produce stable p-ZnO and this conductivity is obtained with N doping [8].

2. Experimental

Undoped p-ZnO thin films deposited onto p-type silicon substrate (100) from a pure zinc metal target (99.99%) and oxygen (99.999%) atmosphere using a home building RF-magnetron sputtering. Previous of the deposition, the silicon wafers were cleaned using 1:H_2O_2 + 4:H_2SO_4 solution (Piranha clean) followed by dip in 10 % HF solution to remove the silicon oxide formed in the H_2SO_4 solution. The pressure was maintained at 10 mTorr, the target-substrate distance was 10 cm, the substrate temperature was maintained constant at about 90 ^{o}C and the ZnO thin films were deposited at different RF power (13.56 MHz) of 50 W, 100 W, 150 W and 200 W. The oxygen flow was kept at 70 sccm and process time was 30 and 60 minutes. The thicknesses measurements obtained by spectral reflectance at 632.8 nm wavelength light source. The ZnO stoichiometry analysed using Rutherford Back Scattering measurements. Conduction induced by white light were obtained from metal (Al)/dielectric/Si 700 μm x 700 μm capacitors that were built at the front side of the wafers. The bottom contact in the wafers backside is 500 nm thickness aluminium films. We use I-V measurements for the resistivity analysis and hot probe measurement for obtain the semiconducting type.

3. Results and Discussion

The sputtering process of Zinc oxide p-type thin films has a relatively high deposition rate, (Figure 1). Ndong [9], in similar process pressure and temperature conditions, obtained deposition rate values from 5.8 nm/min to 8.8 nm/min to 50 W sputtering power. However, Xiong [10] obtained ZnO n-type with deposition rate changing from 33.3 nm/min to 166.6 nm/min at 30 mTorr and 350 ^{O}C. Our results show that is possible to deposit ZnO thin films with reasonable deposition rate in extremely low temperature using different substrates as polymeric ones.

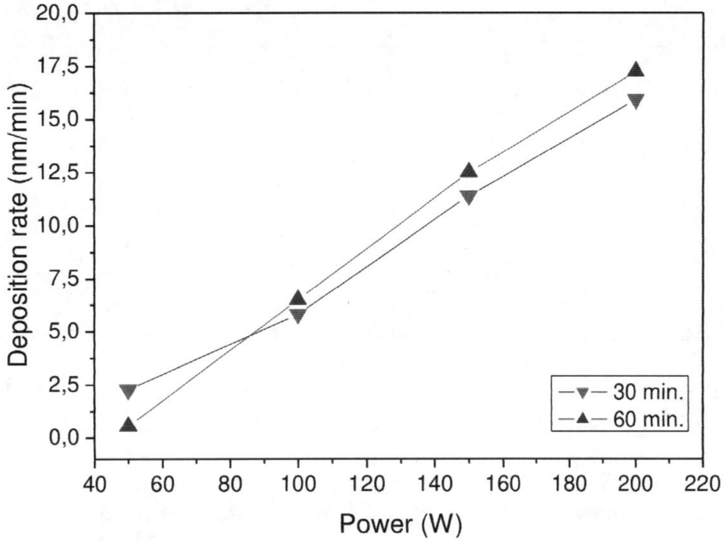

Figure 1 – Zinc oxide deposition rate versus RF power

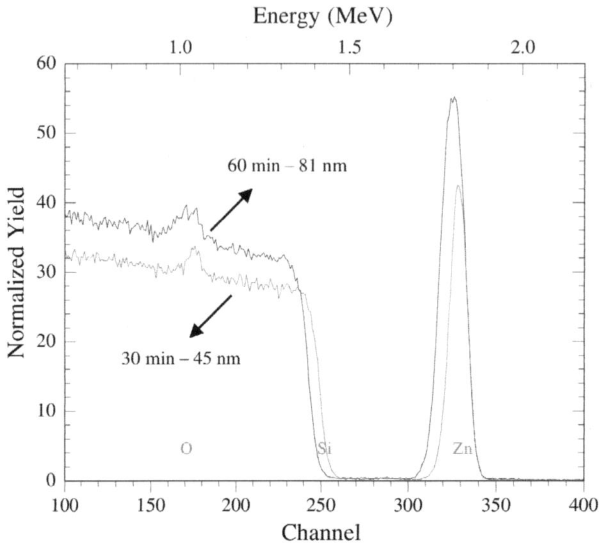

Figure 2 - Zinc oxide RBS spectra for different oxidation times – 100 W

The Zn/O ratio, calculated from RBS spectra, figure 2, changed from 1.03 to 1.12 near to stoichiometric one. Unfortunately, RBS technique was not able to detect any impurities in the samples from the metallic chamber. The intrinsic defects as Frenkel vacancy and Schottky by the zinc excess in the thin films promote the p-type semiconductor characteristics in the ZnO thin films obtained by reactive sputtering, the propriety were observed by hot point probe, all samples shows p type semiconductor characteristics.

Through I-V (current – voltage), and considering breakdown electric field as 1.0 $MVcm^{-1}$, resistivity was calculated, Figure 3. These values are relatively low when compared with thermal silicon oxide (10^{14} to 10^{16} Ω.cm) a common dielectric in silicon technology for transistor devices. Our films shown high resistivity, decreasing which the RF power, the value in the same observed in the literature [11], for undoped ZnO thin films sputtering deposited.

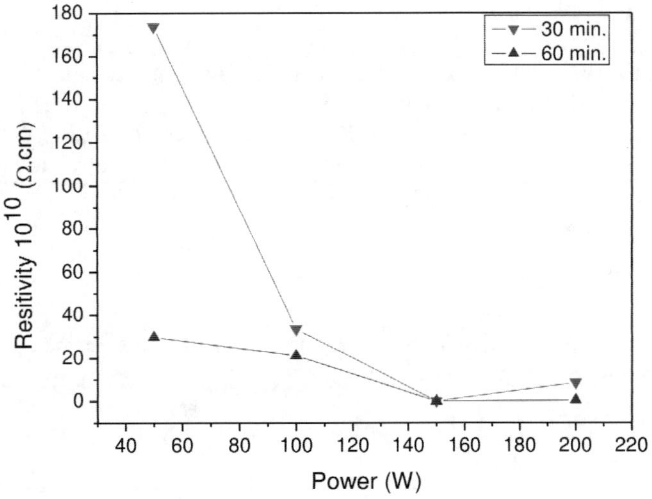

Figure 3– Zinc oxide resistivity versus RF power

Some ZnO films shows sensitivity to white light (Figure 4). When exposed to white light films had increase in their conduction up to 12 times, its suggest the possible application of ZnO thin films as light sensing material in photonics devices. The ZnO samples did not shown any photocurrent characteristic above zero volts.

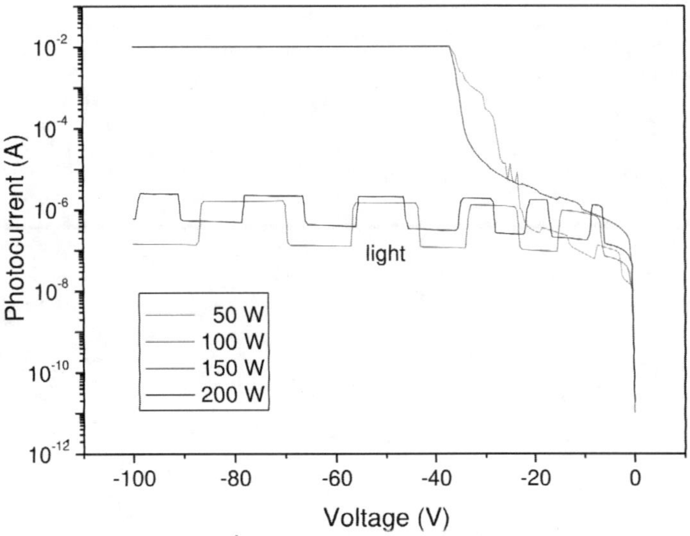

Figure 4 – Zinc oxide photocurrent for different RF power

4. Conclusion

In this work, we deposit nearly stoichiometric p-type ZnO thin films in low temperature by RF-magnetron sputtering for microelectronics and photonics applications. We observe the semiconductor characteristics by the photoelectric effect and p-type characteristics in undoped ZnO thin films. The main influences in semiconductor characteristics of these films were the defects in their structures, native defects and vacancies. The films showed a relative high resistivity and semiconductor characteristics, some samples show photocurrent when exposed to white light , its suggest the possible application of ZnO thin films as light sensing material in photonics devices.

Acknowledgements
The authors acknowledge CNPq for financial support and the Instituto de Fisica da Universidade de São Paulo for the RBS analyses.

References

[1] Godbole B et *al* 2011 *Mater Sci* Appl **2** 643
[2] Nicolay S et *al* 2012 *Sol Energ Mat Sol C* **105** 46
[3] Wenckstern H *et al* 2004 *Appl Phys Lett* **84** 79
[4] Lee J L *et al* 2002 *Thin Solid Films* **420** 112
[5] Jiang M, Tang K andYan X 2012 *J Photonics Energy* **2** 028502-1
[6] Ilican S, Caglar Y and Caglar M 2008 *J Optoelectron Adv M* **10** 2578
[7] Kashif M *et al* 2012 *Microelectron Int* **29** 131
[8] Chavillon B *et al* 2012 *J Am Chem Soc* **134** 464
[9] Ndong R O *et al* 2003 *Microelectron J* **34** 1087
[10] Xiong G L *et al* 2002 *Appl Phys Lett* **80** 1195
[11] Jeong W J , *et al* 2003 *Thin Solid Films* **426** 94

Production of Silicon Oxide like Thin Films by the Use of Atmospheric Plasma Torch

E M Ozono[1], E R Fachini[2], M L P Silva[1,3], L F Ruchko[4] and R M O Galvão[4]
[1]Faculty of Technology of São Paulo- FATECSP, SP, Brazil
[2]University of Puerto Rico, Río Piedras Campus, Puerto Rico, USA
[3]PSI-EPUSP, University of São Paulo, SP, Brazil
[4]Institute of Physics of University of São Paulo, SP, Brazil

E-mail: ozono@fatecsp.br

Abstract. The advantages of HMDS (hexamethyldisilazane) APT-plasma films for sensor applications were explored producing films in a three-turn copper coil APT equipment. HMDS was introduced into the argon plasma at four different conditions. Additional flux of oxygen could modulate the presence of organic components in the film, the composition varying from pure inorganic oxides to organo-silane polymers. Oxygen promoted deposition rates as high as 900 nm/min on silicon, acrylic or piezoelectric quartz crystal substrates. Films with a clustered morphology and refractive index of 1.45 were obtained, mainly due to a silicon oxide structure. Raman spectroscopy and XPS data showed the presence of CH_n and amorphous carbon in the inorganic matrix. The films were sensitive to the humidity of the air. The adsorptive capabilities of outstanding films were tested in a Quartz Crystal Microbalance (QCM). The results support that those films can be a useful and simple alternative for the development of sensors.

1. Introduction

It is well known that plasma processing can be used for a) surface modification, changing the surface wettability, adhesion, adsorption, printability, chemical reactivity, bio-compatibility, protection, sensitivity of light, etc. [1] or b) to deposit films of different kinds on a number of substrates. Those films are useful to develop microelectronic devices, as sensors and microTAS (micro Total Analysis System), because sensors require altering of small structures and the treatment of polymer surfaces is an important step in the manufacturing of many microTAS devices [2, 3].

Atmospheric plasma torch (APT) has high deposition rate of films and produces a range of film densities, being versatile compared with other available plasma equipment and processes. For instance, atmospheric plasma spray systems were used to develop NO_2 sensors and negative temperature coefficient (NTC) devices by the insertion of particles in the plasma, resulting films with various porosities [4, 5]. Aerosols also could be introduced in the film during its formation, promoting a higher degree of retention of precursor's molecular structure during the polymerization process. Problems associated with aerosols are high process pressure and the selection of the appropriate precursor which the particular partial pressure that assures high deposition rate without significant particle formation. Particle formation is a drawback on uniformity or, on the other hand, a strategy to

[1] To whom any correspondence should be addressed.

obtain composites. Thus, dense films to particles can be produced using a proper reactant and fine control of the plasma torch parameters. On such context, organosilicon compounds is a convenient precursor in APT processes because they avoid particle generation, being possible to obtain uniform films of high porosity [6].

Production of silica layers from organosilicon compounds is common, being common precursors HMDS (hexamethyldisilazane), HMDSO (hexamethyldisiloxane) [7] and TEOS (tetraethoxysilane) [8]. Nanoparticles of 160-200 nm were obtained with the introduction of HMDS aerosol, mainly components of particles being SiC_4, Si–CH_2–Si and Si–H [9]. As previously mentioned, a transparent HMDS derived film can be deposited on polycarbonate. The film composition presented few nitrogen (~0.3%) and C–H, N–H, Si–C, Si–N species [3]. Guruvenket et al, employing the same precursor, produced silicon carbonitride hard coatings [10]. Tendero et al summarizes the products from HMDS in APT processes as: a) stoichiometric SiO_2 films (used to protect fibers); b) films made of Si-O with carbon impurities (resistant to base solutions); and c) pure silicon films [11].

Volatile organic compounds (VOCs) pollutants must be evaluated in environmental samples where they occur at low concentration in the presence of analytical interferences. Because of that, it is mandatory the pretreatment of the sample with retention step in some part of the microTAS device. In this situation, adsorptive thin films are very important in the development of sensors and/or microTAS for VOCs analysis. Former works presented hydrophobic HMDS polymerized films with good adhesion on several distinct substrates. Those films presented polar and non polar VOCs adsorptive properties [12, 13]. Nonetheless these films were not tested regarding water adsorption and their resistance toward VOCs exposure.

The use of APT to deposit HMDS derivative films is a common procedure. Less usual is the use of this technique to obtain reproducible adsorptive films in a well controlled and well understood system. This work explores the advantage of the easiness of plasma polymerization of organic silicon compounds in a common process APT design resulting in reproducible adsorptive films for sensors and/or microTAS development in a comprehensive way.

2. Experimental

The produced APT equipment was designed to be used for deposition of thin films; therefore, some special characteristics should be provided in order to preserve monomer structure, such as low power, three coaxial tubes to allow easiness of reactant insertion on the inner section, etc. Thus, the thermal plasma is generated with argon gas under normal atmospheric pressure. The inductively coupled plasma torch consists of a three turns RF copper coil placed at the exit of the torch, around of small three concentric quartz tubes of 6, 14, and 18 mm diameters. Argon gas was injected with a flow of 1 l/min through inner channel to compose the plasma core and second flux of 25 l/min in the outer channel so the required stabilizing tangential swirl at plasma edge. Because thermal plasma is difficult to be ionized it needs to apply a spark by Tesla coil to initiate the discharge. The efficiency of power coupling to the plasma torch is modest. Other requirements such as proper trimming of the cables within the matching box and installation of a low-pass filter with electrostatic shielding in order to reduce the reflected power were fulfilled. Many shootings of torch with plasma have been done. The adjustments at inductive torch coil with plasma resulted at 1420 V of voltage and 11.56 A of current amplitude with a good tuning at 23.47 MHz. The phase shift of 74 degree between voltage and current allowed the delivery of 1037 watts of power at coil. That indicated an enthalpy of 331 W power transferred into plasma. Figure 1a schematizes the admittance of the reactants. The carrier gas injects a controlled amount of the reactant HMDS into the plasma core through the inner channel. If cross flow is preferred, the reactant also can be injected through the outer side of torch. For this alternative, the use of air as carrier gas will favour the oxidation of HMDS molecule. If argon is used, the oxidation of HMDS molecule is not favoured. The experiments were carried out positioning the samples at different distances from the top of the glass torch. Figure 1b shows the plasma torch and a sample positioned over its top (inset).

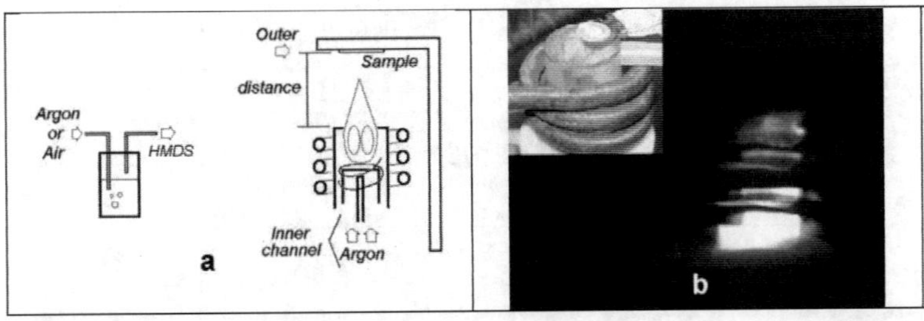

Figure 1: a) schematics of reactants admission into the equipment; b) plasma torch and a substrate positioned to receive the film.

Films were grown on four types of substrates: a) silicon wafers (used in the film's chemical and physical characterization), b) piezoelectric quartz crystal (PQC – 7 MHz, for absorptive capability assessment), c) acrylic (PMMA) (samples used in the chemical resistance evaluation) and d) polyacrilonitrile (PAN) fibers (to verify if the film grows properly and with good adherence on this material). Deionised water and P. A. grade reactants were used, except HMDS (Fluka Chemie GmbH, Switzerland) that was industrial grade.

For all tested substrates, Table 1 summarizes the four conditions used for the film growth and the resulting plasma characteristics. For each condition, 3 samples were prepared at distance of 2, 5 and 10 mm from the torch top to substrate. Since the reactant is inserted on plasma zone by a saturated carrier gas, the HMDS flow was approximately 0.5 g/min for all films that were grown under the plasma conditions during 2 minutes.

Table 1: Different conditions at APT during the growth of the HMDS films.

Condition #	HMDS injection		Carrier gas		Plasma characteristics	
	Inner channel	External channel	Argon	Air*	HMDS molecules and plasma interactions	Oxidation mechanisms
1		x	x		Low	Low
2	x		x		High	Low
3		x		x	Low	High
4	x		x	x	High	High

* Air is admitted outside of the plasma torch via an external channel

The deposition rate and refractive index were determined respectively by profilometry and ellipsometry at two wavelengths (632.8 nm and 830.0 nm). The presence of polar organic functionalities in the films was determined by infrared spectroscopy (FTIR). Raman microscopy was useful to obtain information about non polar organic residues and cluster formation. Additional chemical information came from XPS (x-ray photon electron spectroscopy). Film morphology was inspected by optical and scanning electron microscopy. Water contact angle measurements (goniometer and 4 μL drops) evaluated the hydrophobic character of the films. The PQC setup was described elsewhere [12]. Samples of different relative humidity and VOCs content were used sequentially in order to determine the adsorptive capability of the films. Chemical resistance to VOCs was tested by contact angle measurements with organic liquids.

3. Results and Discussion

Table 2 summarizes the main results obtained by ellipsometry and profilometry measurements. Due to the plasma torch characteristics the thickness is non uniform and some samples are porous; nonetheless, high deposition rates were achieved in all deposition conditions, the highest values correspond to a film with high porosity. Thick films presented numerous clusters and their roughness makes difficult a proper measurement by profilometry. HMDS admission on the inner channel increases ion/molecule reactions and, consequently, the thickness of the films. The high refractive index observed is characteristic of films with low oxygen content. This fact suggests a low oxygen presence in the reactive space even if oxygen is admitted in the system through an external channel (deposition #4). When air is the carrier gas (deposition #3) or a low concentration of HMDS is available during deposition (deposition #1), the film obtained will have high oxygen content with a refractive index similar to silica films (~1.45), these conditions also lead to low deposition rates.

Table 2: Results from ellipsometry and profilometry measurements.

Condition #	Refractive index	Max. thickness (nm)	Characteristics
1	1.3 to 1.4	100	Deposition rate lower than 100 nm/min
2	~2.0	$9.\ 10^3$	Rough and porous films, deposition rate of 1-5 10^3 nm/min
3	1.3 to 1.4	540	Dense films, low deposition rate
4	~2.0	$3.\ 10^3$	Deposition rate up to ~1. 10^3 nm/min

Infrared (FTIR) measurements support the main conclusions obtained by ellipsometry and profilometry measurements; Figure 2 shows typical FTIR spectra. Common features in all FTIR spectra indicate that the films presented the species commonly found in silicon oxide films and HMDS plasma polymerization films. The strongest band corresponds to SiOSi vibration in the 1000-1200 cm^{-1} range. Other species were also found: $SiCH_3$ (~1250 cm^{-1}); CH_n (~ 2950 cm^{-1}); SiC (800 cm^{-1}) and adsorbed water (3400 cm^{-1}). Some samples prepared at a distance of 2 mm from plasma torch with condition #2 showed a weak peak at ~ 2100 cm^{-1}, which corresponds to SiH. This is consistent with low quantity of oxygen and short residence time in the plasma, as discussed by Miettinen [14] in a study of silicon–carbon nanoceramics.

The FTIR SiOSi peaks show different profiles depending on the condition of film growth and they were quite different of the profile obtained by Huang [3] for a film obtained by HMDS plasma polymerization. The Si-O-Si peaks present some resemblance with bands obtained for dense silica films formed from polymerization of TEOS and tetramethylcyclotetrasiloxane (TMCTS), among other precursors [15]. As pointed out by Cui, 1000–1200 cm^{-1} region is typical for a siloxane network, occurring overlapping of peaks at: a) 1075 cm^{-1} (vibration of "fully relaxed SiO_2-network structure with bond angle of 144°"), b) 1100 cm^{-1} (corresponding to "larger angle Si–O–Si bonds"), c) 1150 cm^{-1} (characteristic of "highly symmetric siloxane ring structure") and an additional peak at 1050 cm^{-1} "due to less symmetric and more random network structure" [15]. Thus, if HMDS is exposed to high plasma interaction (#2 deposition), the 1100 cm^{-1} band is privileged, probably owing to the high deposition rate that leads to disordered structure (Figure 2a) and more randomic configuration (Figure 2b) can occur if less reactant is present. The presence of oxygen in the plasma favours a FTIR peak profile similar to SiO_2, as it occurs in TEOS derived plasma films. These conditions induce losses of $SiCH_3$ species, although some weak peak near 1250 cm^{-1} can remain (see Figure 2c), in a situation similar to Cui´s report for TMCTS, which attests the strength of such bond. Water adsorption in films can be enhanced by the presence of SiOH (~940 cm^{-1}) species that promotes hydrogen bonds. Although Huang had observed such species in films made from HMDS in low temperature atmospheric plasmas [3], they were not noticed in FTIR spectra in the present study. Nonetheless, the water presence (broad peak at 3400 cm^{-1}) is a good indicative of adsorptive properties of these films.

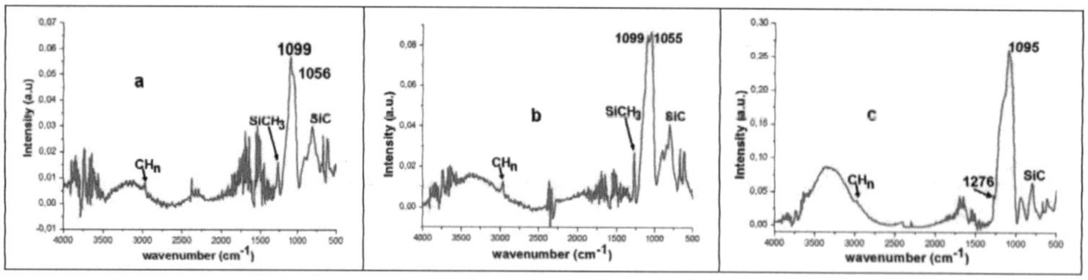

Figure 2: FTIR spectra: (#2 deposition) with a) 5 mm or b) 10 mm from the torch and c) adding oxygen (#4 deposition)

Raman microscopy can probe the small surface features that optical microscopy revealed. Raman microscopy is also capable of evaluating carbon-rich domains. Optical microscopy images show uniform films when weak HMDS-plasma species interaction are favoured (conditions #1 and #3 for the film deposition) as can be seen in Figure 3a. For these samples, weak peaks at ~1450 cm^{-1} that correspond to CH_2 were found in the corresponding Raman spectrum. If strong interactions among

HMDS-plasma species are favoured, features can be observed in the film. If oxygen is absent in the plasma (condition #2), few graphitic domains will be observed (darker points in the centre of Figure 3b). If oxidative plasma is formed (condition #4), the number and size of these graphitic domains increase (see Figure 3c). The frequency and intensity of features increase if the film is formed closer to torch, probably as a consequence of the higher deposition rate (Figure 3, Raman spectrum).

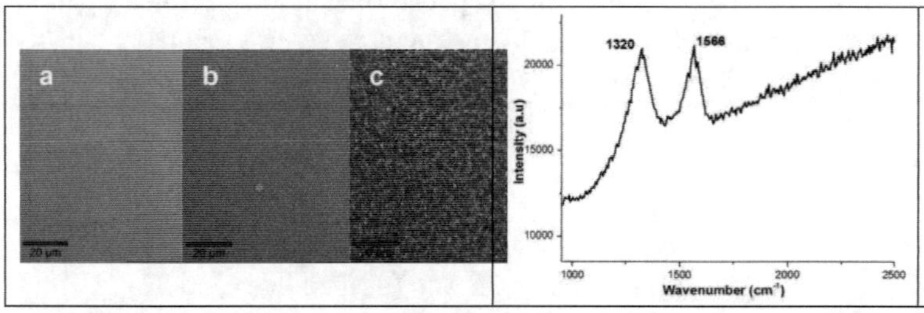

Figure 3: Films formed at substrate-torch 2 mm apart. Optical microscopy of characteristic samples at conditions #1 (a), #2 (b) or #4 (c). Typical Raman spectrum of a film deposited at condition #4.

Water contact angles were approximately 90° for all samples; organic liquids can wet these surfaces. Although these experiments pointed out to hydrophobic/organophilic species at surface, this interpretation could be inaccurate since it could also be due the roughness of the surface, as shown by SEM images, with particles approximately 1 μm long (Figure 4a). To avoid misleading, QCM measurements were performed to evaluate adsorptive properties. Samples were almost insensitive toward VOCs or water. However, samples with highest roughness (condition #4) gave intense response when exposed to a gas stream saturated in water as can be seen in Figure 4b. Probably the high porosity leads to a different phenomenon than adsorption inside the pores, such as clogging. Since these films were resistant to VOCs, they are grown on polymer substrate to test their performance as protective surface layer. Deposition on PMMA was performed without mechanical deformation of the 8-mm-thick substrate. The film was able to prevent acetone-PMMA interaction for few minutes but PMMA surface "cracks" (see detail, Figure 4c). An additional sample was prepared growing the HMDS derivative film in a mat of polymer fibres to test water adsorption in a three dimension system with increased surface area [16]. While PAN micro/nanofibers exposed to argon plasma can be severely destroyed in 15 s, the fibre mat that was protected by the HMDS derivative film showed low damage. After plasma process, a yellowish film that completely covers the fibres was obtained and can be seen in the Figure 4d. Films prepared at 2 and 10 mm apart from torch had very similar XPS spectra peaks, and they presented the same major components at similar concentrations: C: 22-25%, O: 47-44%, F: 1.6-1.5% and Si: 29%. They did not incorporate nitrogen as a film component and both had fluor, probably from a torch environmental contamination.

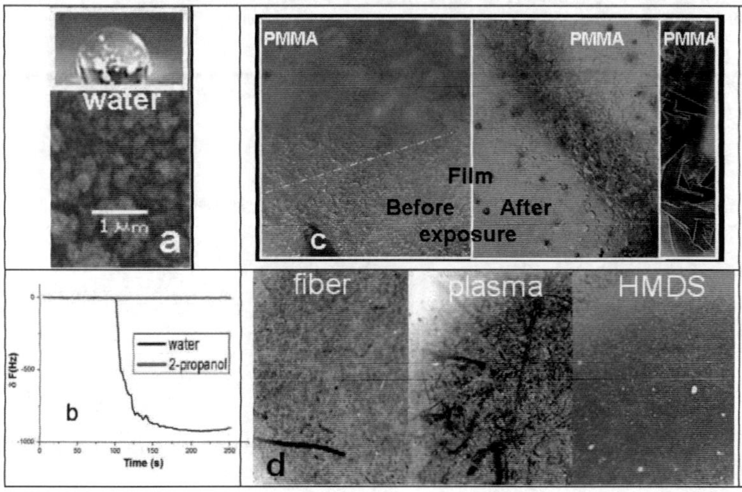

Figure 4: (a) Water contact angle and SEM photo; (b) QCM of a film exposed to air saturated in water or 2-propanol; (c) HMDS film on PMMA exposed to acetone and (c) in sequence from left: fibre on silicon substrate; after plasma treatment without HMDS film protection; PAN fibres mat after 15 s of HMDS derivative film deposition. All samples were approximately 20 mm long.

Moreover, these films presented a more defined granular morphology compared to sample prepared at 5 mm apart of the torch, which structure resembles a "sponge". The last film had no fluor, was enriched in carbon species and impoverished in silicon oxide species when compared with its homologues. It incorporated a significant amount of nitrogen. Its composition was: C: 33%, O: 42%, N: 3% and Si: 22%. The XPS C1s peak was richer than previous films in carbonic species with oxygen and/or silicon (see arrow in the Figure 5). All of this suggests that it was formed at mild conditions in the plasma, with greater retention of the molecular structure of the precursor than its homologues films. The Si 2p peak centred at 102.9 eV vs. 103.2 eV of homologues films confirms this assumption.

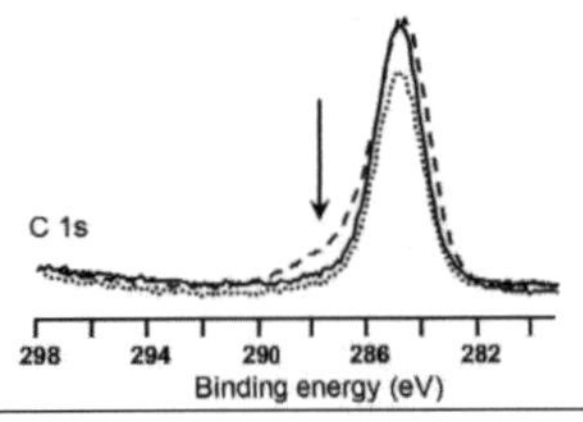

Figure 5 – C1s spectra for three HDMS films prepared at conditions #2 and 2 mm (solid line), 5 mm (dashed line) and 10 mm (dotted line) apart of plasma torch top.

4. Conclusion:
This work establishes the advantages of using HMDS derivative plasma films to construct sensors. If correct values for process parameters are chosen, it is possible to grow a film with a SiO_2 network with organic residues as $Si\text{-}CH_3$. The resulting film demonstrates to have excellent capability to absorb water or to protect polymer surfaces against VOCs attack.

5. Acknowledgments
The authors thank FAPESP and CNPq for financial support.

6. References
[1] Kaczmarek H, Kowalonek J, Szalla A and Sionkowska A 2002 Surface Science. 507–510 883
[2] Oehr C 2003 Nuclear instruments and Methods in Physics Research Section B: Beam interactions with materials and atoms. Ionizing Radiation and Polymers. 208 40
[3] Huang C, Wu S-Y and Chang Y-C 2010 IEEE Transactions on Plasma Science. 38 (5) 1101
[4] Zhang C, Debliquy M, Boudiba A, Liao H and Coddet C 2010 Sensors and Actuators B 144 280
[5] Liang S, Zhang X, Bai Y, Han Z and Yang J 2011 Applied Surface Sci. 257 9825
[6] Massines F and Sarra C 2012 Plasma Processes and Polymers. 9 (11-12) 1041
[7] Alexandrov S 2005 Chem. Vap. Deposition. 11 457
[8] Pappas D 2011 J. Vac. Sci. Technol. A 29 020801
[9] Miettinen M, Johansson M, Suvanto S, Riikonen J, Tapper U, Pakkanen T T, Lehto V P, Jokiniemi J and Lähde A 2011 J. of Nanoparticle Research. 13 (10) 4631
[10] Guruvenket S, Andrie S, Simon M, Johnson K W and Sailer R A 2012 ACS Applied Materials & Interfaces. 4 (10) 5293
[11] Tendero C, Tixier C, Tristant P, Desmaison J and Leprince P 2006 Spectrochimica Acta Part B. 61 (1) 2
[12] Lima, R R, Carvalho R A M, Carvalho A T, Simões E W and Silva M L P 2009 Sensors and Actuators B. 137 185
[13] Lima R R, Carvalho R A M, Silva L M, Simões E W and Silva M L P 2009 Sensors and Actuators B. 137 170
[14] Miettinen M, Johansson M, Suvanto S, Riikonen J, Tapper U, Pakkanen T T, Lehto V-P, Jokiniemi J and Lahde A 2011 J Nanopart Res. 13 4631
[15] Cui L, Ranade A N, Matos M A, Pingree L S, Frot T J, Dubois G and Dauskardt R H 2012 ACS Appl. Mater. Interfaces 4 6587
[16] Silva A N R, Silva M L P, Fachini E R 2013 J. of Physics. Conference Series 421 012013

TiO_2 thin and thick films grown on Si/glass by sputtering of titanium targets in an RF inductively coupled plasma

R Valencia-Alvarado[1,*], A de la Piedad-Beneitez[2], R López-Callejas[1,2], A Mercado-Cabrera[1], R Peña-Eguiluz[1], A E Muñoz-Castro[1], B G Rodríguez-Méndez[1], and J M de la Rosa-Vázquez[3]
[1]Instituto Nacional de Investigaciones Nucleares, Plasma Physics Laboratory, AP 18-1027, 11801, México, D.F., México
[2]Instituto Tecnológico de Toluca, AP 890, Toluca, México
[3]Esime-Zacatenco-IPN, AP 07738, México, D.F., México

*E-mail: raul.valencia@inin.gob.mx

Abstract. TiO_2 thin and thick films were deposited on silicon/glass substrates using RF inductive plasma in continuous wave. The films thickness, as well as phases control, is achieved with a gradual increase in temperature substrates varying supplied RF power or working gas pressure besides deposition time as well. The deposition conditions were: argon 80%/oxygen 20% carefully calibrated mixture of 2 to 7×10^{-2} mbar as working gas pressure range. Deposition time 0.5 to 5 hours, 500 or 600 W RF power at 13.56 MHz frequency and 242-345 °C substrates temperature range. The titanium dioxide deposited on the substrates is grown by sputtering of a titanium target negatively polarized at 3-5 kV DC situated 14 mm in front of such substrates. The plasma reactor is a simple Pyrex-like glass cylindrical vessel of 50 cm long and 20 cm in diameter. Using the before describe plasma parameters we obtained films only anatase and both anatase/rutile phases with stoichiometric different. The films were characterized by X-ray photoelectron spectroscopy (XPS), stylus profilometer, X-ray diffraction (XRD), scanning electron microscopy (SEM) and Raman spectroscopy.

1. Introduction

The titanium dioxide (TiO_2) has been reported as one material for applications in different areas as agent antibacterial, antimicrobial on prostheses and catheters [1-4], as a photocatalytic vehicle in the water purification and in the atmospheric pollutants such as (NO_x) and volatile organic compounds (VOC_s) [5-6]. In such analysis, Blake [7] identified about 600 patents and found about 500 additional references were found in the form of journal and conferences papers. Now today, has been reported applications in the visible light as photocatalytic agent and optoelectronic employment [8].

There are several reports to obtain TiO2 thin films that have been mainly prepared by various methods such as plasma enhanced chemical vapor deposition [9], sputtering [10]. In comparison with other methods, sputtering deposition is one of the widest techniques employed for obtaining large area coatings and strong adhesion. The reactive sputtering technique, for coating nanocrystalline TiO2 thin films, are mainly employed radio frequency (RF) and direct current (DC). In other hand, the reactive DC+RF sputtering technique provide a wide range for to vary the sputtering parameters such as total pressure, partial pressure ratio, DC or RF power, substrate temperature according to the plasma parameters. In these experiments, we observed that the crystal structure and chemical composition of deposited films varied with the employed plasma parameters, as well as the rate of the sputtering from the target.

The objective of this work is to obtain TiO_2 thin or thick films in anatase or rutile phases by controlling the temperature of substrates. This temperature depends on the density of the plasma, which is a function of the RF power and the working gas pressure keeping the rate for Ar and O_2 constant.

2. Experimental details

TiO^2 thin and thick films were deposited on glass slides and silicon wafer, the temperature was kept between 242-345 °C by the action of the plasma alone, that is to say, the temperature substrates depend only plasma density, the gas working pressure and the RF power supply plasma reactor, the deposited films on glass and silicon by the sputtering action were situated 14 mm from titanium target. For this experiment, the target consisted in titanium commercially pure (CpTi) piece cylindrical of 9 mm in diameter and 5 mm long, the dimensions of the substrates were 20×20 mm with glass and 10×10 mm for the silicon. Deposited films were obtained in a home Pyrex-like glass cylindrical vessel of 50 cm long and 20 cm in diameter by sputtering process. The plasma density was controlled by the working gas pressure and RF power supply. In this experiment we employed a calibrated mixture of 4:1 argon/oxygen gas mixture from 2×10^{-2} to 7×10^{-2} mbar preview the vessel was to $<5\times10^{-6}$ mbar pressure base, one RF power supply between 500 and 600 W connected to copper pipe antenna wounded helicoidal cylindrical vessel through an automatic coupling network and one power supply between -3 and -5 kV DC that polarize the target. Figure 1 shows a diagram of the sputtering system preview reported [11].

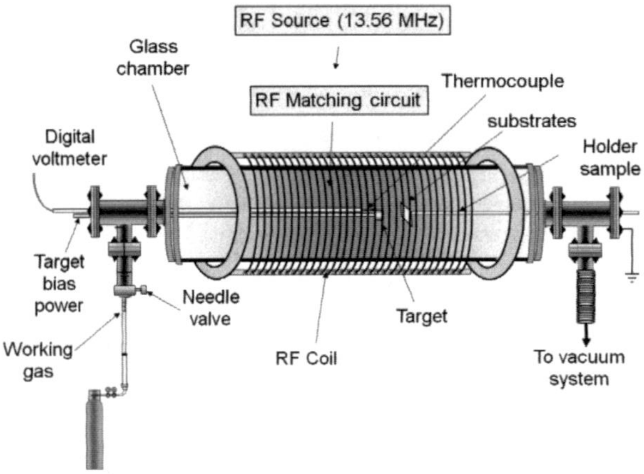

Figure1. Schematic diagram of the experimental sputtering system.

The crystalline structure of the films deposited on silicon wafers were characterized by X-ray photoelectron spectroscopy (XPS) using monochromatized AlKα X-ray source (hv=1486.6 eV), the recorded spectra of all films were referenced to the C1s line of the residual carbon set at 284.6 eV, the thickness of the films were determined by a stylus profilometer, the crystallinity of the films were carried out with an X-ray powder diffractometer (XRD) using CuKα radiation lambda= 1.54056 angstroms from 20° to 90° scan range set applied voltage 35 kV/25 mA and Raman spectroscopy, the surface morphologies, the stoichiometry (titanium:oxygen ratio) and the atomic percentage in the sputtered films analyzed using a scanning electron microscopy (SEM).

3. Results

Figure 2 shows the XPS spectra of the TiO_2 film grown on the silicon substrate. Figure 2(a) shows the peaks of the doublet of Ti2p with energies of 458.03 eV link for $Ti2p_{3/2}$ and 463.6 eV for $Ti2p_{1/2}$ on the surface respectively, the binding energy difference (splitting) between $Ti2p_{1/2}$ and $Ti2p_{3/2}$ is 5.57 eV which indicated that addition of a Ti^{+4} oxidation state titanium. On the other hand, the figure 2(b) shows O1s peak with a binding energy at 529.6 eV of the as-deposited TiO_2 film that is related to oxygen chemically bounded to Ti.

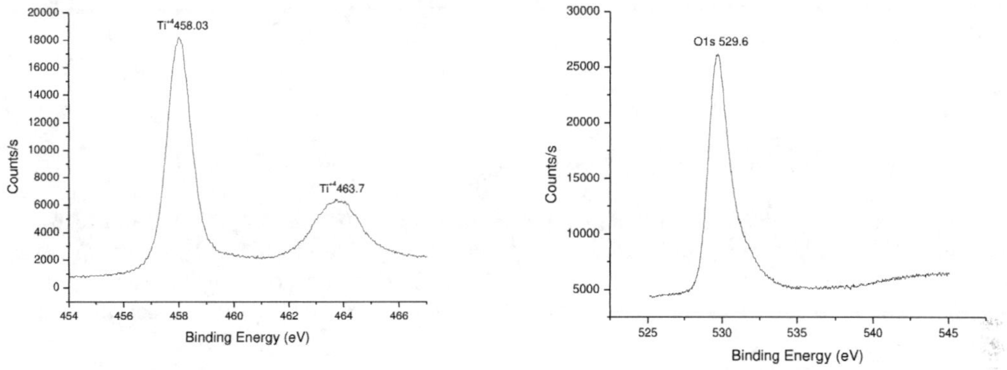

(a) (b)

Figure 2 XPS spectra of (a) Ti2p and (b) O1s on the surface silicon film.

Figure 3 shows XRD spectra of the TiO_2 film deposited at two different working gas pressures. Figure 3(a) shows the crystallinity of the TiO_2 film in the anatase phase deposited at 242 °C on the substrate of silicon in the working gas pressure of 5×10^{-2} mbar at diffractive angles 2θ of 25.4° , 38.1°, 38.9°, 48.3°, 55.3°, 62.6°, 70.3°, 76.2°, and 83.0°. And the other hand, the figure 3(b) shows TiO_2 film when the gas working pressure was 2×10^{-2} mbar with this pressure the substrate reached a temperature of 345 °C, at this temperature, clearly shows the transition from the phase of anatase to rutile phase, in the spectrum are observed both phases at diffractive angles 2θ of 25.3° and 27.4°.

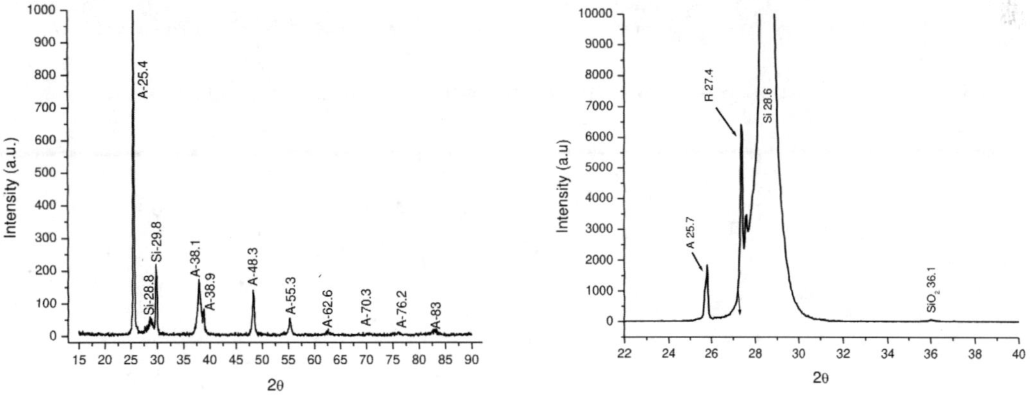

(a) (b)

Figure 3. (a) X-ray diffraction pattern of anatase on silicon and (b) show the transition to rutile phase

Figure 4 shows the different thickness measured by a surface profiler meter of the film deposited on glass substrate, it was variable, in the center of the pattern of deposit thickness was 2500 nm (this graph shows that 0.1 mm is the second ring of deposit which has a thickness of around 1000 nm), at the midpoint of the pattern was 1150 nm and on the periphery of the pattern was around 70 nm, this variation is because the deposited films has a circular patters caused for the magnetic field generated by solenoidal antenna.

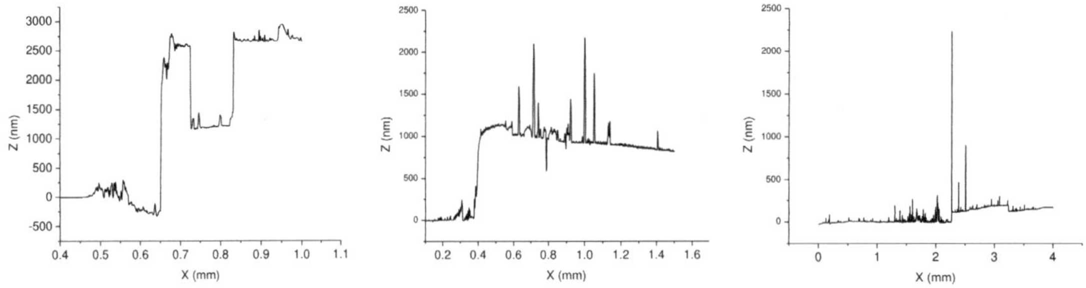

Figure 4 pattern of thickness of the film of the Centre to the Board

Figure 5 shows the micrograph and a photograph of the film in the anatase phase obtained by SEM with an amplification of 30000 on silicon substrate, the grain size was calculated by Sherrer´s formula, whose size was around 80 nm, which is congruent with the sizes observed in the micrograph. The pattern of colors of the photograph of Fig. 5(b) represents the different thicknesses deposited on silicon substrate

(a) (b)

Figure 5. (a) SEM micrograph and (b) normal photograph film on silicon substrate

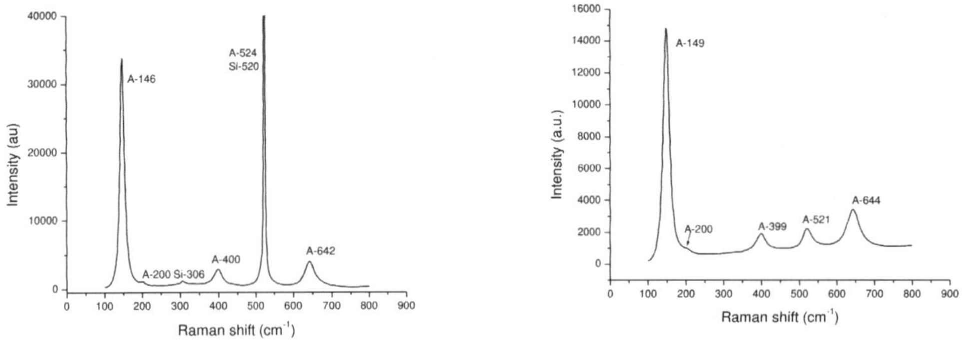

(a) (b)

Figure 6 Raman spectra of the TiO_2 film in: (a) silicon and (b) glass.

Figure 6 shows the graphics acquired by Raman of TiO_2 film in the anatase phase. The Fig. 6(a) and 6(b) Spectra confirm the presence of anatase phase grown on silicon and glass substrates at temperatures below 300 °C. Difference in shift (cm^{-1}) recorded in the graphics is due to a light contrast to the growth of the anatase since silicon and glass substrates differ in their material structure.

4. Conclusions

In this research, we use a plasma reactor inductively coupled to RF generator with which was obtained TiO_2 thin and thick films in anatase and rutile phases by means sputtering of a target of titanium, in this sputtering process, titanium atoms sputtered target and oxygen atoms generated by plasma were deposited on silicon and glass substrates. Atomic concentrations of 60 at.% atoms of titanium and 30 at.% atoms of oxygen obtained by SEM leads to a stoichiometry 2 only in target. In contrast, the atomic concentrations obtained by SEM for the TiO_2 films on silicon and glass substrates were of the order of 75 at.% of oxygen and 25 at.% titanium respectively, it is considered that this rate stoichiometry of 3:1 is due to two reasons: 1) substrates have a percentage of oxygen bigger than target and 2) the applied voltage in the SEM was 20 kV, this rate low to 2 when 15 kV is applied in SEM. Color changes of the films during and the end the oxidation process, the concentric circular patterns suggest the evolution of the composition structure thickness and is related to the deposited film depth. The deposited anatase and rutile phases did not require any additional annealing after the oxidation.

Acknowledgements

This project has been partially founded by CONACyT-México. The authors are grateful to Dr. Manuel Espinoza and Mr. Jorge Pérez for their technical support in the XRD and EDS diagnostic, Dr. Enrique Camps for Raman spectroscopic and Fis. Rafael Basurto for XPS spectroscopic respectively. The technical assistance received from Pedro Ángeles Espinoza, María Teresa Torres M. and Isaías Contreras Villa is much appreciated.

References

[1] Page K, Wilson M and Parkin I P 2009 *J. Mater. Chem.* **19** 3819-3831.
[2] Foster A H, Sheel D W, Sheel P, Evans P, Varghese S, Rutschke N and Yates H M 2010 *J. Photochem. Photobiol.* **A-216** 283-289.
[3] Noimark S, Dunnill Ch W, Wilson M and Parkin I P 2009 *Chem. Rev.* **38** 3435-3448.
[4] Rtimi S, Baghriche O, Pulgarin C, Sanjines R and Kiwi J 2012 *RSC. Adv.* **2** 8591-8595.
[5] Meng F, Song X and Sun Z 2009 *Vacuum* **83** 1147-1151.
[6] Berdahl P and Akbari H 2008 *California Energy Commission, PIER Energy-Related Environmental Research Program* CEC-500-2007-112.
[7] Blake D M 2001 *National Renewable Energy Laboratory Technical Report*, NRET/TP-510-31319.
[8] Fakhouri H 2012 Thin film deposition of pure and doped TiO_2 by RF magnetron sputtering for visible light photocatalytic and optoelectronic applications, these de doctorat, l'Universite Pierre et Marie Curie (Paris-France).
[9] Nakamura M, Makino K, Sirghi L, Aoki T, and Hatanaka Y 2003 *Surf. and Coat. Technol.* **169-170** 699-702.
[10] Tominag K, Ito D, and Miyamoto Y 2006 *Vacuum* **80** 654-657.
[11] Valencia-Alvarado R, López-Callejas R, Barocio S R, Mercado-Cabrera A, Peña-Eguiluz R, Muñoz-Castro A E, de la Piedad-Beneitez A, de la Rosa-Vázquez J M 2010 *Surf. Coat. Technol.* **204** 3078-3081.

TiN Coatings on Titanium Substrates Using Plasma Assisted Ion Implantation

M Cisternas[1], F Mellero[1], M Favre[1,2], H Bhuyan[1,2] and E Wyndham[1]
[1] Instituto de Física, Pontificia Universidad Católica de Chile, Av. Vicuña Mackenna 4860, Santiago, Chile
[2] Centro de Investigación en Nanotecnología y Materiales Avanzados (CIEN-UC), Av. Vicuña Mackenna 4860, Santiago, Chile

E-mail: mncister@uc.cl

Abstract. We present results on producing TiN coatings on titanium substrates, using Plasma-Based Ion Implantation and Deposition (PBII&D). In PBII&D the substrate is immersed into stationary plasma which contains the ion species to be implanted. A negative high voltage pulse is applied to the substrate, which causes the ions in the ion sheath surrounding the substrate to be accelerated across the sheath and implanted into the substrate. We have used a 13.6 MHz, 70 W, radio frequency (RF) plasma, produced with a mixture of 80% N_2 + 20% H_2, at 200 mTorr, with 4 kV, $\approx$10 μs pulses, at a 0.2 kHz rate. During the implantation process the titanium substrate is heated at 400° C range. The resulting TiN coatings are characterized using atomic force microscopy (AFM), x-ray diffraction (XRD), and Vickers hardness test. Optical emission spectroscopy (OES) is used to identify characteristic RF plasma species.

1. Introduction

Titanium is widely used in biomedical application [1, 2]. Because titanium by itself it is not a fully biocompatible material, it requires a biocompatible coating to be suitable for more critical requirements [3, 4]. Besides their well-known properties of hardness, TiN coatings exhibit better biocompatibility features than those of bared titanium. This feature has stimulated research aiming to improve on TiN coatings properties, such as roughness, deepness and adhesion [5, 6]. Different plasma based techniques, such as inductively coupled plasma assisted physical vapor deposition [7], reactive plasma spraying [8], or direct current magnetron sputtering [9], have been used to produce TiN coatings Plasma-Based Ion Implantation and Deposition (PBII&D) is a well-developed technique which allows deep implantation of ions into a substrate [10, 11]. The substrate is immersed into stationary plasma which contains the ion species to be implanted. When a negative high voltage pulse is applied to the substrate, ions in the ion sheath surrounding the substrate are accelerated across the sheath and implanted into the substrate. Successful implantation depends critically on high voltage pulsing parameters (1-100 kV, 1-50 μs, 10-1000 Hz repetition rate). The implantation voltage determines the characteristic depth of the implanted ions, the average implanted ion dose per pulse depends on pulse length, and the repetition rate defines the exposure time required to achieve a given implantation profile. We have previously reported on the synthesis of sub-micron size carbon structures using a combination of 13.6 MHz radio frequency (RF) plasma and PBII&D, operating at 30 W, with 4 kV, 2 μs pulses, at a 2 kHz rate [12]. In order to explore different parameter regimes and aiming to improve on TiN coatings, we have conducted preliminary experiments to produce and characterize TiN coatings on Titanium substrates, using PBII&D techniques in RF plasmas.

2. Experimental set-up

The experiments are conducted using a capacitively coupled RF plasma generator at 13.6 MHz, operating at 70 W, and produced with a mixture of 80% N_2 + 20% H_2, at 200 mTorr working pressure. Nitrogen flow rate was fixed at 13 sccm. The use of a nitrogen-hydrogen mixture is motivated by the fact that at substrate temperatures around 400°, addition of hydrogen in the range 5-50% results in an increased nitride layer, provided that the partial pressure of nitrogen is maintained [13]. As substrate, we used a flat, polished Titanium (TI007945) sample. In a pre-implantation treatment the substrate is exposed to argon plasma at 200 mTorr, for 15 min, in order to improve surface cleanness. During the implantation process the substrate temperatures is maintained at 400°C. Implantation time is kept at 60 min.

PBII&D is performed with an applied negative biasing up to 10 kV, at 0.2 kHz, with $\approx$10 μs pulse width. The high voltage driver is based on a combination of ferrite core pulse transformers and insulated gate bipolar transistors (IGBT). In present experiments the high voltage pulser was used at 40% of the nominal highest voltage output, in order to protect from eventual breakdown to the RF electrodes. Figure 1 shows a characteristic high voltage pulse, as used in the experiments.

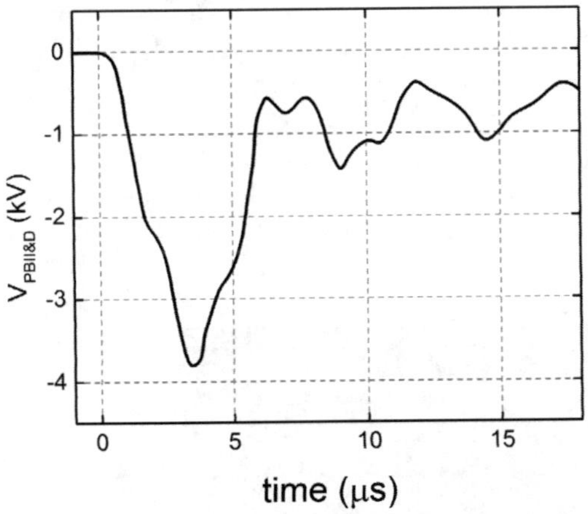

Figure 1. Characteristic high voltage pulse used in the PBII&D process.

The implanted substrates are characterized using atomic force microscopy (AFM), x-ray diffraction (XRD), and Vickers hardness test. Optical emission spectroscopy (OES) is also used to identify dominant species in the RF plasma.

3. Results and Discussion

Figure 2 shows a characteristic emission spectrum of the stationary RF plasma, obtained with a low resolution Ocean Optics PC2000 spectrometer. The strongest emission lines correspond to atomic hydrogen emission, H_α and H_β, together with neutral atomic and molecular nitrogen, and singly ionized nitrogen molecules. Weaker emission from single ionized nitrogen atoms can also be identified. As expected for a low pressure, low power RF plasma, with characteristic electron temperature around 1 eV, emission from neutral species becomes dominant. This fact precludes a clear observation of emission lines corresponding to single ionized nitrogen, as they either coincide with strong molecular bands, or merge with background noise.

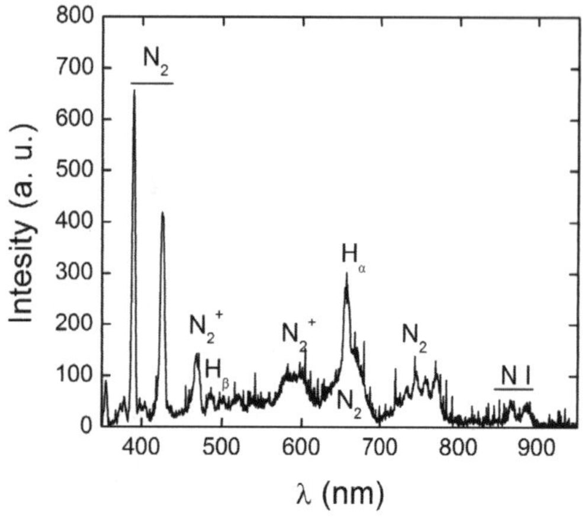

Figure 2. Optical emission spectroscopy of the RF plasma.

The effect on titanium surfaces due to RF plasma exposure, combined with negative high voltage pulsing, was investigated using a NanoWizard 3 NanoScience AFM. Figure 3 shows AFM images of the titanium surface, before and after the PBII&D process. The substrate temperature was kept at 400° during the implantation process. Surface roughness of implanted surfaces is seen to increase by up to a factor of ten, when compared with the unexposed surface. The observed roughness is similar to that reported in previous investigations of titanium nitride synthesized by dynamic nitrogen and titanium PBII&D, using a vacuum arc source for the titanium plasma and a hot filament glow to sustain the nitrogen plasma [14].

Vickers hardness tests at low load, ≈10 gf, indicate that surface hardness is around 3 times that of unexposed substrates, whereas at ≈100 gf it becomes comparable to the unexposed titanium surface. A characteristic implanted layer thickness of around 300 nm is inferred from the Vickers test.

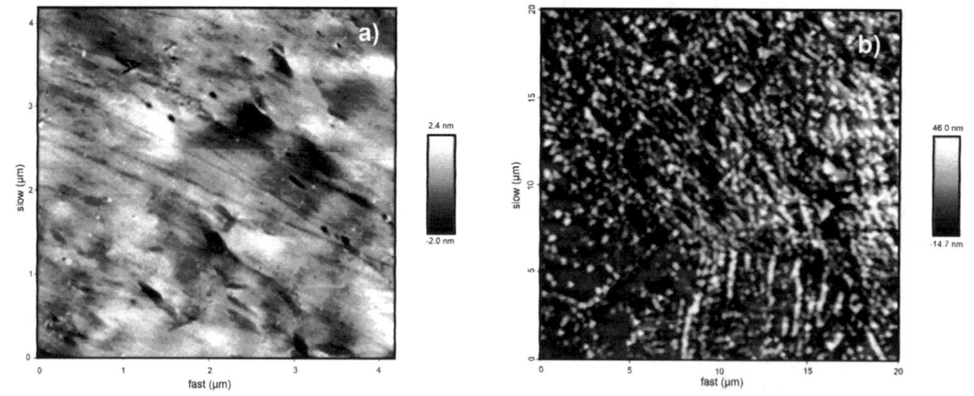

Figure 3. AFM images of, a) Titanium substrate before implantation, and b) substrate surface after PBII&D process.

XRD analysis was performed for different exposed samples, using a Bruker D8 Advance diffractometer. Figure 4 shows the XRD spectrum corresponding to the same exposed sample with an AFM image in figure 3-b. Data analysis of the x-ray diffraction pattern allows to identify distinctive peaks corresponding to TiN and titanium, and also some evidence of titanium oxide. This last, probably due to surface oxidation of previously unexposed substrate, which is not removed by the initial argon plasma treatment. The XRD data hints that a thin surface layer of mixed composition is formed as a result of the PBII&D process.

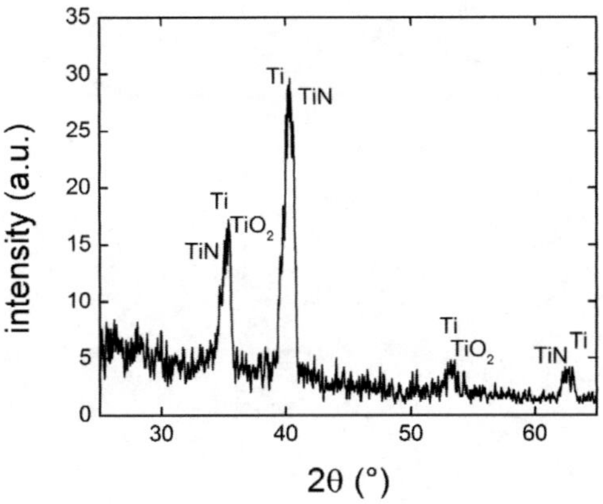

Figure 4. XRD spectrum of exposed sample.

4. Conclusions

Our preliminary results show the feasibility of implanting nitrogen ions onto a titanium substrate using a combination of RF nitrogen plasma and PBII&D techniques. Standard surface science diagnostics indicate that the process results in a sub-micron thickness layer, partially composed of titanium nitride, with improved hardness, as compared with unimplanted samples. Further parametric studies are being performed, in order to optimize the process, together with a more comprehensive characterization of both, the resulting titanium nitride coatings and the RF plasma features.

Acknowledgments

This work has been funded by project FONDECYT no. 1130228 and CONICYT PIA no. ACT1108. M. Cisternas is funded by a postgraduate studies scholarship awarded by CONICYT. F. Mellero acknowledges financial support from project FONDECYT no. 1110380.

References

[1] Liu X, Chu P K and Ding C 2004 *Mat. Sci. Eng.* **R 47** 49
[2] Geetha M, Singh A K, Asokamani R and Gogi A K 2009 *Prog. Mat. Sci.* **54** 397
[3] Ratner D B 2001 *A perspective on Titanium Biocompability (Titanium in Medicine)* eds D M Brunette, P Tengvall, M Textor and P Thomsen (Heidelberg: Springer Berlin) pp 1-12
[4] Auernheimer J, Zukowski D, Dahmen C, Kantlehner M, Enderle A, Goodman S L and Kessler H 2005 *ChemBioChem* **6** 2034
[5] Gordin D M, Gloriant T, Chane-Pane V, Busardo D, Mitran V, Höche D, Vasilescu V, Drob S I and Cimpean A 2012 *J. Mater. Sci.: Mater. Med.* **23** 2953
[6] Jin S, Zhang Y, Wang Q, Zhang D and Zhang S 2013 *Coll. and Surf. B: Biointerfaces* **101** 343
[7] Meng W J and Curtis T J 1997 *J. Elec. Mat.* **26** 1297
[8] Ingo G M, Kaciulis S, Mezzi A, Valente T, Casadei F and Gusmano G 2005 *Electrochimica Acta* **50** 4531
[9] Saikia P, Joseph A, Rane R, Saikia B K and Mukherjee S 2013 *Journal of Theoretical and Applied Physics* **7** 66
[10] Conrad J R 1987 *J. Appl. Phys.* **62** 777
[11] Anders A 2000 *Handbook of Plasma Immersion Ion Implantation and Deposition* (New York: Wiley)
[12] Valderrama E, Favre M, Bhuyan H, Ruiz H M, Wyndham E, Valenzuela J and Chuaqui H 2010 *Surf. Coat. Technol.* **204** 2940

[13] Kumar S, Baldwin M J, Fewell M P, Haydon S C, Short K T, Collins G A and Tendys J 2000 *Surf. Coat. Technol.* **123** 29
[14] Tian X, Wang L, Fu R K and Chu P K 2002 *Mat. Sci. Eng.* **A337** 236

Effect of Ion Irradiation on the Structural Properties and Hardness of a-C:H:Si:O:F Films

Elidiane C. Rangel[1], Nilson C. da Cruz[1], Rita C. C. Rangel[1], Richard Landers[2] and Steven F. Durrant[1]
[1] Laboratório de Plasmas Tecnológicos, Universidade Estadual Paulista - UNESP – Av. Três de Março, 511,
18087-180 Sorocaba, SP, Brazil
[2] Grupo de Física de Superfícies, Departamento de Física Aplicada,
Instituto de Física Gleb Wataghin, Universidade Estadual de Campinas - UNICAMP,
13083-970 Campinas, SP, Brazil

E-mail: steve@sorocaba.unesp.br

Abstract. Amorphous carbon-based thin films, a-C:H:Si:O:F, were obtained by plasma immersion ion implantation and deposition (PIIID) from mixtures of hexamethyldisiloxane, sulfur hexafluoride and argon. For PIIID the sample holder was biased with negative 25 kV pulses at 60 Hz. The main system parameter was the proportion of SF_6 in the reactor feed, R_{SF}. To allow comparison to growth without intentional ion implantation, some films were also grown by plasma enhanced chemical vapor deposition (PECVD). The objectives were to investigate the effects of fluorine incorporation and ion implantation on the film's chemical structure, and principally on the surface contact angle, hardness and friction coefficient. Infra-red and X-ray photo-electron spectroscopic analyses revealed that the films are essentially amorphous and polymer-like, and that fluorine is incorporated for any non-zero value of R_{SF}. Choice of R_{SF} influences film composition and structure but ion implantation also plays a role. Depending on R_{SF}, hydrophilic or hydrophobic films may be produced. Ion implantation is beneficial while fluorine incorporation is detrimental to hardness. For ion implanted films the friction coefficient falls about one third as R_{SF} is increased from 0 to 60%. Films prepared by PIIID without fluorine incorporation present fairly low friction coefficients and hardnesses greater than those of conventional polymers.

1. Introduction

There is at present a growing need for the development of nanoscale devices. The electronics industry fabricates miniaturized components from materials that match thin structures with specific electrical properties [1-3]. The production of microscopic three-dimensional structures, such as pistons, capacitors, transformers and gears, is another promising area with selected material requirements [4-7]. The production of thin films of a specific material is, therefore, a first requisite for such technologies. Despite this, such films or layers are not only used for these ends. Indeed the greatest spectrum of studies in this area deals with surface modification of materials by coating [8-11] Thus the surface properties of the material are modified while the bulk properties remain essentially unchanged.

The coating of computer hard discs is an application for which an extremely fine film is necessary to protect the device against wear but without changing its storage capacity [12, 13]. Another example of applications involves coatings that accelerate/inhibit cellular metabolism and the activity of fungi, which have a decisive role in the development of biomaterials [14-16]. Coating technology continues to be a growth area [17].

Plasma enhanced chemical vapor deposition (PECVD) is a versatile technique for the production of thin films [18]. A key advantage is its ability to produce films simply and cheaply at ambient temperatures. Owing to its low energy consumption and production of clean by-products, the technique is considered environmentally friendly. By altering the chemical composition of the plasma, the applied power or its frequency, or the reactor geometry or substrate temperature, films with a wide range of properties may be produced [18].

A hybrid plasma deposition technique, which permits *in situ* alteration of the film properties is plasma immersion ion implantation and deposition PIIID [19]. In this method, simultaneously with plasma ignition, negative high voltage pulses are applied to the samples, thus attracting high energy positive ions from the discharge. The impacting ions induce a number of effects, including the scission of chemical bonds, thus producing unsaturated groups, the release of volatile group, such as CH_x (x of 1 to 3), and film compaction. Film stoichiometry may also be altered. In addition, the PIIID technique permits adjustment of the bombarding ion intensity during film deposition.

The aims of this work were to determine the effect of ion bombardment during film growth as well as that of fluorine incorporation on the properties of the films produced by PIIID. To allow comparison with the PIIID films, some films were also produced by PECVD. Although the methods are distinct, variation of the system parameters often produces identical tendencies in the behavior of the material properties. The optimization of the treatment parameters is obtained through the analysis of the influence of each parameter on the performance of the material.

2. Experimental

The films were deposited in the system shown schematically in Figure 1. A cylindrical stainless steel vacuum chamber is connected to a pumping system capable of reducing the chamber pressure to about 10^{-3} Pa. Gases are admitted via steel tubing through flowmeters and needle valves. Liquid monomers are retained in glass evaporation cells which are connected to the chamber via needle valves.

A Tokyo Hy-Power radiofrequency (13.56 MHz) power supply (maximum power 300 W) allows the excitation of a discharge between internal, horizontal, circular, stainless-steel electrodes. To produce bombardment by positive ions from the plasma, the lower electrode is connected to a Carl Zeiss UBI2 supply, which produces negative pulses (25 kV, 60 Hz). Greater details of this system are given elsewhere [20].

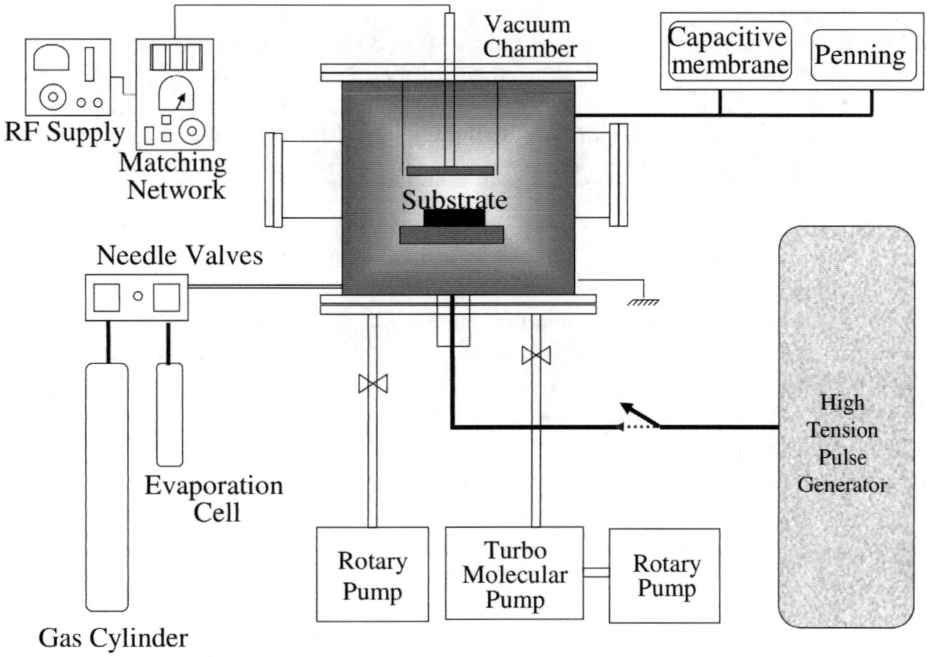

Figure 1 – Details of the deposition system.

Films were deposited from mixtures of hexamethyldisiloxane (HMDSO), sulfur hexafluoride (SF_6) and argon (Ar). Before depositions the system was pumped to the base pressure and HMDSO introduced to produce a pressure of 0.26 Pa, which was maintained constant for all the depositions. The proportion of SF_6 (R_{SF}) in the feed was increased from 0 to 80% while that of Ar was correspondingly decreased, maintaining the total gas pressure at 1.3 Pa. The discharge was excited by the application of rf power to the upper electrode while the substrate holder was polarized with high voltage negative pulses (25 kV, 60 Hz). Deposition times of one hour were used and following each deposition the chamber was filled with nitrogen until atmospheric pressure was reached. The effects of these deposition conditions on the film properties were examined.

Film thickness was measured from well-defined step-heights using a Veeco Dektak 150 profilometer. The molecular structure of films was analyzed by Infra Red Reflectance Absorbance Spectroscopy (IRRAS) using a Bomen MB-101 FT-IR spectrometer. The spectra were taken from samples prepared on aluminum-coated glass slides. X-ray Photoelectron Spectroscopy (XPS) was used to study the chemical composition of the surface of the films deposited onto stainless-steel substrates. The spectra were obtained with a SVW HA100 spectrometer using procedures described previously [21].

The affinity of the films for water was evaluated via measurement of the contact angle, θ, using a Ramé-Hart 100-00 goniometer. Substrates for these films were polished silicon wafers. At least three drops of deionized water of about 2 μl volume were placed on different regions of the sample. The contact angle was measured at ambient temperature on each side of the drop (as seen in front elevation) ten times. The results presented are the arithmetic mean of these measurements (n = 60).

Nanoindentation tests [22] using a Hysitron Triboindenter on films deposited onto polished silicon substrates were undertaken to determine film hardness. The indentation was measured as a function of the normal load applied to the diamond probe in the range from 800 to 10.000 μN. At least eight

indentations per load were made. The hardness was calculated from curves of the indentation versus the applied force, according to the method developed by Oliver and Pharr [23]. Mean values and standard deviations are presented.
Using the nanoindenter's lateral-force device, scratch tests were undertaken on the same samples. At least five scans of 20 μm were taken in different regions of the sample. A constant applied load of 300 μN and a velocity of 1.0 μm s^{-1} were used. The tests were made without lubrification using as a tribological pair the sample and the diamond tip of the indenter, which has a diameter of about 100 nm. Among other parameters, the lateral and normal forces on the tip were determined, permitting calculation of the friction coefficient of the samples.
Topographic images of the sample surfaces were obtained using Scanning Atomic Force Microscopy (SAFM) in the image acquisition mode of the nanoindenter, employing the same samples as those submitted to nanoindentation and hardness tests. A constant force of 2.0 μN was applied to the indenter tip and a scan undertaken over an area of 50 x 50 μm^2. From the images, it was possible to determine the mean roughness [24] (R_a) in regions established as representative of the samples.

3. Results and Discussion

3.1 Thickness, Molecular Structure and Chemical Composition

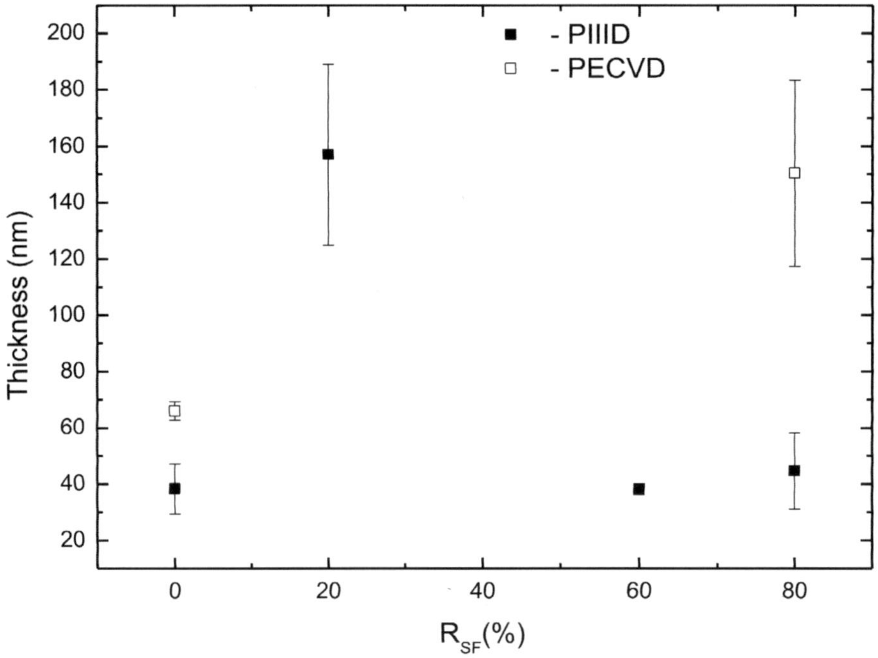

Figure 2 – The film thickness as a function of R_{SF}

Figure 2 shows the thicknesses of the films, which fall in the range 40 to 160 nm. Figure 3 shows the infrared spectra obtained from films deposited at different proportions of SF_6 in the feed to the chamber, R_{SF}, during ionic bombardment and of those produced at R_{SF} of 0 and 80% by conventional PECVD.

The sample produced at 0% SF_6 by PECVD presents absorption bands due to OH (3400 and 957 cm^{-1}), C-H (2870 and 2920 cm^{-1}), C=O and C=C (1600-1700 cm^{-1}) and Si-O-Si (1100-1200 cm^{-1}) groups. The latter peak may also contain a contribution from $SiCH_2Si$ groups, which are readily formed via plama reactions of methylsilyl groups [25-27]. A weak absorption at about 1270 cm^{-1}, related to C-H in CH_3Si groups, is also detected, demonstrating that the HMDSO molecule - $(CH_3)_3SiOSi(CH_3)_3$ – is partially preserved.

The presence of O-H in the structure may be attributed to the oxygen bound in the HMDSO molecule. As oxygen is not linked to hydrogen in this molecule, however, multiple-stage reactions must occur in the plasma phase to form O-H and, therefore, this mechanism is not considered the most important. Residual oxygen in the chamber may contribute more effectively to oxygen incorporation during deposition. In addition, post-deposition reactions between free radicals trapped from the plasma and water vapor and atmospheric oxygen are known to produce hydroxyl groups in the deposited material [28].

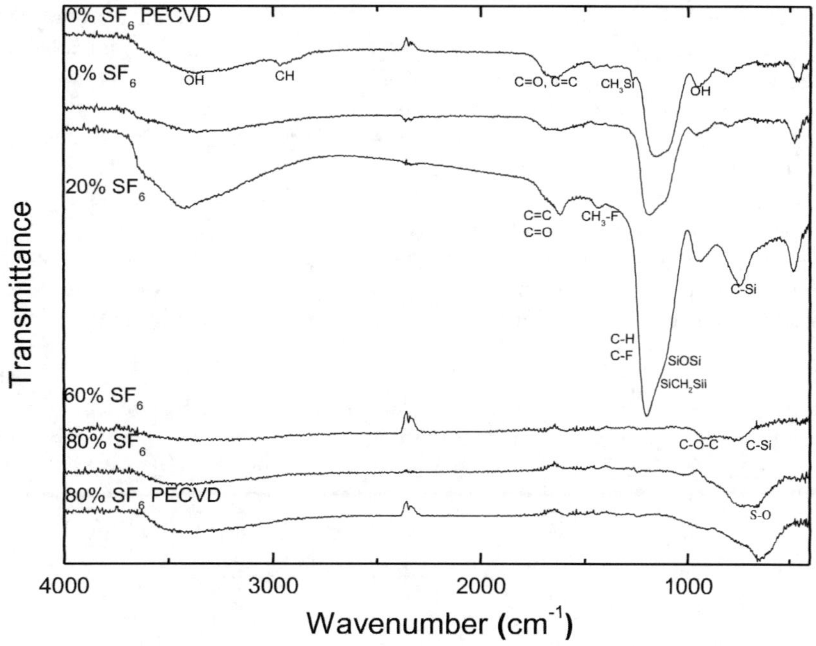

Figure 3 – Transmittance infrared spectra of films produced by PECVD and by PIIID at different R_{SF}.

The detection of absorptions due to C=C and C=O groups is interesting since neither is present in the monomer. These bonds are formed via reactions either in the plasma phase or on the film surface or both, but the constituents are derived from the fragmentation of the monomer molecule. Post-deposition reactions between free radicals and atmospheric oxygen also contribute to the production of C=O.

The spectrum of the film prepared with R_{SF} = 0% by PIIID presents the same absorptions as already mentioned, but with lower intensities in some cases. This effect may be observed by comparing the absorptions at 2970 and 3400 cm^{-1} (due to C-H and O-H, respectively). The fall in the intensity of these absorptions is associated with the different ion bombardment fluences produced in the two methods. In PIIID, the transfer of energy to surface species by collision with fast ions produces, among other effects, the fragmentation of chemical bonds. Species weakly bound to the solid, such as H, are preferentially emitted [29, 30], reducing the concentration of C-H and O-H groups. Additional evidence for this effect is the behavior of the absorption at 1270 cm^{-1}, which derives from vibrations in C-H groups in CH_3Si. When films are prepared by PIIID this band disappears, confirming the greater degree of molecular fragmentation in this process.

Contrary to what is expected for films produced by plasma polymerization of HMDSO without SF_6 in the feed [25-27], no absorbance peaked at around 2100 cm^{-1} and attributable to SiH is seen the the spectrum (film produced by PECVD without SF_6 in the feed). Thus, under the conditions used here, hydrogen is preferentially bound to carbon or lost from the film or both. As already mentioned, the absence of strong absorptions owing to CH groups from all of the spectra, and the tendency of the films grown at high R_{SF} to become 'inorganic', indicate a loss of hydrogen.

The band between 1100 and 1200 cm^{-1}, which appears in the spectra of both films deposited at R_{SF} = 0%, is attributed to the superposition of the absorptions due to CH and SiOSi. The process of dehydrogenation, however, does not affect the intensity of this band as occurs with the absorptions at 2920 and 2870 cm^{-1}, indicating that the contribution of C-H groups to this absorption is negligible.

For R_{SF} = 20%, absorptions in the film spectra appear due to fluorine-containing groups, such as CH_3-F (1436 cm^{-1}) and C-F (1000-1400 cm^{-1}). Atomic fluorine and CF_x (x = 1 to 3) radicals are frequently observed in plasmas containing fluorinated gases [31-33] and are considered film precursors. As the chamber feed does not usually contain compounds in which carbon is bound to fluorine, multiple-stage plasma reactions or plasma/surface reactions or both are necessary to produce these species.

For R_{SF} > 20%, there is a high proportion of atomic fluorine in the plasma which, together with ionic bombardment, contributes to the erosion of the film by etching. Therefore the deposition rate is determined by the balance between film growth and ablation. In PECVD the deposition rate tends to decrease (that is the film thickness decreases for a fixed deposition time) with increasing R_{SF}, which accounts for the absence of absorptions in the spectra of films produced under these conditions. It is worthwhile observing the absorption at 650 cm^{-1} in the two spectra of samples prepared at R_{SF} = 80%, which indicates the presence of sulfur.

Table 1 shows the atomic ratios F/C, O/C and Si/C, determined from XPS spectra of the films deposited with ionic bombardment. The F/C ratio grows with increasing R_{SF}, reaching a maximum of 0.19 at R_{SF} = 80% of SF_6. This result may be the consequence of a decrease in the proportion of carbon, of the increase in the incorporation of fluorine or, more probably, both.

Table 1. Atomic ratios F/C, O/C and Si/C in the PIIID films as a function of the proportion of SF_6 in the feed, R_{SF}.

R_{SF}(%)	F/C	O/C	Si/C
0	-	2.05	-
20	0.02	0.35	0.10
60	0.07	0.23	0.12
80	0.19	0.81	0.04

To clarify this problem, high resolution F1s spectra were deconvoluted, revealing a component at 687.0 eV associated with C-F bonds, and a contribution at 684.0 eV, identified as ionic fluorine in carbon structures [34], that is, free fluorine. The relative contribution (by area) of each component was calculated and the results are presented in Figure 4.

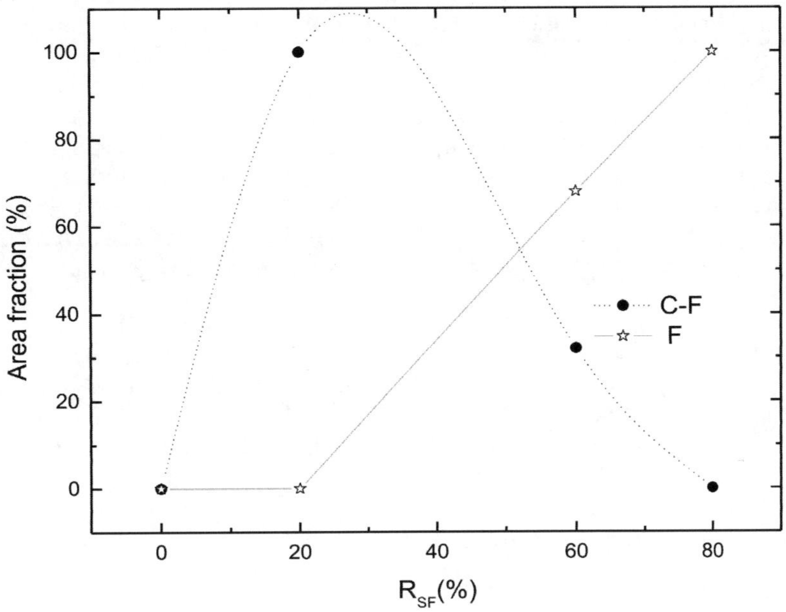

Figure 4 – Contribution of each component of the high resolution F1s peak as a function of R_{SF}.

The proportion of C-F groups is high in the sample prepared at R_{SF} = 20% as shown by some bands in the infrared spectrum of this film. Despite this, when R_{SF} increases, the proportion of C-F decreases in opposition to the increase in the proportion of free fluorine.
The reduction in the quantity of C-F species is explained by the ablation of the film when the proportion of SF_6 is increased, which leads to the removal of carbon and an increase in free fluorine, resulting in an increase in F/C.
The O/C data, also represented in Table 1, reveal the incorporation of oxygen in the deposited material. It is known that when films are produced by PECVD from monomers that do not contain oxygen, some incorporation of oxygen is still observed. The ratio O/C is substantially reduced when R_{SF} is increased from 0 to 20%, but the influence of greater R_{SF} values is less pronounced. This result is attributed to a fall in the concentration of residual free radicals in the films at higher R_{SF}. The intensity of ion bombardment is reduced with the reduced concentration of Ar in the plasma, thus reducing the absorption of oxygen-containing groups.
Nitrogen was also detected, but only in the film deposited without SF_6 in the chamber feed. Here it should be remembered that to reach atmospheric pressure following film deposition, the chamber was filled with nitrogen. Generally, the incorporation of nitrogen occurs to a lesser degree than is observed for oxygen, and its apparent absence at $R_{SF} > 0$ is due to the lower proportion of free radicals in the

samples. As well as the physical effects of ion bombardment, competition between F, H and Si to form bonds with C contributes to a reduction in the concentration of free radicals.
Another interesting behavior is that of Si/C. Silicon was not detected in the film deposited without SF_6 in the feed. This result differs from that obtained by infrared spectroscopy, in which a band arising from SiOSi groups (1100 to 1200 cm^{-1}) was detected. In infrared spectroscopy, however, the beam passes completely through the width of the film, while XPS is a surface analysis technique. Thus the surface and bulk of the film may differ in chemical structure and composition, and this may be due to plasma-surface reactions. When the deposition is undertaken in the absence of F, the relatively high ionic bombardment removes species such as H and Si. This process may also induce the production of free radicals, thus contributing to the increased attachment of oxygen. As the proportion of F increases, ionic bombardment is reduced and the proportion of surface Si increases. At $R_{SF} \geq 80\%$, however, etching of Si increases and the proportion of this species tends to decrease, so that Si/C decreases as F:C correspondingly increases. Thus the presence of Si on the surface is strongly dependent on plasma-surface interactions.
Finally, it is worth noting that sulfur was detected in the samples prepared by PIIID with 60% and 80% SF_6 in the feed. Sulfur incorporation is expected in SF_6 plasmas [35] and was confirmed by the infrared spectra.
Considering the complete set of analyses undertaken it is concluded that the films deposited are amorphous carbon polymers. The material is already fluorinated when $R_{SF} = 20\%$ is employed. At R_{SF} greater than 20%, however, little fluorine is incorporated and there is a reduction in the proportions of C, O and Si, reducing the film deposition rate. Deposition together with ionic bombardment results in surfaces with structures and compositions distinct from those obtained with conventional PECVD.

3.2 Wettability and Roughness

The affinity of the surface of the films for water is illustrated in Figure 5, where the surface contact angle is shown as a function of R_{SF}. The results obtained for films produced by PIIID reveal that θ increases with increasing R_{SF}.

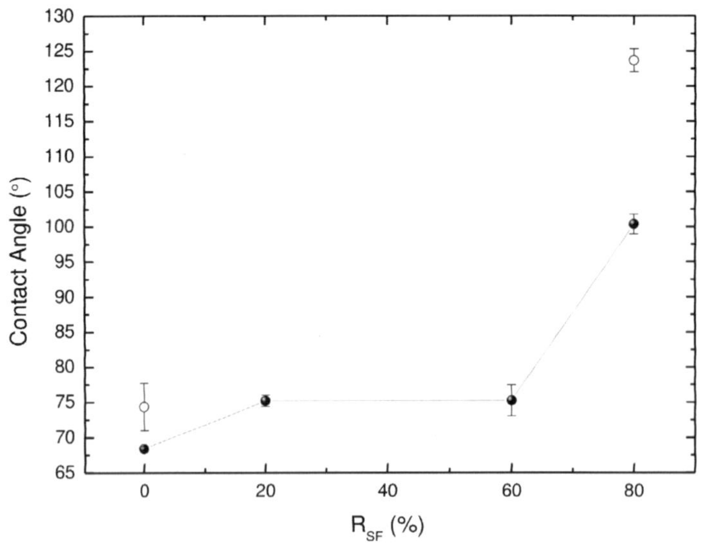

Figure 5 – Surface contact angle of the films produced by PECVD (○) and by PIIID (•) as a function of R_{SF}.

According to the chemical composition, the surface prepared without the addition of fluorine to the discharge presented the greatest proportion of O and N. Such species, connected to carbon, form polar groups that attract water molecules electrostatically. The absence of Si contributes to the hydrophilic character of this sample (Table 1). On the other hand, when R_{SF} = 20%, there is a decrease in the proportion of O that is compensated by the incorporation of F, which in moderate proportions, also contributes to the spread of the drop on the surface. At greater values of R_{SF}, however, the greater incorporation of F and the presence of Si cause the formation of a hydrophobic film with θ of about 100 °.

Although the films deposited at R_{SF} = 80% using the two techniques originated in plasmas with the same chemical composition, their surface wettabilities are distinct owing to ionic bombardment (or its absence).

To complement the interpretation of the wettability, the mean roughness (R_a) of the surfaces was determined, as shown as a function of R_{SF} in Figure 6. The pristine silicon substrate presented a roughness of 4.6 nm. Comparing this value with those obtained for films produced by PIIID, an increase in surface roughness is observed. There are no significant variations in R_a when R_{SF} is increased from 0 to 60%. Only for R_{SF} > 60% is there an increase in surface roughness.

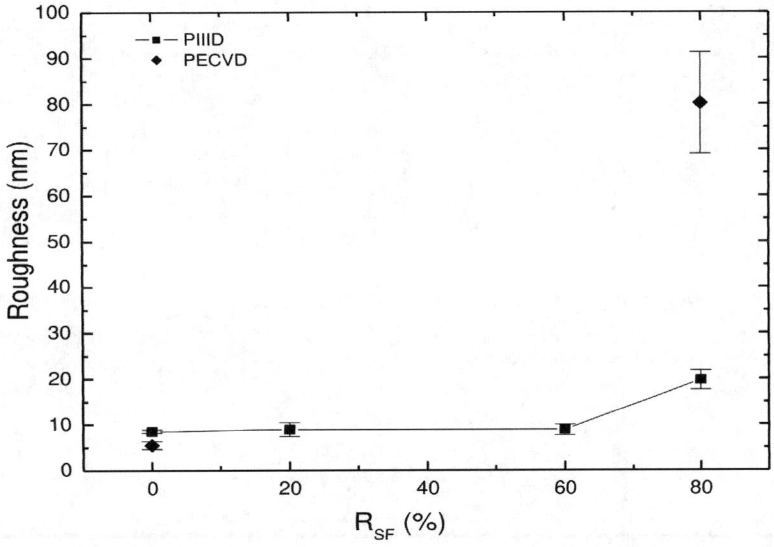

Figure 6 – The roughness of the surfaces coated with films prepared by PECVD and by PIIID at different proportions of SF_6 in the chamber feed. The roughness of the silicon substrate was 4.6 nm.

In contrast, R_a is very similar to that obtained for the silicon substrate when the films are produced without fluorine by PECVD, but increases sharply when R_{SF} is increased to 80%. Figure 7 shows the topographic images of the samples from which the roughness was determined.

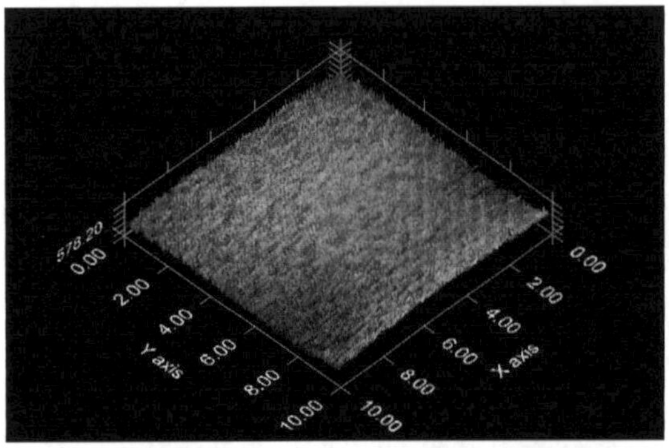

0% SF_6 – PECVD

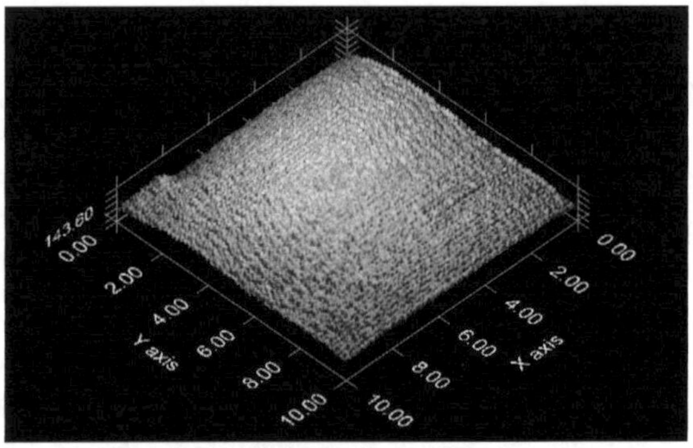

60% SF_6 – PIIID

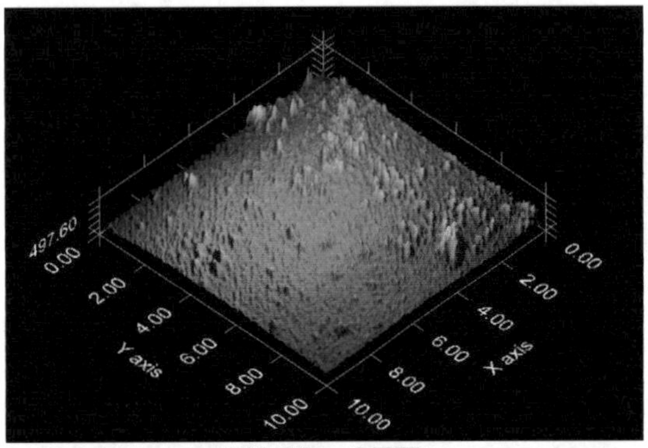

0% SF_6 –PIIID

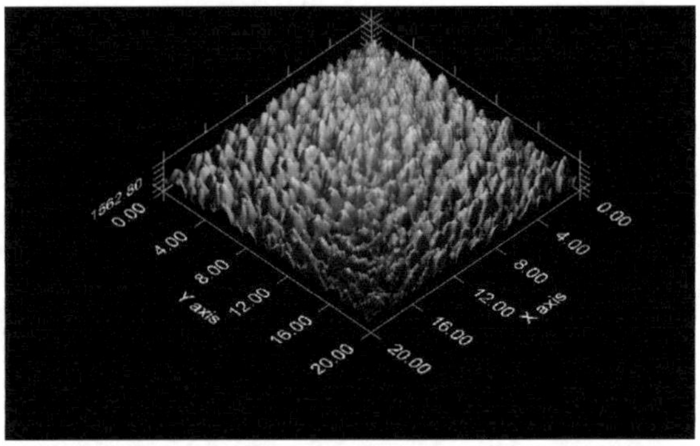

80% SF_6 – PIIID

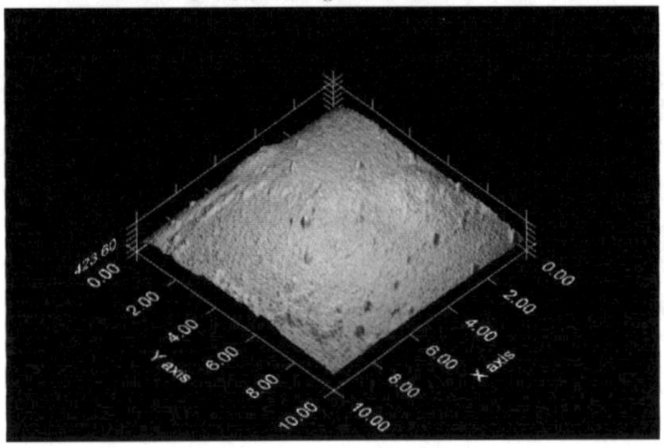

20% SF_6 – PIIID

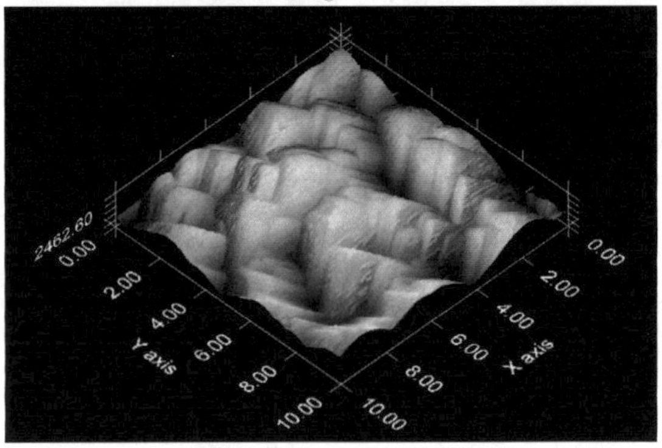

80% SF_6 – PECVD

Figure 7 – Topographic images of the silicon substrates coated with films prepared at different proportions of SF_6 by PECVD and by PIIID. The scales are in μm.

Comparing the images of films obtained with R_{SF} = 0% and R_{SF} = 80% by PECVD, a clear difference is observed. This difference is attributed to the texturization of the silicon surface at the beginning of the deposition by attack by atomic F. Following the deposition of the first few monolayers, this phenomenon ceases and the film follows the topography of the substrate.

On the other hand, when F is present in small proportions in the discharge in depositions with ion bombardment, fluorine ions are formed in the discharge by the rupture of SF_6 molecules, and are also

implanted, decreasing their chemical effect on the surface of the silicon substrate at the beginning of the deposition process. At high R_{SF} (80%), the abundance of F in the plasma induces both ion implantation and ablation of Si. The surface roughness is modified, but not in the same way as observed in samples prepared at the same R_{SF}, but without high energy ionic bombardment. These results demonstrate the effect of the incorporation of F in the discharge and of ionic bombardment on the surface topography. In cases where fluorinated films without texturization are required, PIIID is the best choice. It should be remembered, however, that the effect of F on the texturization of the sample depends on the choice of both the plasma and the substrate.
Comparing the results of Fig. 5 and Fig. 6 it is readily noticed that the slight growth in θ in the 0 to 60% R_{SF} range is a chemical effect of fluorine incorporation since roughness remains unchanged in this region. For the highest R_{SF} value, however, the factor determining θ is roughness.

3.3 Hardness and Friction Coefficient
Table 2 shows the film hardness estimated from curves of hardness as a function of the penetration depth, and taken at about 15% of the film thickness. As the films were all < 200 nm thick, this is a crude calculation. The implanted films, however, show a tendency to soften with greater fluorine incorporation. Implanted films tend also to be harder than PECVD films grown under otherwise identical conditions. The films produced by PECVD with 0% and 80% SF_6 in the feed were measured with a single load applied to the tip. In both cases, the hardness is a few tenths of a GPa, which is characteristic of polymeric materials. These results agree with literature reports, which demonstrate that the inclusion of fluorine in the structure of carbon films deposited in plasmas [36, 37] reduces their hardness. Owing to the thinness of the film produced by PIIID film at R_{SF} = 60% it's hardness could not be determined.

Table 2. Film hardness of the PECVD and PIIID films as determined by nanoindentation

Method	**R_{SF}(%)**	**H(GPa)**
PECVD		
	0	0.8 ± 0.1
	80	0.5 ± 0.5
PIIID		
	0	1.6 ± 0.3
	20	1.5 ± 0.1
	60	-
	80	0.8 ± 0.4

The friction coefficient, μ, determined by scratch tests using the lateral force device of the nanonindenter, is presented as a function of R_{SF} in Figure 8. The films produced by PECVD have friction coefficients of less than 0.1 independently of R_{SF}. From the images of Fig. 7 it may be seen that the topographic differences presented by the samples do not strongly influence μ. It should also be remembered that the diamond tip used as a tribological pair for the film has a diameter of about 100 nm, which makes the dimensions and irregularities of the film prepared at R_{SF} = 80% irrelevant to the interaction.

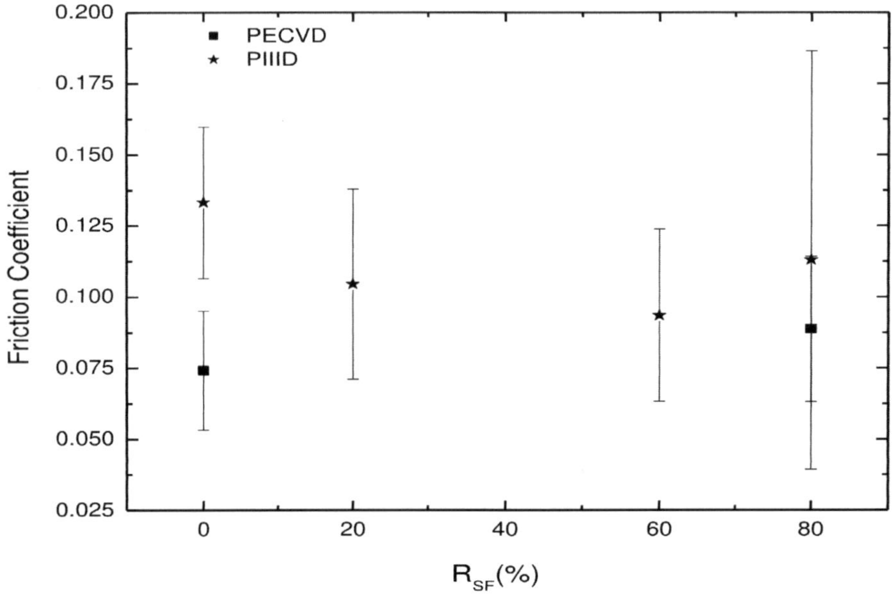

Figure 8 –Friction coefficient of the films produced by PECVD and by PIIID at different R_{SF}. Values obtained with a normal load of 300 μN applied to the tip for a 20 μm scan.

At R_{SF} = 0%, the film deposited with ionic bombardment has a greater value of μ than that produced by PECVD. The friction coefficient is reduced by 30% compared to the film without fluorine when R_{SF} is increased from 0 to 60%. When R_{SF} is further increased, μ tends to increase further.

Comparing the behaviors of the roughnesses (Fig. 6) and of the friction coefficient of the films produced by PIIID, it is observed they are similar only in the range of R_{SF} from 60 to 80%. Otherwise, different tendencies are observed, indicating that the topography is not the dominant factor influencing μ. In this study, the incorporation of a highly electronegative species has a greater influence on the tribology than changes in the surface topography. Depending on the morphology and dimensions of the irregularities, they act together with the electronegative effect of fluorine. This may be seen in the sample deposited at the highest value of R_{SF}, where, despite the increase in the proportion of F incorporated into this sample, μ does not decrease. The tribological behavior of this system thus depends on the balance between these processes.

4. Conclusions

The results presented here show that the chemical composition and structure of the films deposited from plasmas of mixtures of hexamethyldisloxane, sulfur hexafluoride and argon depend on the proportion of gases used and the degree of ionic bombardment induced during film growth. Fluorine was incorporated even when only low proportions of SF_6 were present in the discharge. Contamination of the films with oxygen and nitrogen, to degrees which depend on R_{SF}, was also observed.

Hydrophilic surfaces were obtained by PIIID for R_{SF} = 0, 20 and 60%. Increasing R_{SF} beyond 60%, resulted in hydrophobic films with contact angles of 100°. The same rising tendency was observed in the contact angle of the films prepared by conventional PECVD, but with higher θ values than were observed in the film prepared by PIIID. Thus the thermodynamic properties of the surface may be

controlled via the chemical composition of the discharge, but ionic bombardment is also of great importance.
The hardnesses of non-fluorinated films were increased by ionic bombardment. Moreover, such films possessed hardnesses greater than those typical of similar conventional polymers. While fluorine incorporation (for R_{SF} up to ~60%) tends to slightly improve tribological performance of the diamond/film pair, ionic bombardment has little effect. The minimal improvement in μ due to the incorporation of F does not justify the corresponding loss in mechanical resistance of the material. In summary, ionic bombardment produces some improvement in mechanical resistance of the film while maintaining the friction coefficient at levels acceptable for many applications.

Acknowledgements
We thank the Brazilian Agencies FAPESP and CNPq for financial support.

References

[1] Lakin, K M, 2003, *Microwave Magazine*, IEEE, **4** (4), 61.
[2] Gross, S, Camozzo, D, Di Noto, V, Armelao, and L, Tondello, E, 2007, *European Polymer Journal*, **43** (3), 673.
[3] Di, C, Yu G, Liu, Y, and Zhu, D, 2007, *J. Phys. Chem.* B **111** (51), 14083.
[4] Natarajan, S P, 2007, *IEEE Microwave and Wireless Components Letters*, **17** (2), 858.
[5] Belgacem, B, Calame, F, and Muralt, P, 2007, *J. Electroceram*, **19**,369.
[6] Kimberley, J, Chasiotis, I, and Lambros, J, 2008, *International Journal of Solids and Structures*, **45**, 497.
[7] Saul, C K, Lepienski, C M, Cantão, M P, and Schreiner, WH, 2003, *Sensors and Actuators* A, **107**, 80–84.
[8] Jampala, S N, Sarmadi, M, Manolache, S, and Denes, F S, 2008, *J. Appl. Polym. Sci.*, **107**, 1686.
[9] Holm, A H, Vase, K H, Winther-Jensen, B, Pedersen, SU, and Daasbjerg, K, 2007, *Electrochimica Acta*, **53**, 1680.
[10] Tshabalala, M A, and Sung,L P, *2007 J. Coat. Technol.*, **4**, 483.
[11] Galkowski, M, Malik, M A, Kulesza, P J, Bala, H, Miecznikowski, K, Wlodarczyk, R, Adamczyk, L, and Chojak, M, 2003, *J. Electrochem. Soc.*, **150** (6) B249.
[12] Noda, K, Kawanabe, T, and Naoe, M, 1997, *J. Appl. Phys.*, **81**, 5405.
[13] Kawahara, H, Sakairi, T, Momose, S, Ishihara, Y, and Urai, H, 1996, *Nec Research and Development*, **37** (1), 13.
[14] Kobayashi, Ohgoe, S Y, Ozeki, K, Hirakuri, K, and Aoki, H, 2007, *J. Mat. Sci. Mat. Medicine*, **18** (12), 2263.
[15] Blümmel, J, Perschmann, N, Aydin, D, Drinjakovic, J, Surrey, T, Lopez-Garcia, M, Kessler, H, J, and Spatz, J P, 2007, *Biomaterials,* **28** (32), 4739.
[16] Ai, H, Meng, H, Ichinose, I, Jones, S A, Mills,D K, Lvov, Y M, and Qiao, X, 2003, *Journal of Neuroscience Methods*, **128**, 1.
[17] Gordillo-Vazquez, F J. Herrero, and V J, Tanarro, I, 2007, *Chemical Vapor Deposition*, **13** (6-7), 267.
[18] Yasuda H, in *Plasma Polymerization*, (Academic Press, N.Y., 1985).
[19] Wood. B P, Rej, D J, Anders, A, Brown, I G, Faehl,R J, Malikand, S M, and Munson, C P in Handbook of Plasma Immersion Ion Implantation and Deposition edited by Anders, A (John Wiley and Sons, N.Y., 2000) p. 243.
[20] Rangel, E C, Durrant, S F, Rangel, R C C, Kayama, M E, Landers, R, and Cruz, NC, 2006, *Thin Solid Films*, **515**, 1561.
[21] Durrant, S F, Landers, R, Kleiman, G G, Castro, S G, and Bica de Moraes, M A, 1996, *Thin Solid Films*, **281/282**, 294.

[22] Li, X, and Bhushan, B, 2002, *Mat. Characterization*, **48**, 11.
[23] Oliver W C and Pharr, G M, 1992, *J. Mat. Res.*, **7**(4), 1564.
[24] Petrik, P, Biró, L P, Fried, M, Lohner, T, Berger, R, Schneider, C, Gyulai J, and Ryssel, H, 1998, *Thin Solid Films*, **315,** 186.
[25] Wrobel, A M, Wertheimer, MR, Dib, J, and Schreiber, H P, 1980, *J. Macromol. Sci.- Chem., A***14**, 321.
[26] Wrobel, A M, Klemberg, J E, Wertheimer, M R, and Schreiber, H P, 1981, *J. Macromol. Sci. - Chem., A***15**, 197.
[27] Wrobel, A M, Kryszewski, and M, Gazicki, M, 1983, *J. Macromol. Sci. – Chem., A***20**, 583.
[28] Yasuda, H, Bumgarner, M O, Marsh, H C, and Morosoff, N, 1976, *J. Polym. Sci., Polym. Chem. Ed.,* **14**, 195.
[29] Pivin, J C, and Colombo, P, 1997, *J. Mater. Sci.,* **32**, 6163.
[30] Venkatesan, T, Wolf, T, Allara, D, Wilkens, B J, and Taylor, G N, 1983, *Appl. Phys. Lett.*, **43**, 934.
[31] Durrant,S F, Mota, R P, and Bica de Moraes, M A, 1992, *J. Appl. Phys.*, **71**(1), 448.
[32] Durrant, S F, Rangel, E C, Da Cruz, N C, Castro, S G C, and Bica de Moraes, M A, 1996, *Surf. & Coatings Technol.,* **86**(1-3), 443.
[33] Durrant, S F, Castro, S G C, Bolívar Marinez, L E, Galvão, D S, and Bica de Moraes, M A, 1997, *Thin Solid Films*, **304**(1-2), 149.
[34] Nansé, G, Papirer, E, Fioux, P, Moguet, F, and Tressaud, A, 1997, *Carbon*, **35**(2), 175.
[35] Durrant, S F, Mota, R P, and Bica de Moraes, M A, 1992, *Thin Solid Films*, **220**(1-2), 295-302.
[36] Freire Jr., F L, 2002, *J. Non-Cryst. Sol.*, **304**, 251.
[37] Yao, Zh Q, Yang, P, Huang, N, Sun, H, and Wang, J, 2004, *Appl. Surf. Sci.,* **230**, 172.

Electron Beam Plasma Application for Synthesis of Bioactive Titanium Oxide Coatings

T M Vasilieva[1], I V Sokolov[2] and K V Balakin[1]

[1] Moscow Institute of Physics and Technology, Dolgoprudny, Moscow region, 141700, Russia

[2] Keldysh Research Centre, Moscow, 125438, Russia

E-mail: tmvasilieva@gmail.com

Abstract. Prospective bio-medical applications of the electron-beam plasma (EBP) were experimentally studied. The EBP-treated titanium samples were investigated from the point of view of their bio-compatibility.
The titanium oxide coatings were synthesized in the EBP of oxygen on the surface of plane titanium substrates and on the inner surface of the titanium tubes. The EBB-treatment was able to signifantly improve the surface uniformity and roughness. Titanium oxide TiO_2 (IV) in the rutile form predominated in the coatings composition
The bioactivity of synthesized TiO_2-coatings was characterized by the water contact angle measurements and by the ability to precipitate hydroxyapatite from the model solution which simulated the composition of the body fluid. The studies showed the samples with plasmachemically synthesized TiO_2-coatings to be more hydrophilic than untreated titanium and to precipitate hydroxyapatite on their surface effectively.

1. Introduction

Titanium and titanium alloys are widely used for dental and orthopedic implants production because of their superior mechanical properties such as tensile strength, yield strength, and lightness [1]. Though the metallic titanium implant is covered by a thin native oxide film which stimulates osseointegration, it may corrode away and be separated from the bone by fibrous tissue with immune reactions when the implant is introduced into a bone [2]. Moreover, usually metallic surfaces are not adequately bioactive, and surface treatment is needed to enhance the bioactivity by improving osseointegration of the implant with the bone tissues.

A variety of plasma-assisted techniques have been developed to obtain bioactive titanium surfaces [3-6]. In the present study the titanium oxide layer was synthesized using the electron-beam plasma (EBP), generated by the injection of the electron beam into oxygen media. Under typical conditions of the EBP generation (medium pressure $0,1 < P_m < 10$ kPa and moderate EB power $N_b < 1$ kW) the plasma is strongly non-equilibrium and cold. Being injected into a gas the EB ionizes the gas, excites the gas molecules and is able to cause molecule dissociation. As a result, the EBP composition is complicated and there are a lot of chemically active particles that do not exist under equilibrium conditions. The main advantage of the beam-plasma technologies with respect to the approaches mentioned above are:

- the possibility of the uniform treatment of large-size substrates and bodies with complicated shape;
- shorter treatment time;
- power saving due to the higher gas pressures (in comparison with conventionally used plasmas) at which the EBP-treatment is performed.

The aims of the present study were as follows:

- to experimentally study the plasma-stimulated synthesis of titanium oxides in strongly non-equilibrium low temperature EBP at the surface of the titanium substrates;
- to characterize the morphology and the chemical composition of the formed titanium oxide coatings;
- to prove the bioactivity of the titanium oxides coatings characterizing (*in vitro*) the samples ability to form bone-like hydroxyapatite (the main inorganic component of bones) by its precipitation from a solution which simulate the body fluid.

2. Materials and methods

2.1. Materials

The EBP-stimulated synthesis of titanium oxides was studied at the surface of the plane substrates (size 20×20 mm) and on the inner surface of the thin titanium tube with diameter 15 mm and length 100 mm. The samples were mechanically polished and cleaned with acetone, isopropyl alcohol and deionized water in ultrasonic bath. To uniform the roughness and topography the samples were etched with the mixture of HCl and H_2SO_4 for 30 min at room temperature and washed again with deionized water before the EBP-treatment.

2.2. Characterisation of EBP-treated titanium samples

2.2.1. *Field-emission scanning electron microscopy (SEM).* The surface morphologies were examined using FEI Quanta 200 (FEI Company, USA). The measurements were carried out at the accelerating voltage 5 kV. Three different fields within each sample were randomly chosen, and 3 images of each field were taken at the magnifications ×300, 600, 2000, 3000, 6000, 30 000.

2.1.2. *Raman spectroscopy.* The Raman spectra were recorded using a T64000 - HORIBA Jobin Yvon (Horiba Jobin Yvon IBH Ltd., Japan) equipped with a liquid N_2-cooled CCD detector. The laser excitation line used was the 488 nm of an argon laser. The power of the incident laser beam was about 50 mW on the sample's surface. For focusing of laser spot an X100 objective was used. Typical spectra line width was 0.5 cm^{-1} while the recorded spectra were the average of 3 scans.

2.2.2. *X-ray diffraction (XRD).* XRD was carried out by the sliding beam method using ARL X'TRA powder diffractometer (Thermo Fisher Scientific, USA) equipped with CoK_α radiation source. The conditions were as follows: the angle of incidence of the primary X-ray beam 5°, accelerating voltage 30 kV, and anode current 10 mA. The XRD pattern was recorded in the range of scattering angles 2θ = 10-130°, scanning intervals 0.1°. The resolution of the interplanar spacing measurements was ±0.0001 nm.

2.2.3. *Contact angle measurements.* The wettability of the plane titanium samples before and after the EBP-treatment was characterized by water static contact angles. The contact angle was measured by the sessile drop method using an optical contact angle-measuring device CAM 101 (KSV Instruments, Finland). The contact angle measurements were performed after the EBP-treatment and then the wettability degradation was studied after 24, 48 hours and 1, 2 and 3 weeks.

2.3. Bioactivity characterization of the EBP-treated titanium samples

Bioactivity or the bone bonding capacity of titanium surfaces was quantified by their ability to form bone-like hydroxyapatite (the main inorganic component of bones) *in vitro* [7]. To characterize their bioactivities the EBP-treated samples were soaked in a simulated body fluid solution (SBF) [8] for two weeks.

3. Treatment procedure

For the controllable titanium oxides synthesis the Electron Beam Plasmachemical Reactor (EBPR) described in detail in [9] was used. Figure 1 illustrates the treatment procedure. The focused continuous electron beam (EB) 3 generated by the electron-beam gun 1 that was located in the high vacuum chamber 2 was injected into the working chamber 5 filled with the plasma-generating gas through the injection window 4 [10]. In passing EB through the gas the cloud 8 of the EBP was generated and the sample 6 was inserted into the plasma cloud to be treated. Since all plasma parameters are functions not only of P_m and N_b but of x, y, and z coordinates (z is the axis of the EB injection) the plasma cloud control was required to treat the sample uniformly. The electromagnetic scanning system 10, which is placed inside the working chamber near the injection window, was able to deflect the axis of scattered EB in x and y directions forming a rectangular raster. Varying frequencies and amplitudes of scanning the sample treatment was controlled. To estimate intensities of the fluxes of plasma particles falling on the sample surface the computer simulation of the EBP was carried out for real treatment conditions using the specially developed method [11].

Plane square-shape titanium samples 20x20 mm in size and 2 mm in depth, h, were inserted into the EBPR reaction zone as it is shown in Figure 1a. Each sample had a number of longitudinal and transversal narrow grooves about $(3/4)h$ in depth which notched the entire sample on a set of smaller square segments. The sample was held in the reaction chamber by thin stainless steel wires to minimize the heat leakage from the sample. After treatment the sample was broken along grooves onto separate pieces for further study. Each piece was kept in separate clean glass container with ground glass stopper partially filled with silica gel to prevent the adsorption of organic impurities during storage.

Before treatment the titanium tube was placed inside a cylindrical holder which tightly covered the tube's external surface and also acted as a heat protecting cover. This unit was located in the reaction chamber so that the tube axis coincided with the axis z of the injected EB (figure 1b). The tube was cut on segments immediately after treatment; each segment was kept as described above.

To separate the effect of the plasma-assisted oxidation from the conventional oxide synthesis due to the titanium heating in rarefied oxygen atmosphere special experiment were carried out: the heat-protecting envelope was removed and the water-cooled diaphragm 11 was located in front of the tube localizing the plasma cloud inside the tube only. Properties of inner and outer surfaces of the samples were then studied and compared.

The miniature thermo-sensor 7 was fixed at a certain point (usually at the centre) of the plane substrate to measure the material temperature. A set of thermo-sensors were placed into the tube's holder to monitor the temperature distribution along the tube during the treatment. The local temperatures at various points of both plane and tube samples could be also optically measured by pyrometer Optris LS (Optris GmbH, Germany) to ensure that the temperature distributions were uniform over the samples.

All samples were processed within the temperature range T_s = 250-750 °C. The temperature control was carried out by adjusting the EB current I_b within the range $1 < I_b < 100$ mA at a constant oxygen pressure that was 670 Pa in the described series of the experiments. The optimal treatment time τ was found experimentally and varied from 5 to 15 min depending of the T_s-value that was automatically kept constant during the treatment.

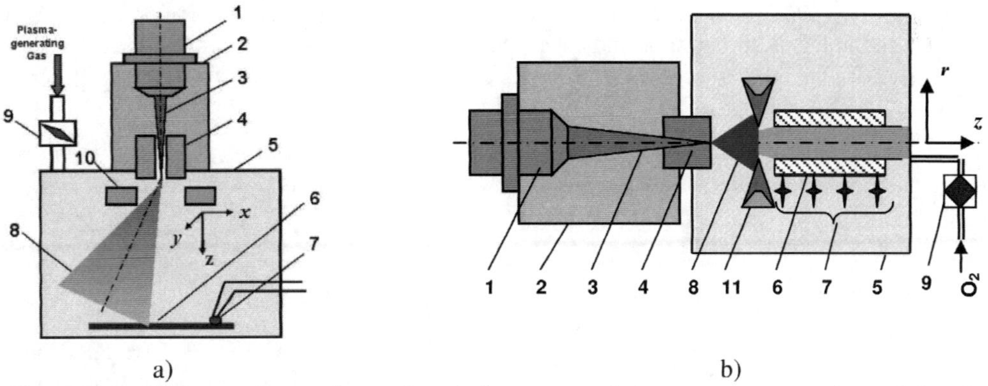

a) b)

Figure 1. The design of electron-beam plasmachemical reactor and the treatment procedure of the plane titanium substrates (a) and titanium tube (b).
1 – electron beam gun; 2 – high vacuum chamber; 3 – EB; 4 – injection window; 5 – working chamber; 6 – plane titanium substrate or titanium tube with heat protecting holder; 7 – temperature sensor(s); 8 – EBP cloud; 9 – gas feeder; 10 – scanning system; 11 – water-cooled diaphragm.

3. The structure and the chemical composition of oxide layer synthesized in EBP

Figure 2 shows the SEM images of plane titanium substrates before and after the EBP-treatment. The surface of original untreated titanium sample had numerous scratches even after mechanical polishing and was coated by the oxide due to the contact with air. This oxide layer was porous with the characteristic pore size ~1 μm and the surface had non-ordered structure (figure 2a). Acid etching improved the surface and formed large-size grains (~ 15 μm in characteristic size) having rigid facets (figure 2b).

The treatment in the oxygen EBP was expected to result in:

- the increase in surface uniformity and improvement of the surface roughness due to the action of fast electrons and heavy plasma particles [12, 13];
- simultaneous synthesis of the titanium oxide layer.

Studies showed that titanium oxide films that were plasmachemically synthesized without any pretreatment in acid consisted of grains with typical size ~10 μm separated by ~1 μm cracks; no pores were observed. The oxide coatings formed in the oxygen EBP on the surfaces pretreated in acidic solution were more uniform in comparison with the above case but they were composed of tightly packed grains of the same typical size ~10 μm and the grain facets became smoother (figures 2c and 2d). The treatment temperature rises from 330 °C to 550 °C amplified this effect: the surface became more uniform with densely packed grains of smaller size (compare figures 2c and 2d).

a) b) c) d)

Figure 2. The SEM images of the titanium plane substrates before and after the EBP-treatment (magnification ×2000): a) orginal titanium sample; b) titanium sample after acidic etching; c) tutanuim sample treated in the EBP of oxygen at 330 °C after preliminary acidic etching; d) tutanium sample treated in the EBP of oxygen 550 °C after preliminary acidic etching.
The samples were treated in EBP at gas pressure 670 Pa, treatment time 15 min

The results of XRD analysis of the inner surface of the titanium tube treated in the oxygen EBP are presented in figure 3. The analysis showed the titanium oxide (IV) in the rutile isoform predominates in the coatings composition. The XRD analysis revealed that the diffraction patterns of two samples treated in the EBP for 5 min at the same oxygen pressure and accelerating voltage at I_b = 18 mA (T_s = 415 °C) and 9 mA (T_s = 275 °C) were not qualitatively different but titanium dioxide content was lower in the second sample. This distinction indicates that the thinner oxide layer was synthesized at lower temperature T_s and, therefore, the oxide layer thickness could be controlled by the temperature of the EBP-stimulated synthesis.

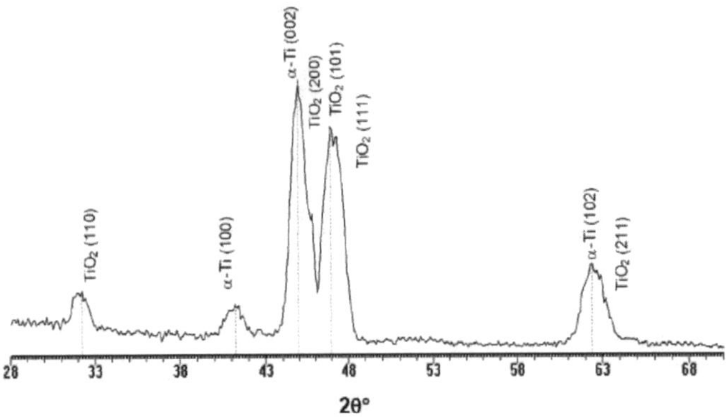

Figure 3. The XRD-analysis of the titanuim tube treated in the EBP of oxygen.
Treatment conditions: gas pressure 670 Pa, EB current 9 mA, temperature 275±5 °C, treatment time 5 min

4. The contact angles measurements and bioactivity characterization of the EBP-treated samples

The measurements showed that the water contact angle of the original titanium samples was only 66.0±2.5 degrees, which is not sufficient to expect high biocompatibility of oxidized surfaces and only samples with specially improved wettability are likely to be more biocompatible. The samples after conventional acidic etching and, especially, the samples with plasmachemically synthesized TiO_2-coatings acquired higher hydrophilicity than untreated titanium (table 1). For all treated samples no time dependence of the contact angle was observed for 48 hours storage but after this period only slight wettability degradation of the plasmachemically treated samples was observed, whereas the wettability of acidic etched samples (without plasmachemical treatment) became equal to that of the original titanium substrate.

Table 1. Water contact angle for untreated titanium substrate and titanium substrates after acidic etching and the EBP-treatment.

Time after the treatment (days)	Sample			
	Original titanium	Acidic etching	Acidic etching and EBP-treatment at 550 °C	EBP-treatment at 550 °C without acidic etching
0	66.0±2.5	21.1±7.6 *	6.9±1.2 *	8.8±1.6 *
1	69.8±2.1	19.3±4.2 *	9.0±1.5 *	15.7±1.9 *
2	67.8±1.5	20.3±5.3 *	17.5±2.3 *	16.4±2.1 *
7	70.0±2.7	29.2±4.9 *	22.8±0.5 *	18.0±1.5 *
21	66.7±1.8	59.3±2.4	32.5±4.2 *, **	20.9±1.5 *, **

All data are expressed as mean ± S.D.
* $p < 0.05$ *vs.* original titanium sample
** $p < 0.05$ *vs.* acidic etched titanium sample

The morphology of hydroxyapatite deposited on various titanium samples after their incubation in the SBF is shown in figure 4. All studied samples precipitated hydroxyapatite from SBF at their surfaces to a greater or lesser extent. The hydroxyapatite deposited at the original titanium sample in a form of discrete clusters (figure 4a). The plasmachemically synthesized TiO_2 improved the hydroxyapatite layer formation at the surface: the layer was more uniform and dense (figure 4b and 4c). The precipitated hydroxyapatite produced the macroporous scaffolds of elongated shape and diameter ~50–100 nm. The majority of the hydroxyapatite particles tended to agglomeration and formed aggregates with diameter of ~3 μm and the hydroxyapatite layer uniformity turned out to improve when the temperature of the preliminary plasmachemical treatment of titanium was higher (compare figure 4b and 4c).

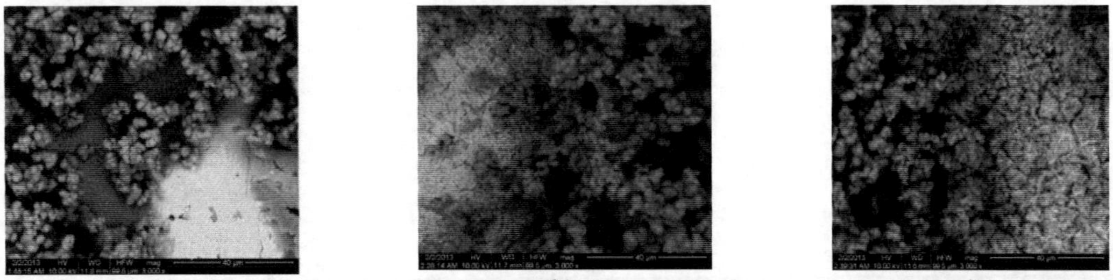

a) b) c)

Figure 4. The SEM images of the hydroxyapatite formed on titanium plane substrates (magnification ×3000): a) control sample; b) sample treated in the EBP at 330 °C; c) sample treated in the EBP at 550 °C. The samples a) and b) were preliminary etched in acid mixture, treated in the EBP of oxygen at pressure 670 Pa for 15 min. All samples were soaked in SBF for two weeks at 37 °C.

4. Conclusions

- The EBP-stimulalted oxides synthesis of titanium allys was experimentally proved. The technique was successfully used to obtain titanium oxide coatings on both plane surfaces and inner surfaces of tubes.

- EBB-treatment was able to improve the surface uniformity and roughness signifantly. The oxide coatings synthesized in the oxygen EBP on the surfaces pretreated in acidic solution were the most uniform and composed of tightly packed grains with typical size ~10 μm. Titanium oxide (IV) in the rutile form predominated in the coatings composition.

- The samples with plasmachemically synthesized TiO_2-coatings were more hydrophilic than untreated titanium and effectively precipitated hydroxyapatite on their surface. Thus, the EBP-stimulated TiO_2 synthesis is the promising technique to produce bioactive coatings on the surface of titanium medical dental and bone implants.

References

[1] Park J B and Lakes R S 1992 *Biomaterials an Introduction* (New York: Plenum Press)
[2] Lin F H, Hsu Y S, Lin S H and Chen T M 2004 *Mater. Chem. Phys.* **87** 24-30
[3] Dudek A 2011 *Arch. Metall. Mater.* **56** 135-140
[4] Boyd A R, Burke G A, Duffy H, Holmberg M, O'Kane C, Meenan B J and Kingshott P 2011 *Mater. Sci. Mater. Med.* **22** 71-84
[5] Katayama H, Katto M and Nakayama T 2009 *Surf. Coat.Technol.* **204** 135-140
[6] Roy M, Balla V K, Bandyopadhyay A and Bose S 2011 *Acta Biomater.* **7** 866-873
[7] Sato K, Kumagai Y and Tanaka J 2000 *J. Biomed. Mater. Res.* **50** 16-20
[8] Kokubo T, Ito S, Huang Z.T, Hatashi T, Sakka S, Kitsugi T and Yamamuro T 1990 *J. Biomed. Mater. Res.* **24** 331-343
[9] Vasilieva T 2010 *IEEE Transac. Plasma Sci.* **38** 1903-1907
[10] Vasilieva T M and Bayandina D V 2010 *Instr. Exp. Tech.* **53** 289-296
[11] Aleksandrov N L, Vasiliev M N, Lysenko N L and Negodaev S S 2011 Electron-beam plasma technologies for simulation of enviromental effects on satelite surface *4-th European Conference for Aerospace Sciences* 4-8 July 2011 S-Peterburg Russia (9 pages on CD)
[12] Aleksandrov N L, Konovalov V P and Son E E 2010 *Encyclopedia of low-temperature plasma* ed E E Son (Moscow: Yanus-K) volume III pp 265-273
[13] Aleksandrov N L, Vasiliev M N, Lysenko N L and A Kh Mahir 2005 *Plasma Physics Reports* **31** 425-435

Copper deposition on fabrics by rf plasma sputtering for medical applications

G Segura[1], P Guzmán[1], P Zuñiga[2], S Chaves [2], Y Barrantes[1], G Navarro[1], J Asenjo[1], S Guadamuz[1], VI Vargas[1] & J Chaves[2].

[1] Plasma Laboratory for Fusion Energy and Applications, Instituto Tecnológico de Costa Rica, Cartago, Costa Rica
[2] Laboratorio Institucional de Nanotecnología, Instituto Tecnológico de Costa Rica, Cartago, Costa Rica

E-mail: geosegu@gmail.com

Abstract. The present work is about preparation and characterization of RF sputtered Cu films on cotton by the usage of a Magnetron Sputter Source and 99.995% purity Cu target at room temperature. Cotton fabric samples of 1, 2 and 4 min of sputtering time at discharge pressure of $1x10^{-2}$ Torr and distance between target and sample of 8 cm were used. The main goal was to qualitatively test the antimicrobial action of copper on fabrics. For that purpose, a reference strain of Escherichia Coli ATCC 35218 that were grown in TSA plates was implemented. Results indicated a decrease in the growth of bacteria by contact with Cu; for fabric samples with longer sputtering presented lower development of *E. coli* colonies. The scope of this research focused on using these new textiles in health field, for example socks can be made with this textile for the treatment of athlete's foot and the use in pajamas, sheets, pillow covers and robes in hospital setting for reducing the spread of microorganisms.

1. Introduction

The plasma technology has been used to modify a huge range of material surfaces. This occurs because studies have been developed to apply plasma for the processing of textiles. Some important characteristics acquired by the implementation of plasma in fabrics are: the acquisition of hydrophilic or hydrophobic character of the material, the increase in dye absorption capacity [13][17], fiber surface cleaning without chemical reagents or wet treatments, as well as the increase in the absorption and adhesion of other materials like polymers and metals on the fabrics that provide specific finishes allowing coatings with fireproof character, with UV protection, with an increase on the surface hardness and antimicrobial property [1].

The recent development of antimicrobial textiles has become very important in the medical field because it's known that the fabrics provide an excellent medium for the adherence, transfer and propagation of microbial species. The effect of microbial growth on textiles produces generation of body odor, affecting human health and wears out textiles; therefore, different agents are used to provide antimicrobial finishes: dyes, polymers like chitosan or polyethylene glycols, also metal and metal salts where silver, copper and mercury compounds are the most effective biocides [4].

Since March, 2008, US Environmental Protection Agency declared copper as a natural and single metallic antimicrobial material. An important characteristic that copper has is its long lasting antimicrobial property, which after only a few hours of exposition it can significantly reduce pathogenic bacteria. Additionally copper is able to donate and accept electrons, thus developing oxidation and reducing reactions that alter the microbial cells. The concentration of copper would determine how many microorganisms are multiplied or eliminated [9][2].

A research [4] has demonstrated the effectiveness of biocidal properties in the copper-oxide impregnated fabrics. Cotton tests showed a 99 percent reduction of organisms for species like *Staphylococcus aureus*, *Escherichia coli*, *Candida albicans* and a 100 percent reduction in the case of a mite named *Dermatophagoides farinae*. Besides the implementation of plasma sputtering techniques on fabrics has resulted in the inactivation of *E. coli* bacteria. In these cases, where the optimal ratio of Cu-loading/cluster size for *E.coli* inactivation was due to the surface and inside Cu-clusters in the fabric and because of the ionic copper species interacting with the bacteria. Also it has been reported that the modification of the cotton by Cu-depositions changes the surface from hydrophilic to hydrophobic with the increase of the sputtering time [5][15].

Thus the main objective of the present study is to qualitatively test the antimicrobial action of copper on cotton fabrics based on the amount of copper deposited with different sputtering samples prepared by RF Magnetron sputtering.

The importance of this research lies in the application of plasma technology to confer antimicrobial properties in textiles. The main impact would be manufacturing antimicrobial fabrics in medical applications. For example, socks can be made with this textile for the treatment of athlete's foot, the use in pajamas, sheets, pillow covers and robes in hospital setting for reducing the spread of microorganisms or on a daily basis in the house to eliminate dust mites in mattresses, quilts, carpets and pillows, thus helping and improving the quality of life in people that suffer asthma and allergies [8].

2. Methodology

2.1. Materials

The fabric used in the experiment is 100% cotton. Cotton fibers are composed 99% of cellulose and non-cellulosic constituents such as: proteins, wax, organic acids, sugar, and others [10]. The cotton samples had an initial size of 10x10 cm and were cleaned by isopropyl alcohol immersion for about 1 hour and were dried in vacuum. For this study, 3 samples were prepared with different sputtering times and a control sample without deposition was used.

2.2. Copper Deposition on Cotton samples

Copper layers were deposited on cotton substrate by RF magnetron Sputtering Gun at room temperature, in a cylinder chamber with an automatic vacuum station. This technique consists on the bombardment of a solid target using energetic particles, mostly ions, applying a radio frequency power supply [6][7][11]. The important parameters established were: the distance between the Cu-target and the cotton substrate was ~8 cm, Argon as gas, a discharge power of 50 W, with a base pressure of $1x10^{-4}$Torr, a discharge pressure of $1x10^{-2}$ Torr, also a Cu-target with 99.995% purity and with 1 inch of diameter, and the deposition times were 1, 2 and 4 min. After the deposition, the edges of cotton samples were eliminated and their sizes shrank to 3x3 cm and finally these samples were stored in sterilized containers. The experimental configuration of the plasma discharge is shown in figure 1.

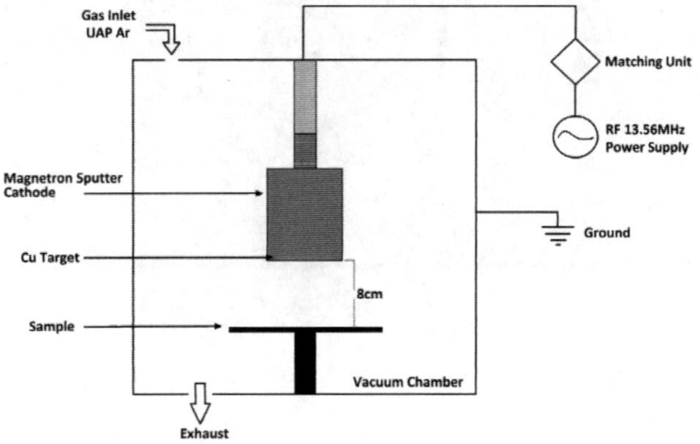

Figure 1. Experimental set up.

2.3. Fiber Characterization Technique
The Scanning Emission Microscopy (SEM) was used to investigate the surface condition and structure of cotton fabrics. This technique consists on the scanning of the sample surface with a beam of electrons and obtaining a three dimensional image [3][14][16].

2.4. Antimicrobial Susceptibility Test
The bacterium used for testing the cotton samples was *Escherichia coli;* it characteristics are: it belongs to the *Enterobacteriaceae* family, it is gram negative, it doesn't form spore, it is motile by means of peritrichous flagella or may be non-motile, facultatively anaerobic, chemo-organotrophic microorganism with an optimum growth temperature of 37°C [20]. Also it is one of the predominant enteric species in the human gut and part of the normal intestinal flora; some of these species provide benefits to the host; however, there is a small group known as pathogenic *E.coli* that can cause severe diarrheal disease in humans [21].

This study used a strain of *E. coli* ATCC 35218. The bacteria were inoculated into a TSA (Tryptic Soy Agar) plate. The cotton fabric samples were placed on a petri dish immediately after the inoculum was done. The provided petri dishes were covered with a lid to prevent contamination and evaporation, and were incubated at 35 °C for 24h. After 24 hours at the incubator the fabric was removed from each plate and the contact area of the fabric was observed. For a better visualization of the bacterial growth after incubation, a stereoscopic microscope (Leica EZ4) was used.

3. Results

3.1. Cotton Fibers
Figure 2 shows cotton fibers without Cu sputtering. The initial cotton structure is braided thus resulting in a nonuniform surface. Also, fiber thicknesses are different with an average of 24,15 um. Braiding can affect the uniformity of the deposited copper layer fabric. But some investigations reported that a rougher surface of the deposited copper fibers improves the adsorption and stabilizes copper ions thus contributing to the inactivation of the bacteria [5][15].

Figure 2. Cotton fibers without Cu sputtering.

3.2. Copper Deposition Cotton

Figure 3 shows sputtered cotton samples for 1, 2 and 4 minutes. Cotton shows light grey color because Cu-content increases with time. The layer color is uniform throughout the sample.

The possible reactions between Ar-plasma and cellulose have been studied. The suggested mechanism for argon plasma induces molecular fragmentation of cellulose and the generation of free radicals through the oxidation process [10]. Therefore this activation process of the cotton fabric may help the adherence of copper ions and the formation of the film [18][19].

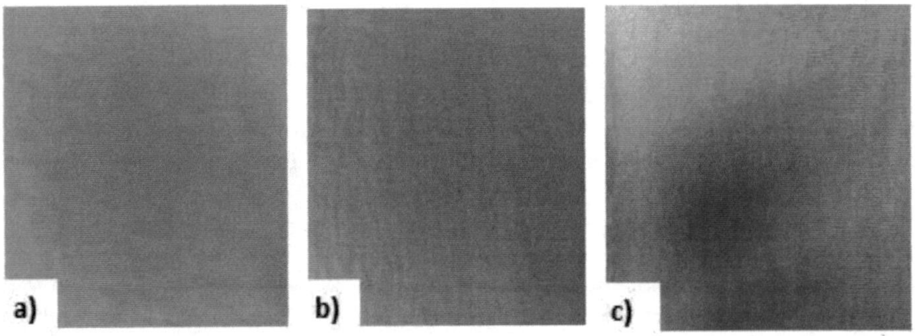

Figure 3. Samples of Cu-cotton sputtered for: a) 1 min, b) 2 min, c) 4 min.

3.3. E.coli Bacterial Inactivation

As shown in figure 4, the bacterial growth appeared as the formation of individual and agglomerated colonies. The picture below shows where the fabric was placed in each test. The fabric was placed on the center for the images on the left whereas the ones on the right show the trace of the edge of the fabric.

As for the pictures on the right, the bacteria did not cross the edge area as the samples show longer time of sputtering. In picture a), growth of bacteria in and out of the edges was observed; in pictures b) and c) there was a bit of bacterial growth beyond the edge, but the concentration of bacteria colonies is focused around the border, while in picture d) there is insignificant bacterial growth within the area.

On the left side images, the brown regions representing lower growth of bacteria decrease as the sputtering time of samples increases. In picture a) there are more brown regions and they are distributed uniformly in the area where there may be bacteria colonies. In picture b) lower brown regions are observed, but there is still a significant group of bacteria colonies. In picture c) and d) this area becomes clearer with lower brown regions that represent the decline in the colony growth.

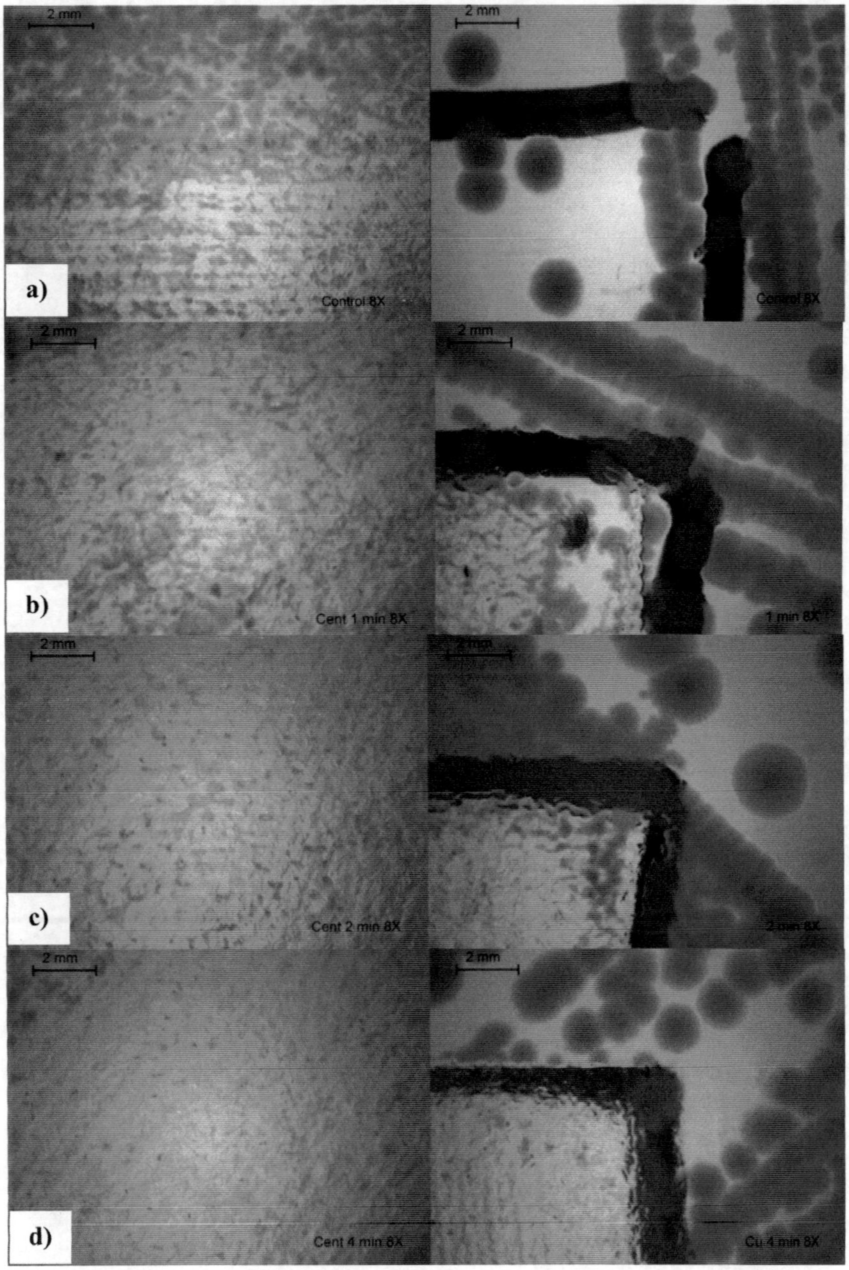

Figure 4. Bacterial growth results in the contact area for each sample:
a) control, b) 1 min, c) 2 min, d) 4 min.

The mechanism by which copper inhibits E.coli is not exactly known but theory explains some possible reactions that may help copper against this bacteria. Copper toxicity to the bacteria is due to the reactions that can be performed by its tendency to alternate their ions, mainly its cuprous Cu(I) and cupric Cu(II) oxidation states [22].

The conditions under which the antimicrobial test is developing affect the inhibition rate and destruction of the bacteria. If the medium is under anaerobic conditions, the contact between the cells and copper surface may cause the reduction of Cu(II) to Cu(I) and the oxidation of Cu(II) to Cu(I), where Cu(I) is more toxic than Cu(II) and stable to this conditions. While if the aerobic condition is presented, the effect of copper is amplified because a redox cycling is produced and leads to the generation of reactive hydroxyl radicals and oxygen species that can damage biomolecules, which is the case studied [22][23][15]. An important aspect to consider is that probably the copper inactivation mechanism is due to the joint action of both ions, not just one.

4. Conclusions

A decrease in the growth of bacteria on contact with Cu was observed; the fabric samples with longer sputtering presented lower development of *E. coli* colonies due to the higher concentration of copper.

The center of the fabric tends to be the best part to analyze bacteria, however, good results in the edge with respect to bacterial growth are presented as well.

The antimicrobial property of copper was evidenced on cotton fabrics, but it was not possible to identify a sample with a sputtering time and optimum concentration of copper to achieve total inactivation. Hence, a subsequent study regarding the quantitative characterization of the observed effect must be applied.

References

[1] Observatorio Industrial del Sector Textil y de la Confección 2010 *Retos del Nuevo Sector Textil-Confección* (España)

[2] Prado J V, Vidal A R and Durán T C 2012 Aplicación de la capacidad bactericida del cobre en la práctica médica *Rev Med Chile* **140** 1325-32

[3] Roaming A and Sandia J 1998 Analytical Transmission Electron Microscopy *ASM Handbook: Materials Charaterization* vol 10, ed American Society for Metals HC. (United States of America)

[4] Gupta D and Bhaumik S June 2007 Antimicrobial treatments for textiles *Indian Journal of Fibre & Textile Research* **32** 254-63

[5] Castro C, Sanjines R, Pulgarin C, Osorio P, Giraldo SA and Kiwi J July 2010 Structure reactivity relations for DC-magnetron sputtered Cu-layers during E. coli inactivation in the dark and under light *Journal of Photochemistry and Photobiology A: Chemistry* **216** 295-302

[6] Depla D, Mahieu S and Greene JE Sputter deposition processes 36

[7] Dietz C 2001 *Desarrollo de sistemas de introducción de muestra en el plasma de microondas* (Departamento de Química Analítica, Universidad Complutense de Madrid)

[8] Gabbay J, Mishal J, Magen E, Zatcoff R, Shemer-Avni Y and Borkow G Copper Oxide Impregnated Textiles with Potent Biocidal Activities 13

[9] Gronemeyer GF Propiedades Antimicrobianas del Cobre *Cobre Salud, Medio Ambiente y Nuevas Tecnologías* ed Worstman HS pp 31-55

[10] Johansson K 2007 Plasma modification of natural cellulosic fibres *Plasma Technologies for textiles* ed R.Shishoo (Cambridge: Woodhead Publishing Limited) pp 247-68

[11] Kiyotaka W, Makoto K and Hideaki A 2004 Sputtering Systems *Thin Film Materials Technology: Sputtering of Compound Materials* (Unites States: William Andrew Inc) pp 135-150

[12] Martín JRS April 2007 Los tejidos inteligentes y el desarrollo tecnólogico de la industria textil *Técnica Industrial* **268** 39-45

[13] Motaghi Z and Shahidi S 2012 Development of polyester-wool fabrics dye ability using plasma sputtering *RMUTP Int. Conf. Textiles & Fashion 2012* (Bangkok Thailand) pp 1-8

[14] Neville A, Mather RR and Wilson JIB 2007 Characterization of plasma-treated textiles *Plasma technologies for textiles* ed R.Shishoo (Cambridge: Woodhead Publishing Limited) pp 301-15

[15] Osorio Vargas P, Sanjines R, Ruales C, Castro C, Pulgarin C, Rengifo Herrera AJ, et al 2011 Antimicrobial Cu-functionalized surfaces prepared by bipolar asymmetric DC-pulsed magnetron sputtering (DCP) *Journal of Photochemistry and Photobiology A: Chemistry* **220** 70-76

[16] Parvinzadeh Gashti M, Alimohammadi F, Song G and Kiumarsi A 2012 Characterization of nanocomposite coatings on textiles: a brief review on Microscopic technology *Current Microscopy Contributions to Advances in Science and Technology* ed Méndez Vilas A (Formatex) pp 1424-37

[17] Shahidi S and Ghoranneviss 2011 M Effect of Plasma on Dyeability of Fabrics *Textile Dyeing* ed Hauser P (Croatia: In Tech)

[18] Vihodceva S, Kukle S and Blums J July 2012 Light Reflection of Cu-sputtering-coated Cotton Fabrics and Effect of Sputtering Time, Acetone Pre-treatment and Fabric Structure *Textiles and Light Industrial Science and Technology* **1** 13-19

[19] Vihodceva S, Kukle S, Blums J and Zommere G 2011 The Effect of the Amount of Deposited Copper on Textile Surface Light Reflection Intensity *Scientific Journal of Riga Technical University, Material Science* **6** 24-29

[20] Fratamico PM and Smith JL 2006 Escherichia coli infections *Foodborne Infections and Intoxications* ed Riemann and Bryan (Elsevier Inc) pp 205-39

[21] FDA 2012 Pathogenic Bacteria: Pathogenic Escherichia coli Group *Bad Bug Book, Foodborne Pathogenic Microorganisms and Natural Toxins* ed Lampel KA

[22] Espírito Santo C, Taudte N, Nies DH and Grass G February 2008 Contribution of Copper Ion Resistance to Survival of Escherichia coli on Metallic Copper Surfaces *Applied and Enviromental Microbiology* 7**4** 977-86

[23] Park HJ, Nguyen TTM, Yoon J and Lee C 2012 Role of Reactive Oxygen Species in Escherichia coli Inactivation by Cupric Ion *Enviromental, Science & Technology* **46** 11299–304

Pulsed laser deposition of carbon nanodots

G Muñoz[1], P Homm[1], F Guzmán[3], H M Ruiz[1], L S Caballero[1], M Favre[1,2], M Flores[3] and S Hevia[1,2]

[1] Instituto de Física, Pontificia Universidad Católica de Chile, Av. Vicuña Mackenna 4860, Santiago, Chile

[2] Centro de Investigación en Nanotecnología y Materiales Avanzados (CIEN-UC), Av. Vicuña Mackenna 4860, Santiago, Chile

[3] Departamento de Física, Facultad de Ciencias Físicas y Matemáticas, Universidad de Chile, Av. Blanco Encalada 2008, Santiago, Chile

E-mail: gamunoz2@uc.cl

Abstract. We present preliminary results on carbon nanodots growth using pulsed laser deposition (PLD), with a graphite target, in a low pressure argon background. In the process thin carbon films are deposited on Porous Alumina Membranes (PAM) positioned over a Silicon substrate. Reproducible carbon nanodots arrays standing on Silicon are obtained by later removal of the PAM. Atomic force microscopy (AFM) was used to characterize the nanodots morphology. The laser carbon plasma was characterized by optical emission spectroscopy (OES), and a fast response, negative biased, Faraday cup was used to measure the characteristic energy and flux of the plasma ions depositing onto the PAM. Reliable Carbon nanodots growth was only observed with Argon background pressures below 18 mTorr

1. Introduction

There is an increasing interest on the possibility of growing nanodots arrays due to potential applications in different fields. In particular, carbon nanodots arrays (CNA) have been used in applications such as a substrate for direct immobilization of DNA[1], electron field emitters[2], or temperature probes[3], among others. Several plasma based techniques have been used to produce nanodots arrays of different characteristics, which include a combination of cathodic arc plasma and anodic alumina pore arrays[4] and template assisted pulsed laser deposition[5]. In this report we present preliminary results on the fabrication of CNA grown on silicon substrates, by using Pulsed Laser Deposition (PLD) and a Porous Alumina Membranes (PAM) as a mask.

2. Experimental set-up

To produce the PAMs, a thin Aluminium film is deposited over Silicon, which under a double anodization process[6] results in homogeneous and highly ordered PAMs with a final thickness between 250 and 500 nm and characteristic diameter around 70 nm. The PLD was carried out in a vacuum chamber, at a base pressure $3 \cdot 10^{-5}$ Torr, using a Nd:YAG laser, 345 mJ, 4.4 ns, at 1.06 μm, focused onto a graphite target, at a fluence around 7 J/cm^2. The PAM was located at 50 mm from the laser spot in order to obtain a homogeneous deposit, covering about one centimeter square. The exposure time ranged from 10 s to 120 s, with the laser operating at 10 Hz. Argon gas background was used, at pressures from 8 to 43 mTorr, in order to investigate the effect of plasma conditions on the resulting nanodots. Atomic force microscopy (AFM) was used to characterize the surface morphology. The laser plasma was characterized by time integrated optical emission spectroscopy (OES) and 20 ns time resolved plasma imaging of visible plasma emission. A fast response, negative biased, Faraday cup was used to measure the characteristic energy and flux of the plasma ions depositing onto the PAM.

3. Results and discussion

Figure 1 shows AFM topographic images of a PAM and the resulting CNA, obtained at base pressure, with 10 s exposure time. The images show a regular pattern of the initial pores and the resulting CNDs.

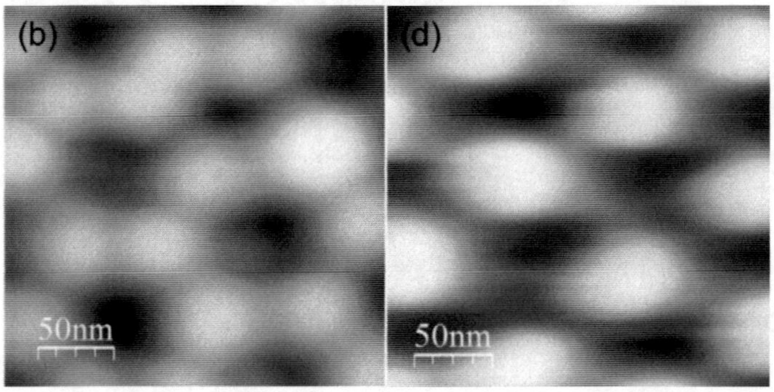

Figure 1. AFM images: (a) corresponds to the PAM used as a mask, and (b) shows the resulting CNA.

It has been found in previous investigations that a low pressure neutral background gas has a strong effect on plasma species composition of laser produced carbon plasma plumes[7]. To explore this effect on CNA growth, we introduced argon in the deposition chamber, at pressures ranging from 8 to 170 mTorr. We found that reproducible CNAs were obtained at base pressure, 8 mTorr and 17 mTorr. In all cases the CNAs showed similar morphological properties, as inferred from SEM and AFM images. As the argon background pressure was increased up to 43 mTorr and beyond, we found no evidence of reproducible and significant nanodots formation.

In order to correlate reproducible carbon nanodots growth with argon background pressure, we investigated the expanding laser carbon plasma plume by using OES and Faraday cup measurements. Figure 2 shows the spectrum of the carbon plasma emission collected from a point on axis, 50 mm from the target surface. The figure shows plasma emission at base pressure, and with argon background in the range from 8 to 170 mTorr. With argon background the spectra are dominated by emission corresponding to vibrational states of the $d_3\Pi_g \rightarrow a_3\Pi_u$ Swan band, of the neutral C_2 molecule[7]. At base pressure the overall plasma emission is orders of magnitude below that with argon background, being the most prominent feature emission from C^+ ions. At higher pressures, where nanodots growth is not observed, the emission intensity increases significantly. This is shown in figure 3, where the integrated emission above continuum of the most prominent C_2 emission bands, $\Delta\nu = 1$ and $\Delta\nu = 0$, collected during the deposition process, from a point close to the PAM surface, is plotted against argon pressure. C_2 emission is seen to reach a maximum around 80 mTorr, and then to decrease. This behavior can be explained in terms of C_2 molecules increasing scattering effects as background pressure increases, and also a quenching effect on molecule formation due to carbon ions collisions with argon atoms. Plasma imaging observations, with 20 ns time resolution, indicate that after reaching the substrate surface, the laser plasma stagnates and remains attached to the surface for characteristic times up to a few microseconds [8].

Figure 4 shows Faraday cup measurements, at different background pressures, obtained on axis, at 50 mm from the graphite target, using a 1.7 cm^2 collecting area. This distance coincides with that where the PAM is located for CNA deposition. As the Faraday cup is negative biased,

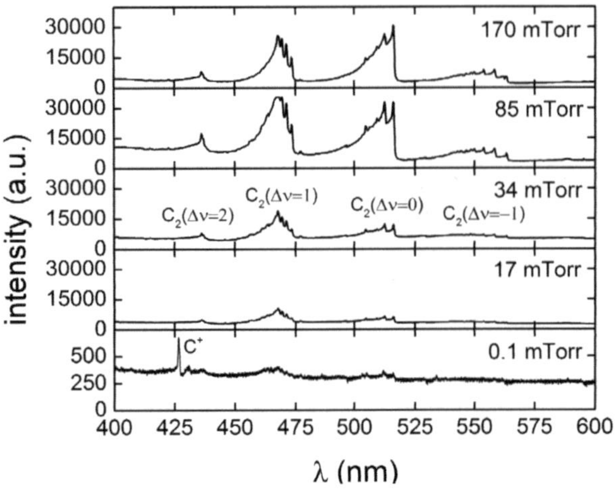

Figure 2. Carbon plasma emission, at different argon background pressures, collected from a 3 mm region located on the plasma plume axis, 20 mm from the graphite target surface. The labels identify emission from argon C_2 molecular bands.

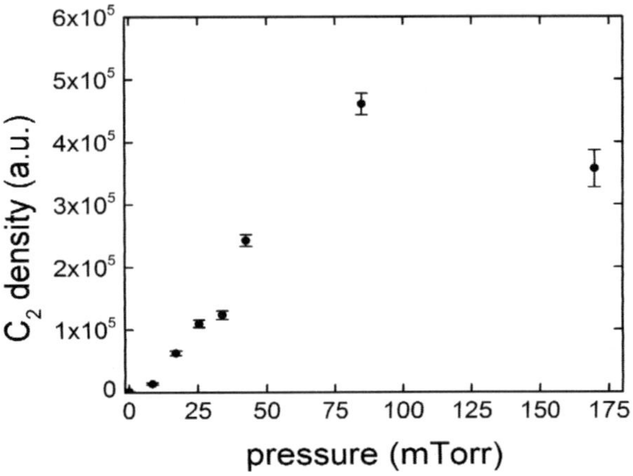

Figure 3. Integrated emission of most prominent C_2 emission bands as a function of argon background pressure.

it also acts as an x-ray diode (XRD), being sensitive to extreme ultraviolet and soft x-ray (XUV) radiation. As a consequence, the sharp spike at around t = 0 corresponds to XUV emission from the initial laser produced plasma. This can be used as a time mark to perform time of flight (TOF) measurements to determine the carbon ions energy spectra. The labels on the figure indicate characteristic energies for single ionized carbon atoms. The total ion dose collected by the Faraday cup varies from $3.4{\cdot}10^{12}$ ions/cm^2 at base pressure, to $2.5{\cdot}10^{12}$ ions/cm^2 and $2.8{\cdot}10^{11}$ ions/cm^2 at 34 and 170 mTorr, respectively.

The main issues that can be inferred from spectroscopic and Faraday cup measurements are a marked increase in C_2 molecules content in the laser plasma, as optical emission is proportional to molecular density, and a decrease in the characteristic energy of the carbon ions in the expanding plasma, both as argon background pressure increases. The trend in increasing C_2 molecules content in the expanding laser carbon plasma and decreasing characteristic carbon ion energy becomes more evident as the argon background pressure is further increased. It is also observed that the expanding laser plasma is composed of several plasma fronts, being C^+ ions the main species in first one, with C_2 molecules becoming dominant in the subsequent fronts[7]. The fact that no reliable carbon nanodots growth is observed at pressures above 17

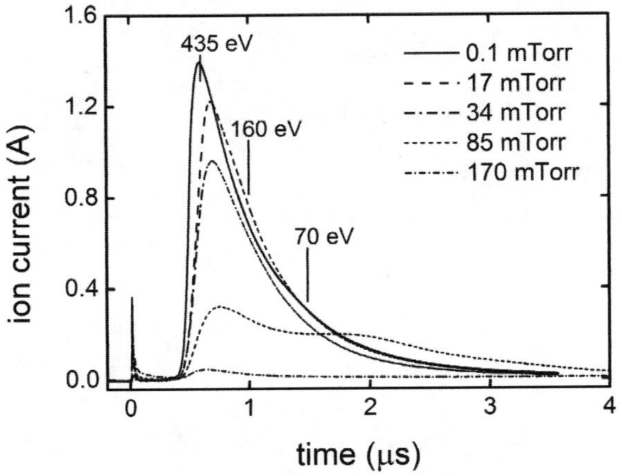

Figure 4. Faraday cup signals collected on axis, at 5 cm from the graphite target, at different argon background pressures. The labels indicate characteristic values of single ionized carbon atoms.

mTorr might be linked to the laser plasma properties. As for the plasma composition, it can be argued that as the plasma content becomes dominated by C_2 molecules the deposition rate increases, up to a point that alumina pores become quickly blocked, thus precluding carbon ions entering and depositing inside the pores. It has also been reported that when a laser produced carbon plasma expands in a low pressure argon background, carbon cluster formation is favored, due to background gas ionization induced by charge transfer collisions with fast carbon ions[9]. It is then expected that as the characteristic cluster size increases, it becomes less likely for them to enter and deposit inside the alumina pores. Additionally, a decrease in the carbon ions characteristic energy results in a reduction of the mean free path for collisions, which in turns increases scattering and radial expansion of the expanding plasma. This is clearly linked to an increase in the argon background pressure. At lower pressures the carbon ions are more ballistic, due to larger mean free path, so their chances to enter inside the pores are expected to be higher.

4. Conclusions

We have shown the feasibility of growing CNAs on a Silicon substrate by depositing a laser produced carbon plasma over a porous alumina membrane. The presence of a low pressure argon background favors reliable CNA growth. However, as the argon background pressure increases further nanodots growth is not observed. This is attributed to enhanced C_2 molecules and carbon clusters formation, resulting from the interaction of the expanding laser carbon plasma and the neutral gas background, which results in partial filling and blocking of the PAM porous. Additional characterizations of the CNAs growth with this technique will be carried out and reported elsewhere.

Acknowledgments

This work has been funded by projects FONDECYT 1110380 and No. 11110352, and CONICYT PIA No. ACT1108. G. Muñoz, P. Homm, F. Guzmán and H. M. Ruiz acknowledge postgraduate studies scholarships from CONICYT.

References

[1] Djenizian T, Balaur E and Schmuki P 2006 *Nanotechnology* **17** 2004
[2] Li S C, Yan P X, Li X C and Chong E M 2010 *Physica E* **42** 1343
[3] Chen P C, Chen Y N, Hsu P C, Shih C C and Chang H T 2013 *Chem. Commun* **49** 1639

[4] Zhang P J, Chen J T, Zhuo R F, Xua L, Lu Q H, Ji X, Yan P X and Wu Z G 2009 *Appl. Surf. Sci.* **255** 4456
[5] Gupta M, Sauer V and Tsui Y Y 2013 *Appl. Phys. A* **110** 817
[6] Hevia S, Homm P, Cortes A, Núñez A, Contreras C, Vera J and Segura R 2012 *Nanoscale Res. Lett.* **7** 342
[7] Ruiz H M, Guzmán F, Favre M, Bhuyan H, Chuaqui and Wyndham E S 2012 *Plasma Sources Sci. Technol.* **21** 034014
[8] Favre M, Caballero L S, Ruiz H M, Guzmán F, Bhuyan H, Veloso F and Wyndham E S 2013 *Proc. 31st ICPIG (July 14-19, 2013, Granada, Spain)* PS3–093
[9] Bailini A, Ossi P M A and Rivolta A 2007 *Appl. Surf. Science* **253** 7682

Emission spectroscopy of an atmospheric pressure plasma jet operated with air at low frequency

L Giuliani[2], J L Gallego[1], F Minotti[2], H Kelly[2] and D Grondona[2]
Instituto de Física del Plasma (INFIP), Consejo Nacional de Investigaciones Científicas y Técnicas (CONICET), Departamento de Física, Facultad de Ciencias Exactas y Naturales, Universidad de Buenos Aires (UBA).
Ciudad Universitaria, Pabellón I, C1428EHA, Buenos Aires, Argentina.
E-mail: leandro@tinfip.lfp.uba.ar

Abstract. Low-temperature, high-pressure plasma jets have an extensive use in plasma biology and plasma medicine, such as pathogen deactivation, wound disinfection, stopping of bleeding without damage of healthy tissue, acceleration of wound healing, control of bio-film proliferation, etc. In this work, a spectroscopic characterization of a typical plasma jet, operated in air at atmospheric pressure, is reported. Within the spectrum of wavelengths from 200 to 450 nm all remarkable emissions of N_2 were monitored. Spectra of the N_2 2nd positive system ($C^3\Pi_u$–$B^3\Pi_g$) emitted in air are the most convenient for plasma diagnostics, since they enable to determine electronic T_e, rotational T_r and vibrational T_v temperatures by fitting the experimental spectra with the simulated ones. We used SPECAIR software for spectral simulation and obtained the best fit with all these temperatures about 3500K. The conclusion that all temperatures are equal, and its relatively high value, is consistent with the results of a previous work, where it was found that the experimentally determined electrical characteristic was consistent with the model of a thermal arc discharge, together with a highly collisional cathode sheet.

1. Introduction

A very active branch in plasma physics is the development of plasma sources able to provide energetic electrons and ions, which produce in turn highly active chemical reactive species, at relatively low gas temperature. This is due to a large number of applications of these sources in plasma biology and plasma medicine, such as pathogen deactivation, wound disinfection, stopping of bleeding without damage of healthy tissue, acceleration of wound healing, control of bio-film proliferation, etc. (see [1] and references therein).

In spite of the existing large amount of experimental work with plasma jets [2-5], the discharge structure and the plasma state are not yet well known. This knowledge is important for the scaling and optimization of the discharge in several potential technological applications.

In a previous work [6], a plasma jet operated with air or argon at atmospheric pressure using the same discharge geometry as used in this work was presented. An electrical study of the discharge, together with a theoretical interpretation, indicated that in some configurations the discharge plasma can be thought of as a low-current arc, in which a relatively large voltage drop occurs mainly in a

[1] Fellow of CONICET.
[2] Researcher of CONICET.

cathodic, collisional sheath. Also, the possibility of turbulence in the gas flow was considered, and linked to the behavior of the external plasma plume for different gas flow rates.

In this work, a spectroscopic characterization of a typical plasma jet, operated in air at atmospheric pressure, is reported. Within the spectrum of wavelengths from 200 to 450 nm all remarkable emissions of N_2 were monitored. Spectra of the N_2 2nd positive system band head ($C^3\Pi_u$–$B^3\Pi_g$) emitted in air were used to determine electronic T_e, rotational T_r and vibrational T_v temperatures by fitting the experimental spectra with the simulated ones.

2. Experimental set-up

A schematic of the experimental device used to generate the atmospheric plasma jet is shown in Figure 1. The device consists in two electrodes with a hole of 1 mm diameter, through which the gas flows. The two electrodes are made of stainless steel disks of 20 mm diameter and 3 mm thickness attached to the surface of centrally perforated dielectric disks (Teflon) of 1 mm in thicknesses. The hole in the center of the dielectric disk is 2 mm in diameter.

The ac power supply is a commercially available transformer for neon light (25 kV, 50 mA, 50 Hz) which is connected to a variable autotransformer (Variac) in order to allow voltage control. Neon light transformers are quite interesting sources for the generation of low current plasma jets, because they provide an intrinsic current limitation of reactive origin (thus avoiding the use of external ballasts in the discharge circuit), combined with a low commercial price.

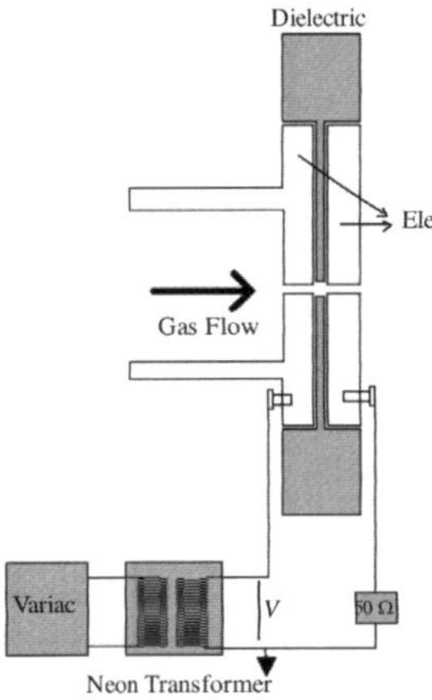

Figure 1. Schematic of the experimental device.

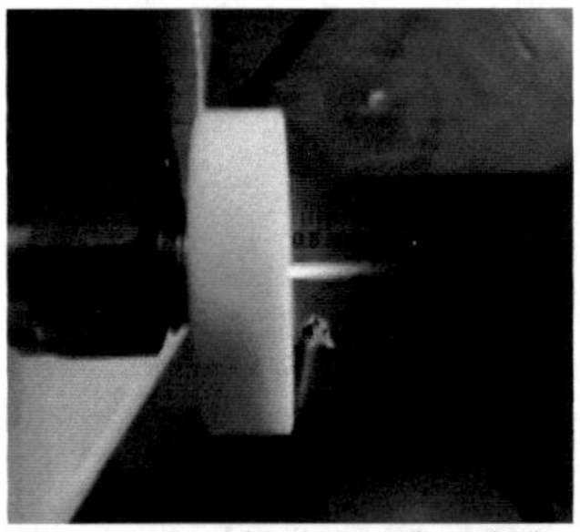

Figure 2. Photograph of the plasma jet.

The voltage V between the electrodes was measured using a digital Tektronix oscilloscope (70 MHz, 1GS/s), the current I was measured from the voltage drop through a 50 Ω resistor, and the air flow was measured using a stainless steel float flow meter.

The spectroscopic measurements were performed with a monochromator (200-500 nm) attached to a photomultiplier (Hamamatsu 1P28) facing the plasma jet. All along this work, the discharge was operated with the power source at 11.2 kV and an air flow of 11.3 l/min.

A photograph of the plasma jet operated with air under the same conditions used for the spectrum measurements is shown in figure 2.

3. Results and analysis

The current and voltage waveforms of the discharge are shown in figure 3.These measurements show that the discharge current was sinusoidal like the transformer voltage and controlled by the transformer impedance (estimated in (57 ± 1) kΩ by a linear regression), being almost independent of the discharge voltage behavior. The voltage between the electrodes shows a quasi-square signal with an amplitude of around 500 V, with a slight concavity opposed to that of the current.

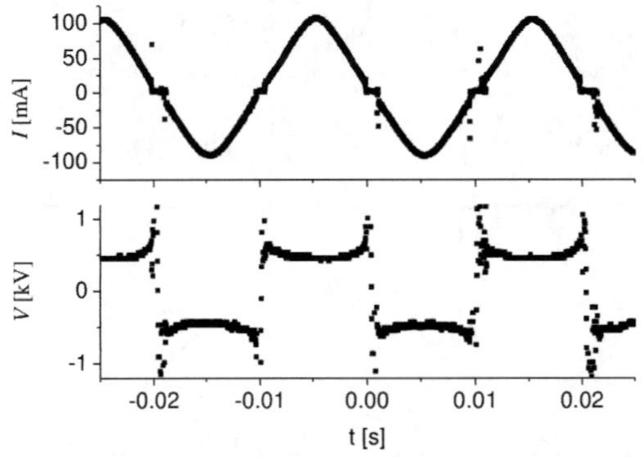

Figure 3. Typical voltage and current waveforms of the discharge.

The emission spectrum of the plasma jet can be seen in figure 4. It was found that the spectrum was dominated by emissions of molecular nitrogen containing N_2 second positive systems. These excited molecules and energetic UV photons play important roles in the sterilizing properties of plasmas. It is worth noting that this spectrum was obtained with the monochromator facing the jet so that most of the light came from the bright spot within the interelectrode region.

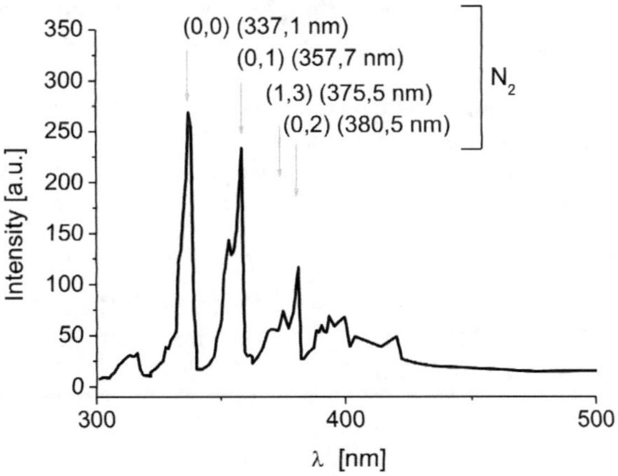

Figure 4. Emission spectrum of the plasma jet with all remarkable emissions of N_2 monitored within the 300-450 nm range.

Experimental spectra of the N_2 2nd positive system band head at 337.1 nm (dots) and SPECAIR software spectral simulations (lines) are shown in figure 5. It was found that the best adjustment is obtained with T_e, T_r and T_v about 3500K. In order to test an alternative discharge mode, an additional simulation is included, corresponding to a streamer discharge with electron temperature of about 1 eV in a cold gas.

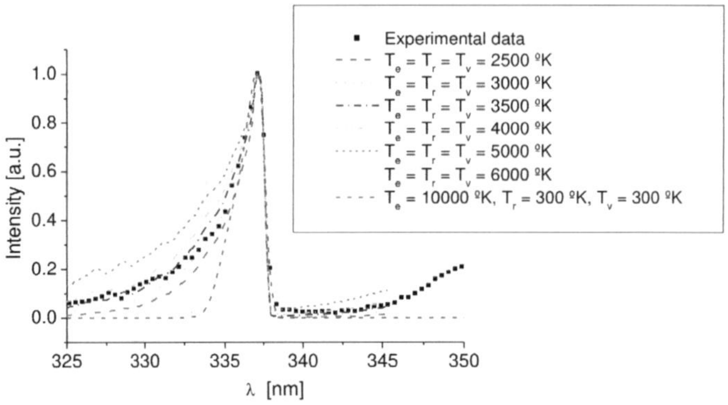

Figure 5. Experimental spectra of the N_2 2nd positive system band head at 337.1 nm (dots) and SPECAIR software spectral simulations (lines).

The result that all temperatures are equal, and their relatively high value, is consistent with the results of a previous work [6], where it was found that the experimentally determined electrical characteristic was consistent with the model of a thermal arc discharge, together with a highly collisional cathode sheet with a relatively large voltage drop.

4. Final remarks

The emission spectrum was mainly dominated by the presence of excited nitrogen species, containing N_2 second positive systems. These excited molecules and energetic UV photons play important roles in the sterilizing properties of plasmas.

The spectroscopic data were consistent with a thermalized plasma with heavy species at high temperature (about 0.3 eV), which is also in agreement with a previous study of the electrical characteristic of the discharge. In contrast, a streamer discharge with higher temperature electrons was not revealed by both studies.

Acknowledgments

This work was supported by grants from the Consejo Nacional de Investigaciones Científicas y Técnicas (CONICET) PIP 11220090100219, Universidad de Buenos Aires (UBA) PID 01/1141 and Agencia Nacional de Promoción Científica y Tecnológica (ANPCyT) PICT 2010No. 0771.

References

[1] Laroussi M 1992 *Plasma Medicine* (Cambridge University Press 2012) Ch. 1.
[2] Hong Y Chand UhmH S 2006 *Appl. Phys. Lett.* **89** 221504.
[3] Park G Y, Park S J, Choi M Y, Koo I G, ByunJ H, Hong J W, SimJ Y, Collins G J and. Lee J K 2012 *Plasma Sources Sci. Technol.* **21** 043001.
[4] Li X, Tao Xand YinY2009 *IEEE Trans. Plasma Sci.* **37** 759.
[5] LiX, TangC, DaiX and YinY 2008 *Plasma Sci. Technol.* **10** 185.
[6] Giuliani L, Xaubet M, Grondona D, Minotti F and Kelly H 2013 *Physics of Plasmas* **20** 063505.

Atmospheric-Pressure Non-Thermal Plasma Jet for biomedical and industrial applications

J Asenjo[1], J. Mora, A.Vargas, L.Brenes, R Montiel, J. Arrieta & VI Vargas[1]

[1] Plasma Laboratory for Fusion Energy and Applications, Instituto Tecnológico de Costa Rica, Cartago, Costa Rica

E-mail: jasca.tec@gmail.com

Abstract. In this work we present the development and evaluation of a low-cost DBD Plasma-JET reactor using Argon as carrier gas, this device is capable of generating a cold plasma plume several centimeters in length making it suitable for use directly in contact with objects and delicate materials, including living tissue.

1. Introduction

Recently Atmospheric-Pressure Non-Thermal Plasmas (APNTP) have become a topic of great interest for a wide range of applications in different branches of industry. In these plasmas the electron temperature is far higher than the temperature of the heavy particles [1]; elastic collisions of the electrons are not effective in contrast to collisions of heavy particles however it can transfer energy to other processes such as ionization, activation or dissociation of molecules This explains the great interest in this type of plasmas.

Over the past 20 years, there has been growing interest in developing this type of devices, particularly for biomedical applications such as bacterial inactivation [2, 3], wound healing [4], dental bleaching [5]; atmospheric plasmas are also of great interest in the industry, improve adhesion for inks, paints, and coatings are examples of plasma applications in order to improve surface properties, changes on the surface energy of any material is one of the purposes of applying plasma [6].

Various types of Atmospheric-Pressure Non-Equilibrium Plasma Jets (APNP-Js) with different configurations have been reported, most of the jets are working with noble gas mixed with a small percentage of reactive gases [1]. According to X. Lu et al. in their work "On atmospheric-pressure non-equilibrium plasma jets and plasma bullets", plasma jets operating with noble gases can be classified into four categories, i.e. Dielectric-Free Electrode (DFE) jets, Dielectric Barrier Discharge (DBD) jets, DBD-like jets and Single Electrode (SE) jets

There are a number of strict requirements that a plasma jet reactor must meet to cover biomedical and industrial applications: low temperature, no risk of arcing, easy to repair, operation at atmospheric pressure, hand-held operation, low ozone generation, etc. [7]. These considerations have been taken into account for the development of reactor

2. Methodology

2.1. Materials

The insulating material used in constructing of the device was Polyoxymethylene (POM), that is a thermoplastic for use in engineering parts that require high rigidity, and excellent dimensional stability, this material is capable of providing the required precision for building small pieces for the reactor.

POM also has also great thermal characteristics: coefficient of linear thermal expansion for most POM resins ranges from 10.4×10^{-5} to 13.5×10^{-5} m/m/°C, the specific heat has been determined as an average over a range from –18 to 100°C; The value is constant at 0.35. Thermal conductivity of the material is in range of 0.30 to 0.37 W/m·k. The melting temperature varies depending on the type of POM from 110 to 170°C degrees [8]. These characteristics make POM a good option for avoid heating problems in the device and also increasing the operating time of it.

The material used for electrical components was cooper.

2.2. Design

Figure 1 shows configuration chosen for reactor, this set-up allows the device work in two working modes depending if plume enters or not in contact with an object. The material for the main electrode and the ground ring was copper; POM was used for all dielectric materials in reactor.

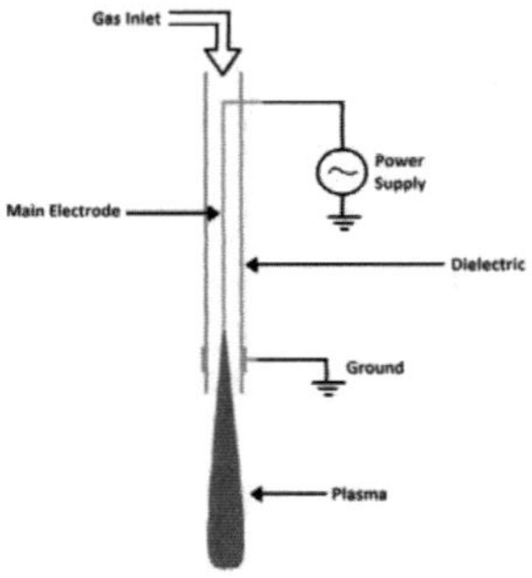

Figure 1. Experimental Set-Up: DBD-like JET

To ensure an easy operation of the device we design a cylindrical reactor with a length of 70.00mm and 25.40mm of diameter. On the top of the reactor there is a gas opening of 4.5mm diameter, at bottom: cooper ground ring with 3.5mm thickness and a dielectric tube of 1.75mm thickness between the electrode and ring. The main electrode consists in a copper wire of 0.8mm thick. Figure 2 shows in detail the design of the reactor and figure 3 shows an explosion view of the real devise and design.

Machining is performed by parts for easy assembly and also to facilitate the replacement of any parts if required. Gas is connected to the device using a tube 1/4OD a common adapter for this diameter was used in order to make the connection and prevent any leakage of gas. Exposed copper cables facilitate electrical connections.

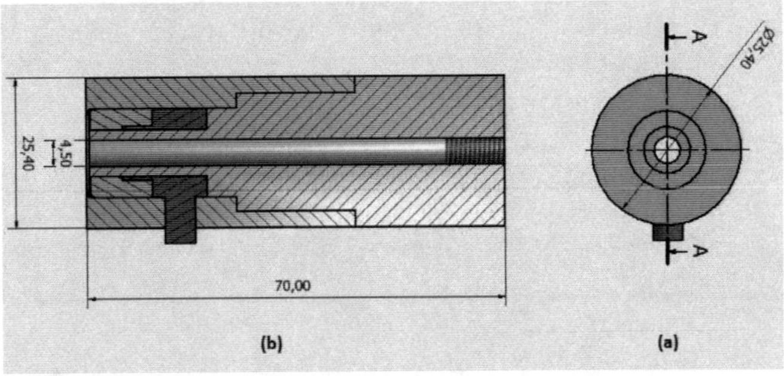

Figure 2. Detailed device design (all units in mm)

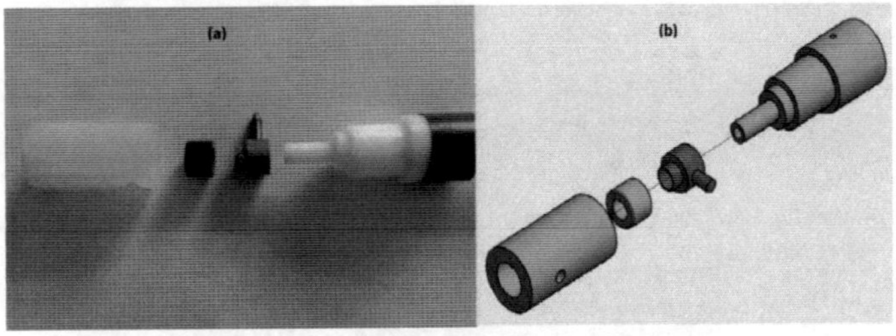

Figure 3. Explosion View (a) Real device (b) Design

Common kHz AC power was used to generate the plasma discharge. In the market there are several low-cost drivers that use a flyback transformer and allows user to control the voltage frequency and duty cycle in order to obtain different characteristics in plasma discharges depending on the application.

3. Results

3.1. Device Testing

For device test was used Argon as gas with flux of 10SCFH. In order to find the optimal discharge conditions according to the capabilities of the power supply, scans were performed at 100V from 20kHz up to 60Khz to locate the optimal discharge frequency which was 25KHz. Then proceeded to vary duty cycles from 50% to 90% for the same purpose, optimal working cycle was 50%

Is important to mention that the optimal point was established as the parameters which give us the longer plume in the reactor however the user can change some parameters depend on the applications needs

Once the optimum values are determined for the frequency and duty cycle several tests at different voltages were performed to observe the behavior of the reactor. Breakdown voltage was determinate at ~72VRMS which was the voltage at we see ionization of gas around main electrode.

Some tests at 100, 150, 200, 250 and 300 VRMS were performed to observe changes in plume length; results in Table 1 and Figure 4 shown us that there are no significant changes in the length of the plume voltages above 200VRMS . No arcs were observed during testing.

Table 1. Plume length at different voltages

Voltage (V_{RMS})	Plume Length (mm)
80V	0
100	18
200	24
250	36
300	39
350	39
400	39

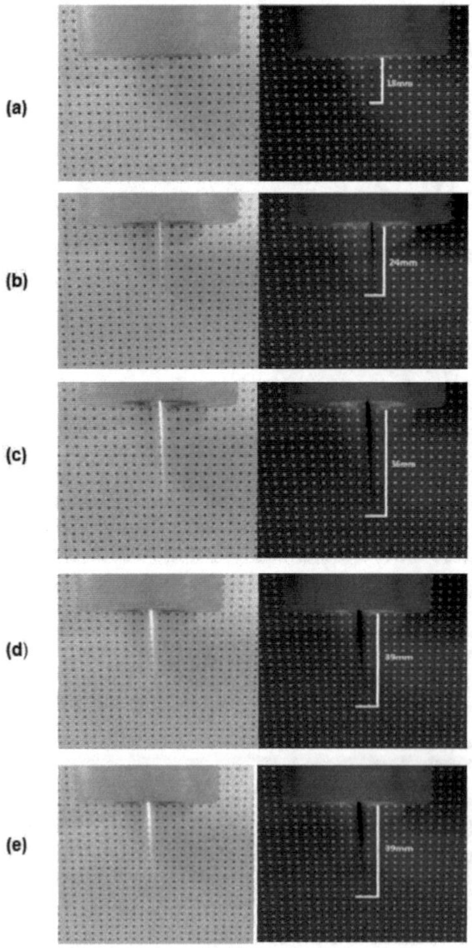

Figure 4. Device Tests (a) 100, (b) 150, (c) 200, (d) 250, & (e) 300 V_{RMS}

The waveform of the output signal was characterized using a digital oscilloscope and setting the power supply at the lowest possible voltage, Figure 5 shows the waveform obtained: the measured frequency was 24.98 kHz with a duty cycle of 50.2% rise time of 9.520 µs and fall time of 9.760 µs. For higher voltages the waveform of the signal remains unchanged.

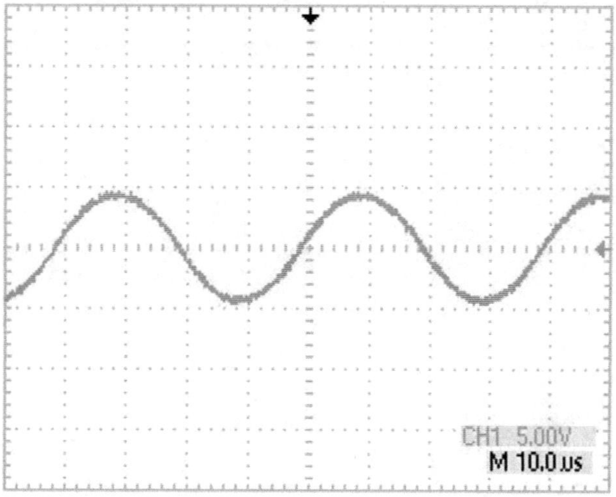

Figure 5. Output waveform from power supply

4. Conclusions

Small size of the reactor makes it suitable for operation in various applications including laminar flow chambers in which it can be used to study the interaction of the plasma with living tissues or bacteria.

Although no change was observed in plume length for voltages above 200VRMS is likely that if there is a change in the temperature of the plasma so in future is necessary to perform some plasma emission analysis to characterize plasma.

The most sensitive part of the device corresponds to the POM dielectric located between the hole for the gas and copper ring due to wear of the materials by the action of plasma, this is indeed the part that must be replaced more often.

References

[1] X. Lu, M. Laroussi and V. Puech, On atmospheric-pressure non-equilibrium plasma jets and plasma bullets, IOPscience 2012 Plasma Sources Sci. Technol. 21 034005 (2012)
[2] Laroussi M 2005 Plasma Process. Polym. 2 391
[3] Deng X T, Shi J J and Kong M G 2006 IEEE Trans. Plasma Sci. 34 1310
[4] Fridman G, Friedman G, Gutsol A, Shekhter A B, Vasilets V N and Fridman A 2008 Plasma Process. Polym. 5 503
[5] Lee H W, Kim G J, Kim J M, Park J K, Lee J K and Kim G C 2009 J. Endod. 35 587
[6] B. Welt, Technical Synopsis of Plasma Surface Treatment, 2009, iopp.org
[7] M.Laurossi and X. Lu, AIP Appl. Phys. Lett. 87, 113902 (2005)

[8] Du Pont ® Delrin Acetal Resin ®, Desing Guide-Module III Retrieved on November 4, 2013 from Website of Du Pont ®: http://plastics.dupont.com/plastics/pdflit/americas/delrin/230323c.pdf

Special thanks to the Workshops of Electromechanical Engineering School at TEC who facilitate the construction of the device

Atmospheric-Pressure Non-thermal Plasma-JET effects on PS and PE surfaces

*a/c J. Arrieta, b/c J. Asenjo, & c I. Vargas a Y. Solis.

[a]School of Engineering& Material Science , Instituto Tecnológico de Costa Rica (ITCR), Cartago, Costa Rica
[b]School of ElectronicEngineering, Instituto Tecnológico de Costa Rica (ITCR), Cartago, Costa Rica
[c]PlasmaLaboratoryforFusionEnergy and Applications, Instituto Tecnológico de Costa Rica (ITCR), Cartago, Costa Rica
E-mail: *jarrietaalfaro@gmail.com

Abstract. The Atmospheric-Pressure Non-Thermal Plasma (APNTP) has become a topic of a great interested for a wide spectrum of applications in different industry branches, including the surface of treatment processes. In this work we evaluate the effect of an argon APNTP exposure to determine changes suffered by a polystyrene (PS) nd polyethylene (PE) polymer surface through RAMAN spectroscopy and SEM. It was determined that the hydrophilic change in energetic terms, i.e. surface activation in the PS and PE polymers is addition of oxygen by surface activation when the samples with jet plasma are exposed with the inert argon gas. It was possible to characterize the hydrophilic shift based on the change in intensity of the spectra.

Introduction

Recently Atmospheric-Pressure Non-Thermal plasmas (APNTP) have become a topic of great interest for a wide range of applications in different branches of industry. In these plasmas the electron temperature is far higher than the temperature of the heavy particles [1]; elastic collisions of the electrons are not effective in terms of it not having a large impact on heavy particles surrounding however electrons can transfer energy to other processes such as ionization, activation or dissociation of molecules.

For over two decades there is a growing interest in the development of devices for the application of this type of plasma discharges in many industrial processes, it is well known also that there is now a wide variety of commercial options to apply treatments of surface activation in different materials with the aim of improving the adhesion properties of certain surfaces.

In the Biomedical field for example, the establishment and quantification of the atomic additions in surfaces (which are carried by the action of the plasma) is very important for the adequate interaction of the cellular matrices in implants and medical devices. In fact, the modification of the original surface through the ion bombardment, corona discharge in air, plasma (nitrogen, argon, axygen), gamma or beta radiation and pre-absorbed surfactants with work gas discharge, is one of the most used methods for the superficial modification of polymeric biomaterials without significantly changing the physicochemical properties and their mechanics [2]

The fundamental physical function of the discharge and the knowledge of basic plasma parameters have been analyzed in several studies. The objective of many studies is to determine the temperature and the concentration of electrons in particular in the active discharge area, however, the chemical dynamics of the process is unknown or has not been extensively studied.

Recently an Atmospheric-Pressure Non-Thermal Plasma-Jet for biomedical and industrial applications has been developed in our laboratory, we evaluate the effect of an argon plasma using this device on a polystyrene (PS) and polyethylene (PE) samples to determine changes suffered in the surface. Some tests RAMAN spectroscopy were performed. It was showed an increase in the intensity of the peaks, which shows a higher surface energy per atom detachment due to exposure to plasma. Furthermore, micofibers in SEM were captured to observe the surface changes of the macro and micro order.

2. Methodology.

2.1. Materials & Experimental Set-Up

Comercial samples of polystyrene and high density polyethylene were used, these samples did not receive any pretreatment prior to application of the plasma discharge. Eight samples of 1cm^2 area were used for each polymer, two unexposed sample was left as a standard, and the others are grouped in pairs that were treated for 1, 2 and 3 minutes.

Figure 1 shows configuration of reactor, this set-up allows the device work in two working modes depending if plume enters or not in contact with an object, the configuration corresponds to a model developed in our laboratory previously [4]

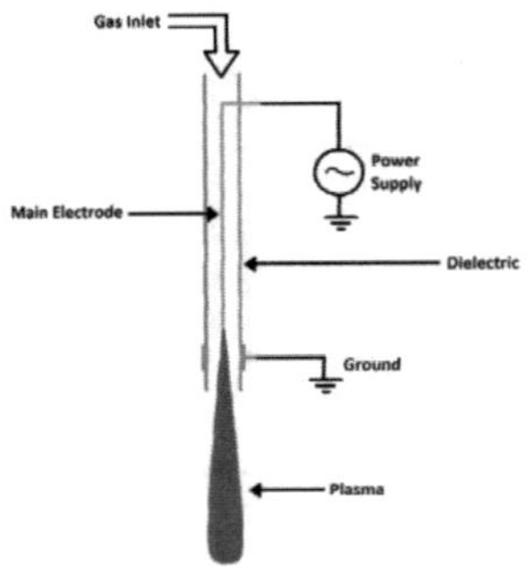

To generate the plasma discharge is used argon as working gas adjusting a flow of 5 l/min, the central electrode of the device is made of copper and glass as dielectric with an aperture of 4.5mm. Using an AC Power Supply set to 25kHz, 300V and a duty cycle of 50%

2.2. RAMAN Spectroscopy.

Raman spectroscopy is obtained by irradiating a sample with a powerful source of visible or infrared laser monochromatic radiation. During irradiation, the spectrum of radiation is scattered at an angle (90 °) anging from 0 to 100 on the intensity of the source is recorded. In these cases, they are useful for the detection of functional groups and fingerprint regions to detect and identify specific compound. Reading the spectrum is done with Weighted intensities for each atomic bond.

2.3. Scanning electron microscope.

It is used as one of the most versatile techniques in the study and analysis of the microscopic characteristics and structural of the solid objects. It works as a superficial study since it does not have a high penetration depth, which is adequate for the surface study.

3. Results

3.1. RAMAN Spectroscopy.

The Raman study is sensible to nanoscale change levels and the addition of functional groups or energy increase in the bound displacement, so it is very important for the analysis of the surface changes, (B. Welt, 2009).
As it is shown in the Figures 3 and 4, in the case of polystyrene, the oxygen addition or interaction by energy changes determine waves between 1300 to 3000, when as showed increase of intensity in spectrum. It is extremely low and does not show a dependence with the time. In the case of polyethylene, it is shown an increase in this intensity and a clear dependence is shown with the exposure time

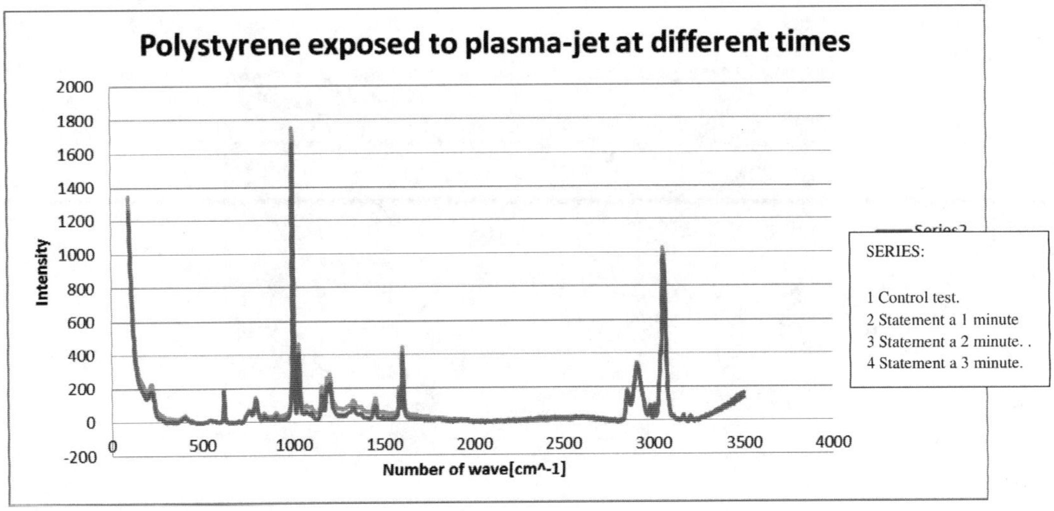

Illustration 2. Graph polystyrene exposed to plasma-jet at different times, (1 min–Blue, 2min-Purpule, 3 min-Green and 4min-Red).

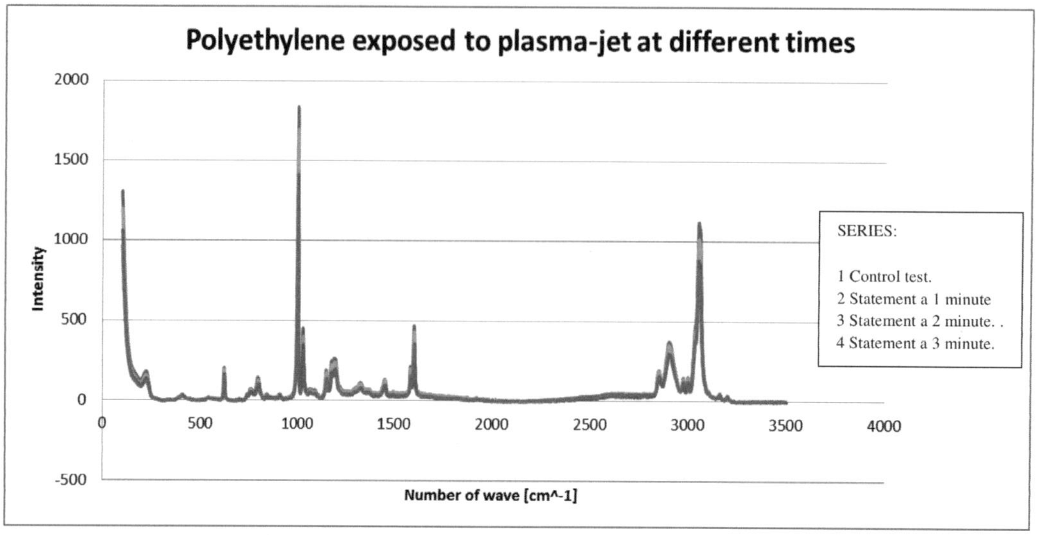

Illustration 3. Graph polyehtylene exposed to plasma-jet at different times, (1 min–Blue, 2min-Purpule, 3 min-Green and 4min-Red).

3.2 SEM

As the Raman analysis also detects changes in roughness it is important to determine every surface change at the macro level [greater than 100 nm], polystyrene was examined only for simplicity and it is observed as shown in Figures 6 and 7 do not change. It can be clearly seen that the same increase in the same area before and after its exposure there are no detectable changes in the microscope

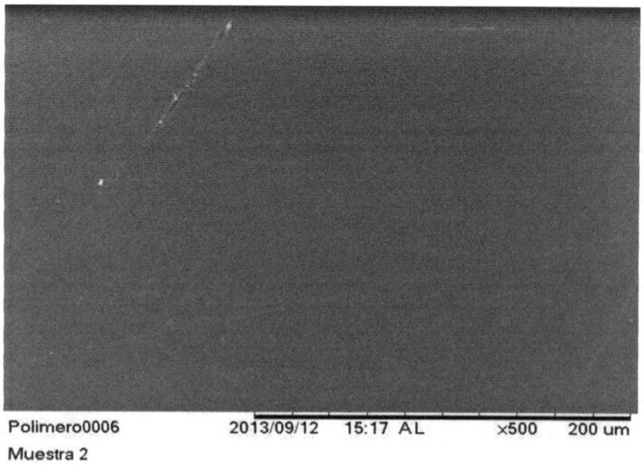

Illustration 4.Control test of polystyrene

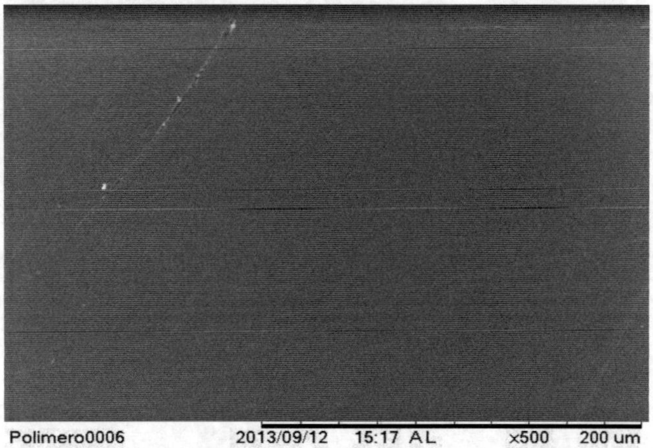

Illustration 5. Polystyrene statement 3 minutes

3.4 Analysis of the results.

The characterizations of the samples were focused in two directions. The analysis macro and micro surface after being exposed to the plasma for 0, 1, 2 and 3 min. The macro-level analysis is the SEM, we can see in Figure 4 and 5, if we compare the polystyrene with the sample exposed for 3 minutes there is change in roughness, which can be seen at 500X. Therefore it can be mentioned that the change in properties of the surface, in this case, the interaction with aqueous or oily solutions, is due to a chemical change in the first layer of atoms, because the Raman study showed an increase in intensity in the peak. The micro-level study confirmed the previous idea, the use of RAMAN spectroscopy could characterize the addition of atomic oxygen to the polymers, and however, it could not quantify the amount of reactant molecules[2]. The theory states that in the range between 1300 and 3000 wave number we found the presence of oxygen in their various functional groups and link availability[3]. In Figure 2, the spectrum [intensity versus wave number] of the polystyrene is observed, we can observe that the increase over time in the range of 1300-3000 was not high enough, and it is held stationary at the time, this is because the energy of benzene is almost saturate , it requires more energy than other functional group because the phenyl structure has a high molecular stability[3]. In the case of Figure 3, we can see peaks in this range because the polyethylene is a simple hydrocarbon chain with one sigma bond if there is a dependence on the exposure time. This is because depending on how much energy the system is submitted, more and more carbon will adhere radicals. in the two spectrums denote increase of intensity in the peaks but not changes in the form. This point showed that materials of this study don´t change your chemical structure, but this energy provide the change of relation with the water.

4. Conclusions

1) The hydrophilic chemical aspect is not physical because no physical, n terms of macrostructure, since evidence was found in the SEM micrograph obtained

2) The results from the RAMAN studies show that the surface changes are product of adittion of more superficial energy ince there is an increase of intensity in the wave numbers between 1300 y 3000.Which is equivalent to the C-O and C=O bonds because this bonds are the most chemically energizable in the molecule. Furthermore, it is seen that the time dependence is only shown with the polyethylene since it lacks the stable and hard to energise since this have one steady state phenyl groups (such in polystyrene). It also contains anoxidizable lineal chain along its structure.

5. References.

1 X. Lu, M. Laroussi and V. Puech, On atmospheric-pressure non-equilibrium plasma jets and **plasma bullets, IOPscience 2012 Plasma Sources Sci. Technol. 21 034005 (2012)**

1.X. Lu, M. Laroussi and V. Puech, 2012 **Plasma Sources SciIOPscience Technol**. 21 034005 (2012)

2. M. Aguilar de Armas, 2002**. *SISTEMAS POLIMÉRICOS HIDROFÍLICOS CON GRUPOS IONIZABLES Y SU APLICACIÓN COMO BIOMATERIALES.*** Madrid, España. UNIVERSIDAD COMPLUTENSE DE MADRID Facultad de Ciencias Químicas
Departamento de Química Orgánica. Tesis Doctoral.

3. A López de Dicastillo, 2011.***Desarrollo y caracterización de polímeros activos hidrofílicos para el envasado de alimentos.*** Valencia, España. Universidad politécnica de valencia. Instituto de Agroquímica y Tecnología de Alimentos (IATA-CSIC). Tesis Doctoral.

4. J Asenjo, J. Mora, A.Vargas, L.Brenes, R Montiel, J. Arrieta & VI Vargas. **Atmospheric-Pressure Non-Thermal Plasma-Jet for biomedical and industrial applications**, in Proceedings of the 14th Latin America Work Shop on Plasma Physics, 27-31 January 2014, San José Costa Rica.

Journal of Physics: Conference Series **591** (2015) 012051 doi:10.1088/1742-6596/591/1/012051

Generation And Applications Of Electron-Beam Plasma Flows

M.N. Vasiliev[1] and Aung Tun Win[2]

[1] Professor, Moscow Institute of Physics and Technology, Dolgoprudny, Moscow region, Russia
[2] Postgraduate student, Moscow Institute of Physics and Technology, Dolgoprudny, Moscow region, Russia

E-mail: csl@mail.mipt.ru

Abstract. Plasma flows generated by continuous or interrupted injection of an electron beam into subsonic or supersonic gaseous streams are considered. Liquid and powder spraying by the electron-beam plasma (EBP) flows is studied as a technique of the aerosol plasma generation. A number of experimental setups generating both free plasma jets and plasma flows in channels are described. Examples of the EBP flows applications for industrial and aerospace technologies are given. The applications are shown to be based on unique properties of the EBP and its stability within very wide ranges of the plasma generation conditions. Some applications of the Hybrid Plasma (HP) generated by combined action of the electron beam (EB) and intermittent gas discharge on flows of gaseous mixtures and aerosols are presented as well.

1. Introduction

The Electron-Beam Plasma (EBP) is generated by injecting an electron beam (EB) into a gaseous medium [1]. To transport the EB from a vacuum chamber, where the EB is originally formed, into a working chamber which is filled with a gas special injection windows (IW) are used. In passing through the gas the EB produces ions and heavy neutral particles in ground and excited states; partially decelerated fast electrons of the original EB and secondary electrons with low and moderate energies are available in the EBP as well. Depending on the EB power N_b and gas pressure P_m plasma temperature T_m varies within a very wide range 300-3000 K. Under typical conditions of the EBP generation ($P_m < 50$ Torr and $N_b \sim 1$ kW) the gas isn't heated highly ($T_m < 10^3$ K), however plasma particles usually reach very high concentrations, i.e. the EBP is strongly non-equilibrium and can be extremely chemically active even at low temperature.

Among peculiarities inherent in the EBP in general [2] the following features should be mentioned specially:

- The EB can be injected into a gas, vapor or their mixtures of any chemical composition;
- The EBP doesn't contract even at higher pressures;
- Being strongly non-equilibrium the EBP is chemically active at low temperatures (down to room temperature and lower);
- When the EB is injected into sub- or supersonic gas flows, as well as into the still gas, the plasma bulk usually remains stable;
- Large size bodies of various shapes can be introduced into the plasma bulk; dispersed additives (liquid drops or solid powders) can be sprayed throughout the plasma bulk too;
- The EB is compatible with other ionizers; the EB can be injected into gas discharge in particular. Combined action of two or more ionizers results in generation of so called Hybrid Plasmas (HP).

The combination of these peculiarities is the basis for development of various setups which can be used for numerous industrial and aerospace applications.

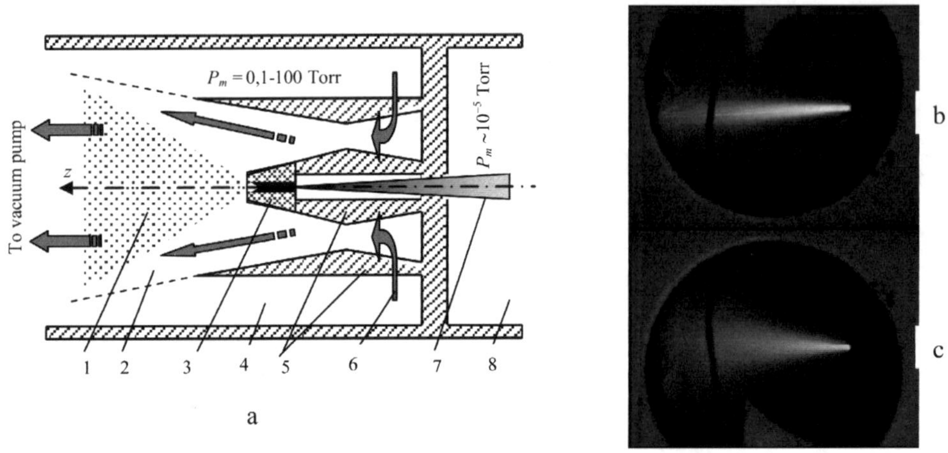

Fig. 1. Free plasma jet generation: a) scheme of the plasma jet generation; b) subsonic plasma jet of air, P_m = 5 Torr; c) unmoving plasma cloud, P_m = 5 Torr (for comparison)
1– zone of the EBP, 2 – gas flow, 3 – IW, 4 – working chamber, 5 – plug nozzle, 6 – gas feeding, 7 – EB, 8 – high-vacuum chamber

2. Techniques of the electron-beam plasma flow generation

Consider free supersonic plasma jet formation in a large-size chamber (see fig. 1) as an example of the EBP flow generation. The focused EB 6 was formed in a high vacuum chamber 8 ($P_m \sim 10^{-5}$ Torr) by an electron gun (the gun is not shown in fig. 1) and then the beam was injected into the working chamber 4 through the IW 4[1]. The IW was placed inside the central body of the plug nozzle 5 which formed supersonic jet 2 of the gas pumped into the prechamber of the nozzle. In passing through the gas the EB excited plasma in zone 1 due to elastic and inelastic processes of the beam-gas interaction. In general, ions and heavy neutral particles in ground and excited states, partially decelerated fast electrons of the original EB and secondary electrons with low and moderate energies are available in the EBP.

The working chamber was filled with a still buffer gas of the same composition as that of the gas being blown or of some other composition. In the latter case the working chamber was equipped with its own gas feeder to maintain the composition of the medium surrounding the jet. The blowing of plasmagenerating and buffer gases was compensated by their evacuation from the working chamber and the value of the gas pressure P_m could be kept constant if the combination of the vacuum pump productivity and total gas discharge was chosen appropriately.

Note that the plug nozzles were used to generate not only supersonic but subsonic plasma jets as well, and the nozzles of other types, e.g. Laval nozzles and non-profiled cylindrical nozzles, turned out to be applicable for the EBP flow generation. All nozzles mentioned above were usually (but not necessarily) combined with injection windows.

3. Generation of the EBP of aerosols

Figure 2a illustrates generation of two-phase EBP flow in free space. The EB was injected through the IW combined with the Laval nozzle, the EB passing along the jet axis through the nozzle throat. As a result, the EBP flow 2 was generated. The liquid was injected into the plasma flow by means of the sprayer 4. We used liquid hydrocarbons and aircraft fuels as the additives to be sprayed. Electrically controlled centrifugal sprayers were used for spraying and pneumatic sprayers could be applied at higher gas flow velocity. The dispersed liquid was mixed with the gas flow and zone 1 of the two-phase EBP was formed. The ratio of the air-liquid mixture components and the shape of the aerosol zone were controlled by the air discharge G_a and by the liquid discharge G_f that can be varied and measured during the experiment.

Plasmas of solid aerosols were generated in a similar way: the mechanical sprayer continuously injected some portions of a powder into the zone of the EBP generation. Powder particles moved together with the plasma flow and then they were caught by a special trap. Though the plasmas of solid aerosols are very

[1] Various IW designs were described in [2] in detail.

promising for applications [3, 7] there is a basic problem in a dusty plasma control since the aerosol particles are able to accumulate ultrahigh electrostatic charge in the EBP [4]. At lower gas pressures (P_m< 1-5 Torr depending on the type of the plasmagenerating gas) the Coulomb forces exceed aerodynamic ones and they can cause instabilities of the plasma flow and even destroy the plasma volume.

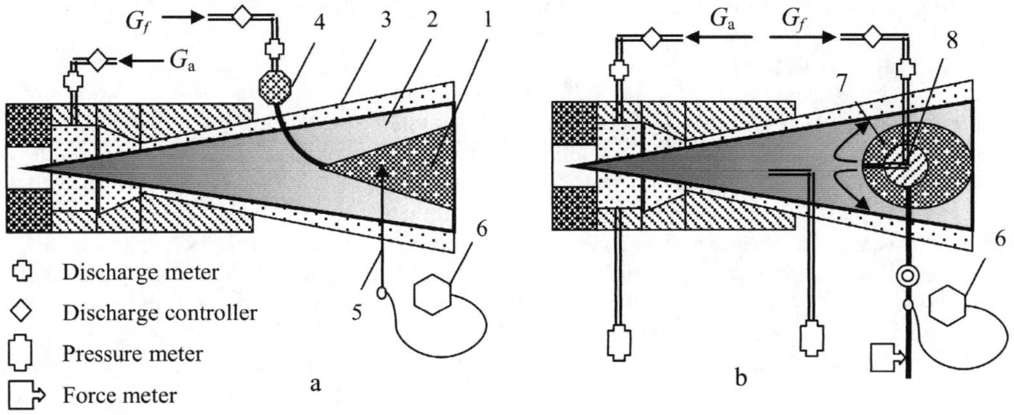

Fig. 2. Aerodynamic experiments in the EBP:
a) generation of the aerosol plasma flows; b) complex aerodynamic experiment.
1 – zone of aerosol EBP, 2 – EBP flow, 3 – gas flow, 4- sprayer, 5 – needle electrode , 6 – high voltage generator, 7 – zone of external combustion, 8 – aerodynamic model

4. Aerodynamic experiments in the EBP

The following experiments were carried out to investigate the EBP influence on:

- ignition and combustion of the aircraft fuels sprayed in free EBP flow;
- ignition and combustion of the aircraft fuels sprayed inside a cylindrical channel;
- external combustion of the air-fuel mixture near the body surface
- drag of bodies of the simplest shape (sphere and cone);
- aerodynamic resistance of long cylindrical channels filled with the flowing EBP.

One of the arrangements of the experiments on plasma-assisted ignition and combustion of air-fuel aerosol is shown in fig. 2a. The aerosol zone 6 was formed inside the EBP flow 2 by pneumatic or centrifugal spraying of the fuel (kerosene) and the air-fuel mixture was ignited by the needle electrode 5 supplied with high voltage from the pulse power source 6. The power source operation frequency was about 100 Hz. Ambient pressure P_m, gas velocity, air and fuel discharges, voltage on the needle electrode and its polarity varied. Experiments showed that usually the air-fuel aerosol:

- could be ignited when the EB was injected into the air flow and it could not be ignited when the EB was turned off;
- stable burning was observed only when the EB was turned on and the igniter was operating.

Thus the EBP assisted the ignition of the air-fuel aerosol and supported the aerosol combustion. The similar behavior of the air-fuel mixture was observed if the aerosol flow was formed in a cylindrical channel that can be considered as a model of a combustion chamber.

Measurements of the friction forces between the EBP flow and the wall of the channel didn't reveal any effects which could be attributed to the plasma processes. Neither plasmachemical modification of the channel surface nor gas ionization and excitation changed the air resistance of the channel. However, the gas heating at higher N_b up to T_m ~10^3 K resulted in the increase of the channel air resistance that correlated with the temperature function of the gas viscosity. The EB injection into supersonic gas jet was found to be able to change the pressure distribution near the surface of the body placed in the stream. This change may be interpreted as the body's drag reduction [5].

Scheme 2b illustrates the experimental study of the body aerodynamic characteristics in the presence of the EBP near the body surface. The model 8 was inserted into the EBP flow on the aerodynamic scale that measured integral force of the flow action on the model. The model had thin channels to inject the gaseous or liquid fuel toward the airflow and high voltage pulses were supplied to the model via the model holder if the fuel ignition was required. Under certain conditions (specially adjusted combination of flow velocity and EB parameters) stable combustion of the fuel occurred and zone 7 of the combined plasma of combustion products was excited around the model surface. The plasma-stimulated external combustion of propane was found to additionally reduce the spherical model down to 20-40% in air flows with Mach number ≈1,5 ($P_m \sim 10$ Torr).

Note that the described experiments were, in fact, experiments with the HP rather than with conventional EBP since the ignition and combustion of the air-fuel mixtures occurred in the gaseous flows excited by combined action of the EB and intermittent high voltage discharge.

5. Materials treatment in the EBP flows

A number of industrial applications of the EBP jets were studied experimentally. In particular, the flows of oxygen and nitrogen plasmas were successfully used to synthesize oxide and nitride layers on the inner surfaces of the metallic tubes [6] and to treat various powders [1]. The same approach proved to be effective to modify physical, chemical and biological properties of polymers and bio-polymers [7, 8].

Figure 3a illustrates the plasma-assisted synthesis of combined nitride-carbide coatings on the inner surface of the titanium tube. Nitrogen and gaseous hydrocarbon were blown into the tube 2 as two separate vortex gas flows 4 and then the nitrogen-hydrocarbon mixture flew along the tube; the EB 5 was injected along the tube axis too. The plasma flow 3 heated the tube and chemically reacted with the material resulting in the Ti-N-C layer formation on the inner tube surface (see fig. 3b). To prevent heat transfer to the ambient gas and to minimize the power consumption for the sample heating the tube was placed inside the heat-protecting envelope 1.

The samples were processed within the temperature range T_s = 350-650 °C. The temperature control was carried out by adjusting the EB current I_b within the range $1 < I_b < 50$ mA at a constant gas mixture pressure that was 10 Torr in the described series of experiments. The optimal treatment time τ was found experimentally and varied from 5 to 15 min depending on the required sample temperature. Local temperatures at various points of the tube were measured by sensors 6; one of them was connected to the feedback of the EB controller to automatically keep the sample temperature constant during treatment. After treatment the tube was cut into fragments to measure the synthesized layer thickness and to study the layer chemical composition. The content of the main chemical elements at various depths under the layer surface is presented in fig. 3c. The figure shows the concentrations of nitrogen and carbon at various depths under the surface. Reaction rates of the surface layer synthesis in the EBP appeared to significantly (several orders of magnitude) exceed those in the gas discharge plasma. Typical formation time of the layer of ~10 μm in thickness was about 10 minutes rather than hours as in conventional industrial reactors with DC- or RF-discharges.

The EBP and HP flows are applicable for the treatment of not only bulky bodies but thin films and even nano-materials. In particular, the experiments with graphene deposited as a thin film on a silicon substrate showed that graphene, being treated in low-speed ammonia HP flow ($P_m \sim 1$ Torr) during $\tau \sim 10$ min, contains about 10% of atomic nitrogen that is a very promising result from the point of view of the graphene functionalization for its further applications.

a b c

Fig. 3. Plasma-assisted synthesis of combined nitride-carbide coatings on the inner surface of the titanium tube: a) scheme of the experiment; b) microsection of the tube fragment after the plasmachemical treatment; c) concentrations of nitrogen and carbon at various depths under the surface of treated sample. 1 – heat-protecting envelope, 2 – titanium tube,43 – EBP flow, 4 – vortex gas flows, 5 – EB, 6 – temperature sensors

6. Conclusions

A number of pilot plasmachemical reactors and plasmadynamic setups based on the generators of the EBP and HP flows had been designed for advanced aerospace and industrial technologies. The reactors were successfully used for the modification of large items, thin films, and powders made of various non-organic, organic and bioorganic materials.

The EBP and HP of liquid aerosols were proved to be promising for the problems of plasma-assisted ignition and combustion of propellants and plasma control of the aerodynamic characteristics of bodies in sub- and supersonic airflows.

References

[1] Vasiliev M N and Aung Tun Win 1992 *AIP. Conf. Proc.* 1482, 519 519-524

[2] Vasiliev M N *Encyclopedia of low-temperature plasma* 2001 **XI** 436-445 (Moscow: Nauka)

[3] Vasiliev M N and Mahir A H *Publications of the Astronomical Observatory of Belgrade* 2011 135-140 (Beograd)

[4] Fortov V E, Gavrikov A V, Petrov O F, Sidorov V S, Vasiliev M N and Vorona N A 2011 *EPL (Europhysics Letters)* **94** No. 5 55001.

[5] Klimov A, Biturin V, Vasiliev M, Vystavkin N and Kuztetsov A 2004 *42-th Aerospace Meeting and Exebition* AIAA 0670 (Reno, Nevada USA)

[6] Vasilieva T M, Lysenko S L and Kukareko V A 2010 *Phys. and Cem. Of Materials Treatment.* No 5 29-36

[7] Vasilieva T M and Mahir A H 2008 *VDI Berihte* Iss. 2027 237-240

[8] Vasilieva T M 2012 *J. Phys.: Conf. Ser.* 012012.

Journal of Physics: Conference Series **591** (2015) 012052 doi:10.1088/1742-6596/591/1/012052

Nitrogen oxides and methane treatment by non-thermal plasma

E Alva[2], M Pacheco[1], A Colín[2], V Sánchez[2], J Pacheco[1], R Valdivia[1], G Soria[3]

[1]*Instituto Nacional de Investigaciones Nucleares, Carretera México-Toluca s/n, La Marquesa, Ocoyoacac, México*
[2]*Universidad Autónoma del Estado de México, Instituto Literario # 100. Col. Centro Toluca, México*
[3]*Instituto Tecnológico de Toluca, Av. Instituto Tecnológico s/n, Metepec, México*

E-mail: marquidia.pacheco@inin.gob.mx

Abstract. Non thermal plasma was used to treat nitrogen oxides (NOx) and methane (CH_4), since they are important constituents of hydrocarbon combustion emissions processes and, both gases, play a key role in the formation of tropospheric ozone. These gases are involved in environmental problems like acid rain and some diseases such as bronchitis and pneumonia. In the case of methane is widely known its importance in the global climate change, and currently accounts for 30% of global warming.
There is a growing concern for methane leaks, associated with a rapid expansion of unconventional oil and gas extraction techniques as well as a large-scale methane release from Arctic because of ice melting and the subsequent methane production of decaying organic matter. Therefore, methane mitigation is a key to avoid dangerous levels of global warming.
The research, here reported, deals about the generation of non-thermal plasma with a double dielectric barrier (2DBD) at atmospheric pressure with alternating current (AC) for NO_X and CH_4 treatment. The degradation efficiencies and their respective power consumption for different reactor configurations (cylindrical and planar) are also reported.
Qualitative and quantitative analysis of gases degradation are reported before and after treatment with cold plasma. Experimental and theoretical results are compared obtaining good removal efficiencies, superior to 90% and to 20% respectively for NO_x and CH_4.

1. Introduction

Nitrogen oxide (NO) and methane (CH_4) are important constituents in emissions from hydrocarbon combustion processes having a direct impact on formation of tropospheric ozone. The methane plays a key role in the global climate change; it is currently responsible for 30% of global warming. Concern about methane has increased due to leakage during extraction of unconventional combustibles, as well as the possibility of a large-scale discharge of methane from the Arctic: as the layer ice melts methane, product of decaying organic matter, is released into the atmosphere. Moreover, methane is a growing source of energy, and its mitigation is then necessary to diminish dangerous levels causing global warming.

NOx refers to a set of emissions mostly composed of nitric oxide (NO), nitric dioxide (NO_2) and nitrous oxide (N_2O), generated in the combustion of any fuel, due to high temperatures and the availability of oxygen and nitrogen. When gases leave the chimney or exhausts cars, NO is oxidized in the atmosphere to NO_2.

Specifically combustion emissions from vehicles have numerous compounds; a characterization of these emissions mainly reports hydrocarbons (methane, ethane, pentane, octane) nitrogen monoxide, nitrogen monoxide, sulfur dioxide, carbon monoxide and carbon dioxide [1].

Regulations around the world are then continuously implemented in order to reduce these gases, European regulations [2, 3] are strictest and the necessity of green technologies becomes critical.

In our laboratory previous work of gas depollution with non-thermal plasma was realized [4-6]. In this work, in order to simplify and to better understand the effect of non-thermal plasma in this kind of emissions; a mixture of 1100ppm NO_x balanced in air and 5000ppm of CH_4 was used. The influence of the hydrocarbon on NOx removal is evidenced with higher removal efficiencies when CH_4 concentration augments.

2. Chemical model

The model considers a gas mixture (NO, helium, air, water vapor, CH_4) and two principal phases in the discharge: the first one corresponds to a very energetic microdischarge head responsible of radicals' formation and, the second phase, the interaction of radicals with toxic compounds.

When electric discharge initiates, a seed electron is formed, gain energy from the applied electric field and lose some of it through collisions with neutral molecules (air and water). These molecules produce reactive species such as free radicals $N^\bullet$, $^\bullet OH$, and $O^\bullet$. They interact, after with NO and CH_4 through thousands of reactions.

Time for noxious molecules degradation is very short, typical time to accomplish the first phase is approximately $1x10^{-8}$ s [7, 8], whereas the interaction of radicals with molecules takes place at around few seconds. The chemical process is outlined in Figure 1.

A more detailed description about theoretical model considerations is depicted by Moreno [Moreno 2008]; in this work the influence of helium in the discharge is taken into account. A very energetic metastable of helium is found at $He(2^3S)$, with approximately 20eV of intern energy and its interaction during the discharge would be then important by Penning effect. However, when water vapor is present in the discharge, this effect is strongly diminished [9] as well as the generation of $O^\bullet$ radicals. Subsequently the influence of this metastable in NOx treatment is not considered in this work.

Furthermore, the influence of He on electrons generation is essential. In several studies the electrons are assumed to be in equilibrium with local field, however this assumption is wrong on the sheath region and, moreover, taking into consideration that gas ionization in atmospheric DBD is significant only over timescale of 1µs [10]; the formation of electrons during the first phase of the microdischarge formation are then calculated in function of electronic temperatures.

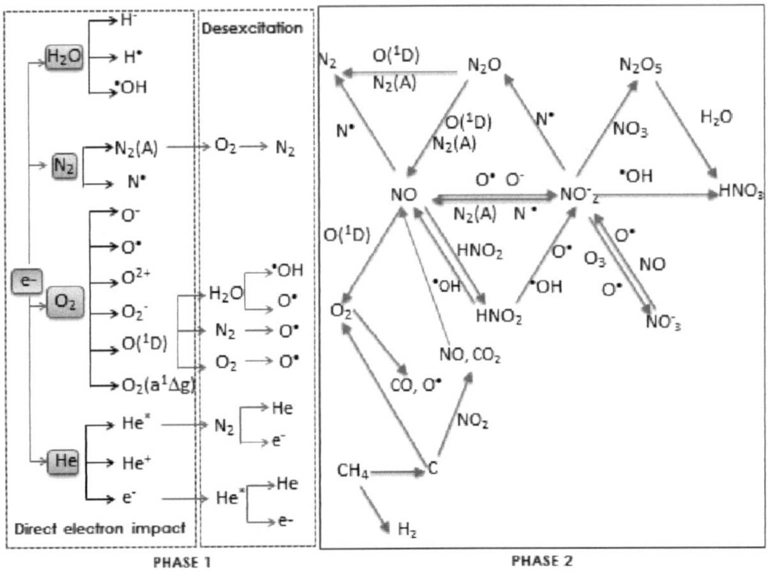

Figure 1. Mechanism of formation of active species in microdischarge head (phase 1) and microdischarge body (phase 2)

3. Experimental setup

To produce the non-thermal plasma, an AC power source was developed. This energy source works with a high frequency series resonant inverter in the range of 20 kHz to 180kHz, able to generate a voltage of 15 kV [11]. The discharge can be sustained at atmospheric pressure using nitrogen or air, as plasma gas. The experimental conditions of this special case were: 36V, 2-5A and 160kHz.

The reactor utilized to treat the gases has a double dielectric barrier configuration. Inlet and outlet gases were analyzed with a PG-250 able to identify and quantify CO, CO_2, NO_x, SO_2 and O_2, a mass spectrometer was also used to identify compounds formed. The experimental system used in the degradation of NO-CH_4 mixture is shown in Figure 2.

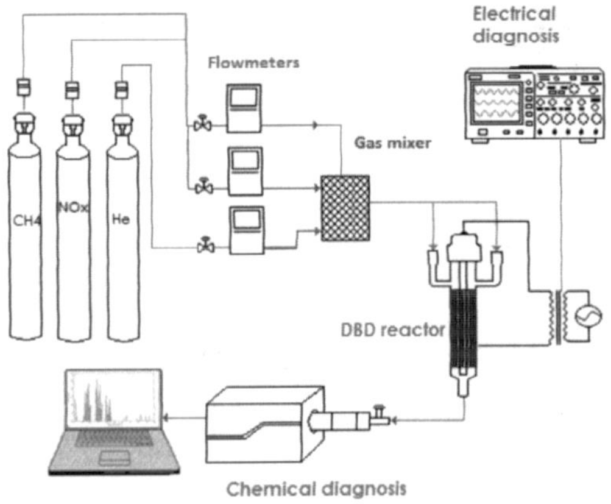

Figure 2. Experimental set up

4. Results

a) Kinetic model

To determine the electronic temperature (T_e), an optical spectroscopy emission study was realized to the plasma, the Boltzmann plot method was employed with NII lines situated at 340.81nm, 391.90nm and 547.80nm. From this, the electronic temperature was deduced obtaining T_e= 0.828eV≅9604.8K. This temperature was used to subsequently calculate coefficients of reactions with helium [10], as detailed in table 1.

Table 1. Reaction coefficients of He reactions

Reaction	Reaction coefficient (cm^3/s)
$e + He \longrightarrow He^* + e$	$1.6277x10^{-19}$
$e + He \longrightarrow He^+ + 2e$	$1.6429x10^{-22}$
$e + He^* \longrightarrow He^+ + 2e$	$3.5503x10^{-10}$
$He^* + He^* \longrightarrow He^+ + He + e$	$2.7x10^{-10}$
$He^* + N_2 \longrightarrow N_2^+ + He + e$	$7x10^{-11}$

The evolution of species formed in the plasma reactor can be appreciated in Figure 3a. A quasi instantaneous conversion of NO to HNO_3 can be observed. The diminution of methane concentration is not so important; it is around 20% as can be observed in figure 3b. An interesting product of methane decomposition is the hydrogen; in figure 3c a concentration of approximately 290ppm is obtained.

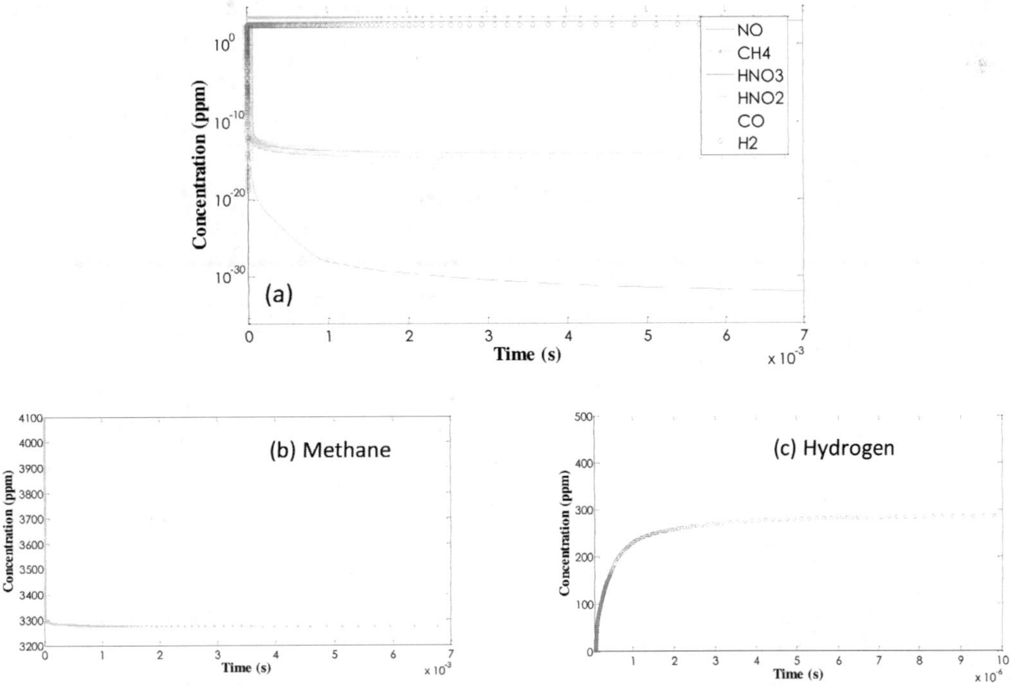

Figure 3 (a) Species concentration in plasma, (b) methane diminution, (c) H_2 production

b) Experimental results

Results from a qualitative analysis are showed in figure 4. A reduction of NO and CH_4 is visible and any acids (i.e HNO_3) cannot be observed.

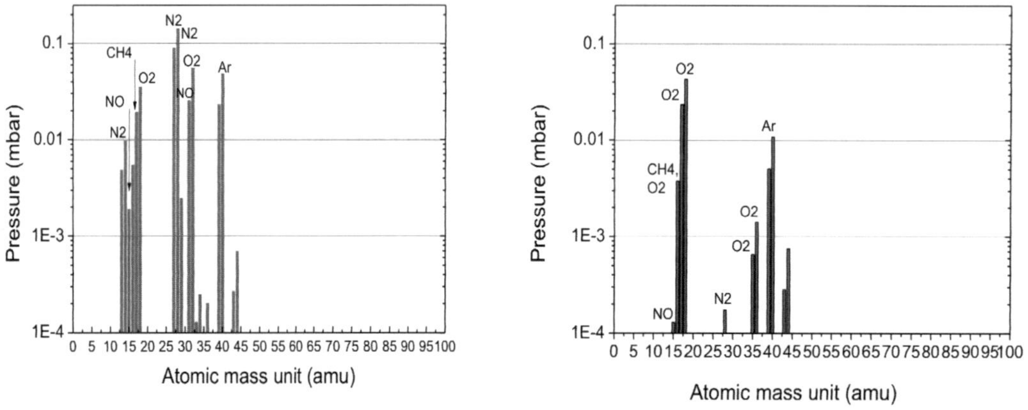

Figure 4. Mass spectra of methane and nitrogen monoxide (a) without plasma treatment (b) with plasma treatment.

The experiences were achieved at 30-35volts and 2-5A; in table 2 the removal efficiencies and the formation of new species are also reported in function of input power. The specific input energy (SIE) is calculated as the relation of the discharge power and the flow rate of feed gas. The NO removal is reported as NOx, an imperceptible, or any, concentration of NO_2 is obtained.

Table 2. Experimental conditions and removal efficiencies and production of CO, CO_2 and H_2

Voltage [V]	I [A]	Pa [W]	SIE [kJ/L]	NO_xremoval (%)	CH_4removal (%)	H_2 [ppm]	CO [ppm]	CO_2 [ppm]
33.99	2.04	69.33	4.55	92.91	21.05	70	650	1000
29.52	2.57	75.86	4.55	95.23	22.30	76	610	1000
34.21	3.32	113.57	6.81	98.31	20.01	72	630	1150
35.00	3.33	116.55	6.99	99.33	24.07	75	600	1200
35.99	5.50	198.00	11.88	100	32.51	90	600	200

From Table 2, the best removal efficiencies at lower SIE of CH_4 and NO_x respectively are 22.3% and 95.23%. At these conditions the production of H_2 and CO are relatively higher, this gas mixture is characteristic for its high energetic capacity [12]. Okumoto et al [13] have obtained CH_4 conversions from 5% to around 30% at SIE under 5kJ/L.

From the comparison between the chemical model and experiences, the prediction about the NO diminution is explained correctly with the mechanism proposed. The generation of H_2 is overestimated in the model; at our experimental conditions the following equations [14] must be studied with more precision:

$$e + CH_4 \rightarrow CH_2 + H_2 + e$$

$$e + CH_4 \rightarrow CH + H_2 + H + e$$

$$CH_3 + H \rightarrow CH_2 + H_2$$

$$H + H \rightarrow H_2$$

5. Conclusions

Results depict the feasibility to non-thermal plasma technology, specifically a 2DBD reactor, for the treatment of automotive toxic gas pollutants. It is important to remark that results obtained from theoretical model suggest the formation of acids (HNO_3, HNO_2), some studies are under development in our laboratory to resolve this problem with a nanostructured carbon filter. $O^{\bullet}$ and $^{\bullet}OH$ radicals mostly generated by O_2 and H_2 molecules play an important role in NO removal. Results obtained from the model, illustrate that in a mixture of NO in humid air, the main path for the NO removal is the oxidation to NO_2 and, soon after, to HNO_3. Concerning the diminution of CH_4 the electron impact is crucial.

Acknowledgements. The presentation of this work was financially supported by COMECYT–Gobierno del Estado de México 2009CO2-131982 and SENER CONACyT 127499 . And to CONACyT scholarship 71674. We are very indebted to Brigida Rojas, Jesús Rosas, E. Loza, A. Salazar, M. Ibañez , M. García and F. Gómez for the experimental and personal support.

References

[1] M. Pacheco, J.Pacheco, R. Valdivia, B. Lefort, A. Santana, Non thermal plasma applied in automotive emissions control, 31st ICPIG, July 14-19, 2013, Granada, Spain (2013)
[2] Kaspar J, Fornasiero P and Hickey N 2003 *Catalysis Today* 77 419-49
[3] Mierlo J, Maggetto G and Lataire P 2006 *Energy Conversion and Management* 47 2748-60
[4] Pacheco M., Pacheco J., Moreno H., Mercado A., Diaz J. y Yousfi M. (2007). DBD-Corona discharge for degradation of toxic gases. Plasma Science and Tech. 9, 682-685.
[5] Moreno H., Pacheco M., Pacheco J., Torres C. y Díaz J. (2007). Modeling and experimental study on nitric oxide removal using DBD-Corona discharge. IEEE Trans Plasma Sci. 35, 1533-1540.
[6] Moreno Saavedra H. (2008). Degradación de gases tóxicos, NOx y SOx, mediante la tecnología DBD-Corona, Tesis de doctorado. Doctorado en Ciencias en Ingeniería Ambiental. Instituto Tecnológico de Toluca, Toluca México, 102 pp.

[7] Y. Kim, W. S. Kang, J. M. Park, S. H. Hong, Y. H. Song, and S. J. Kim, "Experimental and numerical analysis of streamers in pulsed corona and dielectric barrier discharge," *IEEE Trans. Plasma Sci.*, vol. 32, no. 1, pp. 18–24, Feb. 2004.

[8]Kim Y., Hong S H., Cha M. S., Song H. Y. y Kim S. J. (2003). Measurements of Electron Energy by Emission Spectroscopy in Pulsed Corona and Dielectric Barrier Discharge. J. Adv. Oxidation Tech. 6, 17-22.

[9] Stadler K.R. Vidmar R.J. Nersisyan G. y Graham W. G. 2006. Modeling the Chemical Kinetics of High Pressure Glow Discharges in Mixtures of Helium with Air. *Journal Applied Physics*. 99. 093301-1 -093301-8.

[10] X. M. Zhu, M. G. Kong, Electron kinetic effects in atmospheric dielectric-barrier glow discharges, Journal of Applied Physics 97, 083301 2005

[11] J. Pacheco, R. Valdivia, M. Pacheco, J. Ramos, M. Durán, J. Benitez, R. Peña, R. López; "A Universal Resonant Converter for Equilibrium and Nonequilibrium Plasma Discharges", IEEE Transactions on Plasma Science, Vol 32, No. 5, pp 2105-2112, (2004).

[12] J.Pacheco, J. Salazar, R. Valdivia, M.Pacheco, M. Ibañez, G. Soria, J Silva, *An Environmental Application in Acid Gas Cracking with a High Frequency Pulsed Gliding Arc*. IEEE Transactions on Plasma Science, V42(3) 767-773 (2014).

[13] Okumoto, M. Hyun Ha Kim ; Takashima, K. ; Katsura, S. ; Mizuno, A. Reactivity of methane in nonthermal plasma in the presence of oxygen and inert gases at atmospheric pressure Industry Applications, IEEE Transactions on (Volume:37 , Issue: 6) 1618 – 1624 2002

[14] M. Garduño, M. Pacheco, J.Pacheco, R. Valdivia, A. Santana, B. Lefort, N. Estrada,C. Rivera, Hydrogen production from methane conversion in a gliding arc, Journal Of Renewable And Sustainable Energy ISSN: 1941-7012 4, 021202 *(2012)*

Obtention of polymeric membrane fuel cells by low pressure plasma technique: Evaluation of total cell efficiency by function on the amount of platinum and the thickness of the deposited carbon support

A. J. Moreira; N. Ordonez; R. D. Mansano
Electronic System department, Polytechnic School, University of São Paulo, Brazil.

E-mail: adir@lsi.usp.br

Abstract. This work aimed to obtain catalytic support over polymeric membrane building a fuel cell using low pressure plasma technique. For this, polymeric membranes were coated with carbon layer and platinum nanoparticles. The procedures were performed in separate steps in order to obtain firstly carbon layer and catalytic platinum nanoparticles. In the first step, the plasma processes were carried methane in order to obtain carbon layer over the polymeric membrane. At this stage, in order to obtain different thicknesses, were made several processes, reaching a thickness of 0,36μm to 1,4μm. The second step was to get the platinum nanoparticles on the carbon layer. For this, was used a platinum solid source and argon plasma. The study relied primarily on assessing the influence of the carbon layer on the performance of fuel cell. Compared with the commercial processes, it was observed that the results for fuel cells obtained by plasma have a better electric contact on three cell layers (catalyst – electrolyte – reagent). By electrochemical activity test was possible observe increase of reverse voltage of 0.8 volts to 1.24 volts according to increase the thickness of the carbon layer. The same behavior was also observed in the analysis of total efficiency, which was limited to 50% of maximum efficiency of commercial cell due the thickness of the carbon layer deposited during the preparation of this study, indicating a greater thickness with carbon it is possible to achieve the same efficiency of cells better than commercial.

1. Introduction

Fuel cells are electrochemical cells that directly convert the variation of Gibbs free energy of a redox reaction (reduction / oxidation) in electrical energy [1].

Currently, due to increasing demand of energy needed to manage the large amount of technological equipment put on the market, and the growing concern with nature as well as the depletion in renewable, make the fuel cell should be studied more intensively. The subject of intense research is that the problems caused by pollutants such as CO, SO_2 and CO_2 eliminated by fossil fuels are greatly reduced with the use of fuel cell [2].

A fuel cell consists basically of: bipolar plate or interconnector; diffusion electrodes and electrolytes [1, 2].

The hydrogen electrode is the anode or the negative terminal of the fuel cell in which hydrogen enters through channels on the plate [1, 3], and the oxygen electrode or cathode is the positive terminal of the cell. The catalyst turn covers the electrolyte and causes the breakdown of the hydrogen molecule (H_2)

in ions (H^+) and electrons [1, 2, 3]. The electrons generates the electric current and water is generates by ions H^+ and oxygen molecules from cathode.
The main catalytic material used in this system is platinum particles, but their cost is high, resulting in studies to reduce the amount used in fuel cell. One study refers to developing new technologies to reduce the dimensions of these particles, and thus ensure greater contact area with less platinum [2, 3].
Another responsible for the fuel cell performance is the gas diffusion layer (GDL). The GDL is disposed between the catalyst layer and the bipolar plate channels. The main purpose of this layer is to facilitate the reactants flows and products through the channels and interstices, ensuring contact with the bipolar plate. The electrons need to be guided to the catalyst channels of the plates, which are guaranteed by gas diffusion layer. The GDL also serves to maintain a proper draining of water levels, avoiding lose performance [1, 4, 5].
The polymeric ionic membrane is the main component in the PEMFC (Polymeric Membrane Fuel Cells). The most promising PEMFC use perfluorinated ionomer , among which stands out the Nafion ®, first membrane of this type , developed by Du Pont in the 70s [6] .The proton exchange membrane is responsible for conducting protons from the anode to the cathode, electrical isolation between the electrodes and gas barrier between bipolar plates.
The catalysts have properties that vary according to the nanoscale size, improving the catalytic effect when lower is the particle [2, 6]. Catalyzers of Pt / carbon have been widely used as electrode materials for converting chemical energy in electricity through electrochemical reactions. The catalyst consists of small platinum particles (about 50 nm) on a surface of carbon particles. The effective contact area of the catalyst increases due to carbon structure and must be in simultaneous contact with the reactants, with the membrane and the GDL which in turn are in contact with the bipolar plates. The catalyst must be attached to the membrane in order to promote contact with the conducting electrons and protons. The interstices of proton conduction consist of a finely divided carbon particles on the Nafion film [1, 7].
The performance maximization of PEMFC results of increase from active surface of catalyst in electrolyte. However when catalyst area is excessive, increases the catalytic layer thickness, promoting a progressive reduction of the mass transfer. The catalyzer when produced should balance the platinum on carbon, keeping a thinnest layer possible deposited directly on polymer or gas diffusion layer [8, 9]. The distribution of catalytic particles should be such that homogeneous phase region electrolyte - catalyst -reactant is maximized [7].

2. Material

The platinum nanoparticles and carbon support was obtained in a home building parallel plate plasma reactor (figure 1). The plasma was generated by a radio frequency generator at 13.56 MHz. The catalysts were produced using a 99,98% pure platinum wire. The platinum wire was assembled in grid on aluminum rim. The carbon layers were obtained using methane (CH_4) 99,5% of purity. Pure nitrogen gas was used for opening and closing the chamber. *Nafion®* was used as the electrolyte; the sample was 50 x 50 mm area and 1mm thickness. The fuel gases were hydrogen (H_2) 5.0 (99.999%) and oxygen (O_2) 4.0 (99.99%). The membrane was actived using sulfuric acid (H_2SO_4) MOS grade, hydrogen peroxide (H_2O_2) MOS grade and deionized water (18MΩ.cm). The samples were observed by a scanning electron microscope with field emission gun *Nova NanoSEM 400* FEI Company. The catalytic metal was analyzed by X-ray diffractometer *PANalytical XRD Empyream* and the layer thickness was measured by High step meter model 3030 from Dektak - Sloan.

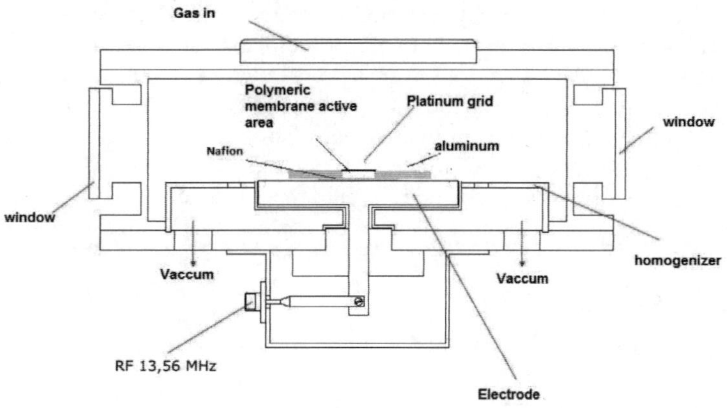

Figure 1: Schematic of plasma reactor showing the platinum grid and aluminum rim.

3. Methodology

The fuel cell was prepared in two steps. The first step was carbon layers deposition in both sides of polymeric membrane and in different thickness. After that, were realized the platinum nanoparticles deposition also in both sides of polymeric membrane. After fuel cells preparation, followed the assembled and efficiency tests. The procedure steps are described bellow:

3.1. Carbon layer deposition

For carbon layer deposition was used methane 99.5 % of purity which was inserted inside of the chamber at low pressure. To fill only the useful area was used an aluminum rim with 100 x 100 mm of size. In the center of this aluminum rim was made an aperture of 30 x 30 mm. The polymeric membrane was positioned on the aluminum electrode (150 mm diameter) and the aluminum rim was placed immediately over it (figure 1). Therefore, besides to fill just the useful area of fuel cell, the aluminum rim kept the polymeric membrane positioned on electrode center, improving the thermic contact from polymeric membrane with the cooled electrode and thus avoiding polymeric membrane damage by high temperature caused by plasma process. The parameters were set up to 300 mTorr of pressure, 200 watts of RF power and the methane gas flow used was 72 sccm. The time for all process was 5 minutes. For the reach different layers thicknesses the processes were made in different steps. For one carbon layer, the time process was 5 minutes in each polymeric membrane side, in other words, ten minutes of total process to both sides. For two carbon layers were made two processes, in other words, was made five minutes of process in each polymeric membrane side and after that, the process was repeated again, totaling twenty minutes of total process deposition. For other carbon layers following the same steps which means the process was repeated three times for three carbon layers, four times to four carbon layer increasing the processes amount until the eight carbon layers. The last process the total time was eighty minutes for eight carbon layers deposition.
The purpose for using only 5 minutes each plasma process was avoid to damage the polymeric membrane because the temperature inside of the chamber caused by plasma effects. The time between each process is enough to decrease the temperature on the electrode and keep to near the room temperature.

3.2. Platinum nanoparticles deposition

The second step was platinum nanoparticles deposition over the carbon layers obtained in previous step.

The platinum nanoparticles were obtained by platinum grid. The same situation was used for reach to platinum nanoparticle just on useful area of fuel cell. Some platinum wires were cut and assembled over an aluminum rim prepared like in carbon layer deposition process, but in this case, the platinum wires were placed on aperture crossing and forming a grid. The purpose of this platinum grid was to fill all useful area of fuel cell using low radio frequency power and lower time process, thus, the amount of platinum nanoparticle deposited on polymeric membrane would be much lower than commercial process, besides the shape and size would be more controlled. In this process the polymeric membrane containing the carbon layers were placed over the electrode and below the platinum grid. After reduce the chamber pressure near to 1 mTorr, was injected inside of the chamber argon 99.995% of purity at flow of 45 sccm and pressure of 500 mTorr. After set up these parameters was controlled the Radio frequency power in 100 watts and so followed the process for 30 seconds each side of polymeric membrane.

3.3. Preparation of polymeric membrane

After the deposition of nanoparticles in the polymeric membrane, the following step was the preparation of the membrane to application in the fuel cell. For the preparation of the cell, chemical cleaning processes were carried out to activate the cell. The following steps were taken: Bath in a hydrogen peroxide solution 3 vol at 80 °C for 1 hour, then three deionized water baths were performed for three hours each at 80 °C and then one sulfuric acid bath also for one hour at 80 °C. Finally more three baths in deionized water for one hour each at 80 °C.

3.4. Tests of electrochemical activities

After all deposition processes and the preparation of polymeric membrane by chemical cleaning, the fuel cells were assembled in a half-cells experimental parker fuel cell in order to verify the total efficiency evaluating mainly the influence of carbon layer thickness. The gases used were hydrogen 99.995% of purity and oxygen 99.995% of purity. In order to keep the membrane humidity was used deionized water at 18 mega ohm. The load was simulated by resistors connected in parallel with the fuel cell. The gases were controlled by mass flow controllers. For comparison test was used a total efficiency commercial cell like reference.

4. Results and discussions

The results of carbon layers thickness measured shown an increase of thickness from 1 to five layers, but after the fifth layer does not have any more increase of thickness (table 1). This can be explained by plasma system build which works like corrosion plasma system. In other words, mean that after a thickness given, the carbon film does not have a good adherence over the previous carbon layer film and then, the plasma corrosion is started over the last carbon layer.

Table 1 Carbon layers thickness comparison

Thickness of carbon layers (μm)		
Commercial fuel cell		130
Produced by plasma technique	1 layer	0.361
	2 layers	0.455
	3 layers	0.578
	4 layers	0.623
	5 layers	1.435
	6 layers	1.368
	7 layers	1.249
	8 layers	1.138

By table 1 is possible to observe that the maximum of carbon layer reached by plasma technique was 1% of commercial fuel cell carbon layer. Although the carbon layer thickness had lower than commercial fuel cell, the electric contact of fuel cell produced by plasma technique was better due the film adherence on the polymeric membrane. By other hand the gas diffusion layer (GDL) was very thin and this hindered the fuel diffusion which decreases the performance of fuel cell.
By figure 2 is possible to see that after five carbon layer, the thickness start to decrease due the corrosion process. By extrapolation, continuing the carbon layer deposition process, each process, the carbon layer would decrease more and more, implying limitation of plasma process in five carbon layers maximum.

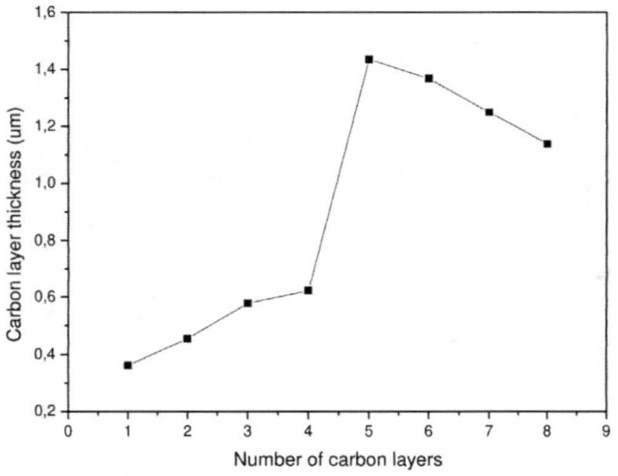

Figure 2: Graphic of carbon layer thickness x number of carbon layers deposited by plasma technique

The platinum nanoparticles obtained by plasma technique shown homogeneous size and shape (figure 3). The shape and size homogenization, besides better distribution over the membrane, improves the performance of fuel cell, so that can be use lower amount of catalizer. Thus the platinum mass deposited on polymeric membrane was 2.16 mg. That mass corresponds to 30 % of total platinum mass used in commercial fuel cell.

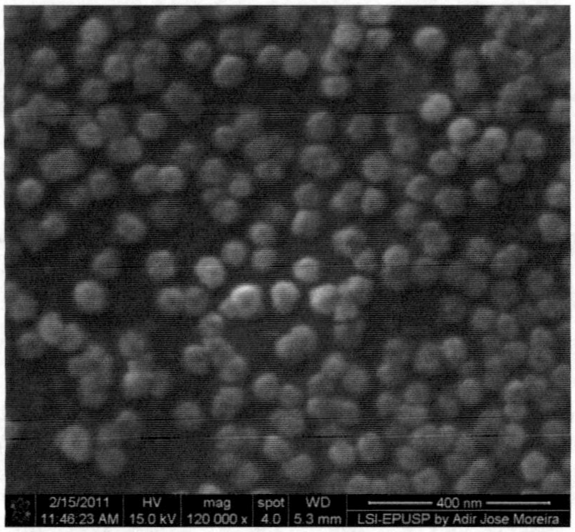

Figure 3: Platinum nanoparticles over carbon film

By the figure 3 can observe that platinum nanoparticles are evenly distributed in the substrate. The nanoparticles size obtained was 50 nm and which is possible to see the uniformity for all particles deposited by plasma technique.
By activity electrochemical test was possible verify the fuel cell efficiency in relation at thickness of the carbon film. Thus was kept the same processes parameters for all platinum nanoparticles deposition, in other words, kept the amount of platinum mass deposited approximately in the same value for all fuel cell produced by plasma technique, changing just the GDL thickness.
So, was to be expecting different results for fuel cells with different carbon layer thickness. This way, when higher the thickness better would the fuel cell performance. Firstly for calculate the total efficiency of fuel cell, was measured the voltage and current (figure 4 and 5).

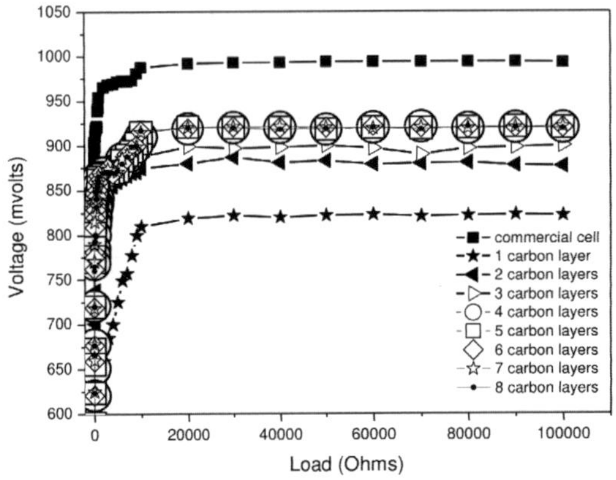

Figure 4: Graphic showing the influence of carbon layer in the voltage

The figure 4 shows the voltage reached by electrochemical activity test. By comparison is possible to see the difference voltages of cells. The commercial cell simulates the conditions of voltage for parameters applied and can see that how much higher the thickness higher the voltage reached.

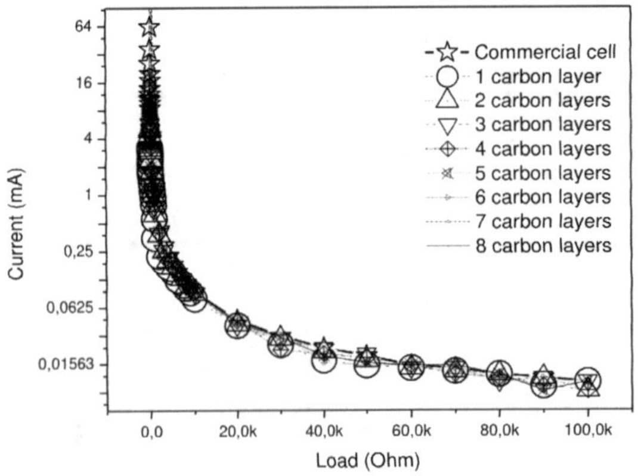

Figure 5: Graphic showing the influence of carbon layer in the current

The evaluation of current shows the operations of cells upon load applied. It is possible to see that the behaviors of the current curves are the same for all cells. Although the curves behaviors of current are the same, the values of current are different, with small difference between the cells for high load and bigger difference when applied small load, reaching around to 62 mA for commercial cell and 20 mA for cell produced by plasma with carbon film higher than 5 layers.
This shows that how much higher the carbon thickness better is the performance of fuel cell.
The performance of fuel cell was compared by polarization curve graphic [figure 6].

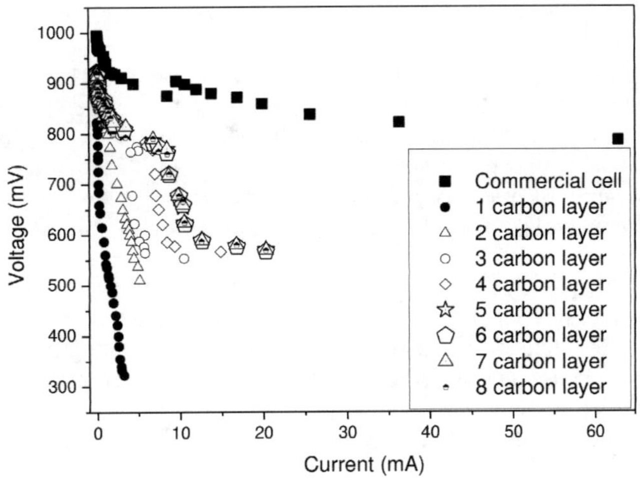

Figure 6: Polarization curve showing the performance of fuel cell according to layer carbon deposited

It's possible to observe in the figure 6 that for the first carbon layers the performance of fuel cell is far below from measured in commercial fuel cell, but as carbon layer thickness increase, the fuel cell performance improves, approaching of commercial fuel cell value and reaching the maximum limit for the fifth carbon layer.
From these graphics was possible calculate the total efficiency of fuel cells [figure7]

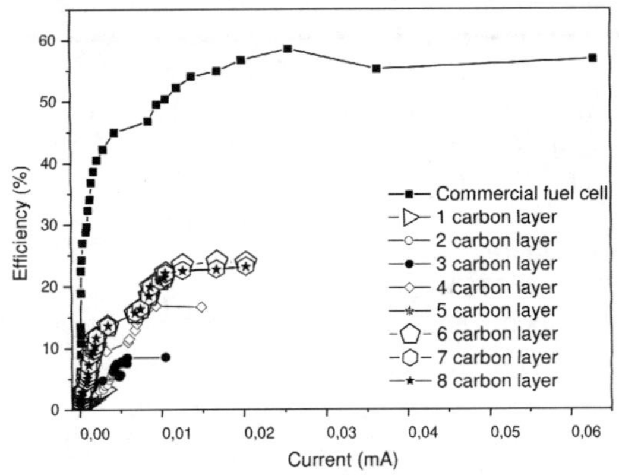

Figure 7: Total efficiency of fuel cell according to layer carbon deposition

The total efficiency was showed in the figure 7 and can be observed the curve behavior is the same for all fuel cell, although measures have been smaller than obtained in commercial fuel cell. By comparison the total efficiency for commercial fuel cell was near to 60 % while the fuel cell produced by plasma technique was near to 30%. Mean to say that using 70% less platinum than commercial fuel cell, was possible achieve almost 50% of total efficiency of a commercial fuel cell even with a thin GDL. The results suggest that with increase of carbon layer thickness the efficiency also increase, this way the cost of fuel cell production decrease because the amount of metal used would be 70 % less.

5. Conclusions

The carbon layer has a great influence on the fuel cell performance, so the higher the carbon layer thickness higher performance. The results obtained in this work show that the carbon layer thickness deposited by plasma although very thin, got a relative efficiency, which is not higher because the limitation of carbon layer thickness deposition by plasma process which occurs due the corrosion on carbon film after some plasma processes. In the other hand the phase region electrolyte - catalyst - reactant was better than commercial fuel cell due the total adherence of carbon film on polymeric membrane.

The possibility of control all processes parameters allows different setups, implying to adjust the deposition process of carbon to increase its thickness in reduced times, avoiding damage to polymeric membrane and increasing the efficiency of the fuel cell, besides, allows to control the size and shape of platinum nanoparticles and consequently decreasing the total mass of platinum applied on fuel cells without lose the efficiency.

This way the results obtained in this work suggest with increase of carbon layer thickness the efficiency also increase, reducing the cost of fuel cell production because the amount of metal used would be 70 % less.

Acknowledgements

The authors thank to FAPESP and CNPQ for the financial support, the Laboratory of Integrated System of Polytechnic School for supplying all support to realize this work. Alexandre Marques Camponnuci, Julio Cesar dos Santos and Rubens de Alcantara Pereira for technical support.

References

[1] Moreira, A.J. Lopera, S., Ordonez, N. Mansano, R.D.; Platinum nanoparticles deposition on polymeric membranes for fuel cell application, Journal of Physics: Conferences Series, 370 (2012), doi: 10.1088/1742-6596/370/1/012030

[2] Moreira, A.J. Ordonez, N. Mansano, R.D.; Low pressure plasma study for platinum nanoparticles synthesis, Nanomater Nanotechnol, 2013, 3:15. doi: 10.5772/56932

[3] Wendt, H.; Götz, M.; Linardi, M.; Fuel cell technology; Química Nova; vol. 23; 2000

[4] Li, W.; Liang, C.; Zhou, W.; Qiu, J.; Zhou, Z.; Sun, G.; Xin, Q.; Preparation and Characterization of Multwalled Carbon Nanotube-Supported Platinum for Cathode Catalysts of Direct Methanol Fuel Cells; Journal Physics Chemistry, B 107; 6292-6299, 2003

[5] Rudge, M.; Büchi, F.; Bipolar elements for PE fuel cell stacks based on the mould to size process of carbon polymer mixtures; Proceedings of the first European PEFC forum, (EFCF) 2001, 299-308

[6] Zawodzinski T., Derouin C., Radzinski S., Sherman R., Smith V., Springer T., Gottesfeld S Water; Uptake by and Transport Through Nafion® 117 Membranes", Journal of the Electrochemical Society, 1993, 140, 1041-1047

[7] Zhigang, Q; Qualification of fuel cell membrane electrode assemblies; Fuel cell research trends pp 3 - 6, 2007

[8] Santis, M.; Investigation of current density in homogeneities in polymer Electrolyte fuel cells, Tesys, Swiss Federal Institute of technology, Zurich; 2006
[9] Reum, M.; Sub millimeter resolved measurement of current density and membrane resistance in polymer electrolyte fuel cells, tesy, Swiss Federal Institute of technology, Zurich, 2008

Journal of Physics: Conference Series **591** (2015) 012054 doi:10.1088/1742-6596/591/1/012054

Patterned growth of carbon nanotubes obtained by high density plasma chemical vapor deposition

A P Mousinho, R D Mansano
Laboratório de Sistemas Integráveis do Departamento de Engenharia de Sistemas Eletrônicos da Escola Politécnica da Universidade de São Paulo, Avenida Professor Luciano Gualberto, 158, trav. 3, São Paulo, SP, Brasil

E-mail: mousinho@lsi.usp.br

Abstract. Patterned growth of carbon nanotubes by chemical vapor deposition represents an assembly approach to place and orient nanotubes at a stage as early as when they are synthesized. In this work, the carbon nanotubes were obtained at room temperature by High Density Plasmas Chemical Vapor Deposition (HDPCVD) system. This CVD system uses a new concept of plasma generation, where a planar coil coupled to an RF system for plasma generation was used with an electrostatic shield for plasma densification. In this mode, high density plasmas are obtained. We also report the patterned growth of carbon nanotubes on full 4-in Si wafers, using pure methane plasmas and iron as precursor material (seed). Photolithography processes were used to pattern the regions on the silicon wafers. The carbon nanotubes were characterized by micro-Raman spectroscopy, the spectra showed very single-walled carbon nanotubes axial vibration modes around 1590 cm^{-1} and radial breathing modes (RBM) around 120-400 cm^{-1}, confirming that high quality of the carbon nanotubes obtained in this work. The carbon nanotubes were analyzed by atomic force microscopy and scanning electron microscopy too. The results showed that is possible obtain high-aligned carbon nanotubes with patterned growth on a silicon wafer with high reproducibility and control.

1. Introduction

Carbon nanotubes (CNTs), with their exceptional anisotropic electrical, thermal and mechanical properties combined with their dimensions – nanometers in diameter and micrometers in length – are ideal candidates for several applications and components in nanotechnology such as nano-wires, transistors, field emitters, reinforced nano-composites and conducting filler in insulating materials[1]. CNTs were discovered in 1991. Remarkable progress has been made in the ensuing years, including the discovery of the two basic types of CNTs (single-wall and multiwall). Great strides have been taken in their synthesis; purification and elucidation of the fundamental physical properties, with important steps are being taken toward realistic practical applications [2]. CNTs are a new form of carbon that shows an equivalent form a bi-dimensional grapheme sheet wrapped as a tube. Single-walled carbon nanotubes (SWCNTs), made of cylindrically rolled grapheme sheets, have attracted a lot of attention due to their interesting and potentially useful electrical and mechanical properties. In particular, regarding their mechanical properties, they have been hailed as the flexible springs of nature. SWCNTs can be conductive and semiconducting. The electrical properties of the CNTs depend on the diameter of the tubes and the chiral angle [3-8]. The CNTs exhibit excellent properties for many applications, including: electrical, electro-optical, optical, optoelectronic, mechanical, electromechanical, magnetic, electromagnetic, chemical, electrochemical, thermal and thermoelectric properties [9]. Additionally, their incredible strength (a consequence of the famously strong carbon bond in the tubular structure) poses further advantages in nano-mechanic applications. CNTs forest is a new class of established materials that can be used for many applications [10]. Carbon nanotubes are generally produced by three main techniques, arc discharge, laser ablation and chemical vapor deposition [8]. Patterned growth of CNTs by chemical vapor deposition represents an assembly approach to place and orient CNTs at a stage as early as when they are synthesized [1]. In this work, we report a method for obtaining patterned growth of CNTs forest at room temperature by High Density Plasmas Chemical Vapor Deposition (HDPCVD) system. We report the patterned growth of CNTs on full 4-in Si wafers, using high-density pure methane plasmas and iron as precursor material (seed). The method for obtaining patterned growth of CNTs included the iron deposition using Magnetron sputtering [11-12], the pattern of regions on the silicon wafers by photolithography processes [12-13] and the patterned growth of CNTs by HDPCVD. The parameters of plasma processes were varied. The CNTs were characterized by Scanning Electron Microscopy, Atomic Force Microscopy, and micro-Raman Spectroscopy [14-17]. The results showed that is possible obtain high-aligned CNTs with patterned growth on a silicon wafer with high reproducibility and control.

2. Materials and Methods

The CNTs were obtained by a HDPCVD system that uses a new concept of plasma generation. In this case, a planar coil coupled to an RF system for plasma generation, and an electrostatic shield for plasma densification are used. In this mode, high-density plasmas are obtained. For the ion acceleration another RF system in planar configuration is used (as seen in Figure 1).

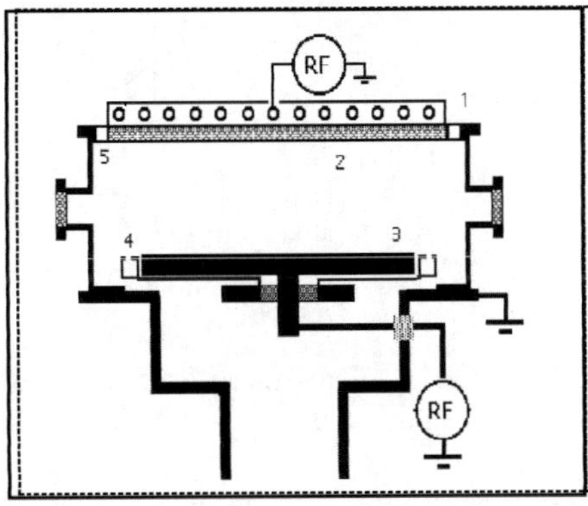

Figure 1. Scheme of the HDPCVD deposition system. 1) planar coil with an electrode shield, 2) pyrex window, 3) electrode, 4) secondary gases distributor and, 5) main gases distributor.

The substrate temperature was not controlled, but measured by a type K thermocouple, which indicated that the temperature never exceed 90 °C. For the CNTs growth, pure methane plasmas processes were used. The HDPCVD system has a main chamber (with diameter of 35 cm), where the electrostatic shield (with diameter of 30 cm) is localized. The electrostatic shield can be heated until 600 °C by an infrared heater. The planar coil has been put in the upper part of the chamber, above a borosilicate window and is covered with an electrode shield. RF power (13.56 MHz) is applied to the planar coil centre and also at the lower electrode (the RF systems are independent). The vacuum system is formed by a turbo-molecular pump (with nominal outflow of 400 l/s) and an auxiliary mechanical pump. The minimum basis pressure that the system can reach is 2.10^{-7} Torr. In the HDPCVD system, there are two gas distributors. The main distributor was constructed in the cover of the chamber, where a small homogenization chamber (secondary distributor of gases) exists. The distribution of the gases in the process chamber has been made by a holed ring. The HDPCVD system has been created to work with gases (six different types) and liquids (two different types) at the same time, thus it is possible to obtain the DLC films with additives (nitrogen, fluorine, oxygen, etc).
The substrates used to obtain the CNTs were three-inch diameter silicon wafers, 380 μm thick and with orientation (100). They were submitted to a Piranha clean, followed by a diluted HF dip. Iron

(precursor material) was deposited on the substrates by Magnetron Sputtering (200 W, 5 mTorr, 20-sccm argon and 30 min). The iron layer was defined using optical photolithography and O_2 plasma treatment. These steps were used to pattern the silicon wafers surface before the CNTs growth.
The CNTs were grown after the samples preparation, with the parameters: 15 mTorr, 250 W (coil power, RF, 13.56 MHz, remote plasma), 40-sccm methane and the deposition time were 3 hours. The structural properties and the diameter distribution of the CNTs were analyzed by Raman spectroscopy. The spectra were collected using a Renishaw micro-Raman 2000 spectrometer on a 40x objective with a photo multiplicator. Unpolarized Raman spectra were acquired at λ = 514.5 nm, the spectral resolution was about 4 – 6 cm^{-1} and the power on the sample was kept well below 1 mW. For the characterization of the SWCNTs structure we used an atomic force microscope (AFM) model SPM 9500J3 (Shimadzu) in tapping mode (10 analyses per sample in different areas of 15 μm x 15 μm and 5 μm x 5 μm). The scanning electron microscope (SEM) used it was a FEI NOVA nano500. The results obtained are shown in this work.

3. Results and Discussion

A typical micro-Raman spectrum obtained for CNTs grown by HDPCVD system is showed in Figure 2.

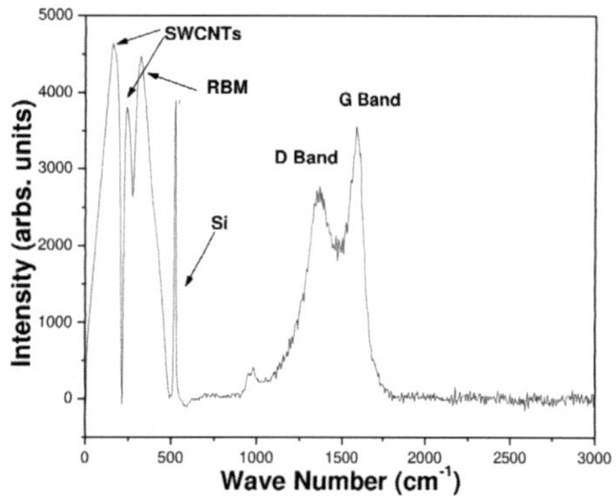

Figure 2. Typical micro-Raman spectrum for CNTs obtained by HDPCVD system.

Using chemical vapor deposition with methane gas for growing the CNTs, complex dissociations occur that could create a large variety of structural characteristics. In this way, observing the Figure 2, we can noticed the presence of four characteristics vibrational modes for a Raman spectrum of a unique CNT (RBM, G band, D band and G' band) and also some new vibrational modes related to CNTs with different diameters. The region between 0 and 600 cm^{-1} is subdivided with some important peaks related to RBM, SWCNTs and CNTs with diameter of 1,8 nm and CNTs with bigger diameters. The micro-Raman spectrum showing details of this region is presented in Figure 3.
The main vibrational modes in CNTs obtained in this work, are centered in ~ 150 cm^{-1}, ~240 cm^{-1} and ~340 cm^{-1}. These peaks are related to the Radial Breathing Mode (RBM). Radial breathing mode (RBM) of CNTs is a low frequency mode, but accounts for the strongest feature observed in the CNT Raman spectrum. For the RBM, all of the carbon atoms in a CNT move in the radial direction synchronously, which generates an effect similar to "breathing" [18]. This mode is unique to CNTs, and is not observed on other carbon systems. Raman measurement of the RBM in CNTs is a standard, straightforward method for precisely determining the diameter of a CNT, distinguishing the CNT chiral-index assignments, or characterizing CNT clusters [18].
In this work, the three peaks that compose the RBM region for Raman spectra obtained for CNTs are related with CNTs with diameter of 1,8 nm and CNTs with larger diameters. The difference in the diameters of the CNTs obtained in this work is related with the segregation of carbon atoms in the iron crystal structure.

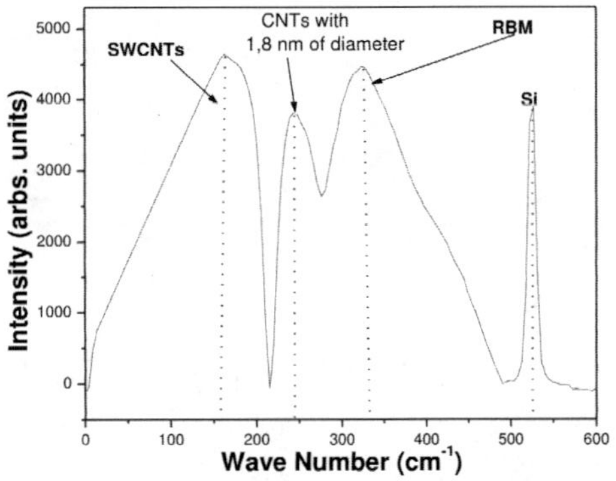

Figure 3. Typical micro-Raman spectrum for CNTs obtained by HDPCVD system, detailing the 0-600 cm^{-1} region.

The presence of iron promotes the high-alignment of CNTs and contributes to the definition of CNTs diameters. Using HDPCVD system is possible control the CNTs diameters. The peak centered at ~150 cm^{-1}, indicates the presence of SWCNTs in the samples. The combination of the peak centered in ~150 cm^{-1} and the G band centered in 1560-1600 cm^{-1} are related with the strong presence of armchair SWCNTs. The presence of many vibrational modes indicates that we have obtained SWCNTs with a large range of the diameters due the presence of the iron in the samples. Besides, there is a presence of other peaks that are related with vibrational modes of the second order. They are related with CNTs with different diameters and other kinds of carbon nanostructure formation in the samples.
Using conventional photolithograph and oxygen plasma treatment we could promote the high-aligned CNTs growth. The iron layer was patterned and the high-aligned CNTs are grown in the regions with iron precursor and in the other regions (without iron), it occur the diamond-like carbon (DLC) formation. Using the photolithography processes we could define the regions where the CNTs would be grown. Thus, we could obtain the patterned growth of CNTs with high-aligned degree. The results could be proved by Scanning Electron Microscopy and Atomic Force Microscopy and they are showed in the next figures.

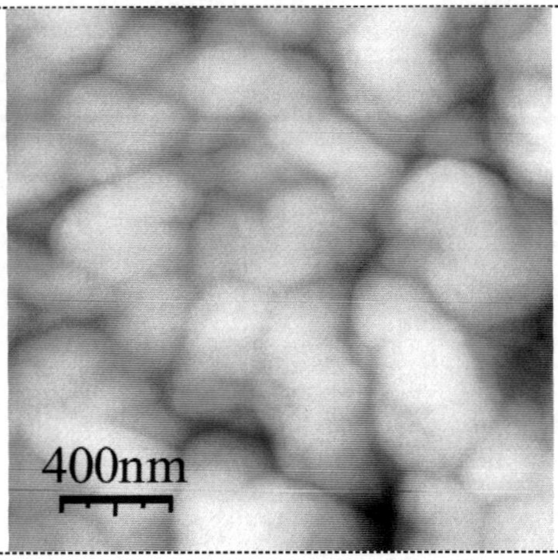

Figure 4. Typical AFM image for CNTs obtained by HDPCVD system.

Figure 5. SEM micrograph of high-aligned SWCNTs obtained by HDPCVD system.

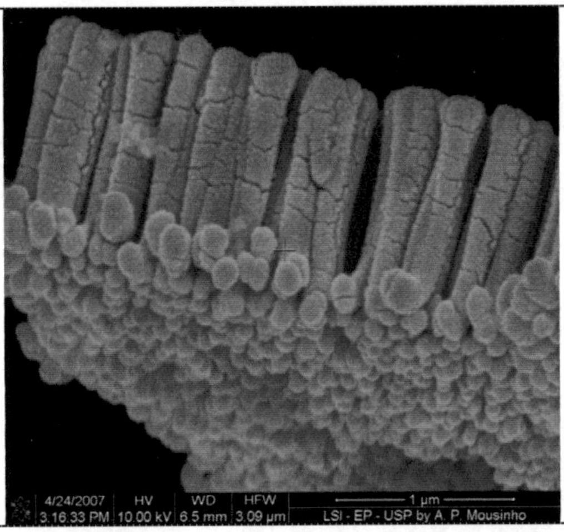

Figure 6. SEM micrograph of high-aligned SWCNTs obtained by HDPCVD system.

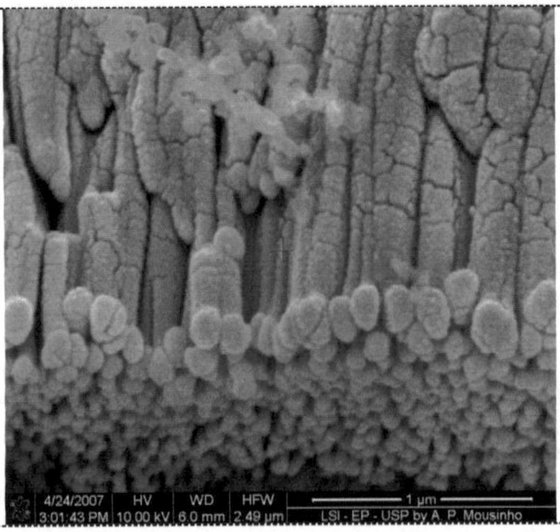

Figure 7. SEM micrograph of high-aligned SWCNTs obtained by HDPCVD system.

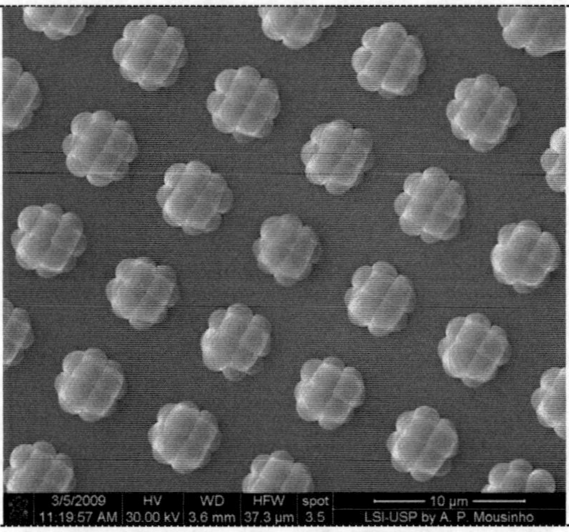

Figure 8. SEM micrograph of patterned growth of high-aligned SWCNTs obtained by HDPCVD system (The "flowers" are islands of CNTs).

Figure 9. SEM micrograph of patterned growth of high-aligned SWCNTs obtained by HDPCVD system (The "squares" are islands of CNTs).

Observing the AFM image and the SEM micrographs we can observe that the CNTs grow up in "clusters" (like a bouquet of flowers). These clusters have high-density of CNTs due the presence of iron in the samples. The presence of iron generates a roughness surfaces in the silicon wafers and promotes locally an increase of carbon-carbon bonds nucleation. With the photolithography processes, we could pattern the CNTs growth.

4. Conclusions

In this work, we have obtained patterned growth of high-aligned carbon nanotubes using pure methane plasmas, iron as precursor material and, photolithography processes and high density plasma chemical vapor deposition techniques at room temperature. The CNTs were analyzed by micro-Raman spectroscopy, SEM and AFM Microscopy. The CNTs obtained in this work showed unique structural properties. With these results is possible obtaining patterned growth of CNTs forest for applications in SWCNT forest-based electronic and optoelectronic devices.

5. Acknowledgements
We gratefully acknowledge the Mr. Nelson Ordonez, PhD Adir José Moreira and Mr. Alexandre Camponucci, for providing technical support and FAPESP and CNPq for financial support.

6. References

[1] Saito R, Dresselhaus G and Dresselhaus M 1998 *Physical Properties of Carbon Nanotubes* (Imperial College, London)
[2] Gogotsi Y 2006 *Carbon Materials* (CRC Press & Francis Group, Boca Raton)
[3] Rana R K, Kolypin Y and Gedanken Y 2001 *Chemical and Physics Letters* **256**
[4] Kurti J 2003 *Molecular Nanostructures* **285**
[5] Antunes E F, Lobo A O and Trava-Airoldi V J 2006 *Carbon* **2202**
[6] Kato T 2003 *Chemical Physical Letters* **422**
[7] Zhang W D and Thong J T L 2002 *Journal of Diamond and Related Materials* **1638**
[8] Vivekchand S R C 2003 *Indian Academic Science* **509**
[9] Mubeen S, Lim J H, Srirangarajan A, Mulchandani A, Deshusses M A and Myung N V 2011 *Electroanalysis* **299**
[10] Mousinho A P, Mansano R D 2011 *Organic Coatings* **326**
[11] Kelly P J and Arnell R D 2000 *Vacuum* **159** 56
[12] Sze S M 1998 *Very Large Scale Integration Technology* (McGraw-Hill Education, Singapore)
[13] Shang H and Liu C 2004 *Journal of Natural Gas Chemical* **95**
[14] Meyyappan M 2004 *Carbon Nanotubes, Science and Applications* (CRC Press & Francis Group, Boca Raton)
[15] Mistra A 2006 *Journal of Diamond and Related Materials* **385**
[16] Camili L, Scarselli M, Gobbo S D, Castrucci P, Nanni F, Gautron E, Lefrant S and Crescenzi D 2011 *Carbon* **1889**
[17] Milijkovic M 2004 *Physical Chemical Technology* **84**
[18] Lei X W, Ni Q Q, Shi J X, Natsuki T 2011 *Nanoscale Research Letters* **492**

Journal of Physics: Conference Series **591** (2015) 012055 doi:10.1088/1742-6596/591/1/012055

SUPER-PARAMAGNETIC NANOPARTICLES SYNTHESIS IN A THERMAL PLASMA REACTOR ASSISTED BY MAGNETIC BOTTLE

R. Cartaya [1], **J. Puerta** [1], **P. Martín** [1,2]

Departamento de Física,Universidad Simón Bolívar [1], Apdo: 8900, Caracas - Venezuela
Universidad de Antofagasta [2], Chile

E-mail: `rene.alejandro.cartaya@gmail.com, jpuerta@usb.ve, pmartin@usb.ve`

Abstract. The present work is a study of the synthesis of super-paramagnetic particles. A preliminary study based on thermodynamic diagrams of Gibbs free energy minimization, was performed with the CSIRO Thermochemical System. In this way, the synthesis of magnetite nanoparticles from precursor powder of ore iron in a thermal reactor, was performed. Then the process was simulated mathematically using magnetohydrodynamic and kinetic equations, in order to predict the synthesis process. A cylindrical reactor assisted by magnetic mirrors was used. The peak intensity of 0,1 tesla (1000 Gauss) was measured at the end of the solenoid. A PlazjetTM 105/15 thermal plasma torch was used. The precursor powder was iron oxide and the plasma gas, nitrogen. The magnetite powder was magnetized whit rare-earth super-magnets, alloy of neodymium-iron boron (NdFeB) grade N-42. The synthesized nanoparticles diameters was measured with a scanning electron microscope LECO and the permanent magnetization with a YOKOGAWA gauss meter, model 325i.Our experimental results show that it is possible the synthesis of super-paramagnetic nanoparticles in thermal plasma reactors.

1. Introduction

The study of plasma has given mankind a great amount of knowledge, and it has been very important in the development of different fields of science, technology and industry with great impact. In this way the plasma has moved to be a scientific curiosity to a fundamental tool for the development of mankind. Plasmas has become now very important in illumination, metal mechanic industry, semiconductors productions, welding, product recycle, medicine and nuclear industry. The plasma technology development has guaranteed great variety of plasma torch configurations and these can be classified as, of transferred and not transferred arch. Considering the last kind of other classification of electrodes and direct current, radiofrequency, hybrids and microwaves. The development of the torches has produced a large impact in the industry on the second period of the XX century, when revolutionized the ship construction. Now the plasma torches have become very important in industrial high technology, mainly in the production of semiconductors, nanorecovery and nanomaterials. The plasma is an state of the matter, characterized to be composed by an ionized gas with several special properties, as to interact with electric and magnetic fields. In this way, the plasma is generated by a DC plasma torch of no transferred arc. The plasma is used to produce nanoparticles and nanomaterials. The new idea is confine the plasma with a magnetic mirror system, in order to increase the

Journal of Physics: Conference Series **591** (2015) 012055 doi:10.1088/1742-6596/591/1/012055

production efficiency.

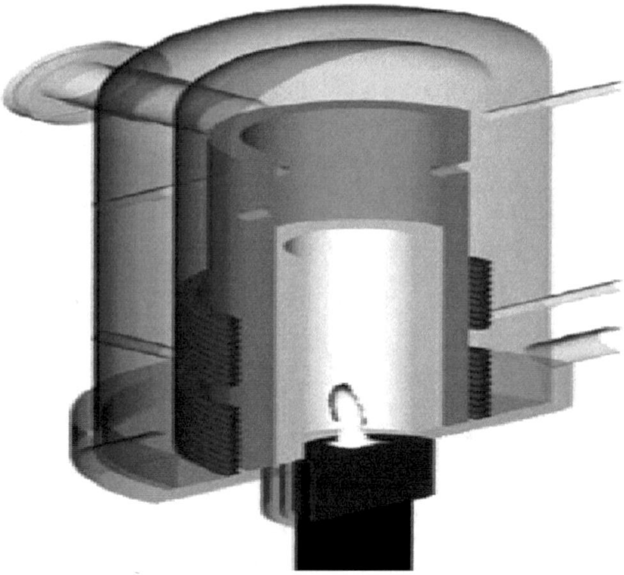

Figure 1: Artistic diagram of a thermal plasma reactor. The plasma torch (A) is shown coupled to the top part of the reactor (B). The coils (C) generating the magnetic field are coaxially inside

2. Aims
To build a thermal plasma reactor with a magnetic confinement system of magnetic mirror type. The reactor is a a non transferred plasma torch to produce nanoparticles, and the process is optimized with the confinement system and the process is optimizes of nanoparticles.

3. Magnetic Confinement Fundamentals
The magnetic field application on the thermal plasma using magnetic confinement will increase the power efficiency of the DC plasma torch. The magnetic fields parallel to a wall reduce wall bombardment of ions and electrons, and therefore reduce impurity contaminations of the plasma atmosphere. The magnetic field furthermore improves uniformity of surface interactions and increases the perpendicularity of ion bombardment on the surface of deposition of nanoparticles and nano tubes.

In this work we are performed experiments with a magnetic mirror confinement configuration, which consist of an axysimmetric magnetic barrier with maximum magnetic field B_{max} in two extremes, and weaker magnetic field B_{min} between extremes in where the plasma is confined.

The fraction F_T of an initially isotropic distribution of particles changes in the uniform field region on the center which remains confined is a function of maximus magnetic field B_{max} and minimum magnetic field B_{min}.

$$F_T = (1 - \frac{B_{min}}{B_{max}})^{1/2} \quad (1)$$

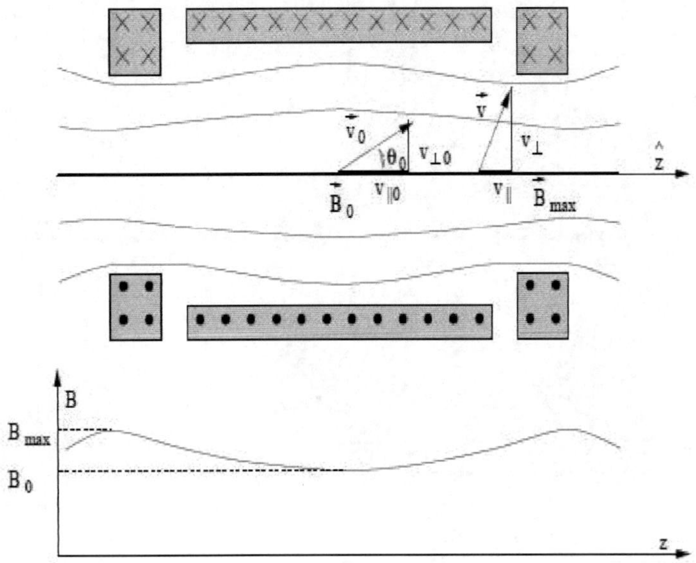

Figure 2: Diagram showing the intensity of the magnetic fields B, as a function of axial distance according to the confinement configuration by magnetic mirror, and its effect on the velocities of the plasma particles

4. Methodology

A confinement system with magnetic mirrors was built which was coupled to the plasma torch is such a way to allow the charge of field intensity B inside the production chamber. To do this a plasma torch was built with autonomous power supply and refrigeration system, as well as a nanoparticles production chamber a confinement system by magnetic mirrors whit variable field intensity B and materials for the nanoparticles production. Once these are produced the nano materials are characterize according to size and other characteristics. First a computational prototype was designed and simulated in agreement with the op experimental operation parameters. Once this was obtained a evaluation of the materials to be used was performed, according to those produced by the local industry taking account of cost and availability. Several simulation were performed in order to get the required parameters. Later, we entrust to national companies the construction of different parts and equipment. Finally the experimental assembly was performed including instrumental and control equipment, measured instruments and refrigeration system. We are now in the process of building the parts of the equipment in order to proceed later to the assembly process and to evaluate the operativity of the experimental of this ensemble. Later we will proceed to couple the thermal plasma reactors and to evaluate the interoperability of the different parts of the systems. Finally, carrying out all the security procedures, we will perform the first experiment with the aim of study the different conditions. After the experiments the study of the results will be performed.

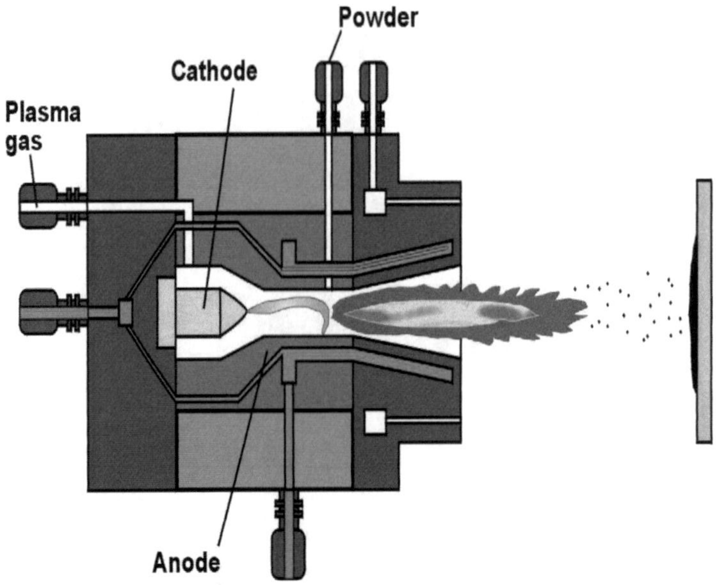

Figure 3: It is shown, schematic representation of DC electrodes torch mouthpiece od the non transferred arch type (A), the refrigeration system (C), the powder injection system (D), the surface for nano particles and nano materials deposition (E), the cathode (F) and the anode (G).

5. Results

The computational model already developed is based for numerical calculations in the finite element method. In this way the temperature profile, velocity density, electric potential and density of the are were obtained for all parameters of the current established. The results obtained in the simulation have been consistent with the theoretical estimations. The experimental phase have not been begin, therefore we do not known if then is agreement between there In the experimental phase we hope to produce nanoparticles, nano and nano recovery with electrodes no-transferred electrodes plasma torch in combination with a confinement system, magnetic mirror type. Our idea is to find the best experimental parameters to increase the efficiency in the production of nanoparticles, nano tubes an nano recoveries. To obtain the best parameters for the introductions of the materials to be processed in liquid state through the use of peristaltic pumps. Referring to the magnetic confinement system, there are two main parameters to be determine during the experimental peocess: average intensity of the magnetic field B, and ratio between the maximum magnetic field Bmax an minimum one, Bmax/Bmin. In reference to the catalyst, there are several options to be consider (Fe), nickel (Ni) and cadmium (Cd).

Figure 4a: It is shown, the results obtained in a previous work realized by the Plasma Physics Group of the University Simon Bolivar in studies on effects of magnetic fields, in this image is shown the results of the nanoparticles formation whithout the effects of magnetic fields.

Figure 4b: Can be seen how the efficiency of nanoparticles productions increases with the magnetic field application.

6. Acknowledgments

Work supported by grant DID-G22, Universidad Simón Bolívar, Caracas, Venezuela, and by great Proyecto ANT/298, Programa MECE2, Universidad de Antofagasta, Chile

Journal of Physics: Conference Series **591** (2015) 012055 doi:10.1088/1742-6596/591/1/012055

7. References

[1] G. Torrente, Estudio numerico y experimental de un reactor de plasma termico asistido por espejos magneticos para la sintesis de ceramicos modificados morfologicamente, Tesis de Doctorado,Universidad Simón Bolívar, Marzo de 2009.

[2] L. Beiras, Modificacion Morfologica a Escala Nanometrica de Materiales Sintetizados con Plasma Termico en Espejos Magneticos, Tesis de Pregrado, Universidad Simón Bolívar, Mayo de 2011.

[3] M.F. Zhukov and I.M. Zasypkin, Thermal Plasma Torches, Design, Characteristics, Applications, Cambridge International Science Publishing, London, 2003.

Journal of Physics: Conference Series **591** (2015) 012056 doi:10.1088/1742-6596/591/1/012056

Contact Glow Discharge Electrolysis in the presence of Organic Waste

Neeraj Sharma, Andres Munoz-Hernandez, Gerardo Diaz and Edbertho Leal-Quiros

School of Engineering, University of California Merced, CA 95343

E-mail: gdiaz@ucmerced.edu

Abstract. The present experimental study was conducted to analyze the interaction between contact glow discharge electrolysis and biomass. For comparison, two sets of tests were conducted, first in the absence of biomass and second in the presence of biomass. It was observed that when limited electrolyte was present in the liquid, the process of normal electrolysis did not transition to contact glow discharge electrolysis. Addition of biomass to the liquid, increased the overall concentration of electrolyte and the process transitioned to contact glow discharge electrolysis after a critical current density was reached near the electrode with smaller surface area. Further, it was observed that most of the energy was consumed in generating steam instead of performing gasification of biomass.

1. Introduction

The field of waste-to-energy conversion using thermal plasma is growing rapidly. A review of current technologies along with performance comparisons of different waste gasification processes based on thermal plasma is presented by Fabry et al. [1].

One of the many useful products of such gasification process is syngas, a mixture of carbon monoxide and hydrogen along with small amounts of other gases. Rutberg et al. [2] evaluated high temperature plasma gasification of wood for the production of syngas. They reported energy conversion ratio of 46.2% approximately, from primary energy of wood with moisture content of 20% (LHV 13.9 $MJkg^{-1}$) into electric.

Huang & Tang [3] conducted thermal plasma pyrolysis of different kinds of organic wastes, varying from plastic and used tire to agricultural residue and medical waste.

Recently Sharma et al. [4] studied contact glow discharge electrolysis as an efficient means to generate steam from liquid waste. In order to prevent the formation of an electric arc between the electrodes, a non-conductive material (gravel) was introduced. In the present study contact glow discharge electrolysis [5] was generated in electrolyte with and without biomass to analyze the feasibility of gasification of biomass with this process.

2. Experimental setup

Experimental setup consisted of a stainless steel cylinder acting as anode and a partially immersed graphite rod acting as cathode as shown in Fig. 1. Tap water with and without biomass was used as electrolyte for the two different experiments.

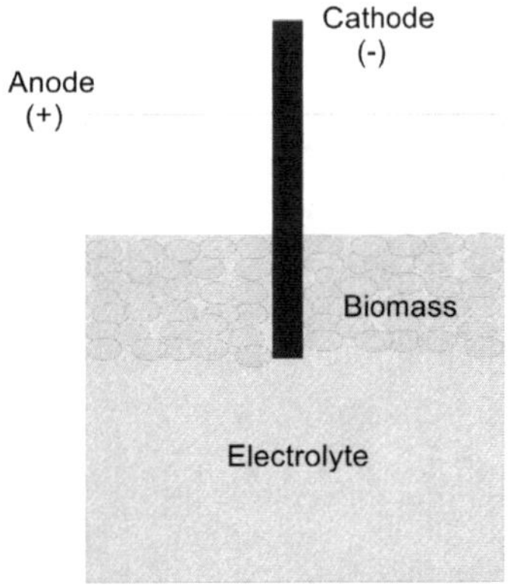

Figure 1. Schematic of the electrolytic cell.

Power was supplied by an ESP-150 DC power supply manufactured by ESAB with control console designed by Foret Plasma Labs. The power supply had a current range between 0 and 150 Amperes and a maximum voltage of 380 Volts. Current output was controlled by a potentiometer knob on the control panel. A National Instruments^{TM} data acquisition system was used along with NI LabVIEW^{TM} to record current, voltage, and temperature with a sampling rate of two seconds. Current values were measured with a Veris direct current sensor model H970HCA with a maximum current range of 200 A and ±3% of accuracy at full scale. Voltages were read directly from a remote control voltage pin on the power supply.

3. Experimental procedure

The stainless steel container acting as anode was filled with tap water until the electrode was submerged to about 5 cm. No biomass was added in the first case. The potentiometer knob at the power supply panel was set to quarter the scale and it remained fixed during the experiment. The potentiometer limited the amount of current passing between the electrodes but the power supply allowed variation of voltages and currents depending on the physical parameters of the process. As no electrolyte was added to the tap water, the amount of ions was relatively low. The cell went into normal electrolysis mode after the power was supplied. Because of Joule heating, water started to evaporate and the effective submerged area of cathode decreased due to the lower liquid level in the container. Because of limited amount of ions, the current density near the cathode never crossed the threshold and the process remained in normal electrolysis mode until the cathode lost contact with water.

In the second case, a measured amount of biomass (dry almond hulls) was added to a fresh refill of tap water to the same level as in the previous case. The biomass added electrolyte to water which was close to room temperature in the beginning of the experiment. The electrolytic cell started in normal electrolysis mode and the electrolyte began to get heated due to Joule heating resulting in a higher rate of evaporation that lowered the electrolyte level, thereby, decreasing the submerged length of the cathode and further increasing the electrical current density near the cathode. At a critical value of the current density and when the electrolyte temperature near the cathode approached its boiling point, normal electrolysis in the cell transitioned to contact glow discharge electrolysis.

4. Results and discussion

4.1. In absence of biomass

At the start of the experiment, tap water was at room temperature. The amount of electrolyte present in tap water was limited since no extra electrolyte was added. As the experiment progressed, the temperature of the tap water increased due to Joule heating. The electrical conductivity of electrolyte increased with temperature [6] and resulted in an increase in the current as shown in Fig. 2 (a). The voltage across the electrode stayed almost constant at 350V as shown in Fig. 2 (b).

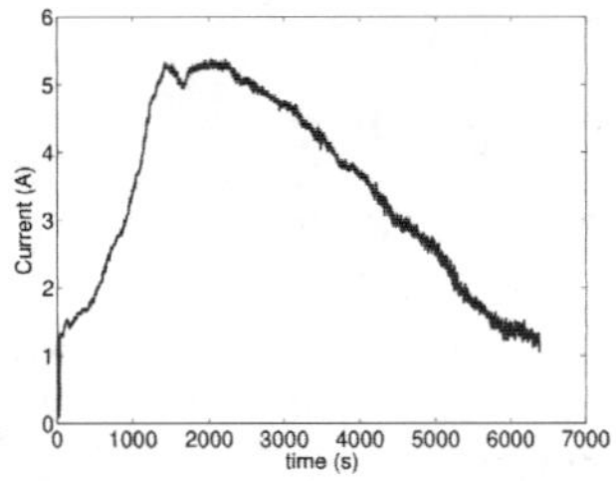

(a) Current

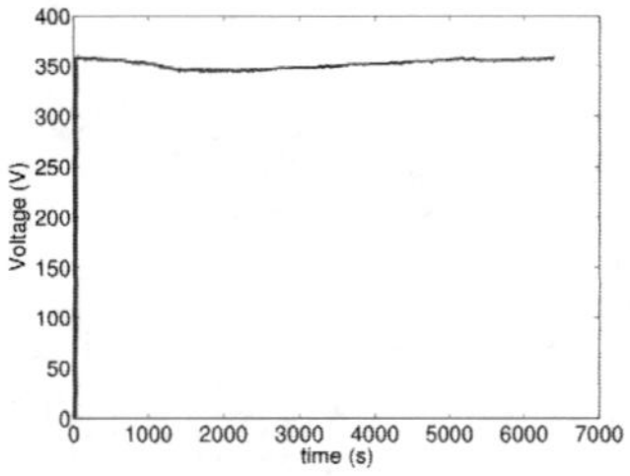

(b) Voltage

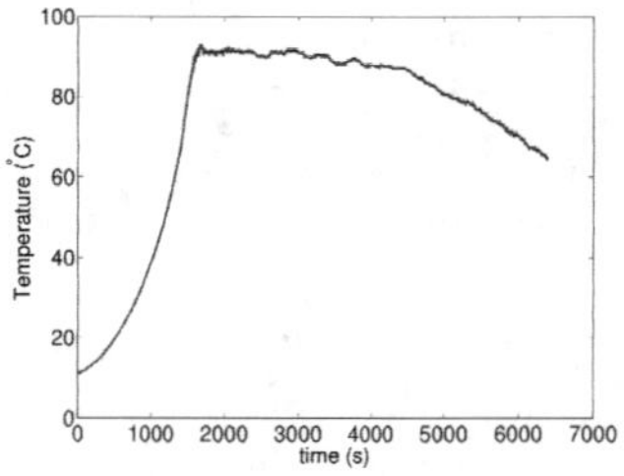

(c) Temperature

Figure 2. Variation of current, voltage, and temperature with time in the absence of biomass.

Joule heating increased the temperature of water until it reached a value near the boiling point of the water as shown in Fig. 2 (c). Current followed the variation in temperature closely and reached a maximum value when temperature reached its maximum. As the length of submerged cathode continued to decrease and the amount of electrolyte present in water was limited, this resulted in further decrease in current until the cathode lost contact with water. With the decrease in current, Joule heating decreased which is reflected in the decrease in temperature after 4500 secs into the test.

4.2. In presence of biomass

Dried almond hulls were added to a fresh refill of tap water to the same level as before. At the start of the experiment, electrolyte was at room temperature. As the experiment progressed, the temperature of the electrolyte started to increase due to Joule heating and biomass added extra electrolyte to the tap water, the electrical conductivity of electrolyte was higher than in the previous case which resulted in overall higher current levels. Electrical conductivity also increased with temperature and resulted in an increase in the current and corresponding decrease in voltage between the electrodes as shown in Figs. 3 (a) and (b).

As the temperature of electrolyte increased, the rate of evaporation increased causing the length of submerged electrode to decrease. This resulted in an increase of electrical current density at the cathode as sufficient electrolyte was present. The normal electrolysis process ceased and contact glow discharge electrolysis was established at a current value of 21 A and when the temperature of the electrolyte near the cathode reached its boiling point as shown in Figs. 3 (c). Formation of vapor film around the cathode increased overall resistance of the cell and resulted in the characteristic drop in current at the transition. No sign of biomass gasification was observed during the glow discharge process but large amounts of steam was generated.

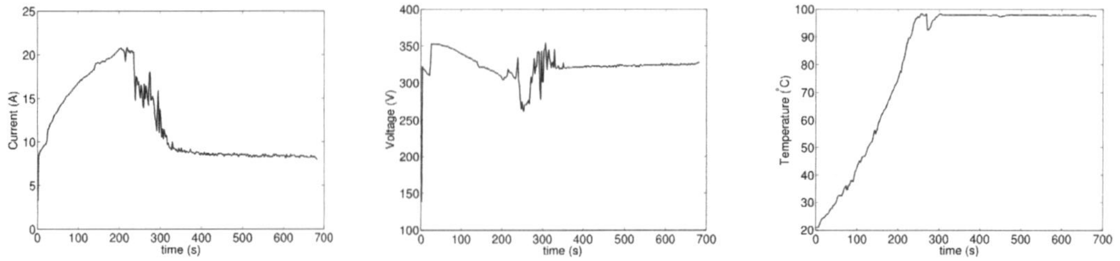

(a) Current (b) Voltage (c) Temperature

Figure 3. Variation of current, voltage, and temperature with time in the presence of biomass.

5. Conclusions

The experiment was designed to test the feasibility of carrying out gasification using contact glow discharge electrolysis in the presence of biomass. Two different cases were tested, with and without biomass. In the absence of biomass, the electrolysis never transitioned to glow discharge because of limited presence of electrolyte. Biomass added electrolyte to tap water and resulted in transition from normal to glow discharge electrolysis. It was observed that most of the heat was absorbed by water to produce steam. Even though biomass was present close to the high temperature, the experiment did not result in any appreciable gasification. In conclusion, the energy dissipated at the cathode was absorbed by the surrounding liquid which was converted into steam. A significantly lower amount of liquid would be needed in order to produce the gasification of the biomass.

Acknowledgments

Acknowledgments The authors would like to acknowledge the support by the California Energy Commission, contract number PIR-08-036 for the development of this work.

References

[1] Fabry F, Rehmet C, Rohani V and Fulcheri L 2013 *Waste and Biomass Valorization* **4** 421
[2] Rutberg Ph G, Bratsev A N, Kuznetsov V A, Popov V E, Ufimtsev A A and Shtengel S V 2011 *Biomass and Bioenergy.* **35** 495
[3] Huang H and Tang L 2007 *Energy Conversion and Management* **48** 1331
[4] Sharma N, Diaz G and Leal-Quiros E 2013 *Electrochimica Acta* **108** 330
[5] Hickling A Ingram M D 1964 *Trans. Faraday Soc.* **60** 783
[6] Sharma N, Diaz G and Leal-Quiros E 2013 *International Journal of Thermal Sciences* **68** 119

Design of a Prototype of Water Purification by Plasma Technology as the Foundation for an Industrial Wastewater Plant

L Barillas

Plasma Laboratory for Fusion Energy and Applications, Instituto Tecnológico de Costa Rica, Central Campus, Cartago, Costa Rica; P.O. Box 159-7050 Cartago, Costa Rica.

laura.plasma.tec@gmail.com

Abstract. In order to mitigate the contamination of water sources due to the spill of sewage without any kind of treatment, mainly generated by the industrial sector; a prototype of water purification by plasma technology has been designed. The prototype will transform liquid water into plasma to eliminate the pathogens from the water, due to their exposure to ultraviolet radiation, electric fields and shock waves, which aid in the destruction of pollutants. The sewage will be accelerated at high speed to convert it into a liquid-gas mixture in order to transform it into plasma, which is achieved when the electrical discharge (of the type dielectric barrier discharge or DBD) is applied to the water by means of high voltage electrodes, from a source of alternating current (AC). Subsequently, the mixture slows down to be return into liquid phase and obtain clean water, all of these without a significantly rise of temperature. The device also has an automatic power control system. Finally, a short feasibility study was conducted in order to use this type of water cleaner in the future as a basis for a treatment plant of industrial waste water, so it comes to replace the current secondary and tertiary treatments used among the industry. It is intended that this new system will be more efficient and cheaper than the current waste water treatments.

1. Motivation

Water availability for human consumption is getting diminished every day. Despite de effort to improve water quality in some regions of the world, the pollution keeps growing; and unfortunately more than 80% of sewage in developing countries is discharged untreated, polluting rivers, lakes and coastal areas [1]. Once these water sources are contaminated, their recovery is almost null, adding that it is estimated that less than 20% of the world's drainage basins exhibit nearly pristine water quality [1].

Based on the statement above, there is a motivation for creating a prototype for water purification by plasma technology as the foundation for an industrial wastewater plant, coming from the industry's need of treating its used waters, which is regulated by law in many countries; in order to protect the environment, the health and common wellbeing. As seen before, wastewater spill without any kind of treatment threatens water sources worldwide, which can affect the human development, but hopefully with the device the treated water will count with such quality that can be reused in other process or even for human consumption.

2. Background on Industrial Wastewater Treatments

Industrial wastewater treatment typically consists of three phases or treatments. The primary treatments are mainly used to eliminate suspended matter, by means of mechanical operations and using chemical additives. The secondary treatments are biological processes, in which depuration of biodegradable organic matter from the wastewater is done by the action of microorganism. Tertiary treatments –which are usually expensive– eliminate the contaminants that last after primary and secondary treatments, such as the dissolved matter. Some of these treatments are stripping, precipitation, reverse osmosis and disinfection by advanced oxidation processes (AOP), which can use combinations of hydrogen peroxide, ultraviolet radiation, photocatalysis and ozone (O_3); focusing on the last one for this paper due to its oxidation potential.

Some of the many advantages of ozone oxidation (ozonation) compare to traditional methods like chlorination are [2]:

- Effective disinfectant without secondary effects.
- No need to store or handle chemicals.
- Ozonation sub products with no adverse effects over health and environment.
- The ozone can safely destroy a vast range of organic pollutants.
- Much more effective killing bacteria, viruses and spores, among others.
- The ozone helps in the removal of color, odor and suspended solid matter.

Based on the analysis of the wastewater treatments commonly used in the industry, an alternative treatment system is proposed with the usage of plasma.

3. Water Purification with Electric Discharges

One way to generate ozone for water purification is by applying an electric discharge on the liquid surface, in such way that it will produce active chemical species (such as OH•, H•, O•, O_3, H_2O_2) [2] capable of disinfecting the water.

The intensity of the electrical fields needed for the electrical discharges in water is lethal for a great variety of microorganisms, and show a synergistic lethal effect when combined with conventional disinfectants such as O_3 and H_2O_2. The electrical discharges in water may also produce ultraviolet (UV) radiation and shock waves, which help in the destruction of pollutants [2].

In the case of water purification by electrical discharges, there are basically three types: Contact Glow Discharge Electrolysis, Dielectric Barrier Discharges (DBD) and Pulsed Corona Discharges. Table 1 shows a summary of the characteristics of the discharges.

Table 1. Summary Table of Electrical Discharges in Water. (Barillas L), ([2])

Discharge type	Electrode characteristics	Voltage	Energy	Other characteristics
Glow Discharge Electrolysis	Thin wire	DC ~ 0.5 kV	100 eV max.	Equilibrium plasma. Anode and cathode can be isolated by a porous glass. Requires cooling system.
Dielectric Barrier Discharges (DBD)	Generally are parallel plates	AC ~ 15 kV	1 – 10 eV	Non-equilibrium plasma. At least one of the electrodes is covered by a dielectric material. No need for cooling system.

Pulsed Corona Discharges	Configurations needle-plate, cylinder-wire (coaxial)	DC ~ 15-100 kV	~ 5 eV	Non-equilibrium plasma. Electrodes can be covered by a thin ceramic layer. Could require cooling system.

4. Design of the prototype

A DBD reactor type was chosen as the model to design, since it gives certain advantages for this case of study (building a prototype later on). The main advantage is that produces a non-thermal plasma that does not rise the temperature significantly; thus, there is no need for a cooling system, making the prototype cheaper and simpler to construct.

First of all, it is considered that the wastewater has gone through previous primary treatments, and that it does not contain solids nor major colloidal suspensions; for this reason, the prototype is considered as a tertiary treatment.

The prototype reactor, showed in **Figure 1**, will transform liquid water into plasma in order to kill the water's pathogens and pollutants, due to their exposure to UV radiation, electrical fields and shock waves. The device will feature a pump and a nozzle which accelerates the sewage to high speed to convert it into a liquid-gas mixture in order to transform it into plasma, which is obtained at the moment of the electrical discharge is applied into the water by means of high voltage electrodes from an AC power supply. Later, the mixture must be decelerated to return it into liquid state and obtain clean water, without a significant temperature rise due to the low pressure in the reactor. Figure 2 shows a general diagram of the prototype.

The main design parameters of the reactor are:

- **Reactor:** Coaxial geometry, copper electrodes, 490 mm long by 120 mm diameter. Outer electrode is hollowed and inner electrode is a solid copper bar. Borosilicate glass (Pyrex®) as dielectric barrier of 2.4 mm thickness.
- **Power Supply:** AC, 120 Vac/60 Hz, 20 kV p-p, 20-70 kHz
- **Nozzle:** Flowrate 10 l/min @ 50 psi
- **Pipelines:** PVC, Ø 1'' SCH 40
- **Pump:** 1 hp, single-phase, 120 Vac/60 Hz; Q = 10-50 l/min, $H_{máx}$ = 52 m
- **Tank:** 48 liters, fiberglass material
- **Control System:** made with relays 120 Vac/10 A
- **Others:** Variable Frequency Drive 0-500 Hz, 115 V, 1 hp. Also a Ø 1'' solenoid valve

The special features of the dynamic (constant flow) coaxial reactor design and the incorporation of a nozzle to create a liquid-gas mixture give a competitive advantage over other DBD reactors based on parallel plates geometries, which is that it provides a greater contact area between the discharge and wastewater (a surround interaction), while the traditional mentioned geometry only interacts with the static superficial layer of the wastewater.

There are especial considerations regarding the power supply, since it was not possible to find a model to characterize it, due to the geometrical configuration of the electrodes (coaxial). Next section will discuss this matter.

4.1. Power Supply Characterization

In order to characterize the power supply needed, a plates or discs electrodes reactor model published in "Characteristics of Dielectric Barrier Discharge Reactor for Material Treatment" by Kostov[3][1] was taken as the base for the current approach.

[1] Model by Kostov (2009) is not fully explained in this paper, for further details please refer to *"Characteristics of Dielectric Barrier Discharge Reactor for Material Treatment"* [3] paper.

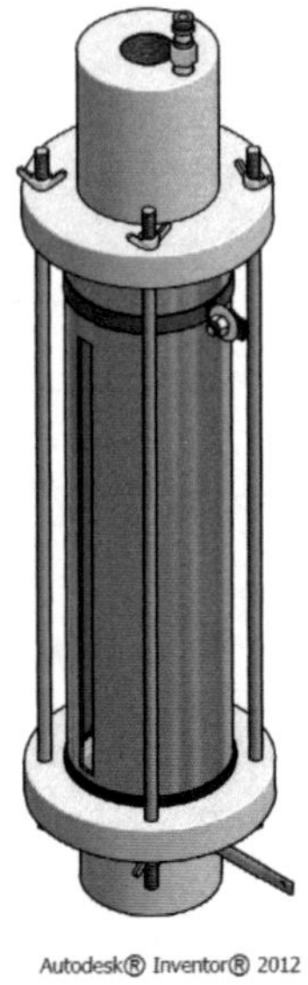

Figure 1. Proposed prototype reactor for water purification (Barillas L)

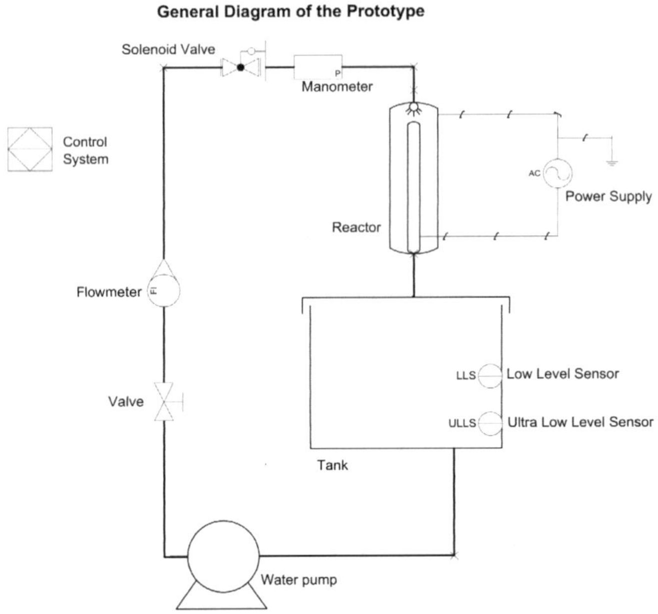

Figure 2. General diagram of the prototype for wastewater purification. (Barillas L)

This model has an equivalent electrical circuit for DBD reactor with parallel plates with two layers of dielectric over each metallic electrode, as shown in Figure 3.

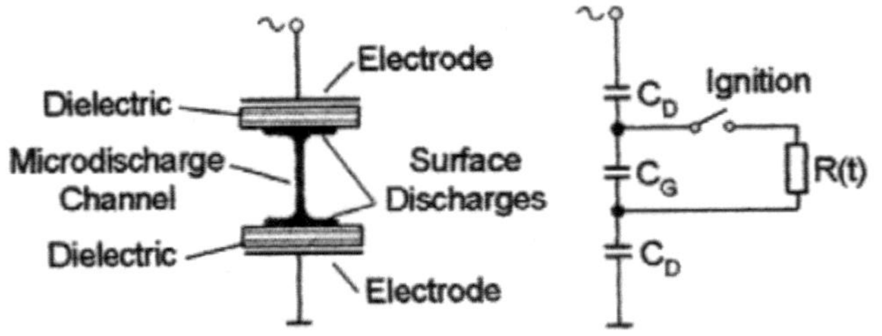

Figure 3. Layout of a parallel plates DBD reactor and its equivalent circuit. ([3])

After a series of equations and analysis of this circuit[1], and assuming that DBD discharge is constituted by various micro-discharges in parallel the plasma resistance can be approximately expressed as $R \cong x\rho/A_{EF}$, where ϱ is the resistivity of a discharge channel, A_{EF} is the total cross section occupied by the micro-discharges, C_D is the capacitance of the dielectric layer and C_G is the capacitance of the discharge gap [3]; then the mean reactor power can be expressed in term of a discharge gap x:

$$\bar{P} = \frac{V^2}{2}\frac{A_{EF}}{x\rho}\left[\left(1 + 4\frac{C_G}{C_D} + 4\frac{C_G^2}{C_D^2}\right) + \left(\frac{4}{\omega^2 C_D^2}\frac{A_{EF}^2}{x^2\rho^2}\right)\right]^{-1}$$

Equation 1

In order to adapt this model to the coaxial geometry reactor, as well as simplify calculations, the cylindrical electrodes will be taken as two sheets, by extending the inner surface of the hollowed electrode (outer electrode), and the surface of the solid electrode. This will result in different areas (A), affecting the term C_D, as $C_{D1} = \frac{k\varepsilon_0 A_1}{d_1}$ and $C_{D2} = \frac{k\varepsilon_0 A_2}{d_2}$; where k is the dielectric constant of the material, d is the thickness of each dielectric barrier, A_1 is the inner area of the cylindrical hollow electrode and A_2 is the area of the surface of the inner electrode.

Thus, Equation 1 must be modified as:

$$\bar{P} = \frac{V^2}{2}\frac{A_{EF}}{x\rho}\left[\left(1 + 4\frac{C_G}{C_{D1}} + 4\frac{C_G^2}{C_{D1}C_{D2}}\right) + \left(\frac{4}{\omega^2 C_{D1}C_{D2}}\frac{A_{EF}^2}{x^2\rho^2}\right)\right]^{-1}$$

Equation 2

With this approach, by having electrodes with different areas, then A_{EF} can be arranged as a trapeze, as shown in Figure 4, indicated in red:

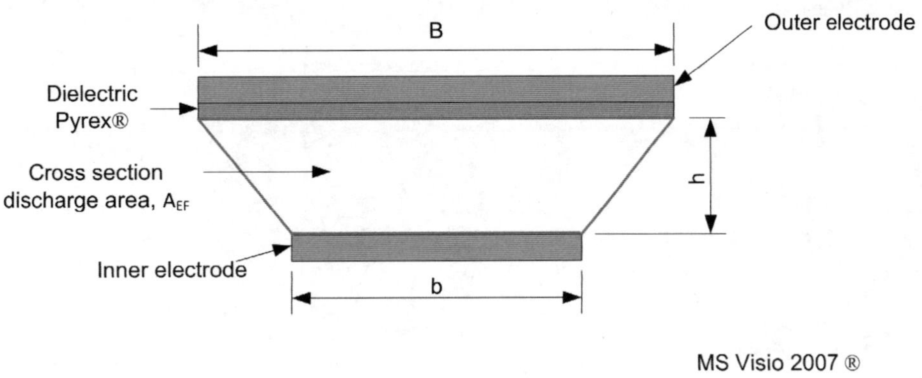

Figure 4. Diagram of the cross section occupied by the micro-discharges, assuming the cylindrical electrodes as sheets to adapt them into the parallel plates electrodes model. (Barillas L)

With this approach, is was possible to build a datasheet in order to determine the voltage (from 1 kV to 20 kV) and frequency (20 kHz to 50 kHz) as functions of power (W). The behaviour of these variables is shown in Graph 1.

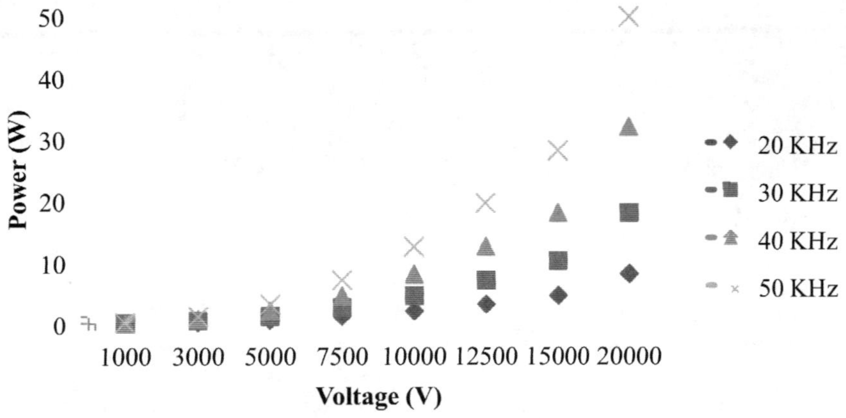

Graph 1. Voltage and frequency as functions of mean power. (Barillas L)

Journal of Physics: Conference Series **591** (2015) 012057 doi:10.1088/1742-6596/591/1/012057

Also, simulations were computed in the AC/DC® Module of COMSOL Multiphysics® Version 4.1, with the electrodes coaxial geometry, in order to determine the optimal values for the power supply. Figure 5 shows the electric potential at different frequencies for the power supply, in a cross section view, around the middle of the reactor length.

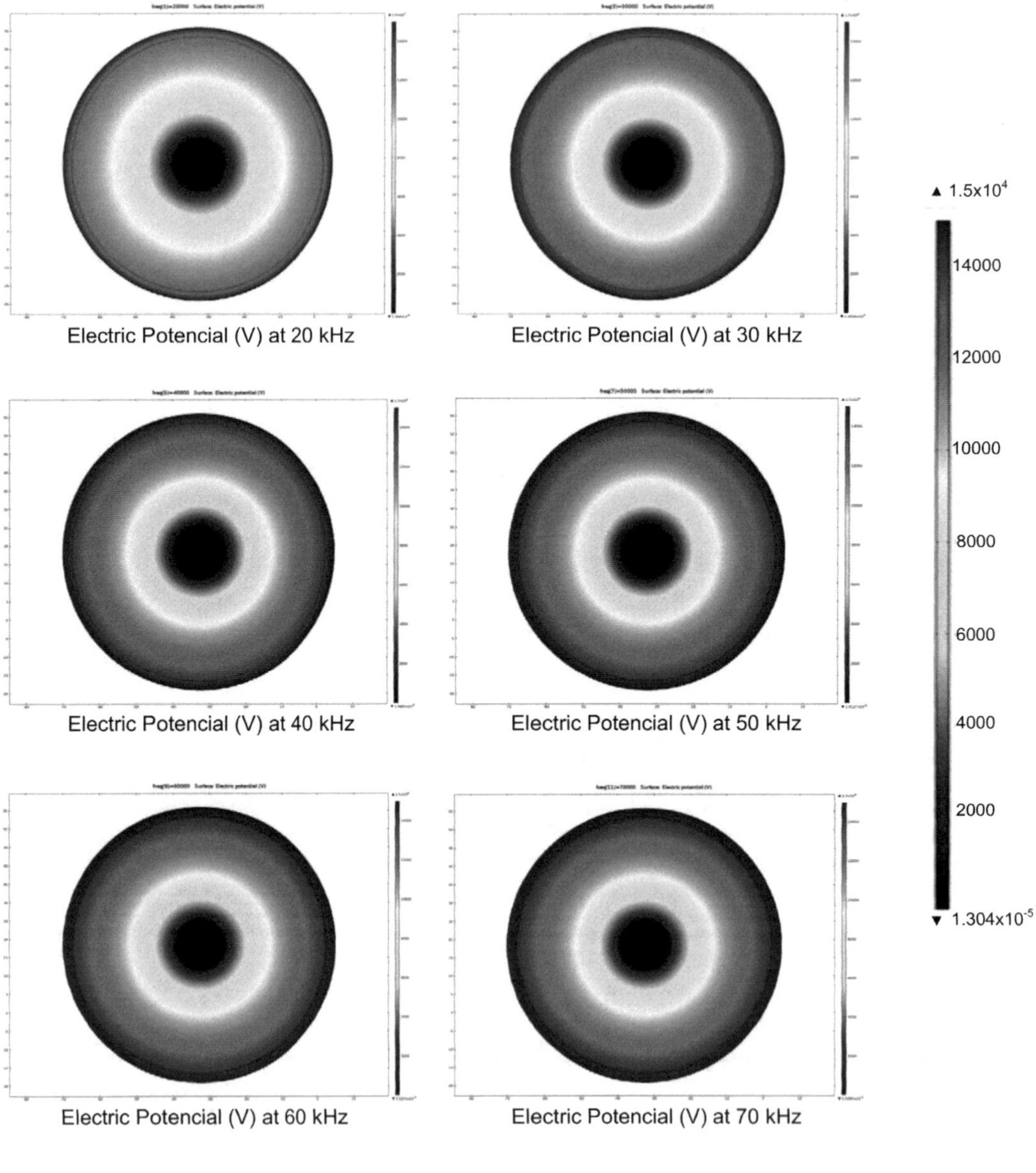

Figure 5. Electric potential at different frequencies for the power supply (20 kHz to 70 kHz), at 15 kV, in a cross section view, around the middle of the reactor length. (Barillas L)

Data from the simulations shows that there is an increase in the intensity of the electric potential towards the centre of the geometry; however, from 50 kHz to 70 kHz there is no significant variation, differently as seen from 20 kHz to 50 kHz. Among other data obtained, in conclusion, it was clear that one can obtain a higher electrical field in the reactor's ionization area from high frequencies and voltages from the AC power supply. After this analysis, a variable power supply (output from 0 to 20 kV peak-peak, 20 kHz to 70 kHz) was chosen with the intention of let the researchers experience with different parameters, in order to find the optimal efficiency parameters.

It is important to mention that many results will only be known in an experimental way, this means, many parameters could change once the prototype is built. Also, the prototype has been design in such way that it could be escalated depending on the demand needs; e.g. if 100 l/min need to be treated, then a 10 coaxial reactor matrix would be set up, since each of them purifies 10 l/min.

5. The Prototype as the Foundation for an Industrial Wastewater Plant.

As explained in section 1, the industrial market shows an obvious need for treating wastewaters, first due to law requirements, and second due to water sources preservation.

The intended use of this plasma water purifier, is to replace a wastewater tertiary treatment, but depending on the effluent quality from the primary treatment, it could even replace the secondary phase in a wastewater plant. Current secondary and tertiary wastewater treatments are expensive and require large installation areas, as well as high maintenance. For example, for activated sludge, great land areas and high cellular retention time are needed, which makes the water take several days to be treated (1 month average).

With the proposed project, the installation areas would be smaller, in such way that the reactor matrix could fit in a machinery (engine) room; and it is pretended that the water is purified almost immediately when passing through the reactors (1 cycle), but even if more cycles are needed, it would only take one day (this is not clear yet due to lack of experimental data).

As an example to characterize the project, it was compared to a traditional wastewater plant that treats 200 m^3/day, with a maximum demand of 277.2 l/min. The primary treatment would remain the same (grills, a grit chamber and a fat and grease trap), and then, the designed device would take place. In this case, for the required demand, a 28 reactor matrix would be needed (it must be remember that each reactor has the capacity of 10 l/min, which is determined by the nozzle flow rate). Figure 6 shows a diagram of the proposed location in a wastewater plant for the designed device.

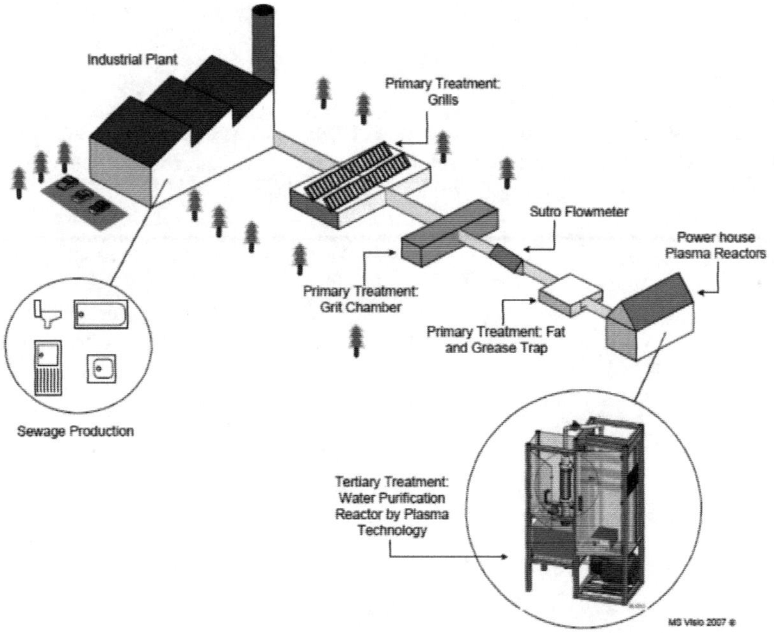

Figure 6. Diagram of the proposed location in a wastewater plant for the designed device. (Barillas L)

Regarding costs, Table 2 shows the possible capitalization costs for the wastewater plant project using the designed water purification reactor by plasma technology.

Table 2. Possible Capitalization Costs for a Plasma Wastewater Plant. (Barillas L)

Specification	Cost (USD)
Prototype reactor	$3,500.00
Traditional activated sludge wastewater plant, with no primary treatment	$160,000.00
Plasma Wastewater Plant, including primary treatment	$130,000.00
Savings	$30.000,00 (20% less)

Summarizing, the competitive advantages of this treatment over traditional treatments, are that it is more effective, efficient, faster, compacter and more economic; aside from the fact that it will not use chemicals, which will make it also safer for people and environment.

6. Conclusions

As mentioned, industries must treat the wastewater in order to ensure the protection of the environment, the health and common wellbeing. It is possible that, due to the high cost and maintenance of traditional treatment plants, people find it difficult to properly treat the used waters. Although experimental data is still pending and some parameters will only be definitive until the device is built and tested, it is intended that with the proposed device, cost, maintenance and space could decrease significantly; which could lead industries (especially from developing countries) to re-consider discharge untreated sewage.

Finally, the ideal state is to offer a whole new cost-improvement, effective, efficient and innovative concept on wastewater plants, also including the primary treatment; for example, by using plasma-synthetized nanostructures that could segregate and retain certain organic/inorganic components from the water, and then, eliminate biodegradable organic matter (like microorganisms, bacteria, virus, etc.) by means of the proposed reactor; in order to obtain a high quality water effluent that can be used in other processes.

Acknowledgment

The author acknowledges the overall help and support from Ivan Vargas, PhD., and Esteban Zamora for his help on the COMSOL Multiphysics® simulations; both from the Plasma Laboratory for Fusion Energy and Applications of the Costa Rica Institute of Technology.

References

[1] World Water Assessment Programme 2009. *The United Nations World Water Development Report 3: Water in a Changing World*. (Paris: UNESCO, and London: Earthscan)
[2] Malik M A, Ghaffar A and Malik S A 2001 *Vols. Plasma Sources Sci. Technol.* **10** 82–91
[3] Kostov K G, Honda R Y, Alves L M S and Kayama M E 2009 *Brazilian J. Phys.* **39** 322–25

From Inert Carbon Dioxide to Fuel Methanol by Activation in Plasma Atmosphere

Alejandro Suarez[1], **Edbertho Leal-Quiros**[1] and **Jorge Gonzalez**[1]
[1] Polytechnic University of Puerto Rico

E-mail: asuarez@pupr.edu; equiros@csufresno.edu

Abstract. The electron-molecules collisions in plasmas are fructiferous field of study, particularly in activation of inert species into extremely active chemical reactants. Several gases such as carbon dioxide, methane, nitrogen, argon, water among others are ionized or simply activated under collisions with electrons creating the complex atmosphere plasmas. Based upon the energy distribution of electrons and the cross section of molecules it is possible to stimulate molecules becoming highly active species. This research explores the assembly of methanol over a surface of Cu-O-Zr catalyst when carbon dioxide reacts with water under the plasma developed into the Electron Cyclotron Resonance device (ECR). The process was continuously monitored by a mass spectrometer and the correspondent results show the formation of methanol. The ECR operated at 10^{-6} torr assisted with a microwave source of 250 Watts of power; the magnetic field was developed with 300 Amp on Helmholtz coils. The mode of operation of plasma was in cusp, and the electron temperature was 6-eV reported by a single Langmuir probe. The mass spectrometry reported the presence of methanol. The catalysts characterization is reported in form of Scanning Electron Microscopy image, a Raman spectroscopy analysis, and the Electron Dispersive Spectroscopy (EDS).

1. Introduction

The incorporation of physical methods to processes is becoming of great importance and creates positive impact in materials and processing technology. Plasma is incorporated easily in chemistry producing sophisticated molecules. The advantage of the plasma processes is to tune finely the budget of energy to activate chemical reactions and produce valuable materials. In this job research the cross sections of carbon dioxide and water molecules with the distribution of energy of the electrons in the plasma become the reference data to stimulate both molecules; several produced radicals adsorbed over proper catalyst produce methanol. This study is of extremely importance since hydrogen atoms coming from water are being incorporated into carbon structure to produce a chemical stable material. It shows that the hydrogen radicals and hydroxyl radicals react in plasma with carbon dioxide consuming completely the water. Studies of mechanisms of reactions producing methanol reveals in a first step the formation of intermediates formate in which two oxygen atoms from carbon are attached to the zirconium metal atoms [1], [2]. The other steps of assembling are not known since it occurs extremely fast. The studying of methanation of carbon dioxide shows to be feasible since its energy of activation is around 1.0 eV, when hydrogen and carbon dioxide react over the surface of Ni (100) [3]. Similar works using ruthenium catalyst found equivalent activation energy [4]. This energy of activation is relatively low if it is compared with the 6 eV transported by electrons in typical ECR machines. The production of methanol is feasible at relatively low energy of activation since methanol has similar structure as methane. The methanol production results more attractive than methane, because methanol at room conditions is liquid and large amount of hydrogen incorporated in carbon

Journal of Physics: Conference Series **591** (2015) 012058 doi:10.1088/1742-6596/591/1/012058

can be transported in a safe manner. This present paper demonstrated the feasibility of methanol production starting with carbon dioxide, water and a proper catalyst when the whole system is confined into a carbon dioxide-plasma. Improved plasma reactor could be operated at pilot level, creating high density plasma for large production of methanol.

2. Energy Distribution and mechanisms of the Chemical reaction

Carbon dioxide and water as polyatomic molecules are active in plasmas depending on their electron-collision excitation cross section, their electrons in these molecules are promoted to high energy electronic states, and even electrons could leave the structures forming ions. The Maxwellian energy of distribution of the electrons in the plasma are good reference to predict the probability of interaction between these electrons and polyatomic molecules. The cross section of both molecules has to overlap finding the threshold energy of activation, as shown below [5], [6].

$$\mathbf{e + CO_2 \longrightarrow CO + O\,(^1s)} \qquad \varepsilon_{threshold} = 3.3eV \quad (1)$$

$$\mathbf{e + H_2O \longrightarrow OH + H} \qquad \varepsilon_{threshold} = 5.1eV \quad (2)$$

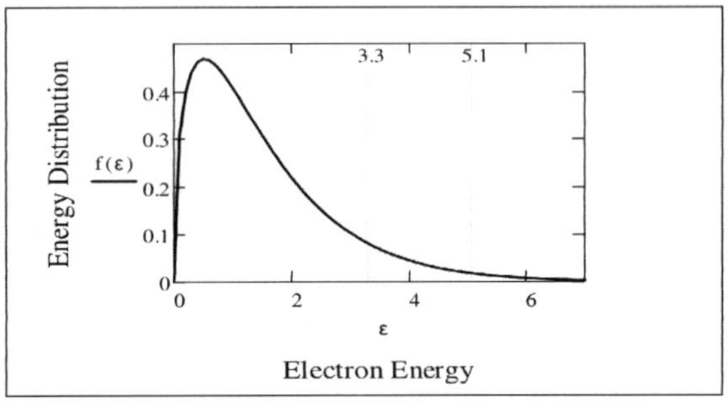

Figure 1. The minimum threshold energy for carbon dioxide and water is 3.3eV and 5.1 eV to interact with electrons in plasma where the typical temperature is about few eV.

The cross sections, of elastic and inelastic collisions are continuously reviewed and updated [1], [2]; for this process the minimum threshold energy related to their cross sections are 3.3 and 5.3 eV. The possible mechanisms of the reaction could be explained as shown below; however the net reaction produces methanol.

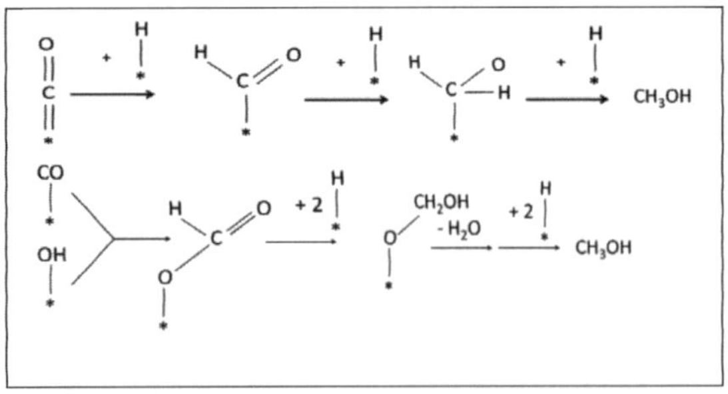

Figure 2. Possible mechanisms of the chemical reaction, the symbol (*) is a possible interaction of species with the solid surface Cu-O-Zr.

3. Methodology

The ECR equipment was used to perform the production of methanol, the equipment reaches a pressure vacuum of 10^{-6} torr. The PUPR Plasma-Cusp Machine required a minimum magnetic field of 0.0875 Tesla for the cyclotron resonance of 2.4 GHz Microwave frequency. Plasma is generated using Electron Cyclotron Resonance Heating (ECRH). This machine could be tuning to generate plasma densities between 10^4 to 10^{12} electrons/cm^3 and electron Temperatures from 0.2eV to 150 eV. For this experiment a microwave power of 250 Watts and 300 Amp of current in the Helmholtz coils that generated the magnetic confinement were used. An electron temperature of 6 eV was measured using a single Langmuir probe. A DME100MS-Dycor Mass Spectrometer is installed in line to monitor the plasma composition. A feedthrough was used to feed water in form of pulses. For each pulse of water a mass spectrum was recorded.

Figure 3. Electron Cyclotron Resonance Plasma Machine at PUPR Polytechnic University of Puerto Rico Plasma Laboratory.[7]

3.1. *Synthesis of the catalyst*

Copper was forced to melt immersed into zirconia, the species were loaded into an alumina tube and the crucible installed into a vacuum chamber was processed at 10^{-3} torr with a stream of hydrogen flowing inside the camber. After melting the material was cleaned with isopropyl alcohol and characterized using Scanning Electron Microscopy (SEM). Figure 4(a) shows the image of Cu-O-Zr by SEM and is clear the uniformly distribution of zirconia spheres over the surface of copper. The testing of EDS shows the presence of Zr and Cu elements shown in Figure 4(b). The Raman spectroscopy indicates the copper oxide formation. Since the atmosphere of the vacuum was hydrogen gas only, then zirconia provided the oxygen to form copper oxide as shown in Figure 5(a) and (b). The catalyst was installed inside the ECR to perform the experiments.

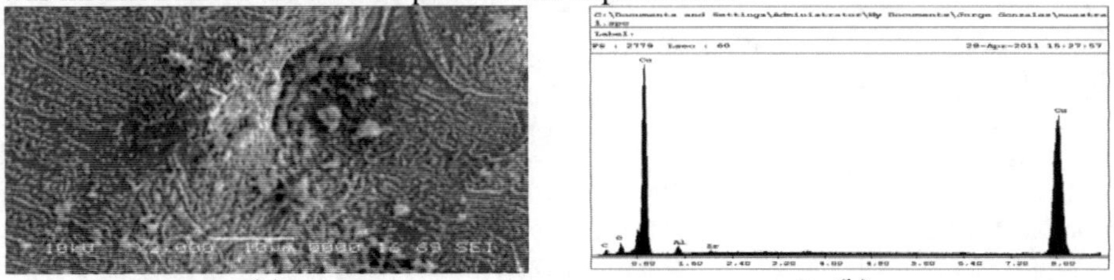

(a) (b)

Figure 4. (a) Uniformly distribution of zirconia over the copper surface and (b) the EDS analysis, showing the presence of Cu and Zr

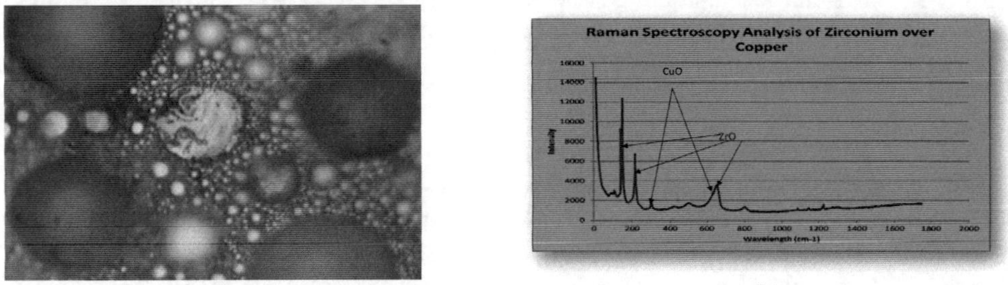

(a) (b)

Figure 5. (a) Optical Microscopy showing the image of the Cu-O-Zr catalyst × 100 and (b) The Raman spectrum shows the presence of copper oxide

4. Experimental Results

The ECR equipment under controlled parameter provided the carbon dioxide plasma in cusp mode for each experiment. The mass spectrums of the plasma composition were reported after each pulse of water and are reported in Figure 6; methanol is formed only under the presence of water. After consuming the water, carbon monoxide is produced into the ECR device. No presence of hydrocarbons was reported in pure carbon dioxide plasma as shown the spectra in Figure 7.

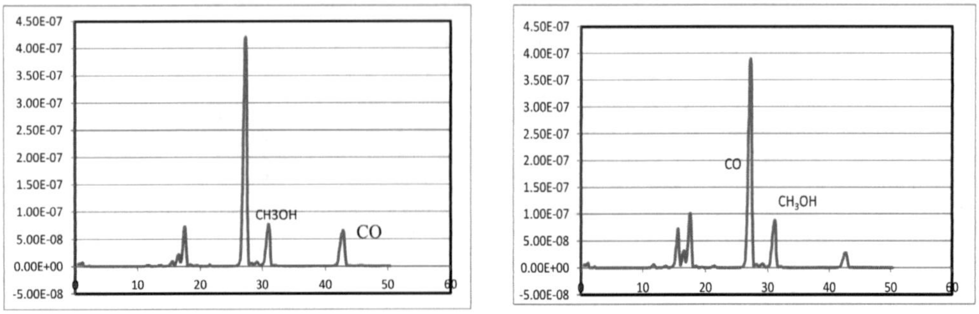

Figure 6. Mass Spectrum of CO_2 plasma after two consecutive pulse of water discharged

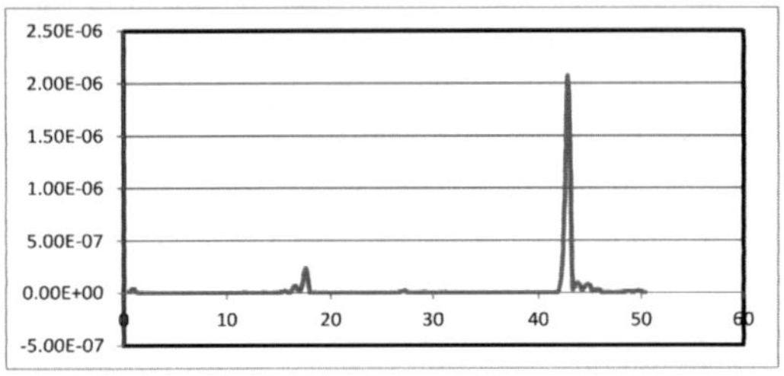

Figure 7. Mass Spectrum of CO_2 Plasma with no pulse of water inside the chamber showing la presence of CO_2.

5. Discussion of Results

Plasma CO_2 is an excellent tool to produce methanol under the presence of Cu-O-Zr catalyst consuming the water. The presence of radicals H and OH produced during the collisions of electron with water molecules produce fast reactions with carbon dioxide. No elemental hydrogen gas is required to be incorporated into carbon structure. Carbon dioxide in excess makes the full consumption of water and carbon monoxide forms as by product due to the limited amount of water during the chemical reaction.

6. Acknowledge

Thanks to Dr. Katyar from University of Puerto Rico- Rio Piedras Campus and Dr. Wilfredo Otaño from University of Puerto Rico-Cayey Campus for the characterization of materials. Thanks also to the personnel of PUPR Plasma Laboratory.

7. References

[1] H. Nakano, I. Nakamura, and T. Fujitani J. Phys. Chem B **105**, 1535-1365 (2001).
[2] W. Wang, S. Wang, X. Ma, and J. Gong, Chem. Soc. Rev., **40**, 3703-3727 (2011)
[3] D. E. Peebles, D. W. Goodmen, and J. M. White, J. Phys. Chem., **87**, 4378-4387 (1983)
[4] P. J. Lunde, F.L. Kesler, Am. Chem. Soc., Div. Fuel Chem., 17:1; Conference: 164th National American Chemical Society meeting, New York, NY, USA, Aug 27, 1972
[5] Yukikazu Itikawa, J. Phys. Ref. Data, **31**, 3 (2002)
[6] Y. Itikawa and N. Mason, J. Phys. Chem. Ref. Data, **34**, 1 (2005)
[7] E. Leal-Quiros and M. Prelas, Rev. Sci. Instrum 61,1708(1990)

Surface analysis of alumina ceramic exposed to shock waves produced by plasma expander

R P Mota[1], E Campos[1,2], C N Santos[1], E F Lucena[2], M Machida[3] and F C L Melo[4]

[1]Faculdade de Engenharia - Campus de Guaratinguetá - UNESP, São Paulo, Brazil

[2]Escola de Especialistas de Aeronáutica (EEAR) - Guaratinguetá, São Paulo, Brazil

[3]Instituto de Física - UNICAMP - Campinas, São Paulo, Brazil

[4]Instituto de Aeronáutica e Espaço -DCTA - São José dos Campos, São Paulo, Brazil

E-mail: rmota@feg.unesp.br

Abstract. Material surface treatment by plasma expander is relatively recent. Plasma expander is based on the inverse pinch effect. The shock waves produced by plasma expander may also promote modifications in ceramic materials exposed to the expander. These modifications are mainly made by ablation phenomenon. This work was intended to verify the shock wave effects on the ionic ceramic samples with high dielectric constant. The alumina ceramic samples were formed by both uniaxial and isostatic pressing methods and sintered at 1650 °C. They were also produced with addition 0.15 wt% of MgO in order to obtain a high densification. The ceramic samples were divided in groups and exposed to 700, 1000 and 1440 pulses during 20 min. These pulses were generated by nitrogen plasma expander at 13.0 Pa and 6 kV. After plasma exposure, there was an increase in roughness parameter values of Al_2O_3 ceramic surface. The treatment by plasma expander did not modify the hydrophilic characteristic of the alumina ceramic samples. The results of hardness test presented no significant changes on hardness mean values.

1. Introduction

This paper aimed to analyze the changes produced by shock waves with nitrogen ions generated by a plasma expander on dense alumina ceramics. The alumina ceramic (Al_2O_3) is an ionic material with a hexagonal crystalline structure, which has high mechanical strength and presents high corrosion and wear resistance. One of the features of high hardness and high elastic modulus of alumina ceramics is due to the ionic bond stability. These ceramics are used in several industry fields. New applications for these materials have emerged in biomaterial industry due to ceramic surface changes such as wettability and morphology [1-5]. Such changes can occur when these materials are exposed to plasma like pinch discharge [6]. Plasma expander is an innovative method for material treatment, based on inverse pinch effect [7-11]. Plasma expander has been used to treat polymer surfaces deposited by plasma [8-10] and in silicon carbide (covalent ceramic) [11].

2. Materials and methods

2.1. Ceramic

The alumina (CT-3000 SG), 0.15 wt% of MgO and 0.12 wt% of CMC were used to produce the ceramics. They were formed by uniaxial pressing (50 MPa) in a stainless steel mold (15 x 15 x 4 mm), followed by isostatic pressing (300 MPa) and sintered at 1650 ºC, with a heating rate of 5 ºC/min. The samples were polished by using diamond paste (granulometry 6 μm, 3 μm and 1 μm).

2.2. Plasma expander

The reactor (Figure 1a and 1b) consists of a glass tube (diameter: 28 cm) covered with two circular brass plates (diameter: 30 cm). In the center, there is a central hole with a conductor bar (diameter: 0.6 cm) isolated by the glass tube, which crosses the bottom plate. The expander has a capacitor (8.5 μF, 20 kV) which is charged at 20 kV by a 20 mA dc power supply. The expansion velocity of the shock wave is measured with a collimated optical fiber coupled to a spectrometer. In the visible spectrum there is a maximum value (approximately 1.3×10^6 cm/s) which decreases to a minimum (0.9×10^6 cm/s) at the edge of the conductive plates.

Figure 1a. Plasma expander system.

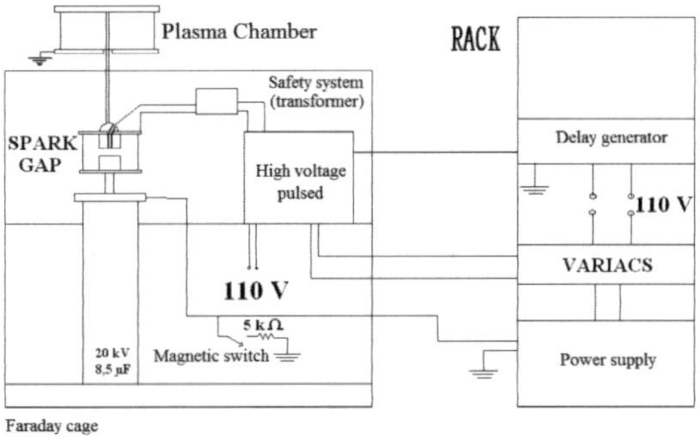

Figure 1b. Plasma expander diagram.

The ceramics were divided in three groups and exposed to 700, 1000 and 1440 pulses during 20 min. These pulses were generated by nitrogen plasma expander at 13.0 Pa and 6 kV. The samples were placed in the reactor (3.0 cm from the center of the conductor). According to [8-10], it was estimated an approximate value of 12 eV for the ion collision energy on the polished surfaces.

2.3. Characterization

All measurements for ceramic characterization were performed before and after the plasma exposure. The apparent porosity was determined according to ASTM C20-00 [12]. In order to determine roughness, it was used a profilometer Mitutoyo - Surftest 301, by choosing Ra (average roughness) and Rt (peak-to-valley roughness) parameters. For Vickers hardness measurements, it was used OttoWolpert-Werke, an equipment with 30 kgf load. The wettability of the sample surface was investigated by goniometer Rame-Hart 300-F1.

The light microscopy was used to monitor the change on the polished surfaces. The "ImageJ", public domain software, developed at the National Institute of Mental Health [13] was used to perform quantitative analyses of the images. The images used for porous and matrix fraction measurements were obtained from polished surfaces by using a Nikon microscope model Epiphot 200. SEM images were captured by Zeiss EVO LS-15 microscope.

3. Results and discussion

Table 1 shows the hardness and porosity results of each ceramic sample group. After the plasma exposure, there were no significant changes in bulk density and apparent porosity and the average values remained nearly the same. Probably, when nitrogen ions collided with the ceramics there wasn't enough energy to go into the material or to remove a large amount of material of the surface.

Table 1 also shows that Vickers hardness mean values decreased very little and standard deviation (dispersion) increased for all sets. These results confirm that nitrogen ions did not cause alterations in the alumina sample bulk.

Table 1. Bulk density, apparent porosity and hardness.

Group	Bulk density [g/cm^3]	Apparent porosity [%]	Vickers Hardness [kgf/mm^2]
		Before plasma expander	
1	3.83 ± 0.01	0.85 ± 0.27	1560 ± 80
2	3.81 ± 0.01	0.84 ± 0.23	1555 ± 80
3	3.81 ± 0.01	0.95 ± 0.23	1557 ± 80
Pulse		**After plasma expander**	
700	3.82 ± 0.02	0.83 ± 0.18	1528 ± 140
1000	3.81 ± 0.01	0.88 ± 0.15	1497 ± 160
1440	3.80 ± 0.01	1.12 ± 0.26	1541 ± 160

Figures 2a and 2b show micrographs obtained from the sample surface (group-1), before and after the plasma exposure, respectively. It can be seen an increase of sample pore number. Furthermore, there pores present a non uniform distribution and irregular morphology after shock wave exposure.

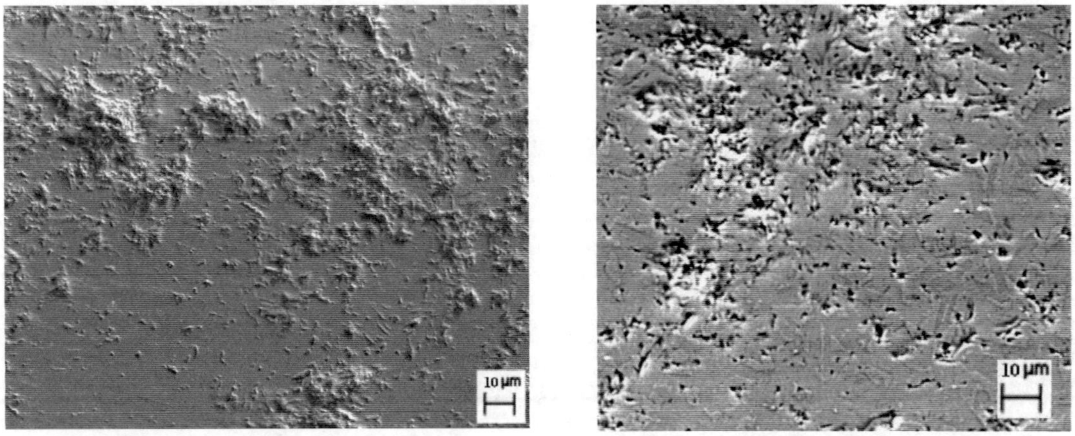

Figure 2a. Surface samples before plasma. **Figure 2b.** Surface samples after plasma.

Table 2 shows roughness, wettability and porous fraction values. In this table, it can be observed that there were significant changes especially in roughness and wettability promoted by the shock waves. Despite the higher roughness values (described by Ra and Rt parameters), the average contact angle remained below 90°, i.e. the ceramic surface maintained the hydrophilic characteristic. The increase of wettability occurred due to the ablation process, which was generated by nitrogen ions of shock waves.

Table 2. Roughness, contact angle and porous fraction measurements.

Group	Ra (μm)	Rt (μm)	Contact angle (°)	Porous fraction (%)	Matrix fraction(%)
		Before plasma expander			
1	0.25 ± 0.06	2.53 ± 0.24	70.2 ± 1.0	21.7 ± 1.5	78.3 ± 1.5
2	0.43 ± 0.10	3.80 ± 1.19	71.3 ± 1.0	25.9 ± 3.2	74.0 ± 3.2
3	0.22 ± 0.06	2.63 ± 0.80	77.4 ± 1.0	20.7 ± 1.6	79.3 ± 1.6
Pulse		**After plasma expander**			
700	0.94 ± 0.14	8.53 ± 1.90	75.5 ± 1.0	24.4 ± 1.5	75.6 ± 1.5
1000	0.79 ± 0.23	5.99 ± 2.22	81.4 ± 1.0	28.7 ± 1.5	71.3 ± 1.5
1440	0.92 ± 0.12	6.34 ± 0.94	86.1 ± 1.0	23.6 ± 0.9	76.4 ± 0.9

4. Conclusions

The porosity measurement, which is based on Archimedes' Principle, and hardness results showed that the sample bulk remained practically unchanged. The plasma expander proved to be a novel surface treatment for promotes changes on Al_2O_3 ceramic surface.

The roughness, contact angle measurements and SEM image analysis indicate that there were changes in the sample surfaces due to momentum transfer from nitrogen ions to ceramic surface. The standard deviations determined after hardness and roughness tests, indicate that the nitrogen ions do not reach the entire ceramic surface with the same intensity.

SEM image analysis and roughness parameters indicate that the sputtering process by nitrogen ions occurs on the entire ceramic surface. However, the increase of the number of plasma pulses did not make the sputtering process more effective, probably due to the surface electric charging.

Acknowledgments

CNPq for financial support and Ms. Marisa Silva (COMAER) for technical assistance.

References

[1] Weidmann G, Lewis P, Reid N 1990 *Materials in Action Series:Structural Materials* (United Kingdon: The Open University)
[2] Barlak M, Piekoszenwski J, Werner Z and Stanislawski J 2009 *Surface and Coatings Technology* **203** 2536-40
[3] Pandiyaraj K N, Selvarajan V, Deshmukh R R and Bousmina M 2008 *Surface and Coatings Technology* **202** 4220-22
[4] Guehennec L L, Soueidan A and Amouriq Y 2007 *Dental Materials* **23** 844-54
[5] Webster T J 2001 *Advances in Chemical Engineering* **27** 125-66
[6] Lee S and Saw S H 2008 *Apllied Physics Letters* **92** 285-90
[7] Anderson O A, Furth H P, Stone J M and Wright R E 1958 *The Physics of Fluids* 1489-94
[8] Rangel E C, Machida M, Durrant S F and Cruz N C 2012 *IEEE Transactions on Plasma Science* **40** 492-96
[9] Rangel E C, Silva P A F and Mota R P 2005 *Thin Solid Films* **473** 259-66
[10] Machida M, Aramaki E A and Moraes M B 2003 *AIP Conf. Proc.* **669** 339-42
[11] Santos C N, Marins E M, Machida M, Campos E, Mota R P, Melo F C L and Hein L R O 2012 *Materials Science Forum* **727-728** 1428-32
[12] ASTM. C20-00 2000 Standard test methods for apparent porosity, water absorption, apparent specific gravity, and bulk density of burned refractory brick and shapes by boiling water
[13] Rasband W S 1997-2012 *ImageJ-U. S. National Institutes of Health*, (Bethesda, Maryland, USA) http://imagej.nih.gov/ij/

PDBD with continuous liquids flows in a discharge reactor

B G Rodríguez-Méndez[1], **D G Gutiérrez-León**[1,2], **J M Belman-Flores**[2], **R López-Callejas**[1], **R Valencia-Alvarado**[1], **A E Muñoz-Castro**[1], **A Mercado-Cabrera**[1], **R Peña-Eguiluz**[1], **A de la Piedad-Beneitez**[3]

[1] Instituto Nacional de Investigaciones Nucleares, Laboratorio de Física de Plasmas, AP 18-1027, 11801, México, D.F., México
[2] Universidad de Guanajuato, División de Ingenierías Campus Irapuato, Salamanca, Comunidad Palo Blanco S/N, CP 36885, México
[3] Instituto Tecnológico de Toluca, AP 890, Toluca, México

e-mail: benjamin.rodriguez@inin.gob.mx

Abstract. This paper presents the design, construction and testing of a cylindrical pulsed dielectric barrier discharge (PDBD) reactor aimed to microbiological elimination of *Escherichia coli* ATCC 8739 bacteria. In the reactor, water flowed continuously and to countercurrent an oxygen gas was injected. The water pumping was carried out with a peristaltic pump type, stainless steel and aluminum constructed, and water was recirculated through norprene tubing. The considered parameters in order to promote energetic efficiency were: the residence time of the water contaminated with bacteria, flow rate of the liquid, shape and material used to build electrodes and dielectric, pressure, and gas injection flow rate. The pulsed power supply parameters are featured by 25-30 kV high voltage, 500 Hz frequency and 30 μs width. The outcome elimination of *E. coli* bacteria at 10^3, 10^4 and 10^6 CFU/mL concentrations reached an efficiency over 0.5 log-order in absence of oxygen; while >2 log-orders when oxygen gas was injected during the process.

Introduction

Water is an indispensable resource for living organisms. In 2002, United Nations Organization declared human right to water as fundamental, it implies have access to water in enough quantity and acceptable quality [1]. Nevertheless, according to the World Health Organization (2013), at global level, approximately 1.1 billion of people do not have access to water sources with an acceptable quality level; 2.4 billion of people do not have some sort of sanitation facility and close to 2 million of people, most of them children under five years old, die every year due to diarrheal disease [2]. In general, the main source of risk is associated with polluted water ingestion, which contains pathogen microorganisms from people and animals feces; therefore, the implementation of an appropriate disinfection process is essential.

Currently, the disinfection process widely implemented in Mexico corresponds to chlorination, in which have been registered responsible compounds to cause cancer such as organochloride compounds and trihalomethanes [3, 4]. Therefore, there is an urgent need to propose, develop and deploy efficient and safe processes as alternatives to chlorination for the elimination of pathogen microorganisms in water.

In recent years a particular interest has grown in the application of pulsed dielectric barrier discharges (PDBD) in water with the aim of achieving synergistic effects (production of reactive chemical species, pulsed electric fields, UV radiation and shock waves), in reduced conditions of time and under the implementation of a safe procedure for the microorganisms elimination in water. Currently, research works have been reported [5-12]; nevertheless, these investigations have been carried out maintaining a certain volume of liquid in a state of rest.

This work aims to evaluate the elimination efficiency of *Escherichia coli* (*E. coli*) bacteria ATCC 8739, which is widely used as indicator microorganism of fecal contamination in water, through a designed pulsed dielectric barrier discharge reactor coupled to a hydraulic system with a total treatment capacity of 500 mL, operating in recirculation mode under steady and laminar flow conditions at 293.15 K and 103 kPa.

2. Experimental setup and microbiological procedure

The experimental setup that was employed in antimicrobial treatment experiments is presented in figure 1. The following subsections are below described in detail:

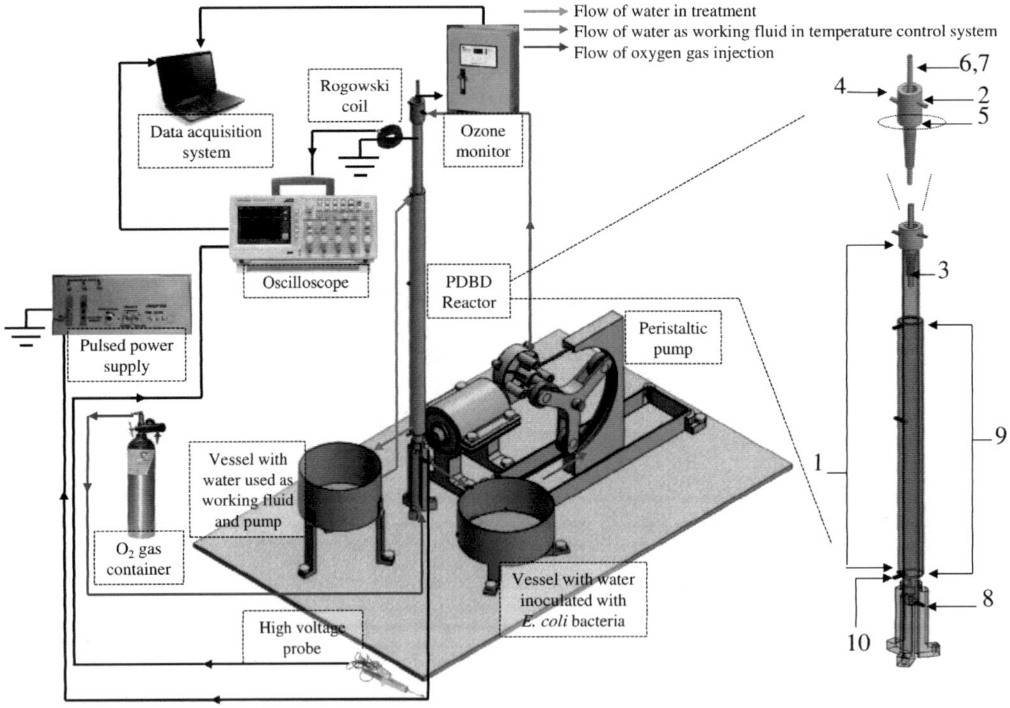

Figure 1. Experimental setup configuration.

2.1 Description of Hydraulic Setup

An experimental hydraulic setup was designed in order to operate in water recirculation, with a total capacity of 500 mL volume for the microbiological elimination of *E. coli* bacteria (see figure 1). The operating conditions were established under steady and laminar flow at 293.15 K of temperature and 103 kPa of pressure.

The experiment was started putting inside a vessel 500 mL of distilled water inoculated at 10^3, 10^4 and 10^6 *E. coli* cells mL^{-1} concentrations. Once started the application of discharges, a peristaltic pump type operating at 1.2 mL s^{-1} was used for the liquid movement from the vessel with water inoculated with *E. coli* bacteria at the top of the reactor, where through an adapted piece, the liquid fluid was distributed inside the chamber to return to the first vessel and receive the treatment again.

2.2 Description of PDBD Reactor

The PDBD reactor is biased by a pulsed power supply, which is featured by pulse height 1-30 kV, frequency 100-2000 Hz and pulse width 0-30 μs; these parameters can be adjusted by the user. The pulse voltage and current signals from the PDBD reactor were picked up by a high voltage probe (Tektronix P6015A, 1000× 3.0 pF and 100 MΩ) and a Rogowski coil (Stangenes 2-0.1 W) (see figure 1). The signals were monitored with a digital oscilloscope (Tektronix TDS2024, 200 MHz-2 GS/s).

The cylindrical geometry PDBD reactor with a volume 38.0 mL constructed of stainless steel 316L (see figure 1) in order to avoid corrosion problems due to the production of reactive oxygen species during the discharge process; it also was used as cathode (see figure 1, point 1). According to figure 1, at the top of reactor vessel was inserted a nylamid piece with a hole, which provides the inlet of liquid, which comes from the vessel with water inoculated with *E. coli* bacteria (point 2), distribute the liquid flow inside the reactor vessel (point 3); enable the outlet of gaseous chemical species generated during the elimination of *E. coli* bacteria (point 4) and, with an adaptation attached at the bottom (point 5), give an adequate and rigid support to dielectric and anode of the system (point 6). The selected material in order to construct the anode corresponded to a tungsten wire of 300 mm length, it was inserted coaxially to the reactor vessel and covered with an alumina hollow rod of 2.44 mm diameter and 350 mm length (point 5), the latter being the dielectric barrier (point 7). The outlet of liquid in treatment (point 8) was carried out by the lateral and inferior side of the reactor vessel. The treated liquid is received in the vessel that contains water inoculated with *E. coli* bacteria; here it is taken the sample to be analyzed microbiologically.

Due to temperature could influence in the elimination process of microorganisms, it is necessary an additional system to control it. Therefore, to maintain a constant temperature during the experiments, in the outside of the PDBD reactor vessel was coupled a cooling system (point 9), which consisted of an isolated jacket with water as working fluid, to avoid bacterial contamination was manufactured with PVC of diameter 25.4 mm and operated at recirculation mode. Additionally, the reactor vessel had a lateral adaptation (10), whose function was allow the oxygen gas to flow at 8.3 mL s^{-1} inside it at countercurrent with the liquid flow.

2.3 Description of Microbiological Procedure

2.3.1 Bacterial growth and preparation. As before mentioned, the effectiveness bacterial elimination in water was determined using the strain of *E. coli* ATCC 8739 from the American Type Culture Collection bacteria, as the selected indicator microorganism to determine the process efficiency. For this, a strain was inoculated in 5 mL of Luria-Bertani (LB) broth and incubated for 86,400 s at 310.15 K, being continuously stirred (overnight culture). The cells were harvested by centrifugation (twice for 600 s at 5000 rpm) and then it adds 5 mL of distilled and sterile water. In order to determine the obtained initial concentration of cells after overnight culture, an aliquot of 100 μL at 1:10^2 dilution is considered, and the count of cells was performed by a phase contrast microscope and a Neubauer chamber. The concentration in cells per milliliter (cells mL^{-1}) was recorded and cells density at stationary phase was approximately 10^7 cells mL^{-1}. The original concentration can also be approached multiplying it by the dilution factor, leading to ~10^9 cells mL^{-1}. The concentration for each experiment was obtained following this method and adjusted by standard progressive dilutions.

2.3.2 Microbiological Analysis. Analysis in triplicate taken from both, treated (taken at the end of every hydraulic cycle) and untreated samples at 10^3, 10^4 and 10^6 *E. coli* cells mL^{-1} concentrations were compounded by inoculating 0.1mL in each Petri dish (90 mm diameter) containing LB agar through a spread-plate technique. Once labeled, the inoculated dishes were incubated overnight at 310.15 K for 86,400 s. After incubation period, the number of colonies from the initial cell concentration and the treated suspension was estimated as the number of surviving cells assuming that every viable bacteria results in the formation of a colony. Colony forming units (CFU) were enumerated, and concentrations per unit volume (CFU mL^{-1}) were determined by direct plate counting method.

2.4 Ozone Concentration Measurements

Ozone is one of the reactive chemical species generated by plasma in air/oxygen presence, which is able to produce important biological effects. By means of a probe connected to an ozone monitor (model 460L Teledyne Instruments), it is possible measure the concentration (in ppm) of residual ozone generated during the discharge at the outlet of gaseous chemical species (see figure 1, point 4). The obtained data was processed on computer software in order to get a graphical interface where ppm concentration versus time is given.

2.5 Scanning Electron Microscopy (SEM) applied to norprene tubing

15 mL in volume of distilled water were introduced inside the reactor vessel and PDBD were applied during 600 s. Then, the water was transferred to a glass and a piece of norprene tubing was inserted inside until a period of 6 days was completed. SEM analysis was carried out in triplicate using a Scanning Electron Microscope JSM-5900 at 20 kV and magnifications ×500 with the aim to determine if norprene is an acceptable material to be used as transport medium for fluids when PDBD process is applied.

3. Results and Discussion

The representative behavior of a discharge limited by a dielectric material is shown in figure 2, through voltage and current waveforms developed by a PDBD reactor during the treatment applied to inoculated water with *E. coli* bacteria. According to Eliasson and Kogelschatz [13] and Rodríguez *et al.* [14], once ionization occurs in the discharge gap, the charge is accumulated in the dielectric causing a limited streamer formation which is reflected in the formation of peaks, entailing to the reduced amount of transported charge into the medium as well as to the entire electrode area. The energy consumption per pulse during the experiment was obtained through the time-integral of the voltage-current product from the waveforms mentioned before, and it corresponds to 134.3 mJ pulse^{-1}; this is bigger than those reported by Hernández *et al.* [11], Rodríguez *et al.* [12], which corresponds to ~100 mJ pulse^{-1}. This variation could be attributed to differences of environmental conditions under experiments were carried out, geometry and internal surfaces of the electrodes and dielectric, and gas quality.

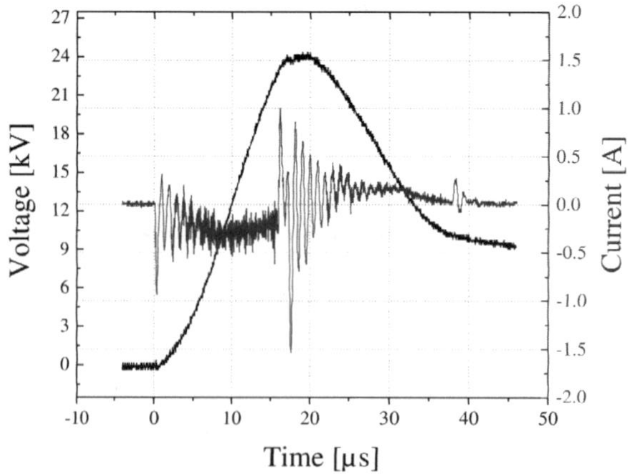

Figure 2. Voltage and current waveforms delivered by the PDBD reactor.

Results of the elimination efficiency of *E. coli* bacteria in a recirculation system, where a PDBD reactor was coupled operating under steady and laminar flow conditions at 293.15 K and 103 kPa, carried out in presence and absence of oxygen gas are presented in figure 3. The required time for every fluid cycle corresponds to 419 s.

In figure 3 are depicted the results obtained in the bounded experimental period of time, at 10^3 CFU mL^{-1} *E. coli* concentration. A reduction of >2 log-orders was attained in presence of oxygen gas and recirculating the fluid 12 times in the system; when the process occurs without gas flow, only <1 log-order bacterial concentration reduction was obtained. In the latter case, 2 additional fluid cycles were applied, obtaining a slight decrease of 1 log-order with respect to the initial concentration. In the case of 10^4 CFU mL^{-1} *E. coli* concentration, with the treatment applied during 18 cycles without oxygen, a reduction of <2 log-orders was obtained; this result was increased considerably at >3 log-orders through oxygen gas injection inside the system. When the treatment was applied at 10^6 CFU mL^{-1} *E. coli* concentration, it was registered a reduction over 3 log-orders in presence of oxygen in 22 fluid cycles; while the gas flow was absent, the elimination efficiency decreased to ~0.5 log-order in the same period of treatment. Therefore, as bacteria concentration increases, the number of required fluid cycles is higher; when the oxygen gas injected inside the system, log-orders decrease of *E. coli* concentrations is more significant, this effect is especially more evident at high concentrations.

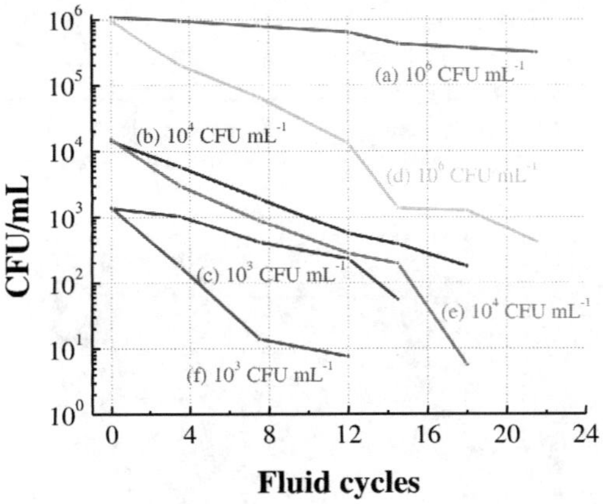

Figure 3. Development of *E. coli* bacteria elimination process under the following conditions: (a), (b) and (c) without oxygen gas flow; (d), (e) and (f) with oxygen injection.

The established experimental conditions in this work involve the oxygen gas injection inside the system in order to promote a mixture of liquid-gas fluids inside the reactor vessel and benefit the production of ozone. Figure 4 displays obtained behaviors of different oxygen gas flows injected inside the reactor. In all cases, it was observed that ozone concentration increased significantly during the first seconds of the process, later a constant tendency of the ozone concentration with respect to time occurred, indicating a saturation. At 8.3 mL s^{-1}, the ozone concentration is higher than the other proposed oxygen flows; it indicates that over >8.3 mL s^{-1}, the ozone concentration drops considerably due to an excess of oxygen gas, which was not used adequately to benefit the production of ozone.

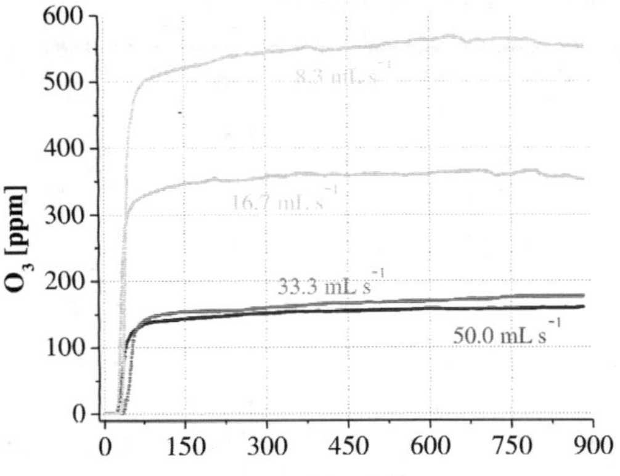

Figure 4. Evolution of ozone concentration versus time of oxygen gas injection into the PDBD.

According to the ozone monitoring during the treatment process applied to water without oxygen injection, it was generated 9.2 ppm while with oxygen 114.4 ppm. Increments in the efficiency of elimination process of *E. coli* bacteria by the oxygen injection at 10^3, 10^4 and 10^6 CFU mL^{-1} concentrations are in the range of 3.5% to 28.5%. It could be attributed to the bacteria-discharge interaction during the included steps in recirculation process such as: flowing of the fluids at countercurrent mode, liquid flow sparging inside the reactor vessel and generating bubbles during the gas flow injection; some of the latter conditions mentioned are in agreement with those proposed by Glaze [16] in order to benefit the mass transfer or reactive chemical species such as ozone into the aqueous phase. Additionally, these results are in accordance with observations performed by Chen *et al.* [17] and Hernández *et al.* [11], who emphasized on the importance of ozone existence as an important chemical species able to cause damage in cytoplasmic membrane deleting the bacteriological activity.

Due to the temperature is considered a factor through the process efficiency could be modified, a cooling system was added to the reactor to maintain a stable temperature (293.15 K) during the treatment; it is in accordance with Gottschalk *et al.* [18], Ingraham [19] who verify that an increase in the temperature involve a decrease in the solubility of ozone and modifies its reaction rate.

Another important factor evaluated in this study corresponded to the norprene tubing used during the process for the liquid transport. Respectively, in table 1 and figure 5 are showing the analytical results and micrographies obtained by SEM before and after the treatment applied to water to eliminate *E. coli* bacteria.

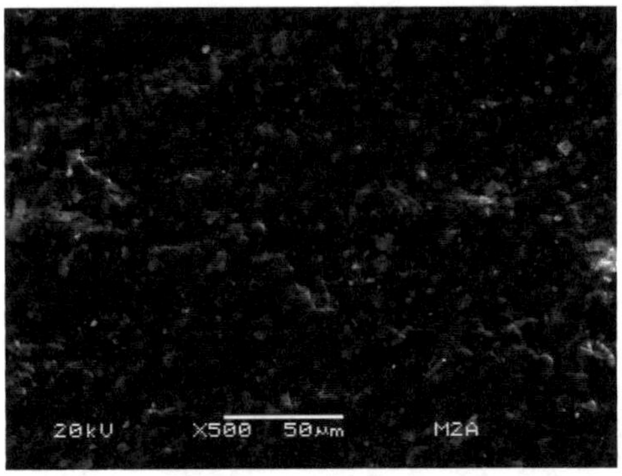

a

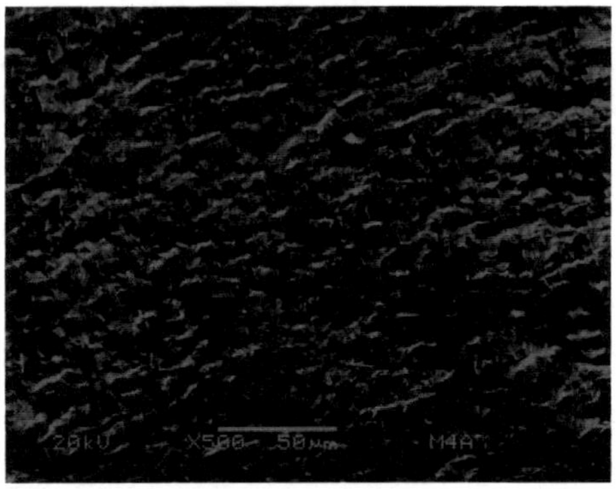

b

Figure 5. Norprene tubing images: (a) before and (b) after the treatment.

Table 1. Atomic percentages of the identified elements before and after elimination process of *E. coli* bacteria in water.

Element	Atomic %	
	Before	After
C	86.49	86.55
O	11.25	13.39
Cl	0.11	0.06
Al	0.22	
Si	0.97	
Na	0.22	

According to the results, norprene tubing exhibited a slight variation in its composition when the elimination process of *E. coli* bacteria was applied; in the case of carbon and oxygen, it was observed an increase of 0.06% and 2.14%, respectively. On the contrary, identified elements in smaller percentages, such as: chlorine, aluminium, silicon and sodium decrease significantly; the last three elements were reduced so that they could not be detected by SEM. Figure 5(a) shows a micrography of norprene tubing before the treatment application; in it is observed a rugose structure due to many disarranged pellets putting on it, which after the treatment (see figure 5(b)), adopted a rigid structure without visible pellets over its surface and an appearance of humid film. In agreement with Askeland and Phulé [20], to be norprene tubing a polymeric material, there exist dissolution of its components in water during a prolonged exposure, carrying to the loss of them and weakening its structure.

Attention should be paid to the dissolution of chlorine in water, which contributes to the production of chlorine, hexachlorobenzene, heptachlor and heptachlor epoxide and methoxychlor, currently regulated by official laws due to their health risk. On the other hand, little increments of carbon and oxygen, indicate a phenomenon of deposition on the tubing internal surface, as a result of the reactions during the process. Additionally, norprene tubing surface could allow lodging and development of microorganisms from one experiment to other; nevertheless, this possibility is eliminated due to it is an autoclavable material.

Conclusions

A PDBD reactor coupled at a recirculation system, which includes a peristaltic pump type and a cooling system, was designed and constructed to operate under steady and laminar flow conditions at 293.15 K and 103 kPa, 25-30 kV high voltage, 500 Hz frequency and 30 µs width, with the aim to eliminate *E. coli* bacteria at 10^3, 10^4 and 10^6 CFU ml^{-1} concentrations found in water contaminated with those microorganims. Results indicate an elimination efficiency over 0.5 log-order when the process was carried out in absence of oxygen; while >2 log-orders when oxygen gas was injected inside the reactor during the treatment. As mentioned above, when bacteria concentration increases, the number of required fluid cycles is higher; when the oxygen gas injection is carried out inside the system, log-orders decrease of *E. coli* concentrations is more significant. Along the recirculation process, experimental conditions including: flowing of the fluids at countercurrent mode, liquid flow sparging inside the reactor vessel and generating bubbles during the gas flow injection, promote an increase on the efficiency of the system. In the case of ozone production by oxygen gas, the efficiency increased in the range 3.5% - 28.5% when injected 8.3 mL s^{-1}, flow which was determined in this study as the optimum to produce ozone that contributes to eliminate *E. coli* bacteria.

Additionally, in this work was determined that norprene is an acceptable material to be used as transport medium for the fluids, when PDBD process is applied to water in order to eliminate *E. coli* bacteria; above mentioned is due to small detected changes by SEM on its atomic percentages corresponding to main components such as carbon (+0.06%), oxygen (+2.14%) and chlorine (-0.05%) and, on its internal surface, which after the treatment showed a more rigid surface. Additionally, norprene tubing is autoclavable; therefore it is possible avoiding the lodging and development of microorganisms inside it between experiments.

Acknowledgments

This work has received financial support from CONACYT, México. The authors are obliged to P. Angeles Espinosa, I. Contreras Villa and M T Torres Martínez for their technical support.

References

[1] United Nations University 2013 Water security & the global water agenda. A un-water analytical brief http://www.zaragoza.es/contenidos/medioambiente/onu/999-eng.pdf

[2] World Health Organization 2014 Water supply, sanitation and hygiene development http://www.who.int/water_sanitation_health/hygiene/en/

[3] Navarro I, Jiménez B, Maya C. and Lucario E S 2007 Assessment of Potential Cancer Risks from THMs in Water Supply at Mexican Rural Communities *Proceedings of the International Conference on Water Management and Technology Applications in Developing Countries, Kuala Lumpur*

[4] Núñez M A, Garmiño R, García G, Garza J, Hernández J L, Cárdenas M L, Martínez S J and Moreno S 2005 Análisis de trihalometanos totales en agua potable del área metropolitana de Monterrey. *A. N. L. Ciencia UANL* **8-3** 358-361

[5] Tanino M, Xilu W, Takashima K, Katsura S and Mizuno A 2007 Sterilization using dieletric barrier discharge at atmospheric pressure *Int. J. of Plasma Environ. Sci. & Tech.* **1-1** 102-107

[6] Kirkpatrick M J, Dobet B and Odic E 2007 Atmospheric pressure humid argon DBD plasma for the application of sterilization-Measurement and simulation of hydrogen, oxygen, and hydrogen peroxide formation *Int. J. of Plasma Environ. Sci. & Tech.* **1-1** 96-101

[7] Yasuda H, Hashimoto M, Masudur M, Takashima K and Mizuno A 2008 States of biological components in bacteria and bacteriophages during inactivation by atmospheric dielectric barrier discharges *Plasma Process Polym* **5-6** 615-621

[8] Subedi D P, Tyata R B, Khadgi A, C and Wong S 2009 Treatment of water by dielectric barrier discharge *J. of Sci. and Tech. in the Tropics* **5-2** 117-123

[9] Dobrynin D, Fridman G, Mukhin Y V, Wynosky M A, Rieger J, Fest R, Gutsol A F and Fridman A 2010 Cold plasma inactivation of *Bacillus cereus* and *Bacillus anthracis* (Anthrax) spores *IEEE Trans. on Plasma Sci.* **38-8** 1878-1884

[10] Quiroz-Velázquez V E, López-Callejas R, Rodríguez-Méndez B G, Peña-Eguiluz R, Mercado-Cabrera A, Valencia-Alvarado R, Hernández-Arías A N, Barocio S R, Muñoz-Castro A E and de la Piedad-Beneitez A 2013 Pulsed power supply and coaxial reactor applied to *E. coli* elimination in water by PDBD *Rev. Int. Contam. Ambie.* **29-2** 25-31

[11] Hernández-Arías A N, Rodríguez-Méndez B G, López-Callejas R, Alcántara-Díaz D, Valencia-Alvarado R, Mercado-Cabrera A, Peña-Eguiluz R, Muñoz-Castro A E, Barocio S R and de la Piedad-Beneitez A 2012 Inactivation of *Escherichia coli* in water by pulsed dielectric barrier discharge in coaxial reactor *J. of Water Health* **10-3** 371-379.

[12] Rodríguez-Méndez B G, Hernández-Arías A N, López- Callejas R, Valencia-Alvarado R, Mercado-Cabrera A, Peña-Eguiluz R, Barocio S R, Muñoz-Castro A E and de la Piedad-Beneitez A 2013 Gas flow effect on *E. coli* and *B. subtilis* bacteria inactivation in water using a pulsed dielectric barrier discharge *IEEE Trans. on Plasma Sci.* **41-1** 147-154

[13] Eliasson B and Kogelschatz U 1991 Nonequilibrium volume plasma chemical processing *IEEE Trans. on Plasma Sci.* **19-6** 1063-1077

[14] Rodríguez-Méndez B G, López-Callejas R, Peña-Eguiluz R, Mercado-Cabrera A, Valencia-Alvarado R, Barocio S R, de la Piedad-Beneitez A, Benítez-Read J S, Pacheco-Sotelo, J O 2008 Instrumentation for pulsed corona discharge generation applied to water *IEEE Trans. on Plasma Sci.* **36-1** 185-191

[15] Traylor M J, Pavlovich M J, Karim S, Hait P, Sakiyama Y, Clark D S and Graves D B 2011 Long-term antibacterial efficacy of air plasma-activated water *J. Phys. D: Appl. Phys.* **44-47** 472001

[16] Glaze W 1987 Drinking-water treatment with ozone *Environ. Sci. Technol.* **21-3** 224–230

[17] Chen C W, Lee H M and Chang M B 2009 Influence of pH on inactivation of aquatic microorganism with a gas-liquid pulse electrical discharge *J. of Electrost.* **67-4** 703-08

[18] Gottschalk C, Libra J A and Saupe A 2000 *Ozonation of water and waste water. A practical guide to understanding ozone and its application* WILEY-VCH Verlag GmbH, D-69469 Weinheim p 119

[19] Igraham J 1987 *Effect of water temperature, pH, water activity, and pressure on growth Escherichia coli and Salmonella Typhimurium. Cellular and molecular biology* Neidhardt F C, Ingraham J L, Low K B, Magasanik B, Schaechter M, Umbarger E Washington DC American Society for Microbiology 2 p 1543-1544

[20] Askeland D R and Phulé P P 2004 *The science and engineering of materials* 4th ed. *CENGAGE Learning* p 681-689

Journal of Physics: Conference Series **591** (2015) 012061 doi:10.1088/1742-6596/591/1/012061

Three-electrode plasma reactor for the removal of toxic gases

J L Gallego[1], L Giuliani[2], D Grondona[2] and F Minotti[2]
Instituto de Física del Plasma (INFIP), Consejo Nacional de Investigaciones Científicas y Técnicas (CONICET), Departamento de Física, Facultad de Ciencias Exactas y Naturales, Universidad de Buenos Aires (UBA).
Ciudad Universitaria, Pabellón I, C1428EHA, Buenos Aires, Argentina.

E-mail: jlgallego@tinfip.lfp.uba.ar

Abstract. Electrical and spectroscopic measurement for the characterization of a novel three-electrode plasma reactor for the treatment of toxic gases is presented. The three-electrode discharge consists in a dielectric barrier discharge (DBD) combined with a corona discharge (CD). The DBD is generated by applying an alternating high voltage signal between two circular aluminium plate electrodes attached to opposite sides of a disk made of dielectric material. The CD is generated applying a continuous negative high voltage to an external cylindrical mesh electrode, coaxial with the DBD electrode system. The gap between the edge of the DBD system and the mesh electrode is approximately 20 mm wide. Up to five DBD electrode systems can be connected in parallel inside the reactor, axially separated from each other by 30 mm. The electrical characterization consisted in the measurement of the current between the DBD system and the external mesh, and the voltages of the electrodes. In order to understand the dynamics of the streamers, a theoretical determination of the laplacian electric field generated by the biased electrodes was done. Optical emission spectroscopy was performed in the range of wavelengths 280-480 nm, containing the typical spectral bands 2nd positive and 1st negative systems of molecular nitrogen.

1. Introduction

The emission of gases containing pollutants by industries, automobiles, and others are a big environmental problem in today's world. To remedy this problem, many techniques have been studied by several scientific teams. Techniques employing plasmas at atmospheric pressure have been very successful in removing toxic molecules present in polluted gases. Some of them are DBD and CD [1, 2]. Plasmas formed in these discharges are non-thermal plasmas in which energetic electrons collide with gas molecules and produce a rich variety of chemical products that undergo reactions capable of transforming harmful substances into non-hazardous products.

In a previous work [3], the development of a coaxial three-electrode plasma curtain discharge was presented. The discharge was based on the combination of a DBD with a CD in a three-electrode system with cylindrical geometry; basically, it resulted from the 'stretching' of a pure DBD discharge by the action of a CD discharge generated between the active electrode of the DBD and a remote third electrode. This discharge configuration, with double DBD electrode systems connected in parallel,

[1] Fellow of CONICET.
[2] Researcher of CONICET.

Journal of Physics: Conference Series **591** (2015) 012061 doi:10.1088/1742-6596/591/1/012061

produces almost double average current than that corresponding to the single one (for the same voltage values), indicating that in the stacked arrangement, both discharge systems develop independently.

In this work, we present a three-electrode plasma reactor for the removal of toxic gases. Electrical characterization was carried out, with air flowing in the reactor, by measuring the interelectrode current. In order to understand the dynamics of the streamers a theoretical determination of the laplacian electric field generated by the biased electrodes was done. Also, a plasma emission spectrum of air was obtained to observe the different species produced in the plasma.

2. Experimental

A schematic of the experimental setup is shown in Figure 1. The three-electrode reactor arrangement consists in two disks of adhesive aluminium tape of 50 μm thickness and 19 (electrode 1) and 17 (electrode 2) mm of radius, flush mounted at both sides of a poly-methyl methacrylate dielectric disk of 40 mm diameter and 2 mm width. One of the disks is connected to an ac power supply and the other to ground by wires placed inside two insulating tubes (10 mm in diameter) which pass along the electrode system. A third electrode (electrode 3), consisting in a steel mesh connected to a dc power supply is attached to the inner wall of a poly-methyl methacrylate dielectric cylindrical tube with 80 mm inner diameter, 240 mm length and with a wall thickness of 5 mm. This tube surrounds the central electrode arrangement and the distance between the edges of the central electrodes and the third electrode is 20 mm. Two plates seal the ends of the tube. Up to five DBD electrode systems can be connected in parallel inside the reactor. The gas, under pressure, goes through the reactor. The gas inlet is placed at one plate and the gas outlet in the other plate.

For our circuit, the optimum excitation ac frequency was 5.3 kHz. The ac voltage amplitude (Vac) was 6 kV and Vdc was -12 kV. The Vac was measured using a high voltage probe (1000 X / 3.0 pF/100MΩ). Current measurements were inferred from the voltage drop through a 50 Ω resistance. These electrical signals were registered with a four-channel digitizing oscilloscope with a bandwidth of 60 MHz and 1 Gs/s of sampling rate. The air flow was measured using a stainless steel float flow meter in the range 0 -17 l/min. Spectroscopy measurements were performed with a monochromator (280-480 nm) attached to a photomultiplier (Hamamatsu 1P28). Figure 2 shows a photograph of the reactor.

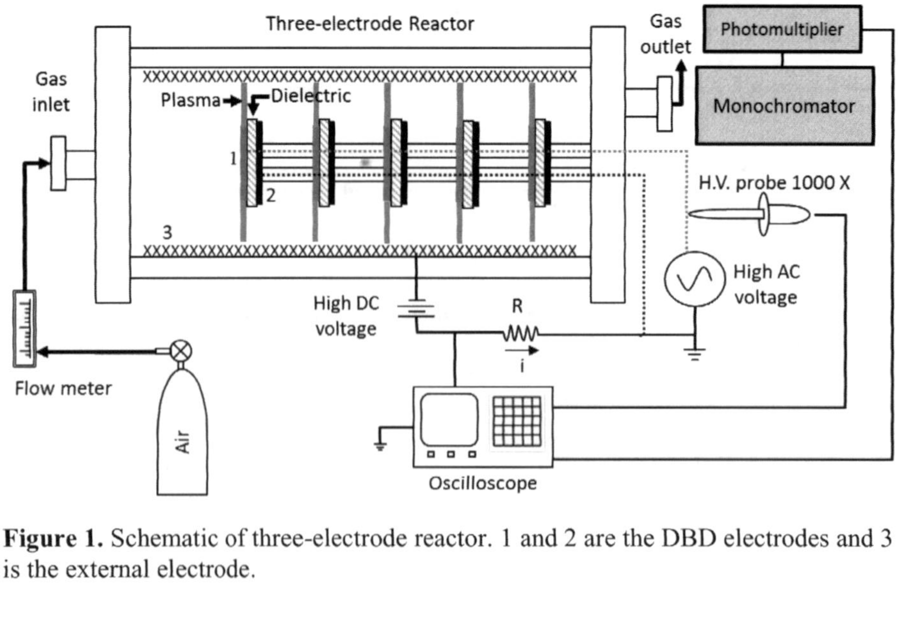

Figure 1. Schematic of three-electrode reactor. 1 and 2 are the DBD electrodes and 3 is the external electrode.

Figure 2. Photograph of the plasma reactor.

3. Results

These In Figure 3 Typical signals of the voltage Vac (a) and the interelectrode current I (b) for five DBD electrode systems without air flow, are presented. Pulses superposed on the capacitive current indicate the presence of streamers crossing the electrode gap. In our configuration, the ignition of the streamers starts at Vac = 2 kV, and the extinction of the streamer occurs when the peak value Vac = 6 kV is reached. To derive some meaningful information on the discharge characteristics the interelectrode current is presented as an average value (Im) by setting the oscilloscope in the average acquisition mode, an average waveform of the current signals over 128 samples were acquired. Then, to ignore the reactive component of the current these statistical-averaged signals were time averaged over one period of the AC voltage to finally obtain the average current values. According to this described procedure the uncertainty in the average current values could be estimated to be around 10%., due to statistical fluctuations.

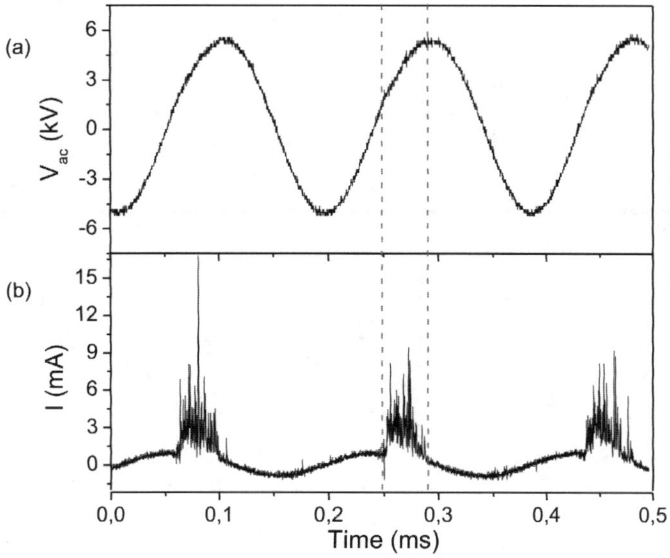

Figure 3. Typical signals of the voltage V_{ac} (a) and interelectrode current (b).

In Figure 4, the average interelectrode current, as a function of air flow values (with different number of DBD electrode systems connected in parallel as a parameter) is presented.

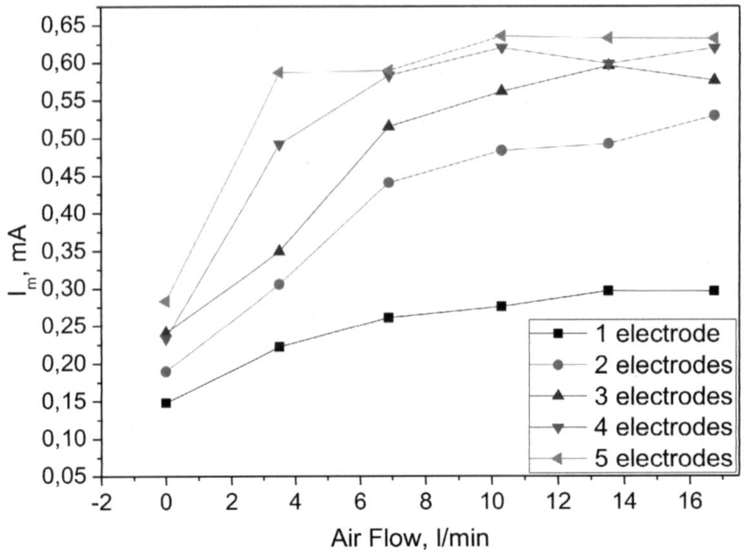

Figure 4. Current as a function of the air flow for different number of DBD electrode systems for V_{ac} = 12 kV pp, V_{dc}= -12 kV.

In Figure 5 an emission spectrum of the discharge in air is presented. A typical emission lines of nitrogen was found in the wavelength range 280 – 480 nm. The relevant lines correspond to the second positive system of the nitrogen molecule.

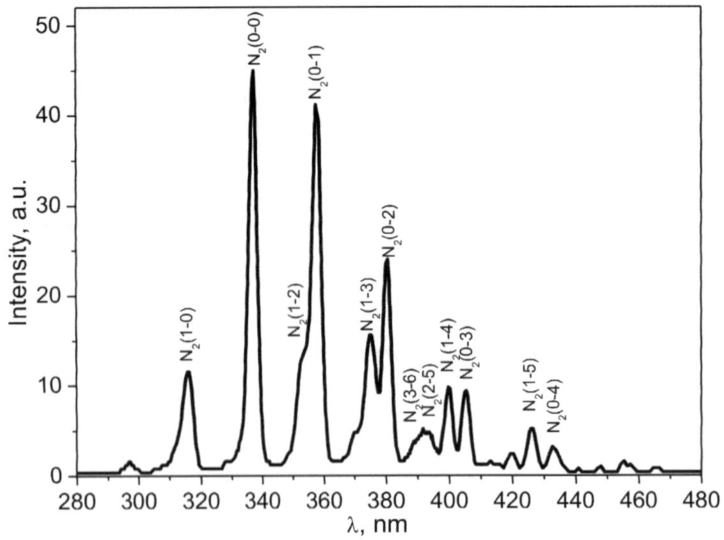

Figure 5. Typical emission spectra,

4. Electrical model

To understand the dynamics of the streamers, the electric field E was evaluated using the analytical expression of the potential for a surface charge density inside a cylinder at a fixed potential [4].

$$\phi(r,z)=\frac{1}{\pi\varepsilon_0}\int_0^{\infty}\left[\int_0^{a_d}\sigma(r')r'I_0(kr')dr'\right]\cos[k(z-z_d)]\left[K_0(kr)-I_0(kr)\frac{K_0(ka)}{I_0(ka)}\right]dk \quad (1)$$

where a is the radius of the external electrode, a_d the radius of the surface charge distribution, located at the axial position z_d, ε_0 is the permittivity of the vacuum, K_0 and I_0 the modified Bessel functions of order 0. Expression (1) is valid for any axial coordinate z, and for radial coordinates $a_d \le r \le a$. For a single metallic disk of radius a_d lying on the surface of a dielectric of relative permittivity ε_r, the total surface distribution of electric charge, including the charge on the disk plus the polarization charges on the dielectric surface, is given by [5]

$$\sigma(r)=\frac{q_d}{2\pi\varepsilon_r a_d^2\sqrt{1-r^2/a_d^2}} \quad (2)$$

where q_d is the electric charge in the disk only. Use of the expression (2) in (1) results in the analytical expression for the potential due to a disk on a dielectric surface, in which the integral is to be done numerically. The total potential was obtained as the superposition of the above expression for each DBD electrode. This expression, when evaluated at each disk position, allows to relate the disk charges with the disk voltages (which are the actual inputs), and the electric field is finally obtained by numerical spatial derivation.

In Figures 5 and 6 the potential and electric field as a function of r for different Vac and external voltage Vdc = -12 kV is presented. According to Raizer [6], to ensure the streamer propagation, the electric field value necessary for the streamer propagation across the air at atmospheric pressure is 4 kV/cm approximately. In Figure 6 can be observed that this value is obtained for Vac values of the figure 3 at which the streamers are generated.

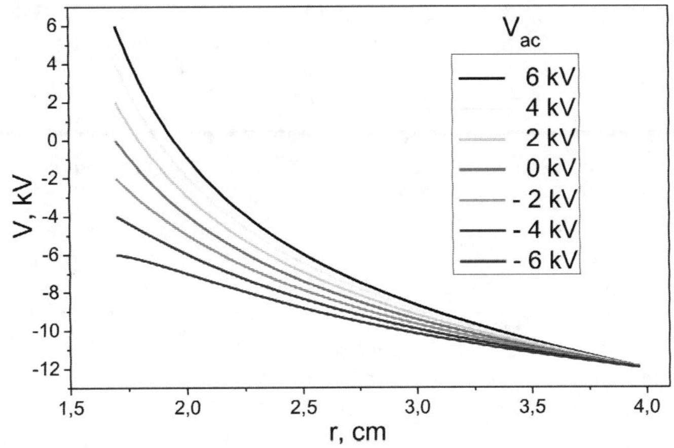

Figure 5. Potential as a function of radius r for different Vac and. Vdc = -12 kV.

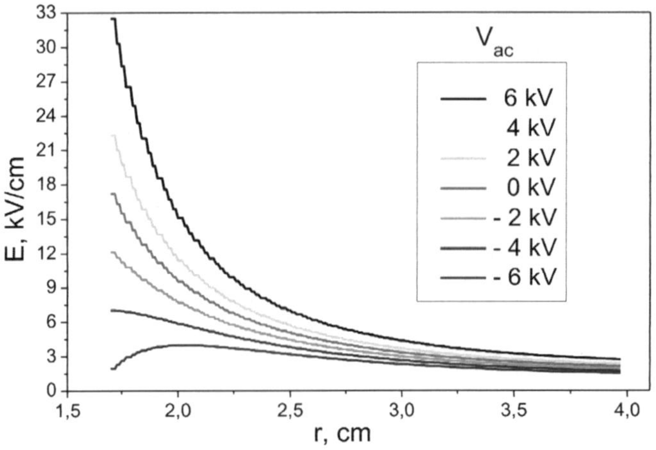

Figure 6. Electric field as a function of radius *r* for different Vac and Vdc = -12 kV.

5. Final Remarks

We have presented a novel three-electrode plasma reactor for the removal of toxic gases. The discharge is stable in a wide range of air flows and for different number of DBD electrodes systems connected in parallel. The main advantage of the geometry of the reactor is that a large plasma volume can be established and presents a natural boundary for the gas flow.

For the applied voltages it was found that the calculated values of the electric field are sufficient to sustain the streamer propagation along the interelectrode gap only during a part of the positive cycle of the Vac voltage in accordance with the experimental observation.

The obtained spectra show the presence of excited molecular N_2, indicating the presence of energetic electrons generated by the streamers crossing the interelectrode gap.

Acknowledgments

This work was supported by grants from the Consejo Nacional de Investigaciones Científicas y Técnicas (CONICET) PIP 11220090100219, Universidad de Buenos Aires (UBA) PID 01/1141 and Agencia Nacional de Promoción Científica y Tecnológica (ANPCyT) PICT 2010 No. 0771.

References

[1] Pacheco-Pacheco M, Pacheco-Sotelo J, Moreno-Saavedra H, Diaz-Gomez J A, Mercado-Cabrera A, Yousfi M 2007 DBD-Corona Discharge for Degradation of Toxic Gases *Plasma Science and Technology* **9** No.6 682-685.

[2] Shang K, Zhuo Y, and Wu Y 2009 Simultaneous Removal of SO_2/NOx by Corona Discharge Plasma *ICEET 09 International Conference on Energy and Environment Technology* **3** 106-109.

[3] Grondona D, Allen P, and Kelly H 2011Development of a Coaxial-Stacked Trielectrode Plasma Curtain *IEEE Trans. Plasma Sci.* **39**, 1466-1469.

[4] Hernandes J A and Assis A K T 2005 Electric potential due to an infinite conducting cylinder with internal or external point charge *J. Electrost.* **63** 1115–1131.

[5] Liang, C H, Li L and Zhi H Q 2004 Asymptotic closed form for the capacitance of an arbitrarily shaped conducting plate *IEEE Proc.-Microw. Antennas* **151** No. 3 217–220.

[6] Raizer Y P 1991 *Gas Discharge Physics* Springer-Verlag Berlin Heidelberg .355.

Impulse Three Phase Power Supply Used for a Gliding Plasma Discharge

J A Salazar -Torres[1], J Pacheco-Sotelo[2], R Valdivia-Barrientos[2], M Pacheco-Pacheco[2], F Ramos-Flores[2], G Soria-Arguello[1] and M Ibañez -Olvera[1]

[1] Instituto Tecnológico de Toluca, Av. Instituto Tecnológico s/n, Metepec, CP 52140, México

[2] Instituto Nacional de Investigaciones Nucleares, Carretera México-Toluca s/n, Ocoyoacac, CP 52750, México

E-mail: jast2007@hotmail.com

Abstract. Power sources used for generating plasma have different configurations depending on the particular application; the aim here comprises the maximum energy transfer to the plasma discharge reaching. This work shows the performance and versatility of a simple impulse phase power source, applied to gliding arc plasma discharge. It is capable of changing the operating frequency from 5 kHz up to 150 kHz and the duty cycle from 1% to 33 % in all three phases, each one connected to three divergent tungsten electrodes. This allows a soft start plasma ignition until the full load is reached. This converter uses a sequential logic circuits composed by flip-flops, gates drivers, IGBT's and high voltage ferrite transformers. These features facilitate the maximum energy transfer to the plasma without using more complex electronic structures. The effect of frequency, duty cycle, voltage and current wave form signals is here described. This power supply has the adaptability to work whit different type of gas such as Argon, Helium, Air and Nitrogen. A Matlab Simulink simulation validates the experimental results. The main features and advantages of this configuration are also defined.

1. Introduction

One of the most important elements in a plasma discharge application is indubitably the power supply, since it must be capable to provide the necessary energy to ionize the gas leading the plasma discharge, in addition to guarantee the maximal power transfer to the load. There are several different power supply configurations, some of them use inverters, resonant circuits, and electronic converters supported by complex systems [1-4]. The impulse power supply here presented, unlike other systems, is able to modify the duty cycle in three or more phases providing a soft start plasma ignition. The operation principle is based in a sequential logic circuits which drives three IGB Transistors connected in common emitter configuration, which in turn drives three high frequency high voltage transformers. The gliding arc starts at the shortest gap (1.5mm) between divergent tungsten electrodes. In this region, due to thermal ionization of neutral particles, a thermal equilibrium between ions and electrons is obtained. Because of the axial gas flux entering into the plasma reactor, the electric arc glides along the electrodes and the discharge length grows until a critical length, determined by the power supply and gas flux; as the arc length expands, the heat dissipation increases. This results in a lowered temperature, and the plasma tends to be in non-thermal plasma discharge [5]. Beyond this point, the energy provided by the electrical supply can no longer sustain the equilibrium phase in the gliding arc and the electrical conductivity of plasma channel decreases until the discharge extinction; then, a restrike appears in the shortest gap, leading to a new periodic cycle, therefore, the discharge regime changes alternatively from thermal to a non-thermal equilibrium medium. Both regions exist in the same chamber reactor, but depending on the location, different kinetics schemes can interact. As an example, according to several authors non-thermal regions were more appropriated for greenhouse-gas treatment [6]. In a sequential process, the gliding arcs have lifetime's ranges from 1to 250ms depending on the gas flow, gas nature and on the electrical field applied in diverging electrodes. Since the arc discharge incessantly glides, neither electrode deterioration nor erosion exists, increasing the electrodes lifetime. This phenomenon has been exploited and has been used for the treatment of pollutants [7-8] and greenhouse gases [9-10]. Recently, the power supply here described has been used for acid gas (H_2S and CO_2) degradation, obtaining high conversion rates and selectivity as well as the specific energy (SE) required during decomposition processes required to generate a mole of CO and H_2 (syngas). Another important factor is the specific energy consumption (SEC) applied to the gas mixture entering to the plasma reactor (H_2S and CO_2), the experimental results of SE (17.62 kJ mol^{-1} and SEC (30.02 $kJmol^{-1}$), as well as the CO_2 and H_2S conversion (29.41 % and 80.44% respectively) and finally the reforming system effectiveness (ECE = 59.05%) were reported elsewhere in [11]. Taking this into consideration, the SE, SEC, and ECE results obtained, place the high frequency pulse gliding arc discharge as one of the most efficient methods for the syngas production.

2. Experimental Setup

The general experimental setup is shown in Figure 1, it consists mainly of four sections: the first section is constituted by the working gas used (helium, argon, air or nitrogen) and the control gas instrumentation, as gas flow controller, mixer, and control valves required for the gas ministration into the plasma reactor. The second section is precisely the gliding arc reactor, composed by a quartz tube of 6cm diameter and 25cm length, inside the quartz tube laid three divergent tungsten electrodes connected to the output high voltage transformer. These transformers (third section) generate the voltage needed to reach the breakdown voltage aforementioned. The fourth section is composed by digital instrumentation to record the necessary experimental data.

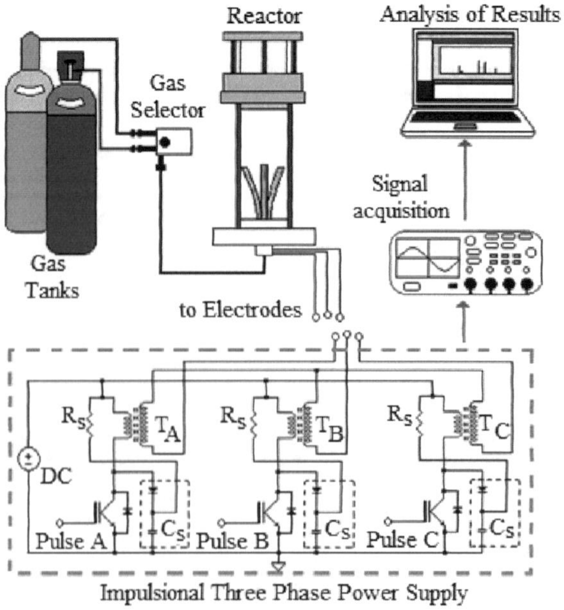

Figure 1. Experimental Setup.

The voltage required for generation of a gliding arc discharge depends on several factors, the nature of gas treated, the electrode material (tungsten), the distance between electrodes and the pressure at which it works. The breakdown voltage of the gas or gases involved in the discharge, can be calculated using the equation of Paschen's law [12] as follows:

$$V_{pd} = \frac{Bpd}{\ln(Apd) - \ln\left[\ln\left(1 + \frac{1}{\gamma}\right)\right]} \quad (1)$$

Where V_{pd} is the breakdown voltage, p is the pressure, d is the distance between the tungsten electrodes; A and B are constants that depend on the gas composition and finally γ is the emission coefficient rate of electrons [13].

According the physical disposition of tungsten electrodes (Figure 2a) the electric field could be obtained (Figure 2b) by using equation (1), the tungsten emission coefficient Van der Wall constants can be obtained in references [13] and [14] respectively.

The experimental Paschen curves are described in figure 2c for the different gas used. In the case of He gas, the work area shows the region were plasma ignition arises, compared with the other gases (Ar, Air and N_2), helium has the lowest V_{pd} (circa 1kV), once the arc electric strikes, it glides along the electrodes and the arc voltage increases to a level that is not more sufficient to sustain the discharge (more than 4kV) and the arc blows-down, a new cycle recommences in the shortest distance between electrodes. Even if the air and N_2 need a most high breakdown voltage, the impulse power supplied is designed to work with these gases.

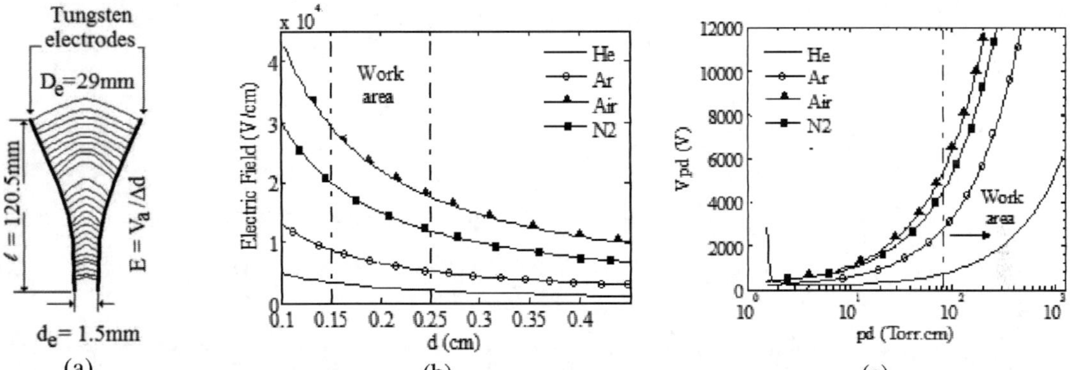

Figure 2. Experimental results: (a) physical specifications, (b) Electric field, (c) Breakdown voltage.

3. Impulse power supply- detailed description

The impulse three phase supply is illustrated in figure 3 just for a single phase; it is mainly composed of four stages: oscillator, sequential logic section, isolation section and power section. The PWM IC TL598 oscillator is responsible for generating the clock pulses, it can also varying the frequency in the order of 15 to 450 KHz and modify the duty cycle of 1% to 33%.

The second stage is a sequential logic circuit composed by a frequency divider (CI 74LS76 JK Flip-Flops) working in MOD8 to generate symmetrical and synchronized three signal pulses with a phase of 120° relative to one other. The next section is an isolation stage ensuring to minimize the possible noise interference from the power stage to the control stage. It consists of three optocouplers TLP250 IC and three Drivers (CI TC4422A) to ensure the required current to the pulse of each IGBT's gate. The fourth and final stage is the power section; composed by three IGBTs (SKM300GA123D) linked in common emitter configuration, and three high-gain (1:150) high frequency power transformers connected in delta configuration to the converged tungsten electrodes. Each transformer has 5KVA of power capacity.

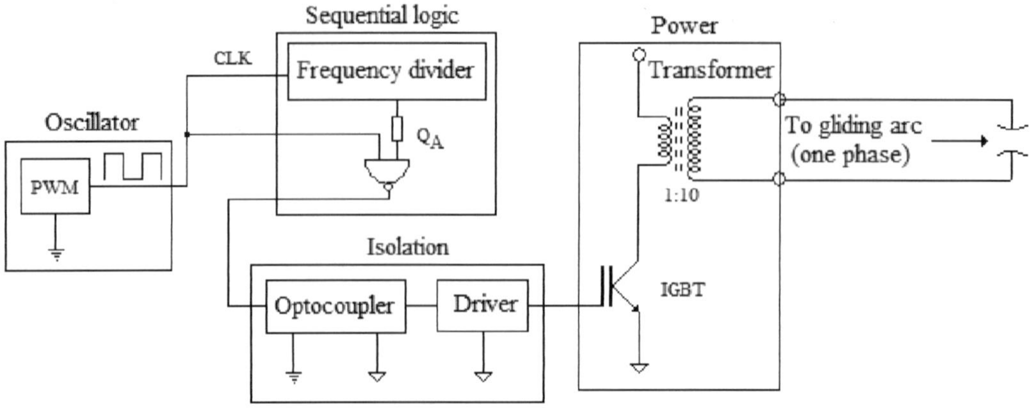

Figure 3. Diagram of the stages of digital control impulse for a single phase.

4. Analysis and results

In the oscillograme of figure 4 it can be seen that the pulses obtained at the gate of the IGBT 's, are shifted 120° relative to one other. Fig 4(a) shows 1% of duty cycle, while figure 4(b) has 33% of Duty Cycle. It is worth to mention that some dead time exist between each signal commutation, to avoid short circuit between phases. The control section works whit 15VDC.

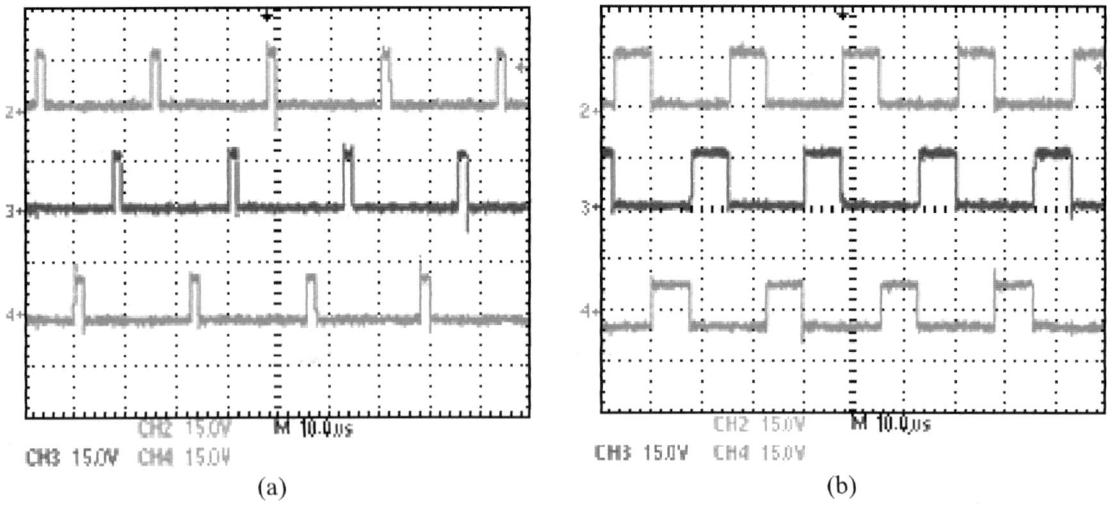

Figure 4. 15VDC gate pulses to (a) 1%,(b) 33%.

In the figure 5 shows the result of this experiment, where both the input power and the output power is measured to determine which ranges of frequency is optimized for maximum efficiency of the system in Helium and Argon plasma discharges.

In the case of Argon discharge (Figure 5a), the optimal frequency correspond to range from 5 kHz to 10 kHz. Note that after this range, the input power P_{in}, is increased considerably whit a relatively low output power P_{out}, consequently the efficiency decreases, therefore after this range it is not adequate to work in the case of argon discharge. Similarly, in the case of Helium discharge the most adequate operation range lies between 5kHz to about 27kHz.

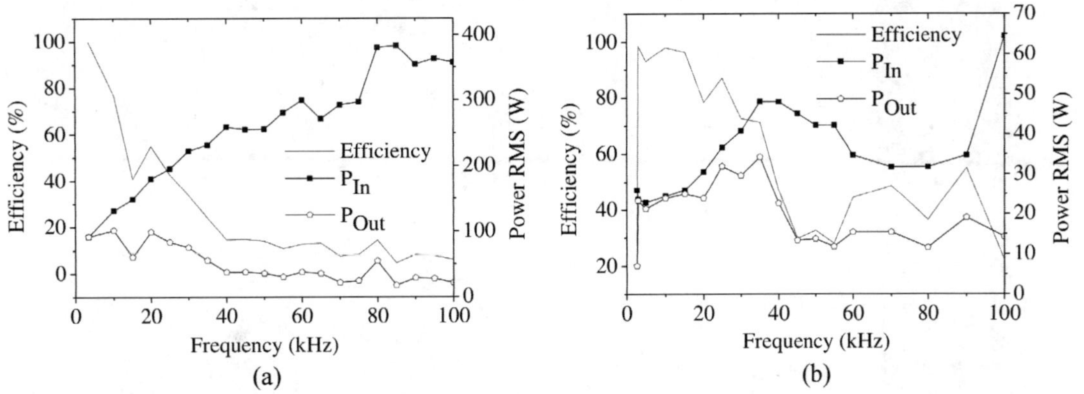

Figure 5. Effect of the frequency in: (a) Argon discharge, (b) Helium discharge

Figure 6 shows the electrical signals of instantaneous and RMS power derived from voltage and current which were obtained experimentally for different working gases: Figure 6a for Helium, Figure 6b for Argon, Figure 6c for Air, and Figure 6d for Nitrogen. In the same order, it can be observed that Helium needs only 75W of RMS power, while Argon needs 120W, Air requires 230W and Nitrogen involves 400W to work in stable conditions, this corresponds precisely for each gas conductance as was foreseen in Paschen curves in Figures 2b and 2c. Being precisely the nitrogen gas having the highest impedance and observing in the center column of Figure 6, it can be appreciated the evolution of the instantaneous change in impedance as the electric arc slides along the electrodes, in all cases the impedance shows a reduced value just when the discharge is well established, after that, the impedance has an exponential increase as the arc length grows until the electric arc extinction occurs and a new cycle begins. The nitrogen presents a more nonlinear behavior because some impedance variations are observed (from 29 to 38 μs) before the arc extinction, this is very possibly due to the nitrogen heat conductivity, which also varies depending on the electric arc temperature. From the instantaneous electrical signals it is possible to characterize the electrical behavior and to determine the impedance discharge for each gas analytical validated by using Matlab Simulink Algorithm in the forward section. This algorithm determines the impedance value, taking into account the resistive, capacitive, and inductive behavior of the plasma simply deduced by the electrical instantaneous signals *V(t)* and *i(t)*, according the following mathematical expressions:

$$R_d = \frac{\int_0^T V(t)i(t)dt}{\int_0^T (i(t))^2 dt} \tag{2}$$

$$L_d = \frac{\int_0^T V(t)\frac{di}{dt}dt}{\int_0^T \left(\frac{di}{dt}\right)^2 dt} \tag{3}$$

$$C_d = \frac{\int_0^T i(t)dt}{V(t)} \tag{4}$$

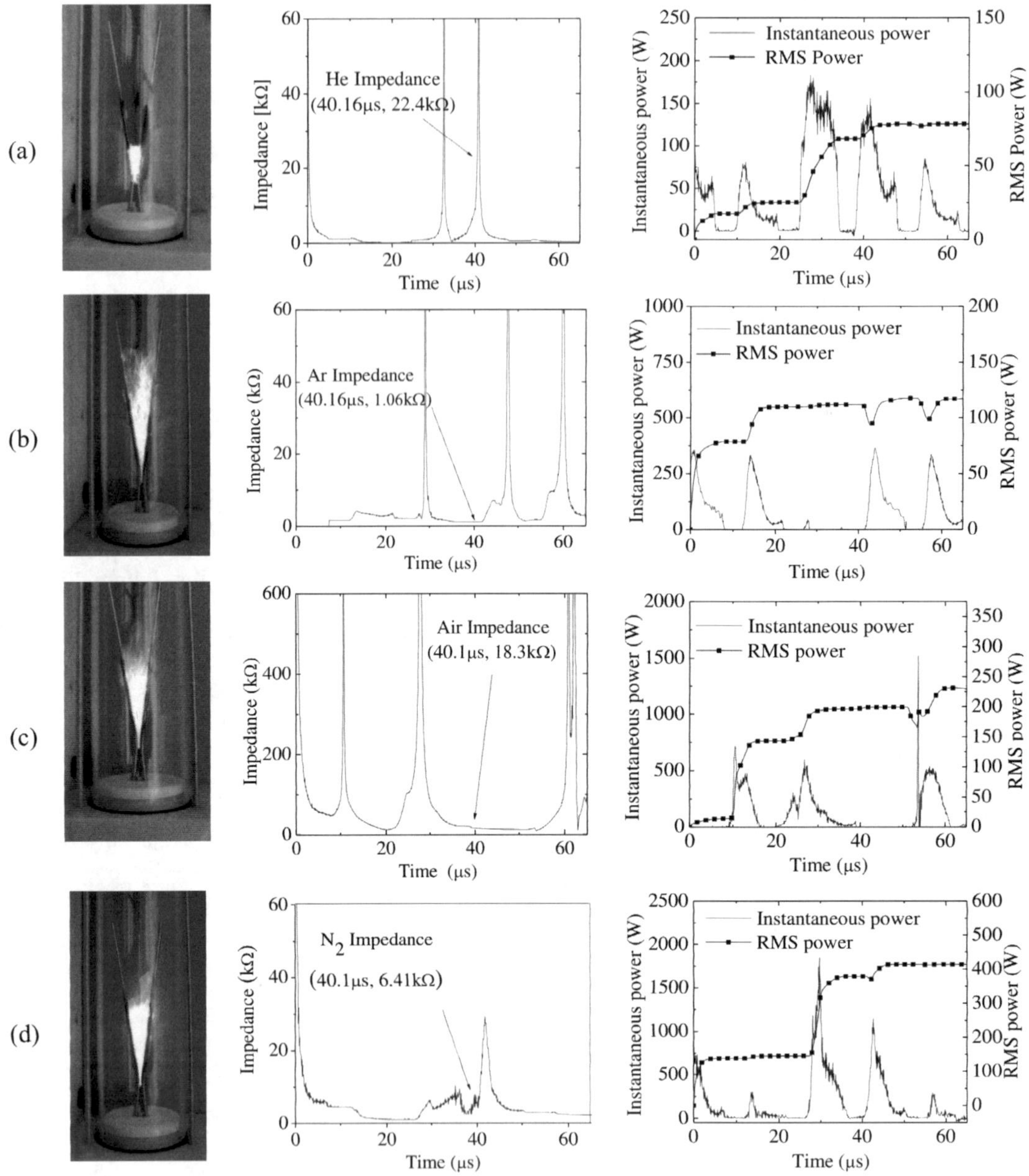

Figure 6. Electric signals and shock characteristics of hybrid plasma with different gases: (a) helium, (b) argon, (c) air and (d) nitrogen

Where R_d is the discharge resistance, L_d is the discharge inductance and C_d is the discharge capacitance, being $V(t)$ the instantaneous voltage applied to the discharge and $i(t)$ the instantaneous discharge current, which in turn correspond to the vectors of the experimentally obtained data voltage and the discharge current respectively.

After obtaining values expressed in equations 2, 3 and 4 one can next to determine the inductive reactance (X_{Ld}) and reactance capacitive (X_{Cd}) to finally find plasma impedance (Z_p), so, we have:

$$X_d = \sqrt{(X_{Cd})^2 + (X_{Ld})^2} = \sqrt{\left(-\frac{1}{2\pi f_s C_d}\right)^2 + (2\pi f_s L_d)^2} \quad (5)$$

Such that:

$$Z_p = \sqrt{(R_d)^2 + (jX_d)^2} \quad (6)$$

Being X_d extremely small, compared with respect to R_d, then, the plasma discharge impedance (Z_p), is purely resistive. In Figure 7, the algorithm which contains the aforementioned mathematical expressions is shown.

Also the power consumption is different, with nitrogen being the more power it consumes, but showing a better sliding, this according to the waveform of the instantaneous power of Figure 6d, where micro manifest -discharges that characterize this behaviour

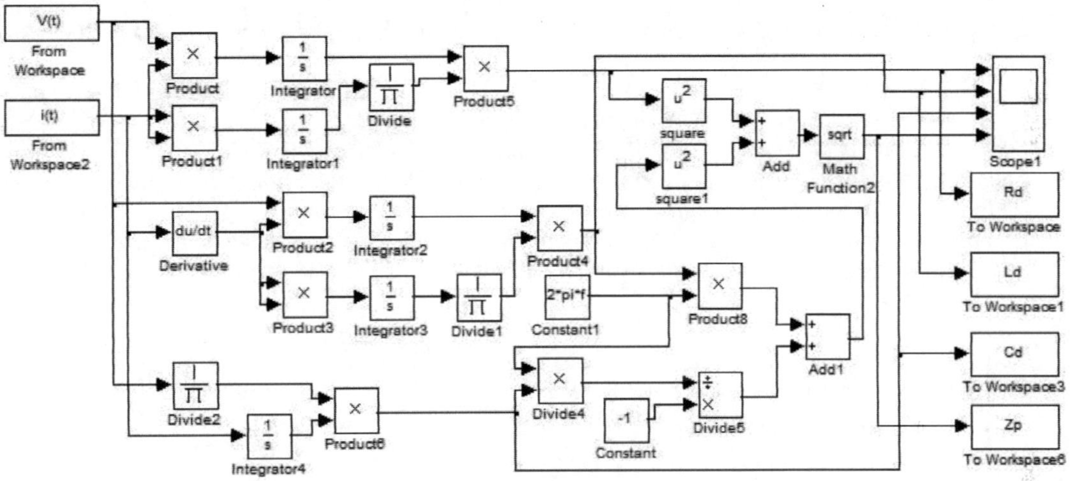

Figure 7. Algorithm in Simulink to calculate the impedance of the plasma.

5. Conclusions

This work shows the performance and versatility of a simple impulse power source, applied to gliding arc plasma discharge. It is capable of changing the operating frequency in a wide range, from 5 kHz up to 150 kHz. As well the duty cycle can be varied from 1% to 33 % in all three phases, besides; a dead time between impulsions avoids short circuit operation in power semiconductors.

Also, the paper describes a very versatile impulsion power supply, adequate to work in plasma discharges for different kind of gases; their principal features were exposed being the wide range of a frequency operation and duty cycle variation for soft start plasma discharge ignition. The gradual soft start avoids components damage in the power supply.

Moreover, this work defines the different impedance in plasma discharge according the gas used. By obtaining the instantaneous current and voltage signals, a data base can be constructed, subsequently, by the Matlab Algorithm application, the determination of the capacitance, inductance, resistance, and consequently impedance plasma discharge can be obtained.

The experimental system can be easily adapted to work with three or more electrodes, and the energy supplied is reduced to the required plasma value, depending on the kind of toxic gases treating. Finally it is found that the frequency and duty cycle are crucial for discharges with high efficiency.

Acknowledgments
The authors are very indebted to the collaborators of the LAP-ININ; F. Ramos Flores, M. Durán García, M. Hidalgo Pérez, G. Soria Arguello. This work was effectuated and financed by ININ, COMECYT and CONACYT.

References
[1] Yueh-Ru Y, Sakugawa T, Namihira T, Takaki K, Minamitami Y and Shimomura M 2007 *IEEE Trans. Dielectric and Electrical Insulated* **14** 1051-64
[2] Jiang C 2012 *Applied Mechanics and Materials.* **130-134** 3381-5
[3] Henryka-Danuta S and Grzegorz- Karol K 2010. *IEEE Region 8 SIBIRCON* 774-9
[4] Baba T, Takeuchi Y, Danuta S and Shin-Ichi A 2012 *Przegląd Elektrotechniczny (Electrical Review)* **NR 6** 88
[5] Pacheco J, García M, Pacheco M, Valdivia R, Rivera C and M Garduño 2012 *Journal of Physics: Conference Series* **370** (6pp)
[6] Ahmeda S, Aitania A, Rahmana F, Al-Dawoodb A, and Al-Muhaish F 2009 *Appl. Catal. A, Gen.*, **359** 1–24
[7] Liang Y, Xiandong L, Wang Y, Shengyong L, and Jian-Hua Y 2010 *J. Phys. Chem.* **114** 260-368
[8] Jian-Hua Y, Zheng P, Sheng-Yong L, Chang-Ming D, Xiao-Dong L, Tong C, Ming-Jiang N and Ke-Fa C 2007 *Journal of Environmental Sciences* **19** 1404–08
[9] Indarto A 2007 *Asian Journal of Water, Environment and Pollution*. **4** 191-4
[10] Zheng B, Jian-Hua Y, Xiaodong L, Yong C and Ke-Fa C 2008 *International Journal of Hydrogen Energy* **33** 5545 –53.
[11] Pacheco-Sotelo J, Salazar-Torres J A, Valdivia-Barrientos R, Pacheco-Pacheco M, Ibañez-Olvera M, Soria-Arguello G and Silva-Rosas J 2014 *IEEE Transactions on Plasma Science* **42** 767-773.
[12] Paschen F 1889 *Analytical Physics* **37** 69.
[13] Braithwaite N J 2000 *Plasma Sources Sci.Technol.* **9** 517-527
[14] Lieberman M A and Lichtenberg A J 2005 *Principles of Plasma Discharges and Materials Processing* ed Wiley Interscience (New Jersey: John Wiley and Sons) pp 545-6

Optical and electrical characterization of C_3H_6O/Ar glow discharge

S. Villalobos[1, 2], F. Castillo[1], O. Flores[1], P.G. Reyes[3], H. Martínez[1],

[1] Laboratorio de Espectroscopia, Instituto de Ciencias Físicas, Universidad Nacional Autónoma de México, Apartado Postal 48-3, 62251 Cuernavaca, Morelos, México
[2]Facultad de Ingeniería, Universidad Nacional Autónoma de México, D.F., México
[3]Facultad de Ciencias, Universidad Autónoma del Estado de México, Toluca, México
ciro@fis.unam.mx

Abstract. A low pressure glow discharge apparatus was used to produce a glow discharge of C_3H_6O/Ar mixture at a total pressure of 2.0 Torr. The emission spectra were measured in the wavelength range of 200 to 1100 nm. The principal species observed were Ar I, C_6H_4O, C_6H_5, CHO, CH_3O, CO_2, CO, H_2O, and H_2. The electron temperature and ion density have been measured using a double Langmuir probe, and the electron temperature and ion density were found in the order of 10 eV and 10^{10} cm^{-3}.

I. - Introduction

Recently, numerous experimental and theoretical works have been conducted to study of plasmas that contains hydrocarbons. This interest was mistreated by the advance of technologies of synthesis of new materials with unique mechanical, electrical, and optical properties [1–3], and also by investigation of processes that occur in space [4], nuclear reactors [5], in detoxification of gas combustion products [6], and in synthesis of composite hydrocarbons [7]. For optimization of available plasma-chemical synthesis technologies and development of new ones, it is necessary to study plasma-chemical reactions and kinetics of nonequilibrium plasma containing hydrocarbons, their fragmentation, and monitor changes in chemical plasma composition.

Plasma is a state that is far from thermodynamic equilibrium; for this reason it is not easy to model chemical processes that occur in gas phase and on adjacent surfaces. Therefore, for verifying the models and results of calculating the processes in plasma, it must be headed to measure its composition immediately during the process (in situ). The situation is complicated by the fact that kinetics and properties of any plasma (laboratory, technological, or natural plasma) are governed by products of stable molecule decay, in particular, by radicals; meanwhile, there are few means for their diagnostics. Radicals and other fragments of molecules in plasma containing hydrocarbons were most frequently studied using optical emission spectroscopy (OES) [8], laser-induced fluorescence (LIF) [9], mass spectroscopy [10], and absorption spectroscopy [2, 11].

Journal of Physics: Conference Series **591** (2015) 012063 doi:10.1088/1742-6596/591/1/012063

Generally, it is not easy to measure the radical concentration in plasma [12]. Direct measurements by (OES) require exact knowledge of electron density and energy along the observation region because, as a rule, only small part of electrons from the high-energy distribution function tail participates in excitation process. The LIF method requires complicated equipment, availability of proper energy levels filled with visible radiation from ground or excited states, and which is most important; the method makes it possible to measure only relative concentrations of radicals. Concerning the absolute values, their usage requires labor-intensive calibrations. Methods based on photon absorption in the visible, IR, and microwave regions were used for measuring absolute concentrations, of CH_3 [13], CH [14], and OH [15]. However, the most frequent method used for measuring radicals and other easily reacting plasma components was mass spectrometry [10].
In the present work, the goal was to explore the OES and to carry out measurements of electron temperature, and ion density in a glow discharge of Ar and C_3H_6O mixture at total pressure of 2.0 Torr.

II. – Experimental setup

The experimental apparatus and technique used to measure the AC plasma emission spectrum is identical to that described in previous work [16, 17]. A brief description is reported here. The system consists of two stainless steel circular plane electrodes, 3 mm thick and 30 mm in diameter. The electrodes are positioned at the centre of the reaction chamber with 10 mm gap spacing. The C_3H_6O and Ar were injected into the reaction chamber through the bottom flange. The same connection was used for the pressure sensor (MKS, Type 947 convection enhanced Pirani). An AC discharge of 60 Hz was used to generate the C_3H_6O-Ar plasma at a pressure of 2.0 Torr. An ultra pure Ar gas and C_3H_6O were used in the course of the measurements. The discharge power supply was maintained at an output of 450 Volts and a current of 3.5 mA (1.6 W). The vacuum chamber was purged with Ar gas several times at a pressure of 1.0 Torr in order to remove the background gas. In order to monitor the plasma discharge, in front of the lateral plane of the electrodes was located an optical fibber (Solarisation-resistant UV and fibber diameter size of 400 μm) that was connected to the entrance aperture of a high-resolution Ocean Optics Inc. Spectrometer Model HR2000CG-UV-NIR. The inlet and outlet slits were 5 μm wide. The data were obtained in a single accumulation of 100 s integration times. The fibber optics assembly was positioned to collect the emission light appearing on the C_3H_6O/Ar plasma. The wavelength accuracy of the spectrometer was calibrated using an Ar Source (Ocean Optics Inc.).
The double Langmuir probe employed in this work consists of a 0.25-mm-radius tungsten wire. The probe was located inside a capillary glass. The tips were extended 2.0 mm beyond the capillary glass with a gap between them of 2 mm [18]. The probe was aligned parallel to the electrodes of the discharge. The applied voltage to the probe was manually scanned from +30 V to -30 V and *vice versa* by a regulated DC power supply. The probe current was monitored by a multimeter. The scanning time for each *I-V* curve was 5-10 minutes. The final *I-V* curve obtained was the result of an average of 6 data scans, from this final curve the electron temperature (T_e) and ion density (n_i) were calculated.

III. - Results

To investigate the plasma properties, the plasma emission spectra were obtained with spectrometer (Ocean Optics) with an optical fibber though the Pyrex glass window. The measurement conditions, such as the integration time of the spectrometer and the position of the optical fibber, were fixed for every measurement.

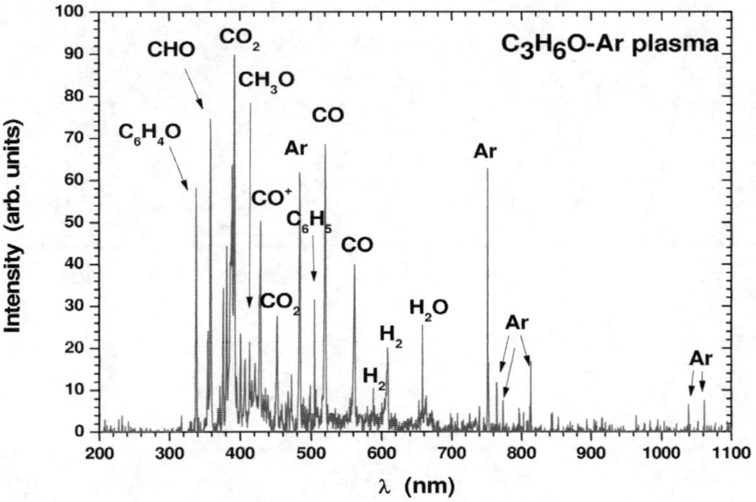

Fig. 1 The OES of C_3H_6O-Ar plasma at 2.0 Torr

A typical OES measurement of glow discharge C_3H_6O-Ar mixture at a pressure of 2.0 Torr is displays in Figure 1, showing the intensities of all observed emission bands and lines. This allowed the analysis of the most luminous area, which corresponds at the centre of discharge. Only the most intense spectral lines and bands of the plasma within the 200-1100 nm range are quoted. The principal species observed were: C_6H_4O at 338.11 nm; CHO at 358.75 nm; CO_2 at 391.2, 452.7 nm; CH_3O at 420.5 nm ; CO^+ at 427.43 nm; C_6H_5 at 504.8 nm; CO at 519.82 and 561.02 nm; H_2 at 588.82 and 608.08 nm; and Ar at 484.78, 751.46, 763.51, 771.37, 1047.01 and 1067.35 nm.

From the measured experimental current-voltage characteristic curve [18] obtained by the double Langmuir probe at a pressure of 2.0 Torr of glow discharge C_3H_6O-Ar mixture, it is possible to get electron temperature (T_e) and ion density (n_i). The T_e and n_i were obtained fitting the theoretical curve to the experimental data. With a correlation coefficient of 0.98, the theoretical curve fit very well to the measured data. That procedure gives an electron temperature of $T_e = (7.31 \pm 0.88)$ eV and ion density of ni = $(1.28 \pm 0.1) \times 10^{10}$ cm^{-3}. Error in the T_e and n_i were discussed in reference [19] and the overall variations in T_e and n_i were found to be 10% and 12% respectively.

In order to analyse the material deposited (MD) on the electrodes, this was collected by mechanical scraping Figure 2, displays Scanning Electron Microscopy (SEM) of the MD on the electrode during 50 h. Overview of the film (knackered) deposited on the electrodes and placed in a graphite film. To determine the surface contact, observations were made in the light microscope. It found that flat faces were in contact with the electrode of copper (Cu), and the rougher faces are exposed to the plasma. It behavior maybe explained on basis the growing of material particles as it's were deposited in the electrode.

Fig. 2 SEM of the material deposited on the electrode.

Different textures of the surfaces of the powders have been observed, a smooth and one rough with granular appearance. The rough surface corresponding to the shelf and it has a granular structure, which it could be formed from the particles generated in the plasma and deposited on the surface of Cu. The particles have a size of 1.3 ± 0.6 microns. From the analysis by EDX of the material deposited can identify the presence of C (92.41%), and O (6.5%). With traces of Si (0.22%), and Cu (0.87%), elements which should be contaminants from the Cu substrate.

IV. - Conclusions

The main species observed in a glow discharge of C_3H_6O-Ar plasma were the molecular bands of CO_2, CHO, CH_3O, CO, C_6H_4O, C_6H_5 and Ar atomic lines. Likewise, by the technique of double Langmuir probe electron temperature and plasma ion density were determined. T_e = (7.31 ± 0.88) eV and n_i = (1.28 ± 0.1) x 10^{10} cm^{-3}. The deposit is mainly composed mostly of carbon and oxygen in less proportion. The processes contribute more acetone to the environment than natural processes conning from of hydrocarbon derivatives (ketones) by its presence in natural (plants, trees, volcanic gases and forest fires) as well as in the exhaust gas of automobiles, smoke snuff and landfill. So, a more complete study of the films formed on the electrodes, give us more information about the processes that occur and that give rise to these films.

Acknowledges

The authors are grateful to H.H. Hinojosa and J. Rangel for technical assistance. This research was partially sponsored by DGAPA IN101613, CONACyT 128714.

References

1. - L. Mechold, J. Ropcke, X. Duten, and A. Rousseau, Plasma Source. Sci. Technol. 10, 52 (2001).
2. - F. Hempel, P.B. Davies, D. Loffhagen, L. Mechold, and J. Ropcke, Plasma Sourc. Sci. Technol. 12, S98 (2003).
3. - N. Mutsukura, Plasma Chem. Plasma Process. 21, 265 (2001).
4.- P. Coll, D. Coscia, M.C. Gazeau, E. De Vanssay, J.C. Guillemin, and F. Raulin, Adv. Space Res. 16, 93 (1995).
5.- F.L. Tabares, D. Tafalla, I. Tanarro, V.J. Herrero, A. Islyaikin, and C. Maffiotte, Plasma Phys. Control. Fusion 44, L1 (2002).
6.- B.M. Penetrante, M.C. Hsiao, J.N. Bardsley, B.T. Merritt, G.E. Vogtlin, A. Kuthi, C.P. Burkhart, and J.R. Bayless, Plasma Sourc. Sci. Technol. 6, 251 (1997).
7.- M. Kareev, M. Sablier, and T. Fujii, J. Phys. Chem. 104, 7218 (2000).
8.- U. Fantz, Plasma Sourc. Sci. Technol. 15 S137 (2006).
9.- L. Magne, S. Pasquiers, V. Edon, F. Jorand, C. Postel, and J. Amorin, J. Phys. D: Appl. Phys. 38, 3446 (2005).
10.- P. Kae-Nune, J. Perrin, J. Guillon, and J. Jolly, Plasma Sourc. Sci. Technol. 4, 250 (1995).
11.- A.K. Ngai, S.T. Persijn, F.J. Harren, H. Verbraak, and H. Linnartz, Appl. Phys. Lett. 90, 081109 (2007).
12.- D.C. Schram, Plasma Sourc. Sci. Techno. 18, 014003 (2009).
13.- A. Welzel, A. Rousseau, P.B. Davies, and J. Ropcke, J. Phys.: Conf. Ser., 86, 012012, (2007).
14.- R. Engeln, K.G. Letourneur, M.G. Boogaarts, M.C. Van de Sanden, and D.C. Schram, Chem. Phys. Lett. 310, 405 (1999).
15.- J. Ropcke, G. Lombardi, A. Rousseau, and P.B. Davies, Plasma Sourc. Sci. Techno. 15, S148 (2006).
16.- H. Martínez, O. Flores, B. Campillo, A. Gomez, L. Salazar, and J.C. Poveda, Radiation Effets & Defects in Solide 167, 582 (2012).
17.- G. García-Cosio, H. Martínez, M. Calixto-Rodriguez, A. Gomez, J. of Quantitative Spectroscopy & Radiative Transfer 112, 2787 (2011).
18. - J.D. Swift and J.R. Schwar, Electric Probes for Plasma Diagnostics, Elsevier, New York, (1991)
19.- L. Salazar-Flores, H. Martinez, A. Guerrero, P.G. Reyes, J. of Physical Science and Application 2, 283 (2012).

Influences of electrode configurations in dual capacitively coupled radio frequency glow discharge plasma

B. Bora[1,2] and L Soto[1,2]
[1]Comisión Chilena de Energía Nuclear, Santiago Chile
[2]Center for Research and Applications in Plasma Physics and Pulsed Power, P4, Chile

E-mail: bbora@cchen.cl

Abstract. Capacitively coupled radio frequency (CCRF) glow discharge plasma is widely studied in the laboratory because of its simpler design and high efficiency for different material processing applications such as thin-film deposition, plasma etching, sputtering of insulating materials etc. A negative dc potential develops between the bulk plasma and the powered electrodes, which is termed as 'self-bias' in RF plasma. This self-bias is generated as a consequences of the geometrical asymmetry of the electrodes, which can be achieved by appropriately design the area of the powered and the grounded electrodes. However, independent control of the dc self-bias in single frequency CCRF plasma is not possible, since the changing in any operating parameters including geometrical asymmetry will also change the plasma parameters. A study on the dual frequency CCRF plasma could be useful in understanding the separate control of the dc self-bias and plasma density, which respectively determine the ion energy and ion flux. In this work, a dual frequency CCRF plasma have been studied on the basis on nonlinear global model to understand the influences of electrode sizes and proper optimization of the CCRF plasma for specific applications.

1. Introduction

In recent years, capacitively coupled radio frequency (CCRF) glow discharge plasma is widely studied in the laboratory because of its simpler design, high efficiency and higher electron temperature for different material processing applications. Bulk plasma is produced in between two parallel electrodes by applying (Radio Frequency) RF voltage to one of the electrodes. The electrodes may be in contact with the discharge or they may be insulated from the discharge by a dielectric. A negative dc potential develops between the bulk plasma and the powered electrodes, which is termed as 'self-bias' in RF plasma. Due to this self-bias the ion energies near the powered electrode reaches up to few hundred eV. This feature makes the CCRF plasma suitable for application in different material processing areas such as thin-film deposition, plasma etching, sputtering of insulating materials etc. However, independent control of the dc self-bias in single frequency CCRF plasma is not possible, since the changing in any operating parameters including geometrical asymmetry also change the plasma parameters. A study on the dual frequency CCRF plasma could be useful in understanding the separate control of the dc self-bias and plasma density, which respectively determine the ion energy and ion flux to the powered electrode. Another aspect of the study on the dual CCRF plasma could be the generation of the dc self-bias in geometrically symmetric CCRF plasma. Presently, many reactive ion etchers use asymmetric plasma chamber to generate a dc self-bias. This requires that the chamber be much larger than the substrate being etched. The chamber could be made smaller in geometrically symmetric CCRF plasma by controlling the ion energy and generating the dc self-bias with the help of dual frequency plasma sources. It is worth to mention here that in our previous studies we have developed a dual frequency CCRF plasma model on the based on nonlinear global model [1] and studied the plasma series resonance heating and influences of frequency in dual frequency CCRF plasma [2, 3]. It have been also studied the geometrical asymmetry effect in 13.56 MHz CCRF plasma for understanding the self-generation of dc self-bias [4]. In this work, a dual frequency CCRF plasma have been studied to understand the influences of geometrical configurations of the electrodes on the plasma heating mechanism.

2. The model and the equations

The schematic of the dual frequency CCRF plasma model used in our study is shown in Fig. 1. It is worth mentioning here that similar kind of sheath model, known as nonlinear global model was used to study the Ohmic and stochastic heating by resonance effects in asymmetric CCRF plasma [1]. The

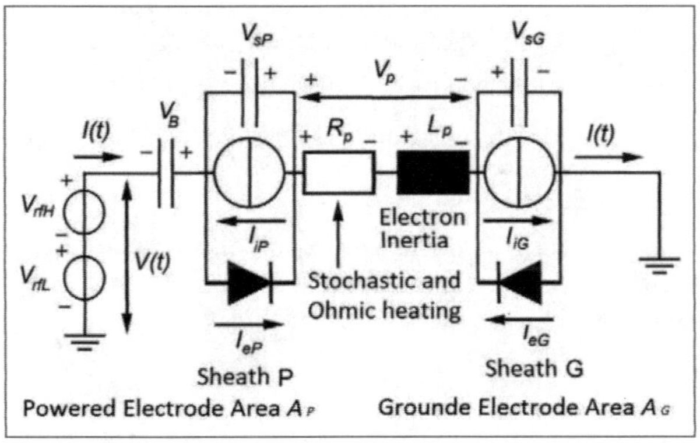

Figure 1. Schematic of the model used to study the influences of electrode geometry in dual frequency CCRF plasma.

nonlinear global model was also used to experimentally measure the plasma series resonance heating and evaluate the plasma parameters from the electrical discharge characteristics [5-9]. It have been also used similar kind of model to study the dual frequency geometrically symmetric CCRF plasma [2, 3]. By assuming the ion density is homogeneous inside the both sheaths, the voltage drop across the sheath at the powered electrode ($V_{S,P}$) and the grounded electrode ($V_{S,G}$) could be correlated with the sheath charges (Q_{SP} and Q_{SG}) by sheath capacitances (C_{SP} and C_{SG}) respectively as,

$$C_{S,P}=\frac{Q_{S,P}}{V_{S,P}}=\left(\frac{2e\varepsilon_0 n_e A_P{}^2}{V_{S,P}}\right)^{1/2} \tag{1a}$$

$$C_{S,G}=\frac{Q_{S,G}}{V_{S,G}}=\left(\frac{2e\varepsilon_0 n_e A_G{}^2}{V_{S,G}}\right)^{1/2} \tag{1b}$$

where, A_P, A_G and n_e are the powered, grounded electrode areas and electron density respectively. The bulk of the plasma is assumed to be quasineutral and homogeneous. The RF current through the bulk of the plasma is carried by electron conduction alone and can be model by Ohm's law by taking into account the acceleration of the electrons by the electric field and their momentum loss due to elastic collisions with the neutrals of the background gas. The collision rate is given by momentum transfer collision frequency (ν_m). It is also known that the stochastic heating is also plays a crucial role in heating of the CCRF plasma at low pressure [1-4]. Thus, considering the Ohmic and stocasting heating, the effective collision frequency can be represented by considering the hard wall model as,

$$\nu_{eff} = \nu_m + v_e / l_p \tag{2}$$

where, $v_e = \left(8eT_e / \pi m \right)^{\frac{1}{2}}$ is the mean thermal speed of the electron [1-4] and l_p is the length of the plasma bulk, m and T_e is the electron mass and electron temperature respectively. Due to the inertia of the electron, the plasma bulk behave like an inductance (L_P) and could be represented as [6-8]

$$L_p = \frac{1}{\omega_{pe}^2 C_0} = \frac{m l_p}{A_p n_e e^2} \tag{3}$$

where, $w_{pe} = \sqrt{n_e e^2 / \varepsilon_0 m}$ is the plasma frequency and $C_0 = \varepsilon_0 A_P / l_p$ is the capacitance of the parallel electrodes.

The electron current inside the sheath at powered and grounded electrode are taken as $I_{e,P} = A_P e n_e v_e \exp\left(-V_{S,P} / T_e \right)$ and $I_{e,G} = A_G e n_e v_e \exp\left(-V_{S,G} / T_e \right)$ respectively. Similarly, the ion currents inside the sheaths at the powered and the grounded electrode are taken as $I_{i,P} = A_P e n_e v_{Bohm}$ and $I_{i,G} = A_G e n_e v_{Bohm}$, with $v_{Bohm} = \left(eT_e / M \right)^{\frac{1}{2}}$ is the Bohm speed and M is the ion mass.

To study the influences of electrodes size in dual frequency CCRF plasma, two voltage waveforms of the angular frequency $\omega_1 = 2\pi f$ and $\omega_2 = 2\omega_1$ with f =13.56 MHz are applied with same amplitude with phase angle ϕ. Thus, the superimposed voltage can be expressed as,

$$V(t) = V_{RF}\left[\cos(\omega_1 t) + \cos(\omega_2 t + \phi)\right] \tag{4}$$

Applying Kirchhoff's law in the model, it can expressed the two sheath charges ($C_{S,P}$, $C_{S,G}$), rf plasma current ($I(t)$) and dc self-bias (V_B) in the following form,

$$\frac{dQ_{S,P}}{dt} = -\left[I(t) + A_{p^{LF}} j_{i0} - A_{p^{LF}} j_{eo} \exp\left(-\frac{Q_{S,P}^2}{2e\varepsilon_0 n_e A_p{}^2 T_e} \right) \right] \tag{5}$$

$$\frac{dQ_{S,G}}{dt} = -\left[-I(t) + A_G j_{i0} - A_G j_{eo} \exp\left(-\frac{Q_{S,G}^2}{2e\varepsilon_0 n_e A_G{}^2 T_e} \right) \right] \tag{6}$$

$$\frac{dI(t)}{dt} = \frac{1}{L_p}\left[V(t) + \frac{Q_{S,P}^2}{2e\varepsilon_0 n_e A_P{}^2} - \frac{Q_{S,G}^2}{2e\varepsilon_0 n_e A_G{}^2} + V_B \right] - \nu_{eff} I(t) \tag{7}$$

$$\frac{dV_B}{dt} = -\frac{1}{C_B} I(t) \tag{8}$$

The equations (5) - (8) are solved with the typical set of plasma parameters, n_e=5x10^{16} m^{-3}, T_e = 1.5ev, l_p=5.7 cm, l_b=0.0027 cm, and ν_m=$K_m n_g$, where K_m = 10^{13} m^{3}s^{-1} is the rate coefficient, n_g (m^{-3}) =3.3x10^{22}

p (Torr) is the neutral gas density, p=2mTorr is the pressure and V_{RF} = 200 volts . The ratio of the electrode geometry (A_G/A_P) varied from 1 to 16, phase angle ϕ from 0 to π. During the computation, the values of the A_P kept constant (with radius 5.7 cm) and varied A_G accordingly. The total dissipation (η_{total}), which is of course the sum of Ohmic and stochastic dissipations are calculated from the rf plasma currents as,

$$\eta_{total} = \left\langle \left(v_m + \frac{v_e}{l_p} \right) L_p I(t)^2 \right\rangle \tag{9}$$

3. Results and discussions

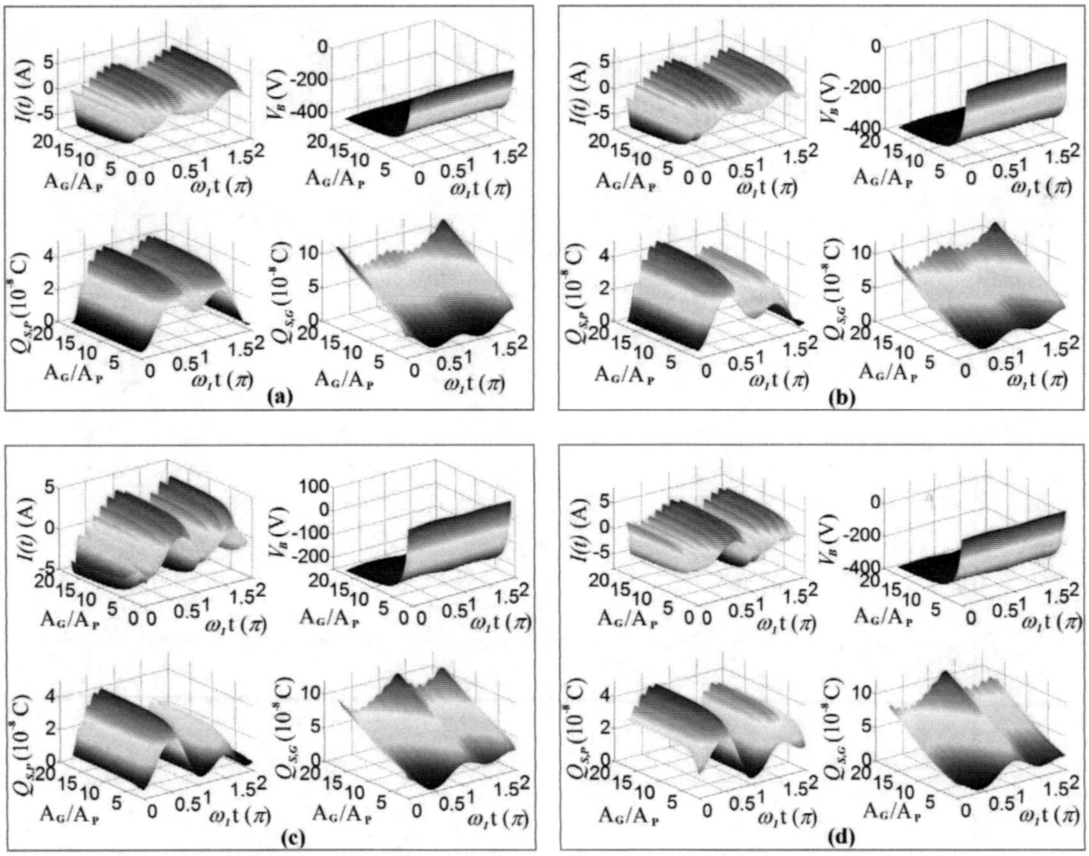

Figure 2. Time evolution rf plasma current, dc self-bias and two sheaths charges for different electrodes area for (a) $\phi = 0$, (b) $\phi = \pi/4$, (c) $\phi = \pi/2$ and (d) $\phi = 3\pi/4$.

The time evolution of the rf plasma current, dc self-bias and the two sheath charges are shown in Fig. 1. It have been observed that distortion in the current waveforms, which is also known as plasma series resonance effect as a results of nonlinearity of the sheaths increase with increasing the ratio of the electrodes area. It have been also observed that the difference in the sheath charges between the two electrode increases significantly with increasing the ratio, which may probably increase the nonlinearity of the sheath charges. Although, significant distortions are present in the rf current and the two sheath charges, no such considerable distortions are observed in the time evolution of the dc self-bias. However, the time dependent of the dc self-bias is observed according with the applied rf voltage waveforms. The dc self-bias is found to increase rapidly with early increase of ratio and looks like saturated after a certain value of the ratio. The magnitude of the dc self-bias is also found to depend on the phase angle between the two applied waveforms.

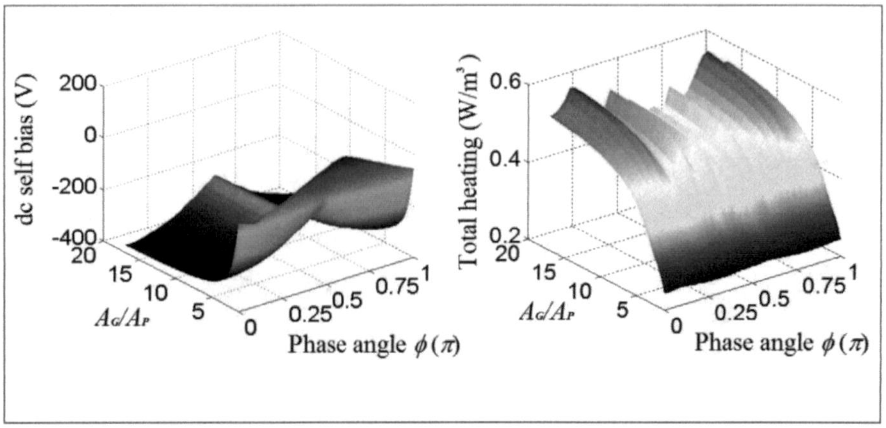

Figure 3. 3 dimensional plot for dc self-bias and total plasma heating (η_{total}) with phase angle (ϕ) and ratio of the electrodes.

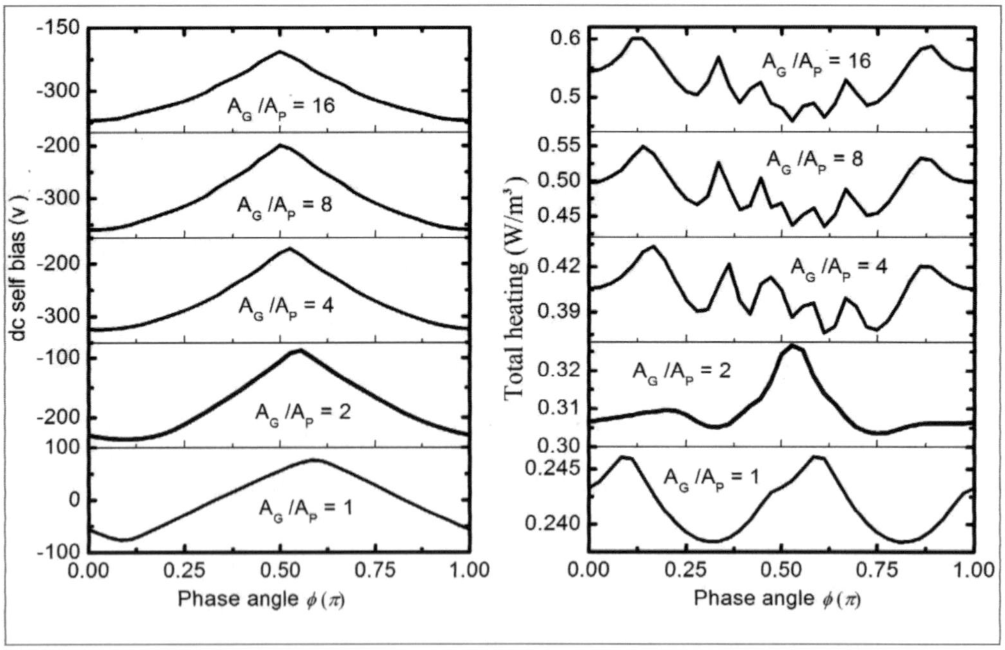

Figure 4. Variation of dc self-bias and total plasma heating (η_{total}) with phase angle (ϕ) for different ratio of the electrodes showing the changes of polarity of the dc self- bias with phase difference for $A_G/A_P = 1$ and local maxima and minima in the variation in total plasma heating.

Figure 3 shows the variation of the time average dc self-bias and total plasma heating with the phase angle and the ratio of the electrodes area. The dc self-bias is found to varied significantly with phase angle depending on the ratio between the electrodes area. For geometrically symmetric (A_G/A_P =1) case, it is observed that the polarity of the dc self-bias could be altered by adjusting the phase angle as shown in Fig. 4. For A_G/A_P =1, or also known as geometrically symmetric case, the variation of dc self-bias is found to well agree with the experimentally observed data [10]. The total heating is also found to increase with increasing ratio of the electrodes area. However, the total heating is found to depend complexly on phase angle showing several local maxima and minima. It is worth to mention here that local maxima and minima on the variation of total heating with driving frequency was also observed due to the excitation of nonlinear plasma series resonance heating [3,11]. Although, it is not conclusive, it is suspected that the plasma series resonance heating may be the reason for local maxima or minima observed in the variation of total plasma heating with phase angle, which may require further studies and understanding.

4. Conclusions

A dual frequency CCRF plasma have been studied on the based on previously developed nonlinear global model to study the influences of electrodes size on the plasma characteristic. It have been observed that the dc self-bias strongly depends on the ratio between the electrodes area. The dc self-bias is also found to depends on the phase angle between the two applied voltage waveforms.

Acknowledgments

Supported by CONICYT grant ACT 1115 and FONDECYT grant 11130048.

References

[1] Mussenbrock T, Brinkmann R P, Lieberman M A, Lichtenberg A J and Kawamura E 2008 *Phys. Rev. Lett.* **101** 085004
[2] Bora B, Bhuyan H, Favre M, Wyndham E and Chuaqui H 2012 *Appl. Phys. Lett.* **100** 094103
[3] Bora B, Bhuyan H, Favre M, Wyndham E and Wong C S 2013 *J. Appl. Phys.* **113** 153301
[4] Bora B and Soto L 2014 *Phys. Plasmas* **21** 083509
[5] Bora B, Bhuyan H, Favre M, Wyndham E, Chuaqui H and Kakati M 2011 *Phys. Plasmas* **18** 103509
[6] Bora B, Bhuyan H, Favre M, Wyndham E, Chuaqui H and Kakati M 2012 *Phys. Plasmas* **19** 064503
[7] Li S Z, Lim J P and Uhm H M 2006 *Phys. Lett. A* **360** 304
[8] Bora B, Bhuyan H, Favre M, Wyndham E and Chuaqui H 2012 *Physics Letters A* **376** 1356
[9] M. A. Lieberman and A. J. Lichtenberg 2005 Principles of Plasma Discharges and Materials Processing, Wiley, New York
[10] Schungel E, Zhang Q Z, Iwashita S, Schulze J, Hou L H, Wang Y N and Czarnetzki U 2011 *J. Phys. D: Appl. Phys.* **44** 285205
[11] Semmler E, Awakowicz P and von Keudell A 2007 *Plasma Sources sci. Technol.* **16** 839

Journal of Physics: Conference Series **591** (2015) 012065 doi:10.1088/1742-6596/591/1/012065

Striations in an ethyl alcohol glow discharge

P.G. Reyes[1]; A. Gómez[1]; C. Torres[1]; H. Martínez[2]; F. Castillo[2], J. Vergara[3]
[1]Laboratorio de Física Avanzada, Facultad de Ciencias, Universidad Autónoma del Estado de México, C.P. 50000, Toluca, Estado de México, México.
[2]Laboratorio de Espectroscopia, Instituto de Ciencias Físicas, Universidad Nacional Autónoma de México, A.P. 48 3, C.P. 62251, Cuernavaca, Morelos, México.
[3]Laboratorio de Análisis y Sustentabilidad Ambiental, Instituto Profesional de la Región Oriente, Universidad Autónoma del Estado de Morelos, C.P. 62715, Xalostoc, Morelos, México.

E-mail: pgrr@uaemex.mx

Abstract. This research shows the behavior of striations in glow discharge generated with high purity ethyl alcohol at a pressure of 0.6 Torr. This paper present the number of striations as a function of the of current and voltage discharge.

Introduction

The formation of alternating dark and bright regions (striations) in the positive column of a glow discharge is a universal property of discharges in tubes [1]. This phenomenon is observed in atomic and molecular gases. The striations can be either stationary or can move. The interest of 3-D stratification in a glow discharge has about ten years [2–6]. It was shown that the spatial luminosity regions for geometry may represent the number of lighting thin closed surfaces nested into each other. The main physical peculiarity of the discharge in this geometry is that recombination processes occur in a volume, contrary to the discharge in a tube, where recombination takes place on the walls. The visual stratifications not observed in simple gases (Ar, N_2, He). The admixture of high molecular gases (acetone, alcohol, benzol, and some others) results in well-pronounced discharge stratification. The parameters of striations (number, geometry spectrum, and intensity) depend on gas pressure, gas composition, current, and geometry of the discharge gap, change of these parameters allows the verification of intensity, color, shape, and topology of a luminosity region. For example, increasing the current leads to the enlargement of striation dimension so that the current density remains the same. The present work studied the striations produced in a glow discharge of ethyl alcohol.

Experimental details

Experimental setup is shown in Fig. 1. The experiments were carried out in an aluminum cylindrical vacuum chamber of 24 cm in length and 21 cm in diameter. The discharge cell consists of two movable parallel electrodes enclosed in the vacuum chamber. The two electrodes were made of planar-copper with5.0 cm in diameter. The electrodes are positioned at the center of the reaction chamber with 10 mm gap spacing. The plasma chamber has a volume of 8.31×10^3 cm^3 and it was pumped down by a vacuum system Varian D302 to a base pressure of 10^{-3}Torr. The ethyl alcohol vapor inlet is controlled by a micro regulator valve with a continuous flow, the alcohol purity has HPLC grade. A DC glow discharge of ethyl alcohol vapor was produced between the two electrodes, keeping a total pressure of 0.6 Torr. The current discharge was between 6.3 and 70.0 mA (measured by a digital Fluke multimeter model 8846A), while the power supply voltage (Spellman SA4) in the range of 470 to830 V. The voltage and current discharges were found as function of the number of striations produced in the glow discharge. In the lateral flange, a quartz window was used to monitor the active species generated in the glow discharge by plasma emission spectroscopy; the spectrum (200–1000 nm) of the emission cell was measured using a spectrometer Ocean Optics HR4000CG-UV-NIR. The spectrometer has a resolution of 0.15 nm.

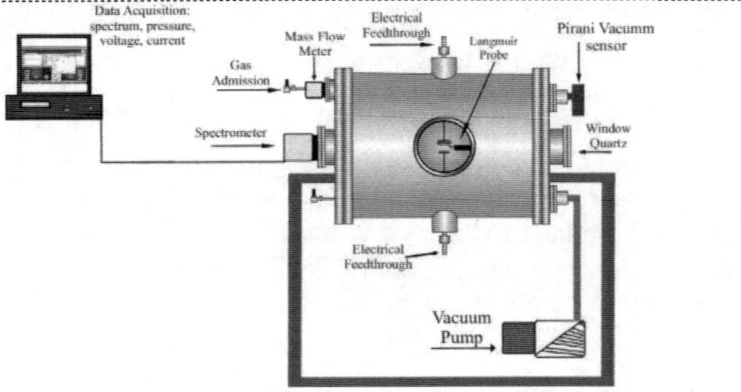

Figure 1. Experimental setup

Results

Optical emission spectroscopy (OES) measurement of anethyl alcohol (CH_3CH_2OH) glow discharge at pressure of 0.6 Torr is display in figure 2.Keeping the pressure and the flow of ethyl alcohol into the reaction chamber and varying the current and voltage of the discharge, it can be observed the formation of striations in the positive column. As the electrical power is incremented the number of striation increase. The electrical power of 2.96 (6.3 mA at 0.47 kV), 8.26 (15.6 mA at 0.53 kV), 20.79 (33.0 mA at 0.63 kV) and 58.1 Watts (70.0 mA at 083 kV) correspond to 1, 2, 3 and 4 striations respectively. This means that the plasma needs more electrical power in the mixture to generate more striations.

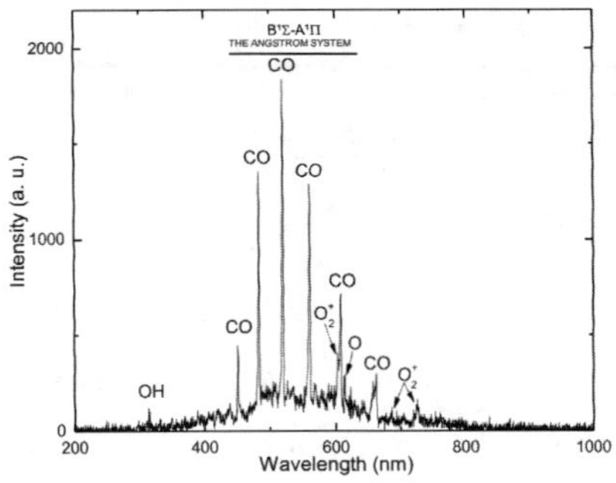

Figure 2. Optical emission spectrum corresponding to an Ethyl alcohol glow discharge, in which it can see the principal species generated.

Figure 2 shows the measurement of OES for the case of the formation of 4 striations.In the case of 1, 2 and 3 striations, the corresponding spectrums have the same bands with only variations in the intensity less than 10%.

The main elements identified in the ethyl alcohol glow discharge are:OHat 312.64 nm ($X^2\Pi$-$A^2\Sigma^+$); CO at 450.55 nm ($d^3\Pi$ -$a^3\Delta$, the Triplet bands); CO at 483.53, 519.82, 561.02, 607.99 and 661.90 nm ($B^1\Sigma$-$A^1\Sigma$, the Angstrom System); O_2^+at 602.27, 685.63 and 723.50 nm ($b^4\Sigma_g^+$-$a^4\Pi_u$,the First Negative System) and O at 613.70 nm (3p5P - $4d^3D°$).From the OES measurements, it can observe that the most intense bands emitted in the ethyl alcohol glow dischargewere O_2^+ and CO [7].The main CO bands are more common found in the range of length wavelength of 412-662 nm [8-9], which correspond to the Fourth positive systemand third positive Angstrom system. In the present experiment, the principal CO band observed by OES measurement was the positive Angstrom system.

Figure 3 displays the power needed to generate the discharge as a function of the number of striations.Fora glow discharge of alcohol at0.6 Torr and power of 58.1 Watt, it is possible to see four striations figure 4d, withwell-defined geometry and intensity. Subsequently, the power was decreased to yield 3, 2 and 1 striations, with powers of 20.79, 8.26, 2.96 Watt, respectively.It can be seen that the electrical power to produce onestriation is approximately 19 times lower than that for producing 4 striations, in this case the voltage and current is kept constant when formed striations. The luminescence is produced in the cathode.

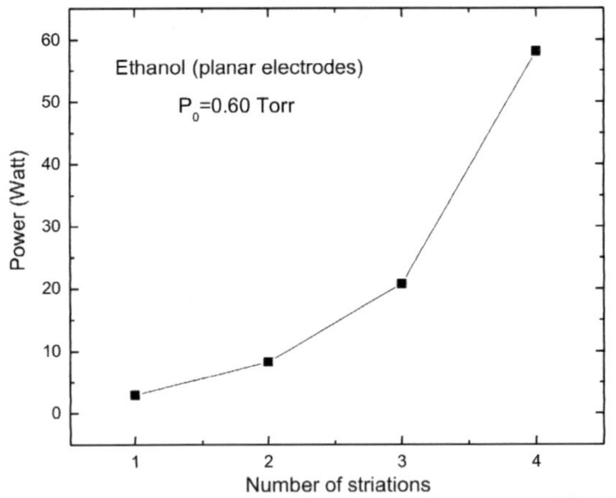

Figure 3. Number of striationsas a function ofelectrical power applied to the discharge.

The striations were stable once they are formed, as it can see in Figure 4.The formation, geometry and color intensity of the images 4cand 4dshow more clearly that the images 4a and 4b. The experiment was a self-sustained glow discharge in all cases shown in figure 4,and it was observed that the parameters measured depend on the composition of the ionized gas and the material of the electrodes, besides the pressure within the reaction chamber. By analysingthe ratio ofthe radius of the striations, it is observed that these are concentric and keep a constant rate depending on the number of striations formed.It found that the relation (r_{n-1}/r_n) where r is the striations radius and n number of striationsis 0.74.

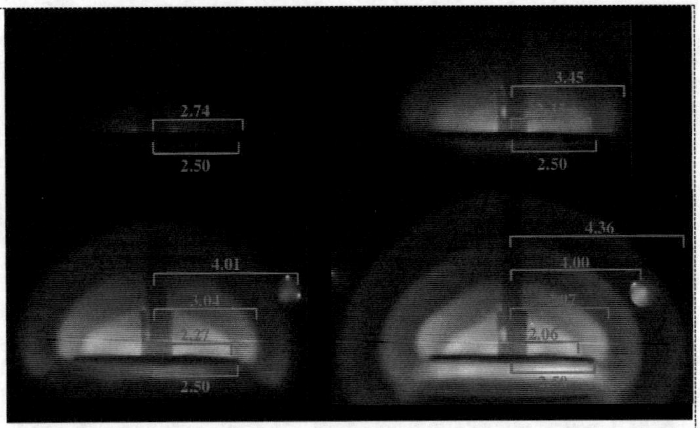

Figure 4. Pictures of striations as afunction of power discharge.

Conclusions

In the present experiment, the number maximum of the striations formed was four. The number of striation grows as the power of the glow discharge increase.The striations only were observed at low pressure. The most intense band observed by OES measurements was CO transitions that correspond to the Angstrom system.It found that the ratio (r_{n-1}/r_n) is 0.74.

Acknowledgements

We are grateful to N. Rodriguez (FC-UAEM) and O. Flores (ICF-UNAM) for their technical assistance. This research was supported by UAEM [3443/2013CHT], DGAPA [IN101613], CONACyT [128714],PROMEP [103.5/13/6626] and PIFI 2010.

References

[1]NovopashinS A, RadchenkoV V and SakhapovS Z 2008 IEEE Trans. Plasma Sci. **36** 998.
[2]Nerushev O A, Novopashin S A, Sukhinin G I and RadchenkoV V 1998 Phys. Rev. E **58** 4897.
[3] Conde L and Leon L 1999 IEEE Trans. Plasma Sci. **27**80.
[4] Conde L, Ibanez L F, and Ferro-FontanC2001 Phys. Rev. E**64**6402.
[5] Dimitriu D G, Aflori M, Ivan L M, Ionita C, and Schrittwieser R W 2007 Plasma Phys. Control Fusion**49**237.
[6] Hoshi Y, Yoshida H, and Tsutsui Y 2002 J. Appl. Phys.**92**5668.
[7] Jing-Liang TAO, Yuan-Quan XIONG 2013 Acta Phys. -Chim. Sin, **29** (1), 205-211.
[8] Krupenie Paul H. 1966 *The Band Spectrum of Carbon Monoxide* (United States of America: National Standard Reference Data Series)
[9] Raymond T. Birge 1926 Phys. Rev. Lett.**28**, 1157–1181.

Electrical and optical characterization of CO_2/He glow discharge

P. G. Reyes[1], C. Torres[1], J. Torres[1], A. Gómez[1], H. Martínez[2], J. Vergara[3]
[1] Laboratorio de Física Avanzada, Facultad de Ciencias, Laboratorio de Física Avanzada, Universidad Autónoma del Estado de México, C.P. 50000, Toluca, Estado de México, México.
[2] Laboratorio de Espectroscopia, Instituto de Ciencias Físicas, Universidad Nacional Autónoma de México, A.P. 48 3, C.P. 62251, Cuernavaca, Morelos, México.
[3] Instituto Profesional de la Región Oriente, Universidad Autónoma del Estado de Morelos, C.P. 62715, Xalostoc, Morelos, México.

E-mail: pgrr@uaemex.mx

Abstract. Glow discharge studies are produced in a atmosphere of CO_2/He mixture at a pressure of 1.5 Torr. The experiment was conducted to observe the optical emission spectroscopy of the plasma, and determined the electron temperature and ion density by the use of a double Langmuir probe. The electron temperature of the mixture was found in the range of 4.54 to 5.37 eV, and the ion density values between 2.15 and 12.00 x 10^9 particles/cm^3. The principal species observed were: CO_2^+, CO^+, CO, C_2, O_2^+, O*, and He*.

1. Introduction

Cold plasmas studies produced at low pressures (few Torr) are of great importance due to their applications in various scientific and technological fields, for example, the characterization of the glow discharge parameters as voltage, current, electron temperature and density, which are essential for their uses in the creation or modification of materials, used in industry; for the simulation of planetary atmospheres[1, 2] and understanding of atomic and molecular processes that occur in such systems. CO_2 is one of the gases that make up our atmosphere, its density is 379 ppm (particles per million) but slight increases in his percentage is one of the responsible of the greenhouse effect [3]. Greenhouse Gas Inventory Report has stated that CO_2 emissions account for 85% of all greenhouse gas emissions with the majority of these emissions created as the result of fossil fuel combustion and the increase in the concentration of CO_2 in the air [3, 4]. Several solutions have been proposed to mitigate the environmental effects of rising CO_2 emissions through both government policy and scientific innovation [5]; direct conversion processes remain the main option for thin films disposition or treatment of materials, with plasma assisted methods, RF plasmas[5, 6], corona discharge [7], dielectric barrier discharge [8] and glow discharges [9]. Due to the damaging effect on the environment than fossil fuels leads, it has been looking for alternative sources of energy waiting for the damage decrease in the near future [10]; why searching for a solution to this problem, so it is important to find physical mechanisms leading to dissociate the main element resulting from the combustion process. This paper presents the electrical and optical characterization of a CO_2/He glow discharge.

2. Experimental details

The schematic diagram of the experimental setup is described in previous work [11, 12]. The discharge cell consists of two movable parallel electrodes enclosed in a stainless steel vacuum chamber. The two electrodes were made of copper disc of 2.5 cm in diameter. The electrodes are positioned at the center of the reaction chamber with 20 mm gap spacing. The plasma chamber had a volume of 1.16×10^4 cm^3 and it was pumped down by a vacuum system (Turbomolecular pump Alcatel ATP80 and mechanical pump Varian DS302) to a base pressure of 10^{-6} Torr. A continuous dynamic flow of CO_2–He gas mixture (ultra-pure gases, Praxair 99.99%) was let in the system through needle valves at the desired pressures. A DC glow discharge was produced between the two electrodes. While keeping the total pressure at 1.5 Torr, the concentration of CO_2 gas in the mixture was done by changing the He partial pressure. To keep a current constant of 10 mA (measured by a digital Fluke multimeter model 8846A), independent of the gas mixture, a ballast resistance (R = 5kΩ) was used, it was done by changing the power supply voltage (Spellman SA4) between 300V (for 100% of He) to 500V (for 100% of CO_2). The discharge voltage in the gas mixture is lower than that in pure carbon dioxide. In a lateral flange, a quartz window was used to monitor the active species generated in the glow discharge by optical emission spectroscopy; the spectrum (200–1100 nm) was measured using a spectrometer Ocean Optics HR4000CG-UV-NIR. The spectrometer has a resolution of 0.15 nm. For the electron temperature and density measurements, the voltage difference applied to the probe was performed by computer control scanning to -30 V to +30 V by a regulated DC power supply. The scanning time for one *I-V* curve was 5-10 minutes. Owing to sputtering or contamination of the probe tips, the total measurement time for one probe was restricted to approximately 5-6 h. The final *I-V* curves obtained were the results of an average of 15 data scans at each probe voltage.

3. Results

From the measured experimental current-voltage characteristic curve (Figure 1) of 33.3%$C0_2$/66.6%He glow discharge plasma obtained by double Langmuir probe at a pressure of 1.5 Torr, it is possible to get electron temperature (T_e) and ion density (n_i) [13].

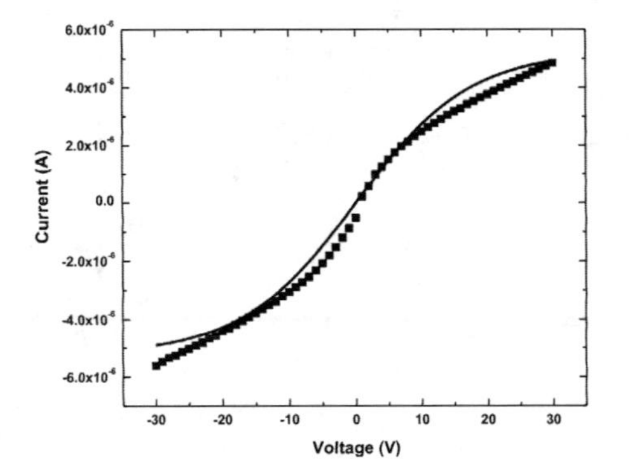

Figure 1. Curve fit of double Langmuir curve of V vs. I.

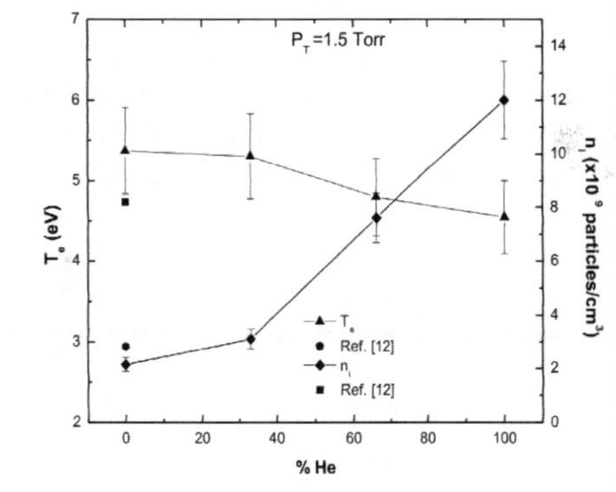

Figure 2. T_e and n_i in a mixture of CO_2/He at different concentrations of He.

Figure 1 shows that the theoretical curve (solid line) fits well to the measured data, with a correlation coefficient of 0.988. That procedure gives an electron temperature (T_e) of the mixture in the range of 4.54 to 5.37 eV, and ion density (n_i) values between 2.15 x 10^9 and 12.00 x 10^9 particles/cm^3, these values are observed in figure 2. Errors in the T_e and n_i were discussed in reference [12] and the overall variations in T_e and n_i were found to be 10% and 12%, respectively.

Figure 2 shows the electron temperature (T_e) and ion density (n_i) in a CO_2/He glow discharge mixture

Journal of Physics: Conference Series **591** (2015) 012066 doi:10.1088/1742-6596/591/1/012066

at different concentration of He gas. The measurement was conducted at concentration of 0 %, 33%, 66% and 100% of He gas. It can see that the presence of CO_2 in the mixture significantly modifies the ion density and electron temperature. The admission of helium to the discharge decreases the value of the electron temperature, while the ion density increases. That could be explained, due that an increase in helium percentage in the mixture can be expected to reduce the electron collision frequency, thereby providing more time for electrons to gain energy from the electric field and thereby accelerating further.

The present result at pure CO_2 discharge is 3.8 more lower than the values obtained recently by Mendez-Martínez et al [12] for ion density, while for electron temperature is 1.8 more higher than the results obtained by Mendez-Martinez et al [12] (see Figure 2).

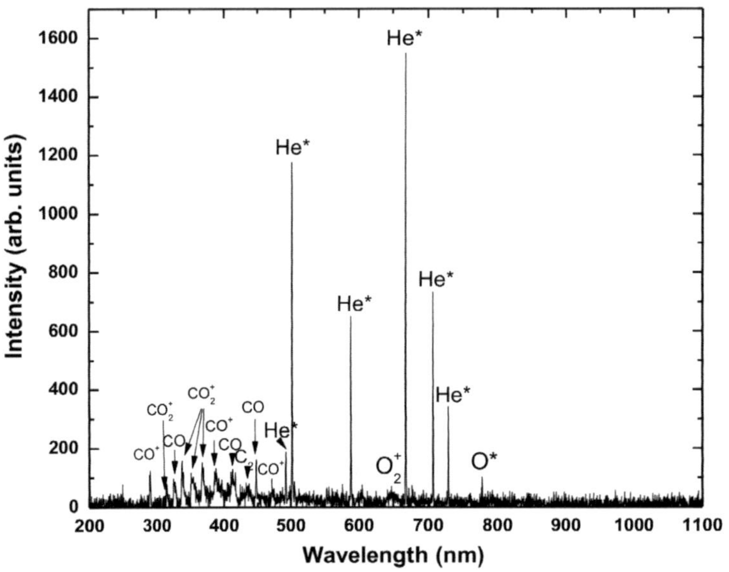

Figure 3. Optical emission spectrum of a mixture CO_2/He.

Figure 3 shows the optical emission spectroscopy of the binary mixture CO_2/He obtained at a total pressure of 1.5 Torr and a concentration of 33.3%$C0_2$/66.6%He. It observed that the more intense emission band correspond to carbon dioxide in the range of 285 to 410 nm; for CO and CO^+ bands between 285 and 475 nm and for helium lines in the range of 490 to 730 nm. It was also possible to identify bands of O_2^+ and lines of O*.

A detailed analysis of emission spectra shows the formation of the following elements: CO_2^+ (288.65, 315.48, 388.27, 354.12, 367.58 and 407.02), CO^+ (289.98, 315.48, 338.37 and 471.53 nm), CO (325.82, 368.11, 412.52 and 447.04 nm), C_2 (435.03 nm), O_2^+ (641.53 nm), O* (777.23 nm), and He* (492.57, 501.90, 587.94, 668.17, 706.82 and 728.44 nm).

The main atomic and molecular processes in the glow discharge produced by the binary mixture of carbon dioxide and helium are mainly due to processes of excitation and ionization by electron impact. So, the formation of these elements can be described as: $CO_2 + e^- \rightarrow CO_2^+ + 2e^-$, ionization; $CO_2 + e^- \rightarrow CO + O^+ + 2e^-$, dissociation; $CO_2 + e^- \rightarrow CO^+ + O^* + 2e^-$, dissociation and excitation; $CO_2 + e^- \rightarrow$

$C + O^+ + O + 2e^-$, dissociative ionization; $C + C \rightarrow C_2$, recombination, and $He + e^- \rightarrow He^* + e^-$, excitation.

4. Conclusions

The species identified by optical emission spectroscopy of CO_2/He glow discharge are: CO^+, CO_2^+, CO, C_2, O_2^+, O* and He*. The electron temperature was found are in the range of 4.54 to 5.37 eV and the ion density between 2.15 x 10^9 and 12.00 x 10^9 particles/cm^3. It observed that the presence of CO_2 in the mixture significantly modifies the ion density and electron temperature. The admission of helium to the discharge decreases the value of the electron temperature, while the ion density increases. That could be explained, due that an increase in helium percentage in the mixture can be expected to reduce the electron collision frequency, thereby providing more time for electrons to gain energy from the electric field and thereby accelerating further.

Acknowledgements
We are grateful to N. Rodriguez (FC-UAEM), O. Flores and F. Castillo (ICF-UNAM) for her technical assistance. This research was supported by UAEM [3443/2013CHT], DGAPA [IN101613], CONACyT [128714], PROMEP [103.5/13/6626] and PIFI 2010.

5. References

[1] Salazar L., Martínez H., Guerrero A. Reyes P. G. 2012 *J. of Physical Science and Applications* **2,** 283

[2] Garcia-Cosio G., Martinez H., Calixto-Rodriguez M., Gomez A. 2011 *J. Quantitative Spectroscopy and Radiative Transfer* **112** 2787–2793

[3] U.S. Environmental Protection Agency 2009 *Inventory of U.S. Greenhouse Gas Emissions and Sinks: 1990–2007*, Washington USA

[4] Knutson T. R., Tuleya R. E, and Kurihara Y. 1998 *Science* **279** 1018-1021

[5] Spencer L. F. and Gallimore A. D., 2011 *Plasma Chem. and Plasma Process.* **31**, 79–89

[6] Savinov S. Y., Lee H., Song H. K. and Na B. 1999 *Ind. Eng. Chem. Res.* **38**, 2540 (RF)

[7] Wen Y. and Jiang X. 2001 *Environ. Sci. Technol.* **21** 595 (pulsed corona)

[8] Li R., Yamaguchi Y., Yin S., Tang Q., Sato T.2004 *Solid State Ionics* **172,** 235. (Barrera dielectrica)

[9] Buser R. G. and Sullivan J.J. 1970 *J. Appl. Phys.* **41** 472 (Glow discharge)

[10] Reyes P. G., Mendez E. F., Osorio-Gonzalez D., Castillo F. and Martínez H. 2008 *Phys. Stat. Sol.* **4**, 907–910

[11] Reyes P. G., Torres C. and Martínez H. 2014 *Radiation Effects and Defects in Solids: Incorporating Plasma Science and Plasma Technology* **169** 285

[12] Mendez-Martínez, E.F., Reyes, P.G., Osorio-Gonzalez D., Castillo F., Martínez H. 2010 *Plasma Sci. Technol.* **12** 314-319

[13] Swift, J. D.; Schwar, J. R. 1991 *Electric Probes for Plasma Diagnostics*, Elsevier: New York USA

Journal of Physics: Conference Series **591** (2015) 012067 doi:10.1088/1742-6596/591/1/012067

Study of N_2/He DC glow discharge

M. Villa[1], O. Flores[2], F. Castillo[2], P. G. Reyes[1], H. Martinez[2*]

[1]Facultad de Ciencias, Universidad Autónoma del Estado de México, Estado de México, México.
[2]Laboratorio de Espectroscopia, Instituto de Ciencias Físicas, Universidad Nacional Autónoma de México, Apartado Postal 48-3, 62251, Cuernavaca, Morelos, México.

[*]E-mail: hm@fis.unam.mx

Abstract. Optical emission spectroscopy and mass spectrometry measurement were used to study a gas mixture glow discharge of He and N_2 at total pressure of 2.0 Torr, a power of 10 W and a flow of 16.5 l/min. The emission bands were measured in the wavelength range of 200 to 1100 nm. The principal species observed were N_2^+ ($B^2\Sigma^+_u \rightarrow X^2\Sigma^+_g$), N_2 ($C^3\Pi_u \rightarrow B^3\Pi_g$), and He, which agree with mass spectrometry measurement.

1. Introduction

Recently, various kinds of plasma have been produced to obtain excited states, active atoms and molecules. The production of active species is very important for a multitude of applications, such as biological sterilization and heating, plasma-assisted ignition and combustion, surface treatment and destruction of harmful substances and pollutants [1-9]. It is well known that an admixture of a molecular gas to a noble gas can substantially change the electrical and spectral characteristics of a glow discharge. The degree and way of these changes depend on the discharge conditions: the sort of noble gas, sort and percentage of the admixture, gas pressure, etc. Here, we restrict ourselves to the analysis of the present state of the art in studying DC glow discharge of a N_2/He mixture.

This paper is intended to acquire knowledge of emission spectroscopy and mass spectrometry in a gas mixture glow discharge of N_2 and Ar, at total pressure of 2.0 Torr. The OES and mass spectrometry dependence on the concentration of the mixture are presented and discussed.

2. Experimental details

The schematic diagram of the experimental setup is identical to that described in previous work [10, 11]. The discharge cell consists of two parallel electrodes enclosed in stainless steel vacuum chamber. The two electrodes were made of stainless steel disc, with 30 mm in diameter. The electrodes are positioned at the center of the reaction chamber with 10 mm gap spacing. The plasma chamber had a volume of 1.16 x 10^4 cm^3 and it was pumped down by a vacuum system (Turbomolecular pump Varian D5302) to a base pressure of 10^{-6}Torr. A continuous dynamic flow of He -N_2 gas mixture (ultra-pure gases, Praxair 99.99%) was let in the system through needle valves at the desired pressures. A DC glow discharge was produced between the two electrodes. While keeping the total pressure of 2.0 Torr, the concentration of He gas in the mixture was done by changing the He partial pressure. To keep a current constant of 10 mA (measured by a digital Fluke multimeter model 8846A), independent of the gas mixture, a ballast resistance (R=5 KΩ) was used, it was done by changing the power supply voltage (Spellman SA4) between 260 V (for 100 % of He) and 250 V (for 100 % of N_2). The discharge voltage in gas mixture is lower than that in pure Helium. A lateral flange was a quartz window, used to monitor the active species generated in the glow discharge by plasma emission spectroscopy; the

Journal of Physics: Conference Series **591** (2015) 012067 doi:10.1088/1742-6596/591/1/012067

optical spectroscopy measurements (OES) were carried out by using a high-resolution Ocean Optics Inc. Spectrometer HR2000CG-UV-NIR and a UV2/OFLV-5 detector, the fiber optic was placed in the aperture entrance. The grating has a spectral response in the range of 200 to 1100 nm.

A quadrupole mass spectrometer (QMS) is connected via cutoff and needle valves to the recipient for process gas analysis by in situ mass spectrometry (MS). The process gas composition and its changes during N_2/He plasma are measured using Faraday cup detection mode at a pressure of 5.0 x 10^{-5} Torr inside the analyzer of the mass spectrometer. Ion currents (I_m) of several mass (m) are recorded simultaneously versus He % in the mixture of the plasma.

In preparation for every MS measurement, the mass spectrometer is pumped down. A final pressure of 1.5 x 10^{-6} Torr is reached afterwards.

3. Results and discussion

OES measurement of glow discharge plasma at a pressure of 2.0 Torr is displayed in Fig. 1. This allowed analysis of the most luminous area, which corresponds to the negative glow near the cathode dark space. Only the most intense spectral lines and bands of the plasma within the 200-1100 range are quoted [12]. The principal species observed were: at 337.13 and 357.69 nm for N_2 ($C^3\Pi_u - B^3\Pi_g$); at 391.44 and 427.81 nm for $N_2^+(B^2\Sigma_u^+ - X^2\Sigma_g^+)$; at 501.57, 587.56 and 667.82 nm for He.

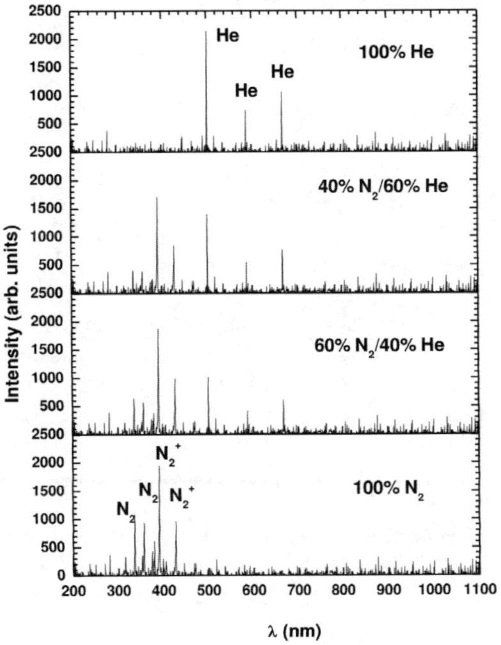

Figure 1. Emission spectra of N_2-He plasma

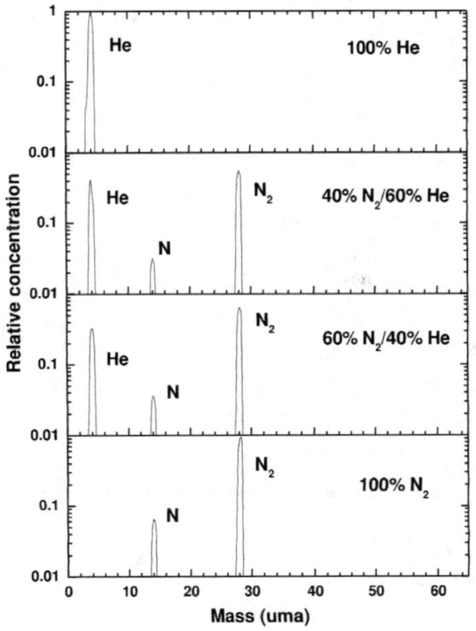

Figure 2. Relative concentration of neutral species.

Figure 2 shows the relative concentrations of three masses as functions of He % in the mixture of the N_2-He plasma at 2.00 Torr. The experimental results display three species of stables neutral, which can be associated to the N_2 molecule and, N and He atoms.

From figures 1 and 2, it can be seen that both shown the same behaviour of the species observed as a function of the He percentage, with the only difference that in the MS measurement, it observes a relatively low intensity corresponding to N atoms, which decreases as the He percentage increases.

Under our experimental conditions the important processes of inelastic electron-atom, electron-molecule and subsequent emission transitions from excited states are

$$N_2(X^2\Sigma_g^+) + e \rightarrow N_2(\mathbf{C}^3\Pi_u) + e \rightarrow N_2(B^3\Pi_g) + h\nu + e \quad (1)$$

$$N_2(X^2\Sigma_g^+) + e \rightarrow N_2^+(B^2\Sigma_u^+) + e \rightarrow N_2^+(X^2\Sigma_g^+) + h\nu + e \quad (2)$$

$$He(1^1S) + e \rightarrow He(3^1P_1) + e \quad (3)$$

$$He(1^1S) + e \rightarrow He(3^3D_{1,2,3}) + e \quad (4)$$

$$He(1^1S) + e \rightarrow He(3^1D_2) + e \quad (5)$$

The excitation energy of the emitting state $N_2(\mathbf{C}^3\Pi_u)$ at 337.13 nm is 11.03 eV and the emission cross sections for electron impact excitation of the nitrogen molecule are in the order of 10^{-22} to 10^{-23} m^2, while the excitation threshold for the $N_2^+(B^2\Sigma_u^+)$ band at 391.44 nm is about 18.75 eV above the ground state of the neutral molecule, and the corresponding cross section for optical emission caused by electron impact approaches 2 x 10^{-22} m^2 for electron energies around 50 eV [13]. Because of its high intensity as can be seen in Figure 1, this band was preferentially used in combination with band from the neutral second positive system for an evaluation based on the corona model [14].

In N_2/He discharges N_2^+ ions are mainly produced by Penning ionization of $N_2(X^2\Sigma_g^+)$ by metastable He atoms. According to data published by Hotop et. al. [15], about 40% of these ions are formed in the state $N_2^+(B^2\Sigma_u^+)$. Production of $N_2(X^2\Sigma_g^+)$ by charge transfer between helium ions and N_2 molecules is not effective under our experimental conditions. The rate constant of this process is high 1.2 x 10^{-9} cm^3/molecule s^{-1} [16] as compared to $\approx 10^{-10}$ cm^3/molecule s^{-1} for Penning ionization process. But the production rate of the He ions is two orders of magnitude smaller than the production of metastable helium atoms. Moreover, it was shown by Collins [17] that N_2^+ is produced in He/N_2 mixtures mostly by Penning ionization process, while charge transfer gives nitrogen atoms in a dissociative or predissociative state, which it can be seen in the MS measurements (see Figure 2).

4. Conclusions

The experimental investigation of cold plasma in which N_2 is predominant can be of great interest for identification of species and processes of likely relevance in multitude of applications, such as biological sterilization and heating, plasma-assisted ignition and combustion, surface treatments and destruction of harmful substances and pollutants.

In this work, we studied by optical emission spectroscopy and mass spectrometry measurements of the plasmas formed in low-pressure (2.0 Torr) glow discharge of N_2 with admixtures of He.
The principal species observed were N_2^+ ($B^2\Sigma^+_u \rightarrow X^2\Sigma^+_g$), N_2 ($C^3\Pi_u \rightarrow B^3\Pi_g$), and He, which agree with MS measurement.
It was shown that N_2^+ is produced in He/N_2 mixtures mostly by Penning ionization process, while charge transfer gives nitrogen ions in a dissociative or predissociative state, which has been corroborated by MS measurements.

Acknowledgments
The authors are thankful to H. H. Hinojosa and J. Rangel for technical assistance. This research was sponsored by DGAPA IN-101613, and CONACyT 128714.

References

[1] E. Marode et al, Plasma Phys. Control. Fusion 51**,** 124002 (2009).
[2] K. V. Kozlov, R. Brandenburg, H. E. Wagner, A. M. Morozov and P. Michel, J. Phys. D: Appl. Phys. 38, 518 (2005).
[3] Y. Zuzeek, S. Bowman, I. Choi, I. V. Adamovich and W. R. Lempert, Proc. Combustion Inst. 33, 3225 (2011).
[4] G. Fridman et. al., Plasma Chem. Plasma Process 27 163 (2007).
[5] N. Chintala, A. Bao, G. Lou and I. V. Adamovich, Combust. Flame 144**,** 744 (2006)
[6] S. M. Starikoskaia, J. Phys. D: Appl. Phys. 39**,** R265 (2006).
[7] I. A. Shkurenkov, Yu A. Mankelevich and T. V. Rakhimova, Phys Rev. E 79, 046406 (2009).
[8] I. A. Shkurenkov, Yu A. Mankelevich and T. V. Rakhimova, Euro. Phys. J. D 61, 95 (2011).
[9] M. Kushner and N. Yu Babaeva, Stud. Health Technol. Inform. 163, 297 (2011).
[10] L. Salazar-Flores, H. Martinez, A. Guerrero-Tapia and P.G. Reyes- Romero, Journal of Physical Science and Application 2, 283 (2012).
[11] G. García-Cosió, H. Martínez, M. Calixto-Rodríguez, A. Gomez, Journal of Quantitative Spectroscopy and Radiative Transfer 112, 2787 (2011).
[12] R. W. B. Pearse, A. G. Gaydon, The identification of molecular spectra, in: University Printing House Cambridge, 1976.
[13] M. Shaw, J. Campos, J. Quant. Spectrosc. and Radiat. Transfer 30, 73 (1983)
[14] I. P. Vinogradov, B. Jettkant, D. Meyer, K. Wiesemann, J. Phys. D 27, 1207 (1994)
[15] H. Hotop, E. Kolb, J. Lorenzen J. Electron Spectrosc. Relat. Phenom. 16, 213 (1979).
[16] S. J. Young, K.P. Horn, J. Chem. Phys. 57, 4835 (1972)
[17] C. B. Collins, W.W. Robertson J. Chem. Phys. 40, 701 (1964)

.

Analysis of green fluorescent protein bioluminescence *in vivo* and *in vitro* using a glow discharge

L Hernández[1], L A Mandujano[1], J Cuevas[1], P G Reyes[2] and D Osorio-González[1,3]

[1] Laboratory of Molecular Biophysics, Faculty of Sciences, Autonomous University of State of Mexico, C.P. 50000, Toluca, State of Mexico, Mexico.

[2] Laboratory of Advanced Physics, Faculty of Sciences, Autonomous University of State of Mexico, C.P. 50000, Toluca, State of Mexico, Mexico.

E-mail: dog@uaemex.mx

Abstract. The discovery of fluorescent proteins has been a revolution in cell biology and related sciences because of their many applications, mainly emphasizing their use as cellular markers. The green fluorescent protein (GFP) is one of the most used as it requires no cofactors to generate fluorescence and retains this property into any organism when it is expressed by recombinant DNA techniques, which is a great advantage. In this work, we analyze the emission spectra of recombinant green fluorescent protein *in vivo* and *in vitro* exposed to a glow discharge plasma of nitrogen in order to relate electron temperature to fluorescence intensity.

1. Introduction

The green fluorescent protein (GFP) is produced by bioluminescent organisms such as the jellyfish *Aequorea victoria*, and the IndoPacific sea anemone relative *Discosoma*; this protein has two excitation maxima at 395 and 470 nm, and a single major emission peak at 509 nm which enable it to exhibit intrinsic fluorescence. The GFP has become the most important marker in cell and molecular biology and a very useful biosensor, since it is used as reporter gene, marker for tumor cells, fusion tag with cloned genes in cellular dynamics studies, or arsenic traces detector in water. In this work, the recombinant GFP was expressed by competent *Escherichia coli* cells and purified at room temperature employing hydrophobic interaction chromatography. Both systems, the transformed cells expressing GFP and the purified recombinant protein, were exposed into a glow discharge produced by a continuous gas flow at different pressures of N_2.

2. Background

The GFP has two excitation peaks, the first near to 395 nm, the second around 475 nm; and also two emission peaks, if excited at the 395 nm peak, the emission is at 508 nm, and on the contrary, if excited at 475 nm, then the broadcast will appear at 503 nm. When this occurs the protein captures more light outside the visible spectrum and the emission becomes visible green light [1-6]. GFP is conformed by a chain of 238 amino acid residues, forming 11 beta strands, organized in a cylinder that

[3] To whom any correspondence should be addressed.

encloses an alpha helix. It contains a special chromophore (constituted by 3 consecutive amino acids: Ser, Tyr and Gly), a chemical group that absorbs and emits light. After synthesis of the protein, three amino acids go through a cyclization and spontaneous oxidation. The carbon atom in the serine's carbonyl forms a covalent linkage to the N atom of the amino group of glycine, followed by removal of water and oxidation of C alfa link – C beta tyrosine to form a double bond. The structure obtained contains a system of conjugated double bonds which gives the property of fluorescence [1, 7-11].

Fluorescent proteins have many applications in science; one of the most common is their use as markers fused to polypeptides or indicators of intracellular chemical - physiological states, in all types of cells, cellular subcompartments and organisms from prokaryotes to mammals. If GFP is attached to other proteins in which the fluorescence depends on the environmental conditions, such compound molecule operates as a sensor in monitoring several cellular complex processes, such as intracellular temperature, the activity and signs displayed in cascades, pH changes, dynamics of second messengers, protein interactions and conformational changes [1,12].

3. Method

3.1. GFP expression and purification

Green fluorescent protein was expressed in *Escherichia coli* competent cells, HB101 K-12 strain. Individual colonies of 1.5 mm in diameter grown in LB solid medium 4%, were transformed with 10 μl of pGLO plasmid at a concentration of 0.08 μg/ml in 1 ml of transformation solution (50 mM $CaCl_2$, pH 6.1) with 1 ml of LB liquid medium 4%. Subsequently, they were cultured on LB solid medium 4%, supplemented with ampicillin (0.1 g/L) and L - arabinose (5 g /L) at 37 ° C for 24 h. The expression was monitored using a long wavelength UV lamp. After 24 h, the transformation efficiency was determined. Individual colonies expressing GFP were grown in tubes containing 2 ml of LB/amp/ara broth [4% LB, ampicillin (0.1 g /L) and L - arabinose (2 g/L)] with vigorous shaking of 200 rpm at 32° C for 24 h . They were then harvested by centrifugation at 13 300 rpm for 5 min and resuspended in 250 μl of Tris - EDTA buffer (TE; 10 mM Tris, 1 mM EDTA, pH 8.0). The cells were lysed by incubation with lysozyme followed by freezing for 24 h. After thawing, they were subjected to centrifugation at 13 300 rpm for 10 min to separate cell debris. GFP was purified at room temperature on columns of hydrophobic interaction chromatography (8 x 40 mm) pre-packed with methyl resins (Macro-Prep®) and pre-equilibrated with 2 ml of equilibration buffer (2M $(NH_4)_2SO_4$/TE , pH 8.0). For each 250 μl of supernatant, there were used 250 μl of binding buffer (4M $(NH_4)_2SO_4$/TE , pH 8.0) and 250 μl of wash buffer (1.3 M $(NH_4)_2SO_4$/TE , pH 8.0), followed by 750 μl of TE as elution buffer. All reagents were obtained from BioRad.

3.2. Spectroscopies

Both systems, transformed cells expressing GFP and the purified recombinant protein in TE buffer, were exposed to a glow discharge produced by a continuous flow of N_2. The experimental setup consisted of a cylindrical chamber with two circular aluminum plate electrodes placed at its center. Prior to discharge, it was held a vacuum of 10^{-3} Torr in order to remove impurities and/or residual gases. In the course of the experiment the pressure was 1 Torr for N_2. To generate the plasma, a potential difference was applied to the electrodes, regulated by a power source which provided direct current (V = 238 V, I = 50 mA).

For analysis *in vivo*, the expressing GFP bacterial colonies on solid medium were mounted in a 1 cm path length quartz cell, using as controls the LB/amp/ara medium without bacteria and LB/amp bacteria. For analysis *in vitro*, 3.1 ml of GFP in TE buffer analysis were placed in a quartz cell of 1 cm path length, using TE buffer as control. Emission spectra were observed for both systems in the wavelength range 200 to 1100 nm, using an Ocean Optics spectrometer HR4000CG with 0.15 nm resolution.

Journal of Physics: Conference Series **591** (2015) 012068 doi:10.1088/1742-6596/591/1/012068

4. Results

Green Fluorescent Protein was successfully expressed using the plasmid pGLO and HB101 K-12 strain of *Escherichia coli*. The efficiency of transformation was on the order of 10^3 transformers per microgram of DNA. The phenotype of the bacterial colonies is consistent with the expected and reported in the literature. In the region of 500-600 nm, it was observed an emission band (*in vivo* and *in vitro*) in which the most intense line is N_2^+. Such band is not typical of a glow discharge of nitrogen at the experimental conditions, so it was attributable to the protein composition, culture medium and/or TE buffer. The fluorescence intensity at 509 nm is evidence that N_2 plasma radiates with sufficient energy to activate the GFP chromophore (figures 1, 2).

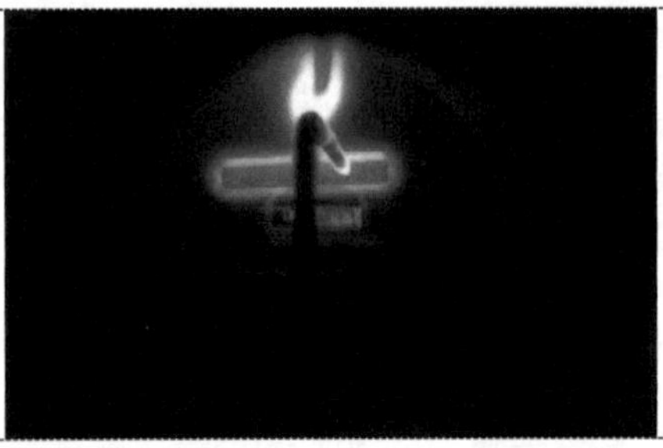

Figure 1. Analysis *in vivo*. Bacteria in culture medium exposed to a nitrogen plasma.

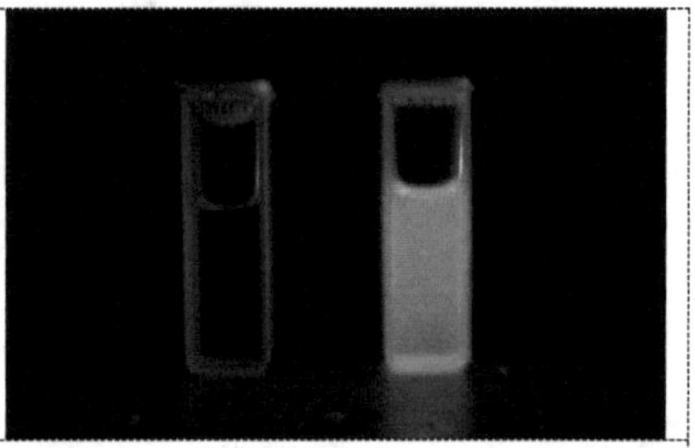

Figure 2. Analysis *in vitro*. Left: Control sample which contains TE buffer. Right: Sample of green fluorescent protein (GFP) dissolved in TE buffer.

The figure 3 shows the emission spectra of GFP *in vivo* (bacterial colonies on solid LB/amp/ara medium) and *in vitro* (TE buffer) in which emission lines and bands corresponds to a typical N_2 plasma generated by a glow discharge, also we can observe the presence of a band in the 500-600 nm range whose species is N_2^+, and a peak at 509 nm corresponding to the fluorescent emission of the GFP.

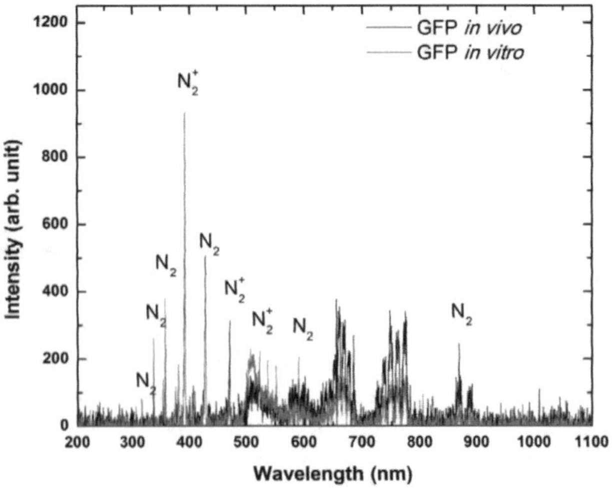

Figure 3. Emission spectra of GFP *in vivo* and *in vitro*.

5. Conclusions

The fluorescence intensity of *in vivo* and *in vitro* systems did not exhibit measurable changes at exposition times into the glow discharge, this is indicative that GFP molecules do not suffer significant changes in their native structure under the experimental conditions.

The experiments showed the advantages of using a glow discharge as external source in order to stimulate the chromophore. We expect that the sample exposition to a more intense plasma will provide favorable conditions to analyze the protein fluorescence anisotropy.

Acknowledgments

The authors wish to thank Autonomous University of State of Mexico for financial support through project ID 3697/2014/CID.

References

[1] Tsien R 1998 The green fluorescent protein *Annu. Rev. Biochem.* **67** 509–44.

[2] Pérez M and Becú-Villalobos D 2009 La proteína verde fluorescente ilumina la biociencia *Medicina, Buenos Aires* **69** 370-4.

[3] Yang F, Moss L and Phillips G 1996 The molecular structure of green fluorescent protein *Nat. Biotechnol.* **14** 1246-51.

[4] Labas Y, Gurskaya N, Yanushevich Y, Fradkov A, Lukyanov K, Lukyanov S and Matz M 2002 Diversity and evolution of the green fluorescent protein family. *Proc. Natl. Acad. Sci.* **99** 4256-61.

[5] Inouye S and Tsuji F 1994 Aequorea green fluorescent protein expression of the gene and fluorescence characteristics of the recombinant protein *FEBS Letters* **341** 277-80.

[6] Ehrmann M, Scheyhing C and Vogel R 2001 *In vitro* stability and expression of green fluorescent protein under high pressure conditions *Lett. Appl. Microbiol.* **32** 230-34.

[7] Prasher D, Eckenrode V, Ward W, Prendergast F and Cormier M 1992 Primary structure of the Aequorea victoria green-fluorescent protein *Gene* **111** 229-33.

[8] Cody C, Prasher D, Westler W, Prendergast F and Ward W 1993 Chemical structure of the hexapeptide chromophore of the Aequorea green-fluorescent protein *Biochem.* **32**:1212-18.

[9] Crudakov D, Matz M, Lukyanov S and Lukyanov K 2010 Fluorescent proteins and their applications in imaging living cells and tissues *Physiol. Rev.* **90** 1103-63.

[10] Zimmer M 2000 Green fluorescent protein (GFP): aplications, structure, and related photophysical behavior. *Chem. Rev.* **102** 759-81.

[11] Rizzo M, Davidson M and Piston D 2010. Fluorescent protein tracking and detection. *Live cell imaging: A Laboratory Manual* ed.Goldman R D, Spector D L and Swedlod J R (New York: Cold Spring Harbor Laboratory Press) pp 3-34

[12] Franco A and Longart M 2009 Aplicaciones de la proteína verde fluorescente (GFP) en la biología celular y en la visualización del sistema nervioso *RET.* **1** 84-96.

Rotating mirror method for measuring the visible and NIR spectrum of plasmas

Luis Bernal[1], **Luis Bilbao**[2]

[1] Department of Physics, University of Mar del Plata, Argentina.

[2] Instituto de Física del Plasma, CONICET-UBA,Ciudad Universitaria, Pab. I, 1428 Buenos Aires, Argentina.

E-mail: bilbao@df.uba.ar

Abstract. We report the construction and operation of an apparatus for measuring optical Doppler shift based on a rotating mirror, and its preliminary application to the measurement of the spectrum of some light sources. We describe preliminary results, and discuss the detection limits and the possibility of using the apparatus for plasma diagnostic.

1. Introduction

In a previous paper [1], we described an apparatus for measuring the Doppler shift using a rotating mirror. Optical Doppler shift experiments are not a simple task since the light source cannot usually be moved in a sufficiently smooth and uniform manner to keep the level of noise well below of that of the signal.

Using a rotating mirror it is possible to overcome many of the noise generating effects. The beam from a light source passes a beam splitter that separates the beam, part is reflected from the advancing side of a rotating mirror; part is reflected from the receding part of a rotating mirror. After reflection the beams are recombined using a beam splitter and measured by a fast photo-detector. Beat between the two beams produces fluctuations in the light intensity at the photo-detector.

Since the spectrum of the measured interference signal is proportional to the spectrum of the original source, then it is possible to use this experiment as an online scanner of the light spectrum of the source.

2. Apparatus

Using a rotating mirror in which one beam is reflected from the advancing side and the other beam is reflected from the receding part of a rotating mirror two effective sources with different velocities can be created. The Doppler beat frequency, f_b, between the two sources is given by [1]

$$f_b = \frac{4\pi dF}{\lambda} \tag{1}$$

where F is the frequency of the rotating mirror, d the beam separation and λ the wavelength of the light. In figure 1 a sample beat pattern of a laser source is shown. The FFT (also shown in figure 1) exhibits a peak at the frequency f_b given by (1).

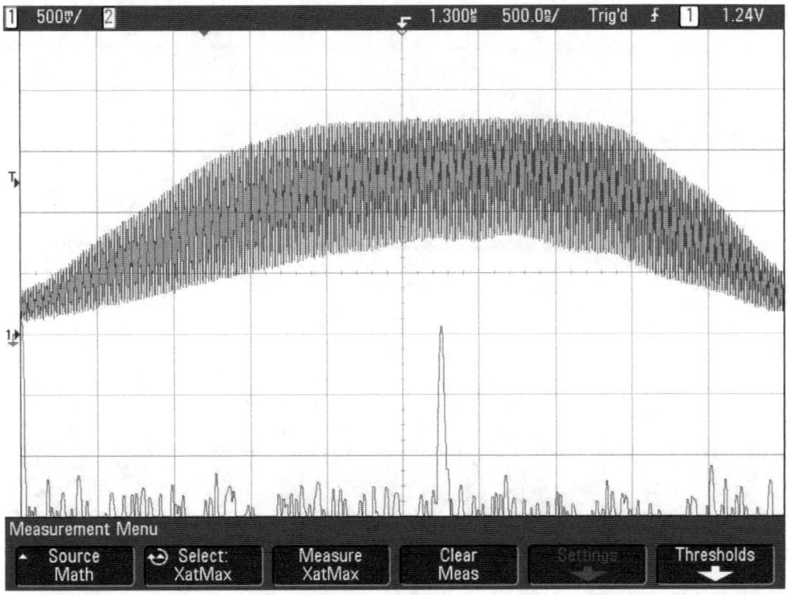

Figure 1. Sample beat pattern of a laser (upper trace, vertical scale 500 mV/div, horizontal scale 500 ns/div) and its FFT transform (lower trace, vertical scale 10 dB/div, horizontal scale 10 MHz/div). From the FFT, the wavelength of the source is obtained using (1).

Note that the beat frequency is a function of the rotational frequency. Therefore, in order to provide a reliable measurement the old motor (used in the previous version) was replaced by an outrunner brushless motor located at the bottom of the mirror. Also an electronic controller was built so the rotational frequency can be varied in a 1 Hz step up to 100 Hz. In figure 2 the new mounting is shown at rest (left) and operating at 50 Hz (right) were the reflected laser beams (separated by 2.4 cm) are also visible. The experimental setup is shown in figure 3, were two sources may simultaneously be used.

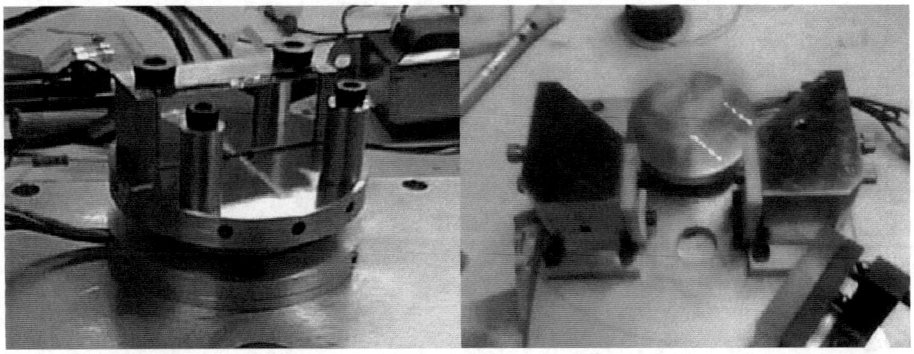

Figure 2. Improved mounting and driver of the rotating mirror (left). The mirror working at 50 Hz (center-right) in between the optical positioners. The green laser beams are also visible.

3. Calibration and preliminary tests

The first idea was to use the apparatus for resolving spectral lines of a given source. Working at 50 Hz and a beam separation of 0.024 m the Doppler beat of the visible spectrum lies between 20 and 40 MHz.

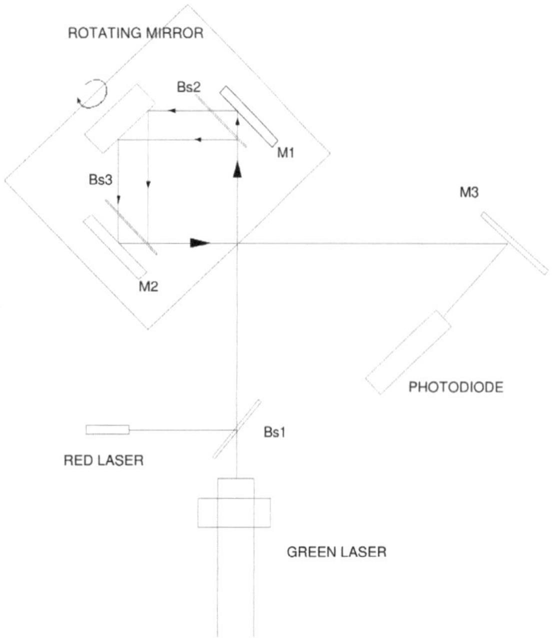

Figure 3. Experimental setup used for calibration. Any of the laser may be replaced by another source.

Using a 500,000 record length, 2 GSa/s digitizer the error of the beat frequency due to the FFT resolution is lower than the error due to the rotational stability that is about 0.1%. Using these values the theoretical resolution will be of about 0.5 nm.

The actual resolution will depend also on the optical components. During these preliminary feasibility tests we have used teaching-grade positioners and optics (ie, from standard teaching kits).

The calibration of the apparatus is performed using two monochromatic sources of known wavelength. The beat spectrum is obtained by means of a FFT transform (a built-in function in most oscilloscopes), and using (1) the corresponding wavelength are obtained. In figure 4 the spectrum of a simultaneous Doppler beat of a red (658.1 nm) and a green lasers (543.5 nm) is shown. The two peaks are clearly resolved in the spectrum and can be used to define the scale for the wavelength. Calibration can also be performed using only a green diode laser since the FFT shows the peaks of the green wavelength (532 nm), the IR pumping wavelength (807.5 nm), and the IR lasing wavelength (1064 nm), see figure 5.

As a preliminary test, a Mercury lamp was used as a source. In figure 6 the Doppler beat and the wavelength spectrum are shown. Acquiring at 100 MSa/s the peaks 576.9+579.0 nm and 546.1 nm are resolved with a resolution of about 2 nm.

4. Conclusions

Using a simple Doppler technique based on the beat between the two beams that reflects off a rotating mirror, it is possible to obtain the line spectrum (or at list a part of it) of a light source with an inexpensive apparatus.

By varying the rotational speed and the beam separation it is possible to scan a wide part of the electromagnetic spectrum. The method is not limited to visible light. The method can be used in any range of the electromagnetic spectrum as long as beam-splitters and mirrors are available.

Work is still in progress. According to these preliminary results a resolution of 0.1 nm will be possible to achieve by an appropriated choice of parameters.

Plasmas are an interesting source of electromagnetic radiation. The present method can be applied to stationary or quasi-stationary plasmas (the maximum time resolution is given by the rotational

Journal of Physics: Conference Series **591** (2015) 012069 doi:10.1088/1742-6596/591/1/012069

frequency) to obtain an online scanner of the light spectrum as a complement to more sophisticated and expensive spectrometers.

5. Acknowledgements

Work partially supported by grant UBA 20020100100866, University of Buenos Aires.

References

[1] L. Bernal and L. Bilbao 2007 *Am. J. Phys.* **75** 216.

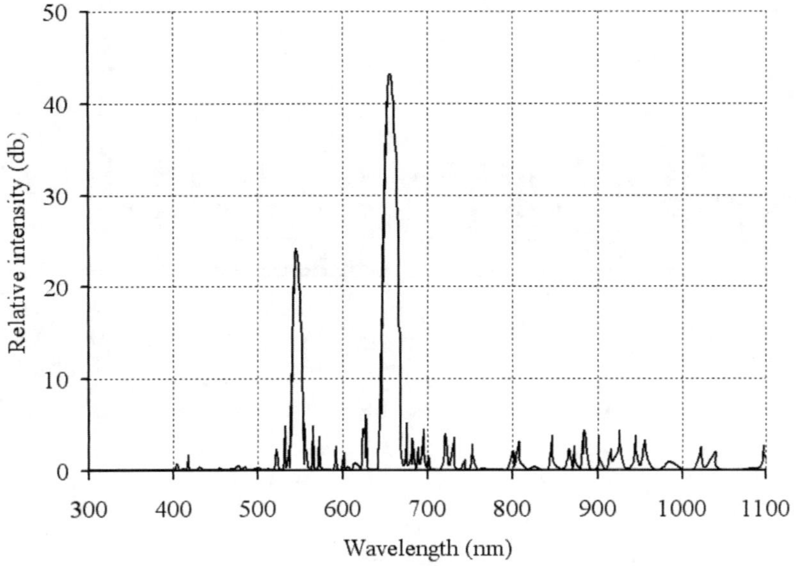

Figure 4. Spectrum of a combined green (543.5 nm, left peak) and a red (658.1 nm) lasers, used for calibration purpose.

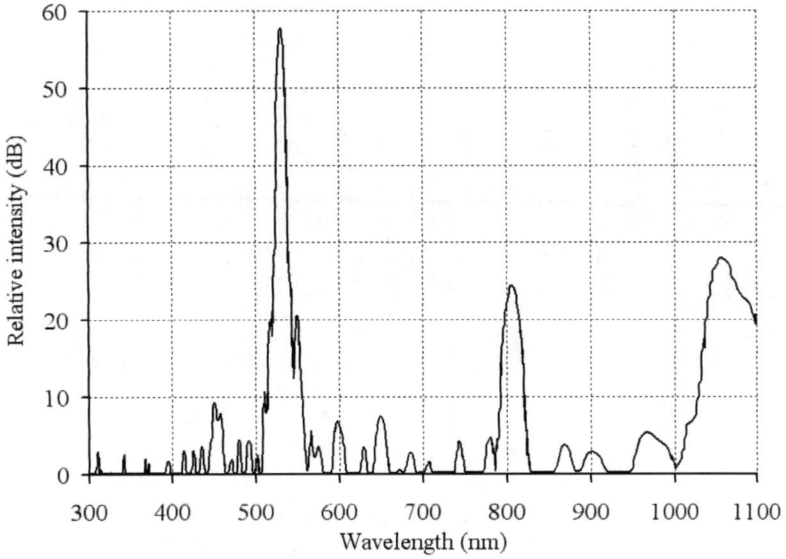

Figure 5. Spectrum of a green diode laser. The peaks (from the left) of the green laser 532 nm, the IR pumping wavelength 807.5 nm, and the IR lasing wavelength 1064 nm are shown.

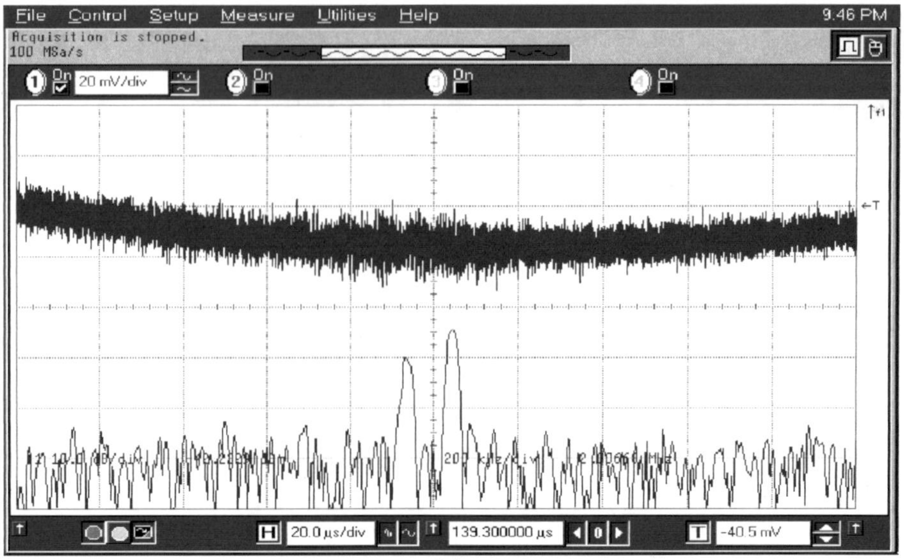

(a)

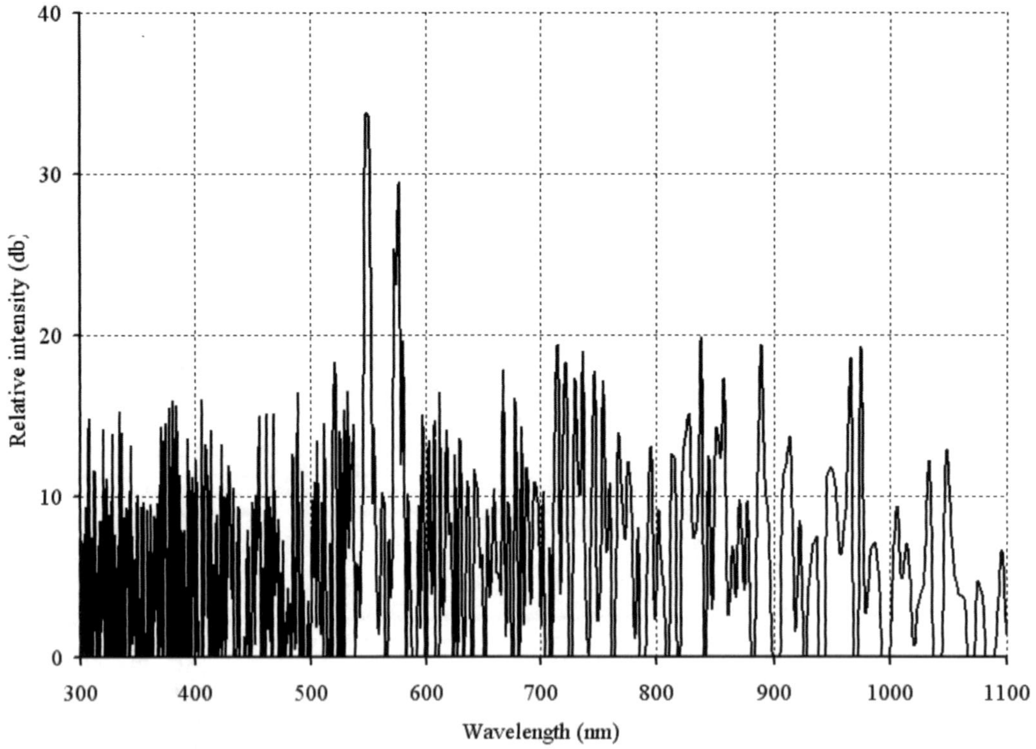

(b)

Figure 6. Spectrum of a Hg lamp. (a) Beat pattern at 100 MSa/s (upper trace, vertical scale 20 mV/div, horizontal scale 20 µs/div) and its correspondent FFT (lower trace, vertical scale 10 dB/div, horizontal scale 200 kHz/div). (b) Spectrum obtained using (1) from the FFT, showing the peaks for 576.9+579.0 nm lines (right peak) and for 546.1 nm line (left peak).